PRINCIPLES AND APPLICATIONS
IN ENGINEERING SERIES

OCCUPATIONAL ERGONOMICS

Principles of Work Design

PRINCIPLES AND APPLICATIONS
IN ENGINEERING SERIES

OCCUPATIONAL ERGONOMICS

Principles of Work Design

EDITED BY

Waldemar Karwowski

University of Louisville
Louisville, Kentucky

William S. Marras

The Ohio State University
Columbus, Ohio

CRC Press
Taylor & Francis Group
Boca Raton London New York

CRC Press is an imprint of the
Taylor & Francis Group, an **informa** business

The material in this book was previously published in *The Occupational Ergonomics Handbook*, W. Karwowski and W.S. Marras, Eds., CRC Press, Boca Raton, FL, 1999.

CRC Press
Taylor & Francis Group
6000 Broken Sound Parkway NW, Suite 300
Boca Raton, FL 33487-2742

First issued in paperback 2019

© 2003 by Taylor & Francis Group, LLC
CRC Press is an imprint of Taylor & Francis Group, an Informa business

No claim to original U.S. Government works

ISBN-13: 978-0-8493-1802-3 (hbk)
ISBN-13: 978-0-367-39531-5 (pbk)

Library of Congress Cataloging-in-Publication Data

Occupational ergonomics : principles of work / edited by Waldemar Karwowski and
    William S. Marras.
        p. cm. — (Principles and applications in engineering)
    Includes bibliographical references and index.
    ISBN 0-8493-1802-5 (alk. paper)
      1. Human engineering. 2. Work design. I. Karwowski, Waldemar, 1953- II. Marras,
    William S. (William Steven), 1952- III. Series.

    TA166.O2565 2003
    620.8'2—dc21

                                                                    2002041777

Library of Congress Card Number 2002041777

Visit the Taylor & Francis Web site at
http://www.taylorandfrancis.com

and the CRC Press Web site at
http://www.crcpress.com

Preface

Ergonomics (or human factors) is defined by the International Ergonomics Association (www.iea.cc) as *the scientific discipline concerned with the understanding of interactions among humans and other elements of a system, and the profession that applies theory, principles, data and methods to design in order to optimize human well-being and overall system performance.* Ergonomists *contribute to the design and evaluation of tasks, jobs, products, environments and systems in order to make them compatible with the needs, abilities and limitations of people.*

Currently, there is substantial and convincing evidence that the proficient application of ergonomics knowledge, in a system context, will help to improve system effectiveness and reliability, increase productivity, reduce employee healthcare costs, and improve the quality of work processes, products and working life for all employees. As ergonomics promotes a holistic approach in which considerations of physical, cognitive, social, organizational, environmental and other relevant factors are taken into account, the professional ergonomists should have a broad understanding of the full scope of the discipline. Development of this book was motivated by the quest to facilitate a wider acceptance of ergonomics as an effective methodology for work-system design aimed at improving the overall quality of life for millions of workers with a variety of needs and expectations.

This book focuses on fundamentals in ergonomics design and evaluation. The volume contains a total of 41 chapters divided into two parts.

Part I covers the background for the discipline and profession of ergonomics and offers an international perspective on ergonomics. It also discusses ergonomics of system design, reviews the sources of ergonomics knowledge, and outlines the principles of certification in professional ergonomics. The development of ergonomics programs and guidance for the reader in searching for relevant ergonomics knowledge on the worldwide web are also discussed.

Part II describes the foundations of ergonomics knowledge, which is presented in three sections including fundamentals of ergonomics, fundamentals of work analysis, and cognitive issues. Section I, *Fundamentals of Ergonomics*, covers the physical, cognitive, informational, and psychosocial aspects of work design and evaluation. This section includes basic knowledge about engineering anthropometry, occupational biomechanics, evaluation of human strength, methods for evaluating working postures and static exertion, physiological job demands, psychosocial work factors, and cognitive issues in work design including design of information devices and controls.

Section II, *Fundamentals of Work Analysis*, contains knowledge about a variety of methods and techniques for ergonomics task analysis and job evaluation. This section also offers information about evaluation of human strength, design of manual lifting tasks, data about push/pull force limits and force exertion in user-product introduction. In addition, methods for evaluation of physical workload due to posture (RULA and OWAS), as well as computer-aided human models and other software-based tools for ergonomic analysis and design are reviewed. Finally, this section outlines the use of physiological instrumentation, quantification of human movement using video-based techniques, and measurement of force and acceleration of human body.

Section III, *Cognitive Environment Issues*, is devoted to human-machine-environment system design and evaluation. The topics covered include human error in complex systems, the speed-accuracy tradeoff in human performance, human-system reliability, techniques for human reliability assessment, safety communication issues, design of warnings, and ergonomics methods in consumer product design.

We hope that this volume will be useful to a large number of professionals, students, and practitioners who strive to improve product and process quality, worker health and safety, and productivity in a variety of industries and businesses. We trust that the knowledge presented in this volume will introduce the reader to the basic principles of ergonomics design and evaluation.

Waldemar Karwowski
University of Louisville

William S. Marras
The Ohio State University

The Editors

Waldemar Karwowski, Ph.D., P.E., C.P.E., is Professor of Industrial Engineering and Director of the Center for Industrial Ergonomics at the University of Louisville, Kentucky. He holds an M.S. (1978) in production management from Technical University of Wroclaw, Poland, and a Ph.D. (1982) in Industrial Engineering from Texas Tech University. He is a Board Certified Professional Ergonomist (CPE). His research, teaching, and consulting activities focus on prevention of low back injury and cumulative trauma disorders, human and safety aspects of advanced manufacturing, fuzzy sets and systems, and theoretical aspects of ergonomics.

Dr. Karwowski currently serves as Secretary-General of the International Ergonomics Association. He is editor of many international journals, including *Human Factors and Ergonomics in Manufacturing,* and *Theoretical Issues in Ergonomics Science,* and consulting editor of *Ergonomics.* He is the author or co-author of more than 200 scientific publications, including 25 books.

Dr. Karwowski is founder and chairman of the International Conference on Human Aspects of Advanced Manufacturing and Hybrid Automation. He was the recipient of the Outstanding Young Engineer of the Year Award, given by the Institute of Industrial Engineering. He was also a Fulbright Scholar at Tampere University of Technology in Finland. He received the President's Award for Outstanding Scholarship, Research, and Creative Activity in the category of Basic and Applied Science at the University of Louisville.

William S. Marras, Ph.D., C.P.E., holds the Honda Endowed Chair in Transportation in the Department of Industrial, Welding, and Systems Engineering at the Ohio State University, Columbus. He is also the director of the biodynamics laboratory and holds appointments in the departments of physical medicine and biomedical engineering. Professor Marras is also the co-director of the Ohio State University Institute for Ergonomics.

Dr. Marras received his Ph.D. in bioengineering and ergonomics from Wayne State University in Detroit, Michigan. His research centers around biomechanical epidemiologic studies, laboratory biomechanic studies, mathematical modeling, and clinical studies of the back and wrist.

His findings have been published in more than 100 refereed journal articles and 12 book chapters. He also holds two patents, including one for the lumbar motion monitor (LMM). His work has also attracted national and international recognition. He has won the prestigious Swedish Volvo Award for Low Back Pain Research and Austria's Vienna Award for Physical Medicine.

Contributors

David C. Alexander
Auburn Engineers, Inc.
Auburn, AL

Susan A.H. Benysh
Purdue University
West Lafayette, IN

Regina Brauchler
University of Stuttgart
Hohenheim, GERMANY

Heiner Bubb
Technische Universität Munchen
Garching, GERMANY

Peter M. Budnick
University of Utah
Salt Lake City, UT

J. Bugajska
Central Institute for Labour
    Protection
Warsaw, POLAND

Pascale Carayon
École des Mines de Nancy
Nancy, FRANCE

Keith Case
Loughborough University
Leicestershire, ENGLAND

Klaus Christoffersen
The Ohio State University
Columbus, OH

David R. Clark
IMSE Department
Kettering University
Flint, MI

David J. Cochran
Department of Industrial
    Engineering
University of Nebraska
Lincoln, NE

E.N. Corlett
Institute for Occupational
    Ergonomics
University of Nottingham
Nottingham
ENGLAND

Brechtje J. Daams
Delft University of Technology
Delft, THE NETHERLANDS

Marjolein Douwes
TNO Prevention & Health
Leiden, THE NETHERLANDS

Colin G. Drury
State University of New York at
    Buffalo
Department of Industrial
    Engineering
Buffalo, NY

J. Dul
TNO Prevention & Health
Leiden, THE NETHERLANDS

Martin T. Freer
Loughborough University
Leicestershire, ENGLAND

Andris Freivalds
The Pennsylvania State University
University Park, PA

Sean Gallagher
National Institute for Occupational
    Safety and Health
Pittsburgh, PA

Anand Gramopadhye
Department of Industrial
    Engineering
Clemson University
Clemson, SC

M. Susan Hallbeck
University of Nebraska
Lincoln, NE

Hal W. Hendrick
University of Southern California
Los Angeles, CA

Sheik N. Imrhan
Industrial Engineering
    Department
University of Texas
Arlington, TX

Toni Ivergard
Institute of Working Life
Ostersund, SWEDEN

Dieter W. Jahns
SynerTech Associates
Bellingham, WA

Jari Järvinen
Motorola Corporation
Delray Beach, FL

Iwona Jastrzebska-Fraczek
Technische Universität Munchen
Garching, GERMANY

Bente Rona Jensen
National Institute of Occupational
    Health
Copenhagen, DENMARK

Steven L. Johnson
Bell Engineering Center
University of Arkansas
Fayetteville, AR

Barry Kirwan
Air Traffic Management
    Development Center
National Air Traffic Services Ltd.
Bournemouth Airport
and University of Birmingham
Christchurch, Dorset
ENGLAND

Danuta Koradecka
Central Institute for Labour
    Protection
Warsaw, POLAND

Karl H.E. Kroemer
Industrial Ergonomics
    Laboratory
Virginia Polytech Institute and
    State University
Blacksburg, VA

Kurt Landau
Institut für Arbeitswissenschaft
Darmstadt, GERMANY

Kenneth A. Laughery
Rice University
Houston, TX

Mark R. Lehto
Purdue University
West Lafayette, IN

Soo-Yee Lim
National Institute for Occupational
    Safety and Health
Washington, DC

Veikko Louhevaara
Institute of Occupational Health
and the University of Kuopio
FINLAND

David R. Lovvoll
Rice University
Houston, TX

Hongzheng Lu
Lucent Technologies
Eatontown, NJ

Holger Luczak
Institute of Industrial Engineering
    and Ergonomics
Aachen, GERMANY

William S. Marras
ISE Department
The Ohio State University
Columbus, OH

Markku Mattila
Tampere University of Technology
Tampere, FINLAND

David Meister
Human Factors Consultant
San Diego, CA

Mathilde C. Miedema
TNO Prevention & Health
Leiden, THE NETHERLANDS

J. Steven Moore
University of Texas
Health Center
Tyler, TX

Gary B. Orr
US Department of Labor
OSHA
Washington, DC

J. Mark Porter
Department of Design and
    Technology
Loughborough University
Leicestershire, ENGLAND

Peter M. Quesada
Department of Mechanical
    Engineering
University of Louisville
Louisville, KY

Robert G. Radwin
Department of Industrial
    Engineering
University of Wisconsin
Madison, WI

Walter Rohmert
Institut für Arbeitswissenschaft
Darmstadt, GERMANY

Christopher Schlick
Institute of Industrial Engineering
    and Ergonomics
Aachen, GERMANY

Joseph Sharit
University of Miami
Coral Gables, FL

Gisela Sjøgaard
National Institute of Occupational
    Health
Copenhagen, DENMARK

Philip J. Smith
The Ohio State University
Columbus, OH

Johannes Springer
Institute of Industrial Engineering
    and Ergonomics
Aachen, GERMANY

Neville A. Stanton
Department of Psychology
University of Southampton
Highfield, Southampton,
    ENGLAND

Terry L. Stentz
University of Nebraska
Lincoln, NE

Brian L. Stonecipher
University of Nebraska
Lincoln, NE

Jatin Thaker
Clemson University
Clemson, SC

Mika Vilkki
Tampere University of Technology
Tampere, FINLAND

Michael S. Wogalter
North Carolina State University
Raleigh, NC

David D. Woods
The Ohio State University
ISE Department
Columbus, OH

Thomas Y. Yen
Department of Industrial
    Engineering
University of Wisconsin
Madison, WI

Mark S. Young
Department of Psychology
University of Southampton
Southampton, ENGLAND

Stephen L. Young
Liberty Mutual Research Center for
    Safety and Health
Hopkinton, MA

Contents

Part I

Background

1

Ergonomics: An International Perspective

Hal W. Hendrick
University of Southern California

1.1 Introduction

In comparison with other disciplines, the science and practice of ergonomics, or human factors as it is also known in North America, is very young. Ergonomics does, however, have at least several distinct roots going back to the early 1900s. For example, in the United Kingdom, we can note the classic work of the Industrial Fatigue Research Board between World Wars I and II. These studies greatly contributed to our understanding of environmental effects on human work performance. In the United States, the development of "scientific management" by Frederick W. Taylor and "industrial psychology" by Hugo Munsterberg and others are among the factors that have contributed to the formation of ergonomics as a distinct discipline. Russia has distinguished histories in psychology, mathematics, and engineering that have helped shape the field of ergonomics. Similar precursors of modern ergonomics can be found in a number of other countries (Hendrick, 1993).

Although its roots go back to the early 1900s, ergonomics, as an identifiable profession, is in only its sixth decade of existence. Since World War II, we have seen the field of ergonomics develop from a few hundred individuals, working on military systems in several industrialized countries, to over 25,000 professional ergonomists working in both industrialized and industrially developing countries throughout the world. These persons, in turn, are supported by thousands of scientists from diverse fields who help develop ergonomically useful data, methods, and related technology. Professional ergonomists now work on a wide variety of systems, ranging from simple hand tools to highly complex equipment, software, and environments. Professor Brian Shackel, the International Ergonomics Association (IEA) historian, has noted the international developmental focus of ergonomics as follows: In the 1950s, military ergonomics; in the 1960s, industrial ergonomics; in the 1970s, ergonomics of consumer goods and services; and in the 1980s, computer ergonomics. The 1990s are proving to be the era of both macro- and cognitive ergonomics, with a major focus on their application to industrial systems. In the 1990s, we also have seen the further maturation of ergonomics into a distinct, stand-alone discipline (Hendrick, 1993).

1.2   The IEA and Its Member Societies

One of the early indications of the development of a clearly definable field of science and practice is the formation of professional societies. The first professional society formed in the field of ergonomics was the Ergonomics Research Society in the United Kingdom in 1949. This society later was renamed The Ergonomics Society, which is how we know it today (IEA, 1996).

The IEA

The idea of forming an international society in ergonomics was first formally put forward at a technical seminar of the European Productivity Agency (EPA) in Leiden, Holland, in 1957. (The EPA, a body working under the auspices of the Organization for European Economic Cooperation, had established a working party on ergonomics in 1955.) The idea was further developed at the meeting of the EPA Steering Committee in Paris, France, in 1958. At a meeting of the Steering Committee in Zurich, Switzerland, in 1959 the name "International Ergonomics Association" was formally adapted. Later in 1959, at an EPA Steering Committee Meeting in Oxford, England, the bylaws drawn up by Etienne Grandjean, Secretary of the Steering Committee, were approved. The first IEA Triennial Congress was held in Stockholm in 1961 (IEA, 1996)

From this beginning, the IEA has steadily grown. Prior to the 1980s, the primary activity of the IEA was to sponsor the triennial congresses. Structurally, the IEA consisted of a Council, comprised of representatives of the federated member societies, and an Executive Committee consisting of the three elected officers: president, secretary general, and treasurer. By 1980, it was clear that ergonomics was entering a phase of rapid growth and expansion internationally. It also was clear that there was a growing need to increase the opportunities for international coordination and exchange of technical and educational information. To meet this need, the IEA began to greatly expand its activities. During the past decade, the IEA developed an expanded committee structure to plan and manage these increased activities; and a series of specialized technical groups which, among other activities, participate in organizing international conferences, symposia, and workshops in their specialty areas.

The IEA now consists of 33 federated and 2 affiliated societies representing over 45 countries. Approximately half of these societies became members between 1991 and 1995. Still other societies, primarily from industrially developing countries, are in the process of forming. Put simply, ergonomics is a young but rapidly expanding science and profession.

The IEA Federated Societies

Most of the 35 IEA federated and affiliated societies now fall into three size groupings: Approximately two thirds have from 30 to 250 members each. Another 20% each have from 450 to 650 members. Three societies fall between 900 and 2000 members; these are the Ergonomics Society (933), the Nordic Ergonomics Society (a federation of the four Nordic country societies, 1497), and the Japan Ergonomics Research Society (1862). The Human Factors and Ergonomics Society in the United States has approximately 4000 members. All told, the IEA federated and affiliated societies are comprised of approximately 14,500 full and associate members (IEA Treasurer's Report, 1995). Including other membership categories (e.g., student members and affiliates) would increase this number to approximately 18,000.

In a recent survey among 25 IEA federated societies (Brown, Noy, and Robertson, 1995; Helander, 1996), the societies estimated that an average of 61.5% of the eligible ergonomists belong to their society. This would indicate that there are over 20,000 ergonomists within the countries represented by the IEA federated societies. Adding the ergonomists living in countries that do not yet have established IEA federated societies would raise this figure to over 25,000.

1.3 The Technology of Ergonomics

A second, and perhaps the most important factor that defines a profession as a unique, stand-alone discipline is its unique *technology*. To the extent that a given professional area, though empirical science, has developed a technology and, as a practice, applies that technology to some end, it distinguishes itself as a unique discipline. Over the five decades of its formal existence, ergonomics has progressively developed and applied its unique technology. That technology may be thought of as *human–system interface technology* (Hendrick, 1991). As a *science*, ergonomics is concerned with developing knowledge about human performance capabilities, limitations, and other characteristics as they relate to the design of the interfaces between people and other system components. As a *practice*, ergonomics concerns the application of human–system interface technology to the *analysis, design,* and *evaluation* of systems to enhance safety, health, comfort, effectiveness, and quality of life. At present, this unique technology has at least four identifiable major components: Human–machine interface technology or *hardware ergonomics,* human–environment interface technology or *environmental ergonomics,* human–software interface technology or *cognitive ergonomics,* and human–organization interface technology or *macroergonomics.* A brief overview of the historical development, present state, and apparent future of each of these four technical areas follows (Hendrick, 1993).

1.4 Ergonomics Technology and Its Application

Hardware Ergonomics or Human–Machine Interface Technology

Originally known as man–machine interface technology, this technology represents the focus of ergonomics during the first three decades of our profession. It primarily concerns the study of human physical and perceptual characteristics, and the application of that knowledge to the analysis, design, and evaluation of controls, displays, and workspace arrangements. It remains today as the single largest aspect of our profession and is likely to remain so in the future.

In the United States, this aspect of ergonomics began with attempts of engineering psychologists to explain why so many military aviation accidents were being attributed to "pilot error" during the early years of World War II. In general, these psychologists found that the real cause was not "human error" but "engineering error." In short, design engineers had failed to adequately take into account "human factors" (i.e., human performance capabilities and limitations) in designing the aircraft human–machine interfaces, and thus, had inadvertently incorporated error-producing characteristics into their designs. The outcome was the birth of the new field of "human factors" in North America. In Europe, several countries (most notably, Germany) were also concerned with human performance capabilities as applied to design (e.g., design of gun sights; white versus red aircraft cockpit lighting). However, it was after the war that the development and application of human–machine interface technology really began in earnest. Faced with the tremendous task of rebuilding their industries, Europe and Japan became concerned with developing and applying a science of work or "ergonomics."

During the 1950s, and 1960s, the primary applications of this rapidly emerging human–machine interface technology were to factory work systems and to transportation systems, with particular emphasis on aircraft design. The primary funding in the U.S. and other countries for the development of human–machine interface technology came from military research and development budgets. Beginning in the late 1960s, we experienced an expansion of hardware ergonomics to many other types of systems. In part, this expansion was a result of the development of computers and related automation of offices and factories. In the future, world competition is likely to be a prime motivator of greater attention to hardware ergonomics in both workstation design (for greater worker effectiveness and quality of production) and product design (as differences in the adequacy of ergonomic design are likely to be the major discriminators between economically successful and unsuccessful products).

Environmental Ergonomics or Human–Environment Interface Technology

The second ergonomics technology is concerned with human capabilities and limitations with respect to the demands imposed by various environmental modalities (e.g., light, heat, noise, vibration, etc.). It is applied to the design of human environments to minimize environmental stress on human performance — including comfort, health, and safety — and to enhance productivity.

One of the main roots of environmental ergonomics can be traced back to the work of the British Industrial Fatigue Research Board in the early 1900s, which carried through the 1930s. In terms of contemporary environmental ergonomics, significant research was begun in the 1960s, at such places as Aston, Loughborough, Wales, and Birmingham universities in the United Kingdom, and various Department of Defense, NASA, and university (e.g., Cornell) research units in the U.S. During this same period, parallel human–environment technology research, development and application was ongoing in other western European countries, Japan, the U.S.S.R., Australia, and elsewhere.

During the last several decades, the importance of understanding the relation of humans to both their natural and man-made environments has gained increasing focus internationally, and a related ergonomics technology is continuously developing. There has even been the development of an ecological approach to human performance modeling as well as to classic ergonomics methods such as task analysis (Vicente, 1990).

Recently, there has been a growing trend toward more stringent ergonomics-related health and safety legislation in both Europe and North America. A progressively increasing international awareness of the importance of ecological issues to human health and effectiveness will ensure the continued expansion of human–environment technology research and application throughout the world.

Cognitive Ergonomics or Human–Software Interface Technology

The third technology is a relatively new development in our profession, having come into being with the development of the silicon chip in the 1960s, and the modern computer revolution which followed. Because this relatively new technology is primarily concerned with how people conceptualize and process information, it is often referred to as "cognitive ergonomics." The major application of this technology is to the design or modify system software to enhance its usability. In the United States, the development of software/cognitive ergonomics has resulted in an increase of approximately 25% in the number of ergonomists, as reflected by the increase in HFES memberships. For example, over 900 of the approximately 4000 HFES members belong to the Society's Computer Systems Technical Group (HFES, 1996). A similar growth appears to have happened in many other countries.

Because of the continuing growth in software technology and application, and the growing realization of the importance of human–software interface technology to effective software design, cognitive ergonomics will continue to be a strong growth area within our profession. The related interest in the technology of artificial intelligence (AI) and its high potential for expert systems development will further fuel the fire under this part of ergonomics development and growth.

Macroergonomics or Human–Organization Interface Technology

Macroergonomics is the newest part of our profession. The central focus of the first three technologies has been the individual operator and operator teams or subsystems. Thus, the primary application of these technologies has been at the microergonomic level. In contrast, because it deals with the overall structure of the work system as it interfaces with the system's people and technology, the human–organization aspect of the human–system interface tends to be *macro* in its focus; hence, it is referred to as "macroergonomics." *Conceptually*, it is a top-down, sociotechnical systems approach to organizational and work systems design, and the design of related human–machine, human–environment, and human–software interfaces (Hendrick, 1986a,b, 1987, 1991).

For many years, organizational factors occasionally have been considered in ergonomics research and practice; but macroergonomics, as a formally recognized area of ergonomics, grew out of a study of future ergonomics needs by the Human Factors Society which was completed in 1980 (Hendrick, 1991). This study noted such factors as (a) rapid changes in technology, (b) changing value systems, (c) demographic shifts (a greying work force), (d) increasing world competition, and the failure of traditional microergonomics to (e) improve overall *system* administrative productivity, (f) reach potentially achievable production quality, and (g) system safety, health, and related quality of worklife goals as underlying drivers of the need for developing and applying a human–organization interface technology.

Since its formal inception a decade ago, macroergonomics has experienced rapid growth throughout the world. Research over the past decade has shown organizational design and management variables to be critical to effective work system design. Anecdotal reports have shown the potential for effective macroergonomic design and follow-through microergonomic design of systems to greatly improve productivity, safety, health, and quality of worklife. For example, in a series of industrial studies in both Japan and the U.S., Nagamachi and Imada (1992), using macroergonomic approaches, achieved reductions in both industrial injuries and motor vehicle accidents of from 76% to over 90%. A field study at L.L. Bean Corporation in the U.S. used a true macroergonomic approach in a total quality management (TQM) change effort in both their production and distribution units, and achieved over 70% reduction in lost time injuries (Rooney, Morency, and Herrick (1993). Ongoing macroergonomic research, such as that on information systems by HUSAT at Loughborough University in the U.K. and on factory automation by Ann Majchrzak and her colleagues at the University of Southern California, offer equally exciting possibilities. Given its potential, macroergonomics should experience an exponential growth during the next several decades.

The Future

Based on the above, all four major technology areas of ergonomics can look forward to a very bright future throughout the world — one characterized by very healthy expansion and growth. In fact, in terms of potential development and growth, the future for ergonomics appears brighter than for any other design-related field except, possibly, biomedical engineering. In any event, it is brighter than for most other design and engineering fields which, while much larger than ergonomics, are likely to see only modest growth in the foreseeable future. The rapid development of computer-based tools in ergonomics will further enhance the discipline's capability and growth.

1.5 Variations Among Countries

Historically, "human factors," as practiced in North America, and "ergonomics," as practiced in Europe and elsewhere, differed noticeably in both their approach and emphasis. Human factors placed relatively more emphasis on the study of human psychological and perceptual characteristics; and on the laboratory research approach to acquiring human factors knowledge and developing human factors technology. Ergonomics, on the other hand, placed more emphasis on the study and application of knowledge about human physical characteristics, including work physiology and biomechanics; and on the field study approach to acquiring ergonomics knowledge and developing ergonomics technology. Historically, the primary application of "human factors" technology was to moving systems, with particular emphasis on aircraft design. In contrast, ergonomics placed more emphasis on the design of factory jobs and work-stations. During the last several decades, both classical "ergonomics" and "human factors" have moved toward one another in terms of emphasis, research methodology, and application. As a result, for well over a decade, those who identify as "human factors professionals" and those who identify as "ergono-mists" have been doing essentially the same kinds of things. In short, the two have become synonymous; and for well over a decade the IEA has formally recognized the two as identical.

Having noted the above, it is still important to mention that national differences *do* exist. Most easily recognizable are national differences in the root professions in which the practicing ergonomists were trained (Hendrick, 1989). These differences, in turn, also reflect proportional differences in the kinds of things ergonomists do in different countries. At the risk of over-generalizing, there appear to be two major categories of root training and related applications that can be found internationally.

In the United States, the U.K., Germany, Japan, and many other countries, the root education of the majority of ergonomists is in industrial, systems, or mechanical *engineering* and industrial or applied experimental *psychology*. Often, ergonomists have studied in both fields. For example, in the United States, ergonomists may have a formal major in one and a minor in the other of these two fields. Ergonomics education historically has been offered as a specialty within both engineering and psychology. In comparison with countries in which this is *not* the general training pattern, a proportionately greater emphasis tends to be given to the design of moving vehicles, command and control systems, and consumer products to improve operational effectiveness, usability, comfort, and safety.

In other countries, such as Australia, Denmark, and Italy, the root education of the majority of ergonomists tends to be medical or medical related. In particular, ergonomists have been trained as physical therapists, occupational therapists, industrial hygienists, or as physicians. In comparison with countries having a predominantly engineering and psychology ergonomics training pattern, a propor-tionately greater emphasis is given to applying ergonomics to the workplace, tending to be more work physiology and biomechanics oriented and having a primary focus on preventing injury in the workplace.

Having noted these differences, it is important to recognize that these differences are *proportional* rather than *absolute*. The full range of educational backgrounds is found in virtually all countries with a developed ergonomics profession. Similarly, the full range of ergonomics applications is found within these countries.

More recently, we are seeing the development internationally of stand-alone professional education programs in ergonomics (Pearson, 1994). Given the maturation of human–system interface technology, this trend toward separate ergonomics professional degree programs should continue. What is common throughout the world is that ergonomics involves the development and application of *human–system interface technology*, and therein lies the identity of our profession internationally. It is this common activity that provides us with our basis for developing internationally harmonized certification criteria and educational guidelines.

1.6   Core Competencies in Ergonomics

One of the major tasks of professional ergonomics societies throughout the world has been, and continues to be, to identify the core competencies required for the practice of ergonomics. This is not surprising, because knowledge of these competencies is essential to determining certification criteria and professional education requirements. For example, a major national study was conducted around 1980 in the U.S. for the Air Force in particular and the Department of Defense (DoD) in general. It is the most extensive study of its kind ever conducted in the field of ergonomics. The purpose of the study was to analyze and evaluate all facets of the development and application of ergonomics (or what then was called human factors engineering) technology to DoD systems, with particular emphasis on how ergonomics is inte-grated into the system development process (Hendrick, 1981).

One of the major purposes of the study was to identify what human factors practitioners actually *do* in applying their technology to the development of a broad spectrum of systems, tools, jobs, facilities, and environments. These data, obtained for literally hundreds of ergonomists, were analyzed to determine what core competencies were actually required to perform the ergonomics tasks. The competencies identified were as follows:

1. Sufficient background in the behavioral sciences to respond to ergonomics questions and issues having psychological or other behavioral implications. Implies the equivalent of a strong undergraduate behavioral science minor.
2. Sufficient background in the physical and biological sciences to appreciate the interface of these disciplines with ergonomics. Implies the equivalent of an undergraduate minor.
3. Sufficient background in engineering to (a) understand design drawings, electrical schematics, test reports, and similar design tools, (b) appreciate engineering design problems and the general engineering process, and (c) communicate effectively with design engineers. Implies formal knowledge of basic engineering concepts at the familiarization level (i.e., at least two undergraduate level courses, or equivalent, in engineering).
4. Ability to evaluate the adequacy of applied ergonomics research and the generalization of the conclusions to operational settings. Requires formal knowledge of the basic statistical methods and principles of experimental design. Implies the equivalent of two introductory level graduate courses in statistics and research design.
5. Ability to (a) evaluate and (b) conduct the various kinds of traditional ergonomics analyses (e.g., functional task, time-line, link). Requires formal training in these techniques.
6. Ability to (a) evaluate and (b) perform classic human–machine integration including workspace arrangement, controls, displays, and instrumentation. Implies formal knowledge of ergonomics human–machine interface technology at the introductory graduate level or equivalent.
7. Ability to apply knowledge of human performance capabilities and limitations under varying environmental conditions in (a) evaluating environmental design and (b) developing environmental design requirements for new or modified systems. Requires formal knowledge of human performance capabilities and limitations in the various physical environments (e.g., noise, vibration, thermal, visual). Implies formal knowledge of environmental ergonomics at the introductory graduate level, or equivalent.
8. Sufficient knowledge of computer modeling, simulation, and design methodology to appreciate their utility in systems development, including ergonomics utilization in function allocation, task time-line analysis, workload analysis, workstation layout evaluation, and human performance simulation (this does not include the ability to actually design the models and simulations). Requires math through calculus and introductory computer science, and knowledge at the familiarization level of measurement, modeling, and simulation because these are applied to ergonomics.
9. Ability to apply knowledge of learning and training methodology to the evaluation of training programs and to instructional systems development (ISD). Requires knowledge of that portion of learning theory and research applicable to training, and of training methodology at the familiarization level.
10. Ability to assist in the development and evaluation of job aids and related hardware. Requires knowledge of the state of the art at the familiarization level.
11. Ability to apply the organizational behavior and motivational principles of work group dynamics, job enrichment and redesign, and related quality of worklife considerations in (a) developing ergonomic system design requirements and (b) evaluating the design of complex systems. Requires formal knowledge of organizational theory and behavior at the introductory graduate level (i.e., what today, somewhat broadened and deepened in scope, is called macroergonomics).
12. Specialized expertise in at least one area of human–system integration technology that goes beyond the introductory graduate level of understanding and application. Requires additional ergonomics course work and a tutorial research or thesis project at the graduate level, and an M.S. degree or equivalent in ergonomics or a closely related academic discipline (e.g., engineering, psychology, safety science, physical therapy).

Note that the above analysis was completed a decade and a half ago. Today, we undoubtedly would add knowledge of software or cognitive ergonomics at the familiarization level to this list.

The above core requirements are just that. They represent what the committee concluded a fully qualified professional ergonomist practitioner, *ideally*, should have. In addition, the project team concluded from its study that at least two years of supervised practice experience was necessary to become a fully qualified professional ergonomist. Although not a formal conclusion of the study, project team members noted that additional unsupervised experience should be required for professional certification, if a certification program were to be implemented.

In reality, very few North American certifiable ergonomics practitioners would meet every single one of the above criteria. However, based upon the first several years' experience of the Board of Certification in Professional Ergonomics (BCPE) in evaluating U.S. and Canadian applicants, they would meet at least seven or eight out of the first eleven, at least minimally, and the twelfth criterion fully. In large part, failure of most ergonomists to meet all of the core requirements is a reflection of their education. Put simply, most have not graduated from true ergonomics programs but from degree programs in related fields with a few ergonomics courses thrown in. In time, with clearly established educational guidelines or accreditation programs reinforced by ergonomics certification programs, professional ergonomics education internationally should become better rounded and the number of true ergonomics graduate degree programs should increase. This will result in many more practicing ergonomists meeting all, or most, of the core criteria.

1.7   Professional Standards

A third indication of the development and maturation of a discipline to a unique, stand-alone status is that of establishing formal standards for professional competency (certification or registration) and education programs (either formal guidelines or accreditation). These kinds of standards protect the public, enhance the stature of the profession, and ensure its continuing viability. In ergonomics, major efforts now are well under way to develop these professional standards and guidelines. Because it is a relatively small profession, it is especially important to harmonize these standards internationally if ergonomics is to have a distinct identity; and the IEA is proactively facilitating this process.

Professional Certification/Registration

Professional registration of ergonomists is not new. For example, the Ergonomics Society in the U.K. has had such a program for a number of years. What is new is the recent, greatly increased interest internationally in professional certification or registration programs, and the development of two major international programs: the Board of Certification in Professional Ergonomics (BCPE), based in North America, and the Center for Registration of European Ergonomists (CREE) in Europe. The BCPE, based on a review of one's professional education, work experience, selected work product(s), and the passing of a three-part, comprehensive basic knowledge and practice written examination, certifies persons from any country as either a "Certified Professional Ergonomist" (CPE) or "Certified Human Factors Professional" (CHFP). The CREE, using similar evaluation criteria but without a written examination, certifies persons as a "European Ergonomist" (Eur. Erg.) who reside or work in the "European Ergonomic Space" (EES). Both certification programs were developed following an extensive review of the literature on ergonomics and ergonomists, and independently arrived at a similar conception of the basic requirements for functioning as a professional ergonomist practitioner. Ergonomists in other areas of the world are also exploring the development of professional certification/registration programs.

Professional Educational Standards and Accreditation

The development of guidelines and standards for professional education programs in ergonomics is also being actively explored by many ergonomics societies and the IEA. Because of the broad number of disciplines in which ergonomics specialization is offered internationally, harmonizing educational standards internationally is proving to be a more difficult problem, but progress is being made. The IEA sponsored a major conference on professional standards in conjunction with financial support from, and hosting by, the Italian Ergonomics Society in Palermo in 1992. This was followed by a second conference in conjunction with the IEA Triennial Congress in Toronto, Canada, in 1994. A major aim of both conferences was to work toward international harmonization of ergonomics education standards. To date, the most developed professional education standards program is the accreditation program of the Human Factors and Ergonomics Society in the United States. This program was developed over approximately a ten-year period. Thus far, approximately 15 of the 60+ human factors/ergonomics degree programs, and specialty programs within related degree programs in the United States have been accredited.

1.8 International Trends in Ergonomics

Employment of Ergonomists

Based on a survey of 25 IEA federated societies, 29% of their members are employed in educational institutions as academics, 27% practice in industry, 15% are researchers, 10% are private consultants, 8% work for the government, and 11% are employed in other occupations (Brown, Noy, and Robertson, 1995; Helander, 1996).

A recent all-member survey of the Human Factors and Ergonomics Society in the United States found the following distribution: 35% work in industry, 23% are consultants, 19% are academicians, and 17% are with the government, including the military services. In terms of primary work focus, 41% do design and development, 32% are in research and/or teaching, 14% in safety assurance, and 10% are in management or administration. Only 2% are in medical rehabilitation.

In terms of educational level, 84% hold an advanced degree (46% Ph.D., 38% masters). Of these, 41% are in psychology, 24% in engineering, and 17% in human factors or ergonomics. (Hendrick, 1996)

Major Problem: Industry, Government, and Public Awareness

The recent IEA survey of 25 member societies found the major problem facing the ergonomics profession was a lack of awareness and recognition by industry and government decision makers and the public in general (Brown, Noy, and Robertson, 1995; Helander, 1996). The recent HFES all-member survey further confirms this and indicates that a more proactive effort is needed by the society to raise the consciousness of industry and government decision makers as to the value and cost benefits of ergonomics (Hendrick, 1996).

Most Important Early Versus Current Applications

Based on the survey of 25 IEA federated societies (Brown, Noy, and Robertson, 1995; Helander, 1996), there has been a significant shift in the most important areas of ergonomics since the societies were founded as compared with today. Table 1.1 shows those comparisons for the five most important areas. Note particularly the emergence of safety, workload, and human computer-interaction (HCI) as among the most important application areas internationally. Implicit in this shift is the resurgence of industrial ergonomics and the impact of computers and automation on human performance and work.

Emerging Areas of Ergonomics

Of equal or greater importance, the IEA survey also identified the emerging areas in ergonomics in the countries represented by the 25 societies surveyed. The most frequently cited of these are shown in Table 1.2 (Brown, Noy, and Robertson, 1995; Helander, 1996). Note particularly the emergence of organizational design, psychosocial work organization, and related ergonomics methodology as, perhaps, the major emerging new area of ergonomics. Reflected here is the growing recognition that a more macro-

TABLE 1.1 The Five Most Important Early and Current Applications
of Ergonomics in 25 Ergonomics Societies

Importance	Early applications	Current applications
1	Anthropometry	Safety
2	Work physiology	Industrial engineering
3	Industrial engineering	Biomechanics
4	Biomechanics	Workload
5	Psychology	Human–computer interaction

TABLE 1.2 Important Emerging Areas of Ergonomics
in 25 Ergonomics Societies Around the World

Topics	Frequency
Methodology to change work organization and design	7
Work-related musculoskeletal disorders	7
Usability testing for consumer electronic goods	6
Human–computer interface software	6
Organizational design and psychosocial work organization	5
Ergonomic design of physical work environment	4
Control room design of nuclear power plants	3
Training ergonomics	3
Mental workload	3
Workforce cost calculation	3
Product liability	2
Road safety and car design	2
Transfer of technology to industrially developing countries	2

ergonomic, systems approach often is needed to effect major improvements in productivity, health, safety, and quality of worklife in complex sociotechnical systems.

Note also the international concern over preventing work-related musculoskeletal disorders, and the continued growth of HCI software design or cognitive ergonomics. Equally important is the emergence and growth of usability testing (and other methods of employee participation) as a major ergonomics methodology. This also is a reflection of another major trend in ergonomics internationally — the shift from technology-centered ergonomics to a more human-centered ergonomics. All of these emerging areas are highly relevant to the science and practice of industrial ergonomics.

1.9 Managing the Development of Ergonomics

As with any maturing organization or profession, as it develops and expands it will experience growing pains. When this happens, more attention must be paid to actively *managing* that growth and development. With respect to managing the development of the ergonomics discipline internationally, at least four major facets have been identified (Hendrick, 1993): (1) the orientation of ergonomics practice, (2) professional standards, (3) development of a more adequate database for predicting human performance, and (4) supporting ergonomics development in industrially developing countries. More recently, the importance of documenting the cost benefits of ergonomics applications and ensuring that this information is shared with organizational and government decision makers has been highlighted (Brown, Noy, and Robertson, 1995; Helander, 1996; Hendrick, 1996).

Orientation of Professional Practice

Historically, the actual practice of our profession has tended to be technology oriented. New system design or modification often is motivated by the desire to exploit new developments in technology. The ergonomist, if consulted at all, is called in after selection of the technology to "fit the human to the machine." In effect, the method of function and task allocation most frequently employed throughout the world is the leftover approach (i.e., whatever cannot be done by the machine is assigned to humans) (Bailey, 1989). When compared to systems designed without any ergonomic attention, this approach has proven effective in reducing accidents and improving human comfort and effectiveness. Not surprisingly, this success has further encouraged continuation of the technology-oriented approach to employing ergonomics in system design.

There are at least two major problems with a technology-oriented approach to ergonomics. First, it is *reactive* rather than *proactive*. The ergonomist's role becomes one of reacting to the system's hardware once it has already been selected or designed. As a result, the ergonomist's potential for enhancing total system performance is severely limited. Second, history repeatedly has shown this approach to lead to *suboptimization* of work system design, including poor utilization of human capabilities and a failure to incorporate well-known intrinsic motivational factors into jobs (e.g., see Hackman and Oldham, 1975); in short, to a relatively dehumanized work system (Hendrick, 1994).

What has become increasingly evident over the past decade is that we must proactively adopt a more *human-centered* approach to ergonomic practice. Whenever possible, we should begin the function analysis process by *justifying* the use of a human, and then selecting or adopting technology as required to *assist* the human in accomplishing system objectives (Bailey, 1989). This kind of approach to our professional practice transforms our role from one of *reactive advisors* to that of *proactive change agents*. As proactive change agents, the effectiveness of ergonomists can be greatly increased, and the end result will be many more humanized and productive work systems than presently exist in any country.

Professional Standards

The second area where proactive management of ergonomics is required is in the area of professional standards for (1) our academic programs (accreditation), (2) ensuring competency in ergonomics practice (certification), and (3) prescribing how we conduct our business (code of practice). As already noted, these kinds of standards mark the maturation of any profession and are well along in development internationally. Most recently, the IEA developed a draft code of practice for use as guidance by national and regional ergonomics societies in developing their own codes.

Development of a More Adequate Data Base

As Meister (1992) has noted, we need to be able to predict human performance, because this is the essence of ergonomics. Meister further notes that "the ability to predict is power. What design engineer could refuse our advice if he knew that it would cost him heavily in terms of the performance of his system? But to achieve this power, to achieve the credibility that will induce the designer to follow our advice, we need numbers. ...To be able to predict, to develop performance standards, we need a database, and this is much more than what they mean when most ergonomists use the term; it is much more than a series of studies plucked from journals. A database means extracting the raw data formally, combining data from various sources, relating the human performance values to the tasks performed, the system in which the performance occurs, performance-shaping factors, etc. Above all it means transforming the raw data into error probabilities and relating these to design variables" (Meister, 1992, pp. 259-260). Because ergonomics is a numerically small discipline, achieving the kind of database Meister envisions will require a *managed*, collaborative systems effort internationally.

A possible first step in this direction was recently taken by the Human Factors Committee of the National Research Council in the U.S. This group has completed a comprehensive project to identify the research needs in human factors/ergonomics. The report of the study's findings has been published by the National Research Council. Hopefully, this report will help provide a basis for establishing and managing an international ergonomics effort to develop a more adequate ergonomics database. The challenge, of course, will be to identify the necessary funding sources and to develop a management structure for this much needed but highly ambitious research effort.

Supporting Ergonomic Development in Industrially Developing Counties

Perhaps the single greatest weakness in the application of the current ergonomics knowledge base is that well over half of the world's population does not benefit from it, or does so in a very limited way. This is particularly critical in the area of industrial ergonomics for improving health and safety.

The IEA has several projects under way to help support ergonomics growth and awareness in industrially developing countries. These include IEA roving ergonomics seminars and the recently published IEA–ILO developed manual, *Ergonomics Checkpoints,* published by the International Labor Organization (ILO). *Ergonomics Checkpoints* contains over 130 common ergonomic applications, including step-by-step instructions and illustrations on how to apply ergonomics principles, guidelines and specifications. It is designed specifically for use with only limited ergonomics training and, in the IEA roving seminars, has proven to be an excellent "textbook" for providing basic ergonomics instruction — particularly for industrial applications.

The IEA effort is admirable but represents only a limited step in managing and enhancing the development of ergonomics in the industrially developing portions of the world. National and regional ergonomics societies, international and national government agencies, etc., also need to be tapped for assistance.

1.10   Conclusion

In summary, the future of ergonomics internationally has never been brighter. If ergonomists around the world properly and proactively manage their profession's development, the profession not only will greatly expand in numbers, but its contribution to improving the human condition will increase by at least several orders of magnitude.

References

Bailey, W. (1989). *Human Performance Engineering,* Second Edition, Prentice-Hall, Englewood Cliffs, NJ.

Board for Certification in Professional Ergonomics (1992). *BCPE* (brochure). Bellingham, WA: Board for Certification in Professional Ergonomics.

Brown, O. Jr., Noy, I, and Robertson, M. (1995). Special survey of IEA federated societies. International Ergonomics Association, c/o Human Factors and Ergonomics Society, Santa Monica, CA.

Hackman, J. R. and Oldham, G. (1975). Development of the job diagnostic survey, *Journal of Applied Psychology,* 159-170.

Helander, M. G. (1996). The human factors profession, in G. Salvendy (Ed.) *Handbook of Human Factors and Ergonomics,* Second Edition (pp. 3-15). New York: Wiley.

Hendrick, H. W. (1996). All member survey, preliminary results. *Human Factors and Ergonomics Society Bulletin, 39,* 1, 4-6.

Hendrick, H. W.(1993). The IEA and international ergonomics: Past, present and future. In O. Brown, Jr. (Ed.), *Ergonomics in Russia, the other independent states, and around the world,* Volume 2 (pp. 1-12). St Petersburg: Russian Ergonomics Association.

Hendrick, H. W. (1991). Human factors in organizational design and management. *Ergonomics, 34,* 743-756.

Hendrick, H. W. (1989). Human factors/ergonomics societies around the world: characteristics and issues. *Human Factors Society Bulletin, 32,* 8-10.

Hendrick, H. W. (1987). Organizational design, in G. Salvendy (Ed.), *Handbook of Human Factors* (pp. 470-494). New York: Wiley.

Hendrick, H. W. (1986a). Macroergonomics: A conceptual model for integrating human factors with organizational design. In O. Brown, Jr. and H. Hendrick (Eds.), *Human Factors in Organizational Design and Management-II*. Amsterdam: North Holland, pp. 467-478.

Hendrick, H. W., (1986b). Macroergonomics: a concept whose time has come, *Human Factors Society Bulletin, 30,* 1-3.

Hendrick, H. W., (1981). Engineering education's response to the need for human factors engineers. Presented during the Professional Session on Human Factors in Engineering Education, *89th ASSE Annual Conference*, Los Angeles, CA.

Human Factors Society (1996). *Directory of the Human Factors Society.* Santa Monica, CA: Human Factors Society.

IEA (1996). The History of the International Ergonomics Association, in *Basic Documents of the International Ergonomics Association*, Utrecht, Holland: IEA, pp. 1-2.

Meister, D. (1992). Some comments on the future of ergonomics. *International Journal of Industrial Ergonomics, 10,* 257-260.

Nagamachi, M. and Imada, A. S. (1992). A macroergonomic approach for improving safety and work design, in *Proceedings of the Human Factors Society 36th Annual Meeting*. Santa Monica, CA: Human Factors Society, pp. 859-861.

Pearson, R. (Ed.) (1994). *International Directory of Educational Programs in Ergonomics/Human Factors* (Third Edition). International Ergonomics Association, c/o Human Factors and Ergonomics Society, Santa Monica, CA.

Rooney, E. F., Morency, R. R., and Herrick, D. R. (1993). Macroergonomics and total quality Management at L. L. Bean: a case study, in R. Nielsen and K. Jorgensen (Eds.), *Advances in Industrial Ergonomics and Safety V.* London: Taylor & Francis, 493-498.

Vicente, K. J. (1990). A few implications of an ecological approach to human factors. *Human Factors Society Bulletin, 30,* 1-4.

2

The Ergonomics of System Design

David Meister
Human Factors Consultant

2.1 Introduction

It is necessary to define what is meant by *system* and by *design* because these terms may mean different things in various contexts. The term *system* in its most general sense is used to describe the entire spectrum of human-machine equipment in which humans interact with physical objects. Systems vary on a scale of complexity, encompassing very complex systems like aircraft or a manufacturing assembly line. The term also describes somewhat less complex equipment like automobiles or the workstations or subsystems that control higher-level system functioning, such as those in the control room of a process control system like a nuclear power plant (NPP) or the bridge of a steamship. Lower on the system scale is the handheld tool or appliance, like the hammer, shovel, or can opener. In this chapter the term system is used to designate all of these, although the differences among them produce significant differences in personnel requirements and performance. The brevity of this chapter does not permit detailed discussion of this and other topics (but see Meister, 1991, for additional details).

The term *design* is also used very broadly to describe the design of everything from an entire company or facility to the simplest tool. System design is the process in which the developmental ergonomist or *human factors engineer* (HFE) performs his/her tasks as part of a design team. System design is not often a concern of the industrial ergonomist or IE, but should be, because system development of some sort always precedes industrial ergonomics and often determines the specific problem the IE must face. If, for example, the designer of a machine tool designs it improperly, the IE may be faced with a high incidence of muscular trauma and/or accidents. Moreover, one of the functions of the IE is to evaluate and redesign a manufacturing facility when one of these preceding conditions arises (Narayan and Rudolph, 1993). The redesign employs some of the same processes as those in the initial design of the primary system. Whatever the specific characteristics of the item being designed or redesigned, the design processes involved are essentially the same, although certain aspects may be emphasized more in one type of system than in another.

The design of the primary system also influences manufacturing processes which are the special purview of the IE (Dockery and Neuman, 1994). These manufacturing processes also involve system design at two levels: first, of any special machine tools needed to perform assembly, and, second, the design of the total manufacturing facility, e.g., the spatial arrangement of the workstations that comprise it. Many machine tools are quite common, but those developed specifically for the production of the primary system will require design, which should be of interest to the IE.

2.2   System Development

System development is the total life cycle of the system being designed, up to the point at which it is released to the customer (see Meister, 1987a). It has three major functions: *analysis* of the design problem, *solution* of the problem, and *testing* of that solution. System development is *iterative* and progressively more detailed (commonly termed "top-down" design), which means that initial solutions are somewhat general and therefore tentative, and are often refined at more detailed levels. Initial design solutions are often revised after testing, at which point more information is gathered. (this is called "bottom-up" design, see Meister, 1991.)

System design always presents a problem, which begins with the specification of design requirements. The problem is, what is the object to be designed supposed to do, and how? The problem is to create hardware and software that will satisfy design requirements. The problem is made more difficult by the fact that there is more than one possible design solution; there is almost always a number of alternative configuration solutions, from among which the design team will have to select one that is the most desirable. Moreover, the solution has multiple, possibly competing dimensions: performance versus cost versus reliability versus operator ease, etc. For example, a system can be more automated, but automation has a higher cost.

Dockery and Neuman (1994) and Marcotte, Mervin, and Lagemann (1995) have demonstrated that the design process involves more than engineering alone. Moreover, there are several players in the design drama: the design engineer (one or more), who has direct responsibility for the finished product; the manufacturing engineer; the HFE and other specialists, who are part of the design team; engineering management, which makes high-level development decisions based on nonengineering criteria, such as costs; and the user, who may play several roles.

The User

One of the newer developments in system design is the increasing involvement of the *user* in the design process. That comes about in several ways. (1) The user, defined as the customer who has ordered the development of the new system, will be asked at the beginning of the project to provide details of the system mission and its implications for operator performance. From a behavioral standpoint, the design specification is almost always seriously deficient, because operator requirements are lacking. The user (company management and its experts) will be asked to describe typical behaviors and standards of operator performance. (2) The user will be asked to present a profile of those lower-level personnel who will eventually use or operate the system — characteristics such as age, gender, intelligence, available skills, and handicaps, all of which may have implications (such as the effects of handicaps) for system design (see Scerbo, 1995). The preceding applies primarily to the design of new or heavily upgraded systems; if the system is only a minor upgrade, much of this information should already be available. (3) The user can be asked to provide subjects to test alternative configurations (Dolan, Wiklund, Logan, and Augaitis, 1995). (4) As the customer for the new system, the user will ultimately be asked to approve the completed system.

It need hardly be said that user involvement in the design process will help to avoid gross errors in the ultimate design. Nonetheless, the customer cannot be asked to participate in daily design decisions. Moreover, some customers have only a vague notion of how the genuinely new system is to be utilized. The HFE should ask for as much information from the user as is available and make the best of what s/he receives.

2.3   Design Analysis

The starting point for design is the requirement *specification*, almost always in written and graphic form, and providing more or less detail. Usually the requirement is much more specific in terms of the engineering qualities it demands than it is for behavioral requirements. The requirement will specify such items as power usage or output, a target range, a maximum fuel consumption. One almost never finds this for behavioral elements of the system, mostly because the customer does not feel there is enough information to be able to specify personnel qualities. Because of this, the first task of the HFE is to analyze the requirement to determine those aspects that may affect or constrain operator performance.

As part of this analysis, the HFE must collect data about any predecessor system (e.g., test results, task analyses, performance data) before analyzing the requirements of the new system. Among the factors that the behavioral analyst must consider are: the degree of automation, which determines in large part what the operator must do; the primary human functions the system emphasizes — physical, perceptual-motor or cognitive — and in what way these will ultimately influence hardware or software design; the conditions under which the system (including the operator) must perform; the number and sources of information to which the operator must attend while operating the system; how rapidly the system functions and how quickly the operator must respond; how forgiving the system will be if the operator makes errors; and any anthropometric constraints. For systems that are in common use and have a long history, such as automobiles, the preceding considerations may be somewhat insignificant, but they become significant if the system is new or highly advanced over any predecessor system (e.g., the Apollo space module or a new-generation fighter aircraft). If the IE is asked to participate in the design of a new or modified manufacturing facility, some of the same questions must be asked. What does the worker have to do? What stresses are likely to be imposed by the manufacturing facility? What requirements in terms of speed, physical activity, communication, decision making, etc. are imposed on the worker?

If the system is being designed for the government, the HFE's design analysis will probably be quite formal. The government will probably insist on a formal human engineering plan. This is part of the various procedures set up by the government in programs such as *Manprint* (Booher, 1990). For a system designed for a commercial customer, or one which is for general use by the public, the formal documentation may be at a minimum.

One factor that impacts the design analysis is the molecularity of the actions performed by the operator. If the operator's actions are largely motoric and repetitive, one can make use of the predetermined time systems (PTS) which describe the time required to perform the action. This is feasible in the design of a manufacturing assembly line, but the design of the primary system usually requires the operator to perform at a much more molar level, involving perceptual processes, analysis of system data, and decision making; for these, PTS are not much good. It would be highly desirable to predict the human error associated with each operator action, but here the available predictive data do not exist, particularly as they relate to cognitive and perceptual-motor processes. It is possible to make use of subject matter experts (SMEs) who can rate the error probability of major actions, and this may be all that one can do.

Depending on the nature of the system, the HFE will be particularly concerned with certain aspects of that system and its operator/user. In the highly automated systems of the upcoming 21st century, such as process control plants, the operator will be a monitor and diagnostician and will work primarily with system status symptomology (see Meister, 1996), and so the HFE will be interested in the nature and amount of information the system provides. The characteristics of the operator are also becoming increasingly important to the HFE. For example, the aging process produces reductions in strength, sight, hearing, and speed of response, and these must be considered in evaluating the adequacy of the design. Use of equipment by the physically and mentally handicapped imposes special demands on design. In the design of a home for paraplegics, for example, anthropometric considerations will be primary. If the user is mentally retarded, mental demands posed by the equipment must be considered (Robertson and Hix, 1994).

The starting point of the design analysis is the determination of what the operator is required to do by the system, to what degree of precision, and how this requirement will affect the probability of successful task accomplishment (e.g., in terms of potential errors and time to accomplish the task). In a manufacturing facility whose outputs are physical, task accomplishment would be measured in terms of quantity and quality of items produced.

The analysis of what the operator must do proceeds in parallel with the development of alternative design concepts. This is because what the operator is required to do is determined by the design concept. The consideration of alternative design configurations is, in its initial stages, almost always free-wheeling discussion ("brainstorming") by the design team. Only after the initial concept is agreed upon, does the engineer develop formal drawings. The HFE has the dual task of contributing to the design discussion and simultaneously analyzing the effect of the alternative design possibilities on operator performance. If, for example, an engineer proposes a design that requires an operator to move 500 pounds without mechanical aids, the HFE will tell him/her that such a requirement is impossible. In actuality, since design engineers ordinarily recognize such disqualifying conditions, most design concepts do not make obvious extraordinary physical or perceptual-motor demands on operators, but the HFE must be alert to less severe demands (such as mental) that, even if not impossible, may cumulatively reduce the effectiveness of personnel responses.

There are formal task analytic (TA) techniques (see Meister, 1985, and Kirwan and Ainsworth, 1992) that decompose required operator actions into their physical, perceptual-motor, and cognitive elements. Whatever the form of the TA, it always breaks required actions down into three elements: the stimulus to which the operator must respond; the operator's internal processes that produce a response; and the physical response. The relationship among these three determines the amount and type of system demand on the operator to which design must respond.

During the initial brainstorming of design concepts the HFE can perform such an analysis only very grossly. It is only when a small number of design configurations is decided upon as viable system candidates that the HFE can engage in a formal analysis and comparison of these alternatives in terms of behavioral factors. The great difficulty is that the design engineers on the team, not being overly concerned with behavioral aspects, are likely to focus precipitously upon a single design configuration without adequate consideration of these aspects. The HFE's job is to slow them down just enough so that behavioral factors can be adequately considered. Since initial design concepts may be relatively gross, the HFE must utilize creative imagination to fill in details.

This chapter cannot describe, even briefly, the available TA procedures because of space limitations. These procedures make it possible to decompose operator actions in minute detail, and to describe these in both verbal and graphic modes in the sequence in which they occur. Initially TA formats were verbal only; since then, verbal/graphic equivalents such as Operational Sequence Diagrams, Decision-Action Diagrams, and cognitive TA have been developed. There are also analytic techniques to determine the workload imposed on the operator, to determine the optimal arrangement of subsystems or machine units in the work area, and to examine the communications or interactions among team members. The length of time it takes the HFE to perform a TA depends on the size and complexity of the system being analyzed, the amount of detail the analysis requires, and the time permitted for the analysis. The TA for a system as complex as the Atlas ICBM, in whose development this author participated, resulted in a series of volumes that almost filled a small room. Engineering, cost and time constraints do not usually permit such an extensive TA, and any TA that can influence design is likely to be rather basic; more detailed ones are more useful in determining training and logistic requirements, as well as operational procedures. These last will not be considered here. TA, like system development as a whole, is an iterative, refining process; as design becomes more detailed, the TA also becomes more detailed.

The culmination of the design analysis process which, depending on system complexity and the novelty of the system, may take days, weeks, or even months, is the selection of a single "best" design configuration and the refinement of the details of that configuration. The selection of the one among several possible configurations the design team will "go with" should ideally be a quantitative decision. There are quantitative techniques for making such a decision (e.g., Sadacca and Root, 1968, Meister, 1985), but these do not consider solely behavioral factors. The latter is only one of the factors the design team and management considers; others are performance output, cost, reliability, manufacturing ease, etc. Moreover, the method of making the design selection may not be formal and quantitative, although it may have quantitative elements. The HFE contributes to the overall decision by providing an analysis of the design alternatives in terms of how each will affect operator performance.

2.4   Design

Some authors divide design into a preliminary design phase, in which analysis is pre-eminent, followed by a detail design phase in which drawings of components and circuitry are produced. These and other categorizations do not mean a great deal in reality, since in every phase there is analysis, design, and test to some degree, and system development as a whole has many feedback loops.

Unless the HFE is also the designer, which is quite rare, s/he acts as the representative of behavioral interests to the design team by evaluating design drawings in terms of their potential impact upon the operator, and by acting as a specialist consultant to the team. The primary task is to review design drawings to ensure that no aspect of the design is likely to lead to inadequate operator performance. One form of review examines the drawing alone by comparing its characteristics with a checklist of characteristics that the design should have. The checklist is likely to be based on such well-known standards as MIL STD 1472 (Department of Defense, 1992) or, in the case of software, Smith and Mosier (1986). This review is performed by the HFE singly and/or with the designer. Another manner of conducting a review is to perform what is called a "walkthrough," which requires both preliminary drawings of the human-machine interface and procedures for system operation. The design team operates the equipment symbolically, examining each step in the procedure with reference to the design drawings to determine whether the operator will be able to perform the step and to note potential problems in that performance. In one form of the walkthrough, someone takes the part of the operator and simulates the action required by motioning at each step to the control on the drawing to be moved and the display to be observed.

The walkthrough is easiest to conduct when system operation is in the form of a step by step procedure but can also be performed even if much of the operator's activity is perceptual (attending to displayed information) and cognitive (analysis of that information). If the operator's role in system operation is to monitor its performance and diagnose its malfunctions, the walkthrough can be couched in terms of the type of malfunction that can arise and how the operator should interpret the symptomology of that malfunction. The focus of such a design review is to determine how difficult or easy it will be to recognize and interpret the symptomology.

As an ergonomics consultant, the HFE is a specialist on the fine points of controls and displays and the human-machine interface generally. The designer *may* ask for recommendations with regard to these; the HFE *should* volunteer his/her opinions.

2.5   Test

Behavioral testing in system development is of two types: that performed to support design (developmental testing) and that performed to verify adequacy of the final design solution (operational testing). Any test may incorporate both aspects, but developmental testing occurs earlier in design, whereas operational testing will always occur toward the close of development (see Meister, 1987b).

It is also necessary to distinguish between research and testing. Research seeks to understand the relationship between system variables and their effect on operator performance. Neither developmental nor operational testing is concerned with such questions, but seeks merely to determine the adequacy of the design and any problems that exist. Only very rarely would system development testing become involved with research on behavioral variables.

Developmental tests are performed to determine (a) which of two or more alternative configurations is best from a behavioral standpoint; (b) if, given a configuration which has already been selected as best, will the operator actually be able to perform effectively with it; (c) what problems, if any, will the operator have with a particular configuration. These questions pertain to both the *operability* and *maintainability* of the system.

It might be supposed that by this time questions of alternative configurations would have been resolved, but these often recur because they are repeated at more detailed levels of design. For example, the individual subsystems to be monitored in a nuclear power plant control room may have been identified early in design, but later the characteristics of the individual workstations in the control room (particularly the information they will present) must be determined; and where alternative configurations of a workstation are possible, testing must decide which is best.

Developmental Testing

The tests that were described previously in the design phase (review of drawings, walkthroughs) were symbolic and conceptual; now actual performance, using subjects, and a physical replication of the system, must be examined. The physical representation of the system is the mock-up or prototype. For some systems, like aircraft, the development of a static or functional mock-up is an accepted means of testing proposed solutions to design problems that arise. This is like a dressmaker's dummy on which changes in dress form can be tried out. In the same manner, if there is a question about which of several design formats or procedures should be accepted, it is possible to test the adequacy of these by requiring subjects to use them under simulated operational conditions. The mock-up can be built of various materials (wood, cardboard, styrofoam) to roughly the same dimensions as the operational equipment will have (machine tolerances are usually not required). The mock-up, although largely static, may have actual controls and displays; these may be instrumented to display operational signals, and the operator may be given the capability of operating controls to produce changes in those displays. Thus, some mock-ups are completely static, others are partially functional, and still others are fully functional (Janousek, 1967). The more functional the mock-up, the more it approaches a simulator, and the more useful are the test results. In attempting to replicate operational performance, procedures used in the mock-up test must also approximate those that will be used operationally. The further along development has proceeded, the closer will the mock-up test replicate operational systems operation. The mock-up test is run as much as possible like the operation of the actual equipment, and data are collected and analyzed just as one would collect operator performance data on actual equipment.

The mock-up test whose purpose is to test the adequacy of software is called "prototyping." This type of test is easier to conduct than that of hardware, because it is easier to write software programs. There are also special software test beds into which the test software can be inserted.

It is easier to conduct a mock-up/prototype test when the system is to be operated in a step-by-step manner. When the system under test has a more flexible, contingent procedure (characteristic of information-driven systems) the instrumentation required to perform the test is more complex and in particular may require the HFE to include various modes of operation and potential malfunctions.

The mock-up or prototype test can be performed in various ways: it can be more or less formal; subjects may be fellow workers or more carefully selected to represent eventual system operators or users; data may be performance measures (e.g., time, errors) or may be judgments made by observers of subject performance. It is presumed that a more formal test with selected subjects, performance data, etc. will be more valid than a less formal test involving only observation and subjective judgment.

What questions can such a test answer? Are required tasks performed successfully (e.g., can a quadriplegic in a motorized wheelchair reach and grasp objects on shelves in a simulated habitat? How long does it take him/her to do so, and with what degree of strain? What significant difficulties are manifested? What changes in design are suggested by the test results? The developmental test enables one to refine the design and also provides assurance that the design selected is the most desirable one.

In attempting to gauge the level of difficulty which the design presents to the operator/user, or to solicit opinions about design adequacy from test subjects, it is necessary to make use of judgmental data, either from the test subject or from an SME observer. This requires the use of subjective test methods, which may be a post-test interview, questionnaire, or rating scale. It is impossible in this chapter to describe how one designs such tools. (Instructions for doing so can be found in Meister, 1985.) Obviously, any questions or judgments regarding difficulties experienced by the test subject or designed to elicit opinions about design adequacy must reflect the system characteristics and operating procedures under test. Examples of interview items, questionnaires, and rating scales can be found in the references provided and can easily be modified to fit the individual test situation. Whether one uses such tools in the developmental test depends on how formal the HFE wishes to be.

Design is iterative because the test results may suggest the need for design revision. However, any performance difficulties found in the test must be sufficiently significant to overcome the inherent resistance of engineers to make potentially expensive modifications in the design.

Operational System Testing (OST)

This has several variations: (1) formal testing in the engineering facility or at a test range, testing which reproduces the operational environment (OE); (2) usability testing conducted in the user's facility during normal operations. The latter is much more formal than the former, because one can have much less control over the OE. The usability test is more often conducted for general commercial systems, such as office furniture and word processors. A third type of OST is what one can call a "redesign test," because it occurs after the system has been accepted and is in operation. This test is performed when difficulties such as excessive muscular trauma or accidents occur (or a need is felt to upgrade performance) in a manufacturing facility.

1. Formal OST

Formal OST immediately precedes the completion of the system and its release to the customer, and verifies that the system complies with the design requirement. To the extent that the customer has specified standards that the OST can demonstrate, it is not difficult to conduct such a test. If the test is to be valid, then the system must be completely operational, that is, it must have not only the same physical characteristics of the system as it will have when fielded operationally, but it must utilize operating and maintenance procedures that would be utilized in routine operations. Subjects who perform in the OST should be either operational personnel themselves or those who have similar characteristics. Note the emphasis on system maintenance as well as operation. In the highly automated systems that are now being developed, the quality of maintenance is as critical as the quality of normal output. Certainly, this is true with regard to operator behavior, since, as was pointed out previously, in such systems the operator is a monitor, diagnostician, and maintenance person when (as is inevitable) the system fails.

If formal operator performance requirements were written into the design specification (although they almost never are), it is possible to compare actual operator performance with the requirement. In the absence of these requirements it may seem as if running an OST would be pointless; but it is in fact quite useful, because, even if formal behavioral standards are not available, SMEs can examine the performance data and comment on whether it represents what should be expected of trained personnel. The performance of operators can be videotaped, and communications (if these are involved) can be recorded.

The same questions that are asked of the mock-up or prototype test are asked also of OST. The criteria for recommending changes as a result of performance discrepancies are much changed, however; only the most serious behavioral inadequacies that threaten system output will warrant consideration. Since the system is about to be released for operational use, proposed changes, except the most minor, are enormously expensive and will be rejected.

2. Usability Testing

In formal OST the system is tested by simulating the OE within the engineering facility; in usability testing the system is transferred to the user and thus becomes part of the OE (Scerbo, 1995). The system must be portable in the sense that it can be taken out of the engineering facility (for example, office equipment). The requirements for usability testing are much the same as those for formal OST. In particular, the user must be willing to use the system for a period of time; where the system requires training of user personnel, these must be given training; and the user must be willing to record the data needed by system developers to evaluate the system.

One does not have the same control over the system in usability testing as one has in formal OST, but whatever performance measures are desired must be collected. If the system is designed to collect performance data automatically, this presents no difficulty; but even a highly automated system may depend on some manual data collection procedures, and the user must be willing to devote a certain amount of time and effort to that collection, e.g., logs. User personnel must be debriefed periodically.

The same questions as those of formal OST are asked in usability testing — Is the system adequate to design requirements? Are any difficulties experienced and changes needed? Is the user satisfied with the system? The degree of user satisfaction with the new system is much more important than in formal OST, because almost all systems in usability testing are commercial systems to be sold generally and in competition with comparable systems. Subjective tools like the interview and the rating scale become especially important.

3. Redesign Testing

IEs are perhaps more familiar with this type of testing, since it is frequently performed in relation to manufacturing but only after performance discrepancies are noted (e.g., high accident rates, muscular trauma incidence). The redesign test (see Schmidt, Petree, and Laughery, 1984, as an example) is essentially a "before and after" test. The situation complained of is analyzed (before) to determine what is causing the problem; then certain changes are introduced (e.g., new office equipment, new machine tools, new procedures); and the system is tested again to determine whether the changes have resolved the problem.

Since redesign testing begins by evaluating the situation, this requires the use of an instrument like task analysis (what is the operator supposed to do?) to determine if the task itself imposes difficulties. Worker operations are observed both directly and by videotaping to aid in diagnosing the situation; the workers themselves are interviewed. The quality of the product produced by the worker is examined, using SMEs to determine whether product quality has been endangered and in what ways.

If the results of this investigation suggest certain design or procedural changes, then the diagnostic phase of the test concludes with redesign and installation of the changes. Following this, performance data are collected (e.g., number of items produced, their quality, worker performance times and errors, number of operator complaints). This is the "after" part of the test. "Before" and "after" data are then systematically compared to determine if the new changes have produced a significant reduction in the problem. Such a comparison should involve statistics of significance of differences (as in Allen and Gerstberger, 1973).

All testing requires substantial periods of time. Management may exert pressure to compress the test period as much as possible, so that the new system can be operational as quickly as possible. Precisely how much time is required depends on the specifics of the new system. The cost in terms of resources, both human and machine, to perform such tests may be high, but without such tests the system will certainly be at risk.

2.6   Conclusions

For reasons of brevity it has been necessary to disregard any consideration of the review of procedures, development of technical manuals, training, and simulator design, any or all of which may become the HFE's responsibility. A detailed list of HFE activities during system development would also include writing analytic and test reports, participating in design reviews, collecting and disseminating data, etc. To be successful, the HFE must be in a continuing relationship with other engineers (e.g., reliability, industrial designers, training) and must be cognizant of difficulties other specialties encounter. Because system development encounters many obstacles involving detours, short cuts, and the repetition of processes, the HFE must be adaptable to accommodate these variations.

The extent to which all the procedures in this chapter are followed depends to a certain extent on the size of the system, equipment, or tool. For example, the design of a new blender may require fewer of these procedures; the design of a control room for a new process control facility will require all of them.

Although industrial ergonomics and HFE logically intersect, there has probably been too little awareness of the other on the part of those who practice each specialty. It would be helpful for each to become more cognizant of the other, because even if the industrial ergonomist does not himself/herself engage in system design, s/he should become more aware of the impact of system design on industrial processes.

Defining Terms

Human factors engineer (HFE): The specialist who, during system development, applies behavioral principles to the analysis, design, and testing of the system.

Specification: The written list of requirements that represent the standards to which the design engineering team performs during system development.

System: Any physical entity operated or used by a human to produce or utilize a desired output.

System design/development: The engineering process by means of which a system is constructed.

User: The management customer who has ordered or will ultimately purchase the system and those employees and the general public who will operate and use the system after it is built.

References

Note: The term "Proceedings" below refers to Proceedings of the annual meetings of the Human Factors and Ergonomics Society.

Allen, T.J. and Gerstberger, P.G. 1973. A field experiment to improve communications in a product engineering department: The non-territorial office, *Human Factors*, 15(5): 487-498.

Booher, H.R. ed. 1990. *Manprint, An Approach to Systems Integration*. Van Nostrand Reinhold, New York.

Casey, S.M., Dick, R.A., and Allen, C.C. 1984. Human factors and performance evaluations of the emergency response information system. *Proceedings*, 225-229.

Clarke, M.M. and Kreifeldt, J.G. 1984. A control room concept for remote maintenance in high radiation areas. *Proceedings*, 230-233.

Department of Defense 1992. MIL-STD-1472. *Human Engineering Design Criteria for Military Systems, Equipment and Facilities*. Washington, D.C.

Dockery, C.A. and Neuman, T. 1994. Ergonomics in product design solves manufacturing problems: Considering users' needs at every stage of the product's life. *Proceedings*, 691-695.

Dolan, W.R., Wiklund, M.E., Logan, R.J., and Augaitis, S. 1995. Participatory design shapes future of telephone handsets. *Proceedings*, 331-335.

Evans, T.E., Jr., Lucaccini, L.F., Hazell, J.W., and Lucas, R.J. 1973. Evaluation of dental hand instruments. *Human Factors*, 15(4): 401-406.

Janousek, J.A. 1970. The use of mock-ups in the design of a deep submergence rescue vehicle. *Human Factors*, 12(1): 63-68.

Kirwan, B. and Ainsworth, L.K., eds. 1992. *A Guide to Task Analysis*. Taylor & Francis, London.

Koch, C.G. and Richardson, R.M.M. 1984. Case study evaluation of color graphics for process control. *Proceedings*, 196-200.

Lowe, B.D., You, H., Bucciaglia, J.D., Gilmore, B.J., and Freivalds, A. 1995. An ergonomic design strategy for the transit bus operator's workspace. *Proceedings*, 1142-1146.

Marcotte, A.J., Mervin, S., and Lagemann, T. 1995. Ergonomics applied to product and process design achieves immediate, measurable cost saving. *Proceedings*, 660-663.

Meister, D. 1985. *Behavioral Analysis and Measurement Methods*. Wiley, New York.

Meister, D. 1987a. Systems design, development, and testing, in *Handbook of Human Factors*, ed., G. Salvendy, pp. 17-42. Wiley, New York.

Meister, D. 1987b. System effectiveness testing, in *Handbook of Human Factors*, ed., G. Salvendy, pp. 1271-1297. Wiley, New York.

Meister, D. 1991. *The Psychology of System Design*. Elsevier, Amsterdam, the Netherlands.

Meister, D. 1996. Human factors test and evaluation in the Twenty-first century, in *Handbook of Test and Evaluation*, eds., T.G. O'Brien and S.G. Charlton, pp. 313-322. Erlbaum, Mahwah, NJ.

Narayan, M. and Rudolph, L. 1993. Ergonomic improvements in a medical device assembly plant: A field study. *Proceedings*, 812-816.

O'Hara, J.M. 1994. Evaluation of complex human-machine systems using HFE guidelines. *Proceedings*, 1008-1012.

Robertson, G.L. and Hix, D. 1994. User interface design guidelines for computer accessibility by mentally retarded adults. *Proceedings*, 300-304.

Sadacca, R. and Root, R.T. 1968. A method of evaluating large numbers of system alternatives. *Human Factors*, 10(1): 5-10.

Scerbo, M.W. 1995. Usability testing, in *Research Techniques in Human Engineering*, ed., J. Weimer, pp. 72-111. Prentice-Hall. Englewood Cliffs, NJ.

Schmidt, J.K., Petree, B.L., and Laughery, K.R., Sr. 1984. The test of a task re-design. *Proceedings*, 829-831.

Smith, S.L. and Mosier, J.N. 1986. *Guidelines for Designing User Interface Software*. Report ESD-TR-86-278. Electronics Systems Division, Hanscom Field, MA.

For Further Information

The Handbook of Human Factors, edited by Gavriel Salvendy, describes every aspect of the discipline. *Research Techniques in Human Engineering*, edited by Jon Weimer, is an excellent applied reference. *Human Factors in Simple and Complex Systems* by Robert W. Proctor and Trisha Van Zandt provides the psychological theory and research which underlies human factors practice.

Further information about various Human Factors specialties can be secured by writing to the Human Factors and Ergonomics Society, Post Office Box 1369, Santa Monica, CA 90406-1369. Telephone 310/394-1811.

3

A Guide to Scientific Sources of Ergonomics Knowledge

Holger Luczak
Institute of Industrial Engineering and Ergonomics
Aachen, Germany

Christopher Schlick
Institute of Industrial Engineering and Ergonomics
Aachen, Germany

Johannes Springer
Institute of Industrial Engineering and Ergonomics
Aachen, Germany

3.1 Generation and Consolidation of Ergonomics Knowledge

Knowledge as a product of scientific discovery grows in the context of people, institutions, organizations, or societies which constitute the special interests of their members. Such a "scientific community" defines a set of rules and procedures to create, publish, discuss, revise, establish, and store information on the one hand, or to reject it on the other. Knowing about these rules can guide the information seeker through the vast amount of scientific information available. Therefore, the ergonomics knowledge generation process can be analyzed separately from the process of knowledge consolidation.

Generation of Ergonomics Knowledge

Analogous to the broad spectrum/scope of ergonomics problems, a similarly broad repertoire of methods is necessary for the generation of knowledge. Beneath specific ergonomics methods, numerous techniques from natural, social, engineering, and human sciences are applied to gather information about "man-at-work." These analytical methods can be assigned to four categories:

- Systematical observations
- Questionnaires, interviews, and surveys
- Physiological measurements
- Physical and chemical measurements

Whereas numerous methods of observing and interviewing are derived from the social sciences, the latter two groups are typical of the approach of natural sciences.

Systematical Observations

Methods of systematical observation can be differentiated according to five criteria (Friedrichs, 1975):

1. Open versus hidden observation: Can the observer (or a technical means, such as a camera or a sensor) be perceived/recognized by the working person?
   If the researcher suspects that the behavior of the working person changes under observation (problem of reactivity of measurement), it is useful to use a hidden method of observation for knowledge acquisition. For ethical reasons, the researcher should inform the working person(s) afterward about the observation, allowing the subject to choose to object to the use of the data gathered. Beneath ethical considerations, numerous legal restrictions have to be taken into account. Thus, hidden observation is a method of minor importance in ergonomics knowledge acquisition.

2. Participatory versus nonparticipatory observation: Does the observer take part in the working situation, or is he or she an external noninvolved person?
   When a researcher doing a field study is employed at a normal workplace in a company, participatory observation is used to avoid disturbance of the working processes and to get more authentic information. In the latter case, one usually combines participatory observation with the hidden approach.

3. Standardized versus nonstandardized observation: Is the observation structure following a standardized scheme or is it more or less explorative and unsystematic?
   The more precisely the hypotheses and goals of the investigation can be formulated and the better the previous knowledge about the object/subject of research available, the more advisable is the use of standardized observation methods due to their better economy in gathering specific information, and in evaluation and interpretation of results.

4. Artificial versus natural situation: Has the situation to be observed been designed solely for the purpose of investigation or does it exist independently of the research project?
   With this differentiation, laboratory and field studies, as well as simulated workplaces (flight simulators, weapon systems simulators, etc.), are taken into account.

5. Self-observation versus observation of the outside world: Is the observer his own subject?
   In ergonomic data acquisition, self-observation is mostly used as one method in conjunction with other methods, for example, an investigated working procedure is performed by the researcher himself to get insights into specific difficulties and problems of execution.
   In the generation of ergonomics knowledge, the open, nonparticipatory observation of the outside world is the predominant method. Open observation does not necessarily mean that the observed persons have to be informed beforehand about the purpose of the investigation. In many cases it seems to be necessary that the real purpose be left unclear to the subjects during the observation in order to avoid results that are influenced voluntarily or involuntarily by the subjects.

Questionnaires, Interviews, and Surveys

These methods can be subdivided into four categories according to the degree of standardization of the question and the respective range of answers:

1. Standardized questions and standardized answers
   The investigation is usually performed in a written manner. Typical representation of this form of knowledge generation is the questionnaire with predetermined answer-categories. The answer-categories can be alternatives (yes/no; right/wrong) or imply a choice on a scale (for example, intensity: no/a bit/somewhat/rather/predominantly/totally; or frequency: never/seldom/sometimes/often).
   A general problem of this type of investigation is the predetermination of all questions and all answer categories beforehand; another problem is the possible misunderstanding of the question by the subject and the coding of possible corrective measures and additional information. Also it is sometimes not obvious, for instance with postal inquiries, who filled in the questionnaire. An advantage is the simple data processing, which can be done automatically by character-recognizing and processing machines.

This type of questionnaire is mostly used in studies of the subjective evaluation of working situations — for instance classifications of stressors and strains by lists of attributes (tiresome, monotonous) to which multistep intensity scales are assigned.

2. Standardized questions and nonstandardized answers

   The investigation is done either as a standardized interview, when a person answers predetermined questions verbally in his/her own words, or in written form with free formulation of answers. The answers can be assigned to categories and classifications afterward. This type of knowledge acquisition seems to be advantageous if the investigator cannot foresee all answer categories. The data processing, however, is more complicated and time consuming.

3. Nonstandardized questions and standardized answers

   This type of investigation is seldom used in the acquisition of ergonomics knowledge. A free-formulated question should be answered by a choice among several figures/drawings/sketches (about alternatives of ergonomic design) or a choice among preformulated statements. Nonstandardized questions are mostly coupled with the verbal presentation/interview.

4. Nonstandardized questions and nonstandardized answers

   This type of investigation, known as free interview or narrative interview is especially used when the prior (*ex ante*) knowledge about the object is scarce and the process of the interview leads to more specific and detailed questions. The evaluation of the information gathered is time consuming and therefore limited to case studies.

In ergonomic knowledge acquisition two other survey techniques are important. *Self-documentation* is used when job items have to be protocolled over a long time. The method can be standardized to different degrees, but it usually pertains to momentary task execution with a time-line protocol. Since the method is very economical for the investigator, it is used mostly in field studies with a broad repertoire of work systems over a long period of time (Frieling and Sonntag, 1987).

The method of *verbal protocols* (thinking aloud) is primarily used in cases in which the mental processes of a person have to be followed in detail (task research, user interface design). Therefore, the structure of cognitive processes is recorded in laboratory settings. The person is asked to vocalize his/her thoughts during task execution. This is usually recorded on tape and categorized afterward according to schemes. Besides the fact that the method is work-intensive, it may also suffer from a person's inability to express appropriately his/her thoughts during a complicated task execution.

Physiological Measurements

A person's state of strain frequently cannot be diagnosed by observation or interview because external signs of strain are difficult to interpret and interviews at short intervals would hinder task execution. Furthermore, persons may want to hide their real status of strain and thus try to give wrong information. Physiological measures (heart rate for example) are considered to be objective, because a person is mostly unable to influence them, and they can be recorded continuously. In addition, physiological measures may demonstrate strain levels which are not perceived by the working person and thus cannot be identified in interviews. The following physiological systems allow measurements for the purpose of ergonomic diagnosis:

- Circulation and respiration system: heart rate, arrhythmia, respiration frequency, blood pressure, etc.
- Motor and limb system: biomechanical variables, electromyogram, tremor, etc.
- Brain and central nervous system: EEG, CNV, evoked potentials, etc.
- Visual system: eye movements, EOG, lid frequency, flicker fusion frequency, etc.
- Dermal system: skin conductance/resistance, EDA, skin con./res. responses.
- Hormonal system: catecholamines, cortisol, etc.
- Metabolic system: respiration volume, O_2/CO_2, energy expenditure, etc.

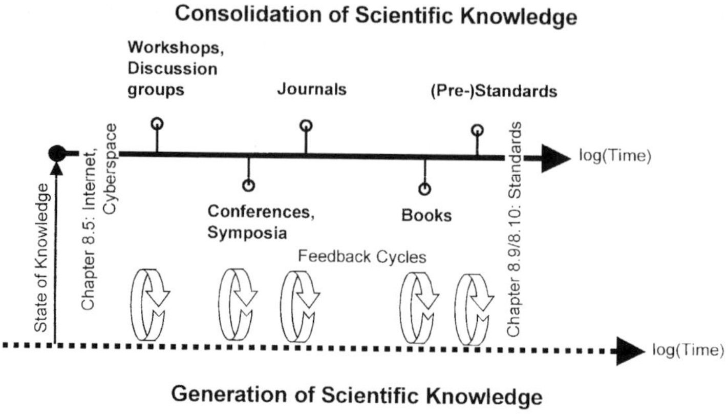

FIGURE 3.1   Time span for the consolidation of scientific knowledge.

In principle, the interpretation of physiological measurements follows two patterns:

- The measured variable directly indicates a bottleneck in the respective organic system. This pattern applies, for example, to heart rate with respect to a heavy muscular work load.
- Variations in organism measurements are interpreted as indicators of a central process. The decrease of flicker fusion frequency, for instance, is seen as an indicator of fatigue.

Physical and Chemical Measures

The physical and chemical methods (beneath the physiological measurements, which are, in principle, physical and chemical measures as well) can be subdivided into those related to the working person and those related to the work environment. The first group consists of methods of time and motion study as well as the analysis of body dimensions and forces. So these measures imply times, distances, forces, and variables derived, like speed, acceleration, and (physical) power. For the description of the work environment techniques of climate, radiation, illumination, vibration, and noise measurement, as well as the quantitative diagnosis of gases, dusts, chemical compositions, etc., are considered.

Consolidation of Ergonomics Knowledge

In order to consolidate the generated ergonomics knowledge there have to be people in a scientific community who discuss, criticize/encourage, transmit, and "store" the information. Knowledge can be differentiated as personal, institutional, organizational, or societal. All these types of knowledge have their own means of distribution and, therefore, need specific guidelines for their use. In this section, scientific knowledge is characterized as consolidated knowledge in the field of ergonomics, which can be acquired by everyone. In principle, that means no barriers hinder peoples' access. The factor "time" structures the consolidation process as illustrated in Figure 3.1.

In the early consolidation phase, different techniques and media are used to discuss ideas and to refine statements and hypotheses. One technique in this phase is the *workshops*. These are often organized in conjunction with conferences and symposia (see next phase) but have a more open and informal character. Another technique is *discussion groups* which are formally or informally organized by special interest groups. (Today these discussion groups are often aided by electronic communication media like the Internet.) The special interest groups are usually established by scientific societies, by scientific projects from a variety of institutions and organizations, or by informal relations among researchers sharing a common interest (for example, the technical groups of the International Ergonomics Association).

After its creation and informal discussion, scientific information is distributed to a broader audience. A characteristic of this phase is that the creator of the information does not exactly know who receives it. In the initial step, a typical way of publishing scientific papers is through *conferences* and *symposia*. Two important examples in the domain of ergonomics are the annual meeting of the Human Factors and Ergonomics Society and the triennial congress of the International Ergonomics Association. Unfortunately, similar events held by scientific societies like the SELF, GfA, NES, etc., not held in the English language do not capture the attention of the worldwide scientific community. In general the whole set of proposed papers has to pass a selection process of a reviewing committee before the researcher can present the results in the conference/symposia "proceedings." Normally, the members of the reviewing committee are well-known experts in a certain domain, which guarantees the scientific quality of the proposals. If the proposal passes the review process and is printed in the proceedings, this presents a good opportunity for discussion and defense of the results of one's research. Today, the presentations — abstracts or long form — are not printed only on paper, but are also often published in an electronic media such as CD-ROM. Due to printing and publication processes, there is a certain time lag between the printed/electronic product and the personal presentation. With reference to the stability of knowledge, this time lag can be either positive or negative: if the material is printed before the conference (preprints), the results of the discussion, the criticism of specific topics, etc., cannot be included in the publication. The results are "older" than the media suggests. If the publication is made after the conference, the results of discussions can be reflected by the "proceedings," including a documentation of the discussion process (protocols, etc.). Thus, the results are more actual than those presented at the conference; but in the worst case the publication is rejected *a posteriori*. Usually the research activities are continued and, depending on the publication time, the scientific information will probably be older than the actual research. This is one reason for using electronic publication provided mainly by the Internet.

The next step of consolidating scientific knowledge is the publication through *national* or *international journals* which are, in some cases, directly linked to a conference. Two examples of primarily basic research-oriented international journals are *Ergonomics* and the *Human Factors*. *Applied Ergonomics* and the *International Journal of Industrial Ergonomics* are examples of international journals with a stronger focus on application-oriented research. In general, every scientific journal has an established review procedure to guarantee the scientific quality standards of the corresponding community, which are reflected by the journal as well. Therefore, the review process is also a stabilization procedure, because the reviewers' comments and proposals are influencing the discussion of the results and the conclusion. Due to publication processes lasting up to two years, the relevancy compared to contemporary research can be rather low. Nevertheless, the published information in journals can be characterized as "state of the current research."

In addition, *scientific books* are primarily written by researchers having a sound knowledge and experience in a specific domain. But economic criteria must also be taken into account, because a publisher will only be willing to publish a book if the interested community (customer base) is large enough to have market potential. With reference to guides to scientific sources of ergonomics knowledge, the most interesting books are *handbooks*. Some good examples of handbooks with a large circulation are the *Handbook of Human Factors* (Salvendy, 1997), the *Handbook of Industrial Engineering* (Salvendy, 1992), the *Handbook of Perception and Human Performance* (Boff et al., 1986), or the *Handbook of Human-Computer Interaction* (Helander et al., 1997). They normally give an excellent overview of a research field for two reasons. First, the research is broadly represented, and second, the scientific quality of the contributions is high because the editor wants to acquire the "scientific capacities" of a research domain for the authorship.

Concerning the time span between creation and establishment of scientific knowledge, the final step of the whole consolidation process is the definition of *standards*. Because standards have the same status as legislation, they are subject to many discussions from different social perspectives: employers and employees, manufacturers and customers, legislation and politics, etc. Before the establishment of standards, the status of the represented scientific knowledge is a draft or prestandard. Critical reflections of standards are usually part of the whole variety of publications such as conference proceedings, journals, and books.

Design-Oriented Compounds of Ergonomics Knowledge

Beneath the presented generic sources, there are also specific compounds of ergonomics knowledge, which are strongly related to the design of work systems. The process of work systems' design can be characterized as a problem-solving task: based on general ergonomic objectives and technological, economical, psychological, social, etc., constraints, the whole task is divided into meaningful subtasks (analysis). This reduction of general task complexity allows the acquisition of existing knowledge for a solution as well as the development of new solutions. When combining partial solutions to the overall solution (synthesis), conflicting partial solutions must be identified and nonconflicting solutions must be elaborated. Since this iterative design process is costly and time consuming, different *compounds* of ergonomics knowledge are available. A compound of ergonomics knowledge is characterized as a generic, design-oriented piece of knowledge in which the different reasons for specific solutions are hidden. In refining more problem-unspecific to more problem-oriented methods, six types of compounds can be differentiated:

- General principles and rules
- Checklists and data lists
- Cases and best practices
- Databases
- Knowledge bases and expert systems
- Problem-oriented tools

Following are some examples of compounds of ergonomics knowledge.

General Principles and Rules

A frequent question in work systems' design is the allocation of functions between man and machine. MABA–MABA rules (man are better at–machines are better at, according to Fitts, 1951) compare the abilities of humans and technical systems in order to decide how functions should be appropriately allocated. But MABA lists are only of general relevance for the evaluation of functions. Different metrics of human and machine, and therefore the incompatibility of the compared functions, are the major criticism of these rules. Nevertheless, they are a general guideline in systems automation.

Due to the general objective of stress reduction, the maximization of efficiency concerning the relation of the input (physiological costs, attention, etc.) and workers' outcome (assembled devices, identified errors, etc.) can be postulated. This general principle of work design implies two different types of design improvements. First, it is possible to achieve the same result, e.g., assembling the underside of a car, with lower physiological costs by rotating the car instead of assembling over head. Second, the designer can prove that additional tasks can be performed with nearly the same, and tolerable, costs, e.g., the use of machines' process times for the handling of additional machines.

Because the organic system has an exponential characteristic of fatigue and recreation, fatigue should be minimized in order to reduce recreation time. The time organization of work should consider this condition when scheduling breaks: more frequent but shorter breaks are better for efficiency than fewer and longer breaks of the same total duration (short breaks rule).

For purposes of occupational safety, engineering rules can be defined to guarantee a safe function of a machine, equipment, etc. (Rohmert and Becker-Biskaborn, 1974). Normally the principles and rules are formulated in relation to design characteristics, e.g., in the cited safety example, the connection principle (connection of safety and handling equipment) or the economics principle (don't disturb working procedures with safety equipment).

Mainly for human information processing tasks and the respective ergonomic systems' design, general principles of compatibility are helpful: stimulus–response, stimulus–stimulus, and response–response compatibility are useful to reduce the required mental capacities of encoding signals (perception), cognition, and carrying out a response (information output). Especially for those tasks with a high degree of decision making (knowledge-based information processing according to Rasmussen, 1983), mental or conceptual compatibility is required to reduce the costs induced by handling the system, e.g., using a computer program, rather than performing the task itself.

Checklists and Data Lists

The complexity of system characteristics to be designed requires aids to support design decisions. Checklists have been developed for more than 30 years (e.g., *Ergonomic System Analysis Checklist*, N.N., 1964) which compound ergonomics knowledge. Ergonomic objectives, ordered by different ergonomic problems, are related to questions the designer can ask for the evaluation of worksystems (e.g., Rohmert, 1974). Because checklists are merely an analytical tool, the most often cited weaknesses are missing guides to improved alternatives, as well as conflicts between different objectives (e.g., Easterby, 1967). It should be stated that, independent of the type of use, the checklists should be used by ergonomic experts only.

Checklists are also used if the conformity of design parameters of a worksystem has to be compared with regulations of standards or legislation. An example is the design of human–computer interfaces: ISO 9214 comprises a list of design rules. Software-ergonomic checklists (e.g., EVADIS, Oppermann et al., 1992) are formulated to enable the quantitative evaluation of existing or planned systems with respect to the standard. As stated above, the most often cited problem in this case is developing appropriate improvements.

Cases and Best Practices

Improving the workplace is sometimes a time- (and knowledge-) consuming challenge. Especially smaller companies, which have neither the staff nor the time for elaborating the "optimal" ergonomic solution, can best use practice cases that are documented through a variety of publications. For example, there are the "ergonomic checkpoints" (ILO, 1996), which provide an easy-to-understand compound of best practice cases. Mostly, the cases are organized according to different design problems, e.g., the height of desks, the use of colors, work organization, etc.

If the system to be designed is based on a set of design elements that are used for a variety of applications, e.g., the user–interface design in human–computer interaction, best practices are normally elaborated with help from different guidelines (see above) or standards (either company standards or public standards). In the human–computer interface domain each standard (X-Windows, Macintosh, etc.) has its own publication of "best practices" (e.g., OSF/Motif, 1993).

Databases and Information Systems

The designer of a work system needs a variety of data, such as geometrical dimensions, performance measurements (e.g., perception parameters) etc., which are usually stored in conventional or electronic databases. Following the latest developments, the hypertext/hypermedia databases are of great interest, especially concerning remote and online access. For example, the military standard MIL-STD 1472D (1991), "Human Engineering Criteria for Military Systems, Equipment, and Facilities," provides the user with criteria, principles, and practices as well as tables and figures for system design.

The more a database consists of guidance functions, the more it becomes an information system (see, for example, Becker-Biskaborn, 1975). The structuring of data, the rules which should be formulated for the use of data, as well as the relationships between the rules (conflicts, dependencies, etc.) are highly important for an efficient use of the information system.

Knowledge Bases and Expert Systems

The application of artificial intelligence techniques in the ergonomic domain has influenced a variety of system developments. In contrast to information systems, the expert system uses a set of rules (knowledge base) in conjunction with inference mechanisms to elaborate design proposals. Because the defined rules are only valid for a limited application domain, the design recommendations can only be used for partial problems of work systems' design.

An example of an expert system is ErgonExpert (Laurig and Rombach, 1989), which produces design recommendations for manual material handling tasks (anthropometric and biomechanical problems). The rules are generated by a model based on anatomy and biomechanics as well as on different empirical investigations, such as fatigue analysis and epidemiological research results.

Problem-Oriented and Rapid Prototyping Tools

For an increasing number of design problems computer-based tools are used. Because the tool uses information for its design purposes the tool itself compounds ergonomics knowledge. CAD-based man models (see, for example, Kroemer, 1988) are examples of tools used for anthropometrical and biomechanical design purposes. Each model is based on anthropometrical databases, storing the human geometry as basic information. Design-oriented information, and therefore ergonomics knowledge, is implemented if the basic information is combined with higher-level aggregates such as the influence of clothing (shoe heights, limited body angles because of clothing, etc.), or measurements of comfort as defined by models like RAMSIS (Tecmath, 1994). Furthermore, the aggregation of basic information to meta-information, like reaching areas, angles of vision, body angles, etc., refers to a compound of ergonomics knowledge.

Another category of design-oriented tools is user interface management systems (UIMS), which are used for rapid prototyping and programming of human–computer interfaces. The UIMS-tool (e.g., Visual Basic) offers a variety of design components like switches, radio buttons, or common dialog boxes for the graphical user interface. Each component has its predefined function and, therefore, can be used by the designer only in a limited application context (e.g., 1–n choice: take radio buttons; m–n choice: take switches, etc.). Because the graphical controls are predefined, the ergonomics knowledge of different functions and dimensions, as well as the question of color, location, relation to other elements, etc., is also compounded.

3.2 Structure of Ergonomic Problems and Empirical Investigation of Scientific Sources

Ordering Model

With regard to the research domain of ergonomics, scientific information and knowledge is acquired from different disciplines such as medicine, physiology, psychology, engineering, economics, sociology, etc. Hence, the sources of ergonomic knowledge are distributed in a variety of congress proceedings, journals, and books. This fact sometimes makes it costly to find the relevant information and, therefore, the utility of structuring aids is potentially high.

An ordering model of ergonomic problems and knowledge by Luczak and Volpert (1987) is oriented toward different levels for describing structures and processes that result when a working person is followed with different analytical approaches for hours, days, or weeks. Considering procedural aspects of work processes, the following levels can be distinguished (Figure 3.2):

V1. Activity of sensomotory automatisms of a person, i. e., elementary operations in sequencing and control of movements
V2. Goal-oriented, consciously controlled action of a person
V3. Motive-related activity of persons, whose concrete results are produced by the sequential and logical arrangement of action
V4. Cooperative work in groups, where the working person has to tune his activity to the activities of other persons
V5. Organization within the company (employers, employees) and between companies to define the roles and orientations to which the working person has to contribute implicitly or explicitly to tasks
V6. Work-oriented political actions, which shall maintain or modify the frame for the parties within the company and which may have severe consequences for any working person

In a structure-oriented form, the levels can be distinguished from the top in the following manner:

S7. Political and societal organization of work.
S6. Forms of industrial relations and organization.
S5. Forms of cooperation in groups and human relations.

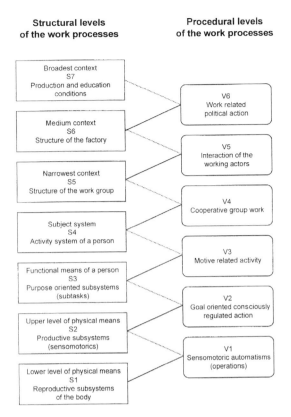

FIGURE 3.2 Structural and procedural levels of the worker process. (From Luczak, H., Volpert, W., 1987, *Arbeitswissenschaft. Kerndefinition–Gegenstandskatalog–Forschungsgebiete*, RKW-Verlag, Eschborn. With permission.)

S4. Forms and types of work and personal activities (individual work).

S3. Subtasks and workplaces.

S2. Operations with tools and working means.

S1. Vegetative systems and environmental factors.

It can easily be seen that these levels have a lot to do with the orientation of work-related scientific disciplines. Thus level S1 is *conditio sine qua non* of occupational physiology and health, but this discipline can widen its approach to levels S2 and S3 as well. Level 7 describes the specialty of political economy and macroeconomics in the sense of labor policy, but excursions to level 6 (industrial relations and organization) and even to level 5 (human relations and group work) may be possible. Levels S2 and S3 may be the focus of industrial engineering (motion study) and ergonomical design, whereas levels S5 and S6 form the core of personnel management and occupational sociology. An integrative approach to work can be assigned to levels S3 to S5.

Empirical Investigation

One of the most helpful sources of ergonomics knowledge is *Ergonomics Abstracts,* edited by The Ergonomics Information Analysis Center at the School of Manufacturing and Mechanical Engineering, University of Birmingham, U.K. The *Ergonomics Abstracts* have been published on paper by Taylor & Francis since 1967. In addition, since 1994 (vol. 26) the scanned abstracts are also available on CD-ROM, which significantly improves the search and retrieval procedure.

Based on the 1996 edition of *Ergonomics Abstracts* an empirical investigation was made with the purpose of an exhausted scanning of relevant sources (abstracts), which are included in conference proceedings, journals, and handbooks. These abstracts were analyzed in relation to different ergonomic aspects as represented by the ordering model introduced in the previous section. The empirical basis stored in the 1996 CD-ROM edition is approximately 44,000 abstracts from the whole variety of ergonomics. Most abstracts are dated no later than 1985. For economic reasons, small congresses and journals with just a small number of scanned abstracts (<10) were excluded from the CD-database, so that approximately 24,000 abstracts were finally investigated.

The classification scheme of the *Ergonomics Abstracts* corresponds with the ordering model. The assignment of classification items of *Ergonomics Abstracts* to the structural levels of the ordering model is shown in Table 3.1 (an additional zero level was added to the ordering model in order to categorize general aspects).

TABLE 3.1   Assignment of the Classification Items of Ergonomics Abstracts (item number in brackets) to the Structural Levels of Work Processes (see Figure 3.2).

---

0. Basics
General (1)

S1. Autonomous Organic Systems & Work Environment
Physiological and Anatomical Aspects (3)
Environment:
   Illumination (29); Noise (30); Vibration (31); Whole Body Movement (32); Climate (33); Atmosphere (34); Altitude, Depth and Space (35); Other Environmental Issues (36)
Health and Safety:
   General Health and Safety (48); Etiology (49); Injury and Illness Prevention (50)

S2. Willfully Steered Organic Systems & Tool/Work Means
Psychological Aspects (2)
Man-Machine Interface Design:
   Choice of Communication Media (10); Input Devices and Controls (21); Visual Displays (22); Auditory Displays (23); Other Modality Displays (24); Display and Control Characteristics (25)
Workplace Design:
   General Workplace Design and Buildings (26); Workstation Design (27); Equipment Design (28)

S3. Tasks & Workplace:
   Visual Communication (8); Auditory and Other Communication Modalities (9); Person-Machine Dialogue Mode (11); System Feedback (12); Error Prevention and Recovery (13); Design of Documentation and Procedures (14); User Control Features (15); Language Design (16); Database Organization and Data Retrieval (17); Programming, Debugging, Editing and Programming Aids (18); Software Performance and Evaluation (19); Software Design, Maintenance and Reliability (20)

S4. Qualification, Motivation and Work Contents & Types of Work
Individual and Task Related Organizational Factors:
   Individual Differences (5); Psychophysiological State Variables (6); Task Related Factors (7)
Job Design:
   Job Attitudes and Job Satisfaction (40); Job Design (41)

S5. Work Group & Cooperative Processes
Group Factors (4)
Group Conflicts and Support:
   Supervision (45); Use of Support (46)

S6. Company & Company Wide Organizational Measures
General System Characteristics:
   General System Features (37); Total System Design and Evaluation (38)
Hours of Work (39)
Payment Systems (42)
Human Resources:
   Selection and Screening (43); Training (44)

S7. Society & Social and Labour Politics
Technological and Ergonomic Change (47)
Social and Labour Politics:
   Trade Unions (52); Employment, Job Security, and Job Sharing (53); Productivity (54); Women at Work (55); Organizational Design (56); Education (57); Law (58); Privacy (59); Family and Home Life (60); Quality of Working Life (61); Political Comment and Ethical Considerations (62)

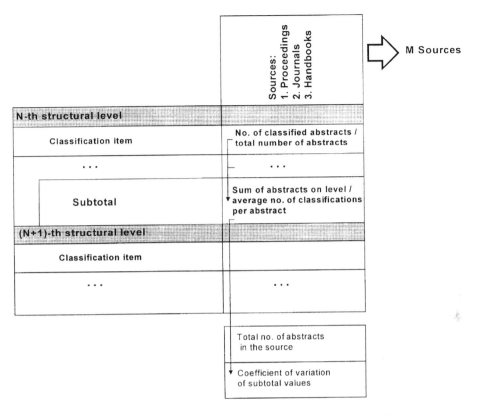

FIGURE 3.3   Calculation scheme of the four "relevance measures."

Based on this systematology, four "relevance measures" were calculated for every classification category (see calculation scheme in Figure 3.3):

1. As a first measure, the total number of scanned abstracts per investigated source was calculated. This was used as a simple indicator of the relevance of the sources in relation to the whole field of ergonomics knowledge. (If the proceedings or journal contained fewer than 10 abstracts, the source was excluded from the calculation.)

2. The number of abstracts which were classified with the corresponding classification item (cf. Table 3.1) divided by the total number of abstracts on the level of the systematology was used as an indicator of the relevance of the source in relation to the specific classification item.

3. A subtotal measure (weighted relevance indicator) was calculated, which equals the sum of the abstracts classified to a whole level of the systematology divided by the average number of classifications per abstract. Thus the subtotal measure is "normalized," because the same abstract might be classified twice or three times with respect to an ordering level, and multiple search results for the same abstract must be taken into account. (Thus, the sum of the subtotal measures over all ordering levels is approximately 100 percent.)

4. The coefficient of variation (CV) of the subtotal measures was calculated. The CV equals the standard deviation of the subtotal measures divided by their average and is therefore an indicator of the degree of specialization of the scientific source as a whole. A large CV indicates a high degree of specialization on few classification items and vice versa.

The results concerning the four measures are shown in Table 3.2 for conference/symposia proceedings, Table 3.3 for journals, and Table 3.4 for handbooks. The titles of the various proceedings which were combined in one heading, as used in Table 3.2, are listed in the appendix.

TABLE 3.2 Results of the Empirical Investigation Concerning Congress/Symposia Proceedings

Ergonomics Abstracts - Classification	Advances in Human-Computer Interaction	Advances in Industrial Ergonomics	Advances in Man-Machine Systems	Analysis, Design and Evaluation of Man-Machine Systems	Computers for Handicapped Persons	Contemporary Ergonomics	Design of Work and Development of Personnel in Advanced Manufacturing	Empirical Studies of Programmers	Environmental Ergonomics	Ergonomics of Hybrid Automated Systems	European Coal and Steel Community	Graphics Interface	HCI
0 Basics													
GENERAL	0.0	10.2	0.0	3.1	0.0	11.5	5.3	0.0	9.2	8.3	19.7	4.2	4.2
Subtotal ¹	*0.0*	*3.7*	*0.0*	*1.1*	*0.0*	*4.4*	*1.9*	*0.0*	*3.7*	*3.4*	*4.3*	*1.6*	*1.6*
S1 Autonomous Organic Systems & Work Environment													
PHYSIOLOGICAL AND ANATOMICAL ASPECTS	0.0	48.1	9.5	2.6	3.3	21.6	5.3	0.0	64.2	3.4	28.9	0.0	4.0
ENVIRONMENT	0.0	12.7	4.8	0.5	0.0	8.5	0.0	0.0	80.8	3.8	78.6	0.0	2.6
HEALTH AND SAFETY	0.0	72.4	14.3	15.9	0.0	33.0	21.1	0.0	62.5	28.3	75.1	0.0	6.5
Subtotal ¹	*0.0*	*48.0*	*8.7*	*6.9*	*1.5*	*23.9*	*9.3*	*0.0*	*83.0*	*14.4*	*40.1*	*0.0*	*5.0*
S2 Wilfully Steered Organic Systems & Tool/Work Means													
PSYCHOLOGICAL ASPECTS	23.5	8.1	76.2	46.7	10.0	22.1	42.1	57.4	0.8	19.2	23.1	16.7	21.3
MAN-MACHINE INTERFACE DESIGN	8.8	5.4	38.1	18.5	96.7	18.3	0.0	1.9	0.8	7.9	20.2	50.0	19.0
WORKPLACE DESIGN	0.0	27.1	0.0	6.7	10.0	25.9	0.0	0.0	3.3	8.7	75.7	2.1	4.2
Subtotal ¹	*11.3*	*14.6*	*34.8*	*26.2*	*53.0*	*25.2*	*14.8*	*19.8*	*2.0*	*14.5*	*26.1*	*26.0*	*17.0*
S3 Tasks & Workplaces													
Subtotal ¹	*55.7*	*1.6*	*24.6*	*26.2*	*25.8*	*12.1*	*3.7*	*54.3*	*0.0*	*10.2*	*5.1*	*59.1*	*36.3*
S4 Qualification, Motivation and Work Content & Types of Work													
INDIVIDUAL AND TASK RELATED ORGANIZATIONAL FACTORS	14.7	23.2	33.3	16.4	0.0	21.2	21.1	13.0	19.2	9.1	17.9	2.1	21.5
JOB DESIGN	0.0	20.5	4.8	13.3	0.0	11.0	31.6	5.6	0.0	30.2	40.5	2.1	11.9
Subtotal ¹	*5.2*	*15.8*	*11.6*	*10.8*	*0.0*	*12.2*	*18.5*	*6.2*	*7.7*	*15.9*	*12.8*	*1.6*	*12.8*
S5 Work Group & Cooperative Processes													
GROUP FACTORS	14.7	14.7	0.0	3.1	0.0	8.7	0.0	51.9	7.5	1.9	1.2	2.1	6.9
GROUP CONFLICTS AND SUPPORT	5.9	1.0	9.5	8.2	0.0	2.8	0.0	3.7	0.0	2.3	0.6	2.1	3.7
Subtotal ¹	*7.2*	*5.7*	*2.9*	*4.1*	*0.0*	*4.4*	*0.0*	*18.5*	*3.0*	*1.7*	*0.4*	*1.6*	*4.0*
S6 Company & Company Wide Organizational Measures													
GENERAL SYSTEM CHARACTERISTICS	55.9	7.2	19.0	50.8	33.3	27.3	31.6	0.0	0.0	34.7	19.7	20.8	35.3
HOURS OF WORK	0.0	4.6	0.0	0.0	0.0	0.8	5.3	0.0	0.8	1.5	1.2	0.0	1.5
PAYMENT SYSTEMS	0.0	0.1	0.0	0.0	0.0	0.0	0.0	0.0	0.0	1.1	0.0	0.0	0.0
HUMAN RESOURCES	2.9	4.4	28.6	5.1	10.0	6.2	31.6	1.9	0.8	5.3	6.4	0.0	3.4
Subtotal ¹	*20.6*	*5.9*	*14.5*	*20.4*	*19.7*	*13.0*	*24.1*	*0.6*	*0.7*	*17.3*	*6.0*	*7.9*	*15.4*
S7 Society & Social and Labour Politics													
TECHNOLOGICAL AND ERGONOMIC CHANGE	0.0	4.6	0.0	7.7	0.0	5.3	26.3	0.0	0.0	31.3	11.0	2.1	10.5
SOCIAL AND LABOUR POLITICS	0.0	8.6	9.5	4.1	0.0	7.6	52.6	1.9	0.0	24.9	12.7	4.2	9.8
Subtotal ¹	*0.0*	*4.8*	*2.9*	*4.3*	*0.0*	*4.9*	*27.8*	*0.6*	*0.0*	*22.7*	*5.2*	*2.4*	*7.8*
Average Number of Classifications per Abstract	2.9	2.8	3.3	2.7	2.2	2.6	2.8	3.0	2.5	2.5	4.6	2.6	2.6
Number of Abstracts:	34	803	21	195	30	792	19	54	120	265	173	48	875
Coefficient of Variation:	1.51	1.22	0.96	0.82	1.55	0.66	0.84	1.50	2.29	0.57	1.10	1.65	0.89

¹ Weighted Relevance (Subtotal Divided by Average Number of Classifications per Abstract)

TOTAL: 11965

Source	Values (grouped)	n	ratio
Work with Display Units	8.4 / **2,7**; 11.6 19.9 56.6 / **28,6**; 22.7 32.7 10.8 / **21,5**; **8,2**; 45.8 16.3 / **20,2**; 10.0 1.6 / **3,8**; 11.2 7.2 0.4 1.6 / **6,6**; 13.5 12.7 8.5 3.1	251	0.77
Visual Search	0.0 / **0,0**; 6.3 4.8 0.0 / **5,2**; 95.2 28.6 1.6 / **58,5**; **11,9**; 25.4 4.8 / **14,1**; 15.9 3.2 / **8,9**; 0.0 0.0 3.2 1.5; 0.0 0.0 0.0 2.1	63	1.55
Vision in Vehicles	4.9 / **1,6**; 2.5 23.3 24.5 / **16,5**; 69.3 70.6 38.7 / **58,7**; **7,3**; 19.0 1.8 / **6,9**; 14.1 0.6 / **4,8**; 3.1 0.6 0.0 8.0 / **3,8**; 0.0 1.2 0.4 3.0	163	1.54
The Ergonomics of Working Postures	2.9 / **1,2**; 97.1 2.9 41.2 / **59,3**; 0.0 2.9 32.4 / **14,8**; **0,0**; 14.7 23.5 / **16,0**; 17.6 0.0 / **7,4**; 0.0 0.0 0.0 0.0; 0.0 2.9 1.2 2.4	34	1.60
The Ergonomics of Manual Work	13.2 / **5,8**; 49.7 6.0 41.7 / **43,0**; 6.6 2.6 25.2 / **15,2**; **0,3**; 31.1 25.8 / **25,1**; 9.3 0.0 / **4,1**; 4.0 2.0 0.0 0.7 / **2,9**; 3.3 4.6 3.5 2.3	151	1.19
SAFE Association	8.4 / **3,6**; 17.6 66.1 82.0 / **70,3**; 3.3 8.4 26.8 / **16,3**; **1,8**; 5.0 1.7 / **2,8**; 2.1 0.0 / **0,9**; 3.8 0.0 0.0 5.4 / **3,9**; 0.0 0.8 0.4 2.4	239	1.91
People and Computers	4.2 / **1,5**; 0.4 0.0 1.1 / **0,5**; 24.3 8.8 1.1 / **12,2**; **54,0**; 10.9 2.8 / **4,9**; 9.2 6.0 / **5,4**; 51.1 0.0 0.0 3.5 / **19,4**; 1.8 4.2 2.1 2.8	284	1.43
National Board/National Institute of/for Occupational Safety and Health	19.2 / **7,3**; 24.3 53.1 69.5 / **55,9**; 6.8 3.4 20.3 / **11,6**; **0,0**; 29.4 11.3 / **15,5**; 10.2 0.0 / **3,9**; 1.7 1.1 0.0 1.7 / **1,7**; 5.6 5.1 4.1 2.6	177	1.46
Man-Computer Interaction Research	0.0 / **0,0**; 3.9 1.3 0.0 / **1,6**; 61.0 10.4 1.3 / **22,5**; **30,9**; 55.8 20.8 / **23,7**; 18.2 9.1 / **8,4**; 20.8 1.3 0.0 9.1 / **9,6**; 7.8 2.6 3.2 3.2	77	0.93
International Ergonomics Association	13.7 / **6,3**; 28.1 10.2 36.4 / **34,2**; 14.3 9.1 18.7 / **19,3**; **4,9**; 20.9 14.6 / **16,3**; 9.9 0.9 / **5,0**; 10.7 2.5 0.1 3.5 / **7,7**; 5.2 8.8 6.4 2.2	1438	0.82
Industrial Ergonomics and Safety	10.0 / **3,8**; 45.3 13.3 69.6 / **48,3**; 9.3 5.0 25.3 / **14,9**; **2,3**; 22.1 19.4 / **15,6**; 13.3 1.2 / **5,5**; 4.8 4.2 0.1 4.2 / **5,0**; 4.9 7.6 4.7 2.7	1078	1.22
IEEE	6.3 / **2,5**; 0.0 6.3 9.4 / **6,2**; 40.6 25.0 3.1 / **27,2**; **34,6**; 0.0 15.6 / **6,2**; 0.0 6.2 / **2,5**; 40.6 0.0 0.0 3.1 / **17,3**; 0.0 9.4 3.7 2.5	32	0.99
Human Computer Interaction - Interact '87 und '90	2.1 / **0,7**; 0.6 0.3 0.0 / **0,3**; 22.6 17.1 0.9 / **14,1**; **52,0**; 16.8 6.1 / **8,0**; 12.8 7.9 / **7,2**; 40.2 0.0 0.3 3.7 / **15,4**; 2.4 4.3 2.3 2.9	328	1.35
Human Factors and Ergonomics Society	11.5 / **4,3**; 12.9 7.1 22.5 / **16,1**; 34.3 28.6 18.2 / **30,8**; **15,0**; 24.0 8.9 / **12,5**; 12.9 4.0 / **6,4**; 19.0 1.1 0.1 10.7 / **11,7**; 2.5 5.9 3.2 2.6	2637	0.71
Human Factors Association	15.1 / **5,2**; 28.8 12.1 68.2 / **37,7**; 15.6 12.8 32.5 / **21,0**; **5,9**; 22.2 18.8 / **14,2**; 13.5 0.7 / **4,9**; 10.8 3.2 0.2 5.5 / **6,8**; 4.1 8.2 4.3 2.9	437	0.94
Human Factors in Telecommunications	9.1 / **3,1**; 0.9 0.9 1.3 / **1,0**; 15.7 73.9 12.6 / **30,5**; **32,5**; 8.3 1.7 / **3,4**; 6.5 8.7 / **5,1**; 56.1 1.3 0.0 2.2 / **20,1**; 3.9 9.1 4.4 3.0	230	1.05
Human Factors in Organizational Design and Management	9.8 / **3,8**; 7.7 4.1 22.2 / **13,2**; 11.3 11.3 12.9 / **13,8**; **5,2**; 20.1 40.7 / **23,6**; 9.8 1.0 / **4,2**; 26.3 1.5 0.0 8.2 / **14,0**; 23.7 33.5 22.2 2.6	194	0.62
Human Factors in Manufacturing	10.0 / **3,8**; 6.9 4.6 16.2 / **10,6**; 13.8 5.4 13.1 / **12,4**; **5,6**; 3.8 30.8 / **13,3**; 4.6 2.3 / **2,7**; 28.5 0.8 2.3 12.3 / **16,8**; 42.3 48.5 34.8 2.6	130	0.82
Human Factors in Information Systems	0.0 / **0,0**; 0.0 0.0 0.0 / **0,0**; 29.2 8.3 0.0 / **14,3**; **23,8**; 33.3 12.5 / **17,5**; 8.3 0.0 / **3,2**; 66.7 0.0 0.0 16.7 / **31,7**; 8.3 16.7 9.5 2.6	24	0.92
Human Factors in Computing Systems	18.6 / **5,7**; 1.2 0.0 0.0 / **0,4**; 30.2 18.6 1.2 / **15,3**; **44,5**; 17.4 2.3 / **6,0**; 24.4 12.8 / **11,4**; 45.3 0.0 0.0 0.0 / **13,9**; 5.8 3.5 2.8 3.3	86	1.12
Human Factors and Power Plants	12.8 / **5,4**; 0.9 3.2 45.7 / **21,0**; 22.8 19.2 5.9 / **20,2**; **14,0**; 8.2 12.3 / **8,7**; 1.8 15.1 / **7,1**; 25.6 2.7 0.0 18.3 / **19,6**; 1.4 8.2 4.0 2.4	219	0.57
Human Factors and Industrial Design	10.8 / **4,1**; 20.6 1.0 21.6 / **16,2**; 10.8 12.7 73.5 / **36,5**; **14,8**; 4.9 2.0 / **2,6**; 16.7 2.9 / **7,4**; 43.1 0.0 0.0 1.0 / **16,6**; 1.0 3.9 1.8 2.7	102	0.92
Human Decision Making	2.2 / **0,9**; 5.8 0.7 8.0 / **6,2**; 65.0 21.9 2.9 / **37,8**; **18,8**; 26.3 8.8 / **14,8**; 5.8 5.5 / **5,5**; 27.7 0.7 0.0 5.8 / **14,5**; 2.2 1.5 1.8 2.4	137	0.97

TABLE 3.3 Results of the Empirical Investigation Concerning International Journals

Ergonomics Abstracts - Classification	Accident Analysis and Prevention	ACM Transactions on Information Systems	Acta Psychologica	American Journal of Physical Medicine & Rehabilitation	American Journal of Psychology	Annals of Occupational Hygiene	Applied Cognitive Psychology	Applied Ergonomics	Applied Occupational and Environmental Hygiene	Australian Journal of Psychology	Australian Safety News	Automatica	Aviation, Space and Environmental Medicine	Behavior and Information Technology
0 Basics														
GENERAL	5.2	2.1	1.4	0.0	0.0	34.9	0.0	11.0	20.0	0.0	16.7	6.3	2.2	6.1
Subtotal	*1.8*	*0.7*	*0.6*	*0.0*	*0.0*	*11.4*	*0.0*	*3.3*	*6.2*	*0.0*	*6.6*	*2.0*	*0.7*	*2.0*
S1 Autonomous Organic Systems & Work Environment														
PHYSIOLOGICAL AND ANATOMICAL ASPECTS	2.2	2.1	6.8	84.2	0.0	7.0	0.0	46.4	36.7	5.9	8.3	0.0	47.7	1.4
ENVIRONMENT	4.4	0.0	0.7	5.3	0.0	72.1	7.2	16.0	56.7	0.0	31.0	0.0	75.6	2.6
HEALTH AND SAFETY	113.3	0.0	0.7	84.2	0.0	120.9	8.7	49.3	113.3	29.4	119.0	6.3	39.6	4.9
Subtotal	*41.3*	*0.7*	*3.7*	*70.2*	*0.0*	*65.2*	*6.0*	*33.7*	*63.9*	*12.0*	*63.0*	*2.0*	*53.8*	*2.9*
S2 Wilfully Steered Organic Systems & Tool/Work Means														
PSYCHOLOGICAL ASPECTS	46.7	21.3	93.2	10.5	89.5	2.3	87.0	17.7	0.0	52.9	2.4	50.0	33.7	28.0
MAN-MACHINE INTERFACE DESIGN	12.6	17.0	13.0	5.3	15.8	2.3	15.9	24.2	0.0	23.5	7.1	18.8	10.0	28.8
WORKPLACE DESIGN	12.6	2.1	0.0	0.0	0.0	4.7	0.0	40.2	46.7	5.9	11.9	0.0	2.6	9.8
Subtotal	*24.7*	*14.1*	*48.3*	*6.4*	*55.6*	*3.0*	*38.8*	*24.7*	*14.4*	*28.0*	*8.5*	*21.6*	*15.3*	*21.8*
S3 Tasks & Workplaces														
Subtotal	*2.3*	*71.1*	*10.9*	*0.0*	*16.7*	*1.5*	*16.4*	*4.8*	*3.1*	*6.0*	*2.4*	*29.4*	*1.3*	*27.7*
S4 Qualification, Motivation and Work Content & Types of Work														
INDIVIDUAL AND TASK RELATED ORGANIZATIONAL FACTORS	30.4	4.3	37.7	26.3	26.3	14.0	40.6	30.6	3.3	94.1	14.3	18.8	46.1	23.3
JOB DESIGN	4.4	10.6	0.0	0.0	0.0	2.3	2.9	18.4	13.3	29.4	6.0	31.3	5.2	18.2
Subtotal	*12.0*	*5.2*	*17.1*	*10.6*	*13.9*	*5.3*	*16.4*	*14.8*	*5.2*	*42.0*	*8.1*	*15.7*	*16.9*	*13.6*
S5 Work Group & Cooperative Processes														
GROUP FACTORS	32.6	2.1	17.8	21.1	21.1	14.0	33.3	24.2	10.0	29.4	1.2	0.0	13.3	17.3
GROUP CONFLICTS AND SUPPORT	2.2	4.3	0.0	0.0	0.0	0.0	11.6	4.1	0.0	0.0	1.2	0.0	0.7	8.1
Subtotal	*12.0*	*0.7*	*8.1*	*8.5*	*11.1*	*4.5*	*16.9*	*8.5*	*3.1*	*10.0*	*0.9*	*0.0*	*4.6*	*8.3*
S6 Company & Company Wide Organizational Measures														
GENERAL SYSTEM CHARACTERISTICS	2.2	10.6	6.2	0.0	0.0	0.0	2.9	12.4	3.3	0.0	1.2	68.8	1.2	35.2
HOURS OF WORK	3.0	0.0	0.7	0.0	0.0	4.7	2.9	3.3	0.0	0.0	6.0	0.0	2.6	0.6
PAYMENT SYSTEMS	0.0	0.0	0.0	0.0	5.3	2.3	0.0	0.7	0.0	0.0	0.0	0.0	0.0	0.3
HUMAN RESOURCES	5.9	0.0	17.1	0.0	0.0	2.3	7.2	6.2	3.3	0.0	2.4	0.0	16.0	7.2
Subtotal	*3.8*	*3.7*	*10.9*	*0.0*	*2.8*	*3.0*	*4.9*	*6.8*	*2.1*	*0.0*	*3.8*	*21.6*	*6.5*	*14.2*
S7 Society & Social and Labour Politics														
TECHNOLOGICAL AND ERGONOMIC CHANGE	0.0	10.6	0.0	5.3	0.0	9.3	0.0	4.3	6.7	0.0	3.6	12.5	0.3	15.0
SOCIAL AND LABOUR POLITICS	5.9	29.8	0.7	5.3	0.0	9.3	1.4	6.7	0.0	5.9	13.1	12.5	2.2	14.1
Subtotal	*2.0*	*3.7*	*0.3*	*4.3*	*0.0*	*6.1*	*0.5*	*3.3*	*2.1*	*2.0*	*6.6*	*7.8*	*0.9*	*9.5*
Average Number of Classifications per Abstract	2.9	2.9	2.2	2.5	1.9	3.1	2.7	3.3	3.2	2.9	2.5	3.2	3.0	3.1
Number of Abstracts:	135	47	146	19	19	43	69	418	30	17	84	16	581	347
Coefficient of Variation:	1.13	1.93	1.24	1.89	1.49	1.72	1.02	0.90	1.69	1.20	1.65	0.89	1.43	0.71

¹ Weighted Relevance (Subtotal Divided by Average Number of Classifications per Abstract)

Journal										N	Index
IEEE Transactions on Systems, Man and Cybernetics	2.0 / **0.8**	5.5, 0.0, 5.5 / **4.5**	65.4, 21.3, 2.8 / **36.2**	**27.8**	13.8, 6.3 / **8.1**	5.9, 5.9 / **4.8**	30.7, 0.0, 0.0, 6.7 / **15.2**	1.2, 5.5 / **2.7**	2.5	254	1.04
IEEE Transactions on Biomedical Engineering	3.4 / **2.0**	62.1, 13.8, 13.8 / **52.0**	13.8, 13.8, 6.9 / **20.0**	**4.0**	20.7, 0.0 / **12.0**	3.4, 0.0 / **2.0**	13.8, 0.0, 0.0, 0.0 / **8.0**	0.0, 0.0 / **0.0**	1.7	29	1.38
Human Relations	7.7 / **2.3**	0.0, 0.0, 17.9 / **5.3**	15.4, 0.0, 5.1 / **6.1**	**0.0**	51.3, 110.3 / **48.1**	12.8, 7.7 / **6.1**	23.1, 0.0, 0.0, 5.1 / **8.4**	7.7, 71.8 / **23.7**	3.4	39	1.28
Human Performance	2.5 / **1.1**	11.3, 5.0, 7.5 / **10.2**	55.0, 8.8, 0.0 / **27.4**	**2.7**	65.0, 23.8 / **38.2**	15.0, 1.3 / **7.0**	1.3, 3.8, 0.0, 18.8 / **10.2**	2.5, 5.0 / **3.2**	2.3	80	1.06
Human Movement Science	0.0 / **0.0**	43.9, 1.0, 1.0 / **25.4**	84.7, 5.1, 1.0 / **50.3**	**5.1**	23.5, 0.0 / **13.0**	10.2, 0.0 / **5.6**	0.0, 0.0, 0.0, 1.0 / **0.6**	0.0, 0.0 / **0.0**	1.8	98	1.40
Human Factors	3.7 / **1.3**	14.6, 7.5, 18.7 / **14.5**	56.1, 41.9, 15.2 / **40.3**	**13.2**	36.0, 3.7 / **14.2**	21.3, 3.5 / **8.8**	5.5, 3.0, 0.2, 10.2 / **6.7**	1.0, 1.6 / **0.9**	2.8	508	1.00
Human-Computer Interaction	4.4 / **1.2**	0.0, 0.0, 2.7 / **0.7**	52.2, 23.9, 0.9 / **21.6**	**37.5**	18.6, 13.3 / **8.9**	27.4, 10.6 / **10.7**	39.8, 0.0, 0.0, 8.0 / **13.4**	1.8, 19.5 / **6.0**	3.6	113	0.97
Health and Safety at Work	56.0 / **21.2**	4.0, 48.0, 104.0 / **59.1**	8.0, 4.0, 24.0 / **13.6**	**0.0**	0.0, 4.0 / **1.5**	0.0, 0.0 / **0.0**	4.0, 0.0, 0.0, 3.0 / **1.5**	0.0, 8.0 / **3.0**	2.6	25	1.63
European Work and Organizational Psychologist	5.6 / **1.7**	0.0, 0.0, 16.7 / **5.2**	5.6, 5.6, 0.0 / **3.4**	**0.0**	44.4, 127.8 / **53.4**	5.6, 5.6 / **3.4**	0.0, 0.0, 0.0, 5.6 / **1.7**	27.8, 72.2 / **31.0**	3.2	18	1.55
European Journal of Cognitive Psychology	0.0 / **0.0**	1102.7, 2.7, 0.0 / **86.7**	94.6, 24.3, 0.0 / **9.3**	**0.8**	21.6, 0.0 / **1.7**	16.2, 2.7 / **1.5**	0.0, 0.0, 0.0, 0.0 / **0.0**	0.0, 0.0 / **0.0**	12.8	37	2.41
European Journal of Applied Physiology and Occupational Physiology	0.5 / **0.2**	96.0, 25.9, 9.2 / **53.9**	3.5, 0.2, 2.1 / **2.4**	**0.0**	53.2, 6.6 / **24.6**	36.2, 0.0 / **14.9**	0.0, 3.1, 0.0, 6.1 / **3.8**	0.0, 0.7 / **0.3**	2.4	425	1.51
Ergonomics in Design	8.7 / **3.5**	10.9, 8.7, 45.7 / **26.3**	13.0, 17.4, 43.5 / **29.8**	**9.6**	10.9, 4.3 / **6.1**	10.9, 0.0 / **15.8**	32.6, 0.0, 0.0, 6.5 / **15.8**	2.2, 8.7 / **4.4**	2.5	46	0.84
Ergonomics	9.8 / **5.1**	0.2, 0.0, 0.3 / **0.3**	26.9, 10.6, 16.6 / **28.4**	**6.0**	44.7, 15.0 / **31.3**	21.3, 0.8 / **11.6**	8.0, 9.8, 0.1, 4.0 / **11.5**	2.5, 8.4 / **5.7**	1.9	1197	0.91
CSERIAC Gateway	45.2 / **23.3**	22.6, 12.9, 16.1 / **26.7**	22.6, 12.9, 6.5 / **21.7**	**6.7**	19.4, 0.0 / **10.0**	3.2, 0.0 / **1.7**	12.9, 0.0, 0.0, 3.2 / **8.3**	0.0, 3.2 / **1.7**	1.9	31	0.80
Computers in Human Behavior	4.2 / **1.4**	3.5, 0.0, 2.8 / **2.2**	34.0, 22.9, 0.0 / **19.7**	**15.9**	77.8, 9.7 / **30.3**	24.3, 3.5 / **9.6**	14.6, 0.7, 0.0, 21.5 / **12.7**	12.5, 11.1 / **8.2**	2.9	144	0.76
Computers & Industrial Engineering	13.4 / **6.1**	25.4, 3.0, 22.4 / **23.1**	10.4, 4.5, 9.0 / **10.9**	**24.5**	11.9, 13.4 / **11.6**	9.0, 9.0 / **8.2**	14.9, 3.0, 0.0, 9.0 / **12.2**	4.5, 3.0 / **3.4**	2.2	67	0.61
Computer Supported Cooperative Work	0.0 / **0.0**	0.0, 0.0, 0.0 / **0.0**	36.8, 26.3, 0.0 / **21.4**	**12.5**	10.5, 57.9 / **23.2**	5.3, 0.0 / **1.8**	21.1, 0.0, 0.0, 0.0 / **7.1**	5.3, 94.7 / **33.9**	2.9	19	1.01
Communications of the ACM	2.6 / **1.0**	1.7, 0.0, 3.2 / **1.9**	10.3, 14.8, 1.3 / **10.3**	**45.1**	6.5, 9.7 / **6.3**	5.8, 0.0 / **2.3**	40.6, 0.0, 0.0, 2.6 / **16.9**	13.5, 27.7 / **16.1**	2.1	155	1.17
Cognitive Science	0.0 / **0.0**	3.1, 0.0, 0.0 / **1.5**	93.8, 3.1, 0.0 / **47.0**	**27.3**	25.0, 0.0 / **12.1**	18.8, 3.1 / **10.6**	3.1, 0.0, 0.0, 0.0 / **1.5**	0.0, 0.0 / **0.0**	2.1	32	1.34
Cognitive Psychology	0.0 / **0.0**	3.7, 0.0, 0.0 / **2.8**	92.6, 7.4, 0.0 / **75.0**	**8.3**	7.4, 0.0 / **5.6**	7.4, 0.0 / **5.6**	0.0, 0.0, 0.0, 3.7 / **2.8**	0.0, 0.0 / **0.0**	1.3	27	2.03
British Journal of Psychology	1.2 / **0.5**	3.6, 7.2, 7.2 / **7.8**	81.9, 4.8, 0.0 / **37.5**	**3.6**	69.9, 4.8 / **32.3**	33.7, 2.4 / **15.6**	1.2, 0.0, 0.0, 0.0 / **0.5**	1.2, 3.6 / **2.1**	2.3	83	1.18
Behavior Research Methods, Instruments & Computers	1.6 / **0.9**	8.8, 0.8, 2.4 / **6.6**	46.4, 24.0, 0.0 / **38.9**	**23.5**	27.2, 0.8 / **15.5**	6.4, 1.6 / **4.4**	5.6, 3.2, 0.0, 8.0 / **9.3**	1.6, 0.0 / **0.9**	1.8	125	1.05

TABLE 3.3 (continued)　Results of the Empirical Investigation Concerning International Journals

Journal	1	2	3	4	5	6	7	8	9	10	11	12	13	14	15	16	17	18	19	20	21	22	23	24	25	26	n	ratio
Industrial Management & Data Systems	6.3	3.2	0.0	0.0	0.0	0.0	6.3	0.0	0.0	3.2	12.9	0.0	6.3	3.2	6.3	0.0	3.2	25.0	0.0	0.0	0.0	12.9	56.3	62.5	61.3	1.9	16	1.62
Information Processing & Management	4.1	1.7	0.0	0.0	0.0	0.0	16.2	8.1	0.0	10.2	61.6	14.9	2.7	7.3	10.8	6.8	7.3	14.9	0.0	0.0	1.4	6.8	5.4	6.8	5.1	2.4	74	1.61
Interacting with Computers	6.1	2.0	0.8	0.8	0.0	0.5	28.0	22.7	1.5	17.0	42.1	10.6	11.4	7.1	8.3	6.1	4.7	51.5	0.0	0.0	3.8	18.0	8.3	18.2	8.6	3.1	132	1.08
International Journal of Aviation Psychology	2.3	0.8	3.4	4.5	11.4	7.1	40.9	29.5	0.0	25.9	12.6	50.0	18.2	25.1	8.0	0.0	2.9	11.4	1.1	0.0	42.0	20.1	1.1	13.6	5.4	2.7	88	0.80
International Journal of Human-Computer Interaction	7.7	2.2	7.7	3.4	12.0	6.7	40.2	27.4	3.4	20.4	25.6	46.2	23.1	20.0	17.1	7.7	7.1	33.3	3.4	0.9	5.1	12.3	11.1	8.5	5.7	3.5	117	0.68
International Journal of Human-Computer Studies	0.0	0.0	0.0	0.0	1.7	0.6	51.7	17.2	0.0	23.7	49.7	15.5	0.0	5.3	15.5	5.2	7.1	31.0	0.0	0.0	1.7	11.2	3.4	3.4	2.4	2.9	58	1.35
International Journal of Human Factors in Manufacturing	9.8	3.0	3.9	2.0	17.6	7.2	19.6	2.9	6.9	9.0	7.5	17.6	68.6	26.4	2.9	5.9	2.7	33.3	3.9	2.9	5.9	14.1	50.0	48.0	30.0	3.3	102	0.83
International Journal of Industrial Ergonomics	20.5	6.2	55.3	22.5	64.5	42.8	8.7	6.0	34.5	14.8	2.3	27.5	29.0	17.0	19.3	1.7	6.3	9.2	6.8	0.2	3.6	6.0	5.6	9.7	4.6	3.3	414	1.06
International Journal of Lighting Research and Technology	0.0	0.0	4.0	112.0	0.0	50.9	20.0	32.0	28.0	35.1	0.0	20.0	0.0	8.8	8.0	0.0	3.5	0.0	0.0	0.0	0.0	0.0	0.0	4.0	1.8	2.3	25	1.56
International Journal of Manpower	2.3	0.8	0.4	0.2	0.8	0.4	38.5	18.1	1.3	19.7	50.1	20.0	3.8	8.1	17.9	7.9	8.8	25.1	0.0	0.0	4.9	10.2	1.1	4.9	2.0	2.9	530	1.32
International Journal of Production Research	6.4	2.9	2.1	0.0	2.1	2.0	31.9	2.1	17.0	23.5	18.6	8.5	17.0	11.8	6.4	2.1	3.9	21.3	0.0	0.0	10.6	14.7	14.9	34.0	22.5	2.2	47	0.70
International Journal of Psychology	10.0	4.5	0.0	20.0	0.0	9.1	50.0	10.0	0.0	27.3	4.5	50.0	0.0	22.7	40.0	0.0	18.2	0.0	0.0	0.0	10.0	4.5	0.0	20.0	9.1	2.2	10	0.72
International Review of Applied Psychology	0.0	0.0	15.4	23.1	23.1	17.8	15.4	0.0	0.0	4.4	0.0	84.6	69.2	44.4	23.1	7.7	8.9	23.1	0.0	0.0	0.0	6.7	0.0	61.5	17.8	3.5	13	1.17
Japanese Journal of Ergonomics	7.1	3.4	25.3	10.1	10.6	22.2	37.1	21.5	16.1	36.1	7.0	23.7	6.3	14.5	13.6	0.3	6.7	10.6	2.7	0.0	1.9	7.4	0.8	4.9	2.8	2.1	367	0.92
Journal of Applied Physiology	0.0	0.0	96.7	48.4	9.0	57.7	2.5	0.0	1.6	1.5	0.0	57.4	0.8	21.8	38.5	0.0	14.4	0.0	0.0	0.0	12.3	4.6	0.0	0.0	0.0	2.7	122	1.59
Journal of Applied Psychology	0.4	0.2	3.5	2.2	14.3	7.8	19.5	2.6	1.7	9.4	0.7	51.5	77.5	50.7	23.4	6.1	11.6	0.0	1.3	3.5	13.4	7.1	1.3	30.7	12.6	2.5	231	1.29
Journal of Educational Multimedia and Hypermedia	0.0	0.0	0.0	0.0	0.0	0.0	42.1	21.1	1.3	26.1	47.8	15.8	5.3	8.7	10.5	0.0	4.3	21.1	0.0	0.0	5.3	10.9	5.3	0.0	2.2	2.4	19	1.33
Journal of Experimental Psychology: Human Perception and Performance	0.0	0.0	13.7	3.4	1.1	11.6	89.7	11.4	1.7	65.5	9.1	14.3	0.0	9.1	7.4	0.0	4.7	0.0	0.0	0.0	0.0	0.0	0.0	0.0	0.0	1.6	175	1.75
Journal of Experimental Psychology: Learning, Memory and Cognition	0.0	0.0	4.7	6.5	0.0	6.2	97.2	6.5	0.0	58.0	3.6	29.0	0.0	16.1	23.4	2.8	14.5	2.8	0.0	0.0	0.0	1.6	0.0	0.0	0.0	1.8	107	1.55
Journal of Human Ergology	3.6	1.3	64.3	16.4	32.9	39.9	15.0	5.0	12.9	11.6	0.3	45.7	17.9	22.4	35.7	10.7	16.3	0.7	7.1	2.9	2.9	4.8	2.1	7.9	3.5	2.8	140	1.08
Journal of the Illuminating Engineering Society	7.9	3.7	5.6	93.3	10.1	51.3	28.1	40.4	16.9	40.2	1.6	3.4	0.0	1.6	2.2	1.1	1.6	0.0	0.0	0.0	0.0	0.0	0.0	0.0	0.0	2.1	89	1.66
Journal of Information Technology	8.0	2.7	0.0	0.0	68.0	23.3	8.0	0.0	0.0	2.7	8.2	0.0	28.0	9.6	0.0	4.0	1.4	28.0	0.0	0.0	4.0	11.0	64.0	56.0	41.1	2.9	25	1.08

Journal of Managerial Psychology	Journal of Occupational Health and Safety - Australia and New Zealand	Journal of Occupational and Organizational Psychology	Journal of Occupational Medicine	Journal of Organizational Behavior	Journal of Organizational Change Management	Journal of Safety Research	Journal of Science of Labour	Lighting Research and Technology	Military Psychology	New Technology, Work and Employment	Occupational Medicine	Perception	Perception & Psychophysics	Perceptual and Motor Skills	Presence	Professional Safety	Psychophysiology	Quarterly Journal of Experimental Psychology	Reliability Engineering and System Safety	Risk Analysis	Safety Science
5,3	9,5	0,0	6,5	1,1	0,0	6,4	9,0	9,3	0,0	0,0	8,6	0,0	0,6	0,8	2,2	16,4	0,0	2,3	11,1	7,1	23,9
1,6	***4,0***	***0,0***	***1,8***	***0,3***	***0,0***	***2,0***	***3,0***	***3,7***	***0,0***	***0,0***	***2,7***	***0,0***	***0,5***	***0,4***	***1,0***	***6,9***	***0,0***	***1,3***	***5,3***	***2,6***	***6,9***
5,3	11,1	0,0	31,5	2,2	0,0	11,7	23,9	4,0	4,5	0,0	22,9	8,0	4,8	18,9	4,3	4,8	97,7	3,1	0,0	3,6	6,8
0,0	12,7	0,0	24,0	0,0	0,0	8,5	35,8	101,3	4,5	1,4	31,4	11,5	8,9	11,5	0,0	15,8	4,7	3,1	0,0	14,3	14,8
47,4	104,8	42,9	138,0	35,2	15,4	134,0	47,8	13,3	4,5	0,0	120,0	0,9	0,0	5,3	0,0	123,0	2,3	0,0	71,1	78,6	167,0
16,1	***53,6***	***13,3***	***53,4***	***9,4***	***6,1***	***49,5***	***36,5***	***47,1***	***5,4***	***0,5***	***54,5***	***13,6***	***10,8***	***15,6***	***1,9***	***60,9***	***41,3***	***3,6***	***34,0***	***34,6***	***54,2***
5,3	3,2	14,3	6,5	16,5	0,0	23,4	12,7	38,7	40,9	0,0	11,4	74,3	72,5	63,8	28,3	9,1	53,5	93,8	51,1	75,0	23,9
0,0	1,6	0,0	2,5	0,0	0,0	10,6	8,2	29,3	6,8	0,0	0,0	19,5	9,9	15,6	69,6	3,0	7,0	13,2	0,0	0,0	1,1
0,0	3,2	0,0	9,0	2,2	0,0	27,7	12,7	25,3	0,0	0,0	5,7	3,5	1,3	2,9	2,2	16,4	0,0	0,8	0,0	3,6	39,8
1,6	***3,3***	***4,4***	***5,0***	***4,7***	***0,0***	***19,8***	***11,4***	***37,0***	***18,8***	***0,0***	***5,4***	***65,1***	***66,0***	***35,9***	***44,2***	***12,1***	***23,9***	***61,8***	***24,5***	***28,2***	***18,6***
0,0	***0,7***	***2,2***	***0,0***	***0,3***	***0,0***	***2,0***	***2,8***	***4,2***	***3,6***	***0,5***	***0,0***	***10,7***	***7,6***	***3,9***	***21,2***	***2,1***	***0,0***	***5,8***	***13,8***	***7,7***	***2,9***
84,2	31,7	85,7	54,0	87,9	30,8	31,9	45,5	9,3	56,8	1,4	34,3	8,0	11,8	65,8	30,4	6,7	72,1	31,8	0,0	17,9	14,8
68,4	7,9	103,6	25,0	125,3	38,5	10,6	22,4	0,0	29,5	63,9	20,0	0,0	0,0	2,5	0,0	5,5	0,0	0,0	2,2	3,6	4,5
46,8	***16,6***	***58,9***	***21,8***	***53,6***	***27,3***	***13,7***	***23,1***	***3,7***	***33,9***	***21,6***	***17,0***	***5,3***	***9,3***	***29,8***	***13,5***	***5,1***	***28,4***	***18,2***	***1,1***	***7,7***	***5,6***
15,8	17,5	10,7	24,0	26,4	15,4	21,3	24,6	5,3	9,1	4,2	14,3	7,1	5,8	25,9	0,0	0,6	7,0	12,4	0,0	14,3	12,5
5,3	1,6	17,9	0,0	6,6	0,0	4,3	0,0	1,3	2,3	4,2	0,0	0,0	0,3	1,2	0,0	2,4	0,0	2,3	11,1	0,0	8,0
6,5	***7,9***	***8,9***	***6,6***	***8,3***	***6,1***	***8,2***	***8,4***	***2,6***	***4,5***	***2,8***	***4,5***	***4,7***	***4,8***	***11,8***	***0,0***	***1,3***	***2,8***	***8,4***	***5,3***	***5,1***	***5,9***
0,0	3,2	0,0	0,5	0,0	0,0	2,1	1,5	4,0	0,0	2,8	0,0	0,0	0,0	0,4	30,4	1,2	0,0	0,0	26,7	10,7	2,3
0,0	1,6	0,0	5,0	5,5	0,0	0,0	19,4	0,0	6,8	6,9	8,6	0,0	0,0	0,8	0,0	1,2	2,3	0,0	0,0	0,0	1,1
0,0	1,6	3,6	0,0	2,2	0,0	1,1	0,0	0,0	0,0	2,8	0,0	0,0	0,0	0,0	0,0	0,0	0,0	0,0	0,0	0,0	0,0
21,1	1,6	7,1	8,0	6,6	0,0	3,2	0,7	0,0	72,7	6,9	11,4	0,9	1,3	3,7	6,5	4,8	4,7	1,6	4,4	0,0	2,3
6,5	***3,3***	***3,3***	***3,7***	***3,6***	***0,0***	***2,0***	***7,4***	***1,6***	***31,3***	***6,4***	***6,3***	***0,6***	***1,0***	***2,2***	***16,3***	***3,1***	***2,8***	***0,9***	***14,9***	***3,8***	***1,6***
0,0	1,6	0,0	4,0	7,7	30,8	0,0	5,2	0,0	0,0	87,5	0,0	0,0	0,0	0,0	2,2	3,6	0,0	0,0	0,0	0,0	4,5
68,4	23,8	28,6	24,0	71,4	123,1	8,5	16,4	0,0	6,8	119,4	31,4	0,0	0,0	0,8	2,2	16,4	2,3	0,0	2,2	28,6	10,2
21,0	***10,6***	***8,9***	***7,7***	***19,9***	***60,6***	***2,7***	***7,4***	***0,0***	***2,7***	***68,3***	***9,8***	***0,0***	***0,0***	***0,4***	***1,9***	***8,5***	***0,9***	***0,0***	***1,1***	***10,3***	***4,2***
3,3	2,4	3,2	3,6	4,0	2,5	3,1	2,9	2,5	2,5	3,0	3,2	1,5	1,3	2,3	2,3	2,4	2,5	1,7	2,1	2,8	3,5
19	63	28	200	91	13	94	134	75	44	72	35	113	313	243	46	165	43	129	45	28	88
1,26	1,39	1,54	1,42	1,42	1,72	1,31	0,93	1,48	1,09	1,90	1,42	1,75	1,76	1,10	1,21	1,59	1,30	1,66	0,95	0,96	1,41

TABLE 3.3 (continued)　　Results of the Empirical Investigation Concerning International Journals

Scandinavian Journal of Work, Environmental & Health	SIGHI Bulletin	Speech Communication	Speech Technology	Travail et Sante	Travail Humain	Vision Research	Work	Work and Stress	Work, Employment & Society	Workplace Ergonomics
10,3	19,8	0,0	2,1	22,0	3,5	0,0	21,2	4,3	0,0	9,1
2,9	*7,8*	*0,0*	*1,6*	*9,7*	*1,3*	*0,0*	*7,4*	*1,0*	*0,0*	*5,0*
42,5	1,0	0,0	0,0	12,0	13,6	15,4	27,3	16,6	0,0	18,2
57,1	0,0	8,2	2,1	20,0	6,0	6,7	3,0	6,4	0,0	0,0
115,3	1,5	0,0	2,1	94,0	30,2	0,0	81,8	52,9	3,8	90,9
59,4	*1,0*	*6,0*	*3,2*	*55,8*	*18,8*	*14,8*	*39,4*	*18,2*	*1,0*	*60,0*
4,6	12,9	19,7	0,0	10,0	44,2	97,9	3,0	19,8	0,0	0,0
0,8	20,8	32,8	41,7	2,0	8,5	15,4	3,0	0,5	0,0	0,0
8,8	2,0	4,9	6,3	14,0	6,0	0,0	27,3	1,6	0,0	45,5
3,9	*14,1*	*41,7*	*36,5*	*11,5*	*22,2*	*75,9*	*11,7*	*5,2*	*0,0*	*25,0*
0,0	*42,2*	*41,7*	*25,4*	*0,0*	*10,2*	*3,4*	*0,0*	*0,0*	*0,0*	*0,0*
47,5	9,9	1,6	0,0	6,0	28,6	2,6	21,2	144,4	15,4	0,0
32,2	4,0	0,0	0,0	8,0	19,1	0,0	36,4	75,4	57,7	9,1
22,0	*5,5*	*1,2*	*0,0*	*6,2*	*18,0*	*1,7*	*20,2*	*52,6*	*19,8*	*5,0*
27,2	7,9	3,3	2,1	10,0	25,1	5,6	18,2	19,3	30,8	0,0
0,4	7,4	0,0	0,0	2,0	4,0	0,0	0,0	1,1	7,7	0,0
7,6	*6,1*	*2,4*	*1,6*	*5,3*	*11,0*	*3,8*	*6,4*	*4,9*	*10,4*	*0,0*
0,0	48,0	8,2	39,6	2,0	8,5	0,0	0,0	0,5	3,8	0,0
7,7	0,0	0,0	0,0	4,0	11,6	0,0	0,0	19,8	11,5	0,0
0,4	0,0	0,0	0,0	0,0	0,5	0,0	0,0	0,0	7,7	0,0
0,4	4,5	1,6	2,1	4,0	5,5	0,5	15,2	1,6	15,4	0,0
2,3	*20,7*	*7,1*	*31,7*	*4,4*	*9,9*	*0,3*	*5,3*	*5,2*	*10,4*	*0,0*
1,1	1,0	0,0	0,0	6,0	3,0	0,0	0,0	2,7	42,3	0,0
5,7	5,9	0,0	0,0	10,0	19,6	0,0	27,3	50,8	173,1	9,1
1,9	*2,7*	*0,0*	*0,0*	*7,1*	*8,5*	*0,0*	*9,6*	*12,8*	*58,3*	*5,0*
3,6	*2,5*	*1,4*	*1,3*	*2,3*	*2,6*	*1,5*	*2,8*	*4,2*	*3,7*	*1,8*
261	202	61	48	50	199	195	33	187	26	11
1,61	*1,09*	*1,46*	*1,27*	*1,43*	*0,54*	*2,09*	*0,98*	*1,38*	*1,59*	*1,67*

TOTAL 11778

With regard to the total number of classified abstracts concerning conferences/symposia (Table 3.2), the annual meeting of the Human Factors and Ergonomics Society (HFES) is the largest source, followed by the triennial congress of the International Ergonomics Association (IEA) second, and the annual International Industrial Ergonomics and Safety Conference (IIESC) third. Although all three conferences have a quite similar profile concerning their weighted relevance (subtotal value according to Table 3.2) of high-order levels (levels S4, S5, S6, and S7), the annual meeting of the HFES has a stronger focus on psychological aspects and man–machine interface design (level S2) as well as on tasks and workplaces (level S3). However, the congress of the IEA and the IIESC emphasize physiological/anatomical aspects and health/safety issues (level S1) as well as general equipment design (level S2). If the CV of the weighted relevance is used as a measure for specialization of the three largest sources, the annual meeting of the HFES covers the broadest range (CV = 0.71), closely followed by the triennial congress of the IEA (CV = 0.82), and finally the IIESC (CV = 1.22). According to the CV, the most specialized periodical conferences are the International Conference on Environmental Ergonomics (CV = 2.29) and the annual symposiums of the SAFE Association (CV = 1.91), which both strongly focus on level S2.

The international journal with the most classified abstracts is the monthly published *Ergonomics*, which holds more than 10% of all abstracts (Table 3.3). According to the subtotal value *Ergonomics* focuses on individual and task related organizational factors (level S4), psychological factors (level S2), and group factors (level S5). The second largest source is *Aviation, Space and Environmental Medicine*, which specializes in physiological/anatomical aspects and environmental and health/safety issues (level S1).

TABLE 3.4   Results of the Empirical Investigation Concerning Handbooks

Ergonomics Abstracts - Classification		Handbook of Human Factors	Handbook of Human Performance	Handbook of Human-Computer Interaction	Handbook of Industrial Engineering	Handbook of Military Psychology	Handbook of Occupational Safety and Health
0 Basics							
GENERAL		4,4	2,9	15,4	10,0	5,6	12,5
	Subtotal: [1]	2,0	0,7	4,2	3,1	1,4	5,6
S1 Autonomous Organic Systems & Work Environment							
PHYSIOLOGICAL AND ANATOMICAL ASPECTS		10,3	34,3	1,9	5,0	27,8	6,3
ENVIRONMENT		33,8	60,0	1,9	45,0	66,7	6,3
HEALTH AND SAFETY		17,6	22,9	5,8	30,0	22,2	137,5
	Subtotal: [1]	27,5	28,7	2,6	24,6	30,4	66,7
S2 Wilfully Steered Organic Systems & Tool/Work Means							
PSYCHOLOGICAL ASPECTS		27,9	94,3	30,8	25,0	72,2	0,0
MAN-MACHINE INTERFACE DESIGN		20,6	5,7	23,1	10,0	0,0	18,8
WORKPLACE DESIGN		10,3	5,7	9,6	35,0	0,0	6,3
	Subtotal: [1]	26,1	25,9	17,3	21,5	18,8	11,1
S3 Tasks & Workplaces							
	Subtotal: [1]	11,8	0,7	34,0	4,6	1,4	0,0
S4 Qualification, Motivation and Work Content & Types of Work							
INDIVIDUAL AND TASK RELATED ORGANIZATIONAL FACTORS		13,2	154,3	25,0	25,0	94,4	6,3
JOB DESIGN		11,8	2,9	11,5	30,0	5,6	0,0
	Subtotal: [1]	11,1	38,5	9,9	16,9	26,1	2,8
S5 Work Group & Cooperative Processes							
GROUP FACTORS		2,9	17,1	15,4	5,0	5,6	0,0
GROUP CONFLICTS AND SUPPORT		1,5	0,0	11,5	5,0	0,0	0,0
	Subtotal: [1]	2,0	4,2	7,3	3,1	1,4	0,0
S6 Company & Company Wide Organizational Measures							
GENERAL SYSTEM CHARACTERISTICS		20,6	0,0	42,3	20,0	5,6	0,0
HOURS OF WORK		1,5	2,9	0,0	0,0	5,6	0,0
PAYMENT SYSTEMS		1,5	0,0	0,0	10,0	0,0	0,0
HUMAN RESOURCES		10,3	2,9	13,5	15,0	55,6	18,8
	Subtotal: [1]	15,0	1,4	15,2	13,8	17,4	8,3
S7 Society & Social and Labour Politics							
TECHNOLOGICAL AND ERGONOMIC CHANGE		4,4	0,0	9,6	20,0	0,0	0,0
SOCIAL AND LABOUR POLITICS		5,9	0,0	25,0	20,0	11,1	12,5
	Subtotal: [1]	4,6	0,0	9,4	12,3	2,9	5,6
	Average Number of Classifications per Abstract:	2,3	4,1	3,7	3,3	3,8	2,3
	Number of Abstracts:	68	35	52	20	18	16
	Coefficient of Variation:	0,80	1,26	0,80	0,67	0,97	1,78

TOTAL: 234

[1] Weighted Relevance (Subtotal Divided by Average Number of Classifications per Abstract)

Third is the *International Journal of Manpower,* which strongly emphasizes tasks and workplaces on level S3. *Human Factors* is the fourth largest source just a few abstracts behind the third and specializes in psychological aspects of work and man–machine interface design (level S2). Of these four largest sources *Ergonomics* is the least specialized journal with a coefficient of variation of 0.91, followed by *Human Factors* (CV = 1.00), the *International Journal of Manpower* (CV = 1.32), and finally *Aviation, Space and Environmental Medicine* (CV = 1.43). The most specialized sources are the *European Journal of Cognitive Psychology* (CV = 2.41, centered around level S2), *Vision Research* (CV = 2.09, centered around level S3), and *Cognitive Psychology* (CV = 2.03, centered around level S3).

With regard to the investigation of handbooks as sources of ergonomics knowledge, the total number of abstracts is an inappropriate measure because of significant structural differences. Therefore, the coefficients of variation only are discussed. The *Handbook of Industrial Engineering* is the source with the largest diversity (least specialization) of the whole set with a CV of 0.67, followed by the *Handbook of Human Factors* and the *Handbook of Human-Computer Interaction,* both with a CV of 0.80. Although the sources have the same CV, the *Handbook of Human Factors* has a stronger focus on levels S1 and S2 of the systematology and the *Handbook of Human-Computer Interaction* deals primarily with the S2 and S3 levels. The *Handbook of Occupational Safety and Health* is the most specialized handbook (CV = 2.18), as indicated in the title, and is centered around level S1 of the systematology.

3.3 Conclusion and Future Developments

In the preceeding sections, the creation and consolidation process of scientific knowledge was analyzed and different types of scientific sources were discussed. Due to the multidisciplinary character of ergonomics, a systematology considering structural and procedural aspects of work was introduced which served as a framework for an empirical investigation of proceedings, international journals, and handbooks. With regard to future developments of scientific sources of ergonomics knowledge, probably the biggest challenge to the established publication procedures is the development of electronic information and communication media, especially the Internet. The Internet has the potential to accelerate the creation, consolidation, and dissemination process significantly, because of online, worldwide information access and sophisticated content-oriented search methods which overcome the limitations of traditional paper-based publications. Nevertheless, the primary goal of scientific discovery should be quality rather than "volume" or "time to market," independent of the publication media used. Hence, the definition, implementation, and control of appropriate quality-ensuring procedures will become even more important in a world of fast-growing knowledge sources. Within this context, the efforts of the International Ergonomics Association, which advertises an IEA Press label guaranteeing high-quality contents, should be taken into account,.

Acknowledgments

The authors would like to thank Claudia Peters and Ralf Hunecke for the preparation of the empirical data. In addition, we owe special thanks to Taylor & Francis for their valuable contribution of the *Ergonomics Abstracts* on CD-ROM.

References

Becker-Biskaborn, G.U., 1975, Ergonomische Erkenntnissammlung für den Arbeitsschutz mit Informationssystem, BAU Forschungsbericht Nr. 142, Band I und II, Wirtschaftsverlag NW, Bremerhaven.

Boff, K.R., Kaufman, L., Thomas, J.P. (ed.), 1986, *Handbook of Perception and Human Performance*, John Wiley & Sons, New York.

Easterby, R.S., 1967, Ergonomic checklists: an appraisal, *Ergonomics* 10 (5), pp. 549-556.

Fitts, P.M., 1951, Human engineering for an effective air navigation and traffic control system, National Research Council, Washington, D.C.

Friedrichs, J., 1975, *Methoden Empirischer Sozialforschung*, Rowohlt Verlag, Reinbek.

Frieling, E., Sonntag, K., 1987, *Lehrbuch Arbeitspsychologie*, Verlag Hans Huber, Bern.

Helander, M.G., Landaver, T.K. Prabhu, P.V. (eds.), 1997, *Handbook of Human–Computer Interaction*, second edition, North-Holland, Amsterdam.

ILO, 1996, Ergonomic checkpoints — Practical and easy-to-implement solutions for improving safety, health and working conditions, Geneva, Switzerland.

Kroemer, K.H.E. et al., 1988, Ergonomic Models of Anthropometry, Human Biomechanics, and Operator-Equipment Interfaces, Proceedings of a Workshop, National Academy Press, Washington, D.C.

Laurig, W., Rombach, V., 1989, Expert systems in ergonomics: requirements and an approach, *Ergonomics* 32 (7), pp. 795-811.

Luczak, H., Volpert, W., 1987, *Arbeitswissenschaft. Kerndefinition–Gegenstandskatalog–Forschungsgebiete*, RKW-Verlag, Eschborn.

MIL-STD 1472D, 1991, Human engineering criteria for military systems, equipment and facilities, the hypertext version is provided by CSERIAC, Crew System Ergonomics Analysis Center, Wright-Patterson AFB, OH.

N.N., 1964, Ergonomic system analysis checklist, International Ergonomics Association (editor), Proceedings of the 2nd World Congress of the IEA, Dortmund.

Oppermann, R., Murchner, B., Reiterer, H., Koch, M., 1992, *Softwareergonomische Evaluation (EVADIS II)*, 2nd edition, Walter de Gruyter, Berlin.

OSF/Motif, 1993, *Style Guide — Release 1.2*, Prentice Hall, Englewood Cliffs, NJ.

Rasmussen, J., 1983, Skills, rules, knowledge: signals, signs, and symbols and other distinctions in human performance models, *IEEE Transactions on Systems, Man, and Cybernetics* SMC-13 (3), pp. 257-267.

Rohmert, W., Becker-Biskaborn, G.U., 1974, Ergonomische Prüfliste für den Arbeitsschutz mit Literaturanhang, BAU Forschungsbericht Nr. 116, Wirtschaftsverlag NW, Bremerhaven.

Salvendy, G. (ed.), 1997, *Handbook of Human Factors*, 2nd edition, John Wiley & Sons, New York.

Salvendy, G. (ed.), 1992, *Handbook of Industrial Engineering*, 2nd edition, John Wiley & Sons, New York.

Tecmath GmbH, 1994, RAMSIS-Leistungsbeschreibung und Softwarehandbuch, Tecmath, Kaiserslautern.

Appendix

Advances in Human-Computer Interaction	Advances in Human-Computer Interaction Volume 1 - 4
Advances in Industrial Ergonomics	Advances in Industrial Ergonomics and Safety 1, I, II, III, IV, V, VI
Advances in Man-Machine Systems	Advances in Man-Machine Systems Research, Volume 1, 3, 4, 5
Analysis, Design and Evaluation of Man-Machine Systems	Analysis, Design and Evaluation of Man-Machine Systems 1988, 1989, 1992 Analysis, Design and Evaluation of Man-Machine Systems, Proceedings of the 2nd IFAC/IFIP/IFORS/IEA Conference, Varese, Italy, September 10-12 1985
Computers for Handicapped Persons	Proceedings of the 3rd International Congress on Computers for Handicapped Persons, 1992
Contemporary Ergonomics	Contemporary Ergonomics 1986 - 1995
Design of Work and Development of Personnel in Advanced Manufacturing	Design of Work and Development of Personnel in Advanced Manufacturing
Empirical Studies of Programmers	Empirical Studies of Programmers Empirical Studies of Programmers, Second Workshop Empirical Studies of Programmers, Fourth Workshop Empirical Studies of Programmers, Fifth Workshop
Environmental Ergonomics	Environmental Ergonomics, Edited by I.B. Mekjavic, E.W. Banister and J.B. Morrison, Taylor & Francis, London, 1988 Proceedings of the Fifth International Conference on Environmental Ergonomics, Maastricht, the Netherlands, 1992
Ergonomics of Hybrid Automated Systems	Ergonomics of Hybrid Automated Systems I - III
European Coal and Steel Community	European Coal and Steel Community Ergonomics Action Information Bulletin, 1986 European Coal and Steel Community Ergonomics Action, Luxembourg, Final Reports for various Projects
Graphics Interface	Proceedings of Graphics Interface '88, Edmonton, Canada, 6-10th June 1988 Proceedings of Graphics Interface '90, Halifax, Nova Scotia, 14-18 May 1990 Proceedings of Graphics Interface '91, Calgary, Alberta, 3-7 June 1991 Proceedings of Graphics Interface '92, Vancouver, British Columbia, 11-15 May 1992 Proceedings of Graphics Interface '93, Toronto, Ontario, 19-21 May 1993 Proceedings of Graphics Interface '94, Banff, Alberta, 18-20 May 1994

HCI	HCI '87: Social, Ergonomic and Stress Aspects of Work with Computers/Cognitive Engineering in the Design of Human-Computer Interaction
	HCI '89: Work with Computers: Organizational, Management, Stress and Health Aspects/Designing and Using Human-Computer Interfaces and Knowledge-Based Systems
	HCI '91: Human Aspects in Computing: Volume 1. Design and Use of Interactive Systems and Work with Terminals/Human Aspects in Computing: Volume 2. Design and Use of Interactive Systems and Information Management
	HCI '93: Human-Computer Interaction: Applications and Case Studies/Human-Computer Interaction: Software and Hardware Interface
Human Decision Making	Proceedings of the Fourth European Annual Conference on Human Decision Making and Manual Control, Zeist, the Netherlands, 28-30 May 1984
	Human Decision Making and Manual Control, 5th EAM, 1985
	Proceedings of the 7th Annual Conference on Human Decision Making and Manual Control, Paris, 18-20 October, 1988
	Proceedings of the 10th European Annual Conference on Human Decision Making and Manual Control, Liège, Belgium, 11-13 November 1991
	11th European Annual Conference on Human Decision Making and Manual Control, November 17-19, 1992
	Proceedings of the XII European Annual Conference of Human Decision Making and Manual
Human Factors and Ergonomics Society	Progress for People. Proceedings of the Human Factors Society 29th Annual Meeting, Baltimore, Maryland, September 29 - October 3, 1985, Edited by R.W. Swezey.
	The Human Factors Society, Santa Monica, California,
	Human Factors Society Bulletin, 1986
	Combining Human and Artificial Intelligence: A New Frontier in Human Factors, Proceedings of a Symposium Sponsored by the Metropolitan Chapter of the Human Factors Society, New York, 15 November 1984, Edited by G. Kohl and S.J. Nassau
	Ergonomics and Safety in the Workplace: Proceedings of the Europe Chapter of the Human Factors and Ergonomics Society Annual Meeting in Antwerp, November 1992
	Proceedings of Interface '93, Raleigh, North Carolina, May 5-8, 1993.
	Designing for Diversity. Proceedings of the Human Factors and Ergonomics Society 37th Annual Meeting, Seattle, Washington, October 11-15 1993
	People and Technology in Harmony. Proceedings of the Human Factors and Ergonomics Society 38th Annual Meeting, Nashville, Tennessee, October 24-28 1994
	Human Factors and Ergonomics Society, Santa Monica, California, 1994
	Human Factors and Ergonomics Society, Santa Monica, California, 1995
Human Factors and Industrial Design	Interface 87: Human Implications of Product Design, Proceedings of the 5th Symposium on Human Factors and Industrial Design in Consumer Products, Rochester, New York, May 13-15, 1987
	Interface 91: Proceedings of the 7th Symposium on Human Factors and Industrial Design in Consumer Products, Dayton
Human Factors and Power Plants	Conference Record for 1985 IEEE Third Conference on Human Factors and Power Plants, Monterey, California, 23-27 June 1985
	Conference Record for 1988 IEEE Fourth Conference on Human Factors and Power Plants, Monterey, California, June 5-9, 1988
	Conference Record for the 1992 Fifth Conference on Human Factors and Power Plants, Monterey, California, June 7-11, 1992
Human Factors in Computing Systems	Human Factors in Computing Systems III und IV
Human Factors in Information Systems	Human Factors in Information Systems: An Organizational Perspective, New Jersey 1991
Human Factors in Manufacturing	Proceedings of the 2nd International Conference on Human Factors in Manufacturing and 4th IAO Conference, Stuttgart, West Germany, 11-13 June 1985
	Proceedings of the 3rd International Conference on Human Factors in Manufacturing, Stratford-upon Avon, 4-6 November 1986
	Proceedings of the International Ergonomics Association Conference on Human Factors in Design for Manufacturability and Process Planning, Honolulu, Hawaii, 9-11 August, 1990
	Organization and Management of Advanced Manufacturing

Human Factors in Telecommunications	Proceedings of the 12th International Symposium of the Human Factors in Telecommunications, The Hague, the Netherlands, May 24-27, 1988
	Proceedings of the Tenth International Symposium on Human Factors in Telecommunications, Helsinki, Finland, 6-10 June 1983
	Proceedings of the 11th International Symposium on Human Factors in Telecommunications, Cesson Sevigne, France, 9-13 September 1985
	Proceedings of the 13th International Symposium on Human Factors in Telecommunications, Torino, 10-14 September 1990
	Supplement to the Proceedings of the 13th International Symposium on Human Factors in Telecommunications, Torino, 10-14 September 1990
	Proceedings of the 14th International Symposium on Human Factors in Telecommunications, Darmstadt, Germany, May 11-14, 1993
Human Factors in Organizational Design and Management	Human Factors in Organizational Design and Management, II - IV
Human Factors Association	Proceedings of the Human Factors Association of Canada, 18th Annual Meeting, Hull, Quebec, 27-28 September 1985
	Proceedings of the 19th Annual Meeting of the Human Factors Association of Canada, Richmond (Vancouver), British Columbia, August 22-23, 1986
	Proceedings of the 20th Annual Conference of the Human Factors Association of Canada, Montreal, Quebec, October 14-17, 1987
	Proceedings of the Human Factors Association of Canada 21st Annual Conference, Edmonton, Alberta, 14-16 September 1988
	Proceedings of the Human Factors Association of Canada 22nd Annual Conference, Toronto, Ontario, November 26-29, 1989
Human-Computer Interaction - Interact	Human-Computer Interaction - Interact '87 and '90
IEEE	IEEE Proceedings of the international Conference on Cybernetics and Society, Tucson, Arizona, November 1985
	IEEE Virtual Reality Annual Symposium, Seattle, Washington, September 1993
Industrial Ergonomics and Safety	Trends in Ergonomics/Human Factors III, Proceedings of the Annual International Industrial Ergonomics and Safety Conference Held in Louisville, Kentucky, U.S.A., 12-14 June 1986
	Trends in Ergonomics/Human Factors IV, Proceedings of the Annual International Industrial Ergonomics and Safety Conference Held in Miami, Florida, U.S.A., 9-12 June 1987
	Advances in Ergonomics and Safety I-VI, London 1989-1994
International Ergonomics Association	Proceedings of the Ninth Congress of the International Ergonomics Association, Bournemouth, 2-6 September 1985
	Ergonomics International 88, Proceedings of the 10th Congress of the International Ergonomics Association, Sydney, Australia, 1-5 August 1988
	Designing for Everyone: Proceedings of the 11th Congress of the International Ergonomics Association, Paris, 1991, Volume 1
	Proceedings of the 12th Triennial Congress of the International Ergonomics Association, Toronto, Canada, August 15-19, 1994, Volume 1
Man-Computer Interaction Research	Man-Computer Interaction Research - Macinter - I
	Man-Computer Interaction Research - Macinter - II
National Board/National Institute of/for Occupational Safety and Health	National Board of Occupational Safety and Health
	National Institute for Occupational Safety and Health
People and Computers	People and Computers III - IX, Cambridge University Press
	People and Computers: Designing for Usability
	People and Computers: Designing the Interface

SAFE Association	Proceedings of the 23rd Annual Symposium of the SAFE Association, Las Vegas, Nevada, December 1-5, 1985
	Proceedings of the 24th Annual Symposium of the SAFE Association, San Antonio, Texas, December 11-13, 1986
	Proceedings of the 25th Annual Symposium of the SAFE Association, Las Vegas, Nevada, November 16-19, 1987
	Proceedings of the Twenty-Sixth Annual Symposium of the SAFE Association, Las Vegas, Nevada, December 5-8, 1988
	Proceedings of the 27th Annual Symposium of the SAFE Association, New Orleans, Louisiana, December 5-8, 1989
	Proceedings of the Twenty-Ninth Annual Symposium of the SAFE Association, Las Vegas, Nevada, November 11-13, 1991
	Proceedings of the 30th Annual Symposium of the SAFE Association, Las Vegas, Nevada, November 2-4, 1992
	Proceedings of the 31st Annual Symposium of the SAFE Association, Las Vegas, Nevada, November 8-10, 1993
The Ergonomics of Manual Work	The Ergonomics of Manual Work
The Ergonomics of Working Postures	The Ergonomics of Working Postures
Vision in Vehicles	Vision in Vehicles, Proceedings of the Conference on Vision in Vehicles, Nottingham, 9-13 September 1985
	Vision in Vehicles - II, Amsterdam 1988
	Vision in Vehicles - III, Amsterdam 1991
	Vision in Vehicles - IV, Amsterdam 1993
Visual Search	Visual Search, London, 1990
	Visual Search 2, London, 1993
Work with Display Units	Work with Display Units 86, Amsterdam, 1987
	Work with Display Units 89, Amsterdam, 1990
	Work with Display Units 92, Amsterdam, 1993

4

A Guide to Certification in Professional Ergonomics

Dieter W. Jahns
*Board of Certification in
Professional Ergonomics*

4.1   Introduction

Some form of quality assurance efforts are natural to most professions. These generally involve development of credentialing for educational programs and/or of individuals. Three types of processes are most common: *Accreditation* is established for the regulation of instructional programs. It is voluntary and generally developed and administered by an association of professionals within the field. *Certification* involves a voluntary process of evaluation and measurement of individuals which can then indicate whether they have achieved a professional level of qualifications as judged by professional peers. It is developed and administered by a professional association or a group specifically established for professional development purposes. *Licensure*, while it does credential individuals, is a mandatory process and is administered by a political or governing body. When laws are implemented "to protect the public" from unprofessional practices, it becomes illegal to practice one's profession without a license. Thus, these processes are distinguishable by three aspects: (a) the recipient of the credential, (b) the credentialing body, and (c) the degree of volunteerism involved in obtaining the credential (Jahns, 1991).

Slappendel (1994) reviewed nine ergonomics certification/registration programs in operation around the world. Her findings are summarized in Tables 4.1 and 4.2. The IEA is currently working on policies and procedures to endorse and harmonize credentialing organizations on an international scale. Since IEA (International Ergonomics Association) Federated Societies are more oriented towards information dissemination, and not so much toward control of the profession as a guild structure, there is an increasing trend for cooperative, yet independent credentialing agencies. In "open-market" societies there are also opportunities for sham operators, which makes a supervisory role by IEA Federated Societies desirable. This is happening. For example, the Human Factors Association of Canada (HFAC/ACE) recognizes BCPE (Board of Certification in Professional Ergonomics) as a valid and reliable certification organization operating simultaneously with its own efforts to develop certification processes and criteria for Canadian ergonomists. BCPE has also become a consultant to efforts in South Africa and Japan on an informal basis.

Similarly, in Europe, CREE (Centre for Registration of European Ergonomists) works with the ergonomics societies of member countries in the European Union in evaluating and registering applicants

TABLE 4.1   Certification of the Ergonomist: Programs in Operation as of May 1994*

Certification/Registration Authority		Designation	Acronyms
Non-Society	Board of Certification in Professional Ergonomics (BCPE) U.S.A.	Certified Professional Ergonomist Certified Human Factors Professional Ergonomist in Training: Associate Ergonomics Professional Associate Human Factors Professional	CPE CHFP AEP AHFP
	Centre for Registration of European Ergonomists (CREE) The Netherlands	European Ergonomist	EurErg
	Stichting Registratie ergonomen (SRe) Netherlands	Registered Ergonomist	R.e.
Society	Professional Affairs Board PAB) of The Ergonomics Society, U.K.	Registered Member of the Ergonomics Society (Professional Member) Fellow of the Ergonomics Society Practitioner of the Professional Register	M.Erg.S. F.Erg.S.
	Professional Affairs Board (PAB) of the Ergonomics Society of Australia	Certified Professional Member	C.Erg.
	Membership Subcommittee of the New Zealand Ergonomics Society	Professional Member	M.NZ.Erg.S

* Programs are also in operation in France, Belgium, and Sweden, but information on these was unavailable at the time of writing.

Source: Slappendel, C. 1994. Harmonising the different approaches to the certification of the ergonomist. *Proceedings of the 12th Triennial Congress of the IEA*, Toronto, ON, Canada.

TABLE 4.2   Criteria Applied in Certification Programmes

Designation	Criteria	Recertification
Certified Professional Ergonomist/Certified Human Factors Professional (BCPE)	Masters degree in ergonomics (human factors) or equivalent, *plus* 4 years of full-time professional practice in ergonomics with emphasis on ergonomic design, *plus* submission of a work product, *plus* a passing score on a written certification examination	Not required yet; annual renewal fee
European Ergonomist (CREE)	At least 3 years of academic formation in any field of which the total amount of education in ergonomics is at least 1 year, *plus* at least 1 year of training, plus at least 2 years of experience	Registration is for a 5-year period
Registered Ergonomist (SRe)	Not specified, but are in line with CREE criteria	Every 3 years
Registered Member of The Ergonomics Society (a.k.a. Professional Member)	At least 3 years (or part-time equivalent) in the practice of ergonomics, and/or teaching and/or research of ergonomics relevance since admission to the Society, *plus* evidence of academic achievements	Not required
Fellow of The Ergonomics Society	Registered Member for at least 6 years plus significant contribution to the practice of, teaching of, and/or research in ergonomics for a period of 10 years since becoming an Ordinary Member *plus* substantial contribution to the activities of the Society	Not required
Practitioner on the Professional Register of The Ergonomics Society	Must be a Registered Member of the Society *plus* a minimum of 3 years in active practice during the preceding year	Every 3 years
Certified professional member of the Ergonomics Society of Australia	A suitable qualification *plus* 3 years full-time equivalent experience in the practice of ergonomics	Required
Professional member of the New Zealand Ergonomics Society	A tertiary qualification in ergonomics, or a qualification of which ergonomics made up a substantial portion of the course content, *plus* experience in the practice of ergonomics, or teaching or research of ergonomics relevance	Not required

Source: Slappendel, C. 1994. Harmonising the different approaches to the certification of the ergonomist. *Proceedings of the 12th Triennial Congress of the IEA*, Toronto, ON, Canada.

for the "Eur. Erg." designation. The BCPE and CREE have a reciprocity agreement in place. As CREE President E. N. Corlett (1996) wrote: "Our policy at the moment is to be linked with only one registering body in each country. Because of our constitution, this body has to have certain requirements, as laid out in the European Standard 45013 to which we adhere. We have confirmed that BCPE fulfills these requirements."

As a member of the U.S.-based National Organization for Competency Assurance (NOCA), the BCPE has used the following checklist (provided by the National Commission for Certifying Agencies) for tracking its program relative to accepted criteria:

1. Agency's purpose must be certification of individuals. (Yes)
2. Agency is nongovernmental. (Yes)
3. Agency is national in scope. (Yes — international)
4. Agency is administratively independent. (Yes)
5. Certificants are represented on governing board. (Yes)
6. Governing body is selected from certificants (Yes)
7. Governing body does not select its successors. (No)
8. Governing body has public member. (No)
9. Agency is separate from any accrediting body. (Yes)
10. Agency has completed two national examination administrations. (Yes)
11. Agency has adequate funding. (Yes — barely)
12. Agency has adequately trained staff. (Yes)
13. Certificant evaluation is fair and based on appropriate knowledge and skills. (Yes)
14. Agency reviews evaluation process periodically. (Yes)
15. Agency has adequate examination security. (Yes)
16. Agency sets pass/fail levels appropriately. (Yes)
17. Agency maintains statistical data on the assessment methods. (Yes)
18. Agency publishes certification criteria and descriptive information. (Yes)
19. Agency has nondiscrimination policy. (Yes)
20. Agency has test examination accommodation policy. (Yes)
21. Agency documents uniform policies. (Yes)
22. Agency periodically reviews certification criteria and examinations to show uniform enforcement. (Yes)
23. Agency publicizes process and results nationally. (Yes — internationally)
24. Agency offers examinations in multiple locations. (Yes)
25. Agency has policy on alternate eligibility options. (Yes)
26. Agency notifies examinees promptly of scores, and failures are given specific deficiencies. (Yes)
27. Agency keeps examination scores confidential. (Yes)
28. Examinees may challenge examination results. (No)
29. Agency has policy giving grounds for refusing eligibility. (Yes)
30. Agency maintains a register of certificants. (Yes)
31. Agency maintains an ethics and discipline code. (No)
32. Agency has recertification procedure. (Not yet)

The candidates for certification usually follow the pathways shown in Figure 4.1 (solid lines) by contacting either the certification agency directly or by inquiring of one of the IEA Federated Societies, which then coordinates the certification procedures. Both BCPE and CREE have highly coordinated information exchanges (dashed lines in Figure 4.1) with the IEA and selected, regionally active Federated Societies to harmonize the professional development of ergonomists. Interested readers can contact the organizations listed in "For Further Information" at the end of this chapter. A general overview of BCPE certification criteria and procedures is given next.

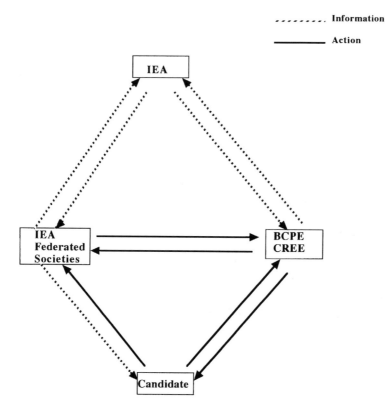

FIGURE 4.1   Communication and actions among ergonomics societies and certification agencies.

4.2   BCPE Certification Requirements

Criteria for Certification

1. CPE/CHFP

   The BCPE job/task analyses led to the following minimum criteria for certification:

   - a master's degree in ergonomics or human factors, or an equivalent educational background in the life sciences, engineering sciences, *and* behavioral sciences to comprise a professional level of ergonomics education, and
   - four years of full-time professional practice as an ergonomist practitioner with emphasis on design involvement (derived from ergonomic analysis and/or ergonomic testing/evaluation), and
   - documentation of education, employment history, and ergonomic project involvement by means of the BCPE "Application for Certification," and
   - a passing score on the BCPE written examination, and
   - payment of all fees levied by the BCPE for processing and maintenance of certification, in accordance with the following schedule:

            Application Fee: U.S. $10.00

            Processing and examination fee: U.S. $290.00

            Annual certificate maintenance fee: U.S. $100.00

2. AEP/AHFP

On March 26, 1995, the BCPE created a new "Associate" category of certification. As of January 1996, a person can be certified by the BCPE as an "Associate Ergonomics Professional" or an "Associate Human Factors Professional" if he or she:
- meets the education requirements for BCPE certification (M.S. in human factors/ergonomics or related field),
- has passed Part-I (on "Basic Knowledge" of human factors/ergonomics) of the BCPE certification examination, and
- is currently working toward fulfilling the BCPE requirement of four years' practical experience as a human factors and ergonomics professional.

Thus, the AEP/AHFP designation is a preprofessional category, while the candidate gains experience. A person can take the Basic Knowledge portion (Part-I) of the BCPE exam immediately after fulfilling the education requirement. Parts-II and III of the exam may be taken after fulfilling the other BCPE requirements. *A person who has graduated from a human factors/ergonomics degree program accredited by an IEA Federated Society (e.g., HFES), will not have to take Part-I of the exam.* BCPE established the "Associate" category to create a path by which individuals could achieve professional certification in progressive steps. The Board also wanted to link this path with accredited educational degree programs by giving some preference (by waiving Part-I of the exam) to applicants who have graduated from programs accredited by IEA Federated Societies. This linkage is an important step in strengthening the human factors and ergonomics profession.

Fees associated with the AEP/AHFP designations are:

Application fee: U.S. $10.00
Processing and examination: U.S. $100.00
Annual AEP/AHFP Maintenance fee: U.S. $60.00

3. CEA

As of July 1998, the new designation CEA (Certified Ergonomics Associate) is available to individuals who:
- have a Bachelor's degree, or equivalent education, in ergonomics (or in one of the scientific disciplines closely related to ergonomics)
- have at least 200 contact hours of ergonomics training commensurate with the "Ergonomist Formation Model (EFM)" adopted by BCPE
- have at least two years of full-time equivalent practice in ergonomics
- obtain a satisfactory score on the BCPE, two-part, multiple-choice answers examination on ergonomics foundations and ergonomics practice methods

These individuals, once obtaining their CEA designation, are expected to be involved in more limited activities of system evaluation and intervention than are CPEs, who have a broader and scientifically deeper involvement with the multidimensionality of work systems development. The CEAs are most likely to use well-established ergonomics principles, methods, and tools in an "intervention strategy" and as an adjunct to reactive workplace issues deriving from worker's compensation insurance, labor negotiations, regulatory compliance audits, and/or standardized ergonomics guidelines.

Fees associated with the CEA designation are:

Application fee: U.S. $10.00
Processing and examination: U.S. $190.00
Annual CEA Maintenance fee: U.S. $75.00

Procedures for Certification

1. The candidate requests application materials by sending a U.S. $10.00 check to: BCPE, P.O. Box 2811, Bellingham, WA 98227-2811. Materials consist of four pages of instructions and seven pages of forms to be filled out by the applicant:
   Section A— Personal data
   Section B— Academic qualifications
   Section C— Employment history
   Section D— Work experience (in ergonomic analysis, design, and testing/evaluation)
   Section E— Work product description (CPE/CHFP only)
   Section F— Signature and payment (US $290.00) record
2. The candidate completes the application and submits it with an official academic degree transcript and a work product (article/technical report/project description/patent application, etc.).
3. The review panel evaluates all submitted materials and makes recommendations to the Board as to whether or not the applicant qualifies to take the written examination.
4. The candidate becomes certified by taking and passing the written examination.

Examination Administration

Applicants who have demonstrated eligibility for the examination will be notified regarding the date and location of the next examination approximately two months before the test date. The examination, requiring a full day, will generally be scheduled for the spring and fall, and will probably be offered as an adjunct to the meetings of ergonomics-related professional societies and associations. Qualified applicants needing accommodations in compliance with the Americans with Disabilities Act (ADA) are asked to specify their accommodation needs to the BCPE prior to signing up for the examination.

Scoring Methods

A panel of prior BCPE certificants with expertise in psychometrics determines the method of establishing passing scores to be used for the examination. Passing scores are established to ensure the applicant's mastery of the knowledge and skills required for professional ergonomics practice. The BCPE will periodically review, evaluate, and, as necessary, revise the examination and scoring to assure that valid and reliable measures of requisite performance capability for ergonomics practice are maintained.

Retaking the Examination

An applicant who does not pass the examination may retake it at the next regularly scheduled examination date and place. A reduced fee will be applied to retakes, and the retake must take place within two years of the original application date.

BCPE Examination: Approximate Weighting of Subject Areas

1. Methods and Techniques (M&T)   30%
2. Design of Human–Machine Interface (DHMI)   25%
3. Humans as Systems Components (Capabilities/Limitations) (HSC)   25%
4. Systems Design and Organization (SDO)   15%
5. Professional Practice (PP)   5%

References

Jahns, D. W. 1991. Certification of professional ergonomists: a status report. *Proceedings 24th Annual Conference of the HFAC/ACE*, Vancouver, B.C., Canada.

Corlett, E. N. 1996. Personal communication, letter dated 12/11/1996.

Slappendel, C. 1994. Harmonising the different approaches to the certification of the ergonomist. *Proceedings of the 12th Triennial Congress of the IEA*, Toronto, ON, Canada.

For Further Information

The Board of Certification in Professional Ergonomics (BCPE)

P. O. Box 2811
Bellingham, WA 98227-2811 U.S. A.
Telephone: 360-671-7601
Fax: 360-671-7681
e-mail: BCPEHQ@aol.com
WWW Home Page: http://bcpe.org

Center for Registration of European Ergonomists (CREE)

Franz Witt, Administrative Director
Aldrinaweg 3
NL-9892 PE Feerwerd
Netherlands
Telephone: 1-31-59-411971
Fax: 1-31-50-134104

National Commission for Certifying Agencies

(A Division of NOCA)
1200 19th Street NW #300
Washington D. C. 20036-2401
Telephone: 202-857-1165
Fax: 202-223-4579

5

Effective Use of the World Wide Web

Peter M. Budnick

ErgoWeb Inc.
University of Utah

5.1 Introduction

The Internet, and the World Wide Web in particular, are revolutionizing the way people seek and access information. For a growing number of people, research begins with the Web, "surfing" around the world in search of the electronic equivalent of a library book — or the electronic equivalent of an art gallery; or, the on-line version of a favorite software program; or, a live or bulletin-board style discussion with others. Or — and this may be the ultimate beauty and utility of the Web — you may find all these elements intertwined into one system through the effective use of "hypermedia," the interactive multimedia capabilities that make up the Web. Using the Web, specialists are communicating and sharing information and ideas faster and more efficiently than ever before, breaking down the traditional barriers of distance and borders, and the information produced by specialists is being distributed to a wider population than ever before.

To traditionalists and hard-copy purists, this may seem a disturbing trend. To this author, it is an exciting human progression, and one that will redefine the level and availability of knowledge around the world. A traditionalist may argue that the speed and ease with which information is distributed in the electronic environment makes it possible for dubious or untested theories to be widely broadcast and accepted as fact by the untrained mind. It is true that this may occur, however, the benefits that improved communications bring far outweigh the potential downsides or misuse of the medium. Further, in disciplines like occupational biomechanics or ergonomics, releasing the knowledge from the world of research to the everyday world of the working person may be the key to practical success. Theorizing and academic debate will continue in the electronic medium, but only a medium like the Web can take the results of such specific bodies of knowledge and bring them to the people who need to know, transcending national, cultural, and corporate borders. It is the understanding and application of ergonomic principles in the everyday working life of individuals that will make work more productive, safer, more fulfilling, more dignified, and ultimately better for both employers and employees.

The Web is constantly evolving, and information, service, and product providers appear and disappear each day. On one hand, it can seem a disorganized environment, bringing equal parts of annoyance and satisfaction. However, as the Internet has proved, this chaotic, even anarchistic environment will self-organize through voluntary persuasion and innovation on behalf of the interested participants. What we know today as the Web was preceded by a number of voluntary Internet protocols that took a set of mere communication links and developed highly organized systems, including e-mail, Telnet, FTP, Gopher, and Usenet. (See the glossary of terms at the end of this chapter if these terms are new to you.) E-mail formed the basis for personal communication; Telnet provided a common standard for communicating between different computer hardware locations; FTP (File Transfer Protocol) provided a standard method to send or retrieve electronic files, such as data sets, text files, and digitally stored images; Gopher provided an organized and easily accessed indexing method for larger stores of information; and Usenet organized millions of unrelated public e-mail discussions into thousands of topic-specific discussions, allowing you to pick and choose only the information you desire to monitor, or to which you wish to contribute. These advancements were all developed and implemented voluntarily.

The Web was a great leap forward in the organization and accessibility of the Internet. Yet, as captured in the name "Web," this new communication protocol is more of an interface advancement than it is an organizational advancement. Anyone who has "surfed" for any amount of time will attest to the fact that it is easy to rush forward (in terms of the excitement of a new frontier) and get lost (in terms of not finding what you set out to find) in this "Web." It is now up to individuals and organizations to organize the knowledge on the Web so that you can locate what you need, when you need it.

It is the distributed design of the Internet communication network that lends to the distributed nature of the Web. The Internet was intentionally designed to be a system of numerous and redundant links between computer hardware spread across wide geographic locations. This design provided a nearly fail safe communications network no matter how large the local disturbance. In summary, when a message is sent across the Internet, it is disassembled into smaller "packets," which are in turn redundantly sent along a variety of different routes through the network. If a "packet" encounters a break in the system, it gets rerouted around the disturbance. At the message destination, the arriving "packets" are reassembled into the original message. So, there is no central path through which Internet traffic passes. (At least not at this date, in the United States. Conceivably, some will design networking so that all traffic passes through central locations which can be more easily monitored and controlled.) Thus, there is no central authority nor location at which to intervene and organize the vast amounts of knowledge available on the Web.

Individuals and organizations are quickly organizing information on their own, however. Such Web sites form the building blocks for a vast body of distributed knowledge, methodologies, etc., yet they exist independently and are often unknown to each other. The Web provides the protocol necessary for these dispersed sites to easily "link" to each other, allowing the Web user to move easily from one source to another. Thus, at the lowest level, a site is organized locally in some way. The next level of organization involves linking from one site to another. This may be anonymous, or it may be a reciprocal arrangement with mutual links. In this way, a group of sites may form a level of organization among themselves. This is good for the user who happens to locate a site that either contains the information sought, or links to others that do, but it still presents a barrier to the user that does not have a starting place in this Web. "Search engines," or meta-indexing sites are the organizational solution that has risen on the Web to help users locate and focus on the information they seek.

This chapter seeks to provide you with a sense of how the Web and the Internet work, but more important, the author hopes to show how you can successfully utilize this excellent resource. The chapter contains no list of Web sites dealing with occupational biomechanics or ergonomics, since such a list would likely be obsolete by the time this is published. Also, you will find no graphics reproduced from the Web for this chapter, because a black and white book format cannot capture the vivid colors and animation frequently encountered on the Web. After reading this chapter, though, you should know how to locate them on your own by using the Web effectively and responsibly.

To demonstrate the effective use of the Internet and the Web in research, this author has deliberately performed all research for this chapter using only the Web. This brings one of the many issues of protocol and standardization to mind: how should one reference an electronic site? Often, an author's name will not appear on a Web page, nor will a date, nor will the location of the publisher. Further, if it is a large compilation of information, there will rarely be page numbers or any other traditional means to identify a place relative to the larger collection. So, the first task is to settle on a referencing style that will help you find the information referred to here, if it still exists when you read this (see the section titled "Finding

What you Need on the Web"). Walker (1996) recommends the following general style for Web sites (she also addresses referencing styles for other Internet protocols):

> Author's Last Name, First Name. "Title of Work." Title of Complete Work.
> [protocol and address] [path] (date of message or visit).

Applying this style to the Walker (1996) page referenced here,

> Walker, J.R., 1996, "MLA-Style Citations of Electronic Sources,"
> http://www.cas.usf.edu/english/walker/mla.html, (19 Jan 1997).

Next, it is instructive to review some of the terms that are commonly used when discussing Internet-based materials. Therefore, the author has compiled a glossary of terms at the end of this chapter that you may want to refer to while reading.

5.2 Copyright Issues

The proliferation of materials on the Internet and World Wide Web has forced many to revisit the issues surrounding intellectual property rights. The Internet makes widespread distribution of multimedia (text, images, video, and sound) easier than ever before. Further, the culture of the Internet, which, until recently, was populated primarily by academics and researchers, is one of sharing information, making it simple to broadcast information of interest to a huge audience at the mere touch of a few keys on a keyboard. This author will not discuss the philosophical aspects related to greater information exchange here, but some mention of copyright issues and the evolving legal concerns is very important for Web users and developers alike. For the interested reader, the author recommends visiting "Copyright and Intellectual Property Rights for Digital Documents" (Berkeley Digital Library, 1996), a Web Page with a comprehensive set of links related to this issue. You are also encouraged to consult with the legal authorities representing your organization if you have any questions.

Many believe that in order for a document to be copyrighted, it must contain a copyright statement. However, according to Stanford University (1996), "Currently, the author's rights begin when a work is created. Copyrighted works are not limited to those that bear a copyright notice." So, one should always assume an electronic document is copyrighted material, and that permission must be requested and granted from the copyright owner in order to use it in a publication or to distribute it to others. Copyrighted materials may be written documents, photographs, electronic images, video, software, databases, any digital works or works transformed into digital format. Copyrights are protected regardless of the medium in which they are created or reproduced.

There are many instances in which the definitions of words like "publish" and "distribute," for instance, are vague. In such cases, you should consider what is commonly referred to as "fair use." O'Mahoney (1995-1996) proposes the "Fair Use Test," based on the "Fair Use" provision in U.S. copyright law. Briefly, the fair use test requires consideration of four factors:

Factor 1 — Purpose and Character of Use
Factor 2 — Nature of Copyrighted Work
Factor 3 — Relative Amount
Factor 4 — Effect on the Market

Interpretation of these factors, and thus determination of what fair use of a particular media is, is constantly evolving. Stanford University (1996) provides one current interpretation:

- The purpose and nature of the use — If the copy is used for teaching at a nonprofit institution, distributed without charge, and made by a teacher or students acting individually, then the copy is more likely to be considered as fair use. In addition, an interpretation of fair use is more likely if the copy was made spontaneously, for temporary use, not as part of an "anthology," and not as an institutional requirement or suggestion.

- The nature of the copyrighted work — With multimedia material, there are different standards and permissions for different media: a digitized photo from a *National Geographic,* a video clip from *Jaws,* and an audio selection from Peter Gabriel's CD would be treated differently — the selections are not treated as equivalent chunks of digital data.

- The nature and substantiality of the material used — In general, when other criteria are met, the copying of extracts that are "not substantial in length" when compared to the whole of which they are a part may be considered fair use.

- The effect of use on the potential market for or value of the work — In general, any use that supplants or diminishes the normal market for the original work is considered an infringement, but a use does not have to have an effect on the market to be an infringement.

The authors caution that the last factor is the most important consideration.

There are works that are considered to be in the "public domain" and can be copied freely by anyone. These include works by the U.S. government (and presumably other governments), works for which a copyright has expired, and works which the owner has explicitly identified as public domain.

When in doubt about whether something is copyrighted, or whether a particular use falls within the "fair use" provisions, always seek permission from the author or copyright owner.

5.3   User Interface Issues

There are many issues to consider when designing Web-based materials. These include careful consideration of file size, layout of the page for effective viewing, and browser compatibility with selected page formatting codes (i.e., HTML). User interface design goes far beyond the confines of this chapter, but a few things are worth mentioning if you decide to produce Web pages.

File size can define whether your pages will ever be viewed. Web users quickly become impatient while waiting for large files to download over a modem or other slow Internet connection. Therefore, you should take care to reduce the number of images contained in a page, or minimize the file size of the images you do include. Sometimes there will be a trade-off between download time and the desired visual effect of your pages, since graphic files often account for the majority of the transfer delay. (It is also important to have good Internet connections and fast computer hardware and networks, but that too is well beyond the scope of this chapter.)

The purpose of a page will often define the size of the file. For example, if you intend a particular document to be printed by the user, you will want to include the entire document in one continuous page, even though it may take longer to download the file. Alternatively, you might provide a link to a compressed downloadable file formatted in a common word processor format, allowing users to print the formatted document on their own, rather than viewing and printing directly from the Web. Presumably, the user will wait the extra time in order to obtain the full document.

On the other hand, if a page contains intermediate information, such as an index page that is just one step for the user in a search for more specific information, you should take care to design a smaller file set that will download quickly. Readers interested in designing effective Web pages may refer to, for example, Diehn and Katz (1996). Among other suggestions, these authors recommend that developers minimize bandwidth impacts; test graphics on a variety of computers; make sure graphics add value; and restrict navigation pages to one screen.

The look and feel of a Web page is influenced by the capabilities and limitations of the HTML formatting language and the speed limitations related to the use of graphics, as noted above. An additional barrier to achieving a desired look is the difference in browser capability. Since the Web is evolving fast and there are no universally accepted development standards, different browser software will interpret the same HTML code in different ways. That is, what looks "good" in one browser, may look "bad" in another. For this reason, you may want to minimize the use of advanced HTML features that have not yet been incorporated into the browsers commonly available and in use among the general public.

Navigation is another important point in Web site design. Web users often complain of "getting lost." While some of this is under the control of the user and the particular browser being used, page designers can take steps to assist navigation, at least on the pages under their control. There are a number of methods one can use to assist users, but the important thing to keep in mind is that the user should be provided with some referencing method that defines the relationship between pages within a given site, such as a navigation bar that appears on every page, allowing the user to quickly jump to a primary page when desired.

If you plan to develop Web pages, here are a few places you can go to learn the basics and the latest advancements.

- The Netscape® "Creating Net Sites" (1997) maintains a set of HTML guidelines and pointers to other Web sources for design assistance.
- World Wide Web Consortium is an international organization founded in 1994 to develop common standards for the evolution of the World Wide Web. This is a good starting place for many Web topics, but with respect to HTML and Web page style guides, see "HyperText Markup Language (HTML)" (1996) and "Web Style Sheets" (1996).
- "HTML and Style Manuals," provided by "InfoQuest" (1996) at http://www.fptoday.com/htmlandstyle.htm provides a nice index of style guides.

5.4   Finding What You Need on the Web

The first few times "surfing" the Web can be exhilarating (so can the rest, for that matter, if you learn to utilize it effectively). The vast amounts of information, a connection to people around the world, the innovative pioneering use of multimedia, and all the possibilities have been likened to an addiction. Not to fear, however, the thrill can quickly dissipate if you merely bounce from site to site without ever honing in on the materials you seek. As Digital Corporation notes at the AltaVista® Web searching and indexing site,

> "The Web is immense. If you only spent a minute per page and devoted ten hours a day to it, it would take four and a half years to explore a million Web pages, a lifetime to explore just this index."

Frustration can be minimized, though, by the effective use of search engines and Web indexes.

With the explosive growth of the Web, any list of specific sites compiled today may well be obsolete in a very short time. Therefore, it is best to review the search methodologies you should apply when you turn to the Web to locate and gather information. To illustrate, this author will take you through a series of searches performed while preparing this chapter.

There is a growing number of Web sites dedicated to searching for specific information, sites, people, and places on the Web. Generally called "search engines," they may utilize specialized software programs that "crawl" the Web reading documents and organizing them into a master index at the home site (e.g., AltaVista®, Excite®, Lycos®). Others are simply a large categorical index of links submitted by interested parties (e.g., Yahoo®). The search engines may be organized in a browsable hierarchical index, in which the user narrows a search by point-and-click.

For example, setting out to find "ergonomic" sites, one might choose the broad category of "science and engineering," which then brings up a number of subcategories falling under that topic. Eventually, you might expect, you will find the topic you seek. However, a different user with another viewpoint may begin searching for "ergonomic" by looking at the broad category of "workplace health and safety." Whether "ergonomic" will be found through either or both of these paths depends on the sophistication and breadth of the indexing method. By experience, this author believes few, if any, indexes have successfully developed a robust indexing system that can capture the expectations of users arriving with varying viewpoints on the same topic.

Other search engines rely on keyword searching schemes or a combination of keyword searching and hierarchical indexes. Effectively used, keyword search methods can be an excellent way to quickly narrow a search to Web pages of interest. However, without taking the time to learn the advanced search methods

TABLE 5.1   Keyword Search Results for "Ergonomic" and "Ergonomics" on Three Example Search Engines

Keyword	Search Engine	Number of Pages Found
ergonomic	AltaVista®	about 40,000
	Excite®	17,481
	HOTBOT®	29,216
ergonomics	AltaVista®	about 40,000
	Excite®	21,297
	HOTBOT®	36,564

TABLE 5.2   Keyword Search Results for the String "Occupational Biomechanics" (without using the quotation marks) on Three Example Search Engines

Keywords	Search Engine	Number of Pages Found
occupational biomechanics	AltaVista®	about 30,000
	Excite®	134,685
	HOTBOT®	2,341

TABLE 5.3   Keyword Search Results for the String "Occupational Biomechanics" on Three Example Search Engines

Keyword(s)	Search Engine	Number of Pages Found
occupational biomechanics	AltaVista®	about 200
	Excite®	1,139
	HOTBOT®	353

employed by the particular site, a user may spend hours wandering through unrelated pages. To illustrate, the author performed one-word keyword searches on three different keyword-driven search engines. The results are summarized in Table 5.1.

Without reading any instructions, the author then did a search for the two-word string "occupational biomechanics" (without using quotation marks). The results indicated that each search engine treats this string differently, which helps build the case that to successfully utilize the Web, one must become very familiar with the syntax and capabilities of at least one search engine. The results are compiled in Table 5.2.

After reading the instructions, the author learned that the search engines were not searching for the phrase "occupational ergonomics," but were instead searching for the booleans "occupational" *OR* "ergonomics," or "occupational" AND "ergonomics" (neither site was clear regarding this issue). Using the various syntax required by the different search engines, the author then performed the search for the string "occupational biomechanics," and received the results shown in Table 5.3.

This did narrow the search but still produced a large number of selections from which to choose. So, in order to narrow the search further, the author consulted the "advanced search" or "help" links at each search engine and quickly learned that each system uses different syntax and search methodologies, and the learning curve can be steep. Some search engines may be exclusively keyword based, while others will employ proprietary "intelligent" search methods that attempt to look at related concepts, related words, and so forth. The advanced search methods also provided options such as search location (the Web, Usenet, etc.), date limits, boolean operators, results ranking, case sensitivity, and more. These advanced features make it possible to narrow a search to an exact document, or narrow a topical search to a manageable set of pages to peruse.

Spending the time to learn and use search engines will save you significant amounts of time when doing research on the Web. However, search engine designers have a long way to go before they make searching easy for the casual user, and this is an area where ergonomics and human factors experts could make substantial contributions.

Glossary of Terms

Browser: The software program installed on a client computer through which the user accesses the Web. This might be thought of as the "window" to the Web. Netscape Navigator® and Microsoft Internet Explorer® are two example browsers.

CGI: "Common-Gate Interface," a protocol allowing Web servers to communicate with server-based computer programs. For example, if calculations are to be performed, the server may call on a server-based computer program to perform those calculations. Once the calculations are complete, the server sends the results to the client through the network.

Client: The computer (or computer user) that "requests" a document from a remote server.

Domain: An Internet address conforming to the Internet Protocol (IP), such as fictitiouscompany.com. This evolving naming convention varies from country to country. In the U.S., the last three characters will often represent the nature of the organization, such as "com" for commercial enterprises, "edu" for educational institutions, "mil" for military, and so on. Other countries follow this convention as well, but add a country designation at the end. Some countries have developed their own domain-naming methodology, usually ending with a country designation.

Email: Protocols which allow individuals to send and receive electronic messages. Once restricted to text only, the protocols are evolving to include hypermedia capabilities.

FAQ: "Frequently Asked Questions." FAQs are common documents that summarize and answer frequently asked questions regarding a topic or a specific site. FAQs can provide a quick study on the topic of interest and are often an excellent starting place.

FTP: "File Transfer Protocol," another method, like HTTP, defined above, in which computers may communicate across the Internet. (FTP may also be used via the Web.)

Home Page: A page dedicated to one individual, or the primary "front" page of a collection of Web pages.

HTML: "Hypertext Markup Language," the computer language used to format Web pages.

HTTP: "Hypertext Transfer Protocol," the common scheme through which Web servers communicate. Also see FTP, Telnet, Gopher, etc., in this definition list.

Hypermedia: The interconnected media of text, static graphics, animated graphics, video, and sound that are published on the Web.

Java: An evolving computer language (developed by Sun Microsystems) that is designed to operate across different computer operating systems (e.g., UNIX, Windows®, etc.). A Java "applette" (a mini software application) executes through the browser on the client side, as opposed to executing on the server side, like CGI applications.

Listserv: Listservs are software programs used to manage an e-mail discussion or distribution list. Users join "Listservs" when they want to monitor or participate in discussions (usually dedicated to a particular topic) with other Internet users via e-mail.

Navigating: Linking from page to page, and site to site on the Web. "Navigating" generally refers to successful, planned "surfing."

Netiquette: Commonly accepted or expected behavior while participating in Internet-based communications. Interested readers may refer to Rinaldi (1996) for a compilation of netiquette topics relating to many Internet communication protocols, including the Web.

Search Engine: A computer software system that locates and catalogs Web pages. Most search engines provide a keyword searching method for clients to locate and link to the cataloged pages.

Search Index, or Web Index: An index of Web links organized by topic that may or may not have been gathered by spiders, etc.

Server: The computer hardware that "serves" electronic documents on the World Wide Web.

Spider, Robot, Crawler, Worm, etc.: Terms used to describe the computer software programs that search the Web in conjunction with certain types of search engines.

Surfing: Linking from page to page, and site to site on the Web. "Surfing" often refers to a somewhat random path of linking (getting "lost" is not unusual for new users).

URL: An acronym for "Uniform Resource Locator," a draft standard for specifying an object on the Internet. Think of this as an address, such as http:www.fictitiouscompany.com. (See Connolly, 1990, for an in-depth discussion.)

Usenet: Thousands of discussions organized by topic. One can monitor a usenet group dedicated to discussions and information exchange on some topic by reviewing posted submissions from others, or one can participate by "posting" a message to the group.

Web Page: An electronic document formatted in the HTML language and made accessible to the Web through a server.

Web Site: A collection of HTML formatted documents that make up one primary server or collection of servers managed by one individual or organization.

World Wide Web: Hereafter referred to as the "Web," this is a distributed system of linked hypermedia documents spanning the world.

References

AltaVista®, AltaVista Technology, Inc.,
    http://www.altavista.com, (22 Jan 1997).

Connolly, D., 1990 (updated 1996), "Names and Addresses, URIs, URLs, URNs, URCs," World Wide Web Consortium,
    http://www.w3.org/pub/WWW/Addressing/Addressing.html, (19 Jan 1997).

Diehn, M. J., and Katz, M., 1996, "WTCS Workshop: Guidelines for Slim Web Pages," Marketing on the Web: Pre-workshop preparation for participants in the WTCS Web Workshop,
    http://owl.warren-wilson.edu/~mdiehn/wtcs/guide.htm, (22 Jan 1997).

Digital Corporation, 1996, "AltaVista Search: Surprise CyberSpace Jump," AltaVista ™ Search, (20 Jan 1997).

Excite®, 1996, Excite Inc.,
    http://www.excite.com, (22 Jan 1997).

Harnack, A. and G. Kleppinger, "Citing the Sites: MLA-Style Guidelines and Models for Documenting Internet Sources," Version 1.3,
    http://falcon.eku.edu/honors/beyond-mla/#citing_sites, (19 Jan 1997).

HOTBOT®, 1996-97, Hotwired Inc.,
    http://www.hotbot.com, (22 Jan 1997).

InfoQuest®, 1996, "HTML and Style Manuals," Front Page Today,
    http://www.fptoday.com/htmlandstyle.htm, (22 Jan 1997).

Netscape®, 1997, "Creating Net Sites,"
    http://www.netscape.com/assist/net_sites/index.html, (30 Jan 1997).

O'Mahoney, B., 1995-1996, "The Fair Use Test," The Copyright Website,
    http://www.benedict.com/fairtest.htm, (21 Jan 1997).

Rinaldi, A.H., 1996, "The Net: User Guidelines and Netiquette — Index," Florida Atlantic University,
    http://www.fau.edu/rinaldi/net/index.htm, (30 Jan 1997).

Stanford University, 1996, "Main Fair Use Index: Library Copyright Guidelines: Copyright Law: Frequently Asked Questions," Copyright & Fair Use,
    http://fairuse.stanford.edu/library/faq.html, (21 Jan 1997).

Walker, J.R., 1996, "MLA-Style Citations of Electronic Sources,"
    http://www.cas.usf.edu/english/walker/mla.html, (19 Jan 1997).

World Wide Web Consortium, 1996, "HyperText Markup Language (HTML),"
    http://www.w3.org/pub/WWW/MarkUp/

World Wide Web Consortium, 1996, "Web Style Sheets,"
    http://www.w3.org/pub/WWW/Style/, (22 Jan 1997).

Yahoo!®, 1994-97, Yahoo! Inc., http://www.yahoo.com, (22 Jan 1997).

6
Professional Ergonomics Issues

Dieter W. Jahns
Board of Certification in Professional Ergonomics

6.1 Introduction

The education, training and career domains of ergonomics in general, and industrial ergonomics in particular, have a long and diverse history. Different countries, cultures, and economies still view the science and technology of work systems from a variety of perspectives. These sometimes divergent perspectives have made it difficult until recently to achieve a consensus on what the scope of ergonomics is, how people should be educated and trained to conduct ergonomics research and practice ergonomics, and how to measure the competencies of those claiming to be ergonomists. It is rather ironic that while ergonomists routinely do job/task analyses, performance/technology assessments, and work-system designs for career fields ranging from astronauts to warehouse workers, they rarely apply these methods to their own career field. A true case of "The cobbler's children have no shoes"! The goal of this chapter is to summarize the current status of professional issues in ergonomics as portrayed in the literature published by the International Ergonomics Association (IEA) and its Federated Societies. This literature also forms the foundation for certification criteria development by the Board of Certification in Professional Ergonomics (BCPE) in the U.S. and by the Center for Registration of European Ergonomists (CREE) in Europe. Readers wanting a broader and deeper survey of the evolutionary development of ergonomics into a transdisciplinary, unique career field should consult the resources listed in "For Further Information." Particularly useful for achieving an awareness of what ergonomics is all about are the books by Chapanis (1996), Booher (1990), Klemmer (1989), MacLeod (1995), Meister (1997), and Salvendy (1987).

6.2 Education

It has long been postulated that ergonomics is akin to engineering because its subject matter involves the design and performance of humans and their technological tools to accomplish work effectively and efficiently. Three root sciences have vied for influence in shaping the quality of working life since the end of the 19th century and since then in an accelerating industrialized society: (a) the biomedical sciences interested in human fatigue, work physiology, biomechanics/kinesiology and anthropometry in occupational settings, (e.g., Amar, 1914), (b) industrial engineering as promulgated by Frederick W. Taylor in his book *The Principles of Scientific Management*, (Taylor, 1911), and (c) industrial/experimental psychology as exemplified by the work of Hugo Münsterberg in 1912 at Harvard University and later Berlin (e.g., Münsterberg, 1912, 1914). Even today, the graduation certificates of ergonomists may be from one of these "traditional" academic disciplines even though not all graduates from the cognate departments are necessarily educated in ergonomics. The situation is clearer in Europe and some parts of Asia where academic departments with core subjects in ergonomics *per se* exist. Where an ergonomics education program is "housed" can have a significant impact on how and what a student learns about the field. For example, it is unlikely that aviation ergonomists will have been exposed to all of the topics covered in this handbook; similarly, it is unlikely that industrial ergonomists will have in-depth knowledge of the perceptual–motor and cognitive requirements of transportation systems or complex, automated industrial processing plants. Both specialization and diversification have occurred in ergonomics in conjunction with economic, demographic, political, and technological pressures. As jobs have changed with advances in science and technology, so has the influence of those sciences on ergonomics; everything became "work-related." Reich (1992, pp. 174-180) predicts that as human enterprise evolves from an industrial/materialistic base to an information/knowledge base, three job categories will dominate economies: (a) routine production services, (b) in-person services, and (c) symbolic–analytic services. Ergonomics can be viewed as the practice of symbolic-analytic services for the benefit of those people who perform routine production services and in-person services.

In order to provide the most effective services, three levels of education are currently being offered by both public and private vendors of knowledge and skills training (Webb and Stager, 1990):

Awareness education: Provides sufficient education in the structure and application potential of ergonomics to enable students to judge its relevance to their current interests.

Familiarization education: Provides sufficient education in the structure, relevant knowledge, and skills in ergonomics to enable a professional in a different field (such as management, engineering, psychology, human resources, or safety) to interact effectively with an ergonomist.

Professional education: Provides a full understanding of the structure, knowledge, and skills of ergonomics to enable the student, after sufficient practical experience, to practice ergonomics.

Awareness education is most appropriate for workers who are involved with *Participatory Ergonomics* and related organizational and management issues. Normally, a survey course of 40 contact hours specifically oriented toward the job, worker demographics and work system with which attending workers are most familiar will provide a learning experience which is useful. Generalized "one-size-fits-all" pep talks on ergonomics should be avoided, as should building unreasonable expectations of what can be achieved with ergonomic data, methods, and techniques. An excellent resource for awareness education is MacLeod's (1995) book *The Ergonomics Edge*.

Familiarization education should build on awareness education in both breadth and depth. The knowledge base of course participants should be used as a point-of-departure for launching into ergonomics topics and how they differ from (and are related to) their occupational responsibilities. In this regard, the current practice of structuring course topics in accordance with the OSHA (Occupational Safety and Health Administration) agenda and guidelines (as was done for this handbook) is not quite appropriate. For one thing, the goals of ergonomics are

- Reasonable human performance
- Reasonable workload
- Reasonable health maintenance
- Reasonable hazard control and injury-risk management

through optimum human–machine system design, personnel selection, and training. The emphasis is on humane use of technology to achieve good system performance. For another thing, OSHA's agenda is much narrower, reactive, and topically different from ergonomics. "Medical management," for example, is important in work systems, as are safety audits and industrial hygiene surveillance, but none of these falls within the scope of ergonomics as a work-system design profession. Ergonomists and other career

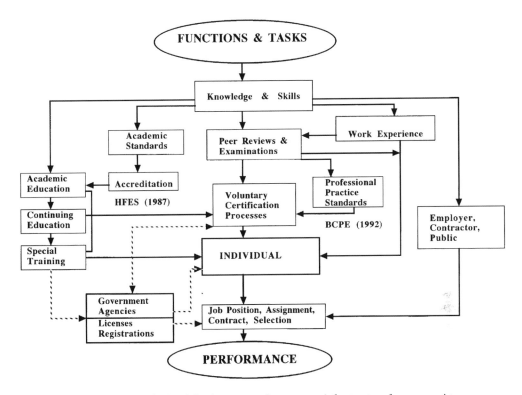

FIGURE 6.1   Professional development and assessment infrastructure for ergonomists.

fields must see their edges clearly (Duncan, 1996) or risk becoming overwhelmed by information overload and/or trivialization of their knowledge and skill base. Being able to drive a car does not mean you can fly an airplane! Try it, and we'll plant daisies on your grave.

Professional education in ergonomics builds on the professional development and assessment model shown in Figure 6.1. The functions and tasks performed by ergonomists have been discussion topics at numerous meetings of related scientific/engineering societies since the 1960s. However, individuals working in the field were often given job titles which reflected the contractual requirements of employers more than academic standards or work experience. "Human Performance Analyst," "Methods Engineer," "Human Factors Specialist," "Human Factors Engineer" and "Engineering Psychologist" have been common titles for ergonomists since the 1920s. Cohesive academic standards for coursework and voluntary certification processes for individuals were slow in coming and still have not been harmonized internationally.

In 1976 a five-day symposium sponsored and financed by the "Special Programme Panel on Human Factors" of the North Atlantic Treaty Organization (NATO) Science Committee brought together 12 experienced university teachers/scientists in ergonomics from seven nations plus four university teachers from other nations who are experts in areas closely related to ergonomics (Bernotat and Hunt, 1977). This group developed recommended curricula structures for six different educational programs in ergonomics:

1. A full program in ergonomics
2. A specialization program in ergonomics for students in engineering
3. An orientation program in ergonomics for students in physiology
4. An orientation program in ergonomics for students in psychology
5. An orientation program in general ergonomics for students in management, engineering, and other disciplines
6. An indoctrination program in general ergonomics for students in physiology, psychology, engineering, and other sciences.

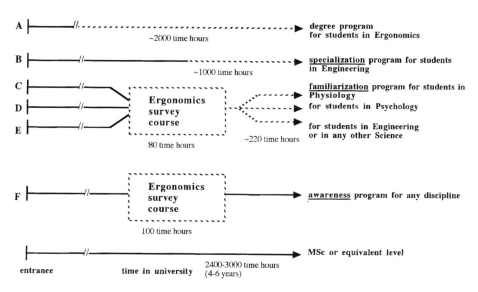

FIGURE 6.2   Recommended curricula structures in ergonomics (see text for details) (From Bernotat, R. and Hunt, D. P. 1977. *University Curricula in Ergonomics.* Wachtberg-Werthhoven, Germany: Forschungsinstitut für Anthropotechnik.)

Figure 6.2 illustrates the academia time history of each of the educational programs. The student/teacher contact hours shown are 60-minute *clock* hours, as opposed to normally shorter *lab/lecture* hours. The solid lines indicate that part of the education which is devoted to the specific discipline listed and general university requirements (GURs); the dashed lines indicate concentration on ergonomics. The bottom line of the figure is meant to show the time line from university entrance up to the granting of a degree at the Master-of-Science level. While there are some programs which are offered to undergraduates in pursuit of a Bachelor's degree (or equivalent diploma), most ergonomics programs are graduate degrees as indicated for programs A, B, C, D, and E of Figure 6.2. The recommended curriculum for a "degreed" ergonomist is shown in Table 6.1. Keep in mind that this was 1976. By 1987, the Human Factors and Ergonomics Society had a voluntary accreditation program in place which used many of the 1976 recommendations for establishing academic standards. The Ergonomics Society (U. K.) also reviews academic programs to establish membership status for its applicants. Table 6.2 lists the HFES and ES (U.K.) accredited programs in the U.S. as of December 1996. Strangely, and most likely for political reasons, these are not titled "ergonomics programs" but follow traditional designators of root sciences or academic departments, even where the campus may house an ergonomics research institute. It should be noted that there are many excellent ergonomics academic programs which for a variety of reasons have not yet applied for accreditation. The interested reader should consult the program directories listed in "For Further Information."

Since "education" is one of the three components upon which professional certification is based, both the BCPE and CREE adopted the Ergonomist Formation Model (EFM) released and published by the HETPEP working group in 1992 (Rookmaaker, et al., 1992). Table 6.3 lists the topics of knowledge, skill, and experience currently used in the evaluative policies, practices, and procedures by BCPE. Notice that certification candidates must be educated, trained, and experienced in at least one "Design" topic of those listed in Table 6.3. It is this tie-in to design knowledge and skills that differentiates ergonomics from other "human factors"-oriented professions like psychology, human resource development, allied health-care providers, etc., who "help" humans by means of a treatment/behavior modification philosophy.

The "Ergonomics Approach" is based on the premise that tasks, tools, and talents form a system for productive human work. The relationships among humans, their tools, and their talents need to be empirically studied in order to create a "human-centered technology." The other categories and details listed in Table 6.3 form the core foundation upon which this technology is built.

TABLE 6.1 The Curriculum for Degreed Ergonomists

Summary of the Time Hours

	Lecture (hours)	Practicum (hours)	Total (hours)
Part I. Basic Knowledge	**250**	**157**	**407**
Human anatomy and physiology	95	55	150
Human psychology	105	102	207
Developmental psychology	50	—	50
Part II. Methods and Techniques	**35**	**150**	**185**
Methods of measurement	—	150	150
Selection	10	—	10
Training	15	—	10
Instrumentation	10	—	10
Part III. Application	**215**	**165**	**380**
Ergonomics in design	60	85	145
The physical environment	90	65	155
Accidents and safety	20	—	20
Systems ergonomics	45	15	60
Part IV. Other Basic Subjects	**345**	**330**	**675**
Statistics	75	25	100
Mathematics	50	50	100
Physics	30	65	95
Computation	20	45	65
Networks, time series, stochastic processes	45	45	90
Systems theory and optimization	50	50	100
Engineering	50	50	100
Communications and "public relations"	25	—	25
Part V. Background Subjects	**145**	**—**	**145**
The individual in work organizations	25	—	25
Organizational context of systems design	10	—	10
Project management	10	—	10
Production & process management & control 50	—	50	
Production & process system design	50	—	50

From Bernotat, R. and Hunt, D. P. 1977. *University Curricula in Ergonomics.* Wachtberg-Werthhoven, Germany: Forschungsinstitut für Anthropotechnik.

TABLE 6.2 Accredited Ergonomics/Human Factors Degree Programs in the U.S. as of December 1996

Georgia Institute of Technology (1)
Louisiana State University (2)
New Mexico State University (1)
North Carolina State University (2)
Ohio State University (1)
State University of New York at Buffalo (1,2)
Texas A&M University (1)
University of Central Florida (1)
University of Dayton (1)
University of Illinois at Urbana-Champaign (1)
University of Southern California (1) (Deactivated 1997)
Virginia Polytechnic Institute & State University (1,2)

(1) = Accreditation by Human Factors and Ergonomics Society (U.S. A.)
(2) = "Vetting" by Ergonomics Society (U. K.)
From Ergonomics Society and Human Factors and Ergonomics Society

TABLE 6.3   Ergonomist Formation Model: Topics of Knowledge, Skills and Experience

General Category			Details	
Code	Name	Numbe r	Name	
A	Ergonomics Principles	1	Ergonomics Approach	
		2	Systems Theory	
B	Human Characteristics	1	Anatomy, Demographics, and Physiology	
		2	Human Psychology	
		3	Social and Organizational Aspects	
		4	Physical Environment	
C	Work Analysis and Measurement	1	Statistics and Experimental Design	
		2	Computation and Information Technology	
		3	Instrumentation	
		4	Methods of Measurement and Investigation	
		5	Work Analysis	
D	People and Technology	1	Technology	
		2	Human Reliability	
		3	Health, Safety and Well-Being	
		4	Training and Instruction	
		5	Occupational Hygiene	
		6	Workplace Design	
		7	Information Design	
		8	Work Organization Design	
E	Applications		Projects pursued by the individual	
F	Professional Issues			

From Board of Certification in Professional Ergonomics, June 1995, *Information on Certification Policies, Practices & Procedures,* 3rd Ed., p. 24, BCPE: Bellingham WA.

6.3   Training

One of the most thorough and comprehensive surveys of ergonomics as a career was conducted in 1989 (with results published in 1992) by the U.S. National Research Council's Committee on Human Factors (NRC-HF) (Van Cott and Huey, 1992). This survey revealed that approximately 83% of ergonomic work in the U.S. is centered in six areas: computers (22.3), aerospace (21.6), industrial processes (16.5), health and safety (8.9), communications (8.2), and ground transportation (5.3). The remaining 17% of work is spread over a large variety of other areas, but it is performed by very few ergonomists, often just one or two people. The principal workplaces of ergonomists are shown in Figure 6.3. where the private business category covers both corporate and self-employed ergonomists. It is interesting that this distribution is similar to that of engineering professionals. The nature of the work performed was categorized by the NRC-HF committee into six categories, spanning 52 tasks as derived from unpublished task/job analyses provided by the Human Factors and Ergonomics Society:

1. Systems analysis
2. Risk and error analysis
3. Design support
4. Test and evaluation
5. Instructional systems design
6. Communications

While it is customary in other career fields to receive post-academic training upon entry into the job market from seasoned practitioners and/or supervisors familiar with the career field, such is generally not the case for ergonomists, most likely because there are so few of them to begin with. The NRC-HF

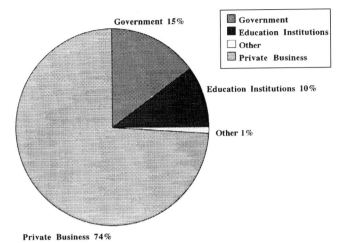

FIGURE 6.3   Principal workplaces of ergonomists. (From Van Cott, H. P. and Huey, B. M. (Eds.), 1992. *Human Factors Specialists' Education and Utilization, Results of a Survey.* Committee on Human Factors, Commission on Behavioral and Social Sciences and Education, National Research Council, Washington, D.C.: National Academy Press.)

committee survey indicated that only 9% of ergonomist supervisors are ergonomists. Thirty-four percent are engineers, with others scattered through business managers, psychologists, industrial designers, etc.

This training dilemma in ergonomics may improve as the field grows, but in the short run, more internships and academia–industry coordinated programs are severely needed.

6.4   Experience

As a scientific approach to work systems development, ergonomics often gets caught in the middle between labor and capital; yet to maintain its neutrality, it is frequently reluctant to take sides in business management, political, and/or regulatory issues. Taylorism (and its reliance on time-and-motion studies to establish labor standards) is at times accused of being too "task focused" and not adequately considering individual differences among workers' capabilities and limitations. Occupational safety and health specialists, sociologists, lawyers, workers' compensation insurers, and industrial/organizational psychologists often see all work and technological innovations as risk factors to the quality of working life. Thus, experience in ergonomics is shaped by how "human factors" in systems development and operation are viewed and dealt with. It does make a difference to the derivation of design solutions whether the human operator and/or maintainer of technological devices is considered to be a "mechanical power source," an "information processor and transducer," an "intelligent controller of processes," a "production cost factor," or the most vital "business asset."

Application scenarios will provide different experiences in how to accommodate human factors in work systems. For example, industrial systems dependent on manual materials handling may create more concern about musculoskeletal issues and thus rely on work physiology, biomechanics, and kinesiology, than transportation or computer systems, which require little physical effort but much information search, decision making, and error-free motoric responses and thus depend on the behavioral/cognitive sciences and linguistics. Because ergonomists work in a variety of settings and situations which are influenced by many different parameters and "stakeholders," they must learn through experience how best to work with other career fields (e.g., engineering, hazard/risk management, project management, labor representatives, system-users/operators, health and safety advocates, etc.).

Unpublished studies conducted by BCPE also determined that ergonomics is practiced at four levels (see Table 6.4). In this regard, the situation is similar to other career fields where the technician or draftsman performs routine engineering tasks; the paralegal assists the attorney; the medical assistant, nurse, or therapist aids the physician; the research/teaching assistant helps the professor. The BCPE started

TABLE 6.4 Levels of Comprehensive Practice in Ergonomics

Purpose:
  To define levels of comprehensive practice for the ergonomics profession.

Rationale:
  The BCPE needs to define levels of comprehensive practice in order to develop measurement instruments that serve to recognize ergonomists progressing in their professional development. Four levels of comprehensive practice are believed to span the profession.

Level of Practice	Background and Experience	What can be expected in terms of the ergonomist's:	
		Capabilities	Areas of Knowledge
Level-1	Has received at least 40 hours of coursework in ergonomics topics taught by qualified ergonomists. Has no experience in independently applying ergonomics knowledge; works under direct supervision from a Level-3 ergonomist or Level-4 ergonomist.	Functions at an introductory level with supervision; uses evaluation devices (checklists, surveys) contracted by qualified ergonomists to collect data and information for making preliminary system analyses.	Has introductory-level knowledge of ergonomics principles and systems theory, characteristics of humans in systems, and methods of work analysis and measurement.
Level-2 BCPE Certification Level at CEA	Has general knowledge of topics in the broad field of ergonomics as received from a baccalaureate degree program in ergonomics or a related field. Alternatively, may have a baccalaureate degree in a nonergonomics discipline and a minimum of 200 hours of continuing education coursework in ergonomics topics of increasing depth and specificity, or one semester (4, 3-hour courses) in ergonomics from an accredited program. Has at least two years of experience in applying ergonomics knowledge.	Works with minimal supervision and/or consultation; conducts basic system analyses using checklists, surveys, questionnaires, videotaping, human performance measurements, etc.; and makes basic recommendations for addressing ergonomics design criteria and specifications. Analysis and design recommendations follow established practices and guidelines (i.e., refers to standards and ensures implementation).	Has general knowledge of ergonomics principles and systems theory, characteristics of humans in systems, and methods of work analysis and measurement. Has basic knowledge and ability to apply basic knowledge of at least one of the following in system design: the relationship between technology and human performance; human reliability; health and safety; training and instruction; occupational hygiene; workplace design; information design; and work organization design. Example systems include consumer products, manufacturing systems, office work, transport systems, process industry; health care systems; and recreation, arts and leisure activities.

Level-3 BCPE Certification Level at CPE/CHFP	Has a working knowledge of all "core" topics comprising the ergonomics profession based on a master's degree or equivalent in ergonomics and at least 4 years of demonstrated experience in applying that knowledge. Alternatively, may have a master's degree in a nonergonomics discipline and a minimum of 320 hours of continuing education coursework in ergonomics topics of increasing depth and specificity, or two semesters (8, 3-hour courses) in ergonomics from an accredited program. Is able to perform independently and to supervise others in ergonomics work.	Works independently; conducts detailed, specific analyses by collecting, analyzing, and interpreting subjective and objective system and human performance data; consolidates and compares data with relevant current research; develops meaningful alternative solutions to identified problems, and proposes practical design solutions specific to the setting; supervises implementation of solutions; employs valid statistical methods to conduct tests/evaluations to verify that identified problems have been appropriately addressed by control measures; clearly communicates recommendations in writing and in oral presentations to all levels of personnel; conducts mission, function, job, and task analyses; supervises, mentors, and trains others in performing necessary support tasks.	Has broad understanding of ergonomics principles and systems theory, characteristics of humans in systems, methods of work analysis and measurement, and issues of importance to the ergonomics profession. Has working knowledge and ability to apply knowledge of at least three of the following in system design: the relationship between technology and human performance; human reliability; health and safety; training and instruction; occupational hygiene; workplace design; information design; and work organization design. Example systems include consumer products, manufacturing systems, office work, transport systems, process industry; health care systems; and recreation, arts, and leisure activities.
Level-4 BCPE Certification Level at CPE/CHFP	Has extensive working knowledge and experience beyond the Level-3 ergonomist with specialization in selected ergonomics application areas or systems. Has at least a master's degree or professional credentials (for example a doctoral or postdoctoral degree) in ergonomics and has received specialty education or training in at least one area of ergonomics. In lieu of specialty training, may have developed recognized expertise through concentrated experience of at least 8 years.	Works independently; performs beyond the capabilities of the Level-3 ergonomist by providing in-depth consultation in general ergonomics or within area of special expertise.	Has extensive understanding of ergonomics principles and systems theory, characteristics of humans in systems, methods of work analysis and measurement, and issues of importance to the ergonomics profession. Has extensive working knowledge and ability to apply knowledge of at least five of the following in system design: the relationship between technology and human performance; human reliability; health and safety; training and instruction; occupational hygiene; workplace design; information design; and work organization design. Example systems include consumer products, manufacturing systems, office work, transport systems, process industry; health care systems; and recreation, arts, and leisure activities.

From Board of Certification in Professional Ergonomics, August 1996, *Fact Sheet: Levels of Comprehensive Practice*, BCPE: Bellingham, WA.

defining the competencies and assessment criteria for the "Certified Ergonomics Associate (CEA)" in 1997. Implementation is scheduled for late 1998. This development is particularly relevant for industrial ergonomics where the demand for practitioners to deal with OSHA issues has outstripped the supply of qualified people at all levels of research, teaching, and practice. The CEA should be able to provide services around Level 2 of Table 6.4. Some people claim that ergonomics is not "rocket science"; they are right! Rocket science is much easier and more orderly than ergonomics. Physics, chemistry, and lots of mathematics are all that rocket scientists need (plus money and a creative vision) to do their job of objective materialism. Ergonomists have to also deal with fuzzy human factors which are subjective in some ways, objective in others, poorly understood, always changing, bounded by ethical and political taboos, and never fully predictable. The job market itself is rapidly changing, as are workforce demographics. The gurus of scientific business management promote new theories faster than they can be implemented and evaluated. Consequently, the budding ergonomist needs about four years of on-the-job experience before he or she can create any sort of stability in systems functioning in unstable equilibrium. Twenty percent of time should be set aside just for tracking changes in technology and their impact on ergonomics. It is up to employers to hire not just experienced ergonomists, but to also provide entry-level experiences. Ergonomics is as vital to business planning as engineering, production, marketing, and financial accounting because ergonomists should build the bridges between human factors and material factors.

6.5 Summary and Conclusions

As was shown in Figure 6.1, a mature career field is established when professional development criteria and assessment standards are in place. In the U.S., accreditation of academic ergonomics programs was started by HFES in 1987. The establishment of professional practice standards and a voluntary certification process for individuals independent of HFES (but in coordination with their resources) was implemented by BCPE in 1992. Hopkin (1994) and Jahns (1996) have predicted that these efforts will impact the infrastructure and further evolutionary development of ergonomics. The IEA is actively involved in harmonizing both professional development and assessment of ergonomists on a global basis, while being sensitive to the unique cultural, demographic, technological, and economic situations of various regions and continents. So long as a core set of criteria and content definitions for ergonomics exists, derivative and suitably modified programs will provide flexibility without weakening the fundamental foundation upon which ergonomics is built.

There are currently more career fields (or occupations) than ergonomists. The U.S. Department of Labor *Dictionary of Occupational Titles* lists over 32,000 occupations; there are an estimated 9,000 to 15,000 practicing ergonomists in the U.S. The challenges ergonomists face are daunting; but then, that is what makes ergonomics an exciting profession.

References

Amar, J. 1914. *Le moteur Humain et les Bases Scientifiques du Travail Professionnel*. Conservatoire Nat. des Arts et Métiers: Paris, France.

Bernotat, R. and Hunt, D. P. 1977. *University Curricula in Ergonomics*. Wachtberg-Werthhoven, Germany: Forschungsinstitut für Anthropotechnik.

Board of Certification in Professional Ergonomics (BCPE). June 1995. *Information on Certification Policies, Practices and Procedures*, 3rd Ed., BCPE: Bellingham WA.

Booher, H.R. (Ed.) 1990. *Manprint: An Approach to Systems Integration*, Van Nostrand Reinhold: New York, NY.

Chapanis, A. 1996. *Human Factors in Systems Engineering*, John Wiley & Sons, Inc.: New York, NY.

Duncan, J. 1996. Do we see our edges clearly?, *The Professional Ergonomist*, Summer 96, BCPE: Bellingham, WA.

Hopkin, C. O. 1994. Accreditation — a mechanism for quality. *Proceedings of 12th Triennial Congress of the IEA*, Vol. 1, 125-127; IEA: Toronto, Canada.

Jahns, D. W. 1996. *A global perspective on ergonomists' education, training and practice assessments for the 21st century.* IEA CybErg Conference: Internet.

Klemmer, E. T., (Ed.) 1989. *Ergonomics: Harness the Power of Human Factors in Your Business.* Ablex: Norwood, NJ.

MacLeod, D. 1995. *The Ergonomics Edge,* Van Nostrand Reinhold: New York, NY.

Meister, D. 1997. *The Practice of Ergonomics.* BCPE: Bellingham, WA.

Münsterberg, H. 1912. *Psychology and Industrial Efficiency,* Harper: New York, NY.

Münsterberg, H. 1914. *Gründzüge der Psychotechnik.* Springer: Leipzig, Germany.

Reich, R. 1992. *The Work of Nations,* Random House: New York, NY.

Rookmaaker, D. P., Hurts, C. M. M., Corlett, E. N., Queinnec, Y., Schwier, W. June 1992. *Towards a European Registration Model for Ergonomists,* Final report of the working group "Harmonising European Training Programs for the Ergonomics Profession" (HETPEP), Leiden, NL.

Salvendy, G. (Ed.) 1987. *Handbook of Human Factors,* John Wiley & Sons: New York, NY.

Taylor, F. W., 1911. *The Principles of Scientific Management,* Harper: New York, NY.

Van Cott, H. P. and Huey, B. M. (Eds.) 1992. *Human Factors Specialists' Education and Utilization, Results of a Survey,* Committee on Human Factors, Commission on Behavioral and Social Sciences and Education, National Research Council, Washington, D.C.: National Academy Press.

Webb, R. and Stager, P. February 1990. *Report of the Human Factors Association/Association Canadienne D'Ergonomie Committee on Professional Education.* HFAC/ACE Unpublished.

For Further Information

Professional issues in ergonomics are most often aired at the technical meetings of related scientific/technical societies and in their meeting proceedings. These are listed below with some auxiliary comments.

International Ergonomics Association (IEA)
c/o HFES
P. O. Box 1369
Santa Monica, CA 90406-1369 U.S.A.
WWW Home Page: http://www.spd.Louisville.edu/~ergonomics/iea/html

The IEA serves as the umbrella agency for 29 Federated Societies in most countries and geographic regions throughout the world. By virtue of membership in a Federated Society, individuals are members of the IEA which only has societies as voting members. The Federated Societies appoint representatives to the IEA.

The official journal of the IEA is *Ergonomics*, published by Taylor & Francis, London, England, which also publishes *Ergonomics International* as a quarterly newsletter (Editor: Dr. Stephan Konz, e-mail: SK@KSU.EDU). The IEA sponsors a triennial congress for dissemination of information on research, methods, and data (2000 San Diego, U.S.). Individuals who want to participate in IEA activities normally do so through technical groups; contact Professor H. Luczak, FIR, Pontdriesch 14/16, 52062 Aachen, GERMANY. e-mail: hluczak@iaw-1.rwth-aachen.de

Human Factors and Ergonomics Society
P.O. Box 1369
Santa Monica, CA 90406-1369 U.S.A.
Telephone: 310-394-1811
Fax: 310-294-2410
e-mail: HFESHQ@aol.com
WWW Home Page: http://hfes.org

HFES was formed in 1957 and currently has over 5,000 members. Its journal is *Human Factors*; it also publishes and distributes many other materials, including *The Directory of Ergonomics/Human Factors Graduate Programs, Consultants Directory, Membership Directory and Yearbook, Meeting Proceedings,* monographs, and brochures. Its annual meeting is usually held in September/October, alternating between east/west U.S. locations.

The Board of Certification in Professional Ergonomics (BCPE)
P. O. Box 2811
Bellingham, WA 98227-2811 U.S. A.
Telephone: 360-671-7601
Fax: 360-671-7681
e-mail: BCPEHQ@aol.com
WWW Home Page: http://bcpe.org

The BCPE is a nonprofit, nonmember certification agency operated by nine directors who establish policies, practices, and procedures administered by a part-time headquarters staff of three people. The BCPE is working with IEA and others to harmonize the teaching and practice of ergonomics. It is a member of the National Organization of Competency Assurance (NOCA) and has a cooperative agreement for "reciprocity" with CREE for the *EurErg* designation operative in the European Union. The BCPE has certified close to 1,000 ergonomists since 1992 and usually offers its certification examinations in the Spring and Fall of each year.

Institute of Industrial Engineers (IIE)
Ergonomics & Work Measurement Division (E & WMD)
25 Technology Park/Atlanta
Norcross, GA 30092 U.S.A.
Contact: Carter J. Kerk
Telephone: 409-862-4149
Fax: 409-847-9005
e-mail: Kerk@zeus.tamu.edu

About 10% of the general IIE membership are active in the E&WM division, which publishes a newsletter and arranges sessions for IIE meetings.

American Industrial Hygiene Association (AIHA)
Ergonomics Committee
2700 Prosperity Avenue, Suite 250
Fairfax, VA 22031 U.S.A.
e-mail: infonet@aiha.org

This is a small committee which addresses environmental and biomechanical factors operating in workplaces from a risk-management perspective.

7

Development of Ergonomics Programs[1]

David C. Alexander
Auburn Engineers, Inc.

Gary B. Orr
USDOL/OSHA

7.1 Introduction

This chapter will describe the development of a "model ergonomics program" for Ergonomics Program Managers. A recent survey[1] of forty ergonomics programs revealed that over half (58%) were floundering, and only 25% were deemed to be successful. "Floundering" was chosen rather than "program failure" and was defined as consuming excessive resources relative to the value provided. In other words, the program could still be operational, but was not expected to be successful in its present form over the long term. Additionally, the survey examined the reasons for floundering and found that, for larger organizations, the managerial aspects of the ergonomics programs were more often the cause than were its technical aspects. The differences are highlighted in Figure 7.1 Clearly, with success rates of less than 50%, there is a critical need for information on the development of effective ergonomics programs.

This chapter provides information that will:

1. Ensure that a comprehensive ergonomics program is developed
2. Ensure that the program is effective and produces the intended results
3. Ensure that the program uses resources effectively
4. Prevent many common program development mistakes
5. Ensure that the program meets both the organization's business needs as well as regulatory/OSHA compliance requirements.

---

[1]Adapted from *The Model Ergonomics Program* © Auburn Engineers, Inc., 1993. Used with permission.

**Ergonomic Technical
Issues**

(Technical skills)

Job analysis
Solving problems
Preventing problems

**Ergonomics Program
Management**

(Managerial skills)

Program Design & Planning
Implementation & Coordination
Program Evaluation

FIGURE 7.1   Ergonomics technical issues vs. ergonomics program management.

This chapter is broken into four sections, each dealing with one major aspect of the ergonomics program. Due to the interaction of the elements, there is some overlap between the sections. The sections are:

- *The Ergonomics Program: An Overview*: This provides an overview of the three major parts of the program: planning, implementing, and evaluating.
- *The Ergonomics Committee*: This covers the structure, responsibilities, logistics, and preparation/training of the ergonomics committee.
- *Activities Necessary to Manage an Ergonomics Program*: These activities include strategic planning, surveillance, tactical planning, ergonomics problem solving, medical management, training, and the prevention of ergonomics problems.
- *Program Management*: This segment provides tools for the management of the program, including the use of measures of success, working the plan, documentation, project evaluation, and audits and assessments.

7.2   The Ergonomics Program: An Overview

This section is based on *An Integrated Plan for an Occupational Ergonomics Program: Your Guide to Developing and Managing an Occupational Ergonomics Program*[2] by Auburn Engineers, Inc. There are three important parts to an effective ergonomics program:

1. Planning
2. Implementation
3. Evaluation

A map of this integrated plan is shown in Figure 7.2.

Planning

Unfortunately, many people fail to plan their ergonomics programs and simply jump into training or project work. While this may seem expedient, it generally leads to floundering and an ineffective program. As a rough rule of thumb, one hour of planning will save forty hours of unproductive and ineffective work. Thus, there are tremendous advantages to planning. Planning consists of two distinct parts: strategic planning and tactical planning.

Strategic planning asks the question, "What do we want our ergonomics program to accomplish?" The answers may be as simple as: reduce our CTD (cumulative trauma disorder) rate by 40%, cut our workers' compensation costs by half, ensure compliance with OSHA regulations, or build a culture which supports

Why Ergonomic Programs Flounder

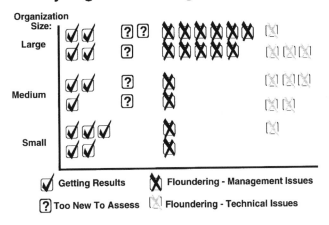

FIGURE 7.2.

Occupational Ergonomics Is Multidimensional!

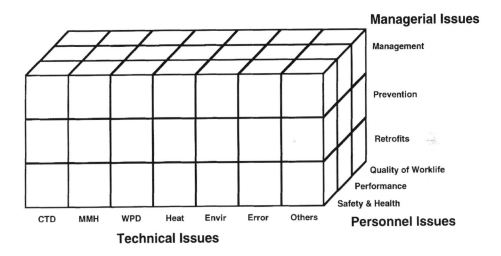

FIGURE 7.3 Possible scope of occupational ergonomics program.

an ergonomics initiative long term. The statement, "If we don't know where we are going, then any path will do," describes the intent of strategic planning.

In addition, strategic planning resolves a number of scope issues as illustrated in Figure 7.3. One helpful way to think about occupational ergonomics is by technical subspecialty, types of problems addressed, and focus of improvement efforts. These are three distinct scales, and as such, should be thought of as three separate dimensions for the ergonomics program. The technical subspecialty dimension determines which aspects of ergonomics will be used. The technical issues include cumulative trauma disorders (CTD), manual material handling (MMH), workplace design (WPD), heat stress, environmental factors such as noise, lighting, thermal comfort, human error issues, and/or other issues. The types of problems addressed are called the managerial issues and include a focus on fixing existing problems (retrofits) and avoiding new problems (prevention), and/or the other managerial and development aspects of the ergonomics program. Finally, the third dimension is the focus of the improvement efforts, which is called the personnel issues. These efforts use ergonomics to simply improve safety and health, improve operating performance such as productivity and quality, and/or improve overall quality of worklife. During the program's design, these dimensions should be explored and an informed choice made regarding the importance of each item.

Tactical planning answers the question, "How do we do that?" Whatever strategic goals are desired, tactical planning is required to determine how to achieve them. Tactical plans provide a list of detailed activities, identify specific financial and human resource requirements, and create a timetable. A written plan is typically one outcome of a sound tactical plan.

Once the strategic and tactical plans are developed, the implementation of the ergonomics program is usually fairly straightforward. Planning allows us to "plan the work and then work the plan." It's simple, and it's effective.

Implementation

Implementation of the ergonomics program usually encompasses four separate, but related, areas:

1. Identifying and resolving current problems (retrofits)
2. Preventing future problems
3. Medical management (prevention and treatment of injured workers)
4. Training

Identifying and resolving ergonomics problems actually consists of three substeps — identifying the ergonomics problems; performing the needed analysis and studies; and developing and implementing solutions. Many organizations use inconsistent methods for identifying ergonomics problems. This substep is necessary for several reasons: it permits problems to be corrected before additional injury/illness occurs, and it permits projects to be prioritized based on risk of injury. Identifying and resolving current problems is often the centerpost of ergonomics activities because they are costly, the injuries place one at risk for OSHA citations, and because injured people (whether assigned to light duty or off work) disrupt production operations. The details of this step are the subject of many technically oriented ergonomics books and short courses.

Preventing future problems is the ability to design equipment, tools, facilities, jobs/tasks, and information so that are ergonomically sound. This step includes both in-house and contract design projects, the purchase of appropriate equipment and tools, and the correct installation of purchased items (e.g., installing purchased equipment at the proper height) and new construction.

Medical management involves the identification and treatment of injured workers. Consistent treatment protocols and appropriate light duty and work conditioning programs are also facets of medical management. Medical management is a particularly important aspect of the treatment of cumulative trauma disorders. Within the medical management area, there are a number of personnel policy issues, such as return to work, light and/or restricted duty, and medical treatment/referral practices.

Training may be required where skills are needed. It is important to realize that training should be performed only to fulfill a need. Training may be required for problem solving, for identification of problems, and for awareness. Some awareness training may be necessary at all levels of the organization. One common error made in implementing ergonomics programs is to do too much training without first determining the need for it. This wastes resources and produces few real results. A second common problem is to perform widespread training too early in the program.

Evaluation

Evaluation should be done at two levels: (1) the project level, and (2) the program level. The project evaluations are done at the completion of each project and evaluate the reduced injury potential, any performance improvements, and costs/benefits.

Program evaluations are time based and are often done semiannually for new programs and annually thereafter. The program evaluations assess progress toward goals and the benefits relative to costs. Following each program evaluation, it is appropriate to review plans and make any necessary adjustments.

7.3   The Ergonomics Committee

An ergonomics steering committee is an effective way to manage the overall implementation of the ergonomics program. The committee should be convened to do this job, and then should be disbanded once the work is complete. This may take from one year to 5 or 6 years, depending on the size of the organization, the exact details of the ergonomics program implemented, and the pace with which the organization chooses to pursue its goals.

Structure

The ergonomics committee for an operating site (a typical site has from 500 to 2,000 people; if it is larger, consider the use of multiple committees) should be comprised of 6 to 10 members. Ideally, there should be representation from loss prevention or safety, medical, human resources, engineering, labor, and production areas. There should be more committee members from production areas than from any other single group. The technical areas may be represented by technical staff or by managers. The production representatives may include labor representatives as well as production management. The production managers on the committee should be at the mid- to senior-management level, and thus will represent significant operating areas of the plant.

The committee can be chaired by the "ergonomist," although strong consideration should be given to having a production manager as chair. Like any committee of this sort, political savvy is important and should take priority over technical skills.

One member of the committee should be the person who will spend the most time with the ergonomics program and its implementation. In general, the more time this person spends with ergonomics, the more that will get done. Realistically, very little will be accomplished if less than 20 to 25% of one person's time is spent with the ergonomics program during its implementation. On the other hand, a dedicated person will do more than twice what a half-time person can accomplish. This person is often called the "site ergonomist," whether or not he or she actually has university training in ergonomics.

Responsibilities

The ergonomics committee is a part-time responsibility for most of its members. It is responsible for planning and managing the ergonomics program, but not necessarily for actually performing all the activities. For example, the committee may identify a need to study and improve a particular job, and then delegate the work to an individual or a problem-solving team. The ergonomics committee is too large and too busy with other activities to really do much detailed problem solving. However, the committee is usually very effective with management functions, such as determining needs, prioritizing activities, obtaining resources, and following up and evaluating the efforts.

Normally, the ergonomics committee will be commissioned by, and will report to, the plant safety committee or a key senior manager. Thus, the role of the committee is similar to that of a middle manager who takes directions from top management, formulates plans, budgets, and schedules, and then oversees the performance of the work. The ergonomics committee, under this scenario, is judged effective when there are plans, ongoing activities, and results. In fact, the "measures of success" are established to measure these aspects of performance.

Logistics

The ergonomics committee meets as needed in order to get the job done. Their time demands are not uniform. The initial time demands are to plan the work and eventually to "work that plan."

Typically, the initial planning (to create the strategic and tactical plans) will take from 8 to 24 hours in a workshop setting, followed by regular meetings of 1 to 2 hours once or twice a month (to implement and manage the plans). The initial tactical plan will provide a list of specific activities, personnel, and resources, on a 6- to 12-month timetable. Subsequent meetings will ensure that the plan is working, and that projects are actually being implemented.

Preparation/Training

The ergonomics committee may need some training to get its job done. There are two types of skill that this committee and its members need: (a) technical skills and (b) program management skills. The technical skills allow the committee to understand ergonomics and ergonomics problem solving. Appropriate technical topics include workplace design, cumulative trauma, manual material handling, heat stress, environmental factors, and human error. Other possible topics include engineering design, maintainability, designing for disabled workers, work shifts and schedules, and job design (macroergonomics).

Program management skills, on the other hand, prepare people for planning, implementation, and evaluation of the ergonomics program. The topics that should be covered include the planning and managing of an ergonomics program, establishing measures, making a program pay off, compliance issues (such as those covered in OSHA's *Ergonomics Program Management Guidelines for Meatpacking Plants*[3]), and building the ergonomics culture.

In closing this section, a simple caveat is necessary: most ergonomics training courses and textbooks deal only with the technical issues and do not provide adequate information on the management of ergonomics programs.

7.4   Activities Necessary to Manage an Ergonomics Program

For some people, the term ergonomics program has come to mean the collection of elements defined by OSHA in the *Ergonomics Program Management Guidelines for Meatpacking Plants*: recognition, evaluation, and control of ergonomics hazards. This is a static model of an ergonomics program, and it fails to recognize all the steps required to initiate an ergonomics program and permit it to mature. This list of activities is broader than the OSHA model and is more business oriented.

The Strategic Plan

The strategic plan will answer the question, "What do we want to do?" Such answers as "We want to do ergonomics" or "We want to implement ergonomics" are *not* appropriate. Ergonomics is a tool that helps to do something, such as reducing workers' compensation costs or reducing the number of lost workday cases. The ergonomics program should be aimed at specific, measurable goals. The strategic plan for ergonomics is best developed by the plant safety committee or a similar group that has a view of overall operations and can see the relative value of different programs (one of their decisions is who should serve on the plant ergonomics committee). However, if strategic planning is not done when the ergonomics committee begins its work, one of its first activities should be to complete the strategic plan and then review it with the plant safety committee or with top management.

To develop the strategic plan, ask the following questions:

1. *What do we want the ergonomics program to do?* This usually becomes the mission, vision, and scope of the ergonomics program.
2. *How do we monitor results?* What data do we measure to demonstrate progress with ergonomics? (These are the "measures of success.")
3. *What are the barriers?* And how can they be overcome?
4. *What policy issues* are likely to be affected?
5. *Who is or should be involved, and what are their roles?* This includes both the ergonomics committee and ergonomics problem-solving groups.

6. *How important is ergonomics* relative to other safety and health issues for our company?
7. What is our general plan?

This information can be documented and shared with the ergonomics committee, the plant safety committee, and management. It should be used to help with program reviews and evaluations.

Surveillance

Some type of surveillance is helpful as the ergonomics program is initiated. With surveillance, one is trying to identify all the ergonomics problems and opportunities — what is wrong and what can be improved? This also will help to determine the magnitude of the ergonomics program and the resources that will be needed. Surveillance does not need to be burdensome or time consuming to be effective. Some things to look for are outlined below:

- *Ergonomics has safety and health implications.* Therefore, one of the first places to look is the OSHA 200 log, workers' compensation, medical records, and restricted work cases. Eventually, as the ergonomics initiative moves from correction to prevention, additional surveillance techniques will be needed to identify ergonomic risk factors and the early warning indicators of potential injuries.
- *Examine traumatic injuries* as well, since their cause is often rooted in an interface problem between the person and some piece of equipment. From these situations, the number and cost of injuries/illness can be determined. Some rough estimates indicate that as many as 25% of traumatic injuries may have an ergonomics root cause.
- *Look for performance problems and cost issues* as well as safety and health concerns. For example, the recruitment, hiring, and training of replacement workers can often cost $2,000 or more. People who work at poorly designed workstations take more breaks, produce less, or turn out lower-quality products. There are a number of ways in which ergonomics can improve job performance, and the ergonomics committee should be alert to these savings.

By determining the overall issues, an organization with an effective ergonomics program may find that ergonomics can easily pay for itself through reduced financial losses.

When looking for ergonomics issues, be sure to look at all aspects of human performance. This is sometimes called the "European Model" of ergonomics because it emphasizes both the physical and cognitive aspects of ergonomics. A useful outline of important ergonomics issues can be found in *The Practice and Management of Industrial Ergonomics,*[4] Chapter 2.

The six primary aspects of ergonomics are:

1. The physical size of people and its implications for the fit of the person at the workplace and within the facility.
2. The cardiovascular system and its limitations on work as measured by work physiology.
3. The major musculoskeletal system and its limitations on manual material handling.
4. The minor musculoskeletal systems and their limitations on fine work, manipulation, and dexterity.
5. The environmental factors, such as lighting, noise, and thermal comfort, and their impact on performance.
6. The cognitive capabilities of people and their impact on processing information and "human error."

Once there is an understanding of all the ergonomics issues, including the troublesome jobs, then a list of the highest-priority jobs should be developed. This list may be known by different names — the top five, the worst ten, or the dirty dozen — and it is simply an effective way to prioritize. The ergonomics committee should be aware of these jobs, and the tactical plans should specifically address when and how they will be resolved. When the ergonomics problems with the initial list of jobs are resolved, a revised list can be prepared. Remember that surveillance is an ongoing activity and will be repeated periodically.

Tactical Planning

While strategic planning is the key to deciding what to do, tactical planning is the key to actually getting things done. Tactical planning is best done in a workshop-type setting with the ergonomics committee. Typically, 8 to 16 hours devoted to tactical planning will really help "jump start" the program. Tactical planning is the link between the goals of the ergonomics program and the specific projects undertaken. Tactical planning will answer these questions:

- What should be done?
- When should it be done?
- Who should do it?
- What are the quality standards?

A practical way to lead the tactical planning is to ask what needs to be done in each of the major implementation areas as well as the major management areas of the ergonomics program. For example, there are projects that help correct existing problems, projects that help prevent future problems, training needs, medical management procedures and protocols, project evaluations, and periodic program reviews. By examining each individual area, lists of needed actions can be developed. These individual lists can then be integrated into a tactical plan covering the upcoming 6 to 12 months, as shown in Table 7.1. Managing the plan then becomes as easy as making monthly assignments and following up on actions, and, of course, making periodic updates to the tactical plan so that it remains current.

Solving Ergonomics Problems

There are several ways in which ergonomics problems can be resolved. The ergonomics committee should carefully consider each problem and decide which approach, from among those discussed below, is best for that situation. Ergonomics problems can be easily resolved by using small group problem-solving processes. An excellent six-step process, which has proven itself time and again, follows:

1. Identify the jobs/tasks at risk and select one to improve
2. Analyze the problem to the extent required
3. Develop alternative solutions
4. Select the most appropriate solution
5. Implement the preferred solution or, if that is not possible, an alternate
6. Follow up to ensure that the problem is resolved

The *use of problem-solving teams* can be both highly effective and enjoyable. The teams usually consist of 4 to 8 people. (For example, a team may consist of 2 to 3 operators from different shifts, a production supervisor, the ergonomist, a mechanic, an engineer, and a nurse.) Some training is needed so that the team can do its job well. The team should use the small group problem solving process outlined above. This author's experience shows that a "quick strike team" (dedicated, full-time work for 2 to 5 days) is preferable to the "quality circle approach" (short, 1- to 2-hour meetings spread over several months) for these appointed teams. Quick strike teams function better, and the projects get done much faster. The team should be able to analyze the problem, develop alternatives, and recommend specific, practical solutions within the 2- to 5-day period. This chapter is not a tutorial on problem-solving teams and additional information can be found in "Using a Quality Action Team to Resolve Ergonomics Problems."

Individuals also can be used to resolve ergonomics problems. These projects may take more elapsed time, but fewer overall labor hours, to resolve a problem. They are particularly good for someone who has some depth of experience and skills and who can talk easily with operating personnel and gather information. Individuals are also good for working on projects directed toward the prevention of ergonomics problems. The person can be a project engineer or any other talented performer.

Additionally, *consultants* can help to resolve ergonomics problems. Consultants are helpful when the organization doesn't have adequate human resources, when the problem is particularly challenging, or

TABLE 7.1   Ergonomics Committee Tactical Plan

May

What?	Will Be Done By Whom?	By When?	Check When Completed
Plant strategy developed			
Ergonomics policy written			
Roles/responsibilities developed			

June

What?	Will Be Done By Whom?	By When?	Check When Completed
First problem solving training			
Initial 2 projects started			
Baseline data collected for measurement systems			

July

What?	Will Be Done By Whom?	By When?	Check When Completed
Initial project recommendations implemented			
Next 2 projects initiated			
Illness investigation procedure tested			

August

What?	Will Be Done By Whom?	By When?	Check When Completed
Project recommendations implemented			
Next 2 projects initiated			
Detailed surveillance completed			
Policy issues identified			

September

What?	Will Be Done By Whom?	By When?	Check When Completed
Project recommendations implemented			
Next 2 projects initiated			
Plans for proactive ergonomic program developed			

October

What?	Will Be Done By Whom?	By When?	Check When Completed
Project recommendations implemented			
Next 2 projects initiated			
Engineering training started			

when it is not cost effective to organize and train a team. A consultant should be able to offer fresh insight, novel solutions, and should show personnel how to resolve problems in a quick and cost-effective manner. A note of caution is necessary — there are many newcomers to the field of ergonomics who can speak the lingo but lack the engineering skills to solve in-plant problems. This problem with consultants will get worse before it gets better. So before you employ a consultant, verify his or her ability and skills — do you want to pay people to learn on your time?

Some ways to recognize unacceptable consultants include:

- A consultant who simply reiterates that you have a problem (you already knew that — that's why you hired him/her)
- A consultant who tells you to automate (you can figure that alternative out on your own)
- A consultant who offers common "textbook solutions" (e.g., suggests that you add lift assist devices for your lifting problems)
- A consultant who offers inappropriate solutions (suggests, for example, that you add footrests to workstations where the only problem is hand–wrist cumulative traumas)

At some point, the *natural work group* should be able to identify and resolve ergonomics problems within its own work areas. The "natural work group" is a work group with its supervisor. This is an important point in the evolution of an ergonomics program because it indicates that ergonomics is becoming part of the culture: ergonomics problem solving is seen as "just part of the job — it's nothing special." Long term, this is the best way to deal with ergonomics problems because ergonomic risk factors will be identified and resolved within the work group before injuries or performance problems occur. Natural work groups do not spontaneously develop problem-solving skills, but after some of their members have worked on problem-solving teams, they can carry back some of their skills and experience. The natural work groups can use either a "quality circle approach" or a "quick strike approach" or some other combination, depending on the time available, the urgency of the problem, and the approaches used to resolve quality and production problems.

Projects should be evaluated after their conclusion. This information will help plan future efforts, and it will provide useful feedback to the team or individual involved, to the ergonomics committee, to the plant safety committee, and to management. Some common factors to evaluate are:

1. The time required to work on the problem relative to the quality of the solutions
2. The effectiveness of the solutions
3. The overall costs (time, materials, equipment, etc.)
4. The overall benefits (injury/illness, lost time, productivity, quality, etc.)
5. The enthusiasm for ergonomics that was generated

Medical Management

This section is not intended to replace sound medical judgment from your healthcare providers. However, it may help the ergonomics committee understand some key areas relative to the medical management aspects of ergonomics. Medical management practices affect many plant policies, such as light duty and work restrictions, work hardening and rehabilitation, and work conditioning prior to full performance on the job. There also may be policy issues relative to medical protocols, treatment practices, and/or record keeping.

One important aspect of medical management is that it complements ergonomics problem solving. In general, there are two ways to deal with overuse and cumulative trauma injuries: (1) fix the problem with ergonomics problem solving, or (2) identify injuries in their early stages and treat the injured workers via medical management. Either approach will impact the frequency and severity rate, although aggressive medical treatment will impact severity far more than frequency. Some organizations choose to emphasize medical management initially, followed by the resolution of the ergonomics problems. The only incorrect strategy is to use medical management exclusively as a long-term strategy without the use of ergonomics problem solving.

When developing the tactical plan for ergonomics, some of the "to do" items may be to examine the pros and cons relative to current policies. For example, an organization may have a current policy of "no light duty or restricted work" that the ergonomics committee wishes to examine. However, since the ergonomics committee is not a policy-making committee, it may choose to have a small team evaluate light duty programs, and prepare a "white paper" which explores the advantages and disadvantages of the current policy and of a proposed new policy. At that point, senior management would be asked to review the document, to develop an understanding of the issues, to enter the debate, and then to make a decision regarding the policy. In general, the ergonomics committee should not recommend medical treatment practices and protocols, although they should verify that any treating physicians are aware of the medical protocols found in the *Ergonomics Program Management Guidelines for Meatpacking Plants* and other OSHA documents.

Training

Training is necessary to ensure that people have appropriate skills. However, when training occurs too early in the process, it can be, and often is, counter-productive. Training needs should be carefully assessed to ensure that people get adequate training, and also to ensure that unnecessary training is not conducted. A training matrix can be used by the ergonomics committee to assess training needs and the organization's ability to provide that training. This training review also provides training objectives for the trainer, and serves as a means to assess the quality of the training. Finally, avoid the temptation to make the training overly complex by using too much technical jargon (ergo-babble), especially for the line organization.

While no training is legally mandated, the *Ergonomics Program Management Guidelines for Meatpacking Plants* suggest the training of workers to spot the early signs of cumulative trauma disorders (much like training on early-warning symptoms for hazardous chemical exposure).

Likewise, training for people exposed to CTDs should include on-the-job activities to reduce CTD exposure (adjusting workplaces, using proper tools, etc.) as well as off-the-job awareness (using a keyboard at home can be just as bad as using a keyboard at work).

Some in-plant training courses to consider are:

- In-depth skill training for safety and health professionals and possibly the ergonomics committee
- Ergonomics problem solving for teams or individuals
- Ergonomics design for engineers
- Awareness training for managers and supervisors
- Operator training on ergonomics (awareness of ergonomics plus job-specific Job Safety Analysis type skills for job protection)

The courses should be customized for the specific training objectives. Thus, some courses will be as short as 30 to 60 minutes, while others can last as long as 20 hours.

Preventing Ergonomics Problems

Prevention of ergonomics problems is necessary in order to avoid having to correct problems in the future. When you build or purchase a piece of poorly designed equipment, you have two bad choices:

1. Live with it
2. Pay to change it yourself

Neither is desirable. Thus, the preferred approach is to ensure that all equipment, tools, and facilities are ergonomically designed prior to purchase and installation.

Two things are necessary for good ergonomics design of equipment, tools, and facilities. First is the expectation that all new designs will be ergonomically sound. Engineers and designers must understand that there is a new requirement for their design work. Second, engineers and designers must have access to usable, up-to-date ergonomics design information (such as the **Ergonomics Design Guidelines**[5]) that allows them to design properly.

People other than engineers and designers are involved in this process, however, and they must be included as well. For example, purchasing agents must specify ergonomics as a feature for all new purchases. Contract engineering firms must be required to pay attention to ergonomics. People who install equipment (anything from setting up a new computer workplace, to the alteration of existing equipment, to the building of a new plant) must all understand and apply ergonomics principles when they make the installation. Good ergonomics design should occur at both the central corporate design staff level and at the facility level.

Some policy issues that are likely to surface when the prevention of ergonomics via design is discussed include:

1. Is a policy on ergonomics design necessary?
2. Is ergonomics design required?
3. How is ergonomics design of purchased items evaluated?
4. Will the organization incur increased design and fabrication costs for sound ergonomics design?
5. How are audits of new designs and finished construction projects performed?

7.5   Program Management

Program management is generally the weakest area when it comes to ergonomics programs. As a result, the information in this section may be the most valuable for the ergonomics committee. Program management is usually not covered in academic programs nor in ergonomics short courses, and many ergonomics program managers simply have to learn these skills on the job.

Measures of Success

You can't manage what you don't measure. Measures of success allow you to determine the results of the ergonomics program. These measures have many benefits including:

- Helping you to set goals and establish priorities
- Telling you whether you are making progress
- Allowing you to judge whether you need additional resources
- Letting you know when you're done

There are many different measures of success that can be used. Each organization should carefully identify a number of possible measures and then select the ones that best fit. A handful of measures (from 6 to 12) reflecting short- and long-term changes, with leading and lagging indicators, works best. If your initial measures aren't working for you, then change them. Periodically, over the life of the ergonomics program the measures should be reviewed and modified. (This is especially important as you move through the stages of ergonomics on your way to an ergonomics culture. For example, early in the program you may want to measure how many people call for ergonomics assistance, but as the program matures, you will want to know how many applications occur without your involvement.)

A few notes about measures are important. There are leading and lagging measures. Lagging measures tell you about things you can no longer control (OSHA Log 200 cases; lost workdays), while leading measures predict what will occur. (First aid cases are a predictor of lost workday cases; number of projects resolved predicts ergonomics risk reduction.)

During the maturing process associated with ergonomics program implementation, the measures will change from activity-based measures (committee established, people trained, surveillance completed, etc.) to outcomes/results-based measures (injuries/illnesses, projects completed, etc.) and finally to systems-based measures (medical management system in place and working; engineering design system in place and working).

TABLE 7.2   Possible Measures of Success for Ergonomics Programs

Measure	When Monitored?	Type of Measure
Number of lost workdays for ergonomics reasons	Quarterly	Outcome; Lagging
Compensation costs for ergonomics reasons	Semi-annually	Outcome; Lagging
OSHA Log 200 cases for ergonomics reasons	Monthly	Outcome; Lagging
First aid cases for ergonomics reasons	Monthly	Outcome; Leading
Progress of ergonomics problem solving efforts	Monthly	Activity; Leading
Progress on milestones listed on the ergonomics committee action plan	Monthly	Activity; Lagging
Number of people trained	Annually	Activity; Leading
Number of projects resolved	Quarterly	Outcome; Leading
Number of ergonomic systems in place	Annually	Systems; Leading

Some examples of measures which can be used are shown in Table 7.2. Remember, though, there are many additional measures that can be used. The measures should monitor information about the objectives you hope to accomplish.

Working the Plan

Working the plan involves tracking data to see how things are going, ensuring that adequate resources are available, and overcoming barriers. It is not an easy job yet, this is precisely what the ergonomics committee is being paid to do over the long term.

With the preplanning of the ergonomics activities, the ergonomics committee can focus on the management of these efforts and spend less of its time on specific projects. A typical meeting agenda should contain these elements:

A. Review of the "Measures"
   • What do the most recent numbers tell us?
   • Are we on target with our plans?
B. Review of the Action Log (assigned projects and activities)
   • What was to be accomplished prior to this meeting?
   • What was actually accomplished?
   • What were the results? Are these results acceptable?
   • If there were problems with the solutions, what were they? Why did they occur? How did you overcome them? Is additional help needed?
C. Review the Action Log (planned projects and activities)
   • Are the upcoming projects still necessary?
   • Who should do them? Do they have time?
   • Assign the project; clarify its scope, verify deadlines, verify the person who is responsible for the project; verify other people involved.

The committee should set up a system to track progress on the tactical ergonomics plan. An action plan format is a good tool for monitoring progress because each action item can be checked off when it's finished. Each monthly meeting should have a review of current projects and activities, followed by the assignment of the new activities that will keep the ergonomics effort moving along.

Another specific tool shown in Table 7.2 can be used to track individual projects. Tracking each project as it progresses through the problem-solving process will identify any weak points in that process. For example, one common problem is noted in the Table 7.3. Failure to fully implement solutions to ergonomics problems is clearly shown when projects routinely reach "select the best solution" but then do not move through "implement solution" in the chart. For example, if the following chart is being reviewed in April, then it is clear that the organization is relatively good at developing solutions but poor at implementing them. Tracking problem-solving data in this way brings these weak points to the surface.

TABLE 7.3   Tracking Ergonomics Projects

Project or Sub-Project Name	Identify & Clarify Project	Analyze Appropriate Data	Develop Alternate Solutions	Select the Best Solution	Implement Solution	Follow Up to Ensure Success
Project # 1	Completed Jan 9	Completed Jan 10	Completed Jan 10	Completed Jan 11	Completed Jan 29	Completed Feb 26
Project # 2	Completed Jan 19	Completed Jan 19	Completed Jan 20	Completed Jan 21		
Project # 3	Completed Feb 6	Completed Feb 6	Completed Feb 13	Completed Feb 14		
Project # 4	Completed Feb 8	Completed Feb 9	Completed Feb 19	Completed Feb 19		

Obtaining resources is another major problem. When cost/benefit analysis is used to justify projects, a concerted effort must be made to determine the full costs and benefits. There are other methods by which to obtain funding for ergonomics projects, and the interested reader should review the chapter on the cost and benefits of ergonomics interventions.

Regarding the issue of overcoming barriers, the four most common major obstacles are:

- Lack of time
- Lack of money
- Too few skills
- Lack of management support

There are, however, some effective means for dealing with each of these problems.[6]

Documentation

Documentation by the ergonomics committee is necessary at several stages of the process. First, the committee should record the tactical plan. The written plan will generate a higher degree of commitment and more action than nonwritten plans.

The committee can prioritize the worst jobs (the dirty dozen) and share the list with management as a means to focus action on the "vital few." Since this list is relatively short, action is much easier to obtain.

When the committee meets, it should keep minutes of the topics discussed and of the progress made. Many committees keep a centralized notebook of minutes, plans, projects, and evaluations.

As projects are assigned, they should be noted, and when solutions are implemented (whether successful or not), they should be recorded. A short videotape of the problem job with narrative describing the problem, which is then followed by a short segment showing the changes made, is a very effective way to document progress. This type of video is also useful in the event of OSHA inspections, as well as when the opportunity arises to share ideas among other company operating sites or even with other companies. The teamworking on a problem can be assigned the duties of documentation for that project.

Project Evaluations

There are two methods that can be used to evaluate ergonomics projects. One is based on reductions in the cumulative stress of the job on the people performing the job, and the other is cost based.

Reduction in cumulative stress:

1. Determine which indicators of excessive stress to use. Some typical indicators are lost workdays, medical claims, OSHA log cases, first aid treatment, pain on the job, and discomfort on the job. In order to evaluate changes, leading indicators (discomfort, pain, and first aid treatments) provide information more quickly than lagging (lost workdays, medical claims, OSHA log cases). In addition, the leading indicators will provide information before an on-the-job injury occurs.

2. Use the appropriate indicators both before and after any changes to the job are made. The baseline data taken before changes are made provide a good comparison for improvements.

Cost-based evaluation of a project:

1. Determine the cost of the changes. This will include the study time for the project team, engineering and maintenance time for alterations, equipment costs, work order costs, downtime (if in a sold out condition), training time, time of supervisors in dealing with the changes, and the time of other staff.
2. Determine the benefits from the project. This will include any reduction in injuries/illnesses, lost workdays, etc., resulting in the following dollar benefits: medical costs, workers' comp costs, costs of replacement workers, overtime costs, training costs, changes in productivity or quality for a less experienced worker, equipment downtime, supervisor's time, etc. In addition, the project may result in enhanced productivity or quality improvements on the job. Finally, there may be less waste or rework on this job.
3. Determine net benefits (the costs subtracted from the benefits) and the benefits/cost ratio (the benefits divided by the costs). These may be compiled for periodic summaries by the ergonomics committee.

Audits and Assessment of the Ergonomics Program

Just as there are audits of the safety program, there is a need for audits of the ergonomics program. Audits are often done by an outside party, while assessments may be conducted by the ergonomics committee itself.

An effective audit tool will have two practical benefits. First, it can provide quantitative information about your ergonomics program. It can tell you where you stand with implementation. A second benefit, however, is that the assessment can give some insight into where you may want to take the program, over what time period, and what specific actions are required to meet those ends. More information on audits and assessments can be found in the chapter titled "Evaluation of Ergonomics Programs."

References

1. *Review of Ergonomics Programs, The Top Ten Reasons Why Ergonomics Programs Fail,* Auburn Engineers, Inc., Auburn, AL 1994.
2. *An Integrated Plan for An Occupational Ergonomics Program, Your Guide to Developing and Managing an Occupational Ergonomics Program,* Auburn Engineers, Inc. Auburn, AL, 1991.
3. *Ergonomics Program Management Guidelines for Meatpacking Plants,* U. S. Department of Labor, Occupational Safety and Health Administration, 1990.
4. *The Practice and Management of Industrial Ergonomics,* David C. Alexander, Prentice-Hall, 1986.
5. *Ergonomic Design Guidelines,* Auburn Engineers, Inc., Auburn, AL, 1990, 1992.
6. *Overcoming the Barriers, Managing Your Ergonomics Program,* Auburn Engineers, Inc., Auburn, AL, 1993.

For Further Information

Assessing Your Ergonomics Program, David C. Alexander, PE, CPE, Auburn Engineers Press, 1995.

Overcoming Four Common Barriers, David C. Alexander, Workplace Ergonomics-Stevens Publishing Co., 1996.

Planning a Successful Ergonomics Program, David C. Alexander, PE, CPE, Workplace Ergonomics-Stevens Publishing Co., 1996.

Framework for Assigning Responsibilities, David C. Alexander, PE, CPE, Workplace Ergonomics-Stevens Publishing Co., 1995.

Why Would You Want An Ergonomics Culture? David C. Alexander, PE, CPE, Workplace Ergonomics-Stevens Publishing Co., 1996.

The Ergonomics Program Report Card, David C. Alexander, PE, CPE, Workplace Ergonomics-Stevens Publishing Co., 1996.

Part II
Foundation of Ergonomics Knowledge

Section I

Fundamentals of Ergonomics

Section I

Introduction

8

Design of Information Devices and Controls

Toni Ivergard
Institute of Working Life
Sweden

8.1 Introduction

This chapter covers the design of traditional information devices, and also the design of devices for communication with computers. At the end of the chapter the design of instructions, forms, and tables will be dealt with.

Instructions on how scales and scale markings on visual instruments should be designed, together with the advantages and disadvantages of different types of visual instruments, are included. It appears, for example, that the common round meter with a moving pointer is the best one for most applications. Where more exact quantitative readings are necessary, the direct-reading digital instrument is best.

A relatively detailed specification for the design of VDU screens is included. The main attribute of VDU screens is their flexibility. They can be used for presenting many different forms of information. However, in control rooms access to a large amount of information simultaneously is often required.

Therefore, it may be necessary to have several VDUs, or to have access to other information devices, such as overview displays as a complement to VDUs.

Methods for producing diagrams, codes, and symbols are also described, and various methods for using color symbols are discussed. The use of colors may have some importance in simplifying the reading of process information. However, the use of colors should be limited, bearing in mind that significant proportions of the population are color-blind. Colors should be used to supply additional information so that the VDU can be read correctly even if all colors disappear.

In man/machine communication, man receives information via the various information devices. In the control room, the information is either visual or auditory, and can be either static or dynamic. Dynamic information is constantly changing, such as the information shown on speed meters, height meters, radar, TV, and temperature and pressure meters. Static information does not change over time and includes road markings, maps, notices, manuals, and any printed or written material.

In this chapter we shall deal primarily with the various types of dynamic information. Static information will also be covered to a certain extent, particularly in the section dealing with VDU screens.

Three main types of information device will be covered in this section:

1. Traditional instruments
2. VDU screens
3. Sound signals

8.2   Traditional Information Devices

Traditional instruments are still the most common forms of information device in the control room. In modern control rooms, however, more and more information is being transferred to VDUs. Traditional instruments may be divided into the following subgroups with regard to their areas of use:

1. Instruments with associated control devices
   Control regulation instruments. The instrument is read and, if necessary, the operator adjusts the machine.
   Instruments for setting-up. Instruments used for making changes in the running conditions.
   Instruments for following (tracking). Instruments usually used in different types of vehicles (e.g., cars, airplanes).
   Instruments for indication. They show such information as "entrance," "way ahead closed," "backwards," etc.
2. Instruments without control devices
   Instruments for quantitative readings. Instruments for qualitative readings.
   Instruments for check readings. Instruments used for detecting and reporting a deviation from a normal value.
   Instruments for comparison. For checking that two machines are producing the same value.

Of the instruments above, those for check readings, control regulation, and for setting up are the most common in industry. The introduction of computer systems for process control, however, has meant that qualitative readings are becoming more common.

Different Types of Visual Instruments

The choice from among the many different types of visual instruments available depends on the information to be presented and how it will be used. Figure 8.1 shows some of the more common types of instrument. Figure 8.1a shows a digital instrument, which displays the various numbers directly. The instrument may have mechanically or electronically generated numbers. Figure 8.1b shows instruments with moving pointers and fixed scales, whereas in Figure 8.1c the instruments have fixed pointers and moving scales.

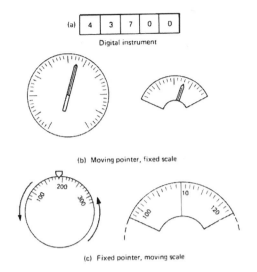

FIGURE 8.1   Various types of instruments.

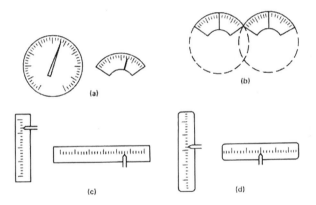

FIGURE 8.2   Varieties of instruments with moving pointers and moving scales.

There are many variants of pointer instruments. Figure 8.2 gives examples of some of these.

1. Round and sector-shaped instruments (Figure 8.2a) are recommended for check and qualitative readings. Round ones are usually better than sector-shaped ones, but take up more space.
2. Vertical and horizontal scales with moving pointers (Figure 8.2b) are also good for check readings. They do not give as much information as that provided by the angle of the pointer in round or sector-shaped instruments. However, this type of instrument takes up little space.
3. Round and sector-shaped instruments with fixed pointers (Figure 8.2c) can be recommended where the whole scale does not need to be seen for quantitative readings which change slowly. Their design allows a relatively long scale to be used without taking up much panel space, but they may need a relatively large area behind the panel.
4. Figure 8.2(d) shows vertical and horizontal instruments with fixed pointers. This type of instrument can also have a very long scale (see Figure 8.3). They cannot be recommended for cases other than where a very long scale is required.

Table 8.1 summarizes the recommended areas of usage for different types of pointer instrument.

In certain cases, it may be desirable to choose other instrument designs. One may wish to use the design to show the function of the instrument. Figure 8.4 gives examples of a round instrument with a

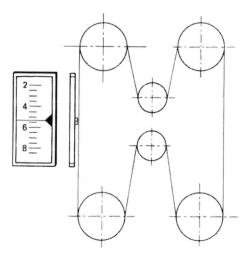

FIGURE 8.3   An instrument that covers a large scale range can be made with a large moving scale and a fixed pointer.

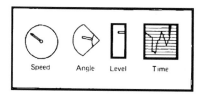

FIGURE 8.4   Form coding through the shape of the instruments.

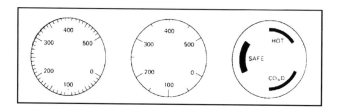

FIGURE 8.5   Different scale complexities.

moving pointer indicating speed. The angle of the material being controlled is shown on a sector-shaped instrument with a moving pointer. Level is shown on a vertical instrument; and time changes are plotted out on paper with a moving pointer; the most recent time is marked at a suitable time interval.

Design of Scales and Markings

The terms to be used in this section are defined thus:

1. *Scales range.* The numerical difference between the lowest and the highest value on the scale.
2. *Numbering interval.* The numerical difference between two successive numbers on the scale.
3. *Scale marking interval.* The numerical difference between two scale markings.

Before choosing a scale for an instrument, one must determine the precision required. Figure 8.5 gives examples of different precisions. It is generally true that the least exact scale, which fulfils system requirements, should be chosen, thus avoiding unnecessary accuracy. If possible, the information should be given in units that require no modification by the operator in order to be used. An example of this

TABLE 8.1 Recommended Areas of Use for Different Types of Pointer Instrument

| | A Moving pointer | | | B1 Fixed pointer | | | B2 Window | | C | D | E |
| | Linear | | | | Linear | | | | Counter | Switch | Lamps |
	Round	horizontal	vertical	Round	horizontal	vertical	Round	Linear			
Without controls:											
Quantitative reading, slow change	o	o	o	o	o	o	o	o	xx	o	o
Quantitative reading, fast change	x	x	o	o	o	o	(x)	(x)	O	o	o
Qualitative reading, direction	(x)	o	x	o	o	o	o	o	o	o	t
Qualitative reading speed	x	o	o	(x)	o	o	o	o	o	o	o
Control/check reading	xx	(x)	(x)	o	o	o	o	o	o	o	o
Comparison, fast	xx	o	o	o	o	o	o	o	xx	o	o
Comparison, slow	(x)	o	o	o	o	o	o	o	o	o	o
Warning with controls:	(x)	o	o	o	o	o	o	o	o	x	xx
Control/check adjustment	xx	(x)	o	o	o	o	o	o	o	o	o
Setting up	x	x	(x)	o	o	o			o		
(Tracking)	Mostly special instruments										
Indicating	xx	o	o	o	o	o	o	o	o	x	x

xx   very good
x   good
(x)   uncertain
o   unsuitable

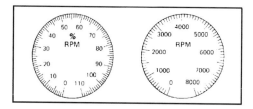

FIGURE 8.6   Comparison of percentage (left) and absolute value (right) scales.

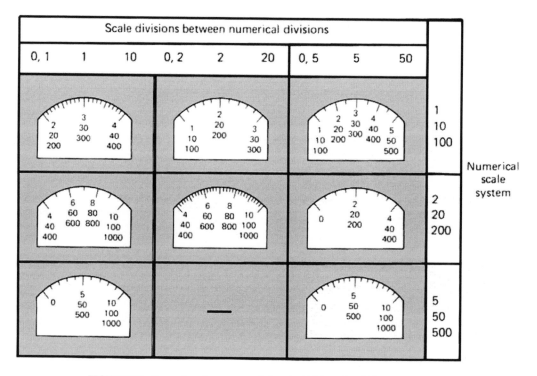

FIGURE 8.7   Examples of recommended scale divisions for different scale sizes.

might be percentage figures being used instead of the actual number of revolutions, so that the operator need not remember different top speeds on different machines (a scale speed is often the same percentage of top speed on all machines). The percentage scale also gives fewer digits to read (see Figure 8.6).

Figure 8.7 shows examples of recommended scale designs for instruments with different scale ranges. On the horizontal axis are the marking intervals, 1, 2, 5 (and 1/10 and 10 times these values). This means that for 2 there are two (or 02 or 20) units between each subsidiary marking. On the vertical axis are the numbering intervals, i.e., the numerical difference between two successive numbers on the scale.

The matrix in Figure 8.7 shows examples of recommended scales and examples of how the markings can be designed. The scale can either be marked with large, medium-sized, or small marks. Certain of the scales use all three sizes. Large markings are always used for the numbers of the scale. The medium-sized and small markings are used for subdivisions, i.e., those divisions that lie between the numbered markings.

There must not be more than nine subdivisions between numerical markings. The scales should be designed so that interpolation between divisions is not necessary. Where there is a lack of space, however, it is better to allow interpolation than to clutter up the instrument with too many markings. If there is room for the scale illustrated in Figure 8.8a, this is the one to choose, otherwise choose the one shown

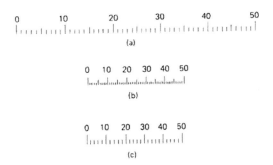

FIGURE 8.8 The physical size of the scale and the number of scale diversions. Scale (b) is not recommended.

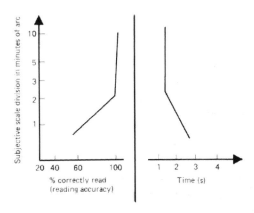

FIGURE 8.9 Effect of subjective scale division on reading accuracy and time. Increasing the subjective scale division to greater than 2 minutes of arc does not increase the reading accuracy or reduce the reading time (From Murrell, K.F.H., 1958, *Fitting the Job to the Worker: A Study of American and European Research into Working Conditions in Industrial Situations,* (Paris: Organisation for European Economic Cooperation–EPA).

in Figure 8.8c. The scale in Figure 8.8b is not suitable. If the alternative in Figure 8.8c is chosen and the scale has to be read off to the nearest whole number, this means having to interpolate one step between markings. The smallest step, which it is necessary to interpolate, is called the *subjective scale division.* There should be zero, two, or five subjective scale divisions for every marked interval.

The basic factor in the reliability of reading an instrument dial is the physical width (measured as an angle of view) of subjective scale divisions (i.e., zero, two, or five subjective scale divisions per marked interval). The scale is subdivided in such a way that the reading tolerance and accuracy one wishes the instrument to have are achieved. Accuracy must not be greater or even less than the allowable pointer error for the instrument in question.

If it is necessary to interpolate fifths in order for the correct degree of accuracy to be obtained, the angle of view must be five times the angle of the subjective scale division, which equals the smallest division which it is necessary to interpolate. If the viewing angle for the subjective scale division is less than a certain critical size, there will be considerable errors in reading it. If, on the other hand, the angle of view of a subjective scale division is greater than this critical size, one cannot expect any further increase in reliability of reading, as shown in Figure 8.9, according to which the width of the critical scale division is two minutes of arc.

The reading reliability over this critical value would be expected to rise to 99% for people with long experience in instrument reading. When designing a scale, therefore, one would start by determining the greatest distance from which the scale is going to have to be read. Based on the distance and the critical angle of two minutes of arc, the smallest size of the subjective scale divisions, which will be needed on a scale, is calculated from the total measuring range the instrument is to have. Knowing that there should be two or five subjective scale divisions between each marked interval, the number of marked intervals can be selected. There should be two, four, or five marked intervals for each numbered interval.

For a scale marked in five subjective scale divisions per marked interval, and which has a total of 100 subjective scale intervals, the total length of the scale is determined according to the reading distance obtained from the following formula:

$$D = 14.4\ L$$

D is the reading distance, and L is the length of the scale. D and L have the same units of length, e.g., cm. If one does not have such a "standard scale" with the above relationship between subjective scale division and marking interval, there is a correction factor, which may be added to the formula:

$$14.4 \times L = \frac{D \times (i \times n)}{100}$$

where i is the number of subjective scale divisions into which each marking interval is to be interpolated, and n is the number of marking intervals.

In practice one is often tied to existing standard instruments. Here one can evaluate the maximum reading distance for the different instruments instead and determine whether this is sufficient for the actual application.

The scale itself should not have long lines joining up the scale markings, but should be open (see Figure 8.10). Figure 8.11 gives the recommended measurements of the scale marking for (a) normal visual conditions and (b) low light levels. The reading distance is approximately 70 cm in this case. The measurements given in Figure 8.11 apply not only to straight instruments, but also to circular and segment-shaped ones.

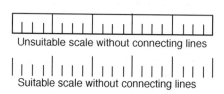

FIGURE 8.10   Examples of suitable and unsuitable scale designs.

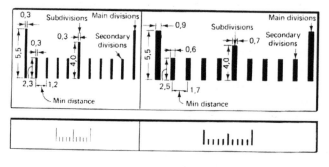

FIGURE 8.11   Recommended design of scales (mm).

The design of the letters and numbers on the scale is important for giving the best possible reading reliability. The following points should be noted especially:

1. The ratio between the thickness of the line and the height of the letter/number for white numbers on a black background should be between 1:10 and 1:20. If the background is dark and the numbers are illuminated from behind, the optimum is between 1:10 and 1:40. The optimum for black letters on a white background lies between 1:6 and 1:8. Black lettering on a white background is normally preferable. For dark adapted and low illumination conditions, white numbers on a black background are preferable (for night viewing in combination with red instrument lighting, the white numbers will be seen as red).
2. The ratio between the height of the numbers/letters and their width should lie between 2:1 and 0:7:1. The optimum is 1.25:1.
3. The appearance of the numbers/letters is important. As a guide, the numbers/letters should be simple in their design. Serifs and too many lines complicate the reading. However, they must not be so simple as to become difficult to understand.
4. Letters/numbers must not be positioned on the scale in such a way that they are shaded by the pointer, and they must be vertical.

The following recommendations are made for the design of the pointer:

1. The pointer must be long enough to reach but not to overlap the scale markings.
2. The pointer should lie as close to the surface of the dial as possible in order to avoid parallax errors.
3. The section of the pointer, which goes from the pivot to the scale markings, should be the same color as the scale markings. The remaining part of the pointer should be the same color as the dial face.
4. On horizontal scales, the pointer should sit on the top of the scale, and for vertical scales it should be on the right-hand side with the numbers on the left.

Sound Signals

Certain types of information are better transmitted as sound signals rather than visual ones. This is the case in the following situations:

1. When the signal is originally acoustic.
2. Warning signals. The advantage is that the operator doesn't need to see the signal in order to detect it; i.e., he or she does not need to look at the instrument constantly.
3. Where the operator lacks training and experience of coded messages.
4. Where two-way communication is required.
5. Where the message concerns something that will occur in the future, e.g., the countdown to the start-up of a process.
6. In stressful situations, where there is a possibility that the operator would forget what a coded message meant.

Tones are preferable in the following situations:

1. For simplicity.
2. Where the operator is trained to understand coded messages.
3. Where rapid action is required.
4. In situations where it is difficult to hear speech (tones can be heard in situations where speech is inaudible).
5. Where it is undesirable or unnecessary for others to understand the message.
6. If the operator's job involves constant talking.
7. In cases where speech could interfere with other speech messages.

Warning signals are without doubt the most common form of sound information.

Sound is an information source which is little used, with the exception of warning signals. Sound signals/information could probably be used to advantage in giving spoken instructions in disaster and other acute situations, for example that there is a fire at a particular location and what action should be taken. In this type of situation one can use a calm voice to give further instructions on necessary actions, such as clear and simple instructions on how to evacuate the building. Various forms of acoustic alarm should be able to be complemented with tones, which would give preliminary information on the type of fault. If there is a hierarchy of alarms, different tones can be used for different groups of alarms. If detailed information is required about the alarms, questions can be entered via a keyboard, or directly by voice. The section on Speech Generation and Recognition later in this chapter includes a description of the design of computer-generated speech, and computers that understand speech.

8.3   Visual Display Units (VDUs)

The information devices traditionally used in control rooms normally have only a single area of application each. The instrument is normally electrically connected directly to some form of measuring device. In certain cases, switches are used that allow the same instrument to be connected directly to several measuring devices. For example, this is usually the case for temperature readings where it is common to have alarms connected to all the measuring points. One then has to turn a special switch in order to be able to read the temperature at all the measuring points.

In the modern computerized process control system, information is collected from many measuring points in the control system. Information is stored and subjected to further processing in the computer, which then sends it on to various receivers, such as the different information devices in the control room. The information devices connected to the computer system are normally the type that can carry out several simultaneous functions, i.e., which can receive different forms of information, such as cathode ray tubes (CRTs).

VDU Design

The collective term for such information devices designed for several different purposes (e.g., for reading temperature, shaft speed, instructions, information on handbook data) is visual display unit (VDU). The most common type of VDU to date is the cathode ray tube (CRT), but there are many other types, such as liquid crystals, plasma displays, and matrices built up of different types of lamp or light-emitting diode (LED).

The important advantage of the VDU is that it can handle most types of presentation although the various technical solutions do have some limitations, e.g., the shape of characters which can be formed on the screen. The pictorial alternative is provided (within the framework of the technical specifications) by programming the computer. It must be stressed, however, that the programming must suit the actual process operator and his or her work and not just reflect how the programmer feels interactive VDUs should be designed.

The latter is unfortunately often the case. This means that many VDU applications in industry have shortcomings or are simply unsuitable and faulty; this does not just mean discomfort and limitations for the operator — it also gives rise to poor performance and an unnecessary number of errors.

There now follow some guidelines on how information on VDU screens should be designed. These recommendations are not final ones that would be applicable to all the different applications and control situations. They must be treated instead as a form of checklist to be used in the design of man–computer communication systems in the control room for process control.

It is also important to stress that it is practically impossible to choose the technical components first, and then expect to get a functional solution. Even if ergonomic requirements are specified for each individual component, there is a risk that the overall solution will be bad. Technical solutions to the design of the interface with the operator can only be produced from detailed descriptions of the job content and analyses of the various work and skill requirements.

Equipment, which can be used for different forms of information presentation, is summarized below, together with some advantages and disadvantages.

Information Presentation Technique/Method Equipment

1. Presentation of set value of a variable or deviation from set value
   Pointer or bar with scale (e.g., thermometer).
   Traditional instrument, plasma or LED forming a bar or equivalent on CRT.
   Moving numbers.
   Dot matrix, 7-segment plasma panel, CRT, etc.
2. Trend or time history of a variable
   Graphics.
   Matrix printer or plotter.
   Line diagram:
   Alt. 1: line printer or plotter.
   Alt. 2: CRT.
3. Relationship between set values and several different values
   Line or bar diagram:
   Alt. 1: CRT or plasma panel.
   Alt. 2: Multi-pen printer or plotter.
4. The way a circuit works light indicators or displayed mnemonics:
   Alt. 1: Lighted buttons, LEDs, rear-projected displays.
   Alt. 2: CRT.
   Alt. 3: Printer.
5. Alarms
   Matrix:
   Alt. 1: Signal/number board.
   Alt. 2: Special reserved CRT.
   Tables in chronological, hierarchical, or random order:
   Alt. 1: CRT reserved for the job.
   Alt. 2: Line printer.
6. Text
   Lighted points in certain groupings:
   Alt. 1: Plasma grouping (usually a 5 × 7 point matrix per symbol).
   Alt. 2: CRT at 25 Hz with interlacing.
   Alt. 3: CRT at 50 Hz frame frequency with no interlacing.
   Lighted line segments:
   CRT with DC positioning (x/y techniques).
7. Diagrams
   Lighted points in a matrix:
   Alt. 1: Plasma grouping.
   Alt. 2: CRT with 25 Hz frame frequency and interlacing.
   Lighted line segments
   CRT with DC positioning.

For the *presentation of the set value,* there are advantages and disadvantages to most types of equipment. The advantage of pointers or thermometer-type bars is that they give a certain indication of the size of the changes as they occur, which is more difficult with a digital presentation. The advantage of a digital presentation is, of course, that it is easy to present a large amount of data at the same time. A numerical presentation, on the other hand, gives a poor understanding of size and quantity. It also makes it difficult to compare several values at the same time.

The ordinary printer or plotter is good for *trends and time histories* of values. However, it is difficult to print out several values at the same time, although good documentation (copies) is provided. The bar chart may mean more difficulty with paper handling, for example, and also requires some form of identification of the individual diagrams. CRTs produce a somewhat poorer picture than graphic printers, but they are more flexible. It is also easy to use a CRT to display different variables in turn. No hard copy can be produced directly from the screen, but this can be obtained using cameras or additional equipment.

In order to present the *relationship between variables and set values* CRTs and plasma panels are flexible and suitable in many cases. Resolution and sharpness are, however, relatively poor in comparison with a multi-pen plotter. The disadvantage of the multi-pen plotter is that the format is often limited. Due to the lack of resolution, the same is also true for the CRT.

In order to see how different *circuits are working*, lighted press-buttons are a good system but impractical if many functions have to be presented at the same time. The CRT can be tabulated for a number of different circuits and functions, and can also be used in conjunction with cursors and light pens.

The more traditional type of display for *alarms* has as its greatest disadvantage its lack of flexibility. It also needs a lot of space, but it does give a good overview and continuous accessibility. The system also works well interactively if the diagram has lighted press-buttons on it, where the buttons function for acknowledgment or corrective actions.

The great advantage in using a CRT is that it is easy to update. It is important, however, to have a CRT reserved specifically for this purpose. In addition, the system is compact and easy to use interactively with a light pen, for example.

If the alarms are presented in tabular form with text, chronologically, hierarchically, or in random order (instead of in a matrix), the greatest disadvantage is that the overview becomes lost. On the other hand, it is possible to include an extra line along with the alarm concerning the action which should be taken. Compared with the CRT, the printer also has the advantage that it produces documentation of the alarm. However, printers are often noisy, and it is often difficult to read the actual line being printed. This application would not require any especially reserved CRT.

The plasma display has a number of advantages in the *presentation of text*. For example, it is very thin; it doesn't cause any flickering effect; and the surface is flat. The contrast on the light surfaces can be high, and it is easy to program. The disadvantages are that only one color can be obtained, and the cost is very high. Different types of CRT are often a more realistic alternative. The great advantage of frame frequencies of 50 to 60 Hz is that the picture is clearer and flicker-free (if the brightness is not set too high). CRTs with DC positioning are considerably more costly, but they give much better contrast and sharpness. The color choice, however, is not so great as on other types of CRTs.

Of the visual display units, the CRT is by far the most common type, and will probably be so for some time to come. CRTs are also used for other purposes, e.g., for surveillance and internal TV. However, with today's technology the CRT can give rise to a number of visual problems. A good printer can be a good alternative if the lack of flexibility is acceptable.

If the recommendations given in the next section are followed, the risk of visual difficulties and eye fatigue will be considerably reduced. Problems may remain, however, for those with vision defects. The operator must then obtain glasses which are made and tested specifically for work with CRTs.

Design of Cathode Ray Tubes (CRTs)

Some brief recommendations on the requirements that should be placed on CRTs, currently the most common form of information presentation aid in the computer control system, are presented. The basic design features of a CRT are that the information shall be *visible* and *easy to read*.

It is important to choose a suitable coding and presentation of information for particular applications. These points are dealt with in the next section. The following recommendations assume a screen with a dark background and light text. This form of CRT demands very careful planning of the lighting. There is much to suggest that screens with dark text on a light background give a better result if one can avoid the problems of flicker, which easily occur on light screens (Berns and Herring 1985).

1. Clarity

   Luminance. Minimum 85 cd/m' for characters (250 ash[1]), with an optimum of 171 cd/m^2 (500 ash).

   Contrast. The optimal character contrast is 94%, but 90% is acceptable.

   Light/dark characters. If flicker can be avoided, dark characters on a light background are preferable. Otherwise, and if other environmental conditions are suitable, light characters on a dark background can be accepted.

   Flicker. For the optimal luminance level of 171 cd/m^2, a frame regeneration rate of at least 50 Hz is required. Certain types of phosphor (e.g., p-20) may need higher frame rates, up to and even over 60 Hz (frequencies of 60 Hz may interfere with mains electrical frequencies of 50 Hz in Europe, and other places).

2. Legibility

   Character size. The height varies between 16 and 27 minutes of arc (visual angle), a width-to-height ratio of 0:75, and a ratio of height to thickness of the strokes making up the characters of between 10:1 and 6:1.

   Character shape. A dot matrix using 9×7 points produces the best symbols. There are many different ways to design letters and numbers, and it is not clear which is best. It is important, however, to choose types which do not cause confusion between letters and numbers as in O and Q, T and Y, 5 and S, I with 2 and K, I with zero 1, O with B and O.

   Resolution. Ten raster lines per character height and more for nonalphanumeric characters.

   Character separation. Fifty percent of character height and 100% between lines.

   Reading distance. For characters of 3.2 mm in height, a reading distance of 69 cm is acceptable. In most practical applications, the reading distance is determined by the height of the characters. The screen should be read from directly in front; angles of more than 30 degrees from the optimum reduce legibility considerably.

   Color. Colors which lie well outside the optimum visual spectrum must be avoided, particularly those in the blue-UV region. Colors in the yellow-green range are good.

   Flashing (e.g., to gain attention). A 3 Hz flashing frequency has no adverse effect on legibility.

   Cursors. A rectangular cursor with a 3 Hz flashing rate is thought to be best.

3. Coding

   Coding improves *performance* considerably, especially for simple information processing tasks.

   Color coding. Best for localization, calculation, and comparison of different information.

   Alphanumeric codes. These are best, and color coding is next best, for recognition and identification of information. Coding is of little or no help in *quantifying* or *size estimation*.

   Components. A code should not comprise more than seven different components (e.g., seven letters or numbers). The fewer components the code uses, the better.

   The *efficiency* of the coding must be seen in relation to the total construction of pictures on the screen.

   Code groups. If possible, different groups of codes, each consisting of components, should not contain the same components.

4. Construction of picture on the screen

   The basic factor in the determination and design of the picture is the function which is to be fulfilled. The following information may be used to provide certain guidelines:

   a. Digital presentation is best for quantitative information.

   b. Pointers on circular scales are best for showing changes.

---

[1]1 ash is the luminance of a white surface illuminated by 1 lux.

c. Vertical and circular scales with moving pointers are good for check readings.
d. Chart recorders are valuable for seeing instrument faults and for obtaining a rapid impression of the system's response.

Diagrams are often preferable to graphs on CRTs. Histograms are especially easy to read. Curves, however, are preferable for reading trends. Several curves can be presented to advantage for comparisons between several variables. Here, semigraphic screens are better, but My screens are to be preferred above all.

Design of tables
a. Whether data should be presented in columns or rows depends on what is more natural for the task in question.
b. Letters are better for identification than numbers.
c. Numbers should be arranged in as few columns as possible.

The design of pictures and the positioning of text within each frame is usually a compromise between several different factors. It is usually very difficult to adhere strictly to the various criteria which are set in designing different frames.

The length of the text also determines the design of frames or pictures because there is only a limited area available, and if the text is very long its length will determine its positioning. In such cases, it may be necessary to abbreviate the text. When doing this, it is of the utmost importance to maintain the comprehensibility of a word as far as possible, even though it is abbreviated.

By placing the same information in the same position in different frames, the possibility of errors by the operators is reduced, together with their search time. This is particularly true for beginners and those who use the terminal infrequently. Consistent positioning of error messages is also very important, because they must always be immediately obvious to the operator, who should not have to search for them.

The most fundamental criteria by which information to be used in a frame should be analyzed are:

1. The importance of the information
2. Its frequency of use
3. The sequence of use

It is clear that even though these criteria have importance on their own, they are very closely connected. One example of this is that the first information to come up on the screen is often both the most important and most frequently used. When the principles for the design of the pictures and the text have been determined, the information should be presented in such a way as to simplify the task of the operator as much as possible.

The CRT-type of display often has poor picture sharpness (focus) at the perimeter caused by the curve on the screen edge, which reduces the clarity of the characters. On many screens, this may mean reducing the amount of information on both the left and the right-hand sides of the screen by 3, 4, or even 5 columns in order to avoid this poor focus area. The effective area of the screen is thus reduced. Another aspect of the design is whether to use every line or every alternate line. No general answer can be given to this; it depends partly on the design of the screen (i.e., the distance between the lines) and partly on the amount of information on the screen.

After determining the above criteria comes the question of how tightly the information should be positioned. The amount of information may be defined as the quantity of information within a defined area. If too much information is displayed, operator performance is decreased (Stewart, 1976; Cakir et al., 1979). Stewart suggests that the search time in seconds is roughly a fifth of the number of alternatives to be searched through. Even if a very experienced operator is able to be selective in searching for information, irrelevant information means that the search time and error frequency will increase, thus reducing the performance of the operator and the system.

Advantages and Disadvantages of VDUs

In older control rooms, each variable was represented by its own instrument, regulator, or control device. These were placed together on a panel. Various ergonomic rules for the design of panels have been produced over the years, with the aim of achieving the best possible operational conditions. The development of VDUs and computers have now provided new forms of information presentation, monitoring, and controlling.

Research by ERGOLAB (Ivergård et al., 1982) has shown that many operators prefer the conventional instrumentation over the computerized versions. They have often described using the computerized alternative as being akin to monitoring the process through a keyhole.

Descriptions have been produced (Stark, 1983) of how people scan a picture for the object they are seeking. It appears that the normal strategy is to fix the gaze on a certain number of points in order to identify the picture; the points chosen on which to set the sight depend on the object. Stark (1983) also found that the experienced viewer will miss some of the points he previously fixed on. Despite jumping over them, he will still be able to make the correct interpretation of the picture. In addition, it will be detected whether there are any changes in the parts of the picture that were skipped over. In other words, people use their peripheral vision for checking whether there are any changes in the picture.

It is likely that a process control room operator will work in a similar fashion in monitoring a VDU or a conventional panel. He fixes on a number of points in order to obtain a picture of the current situation in the process. The operator then updates his mental model of the status of the process. The experienced operator has a considerably higher performance level at monitoring, and probably fixes on considerably fewer points to estimate the status of the whole process. Sometimes the operator only has access to a number of pictures presented on VDU screens. This does not allow parts of the process, other than those currently presented on the screen, to be updated. More conventional instrumentation, where all the information is presented in parallel, gives the operator very different possibilities. By fixing the gaze actively on certain parts and using the peripheral vision for other parts, he or she can continuously update his/her mental model of the status of the whole process.

Given this background information, one can see that there is a natural division of the viewing process into two types: active and passive. When someone fixes their gaze on one or a certain number of points in order to identify an object, this is active vision. In parallel with the active process, passive vision is occurring via the more peripheral parts of the retina.

In active vision, the gaze is turned toward the object and fixed on the central part of the retina where the cones are most dense, i.e., the fovea. In the area around the fovea, the rods dominate. The rods are considerably more sensitive to light and can therefore work under relatively dark conditions. They are also sensitive to movement and changes and are used in pattern recognition. The cones, on the other hand, need more light. The ability to distinguish detail, mainly by the cones, is also thought to increase in proportion to an increase in the light level. There are also cones which have the ability to distinguish colors.

This allows us to draw certain conclusions regarding active and passive vision. Active vision allows the identification of colors and small objects under good lighting conditions. Passive vision permits recognition of patterns and, therefore, especially changes and movements, and passive vision also works if the object is in motion.

In control room work, active vision is used to make detailed readings of a more quantitative nature. It is also used in the identification of color codes and in the detection of small differences in curves or diagrams, for example. Active vision is excellent for VDU viewing. There the operator can call up a particular frame and adjust parameters such as the set value.

Using passive vision, one could tour the control room and identify changes in the process pattern, for example, of lights being lit or extinguished on a panel. Schematic representations of a process with built-in indicators (e.g., signal lamps indicating deviations) are a suitable type of presentation for passive vision. In other words, passive vision is perhaps best applied on a more traditional type of instrument panel.

The choice of presentation method depends very much on the task of the operator. There are three main motives for having an operator in the control room:

1. The operator acts as a supervisor in order to carry out certain standardized routine tasks which, for various reasons, have not been automated. He also has the job of calling for expert help when some unforeseen incident occurs.
2. The operator is himself a qualified expert with the job of carrying out production planning and optimization tasks. The operator deals with simpler, more routine and predictable types of fault. The more serious, unforeseen faults, on the other hand, are passed on to special maintenance experts.
3. The operator is primarily a maintenance-oriented expert who gets information about the process from other sources, especially those which are economic in character. He usually looks after production quality matters himself. The operator is expected to be able to deal with most unforeseen and difficult faults and events in the process.

If the first alternative is chosen, one can determine relatively accurately beforehand what type of information the operator will need in different situations. There are fewer requirements for the more detailed type of overview information. The conventional type of instrumentation therefore provides very little information to this operator and he can largely rely on a number of VDU screens with a predetermined program of frames.

The expert operator who is either production-oriented or maintenance-oriented has a considerably greater need for more detailed, continuous, and parallel presentation of the whole process. If the information is presented on VDUs, most processes would require a large number of VDU screens or extremely large VDU screens. The alternative is to require the operator, even during normal running conditions, to sit down and leaf through all the status frames. He would have to do this in order to update himself actively on the process status and to build up his knowledge of the functioning of the process. Instrumentation of the conventional type offers completely different possibilities for the operator to update his mental picture of the process. Looking at the instrumentation both consciously and unconsciously can do this.

The production-oriented operator needs to be able to see a relatively detailed and dynamic functionally oriented process model. The maintenance-oriented operator, on the other hand, needs to have a more physically oriented model available. The traditional instrument and control panel is a good alternative, but the conventional type of instrumentation is not preferable to a VDU in all processes.

The VDU screen has the great advantage of being flexible. Color monitors with high resolution and detailed pictures also allow the presentation of a large quantity of information in a limited workspace. VDUs are also suitable for presenting different types of information. This may be graphical information (e.g., maps of temperature distributions [isotherms], pressure distributions [isobars], etc.). Even if VDUs cannot always wholly replace conventional instrumentation, they are a necessary complement in modern process control.

Table 8.2 gives a comparison of conventional instrumentation and electronic VDUs. It may be seen that the VDU has many advantageous characteristics, and that the disadvantages are largely of an ergonomic nature. In addition to the characteristics given in Table 8.2, visual problems should be included. These practically always occur when using the CRT-type of visual display. In practice, both VDUs and more conventional instrumentation are required in most cases. When modernizing existing control rooms, it is best to keep the old instrumentation as a reserve and a complement to the VDUs.

When building new control rooms, it is rarely sensible to install both the conventional type of instrumentation and VDUs. On the other hand, it may be a useful complement to the VDUs to produce some form of detailed overview panel. This can schematically and dynamically describe the physical design of the process. It is often desirable, and sometimes necessary, to provide dynamic information on the overview panel. This may, for example, show which valves are open or closed, which pumps are working or not. It may sometimes even be desirable to provide the overview board with quantitative information such as flow rates and levels.

TABLE 8.2 Characteristics of Conventional Electronic Display Devices

Attribute	Conventional	Electronic/Advanced
Nature of presentation	Parallel	Serial
Mode of presentation (digital, analog, etc.)	Fixed	Variable
Availability of information	Continuous	On request
Relationships among items of information	Static	Dynamic
Nature of the interface	Inflexible	Flexible
Incorporation of diagnostic aids	Limited	Readily feasible
Ability to modify or update	Difficult	Easy
Redundancy of displays about control rooms	Costly	Relatively inexpensive
Control room size requirements	Relatively large	Relatively compact
Compatibility between process response and control movement	High	None
Control display feedback	High	Name
Nature of overview information	Detailed	Summarized

Adapted from Seminara, J. L., 1980, *Human Factors Considerations for Advanced Control Board Design,* vol. 4 of *Human Factors Methods for Nuclear Control Room Design* (Palo Alto, CA: Electric Power Research Institute).

The amount of dynamic information provided on the panel depends very much on the type of process. In a continuous, stable process, with relatively few starts and stops, a static overview panel is often sufficient. These may be described as processes with high inertia, where changes in the process variables take place over many minutes, and where a detailed diagram of the principles showing the flow and the interconnections is required. In addition, the most important physical units must be marked on the diagram. A good complement to this form of presentation may be a three-dimensional model of the process.

A dynamic and detailed overview board is required where a complicated network with several alternative connection routes exists. The connections in one part of the network will affect the conditions in another part of the network. A typical example is an electricity supply network, and to a lesser extent a water network. A relatively complex batch process also needs a detailed dynamic overview panel. If the batch process is simple enough to be presented as a picture on a VDU screen, an overview panel is not necessary.

An overview panel is also required in a continuous process with many stops and starts, and where the course of events is relatively rapid. Such processes would have changes occurring within seconds or, at the most, a few minutes. Production-oriented operators will always need a functional overview diagram. This can be suitably presented on a VDU screen. The maintenance-oriented operator has a special need for detail in the overview panel.

8.4   Instructions, Forms, Tables, and Codes

Instructions, notices, and forms give printed information to the operator, often in connection with a product. It is of the utmost importance that this information is presented clearly and concisely and that it is easily read. This is particularly the case when an operator is faced with a new product, one which he rarely uses, or in a situation where time is limited.

The following rules apply:

1. *Letters.* Capitals alone are recommended for general use, but a combination of capitals and lower case (small) letters is allowed; i.e., first letter a capital, followed by lower case letters. The ratio between width and height of the uprights should be between 1:6 and 1:8. The ratio of width to height of letters should be 3:5.
2. *Short messages.* In order to save time and space, the text should be as short as possible, while still maintaining clarity.

3. *Choice of words.* Words and meanings should be kept as simple as possible (see above). Only well-known words should be used. One may use common technical terms for special populations, for example, the use of aeronautical terms when addressing pilots.

4. *Clarity.* Instructions must be short and easily understood. One very simple way to test this is to try out different instructions on different people.

5. *Contrast* in work where dark vision is required, letters should be white or light yellow on a black or other dark (e.g., brown) background. Where no dark vision is required, the letters should be black on a white background. Other colors may also be used for coding purposes, but they should always be chosen with maximum contrast in mind.

6. *Size.* The recommended size of letters depends on the reading distance. At a distance of 70 cm or less, in low lighting conditions, the size should be about 5 mm. In good light, this can be reduced to 2.5 mm.

Diagrams and Tables

The following recommendations apply to diagrams and tables.

1. Diagrams are always better than tables if the shape, variation, or connection between materials is of interest, or if interpolation is necessary. If not, tables are preferable.

2. Simplify the table as much as possible without reducing its accuracy and without the need for interpolations. (An example of this would be where, if the meter marks on the water tank level are not sufficient, several calculations are required to interpret the water level.)

3. Leave at least 4 to 5 mm between columns that are not separated by vertical lines.

4. Where the table columns are long (more than six lines), they should divide into groups of three or four lines. Leave some space between each group.

5. Diagrams should be drawn so that the numbered axis lines are darker than the unnumbered ones. Where only every tenth line is numbered, the fifth line should also be denser than the others, but less dense than the numbered ones.

6. Avoid combining too many parameters in the same diagram — there should not be more than three. If more parameters are shown, more diagrams must be used.

Codes and Symbols

Different sorts of traditional symbols are used to convey a particular meaning. For example the cross represents Christianity for a large proportion of humanity. The alphabet can be used to build up different definitive meanings using various systematic rules. It is important in the construction of an artificial language that all the users agree on the rules, which are to be applied to it. This is the case both in more complex languages (e.g., those which resemble the natural ones) and in the simpler ones, such as those used by machines of different types or as explanations on VDU screens. The assumptions necessary before a language can be built are, first, that everyone agrees what the different characters mean and, second, that there are certain predetermined semantics. There must also be a grammar, which specifies the way in which the different symbols can be combined. Finally, the symbols must be designed in such a way that other people can understand them. As far as visual symbols are concerned, they must be able to be clearly seen and understood in the situation in which they are designed for use.

Languages may be termed natural or artificial. Natural languages are often learned in childhood or after very comprehensive training and/or education. More complex artificial languages, which are to be used in a way similar to the natural languages, also require a very comprehensive education and training. In the design of languages for use in different technical applications in industry, special stress must be put on the need to make the rules simple so that as far as possible they are self-evident.

Even if one can set out such very simple rules for the language, it is still necessary for the rules to be put down in writing so that everyone is agreed on which rules are valid. A semantic, then a grammar, and finally a number of viewpoints on the visual detachability of symbolic languages that can be used on VDU screens are presented.

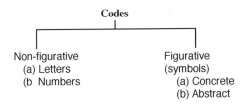

FIGURE 8.12   Categorization of codes.

Semantics

Semantics specify what the different characters mean. In this case, the semantics are summarized in Figure 8.12. The notion of codes is used here as an overall concept covering all types of symbol/character. We are conventionally accustomed to two types of codes:

1. Nonfigurative codes, which consist of several small elements. The individual elements have no meaning in themselves, but when combined they have common unambiguous meanings. The most common form of these nonfigurative codes is numbers and letters.
2. There are also figurative codes (symbols), which are those designed in such a way as to have a meaning themselves without needing to be combined with other symbols. Such figurative codes may be either concrete or abstract. The concrete codes attempt to imitate what they symbolize (e.g., a pedestrian crossing sign is represented by a stylized drawing of a walking person), while the abstract codes symbolize an abstract concept (e.g., Christianity, represented by a cross).

Both figurative and nonfigurative codes are used in process industry applications. The figurative concrete codes should try to resemble the apparatus and machines they represent, and the abstract codes should be used to represent the actual events occurring during the process. Nonfigurative codes, usually in the form of letter and number abbreviations, are also used. Numbers are used both for identification and quantification, although it is preferable to use two different number series for these purposes, for example, Roman numerals for identification and Arabic (ordinary) numbers for quantification.

Grammar

The grammar to be used in this connection must be very simple. It should consist only of:

1. Nouns
2. Adjectives
3. Verbs

Nouns should be used to specify different physical objects such as generators, motors, transformers, and switches. The verb is used to give the condition of the noun, e.g., on, off, running, open, closed. The adjective is used either for specifying which machine/unit is under discussion, or for giving its characteristics, e.g., DC, AC, size. Concrete symbols are suitable for nouns, and abstract symbols are best used for verbs (arrows, colors, etc.). Nonfigurative codes such as abbreviations or numbers are best used for adjectives. Figure 8.13 shows the symbol for a lathe chuck. It has a concrete symbol to specify the noun, arrows are used to denote whether the chuck is opening or closing, and numbers and letters specify which lathe the chuck belongs to.

Comprehensibility of Codes

We shall only look at the ease of understanding of visual codes in this section. It is, however, important to remember that sound codes can be just as useful in many instances as visual codes. One advantage of sound codes is that it is not necessary to be in a particular workplace in order to notice them, and also faster reactions may be expected to sound signals. Hearing can be said to be the dominant sense in terms

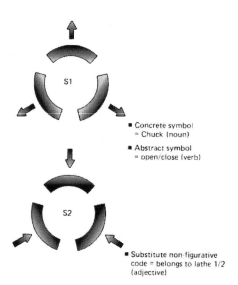

■ Concrete symbol
= Chuck (noun)

■ Abstract symbol
= open/close (verb)

■ Substitute non-figurative
code = belongs to lathe 1/2
(adjective)

FIGURE 8.13   The symbol for a lath chuck.

of noticing, recognizing, and identifying patterns. The ability of a musician to recognize very small variations in a piece of music (which is a pattern presented in the form of sound) is outstanding. The amount of information and the total number of possible different messages that could be presented in this way are immense. Sound also has the advantage that it is not necessary to focus attention on the equipment supplying the information, and one is free to move around while listening for the information.

Another information channel which gives almost the same degree of freedom is the sense of touch. Experiments have been done on producing a language for the deaf that is transmitted through touch. There has been an attempt, in Ivergård (1982) for example, to produce a "hearing glove"; this has an instrument mounted against each finger which produces pressure at different frequencies. Sound, for example from someone talking, is converted electronically and transferred to this hearing glove. It is known that at least four different frequencies can be sensed, so by this method one can transfer at least 4×4 different symbols (information units). Similar methods have also been used by placing the vibration devices on the chest and other parts of the body.

The sense of touch is not only useful as an information transmission channel for the deaf, but it can also be used by people who are overloaded with visual or sound information. The large number of instruments they have to read, for example, often overloads pilots. If we wish to give the pilot even more information, vibrators could be used to transfer information through the sense of touch instead. Like hearing, touch is better than sight for use as a warning signal; while one can close the eyes to remove the sense of vision, the sense of hearing or feeling cannot be shut off. Four variables can be used in the transferal of information by touch: the positioning, frequency, amplitude, and variation of the stimulus.

The visibility of visual codes will not be discussed in detail here except to say that they must be (a) large enough, (b) bright enough, and (c) have sufficient contrast, in order to be visible.

Experience from Gestalt psychology can be useful regarding the suitability of codes and for general rules on their design. In addition, there are a number of perceptual psychological grounds, which form the basis of the number of elements that can be used in different codes in order to avoid confusion. In particular, there is a body of specialist knowledge on the design of letters and numbers.

Table 8.3 summarizes the number of possible variants that can be obtained with different types of stimuli for designing codes. It is important to remember that the estimations given are only approximations. The maximum number only refers to conditions where the person performing the reading has special education or training for that particular form of stimulus. In most normal applications, e.g., work on VDUs, the recommended number of alternatives must be used. Table 8.3 also gives the number of

TABLE 8.3 Numbers of Recognizable Variations of Different Stimuli

Notes	Stimuli	Maximum	Recommended
Colors			
Lamps	10	3	Good
Surfaces	50	9	Good
Design			
Letters and numbers	∞	?	
Geometrical	15	5	
Figures/diagrams	30	10	Good
Size			
Surfaces	6	3	Satisfactory
Length	6	3	Satisfactory
Lightness	4	2	Poor
Frequency (flashing)	4	2	Poor
Slope			
Direction of pointer on dial	24	12	Good
Sound			
Frequency	(Large)	5	Good
Loudness	(Large)	2	Satisfactory

	1	2	3	
Violet	Blue	Green	Yellow	Red
1	2	3	4	5	6	7	8	9

absolutely recognizable units of a particular stimulus. If comparisons are possible, the number of recognizable units becomes considerably greater.

Table 8.4 identifies letters and numbers that are easily confused with each other, and also those that are difficult to read. The risk of confusion on VDU screens is especially great, because the letters and numbers are built in a fairly simple and limited manner.

Figure 8.14 shows the number of seconds it takes to read a particular number compared with the grouping of the digits. It is clear from Figure 8.14 that groups of three digits are the quickest to read. Suggested forms of grouping are given in Table 8.5.

8.5 Using Color

First, it is necessary to define what is meant by color because various color classification methods exist. The CIE system (Commission Internationale de l'Éclairage) (see Figure 8.15) is probably the best-known color classification system, and this has many advantages when it is used for specifying colors for use on VDU screens. This system is described in Wsyzecki and Stiles (1967).

The colors on a CRT are created using three electronic guns, which are aimed at the front of the CRT. This is covered with a phosphor layer. Each gun produces a different color — red, green and blue. The rays travel through a shadow mask which has many small holes in it, and then meet the front of the screen where there are a large number of symmetrically arranged round phosphor dots (see Figure 8.16). This results in different combinations of the colors red, blue, and green. Because the points are very close together, the eye does not see them as individual points, but as mixtures of color.

TABLE 8.4 Interchanging of Letters and Numbers

These:	()* (zero)	8**	B*	D*	1**	Ø** Capital O	Z*	S	G	N	V	Y
Are mistaken	Ø	B		0 (zero)	1	0 (zero)	2	5	6	W	U	W
for these:	6			Q		D		3	C		Y	V
	D			P								4
	9											T

*Interchanged often
**Interchanged very often
The following difficult to read: 2, 4, 5, 8, 9, N, T, I, Q, X, and K

On VDU Screens

Mutual	One-way	
O and Q	C	read as G
T and Y	D	read as B
5 and 5	H	read as M or N
I and L	J,T	read as I
X and K	K	read as R
1* and 1*	2	read as Z*
	B	read as R. 5 or 8

* These three account for more than 50% of all errors

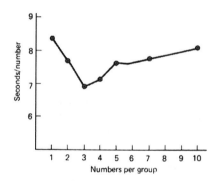

FIGURE 8.14 Average reading time per number for the keying in of 18 to 21 numbers at a time as a function of their grouping.

TABLE 8.5 Examples of How Signs in Codes of Different Lengths Should Be Grouped

Number of digits	Alternative groupings		
7	3, 3, 2	3, 4	3, 2, 2
8	3, 3, 2	2, 2, 2, 2	
9	3, 3, 3		

Rather than a spatial combination of colors, one can have a temporal combination. This means presenting the different colors at different times. Because the succession time of the colors is very short, the eye will not be able to pick out individual colors, and they will be mixed in the proportions in which they were presented. However, the drawback of this type of color presentation is that it gives considerably less well-defined colors, which are difficult to identify.

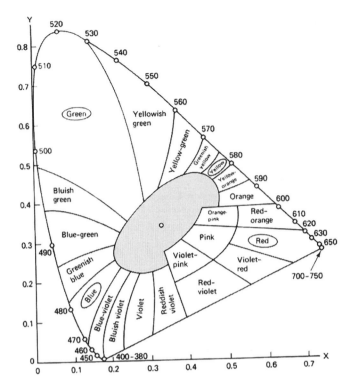

FIGURE 8.15 Chromatic area for common colors shown in the CIE 1931 standard system. The figures on the outer curve (430 to 780) are the wavelength of the light. The colors can be specified by giving their x–y coordinators. The colors which should be used primarily are circled.

In the CIE system, a color mixture will always lie on a line between the two colors of which it is a mixture. If the mixture is made up of three colors, the perceived color will lie within a triangle formed by the lines between pairs of colors from the three. In practice this means that the colors on a CRT will never fall outside the triangle shown in Figure 8.15. This ability to define colors on a CRT fairly simply using the CIE system is its major advantage over other descriptive classification systems. However, an obvious disadvantage of the CIE system is that colors cannot be classified in comprehensive descriptive terms such as color, saturation, brightness, etc. Other descriptive classification systems do have this advantage.

There are a large number of factors, which affect the perceived color. The following are the factors which must be taken into account:

1. Contrast
2. Size
3. Background color
4. Absolute relative discrimination
5. Number of colors
6. Surrounding lighting

When working only with black and white, it is sufficient to use luminance contrast (CR), which is normally defined as:

$$CR = \frac{L_1}{L_2}$$

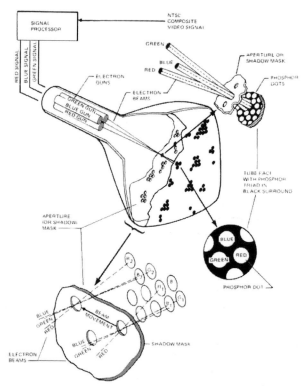

FIGURE 8.16 The colors on a CRT are created by the electronic guns shown in this figure. (From Farrell, R.J. and Booth, J.M., (1975), *Design Handbook for Imagery Interpretation Equipment*, (Seattle: Boeing Aerospace Company). With permission.)

where L_1, is the luminance of the object, and L_2, is the luminance of the surroundings. When color also has to be taken into account, the concept of color contrast is used. A special discrimination index has been worked out which takes into account both luminance and color contrast (see Silverstein, 1982).

Sensing of color is highly dependent on the size and brightness of detail within the picture. Very small details appear less saturated and sometimes change, depending on background color. Also, the ability to discriminate between colors is severely reduced for small pictures, especially along the blue–yellow axis (see Figure 8.15). The background color is also very important because the light from a small object is experienced as though it changes color toward its complementary color in the surroundings.

As the number of colors increases, color discrimination becomes more difficult, and greater demand is made for color coding. Research on the use of colors on VDU screens has shown that a maximum of four to six colors can be used. The number of colors, that can be used is of course highly dependent on whether the discrimination is absolute or relative. In absolute discrimination, the number of colors is further reduced. When color is used for decorative purposes, more colors can be used. The reasons for using colors to any great extent must be powerful ones — it is far easier to produce a poor color picture than a poor black-and-white one.

Comprehensive research has been carried out in recent years into the use of color screens, and the level of knowledge has increased considerably. Unfortunately, some important knowledge is still missing, which makes it difficult to make final recommendations. In particular, the effects of interacting factors are not well understood. For those who wish to learn more about the analytical method used in determining colors on a color screen, Silverstein's (1982) *Human Factors for Colour Display Systems* is to be recommended.

Why should a color display be chosen? In most cases, color displays are just a more costly alternative to the monochrome version. When there is an advantage in using a color display, it does not seem to have an economic basis. It is probable that a color display is *not* preferable to a monochrome display for most types of common office applications. Where there are special office applications needing more complex pictures, or in special applications such as control rooms in the process industries and traffic monitoring situations, color displays are certainly preferable.

Selection of a color display does not depend only on the fact that the operator prefers to have a color screen for aesthetic reasons. The primary reason for choosing a color display is for increased coding potential and the reduction in search time it brings about in complex pictures. It is also important to remember, as demonstrated later, that color contrast increases visibility and thereby decreases the need for luminance contrast. This, too, has many advantages, because color screens can be utilized in rooms with an ordinary level of lighting which fulfils the requirements for other types of visual tasks. Under special visual conditions, for example on planes and ships, the color screen is almost obligatory, although the requirement for well-shielded lights still remains.

Color VDUs drastically increase the requirement for specialist knowledge in the design of the display system and the choice of coding for pictures. It is difficult to make a good picture on a color display. A color picture can easily be made worse than a black-and-white picture if one does not have access to the right knowledge. A poorly designed color display decreases efficiency of performance as compared with a black-and-white screen.

Choice of Colors

A number of general ergonomic rules pertaining to absolute discrimination indicate that for recognizing and naming a color, a maximum of about seven colors is desirable. When it comes to seeing the difference between colors, many different colors and shades can be distinguished, probably well over twenty. Research on color screens has shown, however, that this is unrealistic. One cannot normally have more than three to seven colors on a CRT, depending on a number of different factors (Kinney, 1979; Teichner, 1979; Silverstein and Mayfield, 1981). For an operational color display, where absolute color discrimination is required, it is suitable to have three to four colors, and these should preferably be green, red, blue-green, and possibly a purple-red (plus white or black, depending on the background color used).

Where comparison between colors is possible, six to seven colors can be allowed, preferably red, yellow, green, blue-green (cyan), reddish-blue, and perhaps magenta (plus white or black depending on the background).

Color Screen Character and Symbol Design

It is fairly clear that the requirements will be far more stringent for the design of characters and symbols on a color screen than for those on a monochrome display with a dark background and light text. Research has shown that color sensitivity increases as the color field increases in angle, up to 10 minutes of arc. Small light fields have reduced color saturation, which means that, in practice, light colors tend to be seen as white.

On this basis, characters, symbols, or critical details in the picture which subtend less than 20 minutes of arc should not be used. On larger fields, where greater accuracy in color discrimination is required, this angle should be at least 1°. The lines, which make up the characters, should be thicker on a color screen than on a conventional one. If there is more than one minute of arc between lines, color separation will be distinguishable. The characters on a color screen should not be placed more than 15 minutes of arc from the line of sight. Characters that appear toward the periphery of the screen will not be color-coded to the same accuracy.

Research has also been performed on legibility and luminance contrast of color screens under different environmental lighting conditions (Silverstein, 1982). This has shown that there is no further improvement in legibility when the luminance contrast is increased beyond 5. It should also be remembered that the colors seen on the screen would change depending on the room lighting. The room lighting should therefore not be changed once the color screens are installed.

VDU Screen and Background Requirements

The technical requirements for a color screen are considerably greater than for an ordinary monochrome VDU screen. One must, for example, set considerably greater demands on sharpness of edges and the ability of the guns to send their rays to the correct place at the edges and corners of the screen, so that no convergence problems occur.

It has been shown that increased luminance on a VDU screen increases the color contrast up to a luminance on the screen of 3000 cd/m^2. Increasing the operator's light adaptation level also increases his sensitivity to color (Silverstein, 1982). A light background gives the effect of a higher degree of color saturation. In practice this means that a higher level of environmental light can be used when working with color screens. In many cases it may also be desirable to select a light background color on the VDU screen in order to achieve an even higher level of saturation which, in turn, reduces the risk of both diffused and reflected glare.

Standardization of Colors

There are various conventions that have been specified as standards. There is, for example, an 150 standard (1964) where various safety colors are recommended (see Figure 8.17). There is also an electrical standard (IEC, 1975) describing the recommended use of red, yellow, green, blue, and white colors for signal lamps and buttons. These recommendations are summarized in Table 8.6

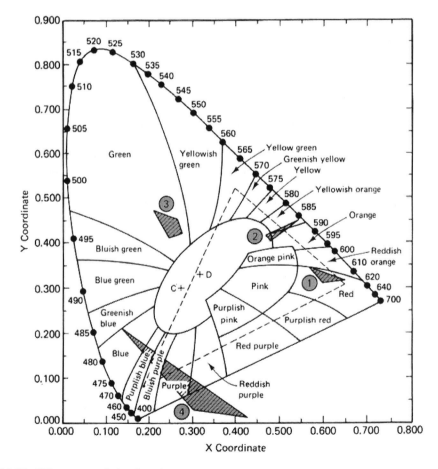

FIGURE 8.17   ISO recommendation for safety colors (R408, Dec. 1964) (1) safety red, (2) safety yellow, (3) safety green, (4) auxiliary blue.

TABLE 8.6 Examples of the Use of Different Colors for Signal Lamps and Control Buttons (SEN 280801)

	Signal lamps	Control buttons
Red	Danger, alarm	Stop, off, emergency
Yellow		Care, caution
Green	Safety	On, start
Blue	Any application	Any application
White	Any application	Any application

Red is used for danger and alarms; yellow illustrates caution, while green is used to signify safety. Control devices (buttons) use red to denote "Stop," "Off," and "Emergency"; yellow for "Action" (to avoid unwanted changes); green for "On" and "Start." Blue and white can be used for any other instructions as required.

There are, unfortunately, risks of misunderstandings with this electrical standard. A switch in an electrical circuit which is *closed* is marked by a red signal lamp outside the door to the switchroom to signify danger. A switch which is open is marked, according to Table 8.6, with red to indicate its "off" condition, i.e., the electrical circuit is "open." So a switch can be marked with red both when it is closed and when it is open! This example illustrates the difficulty of using colors to carry information. Colors must always be complemented with other information in order to avoid misunderstandings. Colors should only be used as a complement and to simplify readings.

In the IEC (1975) standard, it is suggested that in technical applications within the process industries the following practices should be adopted for the use of colors:

Red	Abnormal conditions in variables, and warnings
Green	Static plant information (e.g., pumps, lines, valves, etc.)
Blue-green (cyan)	Alphanumeric information and scales (e.g., legends)
Yellow, orange	Variables (e.g., histograms, curves, etc.)
Violet/red (magenta)	A reserve color to replace red if the red gun fails.

8.6 Speech Generation and Recognition

Language is probably the most distinguishable ability that man possesses. Man/machine systems have long included written language (e.g., in alphanumeric displays and keyboards), but spoken language has only been used for communication between people. Automatic speech generation and recognition by machine now offers an alternative to other forms of input and output to computers. An interactive speech system consists of speech recognition devices for control or information input and speech generation devices as a form of information display.

Automatic speech technology is of great interest for the future. Simpson et al. (1985) gives a comprehensive review of *system design for speech recognition and generation*. From the human factors/ergonomic perspective, a speech recognition system consists of a human speaker, recognition algorithms, and a device that responds appropriately to the recognized speech:

Human speech ~ Recognition algorithms ~ Response device

A speech generation system is the mirror image of a recognition system. It consists of a device to generate messages in the form of symbol strings, a speech generation algorithm to convert the symbol strings to an acoustic imitation of human speech, and a human listener. A speech generation system operates within the context of the user's working environment:

Message generator ~ Transformation of message codes to acoustic
imitation of voice ~ Human listener

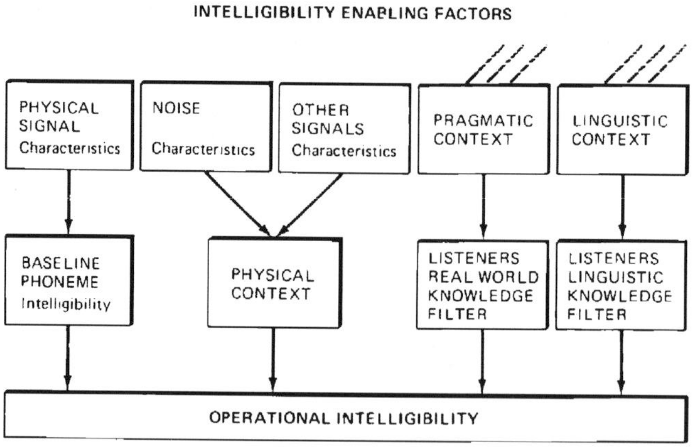

INTELLIGIBILITY ENABLING FACTORS

FIGURE 8.18   Factors that contribute to operational intelligibility (after Simpson et al., 1985). (Redrawn from *Human Factors*, 23, p. 131. © The Human Factors Society, Inc., and reproduced by permission.)

There are many factors that contribute to the operational intelligibility of speech. Simpson et al. (1985) proposed a model for operation intelligibility (Figure 8.18).

It is important to note that intelligibility and human speech are not necessarily correlated. A radio announcer may sound natural despite a background of static noise, but may have low intelligibility. Conversely, synthesized speech warning messages in an aircraft cockpit may sound mechanical, but pilots consider them to be more intelligible than messages received over the aircraft radio.

There are still no comprehensive design guidelines in this area, but some general points can be made. Although the ergonomics/human factors literature includes reports of research that supports certain principles of speech design, this knowledge has not yet been formulated into design guidelines. Human factors methodology is sufficiently well-developed to permit comparison of task-specific speech systems experimentally, but the tools required for producing generic design guidelines for speech systems are not yet available. In the short term, simulation of speech system capabilities in conjunction with the development of improved system performance should prove a productive methodology to achieve these aims.

Speech generation algorithms seem to be more advanced than speech recognition algorithms. Reasonably intelligible text-to-speech from standard English spelling is now available commercially. The recognition counterpart, speech-to-text (not to be confused with speaker identification systems), will probably not be commercially available in the near future.

In the short term, the current recognition algorithms appear adequate for use in favorable environments characterized by low-to-moderate noise levels, and for applications that only require small vocabularies and that do not place the operator under stress. Great caution must be exercised in the use of current technology in stressful situations.

Speech generation algorithms, on the other hand, have demonstrated acceptable performance even under conditions of severe noise and high workload. This technology is sufficiently advanced to be applied appropriately, with careful attention being paid to the integration of ergonomics. Simpson and Williams (1980) have studied the use of synthesized voice for cockpit warnings. They concluded that voice warnings do not need to be preceded by an alerting tone, but can be recommended for practical use.

8.7   Design of Controls

This section covers the design of the more traditional controls and the design of specific controls for communication with computers. Traditional control panels have the advantage that they give feedback to the operator of the maneuvers, that have been carried out. Examples of the advantages and disadvantages of each type of control are presented with design recommendations for these controls.

Keyboards are the traditional input devices used in data processing (DP) applications to communicate with computers. In the process control situation, it can often be advantageous to use other types of controls such as multi-way joysticks or light pens. The advantages and disadvantages of each type of control are described.

8.8 Functional Aspects of Controls

The control device is the means by which information on a decision made by man is transferred to the machine. The decision may, for example, be taken on the basis of previously read information devices or on the basis of information from other sources, or from some form of cognitive process.

Functionally, controls may be divided into the following categories:

1. Switching on/off, start or stop.
2. Increase and reduction (quantitative changes).
3. Spatial control (e.g., continuous control upward, downward, to the left or right).
4. Symbol/character production (e.g., alphanumeric keyboards).
5. Special tasks (e.g., producing sound or speech).
6. Multi-function (e.g., controls for communicating with computers).

Examples of control type 1 include the starting or stopping of motors, or switching lamps on or off. Type 2 may consist of an accelerator pedal to increase and reduce the flow of fuel to the engine. Traditionally, the best-known example of spatial control (3) is the steering of a car. Examples of character production (4) include typewriting and telegraphy. Different forms of control 5 are used for the production of sound. Of special interest here are the machines that are beginning to appear for the production and transmission of speech (as discussed in the previous section).

Of particular importance in control room design are the types of control 6 used in conjunction with computers. Controls operated by hand are of particular use where great accuracy is required in the control movement. Hands are considerably better at carrying out precision movements than feet. Where a very high degree of accuracy of movement is required, it is best for only the fingers to be used.

Other alternatives are also possible for the design of special controls. For example, if one has a large crank, or two cranks coupled in parallel which have to be controlled by both hands, very fine control movements can be made.

Because the power available from the leg is considerably greater than that from the hands, foot controls are suitable for maneuvering over long periods or continuously. Foot controls are also valuable where very large pressures are needed, because the body weight can be added to the force of the strong leg musculature. It may also be necessary to use foot control devices when the hands are occupied in other tasks. However, it should be noted that it may be necessary to use hand controls when there is insufficient space to accommodate foot controls.

When designing traditional types of controls, it is possible to design them in such a way that they naturally represent the changes one wishes to bring about in the process. For example, a lever which is pushed forward may determine the forward direction of movement of a digger bucket. Or the flow in a pipe can be stopped by turning a knob that lies on a line drawn on the panel. In this way the design of the control increases the understanding of the current state of the process.

For communication with computers, the keyboard is often chosen for carrying out all the different control functions. Technically, it is often easy to connect a keyboard to a computer system. Other control devices also exist for communicating with computers, such as light pens. However, a particular failing of this type of multipurpose control device is that the control movements in themselves have no natural analogy with the changes that they aim to bring about in the process.

8.9   Anatomical and Anthropometric Aspects of Control Design

Some of the principal anatomical and anthropometric aspects will be considered. For detailed specifications, some excellent handbooks are recommended: for example, Morgan et al., 1963 and Grandjean (1988). One important limitation of the recommendations available today is that they are based on the Caucasian races. For the Japanese population, for example, the measurements must be adapted for their proportionally shorter leg lengths.

The following rules can be applied in the design of all types of control:

1. The maximum strength, speed, precision, or body movement required to operate a control must not exceed the ability of any possible operator.
2. The number of controls must be kept to a minimum.
3. Control movements that are natural for the operator are the best and the least tiring.
4. Control movements must be as short as possible, while still maintaining the requirement for "feel."
5. The controls must have enough resistance to prevent their activation by mistake. For controls that are only used occasionally and for short periods, the resistance should be about half the maximum strength of the operator. Controls that are used for longer periods must have a much lower resistance.
6. The control must be designed to cope with misuse. In panic or emergency situations, very great forces are often applied and the control must be able to withstand these.
7. The control must give feedback so that the operator knows when it has been activated, even when it has been done by mistake.
8. The control must be designed so that the hand/foot does not slide off or lose its grip.

Table 8.7 gives a summary of the areas of use and the design recommendations for different controls. The controls are discussed in more detail later. Figure 8.19a–c gives the optimal areas for the different controls.

Press-Buttons and Keys

These are suitable for starting and stopping and for switching on or off. This type of control is also suitable for foot control, where it should be operated by the "ball" of the foot. The following recommendations apply to both hand- and foot-operated controls:

1. The resistance of the push-button should increase gradually, and then disappear suddenly to indicate that the button has been activated.
2. The top of the button should have a high coefficient of friction to stop the fingers/feet from sliding off (see Figure 8.20). Where press-buttons are to be activated by the fingers, the concave form is preferable.
3. In order to indicate that the button has been activated, a sound should be emitted if the workplace has low light levels.

Table 8.8 gives detailed recommendations for push buttons.

Toggle Switches

Toggle switches can be used to show two or three positions. Where there are three positions, one should be up, the middle one straight out, and the other one downward. Toggle switches take up very little room. The following recommendations also apply to toggle switches:

1. A sound should be heard to indicate activation of the switch.
2. If a number of switches are used, they should be placed in a horizontal row. Vertical positioning requires more space in order to avoid accidental operation.

Table 8.9 gives detailed design recommendations for toggle switches.

TABLE 8.7 Recommendations for Controls

	Stepwise Adjustments				Continuous adjustments				
	Rotary switch	Hand push-button	Foot press-button	Toggle switch	Small wheel	Wheel	Crank	Pedal	Lever
Large forces can be developed					No	Yes	No	Yes	Yes
Time constraint for adjustment	Medium	Fast	Fast	Very fast					
Recommended number of positions	3–24	2	2	2–3					
Space requirements for placing and using	Medium	Small	Large	Small	Small to medium	Large	Medium to large	Large	Medium to large
Activation by accident	Small	Medium	Large	Medium	Medium	Large	Medium	Medium	Large
Limits of control movement	270°	3.2 × 38 mm	12.7 × 100 mm	120°	None	±60°	None	Small[1]	±45°
Legibility	Good	Acceptable	Bad	Acceptable	Bad	Bad to acceptable	Acceptable	Bad	Good
Visual identification of control position	Acceptable	Bad[2]	Bad	Acceptable	Acceptable[3]	Acceptable	Bad	Bad	Acceptable
Checking control position on panel together	Good	Bad	Bad	Good	Good	Bad	Bad	Bad	Good
Usability as part of a combination of controls	Good	Good	Bad	Good	Good	Good	Bad	Bad	Good

[1] The exception is "cycle" pedals, which have no limit
[2] The exception is when the control is back-lit and the light goes off when the control is activated
[3] Only usable when control cannot be turned more than one revolution Round wheels/knobs must be marked

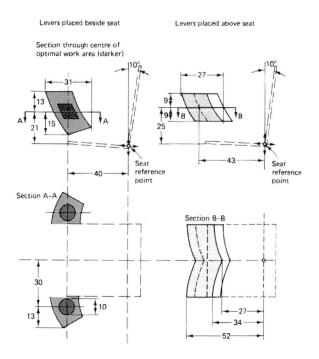

FIGURE 8.19   Preferred vertical surface areas and limits for different classes of manual controls. (Modified from McCormick, E.J. and Sanders, M.S. 1982, *Human Factors in Engineering and Design,* 5th edition (New York: McGraw-Hill.).

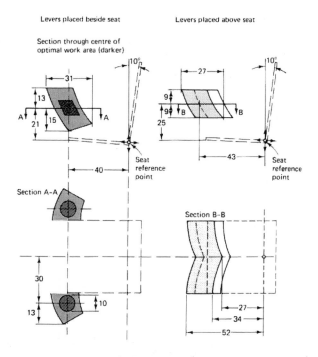

FIGURE 8.19(c)   Optimal working areas for hands moving controls.

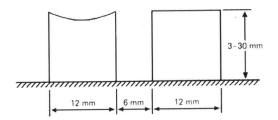

FIGURE 8.20   Press-buttons should be designed so that the fingers will not slide off them. The surface can be made concave or with some form of increased friction.

TABLE 8.8   Recommendations for Design of Press-Buttons

| | Diameter (mm) | | Travel (mm) | | Resistance | | Distance between push buttons (mm) | |
	Min	Max	Min	Max	Min	Max	Min	Max
Finger								
One finger at random	13		3.2	38	280 g	1130 g	13	50
One finger in order							6.5	25
Different fingers in random order					140 g	560 g	13	13
Thumb, nail	19		3.2	38				
Foot								
Normal	13		13		1.8–4.5 kg			
Heavy shoes		25				9 kg		
Stretching ankle				64				
Leg movement				100				

TABLE 8.9   Recommendation for Design of Toggle Switches

Variables	Minimum	Maximum
Size (mm)		
Toggle switch		
for fingers	3.2	25
for hands	13	50
Travel (degrees)		
between positions	30°	
total travel		120°
Resistance (g)	280	1130
Number of positions	2	3

	Minimum	Desirable
Distance between control (mm)		
One finger — random	19	50
One finger — in order	13	25
Different fingers random or in order	16	19

Rotary Switches

These can be divided into two categories — cylindrical and winged. The primary difference between these is that the winged version has a pair of "wings" above the cylindrical part. The wings function both as a positional marker and as a finger grip. Rotary switches may have from 3 to 24 different positions. They require a relatively large amount of space, because the whole hand has to have room to turn around

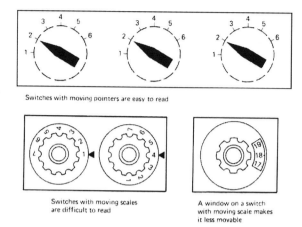

FIGURE 8.21   Design factors for rotary switches.

the switch. However, where multiple position switches are used, they take up less space than the number of push buttons or toggle switches required to fulfill the same function. Rotary switches can either have a fixed scale and moving pointer or a moving scale and fixed pointer. A variant on the moving scale is to have a window, which only shows a small part of the scale. Various models are shown in Figure 8.21.

The following recommendations apply to rotary switches:

1. In most applications, rotary switches should have a fixed scale and moving pointer.
2. There should be a detent in every position.
3. The turning resistance should steadily increase and then suddenly decrease as the next position is approached.
4. Cylindrical switches (knobs) should not be used if the resistance has to be high. In these cases, wing knobs are preferable.
5. Where only a few positions (2 to 5) are needed, they should be separated by 30 to 40 degrees.
6. Where fewer than 24 positions are used, the beginning and end of the scale should be separated by a greater space than between the different positions.
7. Where the workplace has low lighting levels, a sound should be made to denote that the switch has been activated. In these cases, there should also be a definite stop position at the beginning of the scale, so that the positions can be counted out.
8. The scale should always increase clockwise.
9. The hand should not shield the scale.
10. The surface of the switch should have a high coefficient of friction so that the hand does not slip.
11. The distance between panel and knob should be at least 3 mm.
12. The maximum amount of slope on the sides of the knob should be 5 degrees.

Levers

Levers are activated either by the whole hand or just by the fingers. In general, where fine control is needed, only the fingers should be used. The following recommendations apply to levers:

1. The maximum resistance (force) for push–pull movements with one hand, with the control placed centrally in front of the body, is between 12 and 22 kg, depending on how far from the body the control is positioned.
2. The maximum resistance for push–pull movements for two hands is double that for one hand.
3. The maximum resistance for one hand moving in the left–right direction is about 9 kg, and is considerably lower in the opposite direction.
4. The maximum resistance for two-handed movements in the left–right direction is about 13 kg.

5. The lever movement should never be greater than the arm's reach without moving the body.
6. Where precision is required, a supporting surface should be provided for the part of the body used; an elbow rest for large hand movements and a hand rest for finger movements.
7. When levers are used for step-wise control (e.g., gear levers), the distance between positions should be one third of the length of the lever.
8. Where the lever also acts as a visual indicator, the distance between positions can be reduced. The critical distance is then the operator's ability to see the markings.
9. The surface of the lever handle should have a high friction coefficient, so that the hand does not slip.

Cranks

These are suited to continuous control where there are high demands for speed. Cranks can be used for both fine and coarse control depending on the degree of gearing selected. The following recommendations apply to cranks:

1. Cranks are preferable to wheels where two or more revolutions are to be made.
2. For small cranks less than 8 cm in radius, the resistance should be at least 9 N and a maximum of 22 N when rapid movement is required.
3. Large cranks of 125–200 cm radius should have a resistance between 22 and 45 N.
4. Large cranks should be used when precision is required (accuracy between a half to one revolution), with the resistance between 10 and 35 N.
5. The handle should have a high surface friction to prevent the hand slipping.

Wheels

Wheels are used for two-handed operations. Identification of the position is very important if the wheel can be rotated through several revolutions. In addition, the following recommendations apply:

1. The turning angle should not exceed + 60 degrees from the zero position.
2. The diameter of the ring forming the outside of the wheel should be between 18 and 50 mm, and should increase as the size of the wheel increases.
3. The wheel should have a high surface friction so that the hand does not slip.

Table 8.10 shows the relative advantages of different forms of control device for computerized process systems for four common tasks.

8.10   Controls for Communication with Computers

The traditional controls in administrative computer systems are various different types of keyboard. There is often a numerical keyboard as well as the traditional typewriter keyboard. These types of keyboard have been tested over a long period of time and can be well specified. They are also thought to be well suited to most forms of administrative computer system.

For more specialized application, for example, computer systems for control and monitoring of process industries, the situation is often very different. The requirement for control devices is unique to each type of process industry, depending on the process to be controlled and the type of computer system installed. It is thus impossible to give any specific guidelines for the control devices on computerized control systems. However, some overall guidelines may be given. The advantages and disadvantages of different control devices for computerized systems will be examined. Many of the devices are new and have not yet been subjected to ergonomic evaluation. It is therefore difficult to give detailed guidelines for all controls. Finally, some more detailed design recommendations for the different types of keyboard will be presented.

TABLE 8.10   Relative Advantages of Different Forms of Control Device for Computerized Process Systems for Four Common Tasks

	Task			
	1 Numeric data	2 Alphabetic data	3 Position cursor	4 Graphic information
Control				
Fixed-function keyboard	X	XX		
Variable-function keyboard	XX			
Lever			. (X)	
Wheel			(X)	
Light pen			X	
Electronic data board				X
Touch screen			X	(X)
Mouse			X	(X)
Joystick (Track ball)			(X)	

(X), may sometimes be usable; X, usable; XX, very good; —, not recommended. Empty columns mean that the advantages and disadvantages of the application are not known.

Advantages and Disadvantages of Different Controls for Computers

The more traditional types of control can of course also be used for computerized systems. Controls, which have been produced specifically for communicating with computers, include:

1. Keyboard with predetermined functions for various keys
2. Keyboard with variable functions for the keys
3. Light pen
4. Touch screen
5. Electronic data board
6. Voice identification
7. Trackball and joystick (multi-position lever)
8. Mouse

Keyboards with Predetermined Functions for the Keys

Keyboards with predetermined functions for all keys normally have two main parts — an alphanumeric or a numeric part and a function key part. The traditional keyboard — numeric or alphanumeric — will be discussed later. The function key part has different keys for different predetermined tasks, such as starting, stopping, process a, b, c, etc. The keyboard works by the operator pressing the keys in a certain order, which he or she either remembers, or with the aid of some form of crib-sheet. The sequence in which the keys are to be pressed is thus often predetermined both by the system and by the design of the keyboard.

This type of keyboard is characterized by the need for a large number of keys, usually one per function. Where there are many functions and several subfunctions within every main function, problems arise with grouping the keys in the proper way and in positioning the keys in a mutually logical way, which is consistent in terms of movements.

It is unusual to be successful with this at the first attempt; the keyboard will need to be redesigned when it has been operational for long enough for the designer to build up enough experience with it to determine its optimal design. Making changes to the keyboard are often costly, but if it is not redesigned at a later stage, it means that large and frequent arm movements become tiresome and time consuming. The advantage of this type of keyboard is that it needs relatively little computer programming and, to a certain extent, a standard board can be used, at least for the alphanumeric part.

The alternative to having a large number of function keys is to have just a few, and to use particular codes instead, which can be entered numerically or alphanumerically. This type of keyboard is best when the operator is spending a large part of his working time at the keyboard. However, this is relatively uncommon in process industries.

Keyboards with Variable Functions for the Keys

Keyboards on which there is a variety of functions for each of the different keys are relatively uncommon but often exist as part of the more traditional keyboard (e.g., the top row of keys on the keyboard). Keyboards with a variety of functions per key are often particularly useful in process industries. A common form is to have a row of unmarked keys under the monitor screen, and to have squares representing the different keys directly above them on the screen. Depending on the picture being shown on the screen, text appears in different windows showing the functions that the keys have for each frame. There are more advanced systems for this type of keyboard in which there are several rows of unmarked keys and parallel pictures are projected down from the screen onto the keyboard using an arrangement of mirrors in order to show the current function of the keys. Because considerably fewer keys are used on this type of keyboard, fewer hand and arm movements are required by the operator, and the risk of errors occurring is also reduced.

Another application for this type of keyboard is to build lights into the keys. The relevant keys light up for each particular function. The lights in the keys are lit or extinguished when particular keys are pressed, depending on the sequence of operations required. In this way the operator is guided through the correct operation sequence.

Nonilluminated keys are then disconnected from the system. The risk of errors occurring with this type of system is very small, and work on this keyboard is also faster, particularly if the operator is not accustomed to the work. It is important, however, that if the lamps in the keys break, a warning signal must be produced.

There are also applications where keys can be pressed with different pressures. A light pressure on the key causes the function associated with that key to be written out on the screen, and the action is taken if the key is pressed harder. If it is fully depressed a signal is sent to the computer dictating changes to be made to the process. The operator can also receive new information on the screen that informs him which new keys can be used.

Depending on its design, the keyboard can be preprogrammed to lead the operator naturally through the work. This type of programming of the keyboard functions may be an advantage for especially important types of operations, in which errors could have serious consequences. A major disadvantage from the operators' perspective is that they may feel their work is being too highly controlled. Another disadvantage with this type of keyboard is that it requires a lot of programming, and this takes up a large part of the computer's capacity. An advantage is that the hardware does not need to be changed (rebuilding or extending the keyboard, etc.) to any great extent even if a major change is to be made in the function of the control. In other words, this form of control is very flexible.

Light Pens

The light pen consists of a photocell that senses the light radiated from the phosphor on a CRT screen. The light pen reacts every time a pixel on the screen is lit up by the electron ray within the tube. The signal passes from the light pen to the computer, which at the same time receives information on where and when the spot passes different places on the screen. In this way, the computer can identify where the pen is on the screen.

The light pen can be used for pointing to the parts of the screen one wishes to know more about. It can also be used to activate different functions. If, for example, one pointed to a valve and at the same time pressed a button on the side of the pen, this may cause the valve to close. The light pen is suitable for moving cursors on a screen. However, it is difficult to see any operational advantages of light pens over other controls.

If a light pen is used over a long period, it is necessary to have a specially designed armrest to prevent discomfort. The light pen has to come close up against the screen, which means that it is impossible to have any form of reflection shield or filter on the screen, and this can give rise to visual problems. Positioning of the light pen must be exact, which makes considerable demands on vision and also contributes to bad working posture in many instances.

Touch Screens

Touch screens involve moving the finger, a pen, a pointer or some other object within an active matrix placed over the screen This active matrix may be designed in several ways. It could be composed of a thin metal net, for example, on which an electrical circuit is made when it is touched. Electrical bridges and infrared beams can also be used to determine touch on the screen. Another type of touch screen is based on the use of a transparent material, which senses the pressure of the touch on the screen. Special measurement bridges are used to determine how the pressure field is distributed over the screen, and the position of the touch is deduced from this.

Functionally, the touch screen is very similar to the light pen and has similar advantages and disadvantages, although an additional disadvantage is that the screen becomes dirty. An advantage is that it is sometimes faster to point with the finger than with a light pen, however, the technical reliability of the touch screen is usually considered to be lower than that of the light pen.

There is limited research to show that the touch screen can be used for numeric and alphabetical data where there are a limited number of functions, as may be the case in the process industries. If it is necessary to send a large number of different types of words and information to the computer, the traditional keyboard is preferable.

Electronic Data Boards

Electronic data boards consist of a rectangular plate, which represents the surface of the screen. Some form of electric field is created over the plate. When a sensor is run over the board's surface, it "senses" its position on the board. One common form of board is placed directly onto the screen, and in this case, functions very like a light pen or touch screen. Another form of board is placed beside the screen, and one can work with a transparency of a picture. One of the advantages of the electronic data board is that one can very quickly make drawings or change them.

Voice Identification Instruments

As discussed earlier, instruments for voice recognition and identification have been connected to computers for a long time. Recognition of speech, however, is much more difficult. There are many apparent advantages of this type of device. It is, for example, very flexible and requires no special motor skill. The problems are that the equipment available today requires specially trained operators who have to use a very limited vocabulary and have to speak at a particular speed. In the future, however, this type of control may well be more widely applicable. Its present applications are primarily for different forms of emergency and alarm situations. There are many interesting development possibilities for this form of control device within the process industries. Development should progress in such a way that natural words and sentences will be able to be used directly. In an emergency situation, which requires immediate response, the operator should be able to shout "Stop" to control the process if he or she considers this the appropriate action for the situation.

The Trackball

The trackball is a mounted sphere that can be rotated in all directions and can be placed on a table or special fixture. The ball is usually used for moving the cursor on a screen. The cursor moves a certain distance (x/y directions) or with a speed proportional to the movement of the ball.

The Joystick

The joystick, a lever movable in all directions, has a function similar to the trackball.

The Mouse

The mouse is a small device with wheels or a ball mounted on the underside. If the mouse is moved to the left or right, this represents a corresponding movement on the screen. The mouse is especially suited for moving the cursor and for transferring graphic information.

There is no conclusive evidence to produce recommendations for the use of the trackball, mouse, or joystick. In practice, most people seem to prefer the mouse if one has access to a free table surface, otherwise the trackball is generally preferred.

Other Traditional Computer Controls

There are also many traditional types of controls, for example small wheels, levers, or joysticks (control levers that can be moved in two or three dimensions). These more traditional types of control devices are usually used for moving the cursor on the screen. However, in the future there will be a need for new types of controls that better suit the computer applications within the process industries. Traditional control panels, as simulated on a CRT, can sometimes be an improvement on a keyboard, giving direct feedback of different control settings.

The Keyboard

The keyboard is still the most common computer input device. The design of the keyboard has a significant effect on the operator's performance in terms of speed and accuracy. The most common keyboard layout is the QWERTY layout. Where two hands are used on the keyboard, 57% of the workload is on the left hand, even though 80% of the population is right-handed. This is advantageous for the type of job in which the right hand alternates between handwriting and typing. It is also important to have one standard keyboard layout. Although the QWERTY layout is not the most efficient (it is said to have been designed for slowness, so that early mechanical typewriters did not become overloaded), it is now the best compromise because it has become the standard keyboard.

Keyboards are usually designed so that the alphanumeric section is in the center, with the cursor, editing keys and numeric keypad to the right. Function keys may be placed anywhere on the keyboard, but in order to give an aesthetically pleasing design they are often placed on the left-hand side. Lateral hand movements also require less energy than longitudinal (front to back) ones.

There are no specific recommendations for keyboard layout, as their design is extremely sensitive to the task being carried out. However, a degree of flexibility must be incorporated in their design to cater to all variations in user requirements. One solution to this would be to develop a modular keyboard consisting of several units. Each unit would be made up of a different set of keys. The units could be arranged in the desired layout based on the results of the task analysis. However, care must be taken in using a flexible keyboard configuration due to the risk of a negative transfer of training. For example, if an operator carries out a number of different tasks and different keyboards are used for each of the different tasks, then high error rates must be expected.

Chord keyboards are a combination of a keyboard and a coding system. In a similar way to keys on a piano, one can press several keys at the same time. The advantage of this type of keyboard is that key-pressing speed compared with a standard typewriter is considerably greater than 50% faster. There are, however, no special design recommendations for this type of keyboard. In general, it may be said that this type of keyboard needs further study before any firm recommendations can be given regarding suitable areas of use and suitable design.

The numeric keyboard appears in two different designs. The accepted layout is a $3 \times 3 + 1$ key set, but there are two alternatives within this. Adding machines have the 7, 8, and 9 keys on the top row, while push-button telephones have the 1, 2, and 3 keys at the top (see Figure 8.22). In the future, all telephones will use the 123 keypad. Once this happens, it will be recommended that all numeric keyboards are of this design. Uniformity is important, and the user should not have to switch from one keyboard design to the other while working.

**Push-button
telephone layout**   **Adding machine
layout**

FIGURE 8.22   Layout of numerical keyboards.

The height of the keyboard is largely determined by its physical design, for example, electrical contacts and activating mechanism. Thicker keyboards (greater than 30 mm thick) should be lowered into the table surface to ensure a correct user posture. This, unfortunately, does not allow for flexible workplace design. Ideally, keyboards should be as thin as possible (less than 30 mm thick from the desk surface to the top of the second key row (ASDF …) and not need to be lowered into the surface. Recent product development, particularly by ERGOLAB in Sweden, has resulted in keyboards tending to become thinner, and this allows for a more flexible workplace design.

Keyboards can be stepped, sloped, or dished. There is no evidence on the relative advantages of any of these profiles. The most important factor is for the keyboard to be able to be angled between 0 to 15 degrees up at the back and, if the keyboard operator is standing, it is advantageous if it can be raised at the front from 0 to 30 degrees.

The size of the key tops is a compromise between producing enough space for the finger on the key, while at the same time keeping the total size of the keyboard as small as possible. Key tops should be square and 12 to 15 mm in size. This size is quite sufficient for touch typing, but in cases where keyboards are used for other tasks, for example on the shop floor, key sizes can be larger. The spacing of keys is standardized 19.5 mm between key top centers. This is within the ergonomic recommendations of between 18 to 20 mm. The force required for key displacement should be the same on all keys.

For skilled users, the actuating force should be 0.25 to 0.5 N, and the key displacement (travel) 0.8 to 10 mm (from rest to activation of system). For unskilled users, the force should be 1 to 2 N and the displacement 2 to 5 mm. The user requires feedback to indicate that the system has accepted the keystroke. This is an important keyboard characteristic, although the exact requirements vary according to the individual levels of user skill.

In normal typing and other key-pressing tasks, there is kinesthetic (muscle) and tactile (touch) feedback from the actual depression of the key, auditory feedback from the key press and/or activation of the print mechanism, and visual feedback from the keyboard or from the output display. For skilled operators, feedback from the keyboard (sound and pressure change) is of little importance. When learning, and for unskilled operators, this feedback is important. The operator should be able to remove the acoustic feedback.

The color of the keys is not generally regarded as important. A dark keyboard with light lettering is preferable when used in conjunction with light-on-dark image displays, and care should be taken not to cause any distracting reflections on the screen by light key colors. Matte finishes should be used where possible. The recommended reflectance factors for keyboards used in conjunction with negative-image (light on a dark background) VDUs are:

1. The lettering on the keys should be light and clearly defined. Its minimum height should be 25 mm in good lighting conditions. In the case of function keys, certain abbreviations may be required. These must follow a clear, logical pattern, easily identifiable by the operator.
2. Keyboards used with positive image (black on white) VDUs should be lighter in color with darker text. All keytop surfaces should have a matte finish.

Care should be taken when using colors to code various function keys. Attention should not be drawn to a red key or a group of red keys if their importance in the system is minimal. These principles concerning color may also be applied to any information lights found on the keyboard. There are international color standards (IEC 1975) which can apply to both keys and information lights. These standards should be adhered to whenever possible.

References

Berns, T. and Herring, V., 1985, Positive vs. negative image polarity of visual display screens, *ERGOLAB report*, 85:06 (Stockholm: ERGOLAB).

Cakir, A., Hart, D., and Stewart, T., 1979, *The VDT Manual*, (Damstadt: IFRA).

Farrell, R. J. and Booth, J. M., 1975, *Design Handbook for Imagery Interpretation Equipment*, (Seattle: Boeing Aerospace Company).

Grandjean, E., 1988, *Fitting the Task to the Man*, 4th edition (London: Taylor & Francis Ltd.).

IEC, 1975, International Standard, *Colours of Indicator Lights and Push Buttons*, Publ. no. 73 (Paris: International Electrotechnical Commission).

Ivergård, T., 1982, *Information Ergonomics*, (Lund, Sweden: Studentlitteratur).

Kinney, J. S., 1979, The use of color in wide-angle displays, *Proceedings of the Society for Information Display*, **20**, 33-40.

McCormick, E. J. and Sanders, M. S., 1982, *Human Factors in Engineering and Design*, 5th edition (New York: McGraw-Hill).

Morgan, C., Cook, J., Chapanis, A., and Lund, M., 1963, *Human Engineering Guide to Equipment Design*, (New York: McGraw-Hill Book Comp. Inc.).

Murrell, K. F. H., 1958, *Fitting the Job to the Worker. A Study of American and European Research into Working Conditions in Industrial Situations*, (Paris: Organisation for European Economic Co-operation–EPA).

Seminara, J. L., 1980, *Human Factors Considerations for Advanced Control Board Design*, Vol. 4 of *Human Factors Methods for Nuclear Control Room Design* (Palo Alto, CA: Electric Power Research Institute).

Silverstein, L. D. and Maryfield, R. M., 1981, *Color Selection and Verification Testing for Airborne Color CRT Displays*, Proceedings of the 5th Advanced Aircrew Display Symposium, Sept. (Naval Air Test Center).

Silverstein, L., 1982, *Human Factors for Color CRT Displays*, (San Diego, CA: The Society for Information Display).

Simpson, C. A. and Williams, D., 1980, Response time effects of alerting tone and semantic context for synthesized voice cockpit warnings, *Human Factors*, **22**, 319-330.

Simpson, C. A., McCouley, M. E., Poland, E. F., Ruth, J. C., and Williges, B. H., 1985, Systems design for speech recognition and generation, *Human Factors*, **27**, 115-141.

Stark, Lawrence, 1983, Personal Communication (reported in "Study Visit" US-1983 HF — Research in Automation," *Arbetslivsforden*, Stockholm).

Stewart, T. F. M., 1976, Displays & the software interface, *Applied Ergonomics*, **9**, 137-146.

Teichner, W. H., 1979, Color and information coding, *Proceedings of the Society for Information Display*, **20**, 3-9.

Wyszecki, G. and Stiles, W. S., 1967, *Colorscience — Concepts and Methods, Quantitative Data and Formulas*, (New York: John Wiley & Sons Inc.).

9

Engineering Anthropometry

Karl H. E. Kroemer
Virginia Tech

9.1   Overview

People come in a variety of sizes, and their bodies are not assembled in the same proportions. Thus, fitting equipment to suit the body requires careful consideration; design for the statistical "average" will not do. Instead, for each body segment to be fitted, the dimension(s) critical for design must be determined. A minimal or a maximal value, or a range may be critical. Often, a series of such decisions must be made to accommodate body segments or the whole body by clothing, workspace, and equipment.

The following text uses this procedure. It describes the steps involved, provides statistical tools, and supplies anthropometric data. For more detail see Kroemer, Kroemer, and Kroemer-Elbert (1997).

9.2   Terminology

Special terms often used in anthropometry are listed in Table 9.1. Together with the reference planes shown in Figure 9.1, they describe major aspects of anthropometric information used by designers and engineers.

9.3   Designing to Fit the Body

While all humans have heads and trunks, arms and legs, the body parts come in various sizes and are assembled in different proportions. The science of measuring human bodies is called anthropometry. The results of anthropometric surveys are described in statistical terms.

TABLE 9.1 Terms Used in Engineering Anthropometry

Anthropometry —	measure of the human body. The term is derived from the Greek words "anthropos," human and "metrein," measure.
Height —	straight-line, point-to-point vertical measurement.
Breadth —	straight-line, point-to-point horizontal measurement running across the body or a segment.
Depth —	straight-line, point-to-point horizontal measurement running fore-aft the body.
Distance —	straight-line, point-to-point measurement between landmarks on the body.
Curvature —	point-to-point measurement following a contour; this measurement is neither closed nor usually circular.
Circumference —	closed measurement that follows a body contour; hence this measurement usually is not circular.
Reach —	point-to-point measurement following the long axis of the arm or leg.

Terms Related to Body Reference Planes (see Figure 9.1)

Medial or *mid-sagittal* —	cutting body into left and right halves.
Frontal or *coronal* —	cutting body into fore-aft (anterior-posterior) sections.
Transverse —	cutting body into upper/lower (superior-inferior) sections.
Sagittal —	parallel to medial (occasionally used like *medial*).

Anatomical Terms Related to Position

Anterior —	in front of, toward the front of the body.
Posterior —	behind, toward the back of the body.
Ventral —	toward the abdomen (occasionally used like *anterior*)
Dorsal —	toward the back or spine.
Medial —	near or toward the middle.
Lateral —	to the side, away from the middle.
Superior —	above, toward the top.
Inferior —	below, toward the bottom.
*Proximal** —	toward or near the center of the body.
*Distal** —	away from the center of the body.
Superficial —	on or near the surface.
Deep —	away from or below the surface.

* *Proximal* and *distal* usually refer to limbs, where the point of reference is the attachment to the body.

Most body data appear, statistically speaking, in a normal (Gaussian) distribution. Such distribution of data can be described by using the statistical descriptors *mean* (same as *average*), *standard deviation,* and *range*, if the sample size is large enough (see below for more detail). Misunderstanding and misuse have led to the false idea that one could "design for the average"; yet, the mean value is larger than half the data, and smaller than the other half. Consequently, the "average" does not describe the ranges of different statures, arm lengths, or hip breadths. Furthermore, one is unlikely ever to encounter a person who displays mean values in several, many, or all dimensions. The mythical "average person" is nothing but a statistical phantom.

"A pioneer in the field of statistics, Sir Francis Galton (1822–1911) wrote years ago that 'it is difficult to understand why statisticians commonly limit their interests to averages. Their souls seem as dull to the charm of variety as that of a native of one of our flat English counties whose retrospect of Switzerland was that, if its mountains would be thrown into its lakes, two nuisances could be got rid at once.' Basic to virtually all design problems is the fact that mankind is far more like Switzerland than a flat English county, and that, whatever the charms of variety may be, we need statistics to quantify this variety." (Edmund Churchill on page IX-5 of NASA/Webb, 1978.)

Using Percentiles

Most body dimensions are normally distributed. A plot of their individual measures falls inside the well-known bell curve, shown in Figure 9.2. Only a few persons are very short, or very tall, but many cluster

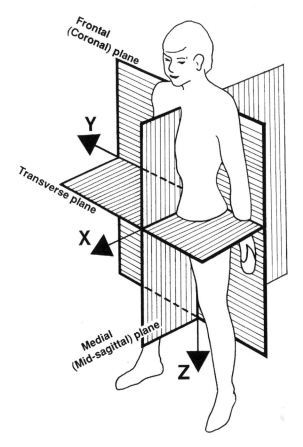

FIGURE 9.1   Reference planes used in conventional anthropometry.

around the center of the distribution (the mean or average). Figure 9.2 shows an approximate distribution of the stature of male Americans; only 2.5% are shorter than approximately 1,620 mm, and another 2.5% are taller than 1,880 mm. In other words: about 95% of all men are in the height range of 1,620 to 1,880 mm, because the 2.5th percentile value is at 1,620 mm and the 97.5th percentile is at 1,880 mm. The 50th percentile is at 1,750 mm. (In a normal — Gaussian — data distribution, mean (m), average, median, and mode coincide with the 50th percentile. The standard deviation (S) describes the peakedness or flatness of the data set. These statistical descriptors are discussed in some detail later in this chapter under "Estimation by Probability Statistics.")

There are two ways to determine given percentile values. One is simply to take a distribution of data, such as shown in Figure 9.2, and determine from the graph (measure, count, or estimate) critical percentile values. This works whether the distribution is normal, skewed, binomial, or in any other form. Fortunately, most anthropometric data are normally distributed, which allows the second, even easier (and usually more exact) approach: to calculate percentile values. This involves the standard deviation, S. If the distribution is flat (the data are widely scattered), the value of S is larger than when the data cluster close to the mean, m.

To calculate a percentile value, you simply multiply the standard deviation S by a factor k, selected from Table 9.2. Then you add the product to the mean m:

$$p = m + k * S \tag{9.1}$$

Steps in Design for Fitting Clothing, Tools, Workstations, and Equipment to the Body

(Kroemer, Kroemer, Kroemer-Elbert 1994)

Step 1: *Select those anthropometric measures that directly relate to defined design dimensions.* Examples are: hand length related to handle size; shoulder and hip breadth related to escape-hatch diameter; head length and breadth related to helmet size; eye height related to the heights of windows and displays; knee height and hip breadth related to the leg room in a console.

Step 2: *For each of these pairings, determine whether the design must fit only one given percentile (minimal or maximal) of the body dimension, or a range along that body dimension.* Examples are: the escape hatch must be big enough to accommodate the largest extreme value of shoulder breadth and hip breadth, considering clothing and equipment worn; the handle size of pliers is probably selected to fit a smallish hand; the leg room of a console must accommodate the tallest knee heights; the height of a seat should be adjustable to fit persons with short and with long lower legs. (How to use and calculate percentiles is explained below.)

Step 3: *Combine all selected design values in a careful drawing, mock-up, or computer model to ascertain that they are compatible.* For example, the required leg-room clearance height needed for sitting persons with long lower legs may be very close to the height of the working surface determined from elbow height.

Step 4: *Determine whether one design will fit all users.* If not, several sizes or adjustment must be provided to fit all users. Examples are one extra-large bed size fits all sleepers; gloves and shoes must come in different sizes; seat heights are adjustable.

If the desired percentile is above the 50th percentile, the factor k has a positive sign and the product $k * S$ is added to the mean m; if the p-value is below average, k is negative and the product $k * S$ is subtracted from the mean. Examples:

1st percentile is at $m–kS$	with $k = -2.33$ (see Table 9.2)
2nd percentile is at $m–kS$	with $k = -2.05$
2.5th percentile is at $m–kS$	with $k = -1.96$
5th percentile is at $m–kS$	with $k = -1.64$
10th percentile is at $m–kS$	with $k = -1.28$
50th percentile is at m	with $k = 0$
60th percentile is at $m+kS$	with $k = 1.28$
95th percentile is at $m+kS$	with $k = 1.64$

Percentiles serve the designer in several ways. First, they help to establish the portion of a user population that will be included in (or excluded from) a specific design solution. For example, a certain product may need to fit everybody who is taller than 5th percentile and smaller than the 60th percentile

TABLE 9.2   Percentile Values and Associated *k* Factors

	BELOW MEAN				ABOVE MEAN		
percentile	factor k	percentile	factor k	percentile	factor k	percentile	factor k
0.001	−4.25	**25**	**−0.67**	**50**	**0**	76	0.71
0.01	−3.72	26	−0.64	51	0.03	77	0.74
0.1	−3.09	27	−0.61	52	0.05	78	0.77
0.5	−2.58	28	−0.58	53	0.08	79	0.81
1	−2.33	29	−0.55	54	0.10	**80**	**0.84**
2	−2.05	**30**	**−0.52**	**55**	**0.13**	81	0.88
2.5	−1.96	31	−0.50	56	0.15	82	0.92
3	−1.88	32	−0.47	57	0.18	83	0.95
4	−1.75	33	−0.44	58	0.20	84	0.99
5	**−1.64**	34	−0.41	59	0.23	**85**	**1.04**
6	−1.55	**35**	**−0.39**	**60**	**0.25**	86	1.08
7	−1.48	36	−0.36	61	0.28	87	1.13
8	−1.41	37	−0.33	62	0.31	88	1.18
9	−1.34	38	−0.31	63	0.33	89	1.23
10	**−1.28**	39	−0.28	64	0.36	**90**	**1.28**
11	−1.23	**40**	**−0.25**	**65**	**0.39**	91	1.34
12	−1.18	41	−0.23	66	0.41	92	1.41
13	−1.13	42	−0.20	67	0.44	93	1.48
14	−1.08	43	−0.18	68	0.47	94	1.55
15	**−1.04**	44	−0.15	69	0.50	**95**	**1.64**
16	−0.99	**45**	**−0.13**	**70**	**0.52**	96	1.75
17	−0.95	46	−0.10	71	0.55	97	1.88
18	−0.92	47	−0.08	72	0.58	98	2.05
19	−0.88	48	−0.05	73	0.61	99	2.33
20	**−0.84**	49	−0.03	74	0.64	99.5	2.58
21	−0.81	**50**	**0**	**75**	**0.67**	99.9	3.09
22	−0.77					99.99	3.72
23	−0.74					99.999	4.26
24	−0.71						

Any percentile value *p* can be calculated from the mean *m* and the standard deviation *s* (normal distribution assumed) by $p = m + ks$.

in hand size or arm reach. Thus, only the 5% having values smaller than the 5th percentile, and the 40% having values larger than the 60th percentile, will not be fitted, while 55% (60% − 5%) of all users will be accommodated.

Second, percentiles are easily used to select subjects for fit tests. For example, if the product needs to be tested, persons having 5th or 60th percentile values in the critical dimensions can be employed for use tests.

Third, any body dimension, design value, or score of a subject can be exactly located. For example, a certain foot length can be described as a given percentile value of that dimension, or a certain seat height can be described as fitting a certain percentile value of lower leg length (e.g., popliteal height), or a test score can be described as falling at a certain percentile value.

Fourth, the use of percentiles helps in the selection of persons to use a given product. For example, if a cockpit of an airplane is designed to fit the 5th to 95th percentiles, one can select cockpit crews whose body measures are at or between the 5th and 95th percentiles in the critical design dimensions.

To Determine a Single (Distinct) Percentile Point

a. Select the desired percentile value
b. Determine the associated *k* value from Table 9.2
c. Calculate the *p* value from $p = m$ plus *k* times *S* (Note that *k*, and hence the product, may be negative.)

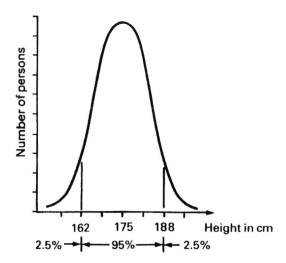

FIGURE 9.2   Distribution of body height (stature) in male Americans, as shown by Kroemer, Kroemer and Kroemer-Elbert (1997) and by Kroemer and Grandjean (1997). About 95% of all men are between 162 and 188 cm tall; about 2.5% are either shorter or taller.

To Determine a Range

1a. Select upper percentile p_{max}
1b. Find related k_{max} value in Table 9.2
1c. Calculate upper percentile value $p_{max} = m + k_{max} * S$.
2a. Select lower percentile p_{min}
    (Note that the two percentile values need not be at the same distance from the 50th p, i.e., the range does not have to be "symmetrical to the mean.")
2b. Find related k_{min} value in Table 9.2
2c. Calculate lower percentile value $p_{min} = m + k_{min} * S$
 3. Determine range $R = p_{max} - p_{min}$.

To Determine Tariffs

A distribution of body dimensions is often divided into certain sections, such as in establishing clothing tariffs. An example is the use of neck circumference to establish selected collar sizes for men's shirts. The first step is to establish the ranges (see above) which shall be covered by the tariff sections. The second step is to associate other body dimensions with the primary one, such as chest circumference, or sleeve length, with collar (neck) circumference. This can become a rather complex procedure, because the combination of body dimensions (and their derived equipment dimensions), depends on correlations among these dimensions, as discussed below. For more information, see McConville's Chapter VIII in NASA/Webb (1978) and the 1975 book by Roebuck, Kroemer, and Thompson.

9.4   Body Postures

To standardize measurements, the body is put into defined static postures:

Standing: the instruction is "stand erect; heels together; rears of heels, buttocks, and shoulders touching a vertical wall; head erect; look straight ahead; arms hang straight down (or upper arms hang, forearms are horizontal and extended forward); fingers extended."

Sitting: on a plane, horizontal, hard surface adjusted in height so that the thighs are horizontal; "sit with lower legs vertical, feet flat on the floor; trunk and head erect; look straight ahead; arms hang straight down (or upper arms hang, forearms horizontal and extended forward); fingers extended."

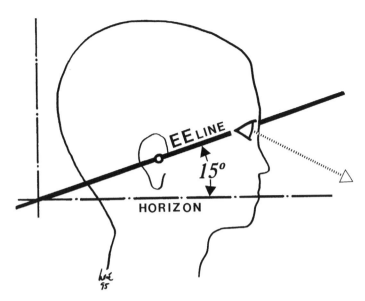

FIGURE 9.3   The ear–eye line serves as a reference to describe head posture and the line-of-sight angle.

Head (including the neck) is held erect (or "upright") when, *in the front view,* the pupils are aligned horizontally, and, *in the side view* the ear–eye line is angled about 15 degrees above the horizon (see Figure 9.3). The ear–eye (EE) line runs through the ear hole and the outside juncture of the eyelids.

People do not stand or sit in these postures naturally. Thus, the dimensions taken on the body in the standardized postures must be converted to reflect real postures. The postures assumed at work or leisure can vary greatly. Therefore, it is impossible to give "conversion factors" that apply to all conditions. The designer has to estimate the corrections that reflect the anticipated postures. Some general guidelines are presented in Table 9.3.

TABLE 9.3   Guidelines for the Conversion of Standard Measuring Postures to Real Work Conditions

Slumped standing or sitting:	deduct 5–10% from appropriate height measurements
Relaxed trunk:	add 5–10% to trunk circumferences and depths
Wearing shoes:	add approximately 25 mm to standing and sitting heights; more for "high heels"
Wearing light clothing:	add about 5% to appropriate dimensions
Wearing heavy clothing:	add 15% or more to appropriate dimensions (Note that mobility may be strongly reduced by heavy clothing.)
Extended reaches:	add 10% or more for strong motions of the trunk
Use of hand tools:	Center of handle is at about 40% hand length, measured from the wrist
Forward bent head (and neck) posture:	ear–eye line close to horizontal
Comfortable seat height:	add or subtract up to 10% to or from standard seat height

Adapted from Kroemer, Kroemer, and Kroemer-Elbert (1997).

9.5   Available Body Size Data

Recent information on body sizes of North American adults, females and males, is presented in Table 9.4. These data were derived from measurements of U.S. Army personnel taken in 1988. In spite of the military sampling bias due to selection, these data are the best available for the total North American adult population. The main reservation is with respect to body weight, which is more variable in the civilian population than in the military. Head, foot, and hand sizes should not differ appreciably between soldiers and civilians. Descriptive data of British, German, and Japanese adults are presented in Tables 9.5, 9.6, and 9.7. Each of these lists contains data, as and if available, measured in similar fashion, as illustrated in Figure 9.4.

TABLE 9.4   Anthropometric Measured Data in mm of U.S. Adults, 19 to 60 Years of Age. The reference numbers of the dimensions are shown in Figure 9.4.

	Men				Women			
Dimension	5th %ile	Mean	95th %ile	SD	5th %ile	Mean	95th %ile	SD
1. Stature [99]	1647	1756	1867	67	1528	1629	1737	64
2. Eye height, standing [D19]	1528	1634	1743	66	1415	1516	1621	63
3. Shoulder height (acromion), standing [2]	1342	1443	1546	62	1241	1334	1432	58
4. Elbow height, standing [D16]	995	1073	1153	48	926	998	1074	45
5. Hip height (trochanter) [107]	853	928	1009	48	789	862	938	45
6. Knuckle height, standing	na	na	na	na	na	na	na	na
7. Fingertip height, standing [D13]	591	653	716	40	551	610	670	36
8. Sitting height [93]	855	914	972	36	795	852	910	35
9. Sitting eye height [49]	735	792	848	34	685	739	794	33
10. Sitting shoulder height (acromion) [3]	549	598	646	30	509	556	604	29
11. Sitting elbow height [48]	184	231	274	27	176	221	264	27
12. Sitting thigh height (clearance) [104]	149	168	190	13	140	160	180	12
13. Sitting knee height [73]	514	559	606	28	474	515	560	26
14. Sitting popliteal height [86]	395	434	476	25	351	389	429	24
15. Shoulder-elbow length [91]	340	369	399	18	308	336	365	17
16. Elbow-fingertip length [54]	448	484	524	23	406	443	483	23
17. Overhead grip reach, sitting [D45]	1221	1310	1401	55	1127	1212	1296	51
18. Overhead grip reach, standing [D42]	1958	2107	2260	92	1808	1947	2094	87
19. Forward grip reach [D21]	693	751	813	37	632	686	744	34
20. Arm length, vertical [D3]	729	790	856	39	662	724	788	38
21. Downward grip reach [D43]	612	666	722	33	557	700	664	33
22. Chest depth [36]	210	243	280	22	209	239	279	21
23. Abdominal depth, sitting [1]	199	236	291	28	185	219	271	26
24. Buttock-knee depth, sitting [26]	569	616	667	30	542	589	640	30
25. Buttock-popliteal depth, sitting [27]	458	500	546	27	440	482	528	27
26. Shoulder breadth (biacromial) [10]	367	397	426	18	333	363	391	17
27. Shoulder breadth (bideltoid) [12]	450	492	535	26	397	433	472	23
28. Hip breadth, sitting [66]	329	367	412	25	343	385	432	27
29. Span [98]	1693	1823	1960	82	1542	1672	1809	81
30. Elbow span	na	na	na	na	na	na	na	na
31. Head length [62]	185	197	209	7	176	187	198	6
32. Head breadth [60]	143	152	161	5	137	144	153	5
33. Hand length [59]	179	194	211	10	165	181	197	10
34. Hand breadth [57]	84	90	98	4	73	79	86	4
35. Foot length [51]	249	270	292	13	224	244	265	12
36. Foot breadth [50]	92	101	110	5	82	90	98	5
37. Weight (kg), estimated by Kroemer	58	78	99	13	39	62	85	14

According to Gordon, Churchill, Clauser, et al. (1989), who used the numbers in brackets.

In both sets of male and female body dimensions the data are presented for the 5th and 95th percentiles. The mean (50th percentile) is also given, as is the standard deviation. This allows us to calculate other than 5th and 95th percentiles by using Equation 1 and the multiplication factors in Table 9.2.

Most populations have not been measured thoroughly and completely. Table 9.8 presents an overview of the most recent anthropometric data on national, ethnic, and geographic populations. It is unfortunate that, in most cases, only few people were measured; their small number *n* makes it unlikely that the statistical descriptors mean *m* and standard deviation *S* truly represent the underlying large population.

Table 9.9 contains general estimates for main regions of the earth. Given the paucity of existing data in 1990 and even today, it is not surprising that the estimates must be applied with great caution; note, for example, that the stature averages estimated for North Americans are several centimeters higher than actually measured, as reported in Table 9.4.

TABLE 9.5 Anthropometric Estimated Data in mm of British Adults, 19 to 35 Years of Age. The reference numbers of the dimensions are shown in Figure 9.4.

		Men				Women			
Dimension		5th %ile	Mean	95th %ile	SD	5th %ile	Mean	95th %ile	SD
1.	Stature	1625	1740	1855	70	1505	1610	1710	62
2.	Eye height, standing	1515	1630	1745	69	1405	1505	1610	61
3.	Shoulder height (acromion), standing	1315	1425	1535	66	1215	1310	1405	58
4.	Elbow height, standing	1005	1090	1180	52	930	1005	1085	46
5.	Hip height (trochanter)	840	920	1000	50	740	810	885	43
6.	Knuckle height, standing	690	755	825	41	660	720	780	36
7.	Fingertip height, standing	590	655	720	38	560	625	685	38
8.	Sitting height	850	910	965	36	795	850	910	35
9.	Sitting eye height	735	790	845	35	685	740	795	33
10.	Sitting shoulder height (acromion)	540	595	645	32	505	555	610	31
11.	Sitting elbow height	195	245	295	31	185	235	280	29
12.	Sitting thigh height (clearance)	135	160	185	15	125	155	180	17
13.	Sitting knee height	490	545	595	32	455	500	540	27
14.	Sitting popliteal height	395	440	490	29	355	400	445	27
15.	Shoulder-elbow length	330	365	395	20	300	330	360	17
16.	Elbow-fingertip length	440	475	510	21	400	430	460	19
17.	Overhead grip reach, sitting	1145	1245	1340	60	1060	1150	1235	53
18.	Overhead grip reach, standing	1925	2060	2190	80	1790	1905	2020	71
19.	Forward grip reach	720	780	835	34	650	705	755	31
20.	Arm length, vertical	720	780	840	36	655	705	760	32
21.	Downward grip reach	610	665	715	32	555	600	650	29
22.	Chest depth	215	250	285	22	210	250	295	27
23.	Abdominal depth, sitting	220	270	325	32	205	255	305	30
24.	Buttock-knee depth, sitting	540	595	645	31	520	570	620	30
25.	Buttock-popliteal depth, sitting	440	495	550	32	435	480	530	30
26.	Shoulder breadth (biacromial)	365	400	430	20	325	355	385	18
27.	Shoulder breadth (bideltoid)	420	465	510	28	355	395	435	24
28.	Hip breadth, sitting	310	360	405	29	310	370	435	38
29.	Span	1655	1790	1925	83	1490	1605	1725	71
30.	Elbow span	865	945	1020	47	780	850	920	43
31.	Head length	180	195	205	8	165	180	190	7
32.	Head breadth	145	155	165	6	135	145	150	6
33.	Hand length	175	190	205	10	160	175	190	9
34.	Hand breadth	80	85	95	5	70	75	85	4
35.	Foot length	240	265	285	14	215	235	255	12
36.	Foot breadth	85	95	110	6	80	90	100	6
37.	Weight (kg)	55	75	94	12	44	63	81	11

According to Pheasant (1986, 1996).

9.6 How to Get Missing Data

Often we design new products for users for whom we do not have exact information about their body size or strength. This is not a great problem if the product is similar to items already in use or if the users are fairly well known to us, such as our colleagues or at least people from our own country. In this case, we can probably take a few measurements on our acquaintances for a "rough guesstimate" of what the needed dimensions might be.

For exact and comprehensive information, however, more than such informal data gathering is necessary. Two avenues are open: one is to conduct a formal anthropometric survey; this is a major enterprise and best done by qualified anthropometrists (Kroemer, 1989; Roebuck, 1995). The other option is to deduce from existing data those that we need to know. There are several engineering approaches to estimate missing data.

TABLE 9.6   Anthropometric Measured Data in mm of East German Adults, 18 to 59 Years of Age. The reference numbers of the dimensions are shown in Figure 9.4.

		Men				Women			
		5th %ile	Mean	95th %ile	SD	5th %ile	Mean	95th %ile	SD
Dimension									
1.	Stature	1607	1715	1825	66	1514	1608	1707	59
2.	Eye height, standing	1498	1601	1705	64	1415	1504	1597	57
3.	Shoulder height (acromion), standing	1320	1414	1512	60	1232	1319	1403	53
4.	Elbow height, standing	na	na	na	na	na	na	na	na
5.	Hip height (trochanter)	na	na	na	na	na	na	na	na
6.	Knuckle height, standing	682	748	819	42	643	703	764	37
7.	Fingertip height, standing	588	652	717	39	557	616	672	35
8.	Sitting height	846	903	958	34	804	854	905	31
9.	Sitting eye height	719	775	831	34	684	733	782	30
10.	Sitting shoulder height (acromion)	552	601	650	31	517	562	609	29
11.	Sitting elbow height	198	244	293	29	190	234	282	28
12.	Sitting thigh height (clearance)	126	151	176	15	125	148	175	15
13.	Sitting knee height	490	531	575	27	458	497	538	24
14.	Sitting popliteal height	410	452	496	26	380	416	455	23
15.	Shoulder-elbow length	na	na	na	na	na	na	na	na
16.	Elbow-fingertip length	432	465	500	20	394	425	556	19
17.	Overhead grip reach, sitting	na	na	na	na	na	na	na	na
18.	Overhead grip reach, standing	1975	2121	2267	89	1843	1973	2103	79
19.	Forward grip reach	704	763	824	37	650	706	767	35
20.	Arm length, vertical	704	762	820	35	650	703	758	33
21.	Downward grip reach	na	na	na	na	na	na	na	na
22.	Chest depth	na	na	na	na	na	na	na	na
23.	Abdominal depth, sitting	na	na	na	na	na	na	na	na
24.	Buttock-knee depth, sitting	560	603	648	27	541	585	630	27
25.	Buttock-popliteal depth, sitting	444	486	527	25	437	479	521	26
26.	Shoulder breadth (biacromial)	365	399	430	20	336	365	393	17
27.	Shoulder breadth (bideltoid)	432	471	510	24	393	437	481	27
28.	Hip breadth, sitting	334	369	406	22	346	401	460	35
29.	Span	1640	1760	1885	75	1503	1616	1735	70
30.	Elbow span	833	895	911	39	757	817	881	38
31.	Head length	179	190	201	7	170	181	191	6
32.	Head breadth	148	158	168	6	141	151	160	6
33.	Hand length	174	189	205	9	161	174	189	9
34.	Hand breadth	81	88	96	5	71	78	85	44
35.	Foot length	243	264	285	13	222	241	260	12
36.	Foot breadth	91	102	113	6	83	93	104	6
37.	Weight (kg)	na	na	na	na	na	na	na	na

According to Fluegel, Greil and Sommer (1986).

Estimation by "Ratio Scaling"

"Ratio scaling" (as used by Pheasant 1986, 1996) is one technique for estimating data from known dimensions. It relies on the assumption that, though people vary greatly in size, they are likely to be similar in proportions. This premise holds true for body components that are related in size to each other, as discussed in detail by Roebuck, Kroemer, and Thomson (1975), by Pheasant (1982), and most recently by Roebuck (1995). For example, many body "lengths" are highly correlated with each other; also, groups of body "breadths" are related, as are "circumferences" as a group. However, not all body lengths (or breadths, or circumferences) are highly correlated with each other, and certainly many lengths are not related highly with breadths, nor with depths or circumferences. Thus, one has to be very careful in deriving one set of data from another.

TABLE 9.7 Anthropometric Measured Data in mm of Japanese Adults, 18-30 Years of Age.
The reference numbers of the dimensions are shown in Figure 9.4.

	Men				Women			
Dimension	5th %ile	Mean	95th %ile	SD	5th %ile	Mean	95th %ile	SD
1. Stature	1599	1688	1777	55	1510	1584	1671	50
2. Eye height, standing	1489	1577	1664	53	1382	1460	1541	49
3. Shoulder height (acromion), standing	1291	1370	1454	50	1208	1279	1367	48
4. Elbow height, standing	970	1035	1098	39	909	967	1028	37
5. Hip height (trochanter)	775	834	899	38	730	787	847	35
6. Knuckle height, standing	na	na	na	na	na	na	na	na
7. Fingertip height, standing	600	644	694	30	563	608	652	27
8. Sitting height	859	910	958	30	810	855	902	28
9. Sitting eye height	741	790	837	29	692	733	778	27
10. Sitting shoulder height (acromion)	549	591	633	26	513	551	588	24
11. Sitting elbow height	216	254	292	23	202	236	269	20
12. Sitting thigh height (clearance)	138	156	176	12	130	143	162	10
13. Sitting knee height	475	509	545	22	442	475	508	20
14. Sitting popliteal height	371	402	434	19	345	372	402	17
15. Shoulder-elbow length	307	337	366	18	289	315	339	15
16. Elbow-fingertip length	418	448	479	18	390	416	445	17
17. Overhead grip reach, sitting	na	na	na	na	na	na	na	na
18. Overhead grip reach, standing	na	na	na	na	na	na	na	na
19. Forward grip reach	na	na	na	na	na	na	na	na
20. Arm length, vertical	na	na	na	na	na	na	na	na
21. Downward grip reach	na	na	na	na	na	na	na	na
22. Chest depth	190	217	246	18	190	215	250	19
23. Abdominal depth, sitting	179	208	245	20	161	188	218	17
24. Buttock-knee depth, sitting	530	567	604	23	511	550	586	22
25. Buttock-popliteal depth, sitting	na	na	na	na	na	na	na	na
26. Shoulder breadth (biacromial)	368	395	423	17	346	367	391	14
27. Shoulder breadth (bideltoid)	na	na	na	na	na	na	na	na
28. Hip breadth, sitting	318	349	380	19	331	358	386	17
29. Span	1591	1690	1795	63	1483	1579	1693	62
30. Elbow span	na	na	na	na	na	na	na	na
31. Head length	178	190	203	7	168	177	187	6
32. Head breadth	152	161	171	6	143	151	160	6
33. Hand length	na	na	na	na	na	na	na	na
34. Hand breadth	79	85	91	4	70	75	81	3
35. Foot length	234	251	269	11	217	232	246	9
36. Foot breadth	97	104	111	5	89	96	103	4
37. Weight (kg)	54	66	80	8	45	54	65	6

According to Kagimoto (1990).

A good rule for ratio scaling is to use only pairings of data that are related to each other with a coefficient of correlation of at least 0.7. (This assures that the variability of the derived information is determined at least by 50% by the variability of the predictor, which derives from the squaring of the correlation coefficient, $0.7^2 = 0.49$.) However, never use ratio scaling if you must assume that the sample from which you want to scale has body proportions different from those of the other set; e.g., many Asian populations have proportionally shorter legs and longer trunks than Europeans or North-Americans.

For sets of highly correlated data, you can establish the estimate E of a ratio scaling factor for a desired dimension (d_y) in the population sample Y

- if you know the value of that dimension in sample X (d_x), and
- if you know the values of a reference dimension D in both samples X and Y (D_x and D_y).

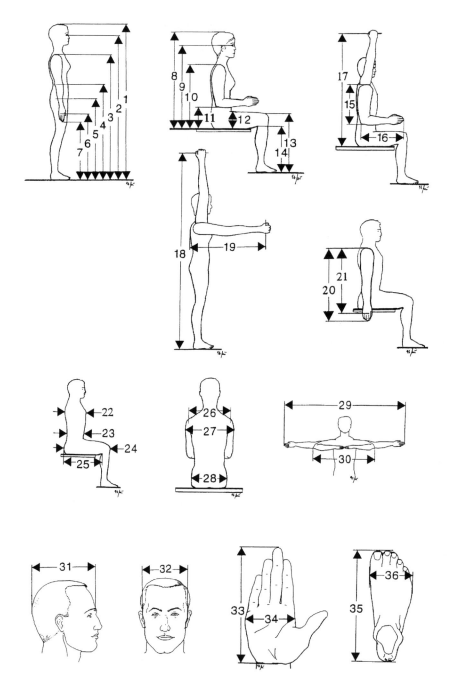

FIGURE 9.4 Illustrations of measured body dimensions.

The following commentary identifies measurements and design implications. The numbers in brackets are the same as used by Gordon et al. (1989), who also provide exact definitions of the anthropometric terms. Most of the listed dimensions are the same as listed by Pheasant (1986, 1996). The data listed in Tables 9.4 through 9.7 have been selected to include both traditional anthropometric sample descriptors (such as stature and weight) and body dimensions and reaches that are relevant to the sizing of workstations and equipment. However, in order to apply this information for engineering purposes, usually the data must be converted to reflect actual positions and motions instead of the "standardized frozen" postures (all body angles at either 0, 90, or 180 degrees) which the subject assumed for the measurements. Furthermore, allotments must be made for clothing, shoes, gloves, and other "real world" conditions. Table 9.3 provides some guidelines.

FIGURE 9.4 (continued)

1. **Stature:**   The vertical distance from the floor to the top of the head, when standing [99]. A main reference for comparing population samples. Relates to the minimal height (clearance) of overhead obstructions. Add height for more clearance, hat, shoes, stride.

2. **Eye height, standing:**   The vertical distance from the floor to the outer corner of the right eye, when standing [D19]. Origin of the visual field. Reference point for the location of vision obstructions and of visual targets such as displays; consider slump and motion of the standing person.

3. **Shoulder height (acromion), standing:**   The vertical distance from the floor to the tip (acromion) of the shoulder, when standing [2]. Starting point for arm length measurements; near the center of rotation of the upper arm (shoulder joint), reference point for hand reaches; consider slump and motion of the standing person.

4. **Elbow height, standing:**   The vertical distance from the floor to the lowest point of the right elbow, when standing, with the elbow flexed at 90 degrees [D16]. Reference point for height and distance of the work area of the hand and for the location of controls and fixtures; consider slump and motion of the standing person.

5. **Hip height (trochanter), standing:**   The vertical distance from the floor to the trochanter landmark on the upper side of the right thigh, when standing [107]. Starting point for leg length measurement; near the center of the hip joint; reference point for leg reaches; consider slump and motion of the standing person.

6. **Knuckle height, standing:**   The vertical distance from the floor to the knuckle (metacarpal bone) of the middle finger of the right hand, when standing. Reference point for lowest location of controls, handles, and handrails; consider slump and motion of the standing person.

7. **Fingertip height, standing:**   The vertical distance from the floor to the tip of the index finger of the right hand, when standing [D13]. Reference point for lowest location of controls, handles, and handrails; consider slump and motion of the standing person.

8. **Sitting height:**   The vertical distance from the sitting surface to the top of the head, when sitting [93]. The vertical distance from the floor to the underside of the thigh directly behind the right knee; when sitting, with the knee flexed at 90 degrees. Relates to the minimal height of overhead obstructions. Add height for more clearance, hat, trunk motion of the seated person.

9. **Sitting eye height:**   The vertical distance from the sitting surface to the outer corner of the right eye, when sitting [49]. Origin of the visual field; reference point for the location of vision obstructions and of visual targets such as displays; consider slump and motion of the seated person.

10. **Sitting shoulder height (acromion):**   The vertical distance from the sitting surface to the tip (acromion) of the shoulder, when sitting [3]. Starting point for arm length measurements; near the center of rotation of the upper arm (shoulder joint), reference point for hand reaches; consider slump and motion of the seated person.

11. **Sitting elbow height:**   The vertical distance from the sitting surface to the lowest point of the right elbow, when sitting, with the elbow flexed at 90 degrees [48]. Reference point for height of an arm rest, of the work area of the hand, and of keyboard and controls; consider slump and motion of the seated person.

12. **Sitting thigh height (clearance):**   The vertical distance from the sitting surface to the highest point on the top of the right thigh, when sitting, with the knee flexed at 90 degrees [104]. Minimal clearance needed between seat pan and the underside of a structure, such as a table.; add clearance for clothing and motions.

13. **Sitting knee height:**   The vertical distance from the floor to the top of the right knee cap, when sitting, with the knees flexed at 90 degrees [73]. Minimal clearance needed below the underside of a structure, such as a table; add height for shoe.

14. **Sitting popliteal height:**   The vertical distance from the floor to the underside of the thigh directly behind the right knee; when sitting, with the knees flexed at 90 degrees [86]. Reference for the height of a seat; add height for shoes, consider movement of the feet.

15. **Shoulder-elbow length:**   The vertical distance from the underside of the right elbow to the right acromion, with the elbow flexed at 90 degrees and the upper arm hanging vertically [91]. A general reference for comparing population samples.

16. **Elbow-fingertip length:**   The distance from the back of the right elbow to the tip of the middle finger, with the elbow flexed at 90 degrees [54]. Reference for fingertip reach when moving the forearm in the elbow.

17. **Overhead grip reach, sitting:**   The vertical distance from the sitting surface to the center of a cylindrical rod firmly held in the palm of the right hand [D45]. Reference for height of overhead controls to be operated by the seated person. Consider ease of motion, reach, and finger/hand/arm strength.

FIGURE 9.4 (continued)

18. **Overhead grip reach, standing**:   The vertical distance from the standing surface to the center of a cylindrical rod firmly held in the palm of the right hand [D42]. Reference for height of overhead controls to be operated by the standing person. Add shoe height. Consider ease of motion, reach, and finger/hand/arm strength.

19. **Forward grip reach**:   The horizontal distance from the back of the right shoulder blade to the center of a cylindrical rod firmly held in the palm of the right hand [D21]. Reference for forward reach distance. Consider ease of motion, reach and finger/hand/arm strength.

20. **Arm length, vertical**:   The vertical distance from the tip of the right middle finger to the right acromion, with the arm hanging vertically [D3]. A general reference for comparing population samples. Reference for the location of controls very low on the side of the operator. Consider ease of motion, reach and finger/hand/arm strength.

21. **Downward grip reach**:   The vertical distance from the right acromion to the center of a cylindrical rod firmly held in the palm of the right hand, with the arm hanging vertically [D43]. Reference for the location of controls low on the side of the operator. Consider ease of motion, reach, and finger/hand/arm strength.

22. **Chest depth**:   The horizontal distance from the back to the right nipple [36]. A general reference for comparing population samples. Reference for the clearance between seat backrest and the location of obstructions in front of the trunk.

23. **Abdominal depth, sitting**:   The horizontal distance from the back to the most protruding point on the abdomen [1]. A general reference for comparing population samples. Reference for the clearance between seat backrest and the location of obstructions in front of the trunk.

24. **Buttock-knee depth, sitting**:   The horizontal distance from the back of the buttocks to the most protruding point on the right knee, when sitting with the knees flexed at 90 degrees [26]. Reference for the clearance between seat backrest and the location of obstructions in front of the knees.

25. **Buttock-popliteal depth, sitting**:   The horizontal distance from the back of the buttocks to back of the right knee just below the thigh, when sitting with the knees flexed at 90 degrees [27]. Reference for the depth of a seat.

26. **Shoulder breadth, biacromial**:   The distance between the right and left acromion [10]. A general reference for comparing population samples. Indication of the distance between the centers of rotation (shoulder joints) of the upper arms.

27. **Shoulder breadth, bideltoid**:   The maximal horizontal breadth across the shoulders between the lateral margins of the right and left deltoid muscles [12]. Reference for the clearance requirement at shoulder level. Add space for ease of motion, tool use.

28. **Hip breadth, sitting**:   The maximal horizontal breadth across the hips or thighs, whatever is greater, when sitting [66]. Reference for seat width. Add space for clothing and ease of motion.

29. **Span**:   The distance between the tips of the middle fingers of the horizontally outstretched arms and hands [98]. Reference for sideway reach.

30. **Elbow span**:   The distance between the tips of the elbows of the horizontally outstretched upper arms with the elbows flexed so that the fingertips of the hands meet in front of the trunk. Reference for "elbow room."

31. **Head length**:   The distance from the glabella (between the browridges) to the most rearward protrusion (the occiput) on the back, in the middle of the skull [62]. A general reference for comparing population samples. Reference for head gear size.

32. **Head breadth**:   The maximal horizontal breadth of the head above the attachment of the ears [60]. A general reference for comparing population samples. Reference for head gear size.

33. **Hand length**:   The length of the right hand between the crease of the wrist and the tip of the middle finger, with the hand flat [59]. A general reference for comparing population samples. Reference for hand tool and gear size. Consider changes due to manipulations, gloves, tool use.

34. **Hand breadth**:   The breadth of the right hand across the knuckles of the four fingers [57]. A general reference for comparing population samples. Reference for hand tool and gear size, and for the opening through which a hand may (or may not) fit. Consider changes due to manipulations, gloves, tool use.

35. **Foot length**:   The maximal length of the right foot, when standing [51]. A general reference for comparing population samples. Reference for shoe and pedal size.

36. **Foot breadth**:   The maximal breadth of the right foot, at right angle to the long axis of the foot, when standing [50]. A general reference for comparing population samples. Reference for shoe size, spacing of pedals.

37. **Weight**:   Nude body weight taken to the nearest tenth of a kilogram. A general reference for comparing population samples. Reference for body size, clothing, strength, health, etc. Add weight for clothing and equipment worn on the body.

TABLE 9.8 Recent Anthropometric Data on National and Ethnic Populations: Averages (and Standard Deviations), in mm but Weight in kg. Contact the author for details on the sources.

	Sample Size N	Stature	Sitting Height	Knee Height, sitting	Weight
Algerian females (Mebarki and Davies, 1990)	666	1576 (56)	795 (50)	487 (36)	61.3 (12.9)
Brazilian males (Ferreira, 1988; cited by Al-Haboubi, 1991)	3076	1699 (67)	—	—	—
Chinese females (Singapore) (Ong, Koh, Phoon, and Low, 1988)	46	1598 (58)	855 (31)	—	—
Chinese females (Taiwan) (Huang and You, 1994)	300	1582 (49)	—	—	51.2 (6.9)
Cantonese males (Evans, 1990)	41	1720 (63)	—	—	60.0 (6.2)
Egyptian females (Moustafa, Davies, Darwich, and Ibraheem, 1987)	4960	1606 (72)	838 (43)	499 (25)	62.6 (4.4)
Indian males (farmers) (Nag, Sebastian, and Mavlankar, 1980)	13	1576 (17)	—	—	44.6 (1.4)
Central Indian male farm workers (Gite and Yadav, 1989)	39	1620 (50)	739 (26)	509 (30)	49.3 (6.0)
South Indian males (workers) (Fernandez and Uppugonduri, 1992)	128	1607 (60)	791 (40)	542 (38)	56.6 (5.1)
Indonesian females	468	1516 (54)	719 (34)	—	—
Indonesian males (Sama'mur, 1985; cited by Intaranont, 1991)	949	1613 (56)	872 (37)	—	—
Iranian female students	74	1597 (58)	861 (36)	488 (23)	56.2 (10.1)
Iranian male students (Mououdi, 1997)	105	1725 (58)	912 (26)	531 (24)	65.7 (10.1)
Irish males (Gallwey and Fitzgibbon, 1991)	164	1731 (58)	911 (30)	508 (28)	73.9 (8.7)
Italian females	753	1610 (64)	850 (34)	495 (30)	58 (8.3)
Italian males (Coniglio, Fubini, Masali, Masiero, Pierlorenzi and Sagone, 1991)	913	1733 (71)	896 (36)	541 (30)	75 (9.6)
Jamaican females	123	1648	832	—	61.4
Jamaican males (Camey, Aghazadeh, and Nye, 1991)	30	1749	856	—	67.6
Korean female workers (Fernandez, Malzahn, Eyada, and Kim, 1989)	101	1580 (57)	833 (32)	460 (22)	53.9 (6.9)
Malay females (Ong, Koh, Phoon, and Low, 1988)	32	1559 (66)	831 (39)	—	—
Saudi-Arabian males (Dairi, 1986; cited by Al-Haboubi, 1991)	1440	1675 (61)	—	—	—
Singapore males (pilot trainees) (Singh, Pen, Lim, and Ong, 1995)	832	1685(53)	894(32)	—	—
Sri Lankan females	287	1523 (59)	774 (22)	—	—
Sri Lankan males (Abeysekera, 1985; cited by Intaranont, 1991)	435	1639 (63)	833 (27)	—	—
Sudanese Males					
Villagers	37*	1687 (63)	—	—	57.1 (7.6)
City dwellers	16*	1704 (72)	—	—	62.3 (13.1)
	48**	1668	—	—	51.3
Soldiers	21*	1735 (71)	—	—	71.1 (8.4)
	104**	1728	—	—	60.0
* (El-Karim, Sukkar, Collins, and Dore, 1981)					
**(Ballal et al., 1982; cited by Intaranont, 1991)					
Thai females	250*	1512 (48)	—	—	—
	711*	1540 (50)	817 (27)	—	—
Thai males	250*	1607 (20)	—	—	—
	1478**	1654 (59)	872 (32)	—	—
* (Intaranont, 1991)					
**(NICE; cited by Intaranont, 1991)					
Turkish females					
Villagers	47	1567 (52)	792 (38)	486 (27)	69.1 (13.8)
City dwellers	53	1563 (55)	786 (35)	471 (25)	65.9 (13.0)
(Goenen, Kalinkara, and Oezgen, 1991)					
Turkish males (soldiers) (Kayis and Oezok, 1991)	5108	1702 (60)	888 (34)	513 (28)	63.3 (7.3)
Vietnamese (American V.)					
Females	30	1559 (61)	—	—	48.6
Males	41	1646 (54)	—	—	58.9
(Imrhan, Nguyen and Nguyen (1993)					
U.S. Midwest workers, with shoes and light clothes					
Females	125	1637 (62)	—	—	64.7 (11.8)
Males	384	1778 (73)	—	—	84.2 (15.5)
(Marras and Kim, 1993)					
U.S. male miners (Kuenzi and Kennedy, 1993)	105	1803 (65)	—	—	89.4 (15.1)

Adapted from Kroemer, Kroemer, and Kroemer-Elbert (1997)

Occupational Ergonomics: Principles of Work Design

TABLE 9.9   Average Anthropometric Data (in mm) Estimated for 20 Regions of the Earth

	Stature		Sitting Height		Knee Height, Sitting	
	Females	Males	Females	Males	Females	Males
NORTH AMERICA	1650	1790	880	930	500	550
LATIN AMERICA						
Indian Population	1480	1620	800	850	445	495
European and Negroid population	1620	1750	860	930	480	540
EUROPE						
North	1690	1810	900	950	500	550
Central	1660	1770	880	940	500	550
East	1630	1750	870	910	510	550
Southeast	1620	1730	860	900	460	535
France	1630	1770	860	930	490	540
Iberia	1600	1710	850	890	480	520
AFRICA						
North	1610	1690	840	870	500	535
West	1530	1670	790	820	480	530
Southeast	1570	1680	820	860	495	540
NEAR EAST	1610	1710	850	890	490	520
INDIA						
North	1540	1670	820	870	490	530
South	1500	1620	800	820	470	510
ASIA						
North	1590	1690	850	900	475	515
Southeast	1530	1630	800	840	460	495
SOUTH CHINA	1520	1660	790	840	460	505
JAPAN	1590	1720	860	920	395	515
AUSTRALIA						
European extraction	1670	1770	880	930	525	570

Adapted from Juergens, Aune, and Pieper (1990).

In this case, you calculate the scaling factor *E* from

$$E = d_x / D_x \qquad (9.2)$$

Since the basic assumption is that the two samples are similar in proportion, the same scaling factor *E* applies to both samples *X* and *Y*:

$$E = d_x / D_x = d_y / D_y \qquad (9.3)$$

with

$$E = d_y / D_y \qquad (9.3a)$$

known, you can calculate

$$d_y = E * D_y \qquad (9.4)$$

in stepwise fashion, as shown in the following:

    Step 1: In population sample *X*, establish a scaling factor *E* between the desired dimension and a known reference dimension. The reference parameter must be common for both population samples; stature is often used. For example, if shoulder height is to be estimated for sample *Y*, and is known in sample *X*, then calculate

$$E = \frac{\text{shoulder height in sample } X}{\text{stature in sample } X} \qquad \text{see (9.2)}$$

Step 2: With E now known, the desired unknown dimension in population sample Y equals E times the reference parameter in sample Y. For example, shoulder height in sample $Y = E *$ stature in sample Y

$$\text{shoulder height in sample } Y = E * (\text{stature in sample } Y) \qquad \text{see (9.4)}$$

The common parameter is often stature because its value is commonly and easily measured. Note, however, that stature is generally related well with other heights, but not necessarily with depths, breadths, circumferences, or weight (as discussed above). Thus, ratio scaling must be done with great caution and careful consideration of the circumstances, especially taking into account statistical correlations.

The technique of ratio scaling has been applied primarily to estimate the mean of a required dimension, and to estimate its standard deviation. For more detail on ratio scaling, read the books by Pheasant (1986, 1996) and Roebuck (1995).

Estimation by Regression Equations

Another way of estimating the relations among dimensions is through regression equations. Most regression equations are bivariate in nature, meaning two variables are involved, and it is presumed that the two variables are linearly related. (That linear relationship is seldom explicitly confirmed.) The general form is

$$y = a + b * x \qquad (9.5)$$

where x is the known mean value and y the predicted mean. The constants a (the "intercept") and b (the "slope") must be determined (known) for the data set of interest. A recent example of this procedure is the estimation of body dimensions of American soldiers by Cheverud, Gordon, Walker, Jacquish, Kohn, Moore, and Yamashita (1990).

If you predict the mean value of y (for any value of x) using the regression equation shown above, you must remember that the actual values of y are scattered about the mean in a normal (Gaussian) probability distribution. The standard error SE of the estimate depends on the correlation r between x and y, and on the standard deviation of y (S_y) according to

$$SE_y = S_y \sqrt{1 - r^2} \qquad (9.6)$$

Roebuck (1995) discussed this concept in some detail, including its extension to develop multivariate regression equations, principal component analyses, and boundary description analyses.

Estimation by Probability Statistics

In most cases, we are unable to measure every person of a user population with respect to body size or strength. If we were able to do so, we could describe the parameters of that total population by the mean (average) μ and standard deviation σ. (The terminology convention in statistics is to use Greek letters to indicate population parameters.) In reality, we can measure only a subgroup (sample), and from its parameters we infer or estimate what the actual population would have yielded. Using roman letters to describe the sample data, we say that

$$m = (\Sigma x)/n \qquad (9.7)$$

where m is the mean (average), x is the individual measurement, and n is the number of measured individuals. The distribution of the data is described by the equation

$$S = \sqrt{\Sigma(x-m)^2/n}$$ (9.8a)

with S called the standard deviation of the sample. If the sample size is small (conventionally, 30 or less) one makes an arbitrary correction by using $(n-1)$ instead of n:

$$S = \sqrt{\Sigma(x-m)^2/n-1}$$ (9.8b)

The smaller n, the larger the standard error SE in sampling. The standard error SE of the mean is determined from

$$SE \text{ of the mean } = S/\sqrt{n}$$ (9.9)

The standard error SE of the standard deviation is determined from

$$SE \text{ of the standard deviation } = S/\sqrt{2n} = 0.71 \ SE \text{ of the mean}$$ (9.10)

As the number n increases, the mean m and the standard deviation S become more reliable estimates of the underlying general population (i.e., of μ and σ).

It is often useful to describe the variability of a sample by dividing the standard deviation S by the mean m (and multiplying the result by 100). This yields the coefficient of variation CV:

$$CV \text{ (in percent)} = 100 \ S/m$$ (9.11)

This expression is independent of the magnitude and of the unit of measurement. Groups of human measurements show characteristic variabilities. Typical coefficients of variation are listed in Table 9.10. This information can be used to judge the reliability of reported data.

Considering the CV is often very helpful when you try to determine the credibility of data published in the literature: unusually large or small CV values indicate that either the distribution of the measured population is indeed different from other populations, or that irregularities in measuring, or in data treatment, or in data reporting occurred. (See the more detailed discussion of possible data variations in the 1997 book by Kroemer, Kroemer, and Kroemer-Elbert, and in the 1995 book by Roebuck.) The CV may also help you make an estimate of the standard deviation of an unknown data set.

TABLE 9.10   Variability of Body Measurements

Variables Measured	CV in %
Body heights (stature, sitting height, elbow height, etc.)	3 to 5
Body breadths (hip, shoulder, etc.)	5 to 9
Body depths (abdominal, chest, etc.)	6 to 9
Reaches	4 to 10
Total body weights	10 to 20
Joint ranges	7 to 30
Muscular static strength	10 to 85

From Kroemer, Kroemer, and Kroemer-Elbert, 1990, *Engineering Physiology: Bases of Human Factors/Ergonomics*, 2nd ed. With permission.

9.7 Combining Anthropometric Data Sets

Occasionally, one must add or subtract anthropometric values; for example, total arm length is the sum of upper and lower arm lengths. If you want to add two measures, such as leg length and torso (with head) length, you generate a new combined distribution, stature. In doing so, you must take into account the covariation *COV* between the two measures of leg and torso: usually (but not always) a taller torso is associated with a taller head. This is mathematically described by the correlation coefficient *r* between the two data sets, *x* and *y*, and their standard deviations, S_x and S_y:

$$COV(x,y) = r_{x,y} * S_y * S_y \tag{9.12}$$

This allows us to calculate the *sum* of the two mean values of the *x* and *y* distributions from

$$m_z = m_x + m_y \tag{9.13}$$

and the estimated standard deviation of *z* from

$$S_z = \left[S_x^2 + S_y^2 + 2r * S_x * S_y \right]^{1/2} \tag{9.14}$$

The *difference z* between two mean values is

$$m_z = m_x - m_y \tag{9.15}$$

and its standard deviation

$$S_2 = \left[S_x^2 + S_y^2 - 2r * S_x * S_y \right]^{1/2} \tag{9.16}$$

Three Examples

EX 1: What is the 95p shoulder-to-fingertip length? You know the mean lower arm LA link length (with the hand) to be 442.9 mm with a standard deviation of 23.4 mm. You also know the mean upper arm UA link length of 335.8 mm and its standard deviation of 17.4 mm.

The multiplication factor of *k* = 1.64 (from Table 9.2) leads you to the 95th percentile. But you cannot calculate the sum of the two 95p lengths because this would disregard their covariance; instead, you calculate the sum of the mean values first:

$$m = m_{LA} + m_{UA} = 442.9 + 335.8 = 778.7 \text{ mm} \qquad \text{see (9.13)}$$

The standard deviation is calculated next, using an assumed coefficient of correlation of 0.4:

$$S = \left[23.4^2 + 17.4^2 + 2 * 0.4 * 23.4 * 18.4 \right]^{1/2} \text{ mm}$$
$$S = 34.6 \text{ mm} \qquad \qquad \text{see (9.14)}$$

The 95p total arm length AL can now be calculated:

$$AL_{95} = 778.7 \text{ mm} + 1.64 * 34.6 \text{ mm} = 835.4 \text{ mm} \qquad \text{see (9.1)}$$

EX 2: What is the average arm (acromion to wrist) length of an American pilot? You know that for a standing pilot the 90th percentile acromial (shoulder) height is 1532.0 mm and the wrist height is

905.6 mm; for the 10th percentile, the values are 1379.5 and 808.6 mm, respectively. The correlation between shoulder and wrist heights is estimated at 0.3. You first calculate the mean 90p and 10p acromion (A) and wrist (W) heights to be able to estimate the standard deviations:

$$m_A = (1532.0 + 1379.5)\,mm/2 = 1455.75\ mm$$

and, with $k = 1.28$ taken from Table 9.2,

$$S_A = (1532.0 - 1455.75)\,mm/1.28 = 59.6\ mm \qquad \text{see (9.1)}$$

$$[\text{or: } S_A = (1455.75 - 1379.5)\,mm/1.28 = 59.6\ mm]$$

Likewise,

$$m_w = (905.6 + 808.6)\,mm/2 = 857.1\ mm$$

$$S_w = (905.6 - 857.1)\,mm/1.28 = 37.9\ mm \qquad \text{see (9.1)}$$

$$[\text{or: } S_w = (857.1 - 808.6)\,mm/1.28 = 37.9\ mm]$$

The average arm length (acromion to wrist, AW) is

$$m_{AW} = m_A - m_W = 1455.75\ mm - 857.1\ mm\ =\ 598.65\ mm \qquad \text{see (9.15)}$$

The standard deviation of the arm length is

$$S_{AW} = \left(59.6^2 + 37.9^2 - 2*0.3*59.6*37.9\right)^{1/2} mm = 60.3\ mm \qquad \text{see (9.16)}$$

EX 3: What is the mass of the head of a 75p Japanese female? The mass of the total body has a mean of 54.0 kg with a standard deviation of 6.0 kg (see Table 9.7). The estimated mass of the head is 6.2% of body mass (from data compiled by Kroemer, Kroemer, and Kroemer-Elbert, 1997). Assume the correlation between total body and head masses to be 1. The mean head mass is

$$\text{mean}_{head} = 0.062 * 54.0\ kg = 3.35\ kg.$$

Given the assumed perfect correlation between head and total body masses, the standard deviation of the head mass may be calculated with

$$E = S_{total\ body}/m_{total\ body} = (6.0/54.0) = 0.11 \qquad \text{see (9.2)}$$

From

$$S_{head} = m_{head} * E = 3.35\ kg * 0.11 \qquad \text{see (9.4)}$$

you calculate

$$S_{head} = 0.37\ kg$$

The mass of a 75th percentile head is (with $k = 0.67$ taken from Table 9.2)

$$\text{mass}_{head\ 75p} = 3.35\ kg + 0.67 * 0.37\ kg = 3.6\ kg \qquad \text{see (9.1)}$$

9.8   The "Normative" Adult vs. "Real Persons"

The "Average Person" Phantom

Without formally stating so, even without consciously being aware of it, we commonly design for a group of "regular" people who are in the 25- to 45-year age bracket; who are of "normal" anthropometry, i.e., have body dimensions such as stature, hand reach, or weight close to the 50th percentile; who are "healthy" in their metabolic, circulatory, and respiratory subsystems; whose nervous control, sensory capabilities, and intelligence are all "near average," and who are able and willing to perform "normally." Thus, by default or for reasons, the normative stereotype of many human factor engineers is the "regular" adult woman or man. In fact, the proverbial "average person" appears only in newspapers, design guidelines, biomechanical models, and textbooks.

This mythical normative adult has become our user prototype to which we compare other subgroups, such as children, temporarily or permanently impaired persons, women during pregnancy, or aging people. Yet, most individuals and whole population subgroups deviate in size, strength, or other performance capabilities from the normative adult. Neither in a statistical sense nor in reality are there persons who are average in most or all respects, and products or processes "designed for the average" fit nobody well (Kroemer, Kroemer, and Kroemer-Elbert, 1994). To achieve ease, efficiency, and safety, it is mandatory to consider the ranges of, the variations in, and the combinations of physiologic and psychologic traits; the foregoing discussions showed ways to accommodate anthropometric variability.

Posture versus Motions

A similarly simplistic approach incorporates the idea of designing for body "posture." In part, this false concept may have been provoked by the standardized erect posture, sitting or standing, utilized in measuring body size, or static strength. Unfortunately, the "upright" posture has been employed as a design model, probably because it is easily visualized and made into a design template. This upright idol was promoted by orthopedists of the late 19th century who translated their postural concerns into the desire for an erect trunk posture, especially when sitting in school or office. Yet, over extended periods of time, the human is unable to maintain any given posture, upright or otherwise. Standing still, immobile sitting, even lying stiffly, quickly become uncomfortable and then, with time, physically impossible to maintain; if enforced by injury or sickness, circulatory and metabolic functions become impaired, bed sores appear. The human body is made to move.

Our bodies are designed for movement especially in the arms, with shoulder and elbow joints providing extensive angular freedom. The strong legs are able to propel the body on the ground, with major motions occurring in the knee and hip joints. Movements of the trunk occur mostly in flexion and extension at the lower back. However, these bending and unbending motions (in the medial plane) are rather limited, and often lead to overexertions, especially if combined with sideways twisting of the torso: low back pain has been reported throughout the history of mankind. Wrist problems have been associated with excessive motion requirements since the early 1700s. Head and neck have limited mobility in bending and twisting. Our thumbs and fingers have limited but finely controlled motion capability.

Ranges of motion (also called mobility or flexibility) depend much on age, health, fitness, training, and skill. Mobility ranges have been measured on dissimilar groups of people with various measuring instructions and techniques; hence, there is much diversity in reported results. However, at least one set of mobility measurements has been taken on groups of 100 females and of 100 males by the same researchers using the same techniques. These data are reported in Table 9.11. Note that the differences in mobility between males and females are negligible in most cases.

Designing to fit motion ranges, instead of fixed postures, is not difficult. The articulations in the human body have varying degrees of freedom for movements. These are shown in Figure 9.5 for major body joints, and the motion ranges are listed in Table 9.11. These maximal ranges were measured on students of physical education, hence, many people will have slightly less mobility than shown. "Convenient" mobility

TABLE 9.11   Comparison of Mobility Data (in degrees) for Females and Males

Joint	Movement	5th Percentile Female	5th Percentile Male	50th Percentile Female	50th Percentile Male	95th Percentile Female	95th Percentile Male	Difference* Female	Difference* Male
Neck	Ventral flexion	34.0	25.0	51.5	43.0	69.0	60.0	+8.5	
	Dorsal flexion	47.5	38.0	70.5	56.5	93.5	74.0	+14.0	
	Right rotation	67.0	56.0	81.0	74.0	95.0	85.0	+7.0	
	Left rotation	64.0	67.5	77.0	77.0	90.0	85.0	NS	
Shoulder	Flexion	169.5	161.0	184.5	178.0	199.5	193.5	+6.5	
	Extension	47.0	41.5	66.0	57.5	85.0	76.0	+8.5	
	Adduction	37.5	36.0	52.5	50.5	67.5	63.0	NS	
	Abduction	106.0	106.0	122.5	123.5	139.0	140.0	NS	
	Medial rotation	94.0	68.5	110.5	95.0	127.0	114.0	+15.5	
	Lateral rotation	19.5	16.0	37.0	31.5	54.5	46.0	+5.5	
Elbow-forearm	Flexion	135.5	122.51	148.0	138.0	160.5	150.0	+10.0	
	Supination	87.0	86.0	108.5	107.5	130.0	135.0	NS	
	Pronation	63.0	42.5	81.0	65.0	99.0	86.5	+16.0	
Wrist	Extension	56.5	47.0	72.0	62.0	87.5	76.0	+10.0	
	Flexion	53.5	50.5	71.5	67.5	89.5	85.0	+4.0	
	Adduction	16.5	14.0	26.5	22.0	36.5	30.0	+4.5	
	Abduction	19.0	22.0	28.0	30.5	37.0	40.0	−2.5	
Hip	Flexion	103.0	95.0	125.0	109.5	147.0	130.0	+15.5	
	Adduction	27.0	15.5	38.5	26.0	50.0	39.0	+12.5	
	Abduction	47.0	38.0	66.0	59.0	85.0	81.0	+7.0	
	Medial rotation (prone)	30.5	30.5	44.5	46.0	58.5	62.5	NS	
	Lateral rotation (prone)	29.0	21.5	45.5	33.0	62.0	46.0	+12.5	
	Medial rotation (sitting)	20.5	18.0	32.0	28.0	43.5	43.0	+4.0	
	Lateral rotation (sitting)	20.5	18.0	33.0	26.5	45.5	37.0	+6.5	
Knee	Flexion (standing)	99.5	87.0	113.5	103.5	127.5	122.0	+10.0	
	Flexion (prone)	116.0	99.5	130.0	117.0	144.0	130.0	+13.0	
	Medial rotation	18.5	14.5	31.5	23.0	44.5	35.0	+8.5	
	Lateral rotation	28.5	21.0	43.5	33.5	58.5	48.0	+10.0	
Ankle	Flexion	13.0	18.0	23.0	29.0	33.0	34.0	−6.0	
	Extension	30.5	21.0	41.0	35.5	51.5	51.5	+5.5	
	Adduction	13.0	15.0	23.5	25.0	34.0	38.0	NS	
	Abduction	11.5	11.0	24.0	19.0	36.5	30.0	+5.0	

* Listed are only differences at the 50th percentile, and if significant ($\alpha < 0.5$).

From Kroemer, Kroemer, and Kroemer-Elbert, 1990, *Engineering Physiology: Bases of Human Factors/Ergonomics,* 2nd ed. With permission.

is somewhere within the range of maximal values shown in Table 9.11, but not always in the middle of the ranges; not seldom, convenient motions are near the limits of mobility. Habits and skill as well as strength requirements may make different ranges preferred.

Design for motions starts by establishing the actual movement ranges. Convenient motions may cluster around the mean of mobility in a body joint, or may be close to the limits of flexibility. For example, a person walking about on a job, or standing, has the knees most of the time nearly extended, that is — in the sagittal view — the knee angles range close to the extreme value of about 180 degrees. The sagittal hip angle (between trunk and thigh) also varies in the neighborhood of 180 degrees. Both angles change to cluster about 90 degrees when sitting. (See Table 9.12.)

Preferred work areas of the hands and feet are in front of the body, within curved envelopes that reflect the mobility of the forearm in the elbow joint, or of the total arm in the shoulder joint; of the lower leg in the knee joint, and of the total leg in the hip joint. Thus, these reach envelopes are often described as

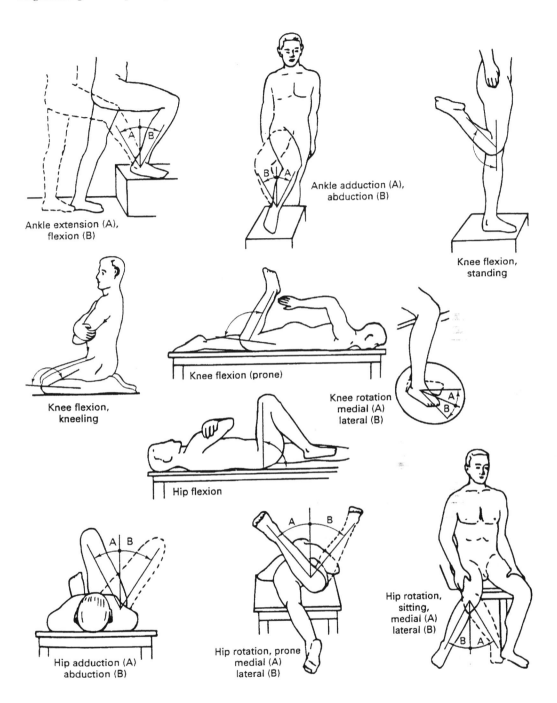

FIGURE 9.5 Maximal displacements in body joints. (From Kroemer, Kroemer, and Kroemer-Elbert, 1990, *Engineering Physiology: Bases of Human Factors/Ergonomics*, 2nd ed. With permission.)

(partial) spheres around the presumed locations of the body joints. However, the preferred ranges within the possible motion zones are different when the main requirements are strength, or speed, or accuracy, or vision — as discussed, in some detail, by Kroemer, Kroemer, and Kroemer-Elbert (1994). Thus, there is not one reach envelope, but different preferred envelopes.

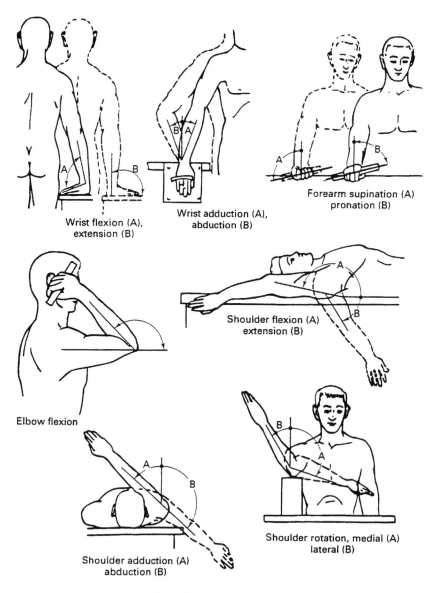

FIGURE 9.5 (continued)

For each job situation, the ergonomic designer determines the dominant requirements of the task; for example, whether the operator

- Works while sitting or walking (standing)
- Performs wide-ranging or specialized work
- Must exert large or small forces
- Executes fast and gross or slow and exact motions
- Needs high or low visual control.

Such circumstances affect the selection of the specific work envelope.

While sitting or moving about, the trunk is normally kept nearly erect, as are the neck and head. In most work situations, the upper arm hangs from the shoulder, while the elbow angle tends to be near 90 degrees; but the wrist is nearly straight. Table 9.12 lists typical body angles at work.

Designing for motion is done in these steps:

Step 1: Select the major body joints involved.

Step 2: *Adjust body dimensions reported for standardized postures* (e.g., Tables 9.4 through 9.9) *to accommodate the real work conditions.* Use Table 9.3 for guidance.

Step 3: *Select appropriate motion ranges in the body joints.* The range can be depicted as the area between two positions, such as knee angles ranging between 60 and 105 degrees; or as a motion envelope, such as circumscribed by combined hand-and–arm movements, or by the clearance envelope under (through, within, beyond) which body parts must fit. Use Table 9.12 for guidance.

Basic work space design faults should be avoided. These include:

1. *Avoid twisted body positions,* especially of the trunk and neck. This results often from bad location of work objects, controls, and displays.
2. *Avoid forward bending of trunk, neck, and head.* This is frequently provoked by improperly positioned controls and visual targets including working surfaces that are too low.
3. *Avoid postures that must be maintained* for long periods of time, especially at the extreme limits of the range of motion. This is particularly important for the wrist and the back.
4. *Avoid holding the arms raised.* This results commonly from locating controls or objects too high, higher than the elbow when the upper arm hangs down. The upper limit for regular manipulation tasks is about chest height.

TABLE 9.12   Mobility Ranges at Work

Angles at	Walking About, Standing	Sitting
Knee	Near extreme stretch: 180 deg. or slightly less	Mostly mid-range: about 90 deg.
Hip (lateral new)	Near extreme stretch: about 180 deg.	Mostly mid-range: about 90 deg.
Shoulder	Mostly mid-range: Upper arm often hanging down	
Elbow	Mostly mid-range: about 90 deg.	
Wrist	Mostly mid-range: about straight	
Neck/Head	Mostly mid-range: about straight	
Back	Near extreme stretch: about erect	

From Kroemer, Kroemer, and Kroemer-Elbert, 1990, *Engineering Physiology: Bases of Human Factors/Ergonomics,* 2nd ed. With permission.

General criteria for work space layout relate to human strength, speed, effort, accuracy, importance, frequency, function and sequence of use, as listed in Table 9.13. Achieving the task while assuring safety for the human, avoiding overuse and unnecessary effort, and assuring ease and efficiency are primary design goals.

9.9 Summary

It is inexcusable to design tasks, tools or workstations for the phantom of "the average person" in a static position. No such person exists, and design for the average fits nobody well. Instead, ranges of body sizes,

TABLE 9.13   Guidelines for Workspace Design

Human Strength —	facilitate extension of strength (work, power) by object location and orientation.
Human Speed —	place items so that they can be reached and manipulated quickly.
Human Effort —	arrange work so that it can be performed with least effort.
Human Accuracy —	select and position objects so that they can be manipulated and seen with ease.
Importance —	the most important items should be in the most accessible locations.
Frequency of use —	the most frequently used items should be in the most accessible locations.
Function —	items with similar functions should be grouped together.
Sequence of use —	items which are commonly used in sequence should be laid out in that sequence.

of motions, and of strengths establish the design criteria. "Designing for function" is easy for the engineer who starts with proper anthropometric information and applies it ergonomically, that is with "ease and efficiency" as the guiding principles.

References

Bhattacharia, A. and McGlothlin, J.D. (Eds.) (1996). *Occupational Ergonomics.* New York, NY: Marcel Dekker.

Cheverud, J., Gordon, C.C., Walker, R.A., Jacquish, C., Kohn, L., Moore, A., and Yamashita, N. (1990). *1988 Anthropometric Survey of U.S. Army Personnel: Correlation Coefficients and Regression Equations.* (Natick TR 90/032-6). Natick, MA: U.S. Army Research, Development and Engineering Center.

Fluegel, F., Greil, H., and Sommer, K. (1986). *Anthropologischer Atlas.* Berlin, Germany: Tribuene.

Gordon, C.C., Churchill, T., Clauser, C.E., Bradtmiller, B., McConville, J.T., Tebbetts, I., and Walker, R.A. (1989). *1988 Anthropometric Survey of U.S. Army Personnel: Summary Statistics Interim Report.* (Natick-TR-89/027). Natick, MA: U.S. Army Natick Research, Development and Engineering Center.

Juergens, H.W., Aune, I.A., and Pieper, U. (1990). *International Data on Anthropometry.* (Occupational Safety and Health Series No. 65). Geneva, Switzerland: International Labour Office.

Kagimoto, Y. (Ed.) (1990). *Anthropometry of JASDF Personnel and Its Applications for Human Engineering.* Tokyo, Japan: Aeromedical Laboratory, Air Development and Test Wing JASDF.

Kroemer, K.H.E. (1983). Engineering anthropometry, in D.J. Oborne and M.M. Gruneberg, (Eds.) *The Physical Environment at Work* (pp. 39-68). London: Wiley.

Kroemer, K.H.E. (1989). Engineering anthropometry. *Ergonomics, 32,* 767-784

Kroemer, K.H.E. and Grandjean, E. (1997). *Fitting the Task to the Human* (5th ed.). London, U.K. and Bristol, PA: Taylor & Francis.

Kroemer, K.H.E., Kroemer, H.B., and Kroemer-Elbert, K.E. (1994). *Ergonomics: How to Design for Ease and Efficiency.* Englewood Cliffs, NJ: Prentice Hall.

Kroemer, K.H.E., Kroemer, H.J., and Kroemer-Elbert, K.E. (1997). *Engineering Physiology. Bases of Human Factors/Ergonomics* (3rd ed.). New York, NY: Van Nostrand Reinhold–Wiley.

NASA/WEBB (Ed.) (1978). *Anthropometric Sourcebook* (3 volumes). (NASA Reference Publication 1024). Houston, TX: NASA (NTIS, Springfield, VA 22161, Order No. 79 11 734).

Pheasant, S. (1986). *Bodyspace: Anthropometry, Ergonomics and Design.* London, U.K.: Taylor & Francis.

Pheasant, S. (1996). *Bodyspace: Anthropometry, Ergonomics and the Design of Work* (2nd ed.). London, UK: Taylor & Francis.

Pheasant, S.T. (1982). A technique for estimating anthropometric data from the parameters of the distribution of stature. *Ergonomics, 25,* 981-992

Roebuck, J.A. (1995). *Anthropometric Methods.* Santa Monica, CA: Human Factors and Ergonomics Society.

Roebuck, J.A., Kroemer, K.H.E., and Thomson, W.G. (1975). *Engineering Anthropometry Methods.* New York, NY: Wiley.

For Further Information

You are likely to encounter more involved considerations if data must be developed that describe composite populations which consist, for example, of subsamples such as a% females and b% males (Kroemer, 1983). Or you may have only mean values of body dimensions, but no information about the associated standard deviations that you must estimate. Also, a large set of design questions arises from link lengths and mass properties of body segments including locations of mass centers and definitions of joint centers, as well as mobility and motions characteristics. Related information and solutions for such challenges can be found in recent publications by Annis (in Bhattacharia and McGlothlin, 1996), Kroemer, Kroemer, and Kroemer-Elbert (1994, 1997), Pheasant (1986, 1996), or Roebuck (1995).

10

Occupational Biomechanics

William S. Marras
The Ohio State University

10.1   Biomechanic Analyses and Ergonomics

Definitions

Biomechanics may be defined as an interdisciplinary field in which information from both the biological sciences and engineering mechanics is used to assess the function of the body. *A major assumption of occupational biomechanics is that the body behaves according to the laws of Newtonian mechanics. By definition, "mechanics is the study of forces and their effects on masses"* (Kroemer, 1987). The function of interest in an occupational ergonomics context is most often a quantitative assessment of mechanical loading that occurs within the musculoskeletal system. The goal of an occupational biomechanics assessment is to quantitatively describe the musculoskeletal loading that occurs during work so that one can derive an appreciation for the degree of risk associated with an occupationally-related task. The characteristic that distinguishes occupational biomechanics analyses from other types of ergonomic analyses is that the comparison is quantitative in nature. The quantitative nature of occupational biomechanics permits ergonomists to address the question of how much exposure to the occupational risk factors is too much exposure.

     The portion of biomechanics dealing with ergonomics issues is often labeled industrial or occupational biomechanics. Chaffin and Andersson (1991) have defined occupational biomechanics as "the study of the physical interaction of workers with their tools, machines, and materials so as to enhance the worker's performance while minimizing the risk of musculoskeletal disorders." This chapter will address occupational biomechanical issues exclusively in this ergonomics framework.

Occupational Biomechanics Approach

In order to effectively address ergonomic issues in the workplace, one must develop an appreciation for the *trade-offs* associated with ergonomics. When one considers biomechanical rationale one finds that it is very difficult to accommodate all parts of the body in an ideal biomechanical environment. It is often the case that in attempting to accommodate one portion of the body the biomechanical situation at another body site is compromised. Therefore, the key to the proper employment of occupational bio-mechanical principles is to be able to consider the appropriate biomechanical trade-offs with various parts of the body associated with different workplace design options. For this reason this chapter will focus on the information required to develop proper biomechanical reasoning when considering a workplace. The chapter will first present and explain a series of key concepts that make up the heart of biomechanical reasoning. From there, these concepts will be applied to different parts of the body. Once this reasoning is developed, we will examine how the various biomechanical concepts must be considered collectively in terms of trade-off when designing a workplace from an ergonomic perspective under realistic conditions. This chapter will demonstrate that one cannot successfully practice ergonomics by simply memorizing a set of "ergonomic rules" (e.g., keep the wrist straight or don't bend from the waist when lifting). These types of rule-based design strategies ultimately result in suboptimizing the workplace ergonomic conditions.

10.2   Biomechanical Concepts

The Load–Tolerance Model

The fundamental concept in the application of occupational biomechanics to ergonomics is that one could design workplaces so that the load imposed on a structure does not exceed the tolerance of the structure. This basic concept is illustrated in Figure 10.1. This figure shows the traditional concept of biomechanical risk in occupational biomechanics. A loading pattern is developed on a body structure that is repeated as the work cycles are repeated during a job. The structure tolerance is also shown in this figure. If the tolerance far exceeds load, then the task is considered safe and the magnitude of the difference between the load and the tolerance is considered the safety margin. As implied in this figure, risk occurs when the imposed load exceeds the tolerance. As many of the industrial tasks become more repetitive, this model is beginning to change. As shown in Figure 10.2, occupational biomechanics logic is beginning to appreciate the fact that, with repetitive loading, the tolerance of the structure of interest may decrease over time to the point where it is more likely that the structure loading will exceed the structure tolerance and result in injury or illness. Thus, occupational biomechanical models and logic are beginning to build systems that consider observations in the workplace, such as cumulative trauma disorders.

Acute versus Cumulative Trauma

It is well recognized that in occupational settings two types of trauma can affect the human body and lead to musculoskeletal disorders. First, *acute* trauma can occur which refers to an application of force that is so large that it exceeds the tolerance of the body structure during an occupational task. Thus, acute trauma is typically associated with large exertions of force that occur infrequently. For example, an acute trauma can occur when a worker is asked to lift an extremely heavy object, as when moving a heavy part. This situation would relate to a peak load pattern that exceeded the load tolerance in Figure 10.1. *Cumulative* trauma, on the other hand, refers to the repeated application of force to a structure that tends to wear it down, thus lowering the structure tolerance to the point where the tolerance is exceeded through a reduction of the tolerance limit. This situation is illustrated in Figure 10.2. Cumu-lative trauma represents more of a "wear and tear" on the structure. This type of trauma is becoming far more common in the workplace because more repetitive jobs are becoming common in industry and are a concern for many ergonomics evaluations.

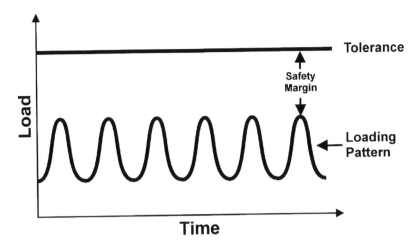

FIGURE 10.1   Traditional concept of biomechanical risk.

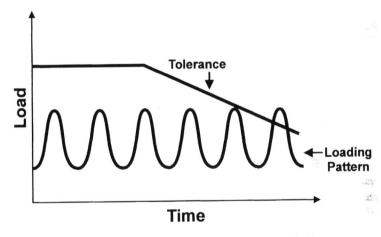

FIGURE 10.2   Realistic scenario of biomechanical risk.

Cumulative trauma can initiate a process that can result in a cycle of reaction of the body's structures that are very difficult to break. This process is illustrated in Figure 10.3. The cumulative trauma process begins by exposing the worker to manual exertions that are either frequent or prolonged. This repetitive or prolonged application of force can affect either the tendons or the muscles of the body. If the tendons are affected, the following sequence occurs. The tendons are subject to mechanical irritation when they are repeatedly exposed to high levels of tension, and groups of tendons may rub against each other. The physiologic reaction to this mechanical irritation can result in inflammation and swelling of the tendon. This swelling will stimulate the activities of the nociceptors surrounding the structure and signal the central control mechanism (brain), via pain perception, that a problem exists. In response to this pain, the body will attempt to control the problem via two mechanisms. First, the muscles surrounding the irritated area will coactivate in an attempt to minimize the motion of the tendon or stiffen the structure. Since motion will further stimulate the nociceptors and result in further pain, motion avoidance is often indicative of the start of a cumulative trauma disorder. Second, in an attempt to reduce the friction occurring within the tendon, the body will increase its production of synovial fluid within the tendon sheath. However, given the limited space available between the tendon and the tendon sheath, the increased production of synovial fluid often exacerbates the problem by further expanding the tendon

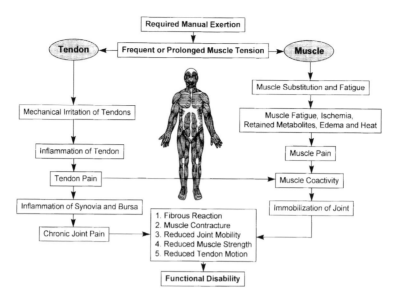

FIGURE 10.3   Sequence of events in cumulative trauma disorders. (Adapted from Chaffin, D.B. and Andersson, G.B. (1991) *Occupational Biomechanics*, John Wiley & Sons, Inc. New York, NY.)

sheath and, thus, further stimulating the surrounding nociceptors. This process results in chronic joint pain and a series of musculoskeletal reactions such as reduced strength, reduced tendon motion, and reduced mobility. Collectively, these reactions result in a functional disability.

A similar process occurs if the muscles are affected by cumulative trauma as opposed to the tendons. Muscles can be easily overloaded when they become fatigued. Fatigue can lower the tolerance to stress and can result in microtrauma to the muscle fibers. This microtrauma typically means the muscle is partially torn and the tear will cause capillaries to rupture and result in swelling, edema, or inflammation near the site of the tear. This process can stimulate nociceptors and result in pain. As with cumulative trauma to the tendons, the body reacts by cocontracting the surrounding musculature and thereby minimizing the motion of the joint. Since the tendons are not involved with cumulative trauma to the muscles, there is no increased production of synovial fluid. However, the end result is the same series of musculoskeletal reactions resulting from tendon irritation (i.e., reduced strength, reduced tendon motion, and reduced mobility). The ultimate result of this process is once again a functional disability.

Even though the stimulus associated with the cumulative trauma process is somewhat similar between tendons and muscles, there is a significant difference in the time required to heal from the damage to a tendon compared to a muscle. The mechanism of repair for both the tendons and muscles is dependent upon blood flow to the damaged structure. Blood provides nutrients for repair and dissipates waste materials. However, the blood supply to a tendon is just a fraction of that supplied to a muscle. Thus, given an equivalent strain to a muscle and a tendon, the muscle will heal rapidly (in about ten days if not reinjured), whereas the tendon could take months to accomplish the same level of repair. For this reason, ergonomists must be particularly vigilant in the assessment of workplaces that could pose a danger to the tendons of the body.

Moments and Levers

Biomechanical loads are *not* defined solely by the magnitude of weight supported by the body. The position of the weight relative to the axis of rotation of the body joint of interest defines the imposed load on the body and is referred to as a *moment*. Thus, a moment is defined as the product of force and distance. For example, a 50 Newton mass held at a horizontal distance of 75 cm (.75 meters) from the shoulder joint imposes a moment of 37.5 Nm (50 N × 0.75 m) on the shoulder joint, whereas the same weight held at a horizontal distance of 25 cm from the shoulder joint imposes a moment or load of only 12.5 Nm (50/n × 0.25m) on the shoulder. Thus, the load on a joint is a function of where the load is held relative to the joint and the mass of the weight held. Load is not simply a function of weight.

As implied by this example, moments are a function of the mechanical lever systems of the body. The musculoskeletal system can be represented by systems of levers, and these lever systems usually form the basis of most biomechanical assessments and models. Three types of lever systems are present in the human body. First-class levers are those that have a fulcrum in the middle of the system, an imposed load on one end of the system, and the restorative or internal load imposed on the opposite end of the system.

As will be discussed later (Figure 10.18), the trunk is an example of a first-class lever. In this example, the spine serves as the fulcrum. As the worker lifts, a moment is imposed anterior to the spine due to the object weight times the distance of the object from the spine. This moment is counterbalanced by the activity of the back musculature, but the mechanical advantage of the back muscles is much less than that of the object lifted. A second-class lever system can be found in the lower extremity. In a second-class lever system, the fulcrum is on one end of the lever, the restorative load is on the other end of the system, and the applied load is between these two. In the body, the lower leg is a good example of this lever system. In this example, the ball of the foot acts as the fulcrum, the load is applied through the tibia or bone of the lower leg. The restorative force is applied through gastrocnemius or calf muscle. In this manner, the muscle activates and causes the body to rotate about the fulcrum or ball of the foot and move the body forward. Finally, a third-class lever system is one where the fulcrum is on one end of the system, the applied load acts at the other end of the system, and the restorative force acts between the two. An example of this system in the human body is the elbow joint and is shown in Figure 10.4.

External and Internal Loading

Two types of forces can load the body during work. *External* loads refer to those forces that are imposed on the body as a result of gravity acting upon an external object being manipulated by the worker. For example, in Figure 10.4a the tool held in the worker's hand is subject to the forces of gravity and imposes a 44.5 N (10 pound) external load at a distance from the joint of 30.5 cm (12 inches) on the elbow joint. However, in order to maintain equilibrium, this external load must be counteracted by an *internal* load that is supplied by the muscles of the body. Figure 10.4a also shows that the internal load (muscle) acts at a distance relative to the elbow joint that is much closer to the fulcrum than the external load (tool). Thus, the internal load or force is at a biomechanical disadvantage and must be much larger (534 N or 120 lbs.) than the external load (44.5 N or 10 lbs.) in order to keep the musculoskeletal system in equilibrium. As shown in this example, it is not unusual for the magnitude of the internal load to be much greater (typically 10 times greater) than the external load. Thus, it is typically the internal loading that contributes most to cumulative trauma of the musculoskeletal system during work. The sum of the external load and the internal load define the total loading experienced at the joint. When evaluating a workstation, the ergonomist must not only consider the externally applied load but must be particularly sensitive to the magnitude of the internal forces that can load the musculoskeletal system.

Factors Affecting Internal Loading

The previous section has discussed the importance of understanding the relationship between the external loads imposed on the body and the internal loads generated by the force-generating mechanisms within the body. The key to proper ergonomic design involves designing workplaces so that the internal loads are minimized. Several properties of the work environment can be manipulated in order to facilitate this goal.

Posture and Length–Strength

The posture assumed when one works can affect the arrangement of the body's leverage system, and thus, can greatly affect the magnitude of the internal load required to support the external load. The

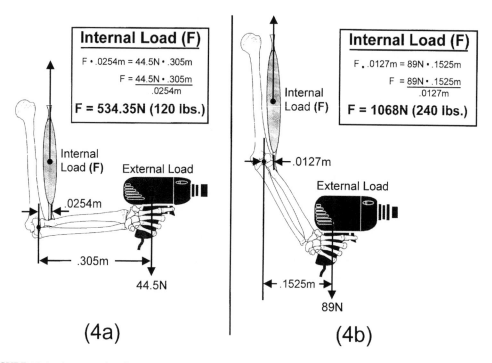

Internal Load (F)

F • .0254m = 44.5N • .305m

$$F = \frac{44.5N \cdot .305m}{.0254m}$$

F = 534.35N (120 lbs.)

Internal Load **(F)**

External Load

.0254m

.305m

44.5N

(4a)

Internal Load (F)

F . .0127m = 89N • .1525m

$$F = \frac{89N \cdot .1525m}{.0127m}$$

F = 1068N (240 lbs.)

Internal Load **(F)**

.0127m

External Load

.1525m

89N

(4b)

FIGURE 10.4 An example of an anatomical third class lever (a) demonstrating how the mechanical advantage changes as the elbow position changes (b).

arrangement of the lever system can influence the magnitude of the external moment (force × distance) imposed on the body as well as dictate the magnitude of the internal forces and the subsequent risk of cumulative trauma. Consider the biomechanical arrangement of the elbow joint that is shown in Figure 10.3. In Figure 10.4a, the mechanical advantage of the internal force generated by the biceps muscle and tendon is minimized by maintaining a posture keeping one's arm bent at a 90° angle. If one palpates the tendon and inserts the index finger between the joint center and the tendon one can gain an appreciation for the internal moment arm distance. One can also appreciate how this internal mechanical advantage can change with posture. With the index finger still inserted between the elbow joint and the tendon if the arm is slowly straightened one can appreciate how the distance between the tendon and the joint center of rotation is significantly reduced. If the moment imposed about the elbow joint is held constant as shown in Figure 10.4b, the mechanical advantage of the internal force generator is significantly reduced. In other words, since the internal moment or distance between the tendon and the joint center is reduced (compared to the situation where the elbow is positioned at a 90° angle), the muscle must produce more force in order to support the external load. This force is transmitted through the tendon and can increase the risk of cumulative trauma. Therefore, the positioning of the mechanical lever system can greatly affect the internal load transmission within the body. The same task can be performed in a variety of ways, but some of these positions are much more costly in terms of loading of the musculoskeletal system than others.

Another important relationship in defining the load on the musculoskeletal system is the length–strength relationship of the muscles. Figure 10.5 shows this relationship. The active portion of this figure refers to structures that actively generate force, such as muscles. The figure indicates that when muscles are close to their resting length they are in a position where they have the greatest capacity to generate force. However, when the muscle length deviates from this resting position, the capacity to generate force is greatly reduced. Hence, when a muscle stretches or becomes very short, the ability to generate force is greatly diminished. Note also, as indicated in Figure 10.5, that passive tissues in the muscle (and also ligaments) can generate tension when muscles are stretched. Thus, the orientation of

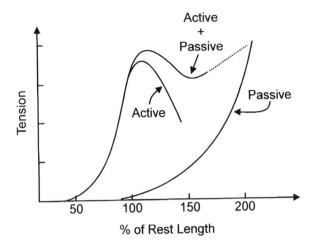

FIGURE 10.5 Length–tension relationship for a human muscle. (Adapted from Basmajian, J.V. and De Luca, C.J. (1985) *Muscles Alive: Their Functions Revealed by Electromyography* (5th edition), Williams and Wilkins, Baltimore, MD.)

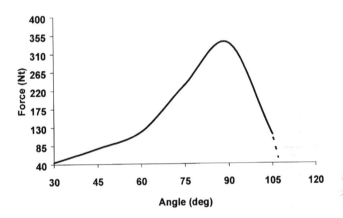

FIGURE 10.6 Length–tension diagram produced by flexion of the forearm in pronation. "Angle" refers to included angle between the longitudinal axes of the forearm and upper arm. The highest parts of the curve indicate the configurations where the biomechanical lever system is most effective. (Adapted from Chaffin, D.B. and Andersson, G.B. (1991) *Occupational Biomechanics*, John Wiley & Sons, Inc. New York, NY.)

the muscle fibers during a task can greatly influence the force available to perform a work task and can therefore influence risk of cumulative trauma. A given tension on a muscle can either tax the muscle greatly or be a minimum burden on it. What might be considered a moderate force for a muscle at the resting length can become the maximum force a muscle can produce when it is in a stretched or contracted position, thus increasing the risk of muscle strain. When this relationship is considered in conjunction with the mechanical load placed on the muscle and tendon via the arrangement of the lever system the position of the joint arrangement becomes a major factor in the design of the work environment. It is typically the case that the length–strength relationship interacts synergistically with the lever system. Figure 10.6 shows the effect of elbow position on the force-generation capability of the elbow. This figure indicates that position can have a dramatic effect on force generation. As already discussed, this position can also have a great effect on internal loading of the joint and the subsequent risk of cumulative trauma.

Force–Velocity

Motion can profoundly influence the ability of a muscle to generate force and load the biomechanical system. Motion can either be a benefit to the biomechanical system if momentum is properly used or it

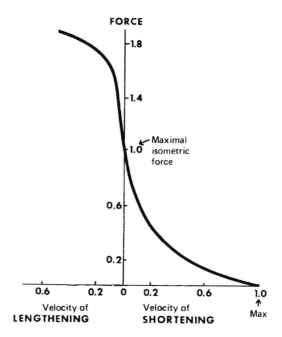

FIGURE 10.7   Influence of velocity upon muscle force. (Adapted from *The Textbook of Work Physiology*, McGraw-Hill, 1977)

can increase the load on the system if the worker is not taking advantage of momentum. The relationship between muscle velocity and force generation is shown in Figure 10.7. This figure shows that, in general, the faster the muscle is moving the greater the reduction in its force capability. As with most of the biomechanical principles mentioned in this chapter, this reduction in muscle capacity can result in the muscle strain occurring at a lower level of external loading and a subsequent increase in the risk of cumulative trauma. In addition, this effect is considered in many dynamic ergonomic biomechanical models.

Strength–Endurance

It is important to realize that strength is a transient factor. A worker may generate a great deal of strength during a one-time exertion. However, if the worker is required to exert large percentages of strength either repeatedly or for a prolonged period of time, the amount of force that the worker can generate is dramatically reduced. Figure 10.8 demonstrates this relationship during an isometric exertion. The broken line in this figure indicates that if a person is asked to generate maximum muscle force, maximum force output is only generated for a very brief period of time. As time increases, strength output decreases exponentially and levels off at about 20% of maximum after about seven minutes. Similar trends occur under repeated dynamic conditions. This indicates that if it is determined that a task requires a large portion of a worker's strength, one must consider how long that portion of the strength is required in order to ensure that the work does not strain the musculoskeletal system.

Rest Time

As has been mentioned several times in this chapter, the risk of cumulative trauma increases when the capacity to exert force is challenged by the external force requirements of the job. Another factor that can affect this strength capacity is rest time. Rest time has a profound effect on the ability to exert force. Figure 10.9 shows how energy for a muscular contraction is regenerated during work. Adenosine triphosphate (ATP) is required to produce a significant muscular contraction. ATP changes to adenosine diphosphate (ADP) once a muscular contraction has occurred. This ADP must be converted to ATP in order to enable another muscular contraction. This conversion can occur with the addition of oxygen to the system. If oxygen is not present, then the system goes into oxygen debt and there is insufficient

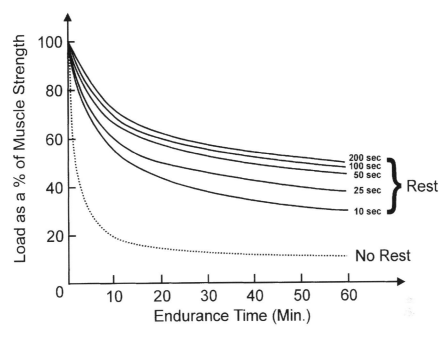

FIGURE 10.8 Forearm flexor muscle endurance times in consecutive static contractions of 2.5 sec duration with varied rest periods. (Adapted from Chaffin, D.B. and Andersson, G.B. (1991) *Occupational Biomechanics*, John Wiley & Sons, Inc. New York, NY.)

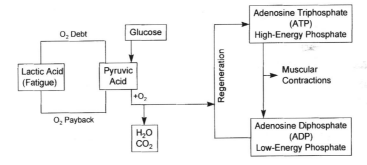

FIGURE 10.9 The body's energy system during work. (Adapted from Grandjean, E. (1982) *Fitting the Task to the Man: An Ergonomic Approach*, Taylor & Francis, Ltd. London.)

ATP for a muscular contraction. Thus, this flow chart indicates that oxygen is a key ingredient in maintaining a high level of muscular exertion. Oxygen is delivered to the target muscles via the blood flow. However, under static exertions the blood flow is reduced and there is a subsequent reduction in the blood flow to the muscle. This restriction of blood flow and subsequent oxygen deficit was responsible for the rapid decrease in force generation over time as shown in Figure 10.8. The solid lines shown in Figure 10.8 show how the force-generation capacity of the muscles increases when different amounts of rest are permitted in a fatiguing exertion. As more and more rest time is permitted, increases in force generation are achieved when more oxygen is delivered to the muscle and more ADP can be converted to ATP. This relationship also shows that any more than about 50 seconds of rest, under these conditions, does not result in a significant increase in force-generation capacity of the muscle. Practically, this indicates that in order to optimize the strength capacity of the worker and minimize the risk of muscle strain, a schedule of frequent and brief rest periods would be more beneficial than lengthy infrequent rest periods.

TABLE 10.1   Tissue Tolerance of the Musculoskeletal System

Structure	Estimated Ultimate Stress (σ_u) (MPa)
Muscle	32–60
Ligament	20
Tendon	60–100
Bone longitudinal loading	
Tension	133
Compression	193
Shear	68
Bone transverse loading	
Tension	51
Compression	133

Adapted from Ozkaya and Nordin, 1991

Load Tolerance

To this point this chapter has considered primarily factors that influence the loads applied to the structures of the body. As mentioned previously, occupational biomechanical analyses must consider not only the loads imposed upon a structure but also the ability of the structure to withstand or tolerate a load during work. This section will briefly review the knowledge base associated with body structure tolerances.

Muscle, Ligament, Tendon, and Bone Capacity

The exact tolerance characteristics of human tissues such as muscles, ligaments, tendons, and bones loaded under various working conditions is difficult to estimate. Tolerances of the structures in the body vary greatly under similar loading conditions. In addition, tolerance depends upon many other factors such as strain rate, age of the structure, frequency of loading, physiologic influences, heredity, conditioning, and many unknown factors. Furthermore, it is not possible to measure these tolerances under human *in vivo* conditions. Therefore, most of the published estimates of tissue tolerance have been derived from various animal and/or theoretical sources.

Muscle and Tendon Strain

Muscle appears to be the structure that has the lowest tolerance in the musculoskeletal system. The ultimate strength of a muscle has been estimated at 32 MPa (Hoy et al., 1990). It is generally believed that the muscle will rupture prior to the tendon in a healthy tendon (Nordin and Frankel, 1989), since tendon stress has been estimated at between 60 and 100 MPa (Nordin and Frankel, 1989; Hoy et al., 1990). Hence, as indicated in Table 10.1, there is a safety margin between the muscle failure point and the failure point of the tendon of about twofold (Nordin and Frankel, 1989) to threefold (Hoy et al., 1990).

Ligament and Bone Tolerance

Ligaments and bone tolerances within the musculoskeletal system have also been estimated. Ultimate ligament stress has been estimated at approximately 20 MPa. The ultimate stress of bone varies depending upon the direction of loading. Bone tolerance can range from as low as 51 MPa in transverse tension to over 190 MPa in longitudinal compression. Table 10.1 also indicates the ultimate stress of bone loaded in different loading conditions.

Disc/Endplate and Vertebrae Tolerance

The mechanism of cumulative trauma in the disc is thought to be related to repeated trauma to the vertebral endplate. The endplate is a very thin (about 1 mm thick) structure that facilitates nutrient flow to the disc fibers (anulus fibrosis). Repeated microfracture of this vertebral endplate is thought to impair the nutrient flow to the disc fibers and thereby lead to atrophy of the fiber and fiber degeneration. It is believed that if one can determine the level at which the endplate experiences a microfracture, one can then minimize the effects of cumulative trauma and disc degeneration within the spine.

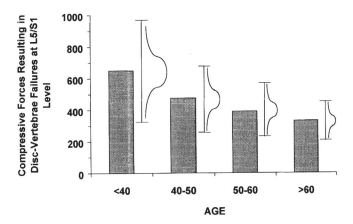

FIGURE 10.10 Mean and range of disc compression failures by age. (Adapted from National Institute for Occupational Safety and Health (NIOSH) (1981) Work practices guide for manual lifting. Department of Health and Human Services (DHHS), National Institute for Occupational Safety and Health (NIOSH), Publication No. 81-122.)

TABLE 10.2 Lumbar Spine Compressive Strength

| Population | n | Strength in kN | |
		Mean	s.d.
Females	132	3.97	1.50
Males	174	5.81	2.58
Total	507	4.96	2.20

Jager et al., 1991

Several studies of disc endplate tolerance have been performed. Figure 10.10 shows the levels of endplate compressive loading tolerance that have been used to establish safe lifting situations at the worksite (NIOSH, 1981). This figure shows the compressive force mean (column value) as well as the compression force distribution (thin line and normal distribution curve) that would result in vertebral endplate failure (microfracture). This figure indicates that for those under 40 years of age endplate microfracture damage begins to occur at about 350 kg (3432 N) of compressive load on the spine. If the compressive load is increased to 650 kg (6375 N), approximately 50% of those exposed to the load will experience vertebral endplate microfracture. When the compressive load on the spine reaches a value of 950 kg (9317 N), almost all of those exposed to the loading will experience a vertebral endplate microfracture. It should also be noted that the tolerance distribution shifts to lower levels with increasing age. In addition, it should be emphasized that this tolerance is based upon compression of the vertebral endplate alone. Shear and torsional forces in combination with compressive loading would be expected to further lower the tolerance of the endplate.

This distribution of risk has been widely used as the tolerance limits of the spine. However, it should be noted that others have identified different limits of vertebral endplate tolerance. Jager, Luttmann, and Laurig (1991) have reviewed 13 studies of spine compressive strength and suggested different compression value limits. Their summary of these spine tolerance limits is shown in Table 10.2. These researchers have also been able to describe the vertebral compressive strength based on an analysis of 262 values collected from 120 samples. They have related the compressive strength of the lumbar spine according to a regression equation:

$$\text{Compressive Strength (kN)} = (7.26 + 1.88\,G) - 0.494 + 0.468\,G) \times A +$$

$$(0.042 + 0.106\,G) \times C - 0.145 \times L - 0.749 \times S$$

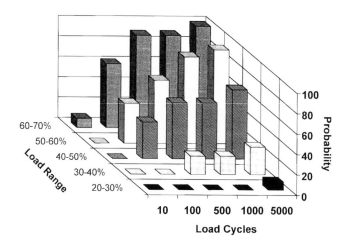

FIGURE 10.11 Probability of a motion segment to be fractured in dependence on the load range and the number of load cycles. (Adapted from Brinckmann, P., Biggemann, M., and Hilweg, D. (1988) Fatigue fracture of human lumbar vertebrae. *Clinical Biomechanics*, 3: Supplement 1, S1-S23.)

where

A = age in decade
G = gender coded as 0 for female or 1 for male
C = cross-sectional area of the vertebrae in cm²
L = the lumbar level unit where 0 is the L5/S1 disc, 1 represents the L5 vertebrae, etc. through 10 which represents the T10/L1 disc
S = the structure of interest where 0 is a disc and 1 is a vertebrae

This analysis suggests that the decrease in strength within a lumbar level is about 0.15 kN of that of the adjacent vertebrae and that the strength of the vertebrae is about 0.8 kN lower than the strength of the discs (Jager et al., 1991). Using this equation, these researchers were able to account for 62% of the variability among the samples.

It has also been suggested that the tolerance limits of the spine varies as a function of frequency of loading (Brinckmann et al.; 1988). Figure 10.11 indicates that the spine tolerance varies as a function of spine load level and frequency of loading.

10.3 The Application of Biomechanics to the Workplace

Biomechanics of Commonly Injured Body Structures

Now that the basic concepts and principles of biomechanics relevant to ergonomics situations have been established we can apply these principles to various work situations. This section will show how one can apply these principles to various regions of the body that are typically affected by occupational tasks.

Shoulder

Shoulder pain is suspected of being one of the most under-recognized musculoskeletal disorders in the workplace. Second only to low back injury and neck pain in clinical frequency and reporting, shoulder region disorders are increasingly being recognized as a major workplace problem by those organizations that have reporting systems sensitive enough to detect such trends. The shoulder is one of the more complex structures of the body with numerous muscles and ligaments crossing the shoulder joint girdle complex. Because of its biomechanical complexity, surgical repair of the shoulder can be problematic. During many shoulder surgeries, it is often necessary to damage much of the surrounding tissue in an

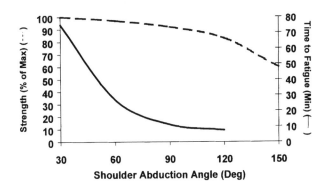

FIGURE 10.12   Shoulder abduction strength and fatigue time as a function of shoulder abducted from the torso. (Adapted from Chaffin, D.B. and Andersson, G.B. (1991) *Occupational Biomechanics*, John Wiley & Sons, Inc. New York, NY.)

attempt to reach the structure in need of repair. Often the target structure is small and difficult to reach. Thus, many times more damage is done to surrounding tissues than the benefit derived to the target tissue. Therefore, the best course of action is to ergonomically design workstations so that risk of initial injury is minimized.

Since the shoulder joint is so biomechanically complex, much of our biomechanical knowledge is derived from empirical evidence. The shoulder represents a statically indeterminate system in that we can typically measure six external moments and forces acting about the point of rotation, yet there are far more internal forces (over 30 muscles and ligaments that must counteract the external moments. Thus, quantitative estimates of shoulder joint loading are rare.

With respect to the shoulder, optimal workplace design is typically defined in terms of preferred posture during work. Shoulder *abduction*, defined as the elevation of the shoulder in the lateral direction, is of concern when work is performed overhead. Figure 10.12 indicates shoulder performance measures in terms of both available strength and perceived fatigue while the shoulder is held in varying degrees of abduction. This figure indicates that the shoulder can produce a considerable amount of strength throughout shoulder abduction angles of between 30 and 90 degrees. However, when we compare fatigue characteristics at these same abduction angles, it is apparent that fatigue increases rapidly as the shoulder is abducted above 30 degrees. Thus, even though strength is not a problem at shoulder abduction angles up to 90 degrees, fatigue becomes the limiting factor. Therefore, the only position of the shoulder that is acceptable from both a strength and fatigue standpoint is a shoulder abduction of, at most, 30 degrees.

Shoulder *flexion* has been examined almost exclusively as a function of fatigue. Chaffin (1973) has shown that even slight shoulder flexion can influence fatigue characteristics of the shoulder musculature. Figures 10.13 and 10.14 indicate the effects of vertical height of the work and horizontal distance, respectively, during shoulder flexion while seated, upon fatiguability of the shoulder musculature. During vertical flexion/extension (Figure 10.13), fatigue occurs more rapidly as the worker's arm becomes more elevated. This trend is most likely due to the fact that the muscles are farther from the neutral position as the shoulder becomes more elevated, thus affecting the length–strength relationship (Figure 10.5) of the shoulder muscles. Figure 10.14 shows that as the horizontal distance between the work and the body is increased, the time to reach significant fatigue is decreased. This trend is due to the fact that, as a load is held further from the body, more of the external moment (force × distance) must be supported by the shoulder. Thus, the shoulder muscles must produce a greater internal force when the load is held farther from the body, and they fatigue more quickly. Elbow supports have been shown to significantly increase the endurance time in these postures. In addition, an elbow support has the effect of changing the biomechanical situation by providing a fulcrum at the elbow. Thus, the axis of rotation becomes the elbow instead of the shoulder, and this makes the external moment much shorter. As shown in Figure 10.15, this not only increases the time one can maintain a posture, but also significantly increases the external load one can hold in the hand.

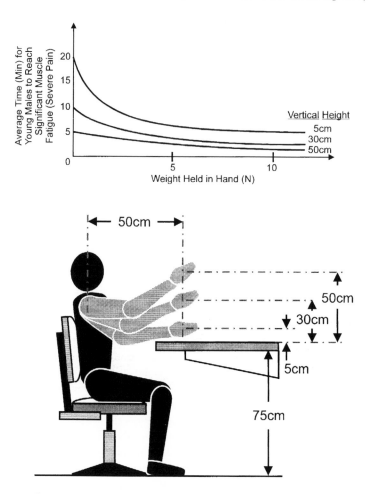

FIGURE 10.13   Expected time to reach significant shoulder muscle fatigue for varied arm flexion postures. (Adapted from Chaffin, D.B. and Andersson, G.B. (1991) *Occupational Biomechanics*, John Wiley & Sons, Inc. New York, NY.)

Neck

Neck disorders can also be associated with sustained work postures. In general, the more upright the posture of the head, the less muscle activity and neck strength is required to maintain the posture. Upright neck postures also have the advantage of reducing the extent of fatigue perceived in the neck region. This relationship is shown in Figure 10.16. This trend indicates that when the head is tilted forward 30 degrees or more from the vertical position, the time to experiencing significant neck fatigue increases rapidly. From a biomechanical standpoint, as the head is flexed forward, the center of mass of the head moves forward relative to the base of support of the head (spine). Therefore, as the head is moved forward, more of a moment is imposed about the spine, which necessitates increased activation of the neck musculature and greater risk probability of fatigue because a static posture is maintained by the neck muscles. When the head is not flexed forward and is relatively upright, the neck can be positioned in such a way that minimal muscle activity is required of the neck muscles and thus fatigue is minimized.

Neck–Shoulder Trade-offs and Work Height

As mentioned earlier, the key to proper ergonomic design of a workplace from a biomechanical standpoint is to consider the biomechanical trade-offs associated with a particular work situation. These trade-offs are necessary because often a situation that is advantageous for one part of the body is disadvantageous for another part of the body. Thus, many biomechanical considerations in the ergonomic design of the workplace require one to consider the various trade-offs and rationales for various design options.

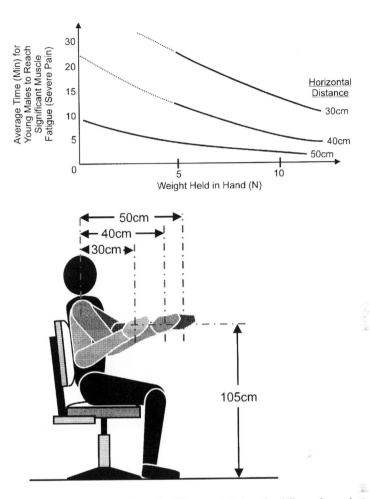

FIGURE 10.14 Expected time to reach significant shoulder muscle fatigue for different forward arm reach postures. (Adapted from Chaffin, D.B. and Andersson, G.B. (1991) *Occupational Biomechanics,* John Wiley & Sons, Inc. New York, NY.)

One of the most common trade-off situations encountered in ergonomic design is the trade-off between accommodating the shoulders and accommodating the neck. This trade-off is often resolved by considering the nature of the work required. Figure 10.17 shows the recommended height of the work as a function of the type of work that is to be performed. Precision work requires visual acuity, which is of prime importance in order for the worker to be able to accomplish the task. If the work is performed at too low a level, the head must be flexed in order to accommodate the visual requirements of the job. This situation could result in significant neck discomfort. Therefore, in this situation the proper work height is dictated by visual acuity requirements, and the work is typically raised to a relatively high level (95 to 110 cm above the floor). This position accommodates the neck but creates a problem for the shoulders because they must be abducted when the work level is high. Thus, a trade-off must be considered. Ideal shoulder posture is sacrificed in order to accommodate the neck because the visual requirements of the job are great. The logic associated with these trade-offs also dictates that the shoulder problems can be minimized by providing wrist or elbow supports at the workplace.

The other extreme of the working height situation involves heavy work. The greatest demand on the worker in heavy work is for a high degree of arm strength, whereas visual requirements in this type of work are typically minimal. Therefore, in this situation, ideal neck posture is typically sacrificed in favor of more favorable shoulder and arm postures. Hence, heavy work is performed at a height of 70 to 90 cm

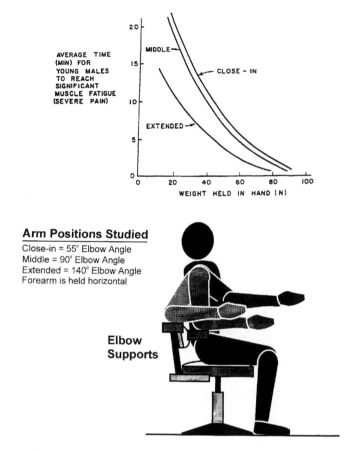

FIGURE 10.15   Expected time to reach significant shoulder and arm muscle fatigue for different arm postures and hand loads with the elbow supported. The greater the reach, the shorter the endurance time. (Adapted from Chaffin, D.B. and Andersson, G.B. (1991) *Occupational Biomechanics*, John Wiley & Sons, Inc. New York, NY.)

above the floor. With the work set at this height, the elbow angles are close to 90 degrees, which maximizes strength (Figure 10.6), and the shoulders are close to 30 degrees of abduction, which minimizes fatigue. In this situation, the neck is not in an optimal position, but the logic dictates that the visual demands of a heavy task would not be substantial and thus the neck should not be flexed for prolonged periods of time. The third work height situation involves light work. Light work is a mix of moderate visual demand with moderate strength requirements. In this situation, work is a compromise between shoulder position and visual accommodation. The height of the work is set at a height between those of the precision work height level and the heavy work height level. This dictates that the work is performed at a level of between 85 and 95 cm off the floor under light work conditions.

The Back

Low back disorders (LBDs) have been labeled as one of the most common and significant musculoskeletal problems in the U.S. that result in substantial amounts of morbidity, disability, and economic loss (Hollbrook et al., 1984; Praemer et al., 1992). Next to the common cold, low back disorders are the most common reason for workers to miss work. Back disorders were responsible for half a billion lost workdays in 1988, with 22 million cases reported that year (Guo, 1993). Among those under 45 years of age, LBD is the leading cause of activity limitations and can affect up to 47% of workers with physically demanding jobs (Andersson, 1991). The prevalence of LBD has also been observed to have increased by 2700% since 1980 (Pope, 1993). The costs associated with LBD are very significant. Estimates of lost wages alone amount to nearly four billion dollars annually (Frymoyer et al., 1983). Recent estimates of the total costs to society from low back pain range from $25 to $95 billion annually (Cats-Baril and Frymoyer, 1991).

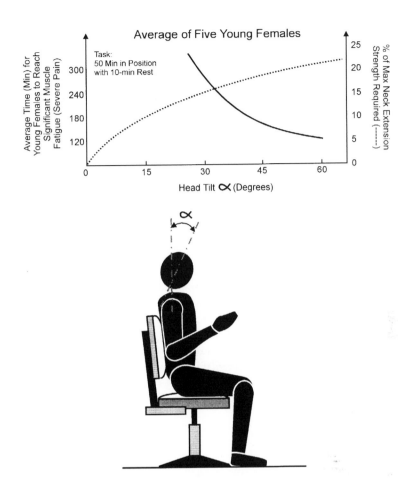

FIGURE 10.16 Neck extensor fatigue and muscle strength required vs. head tilt angle. (Adapted from Chaffin, D.B. and Andersson, G.B. (1991) *Occupational Biomechanics*, John Wiley & Sons, Inc. New York, NY.)

It is clear that the risk of LBD can be associated with industrial work (Andersson, 1981). Thirty percent of occupational injuries in the U.S. are caused by overexertion, lifting, throwing, holding, carrying, pushing, and or pulling objects that weigh 50 pounds or less (National Safety Council, 1989). Twenty percent of all workplace injuries and illnesses are back injuries, which account for up to 40% of compensation costs. Estimates of occupational low back disorder prevalence vary from one to 15% annually, depending upon occupation (Kelsey and White, 1980) and, over a career, can seriously affect 56% of workers (Rowe, 1981).

Manual materials handling (MMH) activities, specifically lifting, dominate occupationally related low back disorder risk. It has been estimated that lifting and MMH account for 50 to 75% of all back injuries (Bigos et al., 1986; Snook, 1989; Spengler et al., 1986). From a biomechanical standpoint, we assume that most serious and costly back pain is discogenic in nature and has a mechanical origin (Nachemson, 1975). Studies have found increased degeneration in the spines of cadaver specimens who had previously been exposed to physically heavy work (Videman et al., 1990). This suggests that occupationally related low back disorders are closely associated with spine loading.

Significance of Moments

The most important concept associated with occupationally related low back disorder risk is that of the external moments imposed about the spine (Marras et al., 1993). As with most body structures, the loading of the trunk is greatly influenced by the external moment imposed about the spine. However, because of the geometric arrangement of the trunk musculature relative to the trunk fulcrum during

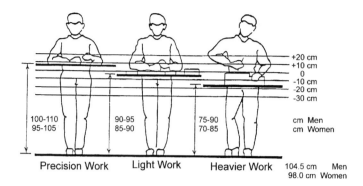

FIGURE 10.17   Recommended heights of bench for standing work. The reference line (+0) is the height of the elbows above the floor (From Grandjean, E. (1982) *Fitting the Task to the Man: An Ergonomic Approach*, Taylor & Francis, Ltd. London. With permission.)

lifting, very large loads can be generated by the muscles and imposed upon the spine. Figure 10.17 shows this biomechanical arrangement of lever system. As indicated here, the back musculature is at a severe biomechanical disadvantage in many manual materials handling situations. Supporting an external load of 222 N (about 50 pounds) at a distance of one meter from the spine imposes a load of 222 Nm of external moment about the spine. However, since the spine-supporting musculature is relatively close to the external load, the trunk musculature must exert extremely large forces (4440 N or 998 lbs.) to simply hold the external load in equilibrium. These internal loads can be far greater if dynamic motion of the body is considered (since force is a product of mass and acceleration). Thus, the most important concept to consider in workplace design from a back protection standpoint is to keep the moment arm at a minimum.

Lifting Style

The external moment concept has major implications for lifting styles, or the best "way" to lift. Since the externally applied moment significantly influences the internal loading, the lifting style is of far less concern compared to the magnitude of the applied moment. Some have suggested that proper lifting involves lifting by "using the legs" as opposed to "stoop" lifting (bending from the waist). However, spine loading has also been found to be a function of anthropometry as well as lifting style. Biomechanical analyses (Park and Chaffin, 1974) have demonstrated that no one lift style is correct for all body types. For this reason, the National Institute of Occupational Safety and Health (NIOSH, 1981) has concluded that lift style need not be a consideration when assessing the risk of occupationally related low back disorder. Some have suggested that the internal moment of the trunk has a greater mechanical advantage when lumbar lordosis is preserved during the lift (McGill, 1986; Anderson and Chaffin, 1986). Thus, from a biomechanical standpoint, the primary indicator of spine loading and, thus, the correct lifting style is whatever style permits the worker to bring the center of mass of the load as close to the spine as possible.

Seated versus Standing Workplaces

Seated workplaces have become more prominent of late, especially with the aging of the workforce and the introduction of service-oriented and data processing jobs. It has been well documented that loads on the lumbar spine are always greater when one is seated compared to standing. This is due to the tendency for the posterior (bony) elements of the spine to form an active load path when one is standing. When one is seated, these elements are disengaged and more of the load passes through the intervertebral disc. Thus, work performed in a seated position puts the worker at greater risk of loading and therefore damaging the disc. Given this situation, it is important to consider the design features of a chair since it may be possible to influence disc loading through chair design. Figure 10.19 shows the results of pressure measurements made in the intervertebral disc of workers as the back angle of the chair and magnitude

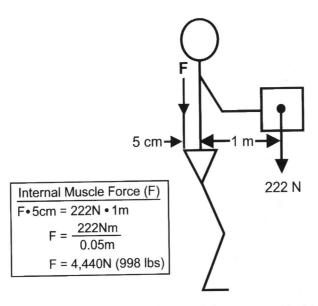

FIGURE 10.18 Internal muscle force required to counterbalance an external load during lifting.

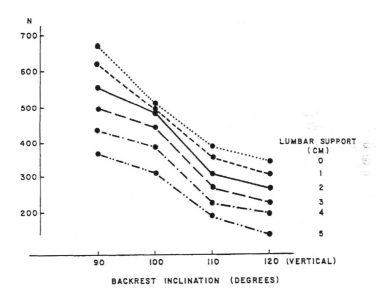

FIGURE 10.19 Disc pressures measured with different backrest inclinations and different size lumbar supports. (From Chaffin, D.B. and Andersson, G.B. (1991) *Occupational Biomechanics,* John Wiley & Sons, Inc. New York, NY. With permission.)

of the lumbar support were varied. This figure indicates that both the seat back angle and lumbar support features have a significant effect on disc pressure. Disc pressure is observed to decrease as the backrest angle is increased. However, increasing the backrest angle in the workplace is often not practical because it also moves the worker away from the work and thereby increases external moment. However, the figure also indicates that increasing lumbar support can also significantly reduce disc pressure. This reduction in pressure is most likely due to the fact that as lumbar curvature (lordosis) is reestablished (with lumbar support), the posterior elements play more of a role in providing an alternative load path, as is the case when standing in the upright position.

Less is known about risk to the low back associated with prolonged standing. It is known that the muscles experience low level static exertions and may be subject to the static overload through the muscle static fatigue process described in Figure 10.9. This fatigue can result in lowered muscle force-generation capacity and can, thus, initiate the cumulative trauma sequence of events (Figure 10.3). It has been demonstrated that this fatigue and cumulative trauma sequence can be minimized by two actions. First, foot rails provide a mechanism to allow relaxation of the large back muscles and, thus, increased blood flow to the muscle. This reduces the static load and fatigue in the muscle by the process described in Figure 10.9. When a leg is lifted and rested on the foot rest, the large back muscles are relaxed on one side of the body and the muscle can be supplied with oxygen. Alternating legs on the foot rest provides a mechanism to minimize back muscle fatigue throughout the day. Second, floor mats have been shown to decrease the fatigue in the back muscles provided that they have proper compression characteristics (Kim et al., 1993). Floor mats are believed to induce body sway which facilitates the pumping of blood through back muscles, thereby minimizing fatigue.

Our knowledge of when standing workplaces are preferable is dictated mainly by work performance criteria. In general, standing workplaces are preferred when: 1) the task requires a high degree of mobility (reaching and monitoring in positions that exceed the reach envelope or when performing tasks at different heights or different locations), 2) precise manual control actions are not required, 3) leg room is not available, (when leg room is not available the moment arm distance between the external load and the back is increased and thus greater internal back muscle force and spinal load result), and 4) heavy weights are handled or large forces are applied. When jobs must accommodate both sitting and standing, it is important to ensure that the positions and orientations of the body, especially the upper extremity, are in the same location under both standing and sitting conditions.

Wrists

The wrist has been of increased interest to ergonomists for the past two decades. The Bureau of Labor Statistics reports that repetitive trauma has increased from 18% of occupational illnesses in 1981 to 63% of occupational illnesses in 1993. Based upon these figures, repetitive trauma has been described as the *fastest growing* occupational problem. Even though these numbers and statements appear alarming, one must acknowledge that occupational illnesses represent 6% of all occupational injuries and illnesses. Furthermore, these figures for illness include illnesses unrelated to musculoskeletal disorders such as noise-induced hearing loss. Thus, the magnitude of the cumulative trauma problem must not be overstated. Nonetheless, there are specific industries (i.e., meat packing, poultry processing, etc.) in which cumulative trauma to the wrist is a major problem, and this problem has reached epidemic proportions within these industries.

Wrist Anatomy and Loading

In order to understand the biomechanics of the wrist and how cumulative trauma occurs in this structure, one must appreciate the anatomy of the upper extremity. Figure 10.20 shows a simplified anatomical drawing of the wrist. This figure shows that few power producing muscles reside in the hand itself. The thenar muscle which activates the thumb is one of the few power-producing muscles in the hand. The vast majority of the power-producing muscles of the hand are located in the forearm. Force is transmitted from these forearm muscles to the fingers through a network of tendons (tendons attach muscles to bone). These tendons originate at the muscles in the forearm, transverse the wrist (with many of them passing through the carpal canal), pass through the hand, and culminate at the fingers. These tendons are secured or "strapped down" at various points along this path with ligaments that keep the tendons in close proximity to the bones. This system results in a hand that is very small and compact yet capable of generating large amounts of force. The price the musculoskeletal system pays for this design is friction. The forearm muscles must transmit force over a very long distance in order to supply internal forces to the fingers. Thus, a great deal of tendon travel must occur, and this tendon travel can result in significant tendon friction under repetitive motion conditions, thereby initiating the events outlined in Figure 10.3. Thus, the key to controlling wrist cumulative trauma is rooted in an understanding of those workplace factors that adversely affect the internal force-generating (muscles) and transmitting (tendons) structures.

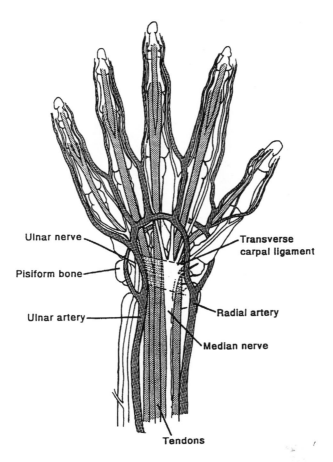

Ulnar nerve

Pisiform bone

Ulnar artery

Transverse carpal ligament

Radial artery

Median nerve

Tendons

FIGURE 10.20   Important anatomical structures in the wrist.

Biomechanical Risk Factors for the Wrist

A number of risk factors for wrist cumulative trauma have been documented in the literature. First, deviated wrist postures are known to reduce the volume of the carpal tunnel and, thus, increase tendon friction. In addition, grip strength is dramatically reduced by deviations in the wrist posture. Figure 10.19 indicates that any deviation from the wrist's neutral position significantly decreases the grip strength of the hand. This reduction in strength is caused by a change in the length–strength relationship (Figure 10.5) of the forearm muscles once the wrist is bent. Hence, the muscles are working at a level that is greater than necessary. This reduced strength potential associated with deviated wrist positions can, therefore, more easily initiate the sequence of events associated with cumulative trauma (Figure 10.3). Thus, deviated wrist postures not only increase tendon travel and friction, but also for a given grip strength requirement, they increase the percentage of muscle activity and relative percentage of muscle force necessary to grip securely.

Second, increased frequency or repetition of the work cycle has been identified as a risk factor for cumulative trauma disorders (Silverstein et al., 1986, 1987). Studies have indicated that increased frequency of wrist motions increases the risk of developing a cumulative trauma disorder. Repeated motions with a cycle time of less than 30 seconds is considered a candidate for cumulative trauma disorder risk (Putz-Anderson, 1984).

Third, the force applied by the hands and fingers during a work cycle has been identified as a risk factor. In general, the greater the force required by the work the greater the risk of CTD. Greater hand forces result in greater tension within the tendons and result in greater tendon friction and tendon travel.

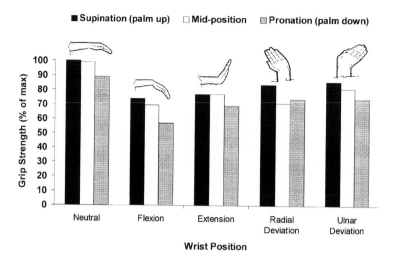

FIGURE 10.21   Grip strength as a function of wrist and forearm position. (Adapted from Sanders, M.S. and McCormick, E.J. (1993) *Human Factors in Engineering and Design*, McGraw-Hill Inc., New York, NY.)

Another factor related to force is wrist acceleration. Industrial surveillance studies have reported that repetitive jobs resulting in greater wrist acceleration are associated with greater cumulative trauma disorder incident rates (Marras and Schoenmarklin, 1993; Schoenmarklin et al., 1994). Since force is a product of mass and acceleration, jobs that increase the angular acceleration of the wrist joint result in greater tension and force transmitted through the tendons. Thus, wrist acceleration can be another mechanism of imposing force on the wrist structures.

Fourth, as shown in Figure 10.20, the anatomy of the hand is such that the median nerve becomes very superficial at the palm. Direct impact to the palm of the hand through pounding or striking an object with the palm, as is often done in assembly work, can directly stimulate the median nerve and initiate symptoms of cumulative trauma even though the work may not be repetitive.

Grip Design

The design of a tool's gripping surface can dramatically affect the activity of the internal force transmission system (tendon travel and tension). The grip opening and shape have a major influence on the available grip strength. Figure 10.21 shows how grip strength capacity changes as a function of the separation distance of the grip opening. This figure indicates that maximum grip strength occurs within a very narrow range of grip openings. If the grip opening deviates from this ideal range by as little as an inch (a couple of centimeters), then grip strength is dramatically reduced. This change in strength is also due to the length–strength relationship of the forearm muscles. Also indicated in Figure 10.22 are the effects of hand anthropometry. The worker's hand size as well as hand preference can influence grip strength and risk. Therefore, proper design of the handles is crucial in ergonomic workplace design.

Handle shape can also affect the strength of the wrist. Figure 10.23 shows how changes in the design of screwdriver handles can affect the maximum force that can be exerted. The biomechanical origin of these differences in strength capacity are most likely related to the length–strength relationship of the forearm muscles as well as contact area with the tool. The handle designs that result in less strength probably permit the wrist to twist or the grip to slip, resulting in a deviation from the ideal length–strength position in the forearm muscles.

Gloves

The use of gloves can significantly influence the generation of grip strength and may play a role in the development of cumulative trauma disorders. When gloves are worn, three effects must be considered. First, the grip strength generated is often reduced. There is typically a 10 to 20% reduction in grip strength when gloves are worn. When using gloves, the coefficient of friction between the hand and the tool can

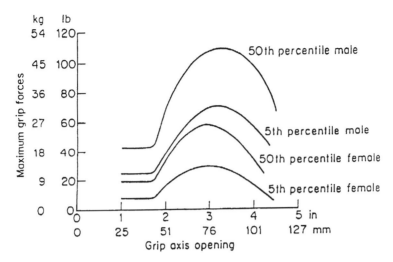

FIGURE 10.22   Grip strength as a function of grip opening and hand anthropometry. (Adapted from Sanders, M.S. and McCormick, E.J. (1993) *Human Factors in Engineering and Design*, McGraw-Hill Inc., New York, NY.)

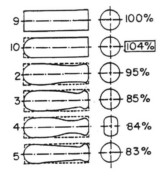

FIGURE 10.23   Maximum force which could be exerted on a screwdriver as a function of handle shape. (From Konz, S.A. (1983) *Work Design: Industrial Ergonomics*, Second Edition, Grid Publishing, Inc., Columbus, OH. With permission.)

be reduced, which in turn permits some slippage of the hand on the tool surface. This slippage can result in a deviation from the ideal muscle length and thus a reduction in available strength. The degree of slippage and the degree of strength lost depends upon how well the gloves fit the hand and the type of material used in the glove. Poorly fitting gloves result in greater strength loss.

Second, when wearing gloves, even though the externally applied force (grip strength) is often reduced, the internal forces are often very large compared to not using a glove. Studies have indicated that, for a given grip strength, the muscle activity is significantly greater when using gloves compared to a bare-handed condition (Sudhakar et al., 1988). Thus, the musculoskeletal system is less efficient when wearing a glove.

Third, the ability to perform a task is generally negatively affected when wearing gloves. Figure 10.24 shows the increase in time required to perform tasks when wearing gloves composed of different materials compared to performing the task bare-handed. The figure indicates that task performance time can increase up to 70% when wearing gloves.

These effects have indicated that there are biomechanical costs associated with glove usage. Less strength capacity is available to the worker, more internal force is generated, and worker productivity is affected. These negative effects of gloves do not mean that gloves should never be worn at work. When

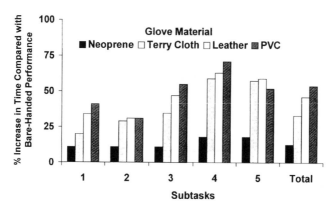

FIGURE 10.24   Performance (time to complete) on a maintenance-type task while wearing gloves constructed of five different material. (From Sanders, M.S. and McCormick, E.J. (1993) *Human Factors in Engineering and Design*, McGraw-Hill Inc., New York, NY. With permission.)

hand protection is needed, gloves should be considered as a potential solution. However, protection should only be provided to the parts of the hand that require it. For example, if the palm of the hand requires protection, fingerless gloves might provide an acceptable solution. If the fingers require protection but there is little risk to the palm of the hand, then grip tape wrapped around the fingers might be considered. In addition, different styles, materials, and sizes of gloves will fit workers differently. Thus, gloves produced by various manufacturers and of different sizes should be available to the worker to minimize the negative effects mentioned above.

Design Guidelines

This discussion has indicated that there are many factors that can affect the biomechanics of the wrist and the subsequent risk of cumulative trauma disorders. This suggests that proper ergonomic design of a work task cannot be accomplished by simply providing the worker with a "ergonomically designed" tool. Since ergonomics is associated with matching the workplace design to the worker's capabilities, it is not possible to design an "ergonomic tool" without considering the workplace design and task requirements simultaneously. What might be an "ergonomic" tool for one work situation may be improper while a worker is assuming another work posture. For example, using an *in-line* tool may keep the wrist straight when inserting a bolt into a horizontal surface. However, if the bolt is to be inserted into a vertical surface a *pistol grip* tool may be more appropriate. Using the in-line tool in this situation (inserting a bolt into a vertical surface) may cause the wrist to be significantly deviated. Hence, there are no ergonomic tools. There are just ergonomic *situations*. What may be an ergonomically correct tool in one situation may not be ergonomically correct in another. Thus, workplace design should be performed with care and trade-offs between different parts of the body must be considered by taking into consideration the various biomechanical trade-offs. Given these considerations, the following components of the workplace should be considered when designing a workplace so that cumulative trauma risk is minimized. First, maintain a neutral wrist posture. Second, minimize tissue compression. Third, avoid actions that repeatedly impose force on the internal structures. Fourth, minimize required wrist accelerations and motions. Fifth, consider the impact of glove use, hand size, and left-handed workers.

10.4   Analysis and Control Measures Used in the Workplace

Several analysis and control measures have been developed to evaluate and control biomechanical loading of body during work. Since low back disorders are often the objective of a biomechanical workplace analysis, most of these analysis methods have focused on spine risk. However, several of the measures also include analyses of risk to other body parts.

Lift Belts

Back support belts or lifting belts have been used with increasing frequency in the workplace. There exists a great deal of controversy as to whether use of these belts is a benefit or a liability during manual materials handling. A review of the literature related to lifting belts offers no clear answer as to the benefits of belt use. Reviews by McGill (1993) and NIOSH (1994) have concluded that there are so few well executed studies that one cannot unequivocally judge the benefits of lifting belts. Therefore, what is known about these devices can be summarized and used as a basis for an informed opinion.

Epidemiological studies have generally been limited in scope and often result in findings that were confounded by other factors such as training, the type of belt used, or the "Hawthorne Effect." Walsh and Schwartz (1990) reported a reduction in low back disorder (LBD) injury rate with the usage of back supports (hard shell corsets) and have recommended that they be used in controlling the risk of low back disorder. However, the data from this study suggest that back supports were only effective for those workers who had previously suffered a low back disorder. Mitchell et al. (1994) retrospectively evaluated injury data associated with belt use over a six year period at Tinker Air Force Base. Over this period, two different types of belts were used. Leather belts were used in the first two years of the study, and Velcro belts were used over the last four years. No relationship between belt usage and back injury could be established, but they did find that those who wore belts suffered more costly injuries once they occurred. Riddle et al. (1992) observed that when workers stopped wearing belts the risk of injury increased. However, this study suffers from small sample size, which makes it difficult to assess the strength of the association. More recently, Straus et al. (1996) in a large prospective study found that lift belts significantly reduced the risk of low back pain in a chain of home improvement stores. However, many unresolved study design questions are also associated with this study. Unfortunately, none of these epidemiologic studies may be considered conclusive because many of them suffer from low participation rates, inadequate observation periods, confounding with training, small sample size, low back-injury rates, improper counterbalancing of experimental treatments, uncontrolled work tasks, reporting bias, and/or previous back injury history (NIOSH, 1994).

Psychophysical studies have attempted to assess whether the magnitude of the weight a person was willing to lift changes when wearing a back belt. McCoy et al. (1988) found that subjects were willing to lift 19% more weight when belts were used but found no difference between belt types. Subjects reported that they preferred the elastic belt. However, this does not suggest that workers would be at lowered risk of back injury because it is not clear that spine tolerance to load would be increased with belt use.

Biomechanically based studies of lifting belts have documented their influence upon trunk motion, trunk muscle activity, and indirect indicators or predictions of trunk loading. The most consistent finding of these studies is that side bending and twisting trunk motion is significantly reduced with belt usage (McGill et al., 1994; Lavender et al., 1995; Lantz and Schultz, 1986). A recent study has indicated that for some subjects there may be a slight reduction in spine loading when lifting asymmetrically without moving the feet. However, this trend was not true for all subjects. Some subjects increased their spine loading under these conditions (Granata et al., 1996). Thus, if belts do have the potential to reduce spine loading, they do not appear to do so in all workers and may increase loading in some workers.

Perhaps the most important reason to be cautious of lifting belts is unrelated to biomechanical loading of the spine. There appear to be physiological reasons to be concerned within the use of lifting belts. One study has shown that lifting belts can significantly increase blood pressure. This could become problematic for workers who have a compromised cardiovascular system.

The brief review indicates that there is a large amount of conflicting evidence as to the benefits or liabilities associated with the use of back belts. A consistent finding among the studies is that if there is a benefit to back belts it is probably for those who have previously experienced a low back disorder. The literature also suggests that belts should only be used for a limited period of time. Until more definitive studies are available, it is prudent to use caution when recommending the use of back belts in a work environment. This includes a screening by an occupational physician who is familiar with the literature so that potential cardiovascular problems can be assessed.

TABLE 10.3   Fmax Table

Period	Average Vertical Location (cm) (in)	
	Standing V > 75 (3)	Stooped V ≤ 75 (3)
1 Hour	18	15
8 Hours	15	12

Reprinted from NIOSH, *Work Practices Guide for Manual Lifting*

1981 NIOSH Lifting Guide

The National Institute for Occupational Safety and Health (NIOSH) has developed two assessment tools or guides to help determine whether a manual materials handling task is safe or risky. The lifting guide was originally developed in 1981 (NIOSH, 1981) and applies to lifting situations in which the lifts are performed in the sagittal plane and to motions that are slow and smooth. Two benchmarks or limits are defined by this guide. The first limit is called the *action limit* (AL) and represents a magnitude of weight in a given lifting situation which would impose a spine load corresponding to the beginning of low back disorder risk along a risk continuum. The AL is associated with the point in Figure 10.10 at which people under 40 years of age just begin to experience a risk of vertebral endplate microfracture (350 kg of compressive load). The weight of an object to be lifted by a worker in a given task is compared to the AL. If the weight of the object is below that of the AL, the job is considered safe. If the weight lifted by the worker is larger than the AL, there is at least some level of risk associated with the task. The general form of the AL is defined according to equation (1).

$$AL = k \, (HF)(VF)(DF)(FF) \qquad\qquad (10.1)$$

where

AL = The action limit in kg or pounds
k = Load constant (40 kg or 90 lbs.) which is the greatest weight a subject could lift if all lifting conditions are optimal.
HF = Horizontal factor defined as the horizontal distance from a point bisecting the ankles to the center of gravity of the load at the lift origin. Defined algebraically as 15/H (metric) or 6/H (U.S. units).
VF = Vertical factor or height of the load at lift origin. Defined algebraically as $(.004) \, |V - 75|$ (metric) or $1 - (.01) \, |V - 30|$ (U.S. units).
DF = Distance factor or the vertical travel distance of the load. Defined algebraically as .7 + 7.5/D (metric) or .7 + 3/D (U.S. units).
FF = Frequency factor or lifting rate defined algebraically as 1 − F/Fmax
F = average frequency of lift, Fmax is shown in Table 10.3

The logic associated with this equation assumes that if the lifting conditions are ideal, a worker could safely hold (and this implies lift) the load constant, k, (40 kg or 90 lbs.). If the lifting conditions are not ideal, the allowable weight is discounted according to the four factors HF, VF, DF, and FF. These four factors are shown in monogram form in Figures 10.25 through 10.28. According to the load discounting associated with these figures, the HF which is associated with the external moment has the most dramatic effect on acceptable lifting conditions. VF and DF are associated with the back muscle's length–strength relationship. FF attempts to account for the cumulative effects of repetitive lifting.

The second benchmark associated with this guide is the *maximum permissible limit* or MPL. The MPL represents the point at which significant risk, defined in part as a significant risk of vertebral endplate microfracture (Figure 10.10). The MPL is associated with a compressive load on the spine of 650 kg,

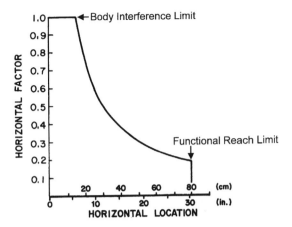

FIGURE 10.25 Horizontal Factor (HF) varies between the body interference limit and the limit of functional reach. (Adapted from National Institute for Occupational Safety and Health (NIOSH) (1981) Work practices guide for manual lifting. Department of Health and Human Services (DHHS), National Institute for Occupational Safety and Health (NIOSH), Publication No. 81-122.)

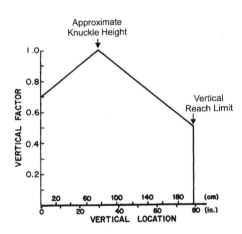

FIGURE 10.26 Vertical Factor (VF) varies both ways from knuckle height. (Adapted from National Institute for Occupational Safety and Health (NIOSH) (1981) Work practices guide for manual lifting. Department of Health and Human Services (DHHS), National Institute for Occupational Safety and Health (NIOSH), Publication No. 81-122.)

which corresponds to a point at which 50% of the people would be expected to suffer a vertebral endplate microfracture. Equation (2) indicates that the MPL is a function of the AL and is defined as follows:

$$MPL = 3 \, (AL) \tag{10.2}$$

The weight that the worker expected to lift in a work situation is compared to the AL and MPL. If the magnitude of weight falls below the AL, the work is considered safe and no adjustments are necessary. If the magnitude of the weight falls above the MPL, then the work is considered risky and engineering changes involving the adjustment of HF, VF, and/or DF are required to reduce the AL and MPL. If the weight falls between the AL and MPL, then either engineering changes or administrative changes, defined as selecting workers who are less likely to be injured or rotating workers, are recommended. The AL and MPL were also indexed to non-biomechanical benchmarks. According to NIOSH (1981), these limits also correspond to strength, energy expenditure, and psychophysical acceptance points.

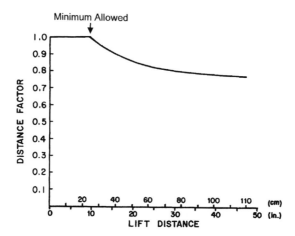

FIGURE 10.27   Distance factor (DF) varies between a minimum vertical distance moved of 25 cm (10 in.) to a maximum distance of 200 cm (80 in). (Adapted from National Institute for Occupational Safety and Health (NIOSH) (1981) Work practices guide for manual lifting. Department of Health and Human Services (DHHS), National Institute for Occupational Safety and Health (NIOSH), Publication No. 81-122.)

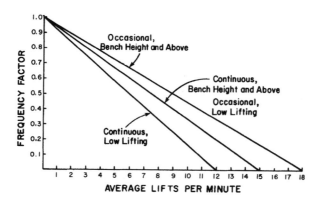

FIGURE 10.28   Frequency Factor (FF) varies with lifts/minute and the F_{max} curve. The F_{max} curve depends upon lifting posture and lifting time. (From National Institute for Occupational Safety and Health (NIOSH) (1981) Work practices guide for manual lifting. Department of Health and Human Services (DHHS), National Institute for Occupational Safety and Health (NIOSH), Publication No. 81-122.)

1991 Revised NIOSH Equation

The 1991 NIOSH revised lifting equation was introduce in order to address those lifting jobs that violate the sagittally symmetric lifting assumption (Waters et al., 1993). The concept of AL and MPL was replaced with a concept of a *lifting index* or LI. The LI is defined in equation (3).

$$LI = \frac{L}{RWL} \qquad (10.3)$$

where

L = Load weight or the weight of the object to be lifted.
RWL = Recommended Weight Limit for the particular lifting situation.
LI = Lifting Index used to estimate relative magnitude of physical stress for a particular job.

TABLE 10.4   Frequency Multiplier Table (FM)

Frequency Lifts/min (F)[‡]	Work Duration					
	≤1 hour		>1 but ≤2 hours		>2 but ≤8 hours	
	V < 30[†]	V ≥ 30	V < 30	V ≥ 30	V < 30	V ≥ 30
≥0.2	1.00	1.00	.95	.95	.85	.85
0.5	.97	.97	.92	.92	.81	.81
1	.94	.94	.88	.88	.75	.75
2	.91	.91	.84	.84	.65	.65
3	.88	.88	.79	.79	.55	.55
4	.84	.84	.72	.72	.45	.45
5	.80	.80	.60	.60	.35	.35
6	.75	.75	.50	.50	.27	.27
7	.70	.70	.42	.42	.22	.22
8	.60	.60	.35	.35	.18	.18
9	.52	.52	.30	.30	.00	.15
10	.45	.45	.26	.26	.00	.13
11	.41	.41	.00	.23	.00	.00
12	.37	.37	.00	.21	.00	.00
13	.00	.34	.00	.00	.00	.00
14	.00	.31	.00	.00	.00	.00
15	.00	.28	.00	.00	.00	.00
>15	.00	.00	.00	.00	.00	.00

[†]Values of V are in inches. [‡]For lifting less frequently than once per 5 minutes, set F = .2 lifts/minute.

Reprinted from NIOSH, *Applications Manual for the Revised NIOSH Lifting Equation.*

If the LI is greater than 1.0, an increased risk for suffering a lifting-related low back disorder exists. The RWL is similar to the 1981 Lifting Guide AL equation (Equation 1) in that it contains factors that discount the allowable load according to the horizontal distance, vertical location of the load, vertical travel distance, and frequency of lift. However, the form of these discounting factors was changed. Moreover, two additional discounting factors have been included. These additional factors include a lift asymmetry factor to account for asymmetric lifting conditions and a coupling factor that accounts for whether or not the load lifted has handles. The RWL is represented algebraically in equations (4) (metric units) and (5) (U.S. units).

$$RWL(kg) = 23(25/H)(1-(0.003|V-75|))(.82+4.5/D))(FM)(1-(0.0032A))(CM) \quad (10.4)$$

$$RWL(lb) = 51(10/H)(1-(0.0075|V-30|))(.82+1.8/D))(FM)(1-(0.0032A))(CM) \quad (10.5)$$

where

H = Horizontal location forward of the midpoint between the ankles at the origin of the lift. If significant control is required at the destination, then H should be measured both at the origin and destination of the lift.

V = Vertical location at the origin of the lift.

D = Vertical travel distance between origin and destination of the lift.

FM = Frequency multiplier shown in Table 10.4.

A = Angle between the midpoint of the ankles and the midpoint between the hands at the origin of the lift.

CM = Coupling multiplier ranked as either good, fair, or poor and described in Table 10.5.

In this revised equation, the load constant has been significantly reduced compared to the 1981 equation. The adjustments for load moment, muscle length–strength relationships, and cumulative loading are still integral parts of this equation. However, these adjustments or discounting factors have

TABLE 10.5   Coupling Multiplier

	Coupling Multiplier	
Coupling Type	V < 30 inches (75 cm)	V ≥ 30 inches (75 cm)
Good	1.00	1.00
Fair	0.95	1.00
Poor	0.90	0.90

Reprinted from NIOSH, *Application Manual for Revised NIOSH Equation, 1994.*

been changed (compared to the 1981 guide) to reflect the most conservative value of the biomechanical, physiological, psychophysical, or strength data upon which they are based. Recent studies report that the 1991 revised equation yields a more conservative (protective) prediction of work-related low back disorder risk (Marras et al., 1997).

2D/3D Static Models

Biomechanically based spine models have been developed to help assess occupationally related manual materials handling tasks. These models assess the task based upon both spine loading criteria as well as through an evaluation of the strength required at the various major body joints in order to perform the task. One of the early static assessment models was developed by Chaffin at the University of Michigan (1969). Both two-dimensional (2D) as well as three-dimensional (3D) static models (Chaffin and Muzaffer, 1991) have been developed to help assess the risk of injury during manual materials handling activities. In both models, the moments imposed upon the various joints of the body due to the object lifted are evaluated assuming that a static posture is representative of the instantaneous loading of the body. These models then compare the imposed moments about each joint with the static strength capacity derived from a working population. The static strength capacity of the major joints assessed by this model have been documented in a database of over 3000 workers. In this manner the proportion of the population capable of performing a particular static exertion is predicted. In addition, the joint that limits the capacity to perform the task can be identified via this method. These models assume that a single equivalent muscle (internal force) supports the external moment about each joint. By considering the contribution of the externally applied load and the internally generated single muscle equivalent, spine compression acting on the lumbar discs is predicted. The predicted compression can then be compared to the tolerance limits of the vertebral endplate (Figure 10.10). The 2D version of this computer model assumes (as does the 1981 NIOSH Lifting Guide) that all lifts occur directly in front of the worker in the sagittal plane. Another important assumption of these models is that no significant motion occurs during the exertion because it is a static model. Figure 10.29 shows the output screen for this computer model where the lifting posture, lifting distances, strength predictions, and spine compression are shown. The 3D version of the computer model works in a similar manner, however, non-sagittal symmetric lifting assumptions are permitted.

Multiple Muscle System Models

One of the significant simplifying assumptions inherent in most static models is that the coactivation of the trunk musculature during a lift is negligible. The trunk is truly a multiple muscle system with many major muscle groups supporting and loading the spine. This can be seen in the cross-section of the trunk shown in Figure 10.30. Studies have shown that there is significant coactivation occurring in many of the major muscle groups in the trunk during realistic *dynamic* lifting (Marras and Mirka, 1993). This coactivation is important because all the trunk muscles have the ability to load the spine. Thus, assumptions regarding single equivalent muscles within the trunk can lead to erroneous conclusions about spine loading during a task. A recent study has indicated that ignoring the coactivation of trunk muscles during

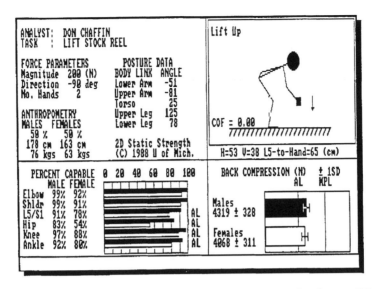

FIGURE 10.29 The 2D Static Strength Prediction Model. (From Chaffin, D.B. and Andersson, G.B. (1991) *Occupational Biomechanics*, John Wiley & Sons, Inc. New York, NY. With permission.)

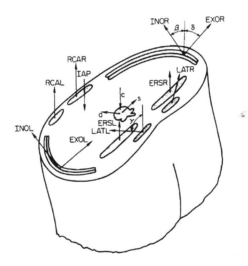

FIGURE 10.30 Cross-sectional view of the human trunk at the lumbrosacral junction. (Adapted from Schultz, A.B. and Andersson, G.B.J. (1981) Analysis of loads on the lumbar spine. *Spine*, 6:76-82.)

dynamic lifting can misrepresent spine loading by 45 to 70% (Granata and Marras, 1995). In an effort to more accurately estimate the loads on the lumbar spine, especially under complex, changing (dynamic) postures, multiple muscle system models of the trunk have been developed. Much of the research in the past decade has been centered around predicting how the multiple trunk muscles coactivation during dynamic lifting.

EMG-Assisted Multiple Muscle System Models

People recruit their muscles in various manners when moving dynamically. For example, when moving slowly the agonist muscle may dominate the muscles' activities during a lift. However, when moving cautiously, asymmetrically, or rapidly there may be a great deal of antagonistic coactivation present. During occupational lifting tasks, these latter dynamic conditions are typically the rule rather than the exception during lifting. For these reasons it has been virtually impossible to predict the instantaneous coactivation and resultant loading on the spine during dynamic trunk exertions. One of the few means to accurately account for the effect of the trunk muscle system coactivation upon spine loading is through the use of biologically assisted models. The most common of these models are electromyographic or EMG-assisted models. These models take into account the individual recruitment patterns of the muscles during a specific lift for a specific individual. By directly monitoring muscle activity, the EMG-assisted model can determine individual muscle force and the subsequent spine loading. These models have been developed and tested under bending and twisting dynamic motion conditions and have been validated (McGill and Norman, 1985; McGill and Norman, 1986; Marras and Reilly, 1988; Reilly and Marras, 1989; Marras and Sommerich, 1991a; 1991b; Granata and Marras, 1993; Marras and Granata, 1995, 1997). Figure 10.31 shows how such models can assess the effects of lifting dynamics upon spine loading. These models are the only ones that can predict the *multidimensional loads* on the lumbar spine under many three-dimensional complex dynamic lifting conditions. The limitation of such models is that it is that they require significant instrumentation of the worker.

Dynamic Motion and Low Back Disorder

As discussed throughout this chapter, it is clear that dynamic activity may increase the risk of low back disorder. However, in order to control this biomechanical situation at the worksite, one must know the type of motion that increases biomechanical load and how much is too much motion from a biomechanical standpoint. These issues were the focus of an industrial retrospective study performed over a six-year period in 68 different industrial environments. Trunk motion and workplace conditions were assessed in workers exposed to high risk of low back disorder jobs and compared to trunk motions and workplace conditions of low risk jobs (Marras et al., 1993, 1995). A trunk goniometer (lumbar motion monitor or LMM) that has been used to document the trunk motion patterns of workers in the workplace is shown in Figure 10.32. Trunk motion and workplace conditions associated with the high-risk and low-risk environments are listed in Table 6. Based on these findings, a five-factor multiple logistic regression model was developed that is capable of discriminating between tasks that indicate probability of high-risk group membership. These factors include: 1) frequency of lifting, 2) load moment (load weight multiplied by the distance of the load from the spine), 3) average twisting velocity (measured by the LMM), 4) maximum sagittal flexion angle through the job cycle (measured by the LMM), and 5) maximum lateral velocity (measured by the LMM). This LMM risk assessment model is the only model capable of assessing the *risk* of three-dimensional trunk motion on the job. This model has been shown to have a high degree of predictability (odds ratio = 10.7) compared to previous attempts to assess work-related low back disorder risk. The advantage of this assessment is that the evaluation provides information about risk that would take years to derive from historical accounts of incidence rates.

10.5   Summary

This chapter has shown that biomechanics provides one of the few means to *quantitatively* consider the implications of workplace design. Biomechanical design is important when a particular job is suspected of imposing large or repetitive forces on a particular structure of the body. It is particularly important to recognize that the internal structures of the body, such as muscles, are the primary loaders of the joint and tendon structures. In order to evaluate the risk of injury from a particular task, one must consider the contribution of both the external loads and internal loads upon the structure. Several quantitative models and assessment methods have been developed that systematically consider the internal loading imposed on the worker due to workplace layout and task requirements. Proper use of these models and methods involves recognizing the limitations and assumptions of each technique so that they are not applied inappropriately. When properly used, these assessments can help assess the risk of work-related injury and illness.

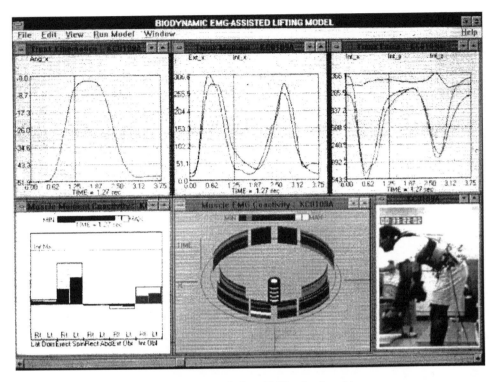

FIGURE 10.31 Windows EMG-assisted model.

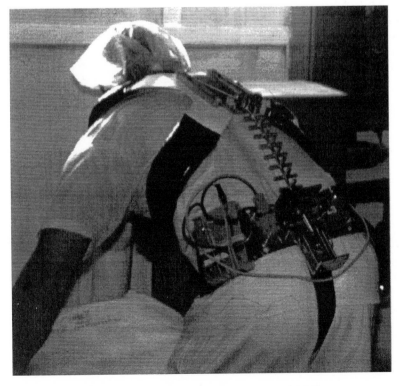

FIGURE 10.32 The lumbar motion monitor (LMM).

TABLE 10.6 Descriptive and t Statistics of the Workplace and Trunk Motion Factors in Each of the Risk Groups

Factors	High Risk (N = 111)				Low Risk (N = 124)				Statistics t
	Mean	SD	Minimum	Maximum	Mean	SD	Minimum	Maximum	
WORKPLACE FACTORS									
Lift rate (lifts/hr)	175.89	8.65	15.30	900.00	118.83	169.09	5.40	1500.00	2.1*
Vertical load location at origin (m)	1.00	0.21	0.38	1.80	1.05	0.27	0.18	2.18	1.4
Vertical load location at destination (m)	1.04	0.22	0.55	1.79	1.15	0.26	0.25	1.88	3.2†
Vertical distance traveled by load (m)	0.23	0.17	0.00	0.76	0.25	0.22	0.00	1.04	0.8
Average weight handled (N)	84.74	79.39	0.45	423.61	29.30	48.87	0.45	280.92	6.4†
Maximum weight handled (N)	104.36	88.81	0.45	423.61	37.15	60.83	0.45	325.51	6.7†
Average horizontal distance between load and L5-S1 (N)	0.66	0.12	0.30	0.99	0.61	0.14	0.33	1.12	2.5*
Maximum horizontal distance between load and L5-S1 (N)	0.76	0.17	0.38	1.24	0.67	0.19	0.33	1.17	3.7†
Average moment (Nm)	55.26	51.41	0.16	258.23	17.70	29.18	0.17	150.72	6.8†
Maximum moment (Nm)	73.65	60.65	0.19	275.90	23.64	38.62	0.17	198.21	7.4†
Job satisfaction	5.96	2.26	1.00	10.00	7.28	1.95	1.00	10.00	4.7†
TRUNK MOTION FACTORS									
Sagittal Plane									
Maximum extension position (°)	-8.30	9.10	-30.82	18.96	-10.19	10.58	-30.00	33.12	3.5†
Maximum flexion position (°)	17.85	16.63	-13.96	45.00	10.37	16.02	-25.23	45.00	1.5
Range of motion (°)	31.50	15.67	7.50	75.00	23.82	14.22	3.99	67.74	3.8†
Average velocity(°/sec)	11.74	8.14	3.27	48.88	6.55	4.28	1.40	35.73	6.0†
Maximum velocity (°/sec)	55.00	38.23	14.20	207.55	38.69	26.52	9.02	193.29	3.7†
Maximum acceleration (°/sec²)	316.73	224.57	80.61	1341.92	226.04	173.88	59.1	1120.10	4.2†
Maximum deceleration (°/sec²)	-92.45	63.55	-514.08	-18.45	-83.32	47.71	-227.12	-4.57	1.2

Lateral Plane									
Maximum left bend(°)	−1.47	6.02	−16.80	24.49	−2.54	5.46	−23.80	13.96	1.4
Maximum right bend (°)	15.60	7.61	3.65	43.11	13.24	6.32	0.34	34.14	2.6*
Range of motion (°)	24.44	9.77	7.10	47.54	21.59	10.34	5.42	62.41	2.2*
Average velocity (°/sec)	10.28	4.54	3.12	33.11	7.15	3.16	2.13	18.86	6.1†
Maximum velocity (°/sec)	46.36	19.12	13.51	119.94	35.45	12.88	11.97	76.25	4.9†
Maximum acceleration (°/sec²)	301.41	166.69	82.64	1030.29	229.29	90.9	66.72	495.88	4.1†
Maximum deceleration (°/sec²)	−103.65	60.31	−376.75	0.00	−106.20	58.27	−294.83	0.00	0.3
Twisting Plane									
Maximum left twist (°)	1.21	9.08	−27.56	29.54	−1.92	5.36	−30.00	11.44	3.2†
Maximum right twist (°)	13.95	8.69	−13.45	30.00	10.83	6.08	−11.20	30.00	2.2*
Range of motion (°)	20.71	10.61	3.28	53.30	17.08	8.13	1.74	38.59	2.9†
Average velocity (°/sec)	8.71	6.61	1.02	34.77	5.44	3.19	0.66	17.44	3.8†
Maximum velocity (°/sec)	46.36	25.61	8.06	136.72	38.04	17.51	5.93	91.97	4.7*
Maximum acceleration (°/sec²)	304.55	175.31	54.48	853.93	269.49	146.65	44.17	940.27	2.9†
Maximum deceleration (°/sec²)	−88.52	70.30	−428.94	−5.84	−100.32	72.40	−325.93	−2.74	1.6*

*Significant at $\alpha \leq 0.05$ (two-sided). †Significant at $\alpha \leq 0.01$ (two-sided).
Adapted from Marras et al., 1993.

References

Adams, M.A. and Dolan, P. (1995) Recent advances in lumbar spinal mechanics and their clinical significance, *Clinical Biomechanics*, 10(1), 3-19.

Anderson, C.K. and Chaffin, D.B. (1986) A biomechanical evaluation of five lifting techniques. *Applied Ergonomics*, 17(1): 2-8.

Andersson, G.B. (1981) Epidemiologic aspects of low back pain in industry. *Spine*, 6: 53-60.

Andersson, G.B. (1991), The epidemiology of spinal disorders, in *The Adult Spine*. Eds. Frymoyer, J.W., Ducker, T.B., Hadler, N.M., Kostuik, J.P., Weinstein, J.N., and Whitecloud, T.S. Raven Press, New York, pp. 107-146.

Basmajian, J.V. and De Luca, C.J. (1985) *Muscles Alive: Their Functions Revealed by Electromyography* (5th edition), Williams and Wilkins, Baltimore, MD.

Bean, J.C., Chaffin, D.B., and Schultz, A.B. (1988) Biomechanical model calculation of muscle forces: A double linear programming method. *J. Biomechanics*, 21 (1) 59-66.

Bigos, S.J., Spengler, D.M., Martin, N.A., Zeh, J., Fisher, L., Nachemson, A., and Wang, M.H. (1986) Back injuries in industry: A retrospective study. II. Injury factors. *Spine*, 11(3), 246-251.

Brinckmann, P., Biggemann, M., and Hilweg, D. (1988) Fatigue fracture of human lumbar vertebrae. *Clinical Biomechanics*, 3: Supplement 1, S1-S23.

Cats-Baril, W. and Frymoyer, J.W. (1991) The economics of spinal disorders, in *The Adult Spine*. Eds. Frymoyer, J.W., Ducker, T.B., Hadler, N.M., Kostuik, J.P., Weinstein, J.N., and Whitecloud, T.S. Raven Press, New York, pp. 85-105.

Chaffin, D.B.(1969) A computerized biomechanical model: development of and use in studying gross body actions. *Journal of Biomechanics*, 2, 429-441.

Chaffin, D.B. and Andersson, G.B. (1991) *Occupational Biomechanics*, John Wiley & Sons, Inc. New York, NY.

Chaffin D.B. and W.H. Baker (1970) A biomechanical model for analysis of symmetric sagittal plane lifting. *AIIE Transactions*, II(1): 16-27.

Chaffin, D.B. (1973) Localized muscle fatigue definition and measurement, *Journal of Occupational Medicine*, 15(4), 346-354.

Chaffin D.B. and Muzaffer, E. (1991) Three-dimensional biomechanical static strength prediction model sensitivity to postural and anthropometric inaccuracies. *IIE Transactions*, 23(3), 215-227.

Frymoyer, J.W., Pope, M.H., Clements, J.H., Wilder, D.G., MacPherson, B., and Ashikaga, T. (1983) Risk factors in low back pain: An epidemiologic survey. J. Bone Joint Surg., 65A, 213-216.

Granata K.P. and Marras, W.S. (1993) An EMG-assisted model of loads on the lumbar spine during asymmetric trunk extensions. *J. Biomechanics*, 26 (12), 1429-1438.

Grandjean, E. (1982) *Fitting the Task to the Man: An Ergonomic Approach*, Taylor & Francis, Ltd. London.

Guo, H.R., (1993) Back pain and U.S. workers (NIOSH report), presented at American Occupational Health Conference, April 29.

Hollbrook, T.L., Grazier, K., Kelsey, J.L., and Stauffer, R.N. (1984) The frequency of occurrence, impact and cost of selected musculoskeletal conditions in the United States, American Academy of Orthopaedic Surgeons, Chicago, IL, pp. 24-45.

Hoy, M.G., Zajac, F.E., and Gordon, M.E. (1990) A musculoskeletal model of the human lower extremity: The effect of muscle, tendon, and moment arm on the moment-angle relationship of the musculotendon actuators at the hip, knee, and ankle. *Journal of Biomechanics*, 23(2): 157-169.

Jager, M., Luttmann, A., and Laurig, W. (1991) Lumbar load during one-handed bricklaying, *International Journal of Industrial Ergonomics*, 8(3), 261-277.

Kelsey, J.L. and White, A.A. III (1980) Epidemiology and impact on low back pain. *Spine*, 5(2), 133-142.

Kelsey, K.L., Githens, P.B., White, A.A. III, Holford, T.R., Walter, S.D., O'Conner, T., Ostfeld, A.M., Weil, U., Southwick, W.O., and Calogero, J.A. (1984) An epidemiologic study of lifting and twisting on the job and risk for acute prolapsed lumbar intervertebral disc. *J Ortho Res.*, 2(1): 61-66.

Konz, S.A. (1983) *Work Design: Industrial Ergonomics*, Second Edition, Grid Publishing, Inc., Columbus, OH.

Kroemer, K.H.E. (1987) Biomechanics of the human body, in *Handbook of Human Factors*, (Ed. Salvendy, G.) John Wiley & Sons, New York, NY.

Lantz, S.A. and Schultz, A.B. (1986) Lumbar spine orthosis wearing. I. Restrictions of gross body motion. *Spine*, 11(8): 834-837.

Lavender, S.A., Thomas, J.S., Chang, D., and Andersson, G.B. (1995) The effects of lifting belts, foot movement and lifting asymmetry on trunk motions. *Human Factors*, (in press).

Marras, W.S. (1988) Predictions of forces acting upon the lumbar spine under isometric and isokinetic conditions: A model — experimental comparison. *Int. J. Ind. Ergonomics*, 3, 19-27.

Marras, W.S. and Granata, K.P. (1994) A biomechanical assessment and model of axial twisting in the thoraco-lumbar spine. *Spine*, 20(13): 1440-1451.

Marras, W.S., Lavender, S.A., Leurgans, S.E., Rajulu, S.L., Allread, W.G., Fathallah, F.A., and Ferguson, S.A. (1993) The role of dynamic three-dimensional trunk motion in occupationally-related low back disorders: The effects of workplace factors, trunk position and trunk motion characteristics on risk of injury. *Spine*, 18 (5), 617-628.

Marras, W.S., Lavender, S.A., Leurgans, S.E., Rajulu, S.L., Allread, W.G., Fathallah, F.A., and Ferguson, S.A. (1995) Biomechanical risk factors and trunk motion. *Ergonomics*, 38(2): 377-410.

Marras, W.S. and Reilly, C.H. (1988) Networks of internal trunk-loading activities under controlled trunk-motion conditions. *Spine*, 13(6), 661-667.

Marras, W.S. and Schoenmarklin, R.W. (1993) Wrist motion in industry, *Ergonomics*, 36(4), 341-351.

Marras, W.S. and Sommerich, C.M. (1991a) A three-dimensional motion model of loads on the lumbar spine: I. Model structure. *Human Factors*, 33 (2), 123-137.

Marras W.S. and Sommerich, C.M. (1991b) A three-dimensional motion model of loads on the lumbar spine: II. Model validation. *Human Factors*, 33 (2), 139-149.

McCoy, M.A., Congleton, W.L., Johnston, W.L., and Jiang, B.C. (1988) The role of lifting belts in manual lifting. *International Journal of Industrial Ergonomics*, 2: 259-256.

McGill, S.M. and Norman, R.W. (1985) Dynamically and statically determined low back moments during lifting. *J. Biomechanics*, 8 (12), 877-885.

McGill, S.M. and Norman, R. (1986) Partitioning the L4-L5 dynamic moment into disc, ligamentous, and muscular components during lifting. *Spine*, 11, 666-678.

McGill, S.M. (1993) Abdominal belts in industry: a position paper on their assets, liabilities and use. *American Industrial Hygiene Association Journal*, 54(12): 752-754.

McGill, S., Seguin, J., and Bennett, G. (1994) Passive stiffness of the torso in flexion, extension, lateral bending, and axial rotation: effects of belt wearing and breath holding. *Spine*, 19(6): 696-704.

Mitchell, L.V., Lawler, F.H., Bowen, D., Mote, W., Asundi, P., and Purswell, J. (1994) Effectiveness and cost-effectiveness of employer-issued back belts in areas of high risk for low back injury. *Journal of Medicine*, 36(1): 90-94.

Mirka G.A. and Marras, W.S. (1993) A stochastic model of trunk muscle coactivation during trunk bending. *Spine*, 18 (11), 1396-1409.

Mundt, D.J., Kelsey, J.L., Golden, A.L. et al. (1993) An epidemiologic study of non-occupational lifting as a risk factor for herniated lumbar intervertebral disc. *Spine*, 18(5): 595-602.

Nachemson, A. (1975) Towards a better understanding of low-back pain: a review of the mechanics of the lumbar disc. *Rheumatology and Rehabilitation*, 14, 129-143.

National Institute for Occupational Safety and Health (NIOSH) (1981) Work practices guide for manual lifting. Department of Health and Human Services (DHHS), National Institute for Occupational Safety and Health (NIOSH), Publication No. 81-122.

National Institute for Occupational Safety and Health (NIOSH) (1994) Workplace Use of Back Belts. Department of Health and Human Services (DHHS), National Institute for Occupational Safety and Health (NIOSH), Publication No. 94-122.

National Safety Council. (1989) *Accident Facts, 1989*, Chicago, IL.

National Safety Council. (1991) *Accident Facts*, Chicago, IL.

National Institute for Occupational Safety and Health (NIOSH), U.S. Department of Health and Human Services, Centers for Disease Control (1973) *The Industrial Environment-Its Evaluation and Control*, U.S. Government Printing Office, Washington, D.C.

Nordin, M. and Frankel, V.M. (1989) *Basic Biomechanics of the Musculoskeletal System*, 2nd edition, Lea and Febiger, Philadelphia, PA, pg 67.

Nussbaum, M.A., Chaffin, D.B., and Martin, B.J. (1995) A back-propagation neural network model of lumbar muscle recruitment during moderate static exertions. *Journal of Biomechanics*, 28(9): 1015-1024.

Ozkaya, N. and Nordin, M. (1991) *Fundamentals of Biomechanics, Equilibrium, Motion and Deformation.* Van Nostrand Reinhold, New York, NY.

Park, K.S. and Chaffin, D.B. (1974) A biomechanical evaluation of two methods of manual load lifting. *AIIE Transactions*, 6(2), 105-113.

Pope, M.H. (1993) Muybridge Lecture, International Society of Biomechanics XIVth Congress, Paris, France, July 5, 1993.

Praemer, A., Furner, S., and Rice, D.P. (1992) *Musculoskeletal Conditions in the United States*, American Academy of Orthopaedic Surgeons, Park Ridge, IL.

Reddell, C.R., Congleton, J.J., Huchingson, R.D., and Montgomery, J.F. (1992) An evaluation of a weight-lifting belt and back injury prevention training class for airline baggage handlers. *Applied Ergonomics*, 23(5): 319-329.

Reilly, C. and Marras, W. (1989) Simulift: A simulation model of the human trunk motion. *Spine*, 14, (1), 5-11.

Rowe, M.L. (1981) Low back disability in industry: an updated position. *Journal of Occupational Medicine*, 13(10), 476-478.

Sanders, M.S. and McCormick, E.J. (1993) *Human Factors in Engineering and Design*, McGraw-Hill Inc., New York, NY.

Schoenmarklin, R.W., Marras, W.S., and Leurgans, S.E. (1994) Industrial wrist motions and risk of cumulative trauma disorders in industry. *Ergonomics*, 37(9), 1449-1459.

Schultz, A.B. and Andersson, G.B.J. (1981) Analysis of loads on the lumbar spine. *Spine*, 6: 76-82.

Schultz, A.B., Andersson, G.B.J., Haderspeck, K., Ortgren, R., Nordin, R., and Bjork, R. (1982a) Analysis and measurement of the lumbar trunk loads in tasks involving bends and twists. *J. Biomechanics*, 15, 669-675.

Silverstein, B.A., Fine, L.J., and Armstrong, T.J. (1986) Hand wrist cumulative trauma disorders in industry. *Journal of Industrial Medicine*, 43: 779-784.

Silverstein, B.A., Fine, L.J., and Armstrong, T.J. (1987) Occupational factors and carpal tunnel syndrome. *American Journal of Industrial Medicine*, 11: 343-358.

Snook, S.H. (1989) The control of low back disability: the role of management. *Manual Materials Handling: Understanding and Preventing Back Trauma*. American Industrial Hygiene Association, Akron, OH.

Spengler, D.M., Bigos, S.J., Martin, B.A., et al. (1986) Back injuries in industry: a retrospective study, I. Overview and costs analysis. *Spine*, 11: 241-245.

Sudhakar, L.R., Schoenmarklin, R.W., Lavender, S.A., and Marras, W.S. (1988) The effects of gloves on grip strength and muscle activity. *Proceedings of the Human Factors Society 32nd Annual Meeting*, October 24 to 28, Anaheim, CA.

UAW International Union (1982) *Strains and Sprains: A Worker's Guide to Job Design*, UAW, Detroit, MI.

Videman, T., Nurminen, M. and Troup, T.D.G. (1990) Lumbar spinal pathology in cadaveric material in relation to history of back pain, occupation, and physical loading. *Spine*, 15(8), 728-740.

Walsh, N.E. and Schwartz, R.K. (1990) The influence of prophylactic orthoses on abdominal strength and low back injury in the workplace. *American Journal of Physical Medicine and Rehabilitation*, 69(5): 245-250.

Waters, T.R., Putz-Anderson, V., Garg, A., and Fine, L.J. (1993) Revised NIOSH equation for the design and evaluation of manual lifting tasks. *Ergonomics*, 36(7): 749-776.

Webster, B.S. and Snook, S.H. (1989) The cost of compensable low back pain, Liberty Mutual internal report.

11

Human Strength
Evaluation

Karl H. E. Kroemer
Virginia Tech

11.1   Overview

Skeletal muscles are able to move body segments with respect to each other against internal and external resistances. Muscle components can shorten dynamically, statically retain their length, or be lengthened. Various methods and techniques are available for assessing muscular strength. The engineering application of data on available body strength requires the determination of whether minimal or maximal exertions, static or dynamic, are critical. Data on body strength are presented for the design of tools, equipment, and work tasks.

11.2   Background and Terminology

Muscular efforts have been of special interest to physiological science; therefore, there is a long tradition of philosophical and experimental approaches and use of terminology. Of particular importance are Newton's three laws:

- The first explains that *unbalanced force acting on a mass changes its motion condition*
- The second states that *force f equals mass m multiplied by acceleration a:*

$$f = m * a.$$

- The third makes it clear that *force exertion requires the presence of an equally large counter force.*

Physiology books published until the middle of the 20th century tended to divide muscle activities into either dynamic efforts lasting for minutes or hours, with work, energy, and endurance as typical topics; or short bursts of contractile exertion. Much research on muscle effort concentrated on the "isometric" condition in which muscle length (and hence body segment position) did not change. (The Greek term *iso* means unchanged or constant, and *metrein* refers to the measure, i.e., length of the muscle.) Consequently, most information on muscle strength was gathered for such static exertion. All other muscle activities were typically called "anisometric," often even falsely labeled "isotonic" or "kinetic," meant to cover all the many possible dynamic muscle uses. Table 11.1 lists and explains terms that correctly describe muscular events.

For the engineer, skeletal muscles are of primary interest because they pull on segments of the human body and generate energy for exertion to outside objects. Skeletal muscles connect two body links across their joint, as shown in Figure 11.1 in some cases muscles cross even two joints. Muscles are usually arranged in "functional pairs" so the contracting muscle is counteracted by its opponent. The muscle, or group of synergistic muscles, pulling in the intended direction, is called agonist (also called protagonist) and the opposite antagonist. Cocontraction, the simultaneous activation of paired opposing muscles, serves to control speed and strength exertion.

There are several hundred skeletal muscles in the human body, known by their Latin names. They are wrapped in connective tissue (fascia) which imbeds nerves and blood vessels. At the ends of the muscle, the tissues combine to form tendons which attach the muscle to bones.

Thousands of individual muscle fibers run, more or less parallel, the length of the muscle. Seen with a microscope, skeletal muscle fibers appear striped (striated): thin and thick, light and dark bands run across the fiber in regular patterns, which repeat along the length of the fiber. One such dark stripe appears to penetrate the fiber like a thick membrane or disc: this is the so-called z-disk (from the German *zwischen*, between). The distance between two adjacent z-lines defines the sarcomere. Its length at rest is approximately 250 Å ($1Å = 10^{-10}$ m), meaning that there are about 40,000 sarcomeres in series within 1 mm of muscle fiber length.

Within each muscle fiber, thread-like myofibrils (from the Greek *mys*, muscle) are arranged by the hundreds or thousands in parallel. Each of these, in turn, consists of bundles of myofilaments. Spaces between them are filled with a network of tubular channels, sacs, and cisterns connected with a larger tubular system in the z-disks, which itself is part of the networks of blood vessels and nerves in the fascia. This is the sarcoplasmic reticulum, the "plumbing and control" system of the muscle. It provides the fluid transport between the cells inside and outside the muscle and also carries chemical and electrical messages.

Two of the myofibrils, myosin and actin, have the ability to slide along each other; this is the source of muscular contraction. Small projections, called cross-bridges, protrude from the myosin filaments toward neighboring actins. The actin filaments are twisted, double-stranded protein molecules, wrapped in a double helix around the myosin molecules. This is the "contracting microstructure" of the muscle.

The only *active* action a muscle can take is to contract; elongation is brought about by external forces that stretch the muscle. According to the "sliding filament theory," contraction is brought about by the heads of adjacent actin rods moving toward each other. This pulls the z-disks closer together: sarcomeres in series (and parallel) shorten, and with them the whole muscle. After a contraction, the muscle returns to its resting length, primarily through a recoiling of its shortened filaments, fibrils, fibers, and other connective tissues. Stretching the muscle beyond its resting length can be done by forces external to the muscle: either by gravity or other force acting from outside the body, or by the action of antagonistic muscle.

You may want to consult books by Asimov (1963), Åstrand and Rodahl (1986), Chaffin and Andersson (1991), Enoka (1988), Kroemer, Kroemer, and Kroemer-Elbert (1994, 1997), Schneck (1990, 1992), and Winter (1990), among others, for more information.

11.3   Relation between Muscle Length and Tension

Stimulation from the central nervous system (CNS) causes the muscle to contract to its smallest possible length which is about 60% of resting length with no external load. In this condition, the actin proteins

TABLE 11.1 Glossary of Muscle Terms

Activation of muscle — *See* contraction.

Co-contraction — Simultaneous contraction of two or more muscles.

Concentric (muscle effort) — Shortening of a muscle against a resistance.

Contraction — Literally, "pulling together" the Z lines delineating the length of a sarcomere, caused by the sliding action of actin and myosin filaments. Contraction develops muscle tension only if the shortening is resisted. *Note: during an isometric "contraction" no change in sarcomere length occurs, and in an eccentric "contraction" the sarcomere is actually lengthened. To avoid such contradiction in terms, it is often better to use the terms activation, effort, or exertion.*

Distal — Away from the center of the body.

Dynamics — A subdivision of mechanics that deals with forces and bodies in motion.

Eccentric (muscle effort) — Lengthening of a resisting muscle by external force.

Fiber — *See* muscle.

Fibril — *See* muscle fibers

Filament — *See* muscle fibers.

Force — As per Newton's Third Law, the product of mass and acceleration; the proper unit is the Newton, with $1 \text{ N} = 1 \text{ kg}$ m s^{-2}. On earth, one kg applies a (weight) force of 9.81 N (1 lb. exerts 4.44. N) to its support. Muscular force is defined as muscle tension multiplied by transmitting cross-sectional area.

Free dynamic — In this context, an experimental condition in which neither displacement and its time derivatives, nor force are manipulated as independent variables.

Iso — A prefix meaning constant or the same.

Isoacceleration — A condition in which the acceleration is kept constant.

Isoforce — A condition in which the muscular force (tension) is constant, i.e., isokinetic. This term is equivalent to isotonic.

Isoinertial — A condition in which muscle moves a constant mass.

Isojerk — A condition in which the time derivative of acceleration, jerk, is kept constant.

Isokinetic — A condition in which muscle tension (force) is kept constant. *See* isoforce and isotonic; compare with isokinematic.

Isokinematic — A condition in which the velocity of muscle shortening (or lengthening) is constant. (Depending on the given biomechanical conditions, this may or may not coincide with a constant angular speed of a body segment about its articulation.) Compare with isokinetic.

Isometric — A condition in which the length of the muscle remains constant.

Isotonic — A condition in which muscle tension (force) is kept constant — *see* isoforce. (In the past, this term was occasionally falsely applied to any condition other than isometric.).

Kinematics — A subdivision of dynamics that deals with the motions of bodies, but not the causing forces.

Kinetics — A subdivision of dynamics that deals with forces applied to masses.

Mechanical advantage — In this context, the lever arm (moment arm, leverage) at which a muscle pulls about a bony articulation.

Mechanics — The branch of physics that deals with forces applied to bodies and their ensuing motions.

Moment — The product of force and the length of the (perpendicular) lever arm at which it acts. Mechanically equivalent to torque.

Motor unit — All muscle filaments under the control of one efferent nerve axon.

Muscle — A bundle of fibers, able to contract or be lengthened. In this context, striated (skeletal) muscle that moves body segments about each other under voluntary control.

Muscle contraction — The result of contractions of motor units distributed through a muscle so that the muscle length is shortened. See contraction.

Muscle fibers — Elements of muscle, containing fibrils, which consist of filaments.

Muscle fibrils — Elements of muscle fibers, containing filaments.

Muscle filaments — Muscle fibril elements, especially actin and myosin (polymerized protein molecules), capable of sliding along each other, thus shortening the muscle and, if doing so against resistance, generating tension.

Muscle force — The product of tension within a muscle multiplied by the transmitting muscle cross-section.

Muscle strength — The ability of a muscle to generate and transmit tension in the direction of its fibers. *See also* body strength.

Muscle tension — The pull within a muscle expressed as force divided by transmitting cross-section.

Myo — A prefix referring to muscle (Greek *mys*, muscle).

Mys — A prefix referring to muscle (Greek *mys*, muscle).

Proximal — Toward the center of the body.

Repetition — Performing the same activity more than once. (One repetition indicates two exertions.)

Statics — A subdivision of mechanics that deals with bodies at rest.

Strength — *See* body strength and muscle strength.

Tension — force divided by the cross-sectional area through which it is transmitted.

Torque — The product of force and the length of the (perpendicular) lever arm at which it acts. Mechanically equivalent to moment.

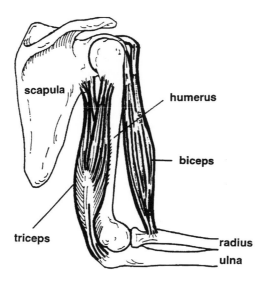

FIGURE 11.1　The biceps muscle reduces the elbow angle as agonist, counteracted by the triceps muscle as antagonist. Note the simplification of the actual conditions in modeling: in addition to the biceps, two other muscles (radialis and brachioradialis) also contribute to flexion about the elbow joint.

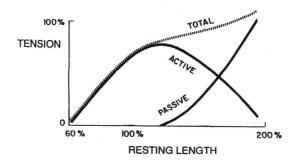

FIGURE 11.2　Active, passive, and total tension within a muscle at different lengths. (From Kroemer, K.H.E., Kroemer, H.J., and Kroemer-Elbert, K.E. (1997). *Engineering Physiology. Bases of Human Factors/Ergonomics* (3rd ed.). New York, NY: Van Nostrand Reinhold. With permission.)

are completely curled around the myosin rods. This is the shortest possible length of the sarcomeres, below which the muscle cannot develop any additional active contraction force.

Near resting length, the cross-bridges between the actin and myosin rods are in an optimal position to generate contact for contractile resistance. If the muscle is elongated further, the actin and myosin fibrils are slid along each other, which reduces the cross-bridge overlap between the protein rods. At about 160% of resting length, so little overlap remains that no active resistance can be developed internally. Thus, the curve of active contractile tension developed within a muscle in isometric twitch is as shown in Figure 11.2 minimal at approximately 60% resting length, rising to about 0.9 at resting length, at unit value at about 120 to 130% of resting length, and then falling back to minimum at about 160% resting length.

The muscle passively resists stretch, like a rubber band. This passive resistance becomes stronger the more the muscle is pulled from its resting length and is strongest near the point of muscle or tendon (attachment) breakage. This is also shown in Figure 11.2 above resting length, the tension in the muscle is the summation of active and passive strains. The summation effect explains why we stretch muscles

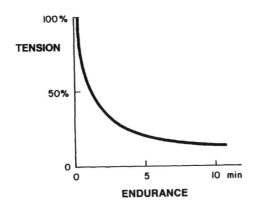

FIGURE 11.3 Muscle exertion and endurance. (From Kroemer, K.H.E., Kroemer, H.J., and Kroemer-Elbert, K.E. (1997). *Engineering Physiology. Bases of Human Factors/Ergonomics* (3rd ed.). New York, NY: Van Nostrand Reinhold. With permission.)

for a strong exertion, like in bringing the arm behind the shoulder before throwing a rock. This "pre-loading" tenses the muscle for a strong exertion.

In engineering terms it is said that muscles exhibit "viscoelastic" qualities. They are viscous in that their behavior depends both on the amount by which they are deformed, and on the rate of deformation. They are elastic in that, after deformation, they return to the original length and shape. These behaviors help to explain why the tension that can be developed isometrically ("statically," especially in a state of eccentric stretch) is the highest possible, while in active shortening (in a "dynamic" concentric movement) muscle tension is decidedly lower.

11.4 Muscle Endurance and Fatigue

Sufficient supply of arterial blood to the muscle and its unimpeded flow through the capillary bed into the venules and veins are crucial because they determine the ability of the contractile and metabolic processes of the muscle to continue. Blood brings needed energy-carriers and oxygen, and it removes metabolic byproducts, particularly lactic acid and potassium, as well as heat, carbon dioxide, and water liberated during metabolism.

The fine blood vessels permeating the muscle are easily compressed by pressure applied to them. A strongly contracting muscle generates pressure within itself, as can be felt by touching a tightened biceps or calf muscle. By this pressure, the muscle compresses its own blood vessels, thus reducing or even shutting off its own blood circulation. The interruption of blood flow through a muscle leads to muscle fatigue within seconds, forcing relaxation. Such fatigue, which occurs slowly when the muscle is not maximally contracting, is experienced painfully when one works overhead with raised arms, e.g., while fastening a screw in the ceiling of a room. Muscle fatigue in the shoulder muscles makes it impossible to keep one's arms raised after only a minute or so, even though nerve impulses from the CNS still arrive at the neuromuscular junctions, and the resulting action potentials continue to spread over the muscle fibers. Muscle fatigue is defined operationally as a "state of reduced physical ability which can be restored by rest."

Figure 11.3 shows the relation between static exertion and muscle endurance schematically: a maximal exertion can be maintained for just a few seconds; 50% of tension is available for about one minute; but less than 20% can be applied for long endurance periods.

11.5   Muscle Tension and Its Internal Transmission to the Point of Application

The term "strength" is often used in reference to any or all of the following:

- The tension *within* a muscle
- The *internal transmission* via body links across joints
- The *external application* of force or torque by a body segment to an outside object.

Confusion can be avoided by distinguishing these aspects and using proper terminology.

Muscle Tension — "Muscle Strength"

Within the muscle, all filament pulls combine to a resultant tension in the muscle. Its magnitude depends mostly on the involved number of parallel muscle fibers, i.e., on the cross-sectional thickness of the muscle. Maximal tensions reported on human skeletal muscle are within the range of 16 to 61 N/cm². Enoka (1988) uses 30 N/cm² as a typical value and calls it "specific (human muscle) tension." If the muscle cross-section area is known (such as from cadaver measurements or from MRI scans), one can calculate a resultant muscle force.

From the muscle, one tendon extends proximally to the origin and, in the opposite direction, another tendon extends outward to the insertion (see Figure 11.4). Like cables, tendons transmit the muscle tension, usually to the surface of a bone, but some end at strong connective tissues, such as in the fingers. The distal tendon may be quite long; for example, the tendons that reach from the muscles in the forearm to the digits of the hand can be 20 cm in length.

Internal Transmission

The tension inside a muscle–tendon unit tries to pull together the origin and the insertion of the muscle–tendon unit across a joint. This generates torque about the articulation with the long bones, the lever arms at which muscle pulls. The torque developed depends, hence, not only on the strength of the muscle, but also on the effective lever arm, or on the pull angle. Figure 11.4 depicts these conditions as

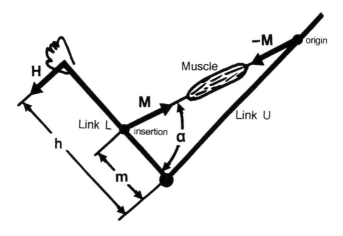

FIGURE 11.4   The muscle–tendon unit exerting pull forces M to links L and U at origin and insertion. Note the simplification in modeling the actual conditions at the elbow: in reality the biceps muscle has its origin proximal to the shoulder joint: two other muscles (radialis and brachioradialis) contribute to flexion about the elbow joint. Torque $T = m * M * \sin \alpha$ must be transmitted across the wrist joint to generate hand force H. (From Kroemer, K.H.E., Kroemer, H.J., and Kroemer-Elbert, K.E. (1997). *Engineering Physiology. Bases of Human Factors/Ergonomics* (3rd ed.). New York, NY: Van Nostrand Reinhold. With permission.)

a simplified scheme of the lower arm link *L* attached to the upper arm link *U*, articulated in the elbow joint. The muscle has its origin at the proximate end of *U*. Its insertion point is at the distance *m* from the joint on the arm link *L*. It generates a muscle force *M*, pulling at the angle α. The torque *T* generated by *M* depends on the lever arm *m* and the pull angle α according to

$$T = m * M * \sin \alpha \qquad (11.1)$$

This torque *T* then counteracts an external force *H*, acting perpendicularly at its lever arm *h*, according to

$$T = m * M * \sin \alpha = h * H \qquad (11.2)$$

External Application — "Body (Segment) Strength"

The final output of the biomechanical system is the torque or force (*H* in Figure 11.4) available at the hand, foot, or other body segment for exertion to a resisting object. This object is usually outside the body, but the resistance may be from an antagonistic muscle. The "body (segment) strength" available for application to an outside object is of primary importance to the engineer, designer, and manager.

The model depicted in Figure 11.4 shows that the amount of force (*H*) available at the body interface with an external object depends on

- Muscle force (*M*)
- Lever arms (*m* and *h*)
- Pull angle (α) which, in turn, depends on the angle between the two links. If all of these are known, the body segment force can be calculated from equation (2)

$$H = (m/h) * M * \sin \alpha \qquad (11.3)$$

Definitions

To help distinguish among muscle tension, its internal transmission, and the exertion to an outside object, it is useful to define terms as follows:

MUSCLE STRENGTH is the maximal tension (or force) that muscle can develop voluntarily between its origin and insertion.

The best word to refer to this is "muscle tension" (in N/mm^2 or N/cm$^{2)}$ but the term strength (force in N) is commonly used. If the variables *m*, *h*, α, and *H* in equation (2) are known, one can solve for the muscle force

$$M = (h/m) * H/\sin \alpha \qquad (11.4)$$

INTERNAL TRANSMISSION is the manner in which muscle tension is transferred inside the body along links and across joint(s) as torque to the point of application to a resisting object.

If several link-joint systems in series constitute the internal path of torque (in Nm or Ncm) transmission, each transfers the arriving torque by the existent ratio of lever arms (*m* and *h* in the example above) until resistance is met, usually the point where the body interfaces with an outside object. This transfer of torques is more complicated under dynamic conditions than in the static case because of changes in muscle functions with motion and because of the effects of accelerations and decelerations on masses.

BODY SEGMENT STRENGTH is the force or torque that can be applied by a body segment to an object external to the body.

The segment is usually identified such as "hand," "elbow," "shoulder," "back," "foot." (Strength in N, torque in Ncm or Nm.)

The original muscle pull (after being internally transmitted to the appropriate body member and transformed in magnitude and direction during that transmission) finally results in a force or torque that can be applied to an outside object: often by hand to a hand tool, or to a handle on a box as in load lifting, or by shoulder or back in pushing or carrying; or by the feet in operating pedals, or in walking or running.

The quality and quantity of the force or torque transmitted to an outside object depends on many biomechanical and physical conditions including

- Body segment employed, e.g., hand or foot
- Type of body object attachment, e.g., by a touch or grasp
- Coupling type, e.g., by friction or interlocking
- Direction of force/torque vector
- Static posture or body motions in dynamic exertion

Within the field of ergonomics (*aka* human factors, *aka* human engineering) *muscle strength* is of particular interest to the engineering physiologist; *internal transmission* is of particular interest to the biomechanist and to the designer because of the implications for body segment posture and motion; and *body segment strength* is of particular interest to the designer of tools, equipment, and work tasks.

Figure 11.5 shows, in the form of a flow diagram, the feedforward of excitation signals from the CNS to muscle to generate tension. EEGs (electroencephalograms) and EMGs (electromyograms) can be recorded and measured, while muscle tension is calculated using biomechanical modeling. Torques are internally transmitted via bone leverages and across articulations to the body segment that applies energy to a resisting object (often a handle or pedal) for task execution. The internal torques are calculated biomechanically, while posture and motion as well as applied force and torque can be measured. Three feedback paths can be identified, although at present they do not provide convenient avenues for measurements. The first is a short reflex loop F_1 originating at interoceptors. The other two loops start at exteroceptors and lead to a comparator where they modify the input to the CNS. F_2 provides kinesthetic signals related to touch, body position, and motion. F_3 relates to sound and vision; its signal can be influenced by the experimenter, e.g., via verbal exhortation or by an instrument showing the applied force.

11.6   Assessment of Body Segment Strength

Generation of muscle strength is a complex procedure of myofilament activation through nervous feedforward and feedback control. It may involve substantial shortening or lengthening of muscle, i.e., a concentric or eccentric effort; or there may be no perceptible change in length, i.e., the effort is isometric. Mechanically, the main distinction between muscle actions is whether they are "dynamic" or "static."

Static Strength

In physiological terms, an isometric muscle contraction generates (usually after some initial sarcomere shortening) the static condition. When there is no change in muscle length during the isometric effort, then involved body segments do not move; in physics terms, all forces acting within the system are in equilibrium, as Newton's First Law requires. Therefore, the physiological "isometric" case is equivalent to the "static" condition in physics.

The static condition is theoretically simple and experimentally well controllable. It allows rather easy measurement of muscular effort. Therefore, most of the information currently available on "human strength" describes the outcomes of static (isometric) testing. Accordingly, most of the tables on body segment strength in this chapter and in other human engineering or physiologic literature contain static data.

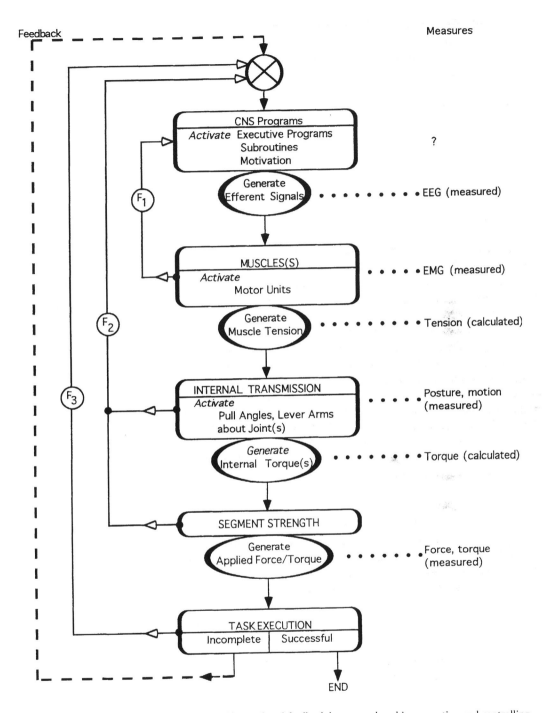

FIGURE 11.5   A conceptual model of the feedforward and feedback loops employed in generating and controlling muscle strength exertion. (Modified from Kroemer 1979, and Kroemer, Kroemer, and Kroemer-Elbert, 1994, 1997.)

Besides the simple convenience of dealing with statics, measurement of isometric strength appears to yield, for most cases, a reasonable estimate of the maximal possible exertion for most slow body link movements, especially if they are excentric. However, the data do not estimate fast exertions well, especially if they are concentric and of the ballistic-impulse type, such as throwing or hammering.

Dynamic Strength

Dynamic muscular efforts are more difficult to describe and control than static contractions. In dynamic activities, muscle length changes and, therefore, involved body segments move. This results in displacement. The amount of travel is relatively small at the muscle but usually amplified along the links of the internal transmission path to the point of application to the outside, for example at the hand or foot.

The time derivatives (velocity, acceleration, and jerk) of displacement are of importance both for the muscular effort (as discussed earlier) and the external effect: for example, change in velocity determines impact and force, as per Newton's Second Law.

Definition and experimental control of dynamic muscle exertions are much more complex tasks than in static testing. Various new classification schemes for independent and dependent experimental variables can be developed. Such a system has been presented for dynamic and static efforts (Kroemer, Marras, McGlothlin, McIntyre, and Nordin 1989; Marras, McGlothlin, McIntyre, Nordin, and Kroemer 1993). It includes the traditional isometric and isoinertial approaches.

Independent variables are those that are purposely manipulated during the experiment in order to assess resulting changes in the dependent variables. For example, if one sets the displacement (muscle length change) to zero — the *isometric* condition — one may either measure the magnitude of force generated, or the number of repetitions that can be performed until force is reduced because of muscular fatigue. This case is described in Table 11.2. Of course, since there is no displacement, its time derivatives, velocity, acceleration, and jerk, are also zero.

Alternatively, one may choose to control velocity as an independent variable, i.e., the rate at which muscle length changes. If velocity is set to a constant value, one speaks of an *isokinematic* muscle strength measurement. (Note that this *isovelocity* condition is often mislabeled "isokinetic.") Time derivatives of constant velocity, acceleration, and jerk, are zero. Mass properties are usually controlled in isokinematic tests. The variables displacement, force, and repetition can either be chosen as dependent variables or they may become controlled independent variables. Most likely, force and/or repetition are chosen as the dependent test variables.

Following the scheme laid out in Table 11.2, one also can devise tests in which acceleration or its time derivative, jerk, are kept constant. These test conditions are theoretically possible, but so far they have not been commonly applied.

For some tests, one sets the amount of muscle tension (force) to a constant value. In such an *isoforce* test, mass properties and displacement (and its time derivatives) are likely to become controlled independent variables, and repetition a dependent variable. This *isotonic* condition is, for practical reasons, often combined with an isometric condition, such as in holding a load motionless.

Note that the term isotonic has often been applied falsely. Some older textbooks described the examples of lifting or lowering of a constant mass (weight) as typical for isotonics. This is physically false for two reasons. The first is that according to Newton's Laws the change from acceleration to deceleration of a mass requires application of changing (not constant) forces. The second fault lies in overlooking the changes that occur in the mechanical conditions (pull angles and lever arms) under which the muscle functions during the activity. Hence, even if there were a constant force to be applied to the external object (which is not the case), the changes in mechanical advantages would result in changes in muscle tonus. It is certainly misleading to label all dynamic activities of muscles isotonic, as is unfortunately done occasionally.

In the *isoinertial* test, the external mass is set to a constant value. In this case, repetition of moving such constant mass (as in lifting) may either be a controlled independent or, more likely, a dependent variable. Also, displacement and its derivatives may become dependent outputs. Force (or torque) applied is likely to be a dependent value, according to Newton's Second Law. (Note that in the isoinertial test the external load is controlled, while in the previously described tests the conditions at the muscle were controlled.)

TABLE 11.2 Techniques to Measure Muscle Performance by Selecting Specific Independent and Dependent Variables

Names of Technique Variables	Isometric (Static) Indep. Dep.	Isovelocity (Dynamic) Indep. Dep.	Isoacceleration (Dynamic) Indep. Dep.	Isojerk (Dynamic) Indep. Dep.	Isoforce (Static or Dynamic) Indep. Dep.	Isoinertia (Static or Dynamic) Indep. Dep.	Free Dynamic Indep. Dep.
Displacement, linear/Angular	Constant* (zero)	C or X	C or X	C or X	C or X	C or X	X
Velocity, linear/angular	0	constant	C or X	C or X	C or X	C or X	X
Acceleration, linear/angular	0	0	constant	C or X	C or X	C or X	X
Jerk, linear/angular	0	0	0	constant	C or X	C or X	X
Force, Torque	C or X	C or X	C or X	C or X	constant	C or X	X
Mass, Moment of Inertia	C	C	C	C	C	constant	C or X
Repetition	C or X	C or X	C or X	C or X	C or X	C or X	C or X

Legend

Indep = independent
Dep   = dependent
C   = variable can be controlled
*   = set to zero
0   = variable is not present (zero)
X   = can be dependent variables

The boxed constant variable provides the descriptive name.

From Kroemer, K.H.E., Kroemer, H.J., and Kroemer-Elbert, K.E. (1997). *Engineering Physiology: Bases of Human Factors/Ergonomics* (3rd ed.). New York, NY: Van Nostrand Reinhold. With permission.

Table 11.2 also contains the most general case of motor performance measurement, labeled "free dynamic." Here the independent variables force and displacement (and its time derivatives) are left to the free choice of the subject. Only mass and repetition are usually controlled but may be used as dependent variables. Force, torque, or some other performance measure is likely to be chosen as a dependent output.

This discussion indicates that dynamic tests indeed require more effort to describe and control than static (isometric) measurements. This complexity explains why, in the past, dynamic measurements other than isokinematic and isoinertial testing have been rarely performed in the laboratory. On the other hand, if one is free to perform as one pleases, such as in the "free dynamic" test common in sports, very little experimental control can be executed. Nevertheless, Table 11.2 shows that it is possible to include both the traditional static and the important dynamic exertions in one systematic matrix of measurements.

11.7 Designing for Body Strength

The engineer or designer wanting to consider operator strength has to make a series of decisions. These include:

- Is the use mostly static or dynamic? If static, information about isometric capabilities, listed below, can be used. If dynamic, other considerations apply in addition, concerning for example physical endurance (circulatory, respiratory, metabolic) capabilities of the operator or prevailing environmental conditions. Physiologic and ergonomic texts (e.g., by Åstrand and Rodahl, 1986; Kroemer, Kroemer, and Kroemer-Elbert, 1994, 1997; Winter 1990) provide such information.

- Is the exertion by hand, by foot, or with other body segments? For each, specific design information is available. If a choice is possible, it must be based on physiologic and ergonomic considerations to achieve the safest, least strenuous, and most efficient performance. In comparison to hand movements over the same distance, foot motions consume more energy, are less accurate and slower, but they are stronger.

- Is a maximal or a minimal strength exertion the critical design factor?

  "Maximal" user strength usually relates to the structural strength of the object, so that a handle or a pedal may not be broken by the strongest operator. The design value is set, with a safety margin, above the highest perceivable strength application.

  "Minimal" user strength is that expected from the weakest operator which still yields the desired result, so that a door handle or brake pedal can be successfully operated or a heavy object be moved.

  A "range" of expected strength exertions is, obviously, that between the considered minimum and maximum. The infamous "average user" strength is usually of no design value (See Chapter 9).

- Most body segment strength data are available for static (isometric) exertions. They provide reasonable guidance also for slow motions, although they are probably too high for concentric motions and a bit too low for eccentric motions. Of the little information available for dynamic strength exertions, much is limited to isokinematic (constant velocity) cases. As a general rule, strength exerted in motion is less than measured in static positions located on the path of motion.

- Measured strength data are often treated, statistically, as if they were normally distributed and reported in terms of averages (means) and standard deviations. This allows the use of common statistical techniques to determine data points of special interest to the designer. In reality, body segment strength data are often in a skewed rather than in a bell-shaped distribution. This is not of great concern, however, because usually the data points of special interest are the extremes. The maximal forces or torques that the equipment must be able to bear without breaking are those above or near the strongest measured data points. The minimal exertions, which even "weak" persons are able to generate, can be identified as given percentile values at the low end of the distribution: often the 5th percentile is selected. This can be done either by calculation or by estimation (See Chapter 9 for details and instructions.)

Designing for Hand Strength

The human hand is able to perform a large variety of activities, ranging from those that require fine control to others that demand large forces. (But the feet and legs are capable of more forceful exertions than the hand, see below.)

One may divide hand tasks in this manner:

- Fine manipulation of objects, with little displacement and force. Examples are writing by hand, assembly of small parts, adjustment of controls.
- Fast movements to an object, requiring moderate accuracy to reach the target but fairly small force exertion there. An example is the movement to a switch and its operation.
- Frequent movements between targets, usually with some accuracy but little force; such as in an assembly task, where parts must be taken from bins and assembled.
- Forceful activities with little or moderate displacement (such as with many assembly or repair activities, for example when turning a hand tool against resistance).
- Forceful activities with large displacements (e.g., when hammering).

Accordingly, there are three major types of requirements: accuracy, displacement, and strength exertion. Design for the first two tasks is described in Chapters 8 through 11 of the book by Kroemer, Kroemer, and Kroemer-Elbert (1994).

Of the digits of the hand, the thumb is the strongest and the little finger the weakest. Gripping and grasping strengths of the whole hand are larger, but depend on the coupling between the hand and the handle (see Figure 11.6.) The forearm can develop fairly large twisting torques. Large force and torque vectors are exertable with the elbow at about a right angle, but the strongest pulling/pushing forces toward/away from the shoulder can be exerted with the extended arm, provided that the trunk can be braced against a solid structure. Torque about the elbow depends on the elbow angle as depicted in Figure 11.6 and, in more detail, in Figure 11.7.

Obviously, forces exerted with the arm and shoulder muscles are largely determined by body posture and body support. Likewise, finger forces depend on the finger joint angles, as listed in Tables 11.3 and 11.4. Table 11.5 provides detailed information about manual force capabilities measured in male students and machinists. Female students developed between 50 and 60% of the digit strength of their male peers, but achieved 80 to 90% in "pinches."

The Use of Tables of Exerted Torques and Forces

There are many sources for data on body strengths that operators can apply (see the tables and figures to follow). While these data indicate "orders of magnitude" of forces and torques, the exact numbers should be viewed with great caution because they were measured on various subject groups of rather small numbers under widely varying circumstances. It is advisable to take body strength measurements on a sample of the intended user population to verify that a new design is operable.

Note that thumb and finger forces, for example, depend decidedly on "skill and training" of the digits as well as the posture of the hand and wrist. Hand forces (and torques) also depend on wrist position, and on arm and shoulder posture. Exertions with arm, leg, and "body" (shoulder, backside) depend much on the posture of the body and on the support provided to the body (i.e., on the "reaction force" in the sense of Newton's Third Law) in terms of friction or bracing against solid structures. Figure 11.9 and Table 11.6 illustrate this: both were derived from the same set of empirical data but extrapolated to show the effects of

- Location of the point of force exertion
- Body posture
- Friction at the feet

on horizontal push (and pull) forces applied by male soldiers.

	Coupling #1.	**Digit Touch:**
		One digit touches an object.
	Coupling #2.	**Palm Touch:**
		Some part of the palm (or hand) touches the object.
	Coupling #3.	**Finger Palmar Grip (Hook Grip):**
		One finger or several fingers hook(s) onto a ridge, or handle. This type of finger action is used where thumb counterforce is not needed.
	Coupling #4.	**Thumb–Fingertip Grip (Tip Pinch):**
		The thumb tip opposes one fingertip.
	Coupling #5.	**Thumb–Finger Palmar Grip (Pad Pinch or Plier Grip):**
		Thumb pad opposes the palmar pad of one finger (or the pads of several fingers) near the tips. This grip evolves easily from coupling #4.
	Coupling #6.	**Thumb–Forefinger Side Grip (Lateral Grip or Side Pinch):**
		Thumb opposes the (radial) side of the forefinger.
	Coupling #7.	**Thumb–Two–Finger Grip (Writing Grip):**
		Thumb and two fingers (often forefinger and middle finger) oppose each other at or near the tips.
	Coupling #8.	**Thumb–Fingertips Enclosure (Disk Grip):**
		Thumb pad and the pads of three or four fingers oppose each other near the tips (object grasped does not touch the palm). This grip evolves easily from coupling #7.
	Coupling #9.	**Finger–Palm Enclosure (Collet Enclosure):**
		Most, or all, of the inner surface of the hand is in contact with the object while enclosing it. This enclosure evolves easily from coupling #8.
	Coupling #10.	**Power Grasp:**
		The total inner hand surfaces is grasping the (often cylindrical) handle which runs parallel to the knuckles and generally protrudes on one or both sides from the hand. This grasp evolves easily from coupling #9.

FIGURE 11.6   Couplings between hand and handle. (Adapted from Coupling the hand with the handle: an improved notation of touch, grip, and grasp, K.H.E. Kroemer, *Human Factors, 28,* 337-339.)

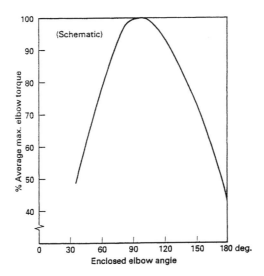

FIGURE 11.7   Relation of elbow angle and elbow torque (From Kroemer, K.H.E., Kroemer, H.B., and Kroemer-Elbert, K.E. (1994) *Ergonomics: How to Design for Ease and Efficiency.* Englewood Cliffs, NJ: Prentice Hall. With permission.)

TABLE 11.3   Mean Forces and Standard Deviations in *N* exerted by 9 Subjects in Fore, Aft, and Down Direction with the Fingertips, Depending on the Angle of the Proximal Interphalangeal Joint (PIP)

| | PIP joint at 30 degrees | | | PIP joint at 60 degrees | | |
Direction:	Fore	Aft	Down	Fore	Aft	Down
DIGIT						
2 Index	5.4 (2.0)	5.5 (2.2)	27.4 (13.0)	5.2 (2.4)	6.8 (2.8)	24.4 (13.6)
2 n	4.8 (2.2)	6.1 (2.2)	21.7 (11.7)	5.6 (2.9)	5.3 (2.1)	25.1 (13.7)
3 Middle	4.8 (2.5)	5.4 (2.4)	24.0 (12.6)	4.2 (1.9)	6.5 (2.2)	21.3 (10.9)
4 Ring	4.3 (2.4)	5.2 (2.0)	19.1 (10.4)	3.7 (1.7)	5.2 (1.9)	19.5 (10.9)
5 Little	4.8 (1.9)	4.1 (1.6)	15.1 (8.0)	3.5 (1.6)	3.5 (2.2)	15.5 (8.5)

n: Nonpreferred hand

From Kroemer, K.H.E., Kroemer, H.B., and Kroemer-Elbert, K.E. (1994) *Ergonomics: How to Design for Ease and Efficiency.* Englewood Cliffs, NJ: Prentice Hall. With permission.

TABLE 11.4   Mean Poke Forces and Standard Deviations in *N* exerted by 30 subjects in the Direction of the Straight Digits.

Digit	10 Male Mechanics	10 Male Students	10 Female Students
1. Thumb	83.8 (25.19) A	46.7 (29.19) C	32.4 (15.36) D
2. Index Finger	60.4 (25.81) B	45.0 (29.99) C	25.4 (9.55) DE
3. Middle Finger	55.9 (31.85) B	41.3 (21.55) C	21.5 (6.46) E

Entries with different letters are significantly different from each other ($p \leq 0.05$).
From Kroemer, K.H.E., Kroemer, H.B., and Kroemer-Elbert, K.E. (1994) *Ergonomics: How to Design for Ease and Efficiency.* Englewood Cliffs, NJ: Prentice Hall. With permission.

TABLE 11.5   Mean Forces and Standard Deviations in *N* Exerted by 21 Male Students* and by 12 Male Machinists

Couplings (see Figure 11.6)	Digit 1 (thumb)	Digit 2 (index)	Digit 3 (middle)	Digit 4 (ring)	Digit 5 (little)	
Push with digit tip in direction of the	91 (39)*	52 (16)*	51 (20)*	35 (12)*	30 (10)*	See also
extended digit ("Poke")	138 (41)	84 (35)	86 (28)	66 (22)	52 (14)	Table 11.4
Digit Touch (Coupling #1)	84 (33)*	43 (14)*	36 (13)*	30 (13)*	25 (10)*	
perpendicular to extended digit.	131 (42)	70 (17)	76 (20)	57 (17)	55 (16)	
Same, but all fingers press on one bar	—	digits 2, 3, 4, 5 combined: 162 (33)				
Tip force (like in typing; angle between	—	30 (12)*	29 (11)*	23 (9)*	19 (7)*	
distal and proximal phalanges about	—	65 (12)	69 (22)	50 (11)	46 (14)	
135 degrees)						
Palm Touch (Coupling #2)	—	—	—	—	—	233 (65)
perpendicular to palm (arm, hand,						
digits extended and horizontal)						
Hook Force exerted with digit tip pad	61 (21)*	49 (17)*	48 (19)*	38 (13)*	34 (10)*	all digits
(Coupling #3, "Scratch")	118 (24)	89 (29)	104 (26)	77 (21)	66 (17)	combined:
						108 (39)*
						252 (63)
Thumb-Fingertip Grip (Coupling #4,	—	1 on 2	1 on 3	1 on 4	1 on 5	
"Tip Pinch")		50 (14)*	53 (14)*	38 (7)*	28 (7)*	
		59 (15)	63 (16)	44 (12)	30 (6)	
Thumb-Finger Palmar Grip (Coupling	1 on 2 and 3	1 on 2	1 on 3	1 on 4	1 on 5	
#5, "Pad Pinch")	85 (16)*	63 (12)*	61 (16)*	41 (12)*	31 (9)*	
	95 (19)	34 (7)	70 (15)	54 (15)	34 (7)	
Thumb-Forefinger Side Grip (Coupling	—	1 on 2	—	—	—	—
#6, "Side Pinch")		98 (13)*				
		112 (16)				
Power Grasp (Coupling #10, "Grip	—	—	—	—	—	318 (61)*
Strength")						366 (53)

From Kroemer, K.H.E., Kroemer, H.B., and Kroemer-Elbert, K.E. (1994) *Ergonomics: How to Design for Ease and Efficiency.* Englewood Cliffs, NJ: Prentice Hall. With permission.

It is obvious that the amount of strength available for exertion to an object outside the body depends on the weakest part in the chain of strength-transmitting body parts. Hand pull force, for example, may be limited by finger strength, or shoulder strength, or low back strength; or it may be limited by the reaction force available to the body, as per Newton's Third Law. Figure 11.9 helps in determining where the "critical body segment "is in that sequence of torques about body joints. Starting at the point of external exertion, e.g., at the hand, one assesses the strength requirements joint-by-joint along the arm, shoulder, and back. Often, the lumbar back area is the "weak link" as evidenced by the large number of low-back pain cases reported in the literature.

To a sitting person, the reaction countering the forces actively exerted through upper body and arms is largely provided by the seat, although some support may be gathered from the floor via the legs. A walking or standing person receives all support from the ground up, of course, and not seldom hip or knee joint strength limits the ability to do hard efforts, such as lifting a load on the back. A slippery surface may make it impossible to push a heavy object sideways with the shoulder; one can experience this in the winter on icy ground when trying to push a car out of the ditch. These examples demonstrate how important is to provide proper body support at seat or ground.

Designing for Foot Strength

If a person stands at work, fairly little force and only infrequent operations of foot controls should be required because, during these exertions, the operator has to stand on the other leg. For a seated operator, however, operation of foot controls is much easier because the body is largely supported by the seat. Thus, the feet can move more freely and, given suitable conditions, can exert large forces and energies.

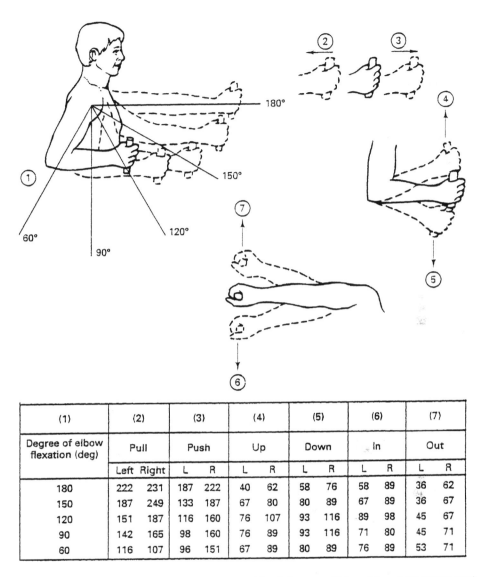

(1)	(2)		(3)		(4)		(5)		(6)		(7)	
Degree of elbow flexation (deg)	Pull		Push		Up		Down		In		Out	
	Left	Right	L	R	L	R	L	R	L	R	L	R
180	222	231	187	222	40	62	58	76	58	89	36	62
150	187	249	133	187	67	80	80	89	67	89	36	67
120	151	187	116	160	76	107	93	116	89	98	45	67
90	142	165	98	160	76	89	93	116	71	80	45	71
60	116	107	96	151	67	89	80	89	76	89	53	71

FIGURE 11.8   Fifth percentile arm forces in N exerted by sitting men. (Adapted from MIL HDBK 759.)

A typical example for such an exertion is pedaling a bicycle: all energy is transmitted from the leg muscles through the feet to the pedals. For normal use, these should be located directly underneath the body, so that the body weight above them provides the reactive force to the force transmitted to the pedal. Placing the pedals forward makes body weight less effective for generation of reaction force to the pedal effort, hence a suitable backrest should be provided against which the buttocks and low back press while the feet push forward on the pedal.

Small forces, such as for the operation of switches, can be generated in nearly all directions with the feet, with the downward or down-and-fore directions preferred. The largest forces can be generated with extended or nearly extended legs: in the downward direction limited by body inertia, in the more forward direction both by inertia and the provision of buttock and back support surfaces. These principles are typically applied in automobiles. For example, operation of a clutch or brake pedal can normally be performed easily with about a right angle at the knee. But if the power-assist system fails, very large

TABLE 11.6   Horizontal Push and Pull Forces in *N* that Male Soldiers Can Exert Intermittently or for Short Periods of Time

Horizontal force*; at least	Applied with**	Condition (μ: coefficient of friction at floor)
100 N push or pull	both hands or one shoulder or the back	with low traction, 0.2 < μ < 0.3
200 N push or pull	both hands or one shoulder or the back	with medium traction, μ ~ 0.6
250 N push	one hand	if braced against a vertical wall 51 to 150 cm from and parallel to the push panel
300 N push or pull	both hands or one shoulder or the back	with high traction, μ > 0.9
500 N push or pull	both hands or one shoulder or the back	if braced against a vertical wall 51 to 180 cm from and parallel to the panel
		or
		if anchoring the feet on a perfectly nonslip ground (like a footrest)
750 N push	the back	if braced against a vertical wall 600 to 110 cm from and parallel to the push panel
		or
		if anchoring the feet on a perfectly nonslip ground (like a footrest)

* May be nearly doubled for two and less than tripled for three operators pushing simultaneously. For the fourth and each additional operator, not more than 75% of their push capability should be added.
** See Figure 11.9 for examples.
Adapted from MIL STD 1472.

forces must be exerted with the feet: in this case, thrusting one's back against a strong backrest and extending the legs are necessary to generate the needed pedal force.

Figures 11.11 through 11.15 provide information about the forces that can be applied with legs and feet to a pedal. Of course, the forces depend on body support and hip and knee angles. The largest forward thrust force can be exerted with the nearly extended legs which leaves very little room to move the foot control further away from the hip.

Of course, the strength that can be exerted with the foot to an object, such as a pedal, depends on the joint strengths that can be transmitted along the "chain" ankle–knee–hip to the seat, which provides the reaction support needed according to Newton's Third Law. This is shown in Figure 11.16. The foot exertion loads all proximal segments according to the prevailing angles and leverarms. Following the diagram outward one sees that bad seat design, frail hip, knee or ankle, or low friction at the coupling of the shoe with the object may all make for a "weak kick."

Information on body strengths has been compiled, e.g., in NASA and U.S Military Standards; by Eastman-Kodak Company (1983, 1986); Kroemer, Kroemer, and Kroemer-Elbert (1994, 1997); Salvendy (1987); Weimer (1993); and Woodson, Tillman, and Tillman (1991). However, caution is necessary when applying these data because they were measured on different populations under varying conditions.

11.8   Summary

Muscle contraction is brought about by active shortening of muscle substructures. Elongation of the muscle is due to external forces. Maximal muscle tension depends on the individual's muscle size and exertion skill.

Prolonged strong contraction leads to muscular fatigue, which hinders the continuation of the effort and finally cuts it off. Hence, maximal voluntary contraction can be maintained for only a few seconds.

In isometric contraction, muscle length remains constant, which establishes a static condition for the body segments affected by the muscle. In an isotonic effort, the muscle tension remains constant, which usually coincides with a static (isometric) effort.

Dynamic activities result from changes in muscle length, which bring about motion of body segments. In an isokinematic effort, speed remains unchanged. In an isoinertial test, the mass properties remain constant.

Human body (segment) strength is measured routinely as the force (or torque) exerted to an instrument external to the body. This is information of great importance to the ergonomic designer/engineer.

	Force-plate[1] height	Distance[2]	Force, N	
			Mean	SD
	50	80	664	177
	50	100	772	216
	50	120	780	165
	70	80	716	162
	70	100	731	233
	70	120	820	138
	90	80	625	147
	90	100	678	195
	90	120	863	141
	Percent of shoulder height		Both hands	
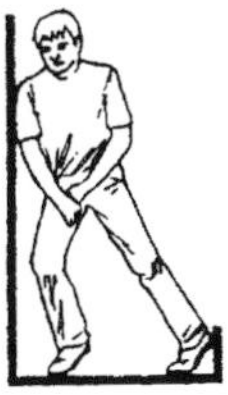	60	70	761	172
	60	80	854	177
	60	90	792	141
	70	60	580	110
	70	70	698	124
	70	80	729	140
	80	60	521	130
	80	70	620	129
	80	80	636	133
	Percent of shoulder height			
	70	70	623	147
	70	80	688	154
	70	90	586	132
	80	70	545	127
	80	80	543	123
	80	90	533	81
	90	70	433	95
	90	80	448	93
	90	90	485	80
	Percent of shoulder height		Both hands	
			Both hands	
Force plate	100 percent of shoulder height	50	581	143
		60	667	160
		70	981	271
		80	1285	398
		90	980	302
		100	646	254
			Preferred hand	
		50	262	67
		60	298	71
		70	360	98
		80	520	142
		90	494	169
		100	427	173
		Percent of thumb-tip reach*		
	100 percent of shoulder height	50	367	136
		60	346	125
		70	519	164
		80	707	190
		90	325	132
		Percent of span**		

[1]Height of the center of the force plate – 20 cm high by 25 cm long – upon which force is applied.
[2]Horizontal distance between the vertical surface of the force plate and the opposing vertical surface (wall or footrest, respectively) against which the subjects brace themselves.

*Thumb-tip reach – distance from backrest to tip of subject's thumb as arm and hand are extended forward.
**Span – the maximal distance between a person's fingertips when arms and hands are extended to each side.

FIGURE 11.9 Mean horizontal push forces and standard deviations in N exerted by standing men with their hands, the shoulder, and the back. (Adapted from NASA STD. 3000 A, 1989.)

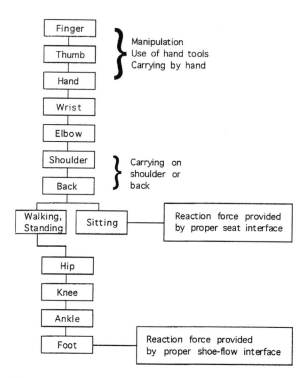

FIGURE 11.10   Determining the critical body segment strength for manipulating and carrying. (From Kroemer, K.H.E., Kroemer, H.J., and Kroemer-Elbert, K.E. (1997). *Engineering Physiology. Bases of Human Factors/Ergonomics* (3rd ed.). New York, NY: Van Nostrand Reinhold. With permission.)

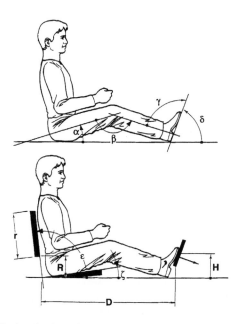

FIGURE 11.11   Conditions affecting the force that can be exerted on a pedal: body angles (upper illustration) and work space dimensions. (From Kroemer, K.H.E., Kroemer, H.B., and Kroemer-Elbert, K.E. (1994). *Ergonomics: How to Design for Ease and Efficiency.* Englewood Cliffs, NJ: Prentice Hall. With permission.)

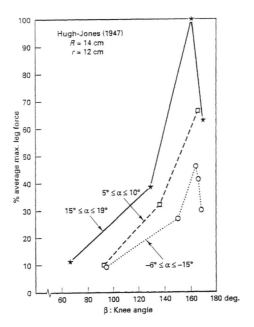

FIGURE 11.12   Effects of thigh angle α and knee angle β (see Figure 11.11) on pedal push force. (From Kroemer, K.H.E., Kroemer, H.B., and Kroemer-Elbert, K.E. (1994) *Ergonomics: How to Design for Ease and Efficiency*. Englewood Cliffs, NJ: Prentice Hall. With permission.)

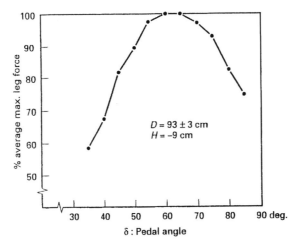

FIGURE 11.13   Effects of ankle (pedal) angle δ (see Figure 11.11) on foot force generated by ankle rotation. (From Kroemer, K.H.E., Kroemer, H.B., and Kroemer-Elbert, K.E. (1994) *Ergonomics: How to Design for Ease and Efficiency*. Englewood Cliffs, NJ: Prentice Hall. With permission.)

   Design of equipment and work tasks for human body segment strength capabilities is done systematically by:

- Determining whether the exertion is static or dynamic
- Establishing with what body part the force or torque is exerted
- Following the chain of strength vectors through the involved body segments to find the "weak link" and to improve and rearrange the conditions, if possible
- Selecting the body strength percentile (minimum and/or maximum) that is critical for the operation.

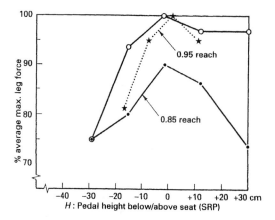

FIGURE 11.14   Effects of pedal height h and leg extension (see Figure 11.11) on pedal push force. (From Kroemer, K.H.E., Kroemer, H.B., and Kroemer-Elbert, K.E. (1994) *Ergonomics: How to Design for Ease and Efficiency.* Englewood Cliffs, NJ: Prentice Hall. With permission.)

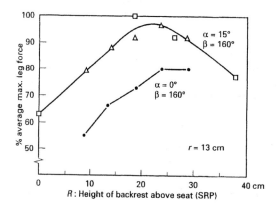

FIGURE 11.15   Effects of backrest height R (see Figure 11.11) on pedal push force. (From Kroemer, K.H.E., Kroemer, H.B., and Kroemer-Elbert, K.E. (1994) *Ergonomics: How to Design for Ease and Efficiency.* Englewood Cliffs, NJ: Prentice Hall. With permission.)

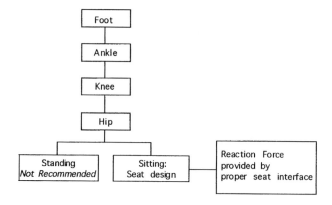

FIGURE 11.16   Determining the "critical body segment strength" for foot operation.

References

Asimov, I. (1963). *The Human Body. Its Structure and Operation.* New York, NY: The New American Library/Signet.

Åstrand, P. O. and Rodahl, K. (1977, 1986). *Textbook of Work Physiology.* (2nd, 3rd ed.). New York, NY: McGraw-Hill.

Chaffin, D.B. and Andersson, G.B.J. (1991). *Occupational Biomechanics* (2nd ed.) New York, NY: Wiley

Enoka, R.M. (1988). *Neuromechanical Basis of Kinesiology.* Champaign IL: Human Kinetics Books.

Eastman-Kodak Company (Ed.) (Vol. 1, 1983; Vol. 2,1986). *Ergonomic Design for People at Work.* New York, NY: Van Nostrand Reinhold.

Kroemer, K.H.E. (1979). A new model of muscle strength regulation, In *Proceedings, Annual Conference of the Human Factors Society,* (pp. 19-20). Santa Monica, CA: Human Factors Society.

Kroemer, K.H.E., Kroemer, H.B., and Kroemer-Elbert, K.E. (1994). *Ergonomics: How to Design for Ease and Efficiency.* Englewood Cliffs, NJ: Prentice Hall.

Kroemer, K.H.E., Kroemer, H.J., and Kroemer-Elbert, K.E. (1997). *Engineering Physiology. Bases of Human Factors/Ergonomics* (3rd ed.). New York, NY: Van Nostrand Reinhold.

Kroemer, K.H.E., Marras, W. S., McGlothlin, J. D., McIntyre, D. R., and Nordin, M. (1989). Assessing Human Dynamic Muscle Strength. (Technical Report, 8-30-89). Blacksburg, VA: Virginia Tech, Industrial Ergonomics Laboratory. Also published (1990) in *International Journal of Industrial Ergonomics, 6,* 199-210.

Marras, W.S., McGlothlin, J.D., McIntyre, D.R., Nordin, M., and Kroemer, K.H.E. (1993). *Dynamic Measures of Low Back Performance.* Fairfax, VA: American Industrial Hygiene Association.

Salvendy, G. (Ed.). (1987). *Handbook of Human Factors.* New York, NY: Wiley.

Schneck, D.J. (1990). *Engineering Principles of Physiologic Function.* New York, NY: New York University Press.

Schneck, D.J. (1992). *Mechanics of Muscle* (2nd ed.) New York, NY: New York University Press.

Weimer, J. (1993). *Handbook of Ergonomic and Human Factors Tables.* Englewood Cliffs, NJ: Prentice Hall.

Winter, D.A. (1990). *Biomechanics and Motor Control of Human Movement. (2nd ed.).* New York, NY: Wiley.

Woodson, W.E., Tillman, B., and Tillman, P. (1991). *Human Factors Design Handbook* (2nd ed.). New York, NY: McGraw-Hill.

12

Methods Based on Maximum Holding Time for Evaluation of Working Postures

Marjolein Douwes
TNO Prevention and Health
The Netherlands

Mathilde C. Miedema
TNO Prevention and Health
The Netherlands

J. Dul
TNO Prevention and Health
The Netherlands

12.1 Introduction

Background

Many work situations require postures which have to be maintained for a long period of time (e.g., machine operation, assembly work, VDU work). Forty-two percent of European workers adopt uncomfortable working postures for more than two hours a day (Paoli, 1992). This is illustrated in Figure 12.1. Depending on the position, of the load, the posture and the duration of holding the posture, there is a risk of acute discomfort and long-term health effects (musculoskeletal disorders) (Keyserling et al., 1988; Kilbom, 1988; Genaidy and Karwowski, 1993; Putz-Anderson and Galinsky, 1993). The percentage of workers on sick leave or unfit for work due to musculoskeletal disorders is enormous. This elicits the development of preventive programs.

Static work load can be diminished by improving the working posture (by optimization of the workplace and the equipment), by reducing the holding time of postures, and by supplying sufficient and

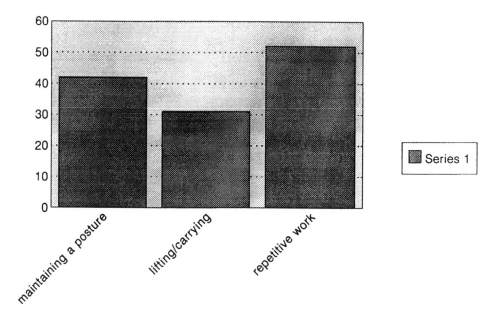

FIGURE 12.1   Percentage of the work force of different countries in the EC who work in uncomfortable postures for more than 2 hours per day (first column), who lift or carry loads for more than 2 hours (second column) or perform repetitive work for more than 2 hours per day (third column). (From Paoli, P. 1992. *First European Survey on the Work Environment: 1991–1992.* European Foundation for the Improvement of Living and Working Conditions Dublin. With permission.)

properly distributed rest pauses. For these measures, standards and evaluation methods need to be developed. Because of the lack of quantitative exposure–effect data, ergonomic recommendations to prevent risks for musculoskeletal disorders due to static load cannot yet be based on long-term health effects. However, acute discomfort can also be considered as an independent evaluation criteria for static postures (Miedema, 1992; Dul et al., 1994; Miedema et al., 1996).

Perceived Discomfort

An indicator of work load is the discomfort perceived by the worker. Static postures can cause strain on interior body structures, such as bones, joints, tendons, muscles, and ligaments. Subjective measurements like registration of body areas affected by discomfort and the intensity of such discomfort can be an indicator of the postural load. Van der Grinten and Smitt (1992) developed a method *localized musculoskeletal discomfort* (LMD) for recording the changing levels of discomfort over the working period. The method uses the 10 points category-ratio scale developed by Borg (1982) for recording the level of discomfort. To indicate the location of discomfort, the body diagram of Corlett and Bishop (1976) is used.

Since discomfort and musculoskeletal disorders are both related to exposure to biomechanical load on the musculoskeletal system (Milner, 1985; Nag, 1991; Putz-Anderson and Galinsky, 1993), reduction of discomfort will presumably contribute to reduction of the risk of musculoskeletal disorders as well (Dul et al., 1994).

Maximum Holding Time

A posture can be maintained for a limited period of time. The *maximum holding time* (MHT) is the maximum time that a posture, with or without external force exertion, can be maintained continuously until maximum discomfort, from a rested state. Because of the endless amount of different combinations of postures and force exertions, data on the MHT are only available for a limited amount of those

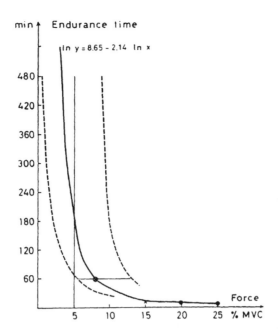

FIGURE 12.2 The relationship between muscle effort and MHT (Sjøgaard, 1986). ● from Rohmert (1960), X from Björkstén and Jonsson (1977), ○ from Hagberg (1981).

combinations. To estimate the MHT for other combinations, the relationship between the MHT and the muscle effort can be used. The muscle effort of a posture is the force (or moment) that is needed to maintain the posture (with or without external force exertion) as a percentage of the maximal force that can be exerted in the same posture.

The relationship between muscle effort and MHT has been studied by many authors. Rohmert (1960) measured MHT for different values of muscle effort for different muscle groups. He found that despite large individual differences in maximal force exertion, equal relative loading (i.e., muscle effort) eliminated differences in MHT. Caldwell (1974) also found little variation in MHT among various levels of MVC when the muscle effort was equal for each individual. The studies of Rohmert (1960), Björkstén and Jonsson (1977), and Sjøgaard (1986) show that the MHT of contractions decreases exponentially as the muscle effort increases (see Figure 12.2). This relationship between muscle effort and MHT can be used to predict the MHT from the muscle effort of a certain posture.

Sjøgaard also found that for low-level static loads (less than 20% MVC) the variation in MHT is large, possibly because of differences between the various muscle groups (e.g., muscle structure and fiber composition).

The Relation Between MHT and Discomfort

The time that a static posture can still be maintained continuously after a period of loading (and resting) is called the Remaining Endurance Capacity (REC) and is expressed as a percentage of the MHT. Experiments have shown that at group level perceived discomfort, as measured with a 10-point rating scale (Borg, 1982), increases linearly in time, independent of the magnitude of MHT (Taksic, 1986; Manenica, 1986; Meijst et al., 1995; see Table 12.1). For example, a discomfort level of 5 after 10 minutes of holding a certain posture, means that the MHT of this posture is 20 minutes.

Maximum Acceptable Level of Discomfort

For standardization purposes agreement should be reached on the level of discomfort that can be considered acceptable. The choice of the maximum acceptable level of discomfort is not a scientific question but a matter of agreement between parties involved. Hagerup and Time (1992) have (arbitrarily)

TABLE 12.1   The linear relationship between the 10-point category-ratio scale for recording discomfort and the percentage of the maximum holding time (MHT).

Time (% MHT)	Remaining Endurance Capacity (% MHT)	Discomfort score (Borg CR-10 scale)	
0%	100%	0	nothing at all
10%	90%	1	very weak (just noticeable)
20%	80%	2	weak (light)
30%	70%	3	moderate
40%	60%	4	somewhat strong
50%	50%	5	strong (heavy)
60%	40%	6	
70%	30%	7	very strong
80%	20%	8	
90%	10%	9	
100%	0%	10	extremely strong (maximal)

proposed a division of the Borg scale into three categories. They consider the mean score of a group of individuals acceptable if this score is 1 to 3. Rose et al. (1992) found that when the subjects were allowed to decide on the duration of a static work task themselves, they stopped to pause at approximately 20% of the MHT. Until we have more data to make a better choice, we use a maximum acceptable mean discomfort level of 2 on the Borg scale (weak discomfort). Because of the linear relationship between discomfort and REC at group level, this implies that the holding time of a continuous static posture should be no more than 20% of the MHT of that posture. The duration of an intermittent exercise until the minimum acceptable mean discomfort score 2 can be considerably longer, depending on the work–rest schedule.

Ergonomic standards are meant to protect a certain percentage (e.g., 95%) of the population. If the maximum acceptable mean level of discomfort would be a score of 2, we estimate that at least 50% of the population will have less than "weak discomfort" (score 2), and 95% of the population will have less than "strong discomfort" (score 5). These estimations are based on the distribution of individual discomfort scores in our experimental data set (Dul et al., 1994).

Methods to Evaluate Static Load

As stated before, muscular fatigue and risk of musculoskeletal disorders can be reduced in a number of ways. The muscle effort during work can be reduced by improving the posture or reducing the external force exertion. The working posture and required external force can be controlled by variables such as the working height, reaching distance, and the force required to operate a machine. When postural and force changes are difficult to establish, sufficient and properly distributed rest pauses can be supplied. The duration and distribution of work and rest periods can be controlled by organizational (time based) factors. Both types of variables can be influenced by designers and manufacturers of machinery and by occupational health and safety personnel.

In this chapter, guidelines and methods are presented to help occupational health practitioners and designers bring muscle fatigue during work to an acceptable level. In the next paragraph an evaluation method for working postures based on hand positions is presented. A subsequent section describes an evaluation method for working postures alternated with rest periods ("The Work–Rest Model for Static Postures"). In the final section, guidelines for static load used in CEN and ISO standards are presented.

12.2   Evaluation Method for Hand Positions

Due to the increment of discomfort in time, holding time (%MHT) can be taken as a measure for making recommendations concerning the maximum duration of static postures. We studied endurance data of experiments found in the literature and ranked the postures (Dul et al., 1993). Based on this ranking we

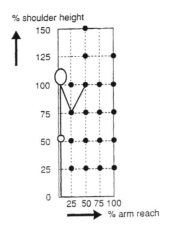

FIGURE 12.3 The 19 postures found in literature are indicated with ●. Posture is defined by the horizontal and vertical hand position. The vertical hand position is expressed as a percentage of the shoulder height and the horizontal hand position as a percentage of the arm reach in the upright standing posture.

developed recommendations for the holding time of static standing postures. This section describes the development and results of a method to evaluate postures on the basis of hand position. In the last part of the paragraph an example of using this method in practice is included.

MHT Data from the Literature

Information was gathered about the MHT of 19 different standing postures which were maintained without rest pauses and without external load. These data were found in seven studies (Corlett and Manenica, 1980; Hagberg, 1981; Boussenna et al., 1982; Milner, 1985; Taksic, 1986; Manenica, 1986; Meijst et al., 1995). All postures were defined by two parameters, i.e., the horizontal distance (% shoulder height) and vertical distance (% arm reach) of the position of the hands with respect to the feet in upright standing posture. Shoulder height (SH) is defined as the distance from acromion to the floor in the upright position. The arm reach (AR) is defined as the maximum distance from the knuckles to the wall when standing upright with the back against the wall and the shoulder in 90° anteflexion. The 19 postures differ in the combination of 25, 50, 75, 100, 125, or 150% SH and 25, 50, 75, or 100% AR, and are shown in Figure 12.3. In all studies the participants were asked to maintain the posture as long as they could. In almost all studies the subjects had to perform a task while holding the posture. These tasks implied television games, spot-tracking, or tapping tasks. While maintaining the posture, location and amount of perceived discomfort was registered. The experiments ended when maximum discomfort was reached (score 10 on a 10-point rating scale; Borg, 1990). The MHTs that were recorded are summarized in Table 12.2.

Description of the Method

As can be seen in Table 12.2, there is much variation in MHT's of similar postures within and between studies. In spite of this variation, a ranking of the 19 postures was made, based on the mean MHT from all available data. These ranked postures are shown in Figure 12.4. The posture 75%SH/50%AR has the highest MHT (35.7 min.), and the posture 25%SH/100%AR has the lowest MHT (2.7 min.). The ranked postures can be arbitrarily classified into 3 groups; "comfortable," "moderate," and "uncomfortable," with relatively large, medium, and small MHT's, respectively. Uncomfortable postures are defined as postures with an MHT smaller than 5 minutes, which implies that maintaining an uncomfortable posture will lead to a relatively quickly increasing feeling of discomfort. All postures with an extremely low or high working height (25% and 150% SH) appear to be uncomfortable postures. According to the classification,

TABLE 12.2   Summary of the data from literature. For each study the postures investigated and their maximum holding times (MHT; in minutes) are given.

Posture SH/AR		Meijst n = 20 10♀/10♂ television game		Corlett n = ? tapping task		Manenica n = 15♀ tapping task		Milner n = 9♂ computer game		Boussenna n = 8♂ spot tracking		Douwes n = 12 6♂/6♀ no task		Hagberg n = 7♀ no task	
		mean	SD	mean	SD	mean	SD	mean	SD	mean	SD	mean	SD	mean	SD
25/25	males														
	females														
	all			4.5											
25/50	males														
	females														
	all			4.0											
25/75	males														
	females														
	all			2.9											
25/100	males									4.41	1.53				
	females					1.57	0.72								
	all			2.0											
50/25	males														
	females														
	all			13.5											
50/50	males	10.89	6.75												
	females	16.52	5.99												
	all	13.71	6.98	8.5											
50/75	males														
	females														
	all			5.0											
50/100	males	8.56	1.67					10.10		5.26	1.66				
	females	7.88	2.92			2.35	1.55	8.12	1.52						
	all	8.22	2.40	3.2				8.73	1.50						
75/25	males														
	females					4.27	2.04								
	all			30.0											
75/50	males	65.5	38.7												
	females	36.5	24.6												
	all	50.9	13.7	20.5											
75/75	males					4.25	1.75								
	females			7.5											
	all														
75/100	males	17.4	11.7							6.35	1.91				
	females	12.0	3.41												
	all	14.7	9.05	4.2											
100/25	males														
	females														
	all			9.0											
100/50	males	38.5	15.7												
	females	16.6	4.7												
	all	26.2	15.1	6.0											

TABLE 12.2 (continued)   Summary of the data from literature. For each study the postures investigated and their maximum holding times (MHT; in minutes) are given.

Posture SH/AR		Meijst n = 20 10♀/10♂ television game		Corlett n = ? tapping task		Manenica n = 15♀ tapping task		Milner n = 9♂ computer game		Boussenna n = 8♂ spot tracking		Douwes n = 12 6♂/6♀ no task		Hagberg n = 7♀ no task	
		mean	SD	mean	SD	mean	SD	mean	SD	mean	SD	mean	SD	mean	SD
100/75	males														
	females														
	all			5.25											
100/100	males	11.42	2.76												
	females	8.45	2.18			3.57	1.38			7.85	2.09				
	all	9.93	2.90												
125/50	males	10.83	3.79									15.3	7.7	21.4	
	females	6.76	2.43			3.26	1.53					9.4	3.8		
	all	8.79	3.78									12.3	6.6		
125/100	males	8.94	2.38												
	females	6.51	3.40												
	all	7.72	3.18												
150/50	males														
	females					3.22	1.93								
	all														

Summary of the literature used in this study: posture is defined as the relative hand position with respect to the feet, i.e., the working height (as a percentage of shoulder height [%SH]) and working distance (as a percentage of arm reach [%AR]). For each study the postures investigated and resulting mean MHT (minutes) for males and females are given.

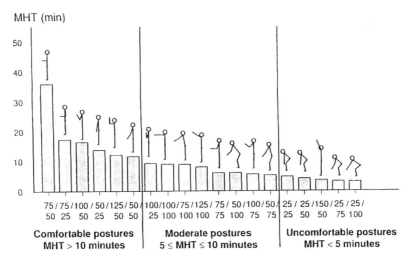

FIGURE 12.4   Ranking of the postures on the basis of the mean values of maximum holding time (MHT) of available data. The recommended holding time is 20% of MHT.

the postures with a combination of a moderate working height (50%, 75%, 100%, and 125% SH) and small working distance (25% and 50% AR) are comfortable postures (MHT longer than 10 minutes). Postures with a moderate working height (50%, 75%, 100%, and 125% SH) and a large working distance (75% and 100% AR) appear to be moderate postures (MHT ≥5 and ≤10 minutes).

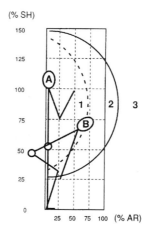

FIGURE 12.5   Recommendations concerning the maximum holding times of static postures. ———— = demarcation line between uncomfortable and moderate postures. ---- = demarcation line between moderate and comfortable postures. 1 = hand positions of comfortable postures with MHT longer than 10 minutes and recommended maximum holding time of 2 minutes. 2 = hand positions of moderate postures with MHT between 5 and 10 minutes and recommended maximum holding time of 1 minute. 3 = Hand positions of uncomfortable postures with MHT less than 5 minutes, which are advised against. Person A adopts a comfortable posture, and person B adopts an uncomfortable posture.

   Holding a comfortable, moderate, or uncomfortable posture for the maximum period of time causes extremely strong (maximal) discomfort in (a part of) the body. By limiting the actual holding time of a posture, discomfort can be limited, even in uncomfortable postures. As mentioned earlier we propose that for a group of individuals the maximum acceptable holding time is 20% of the MHT, corresponding to a discomfort score of 2 of the Borg scale (weak discomfort). To calculate the maximum acceptable holding time of a posture, the MHT has to be divided by 5. To make the recommendations safe for all postures in the three classes, the maximum acceptable holding time valid for each class of hand position corresponds with the lowest maximum acceptable holding time of that class (most uncomfortable posture of that class). Thus, comfortable postures with an MHT of more than 10 minutes are recommended to be maintained 2 minutes maximally. Following the same procedure for a moderate posture the maximum acceptable holding time is 1 minute, and an uncomfortable posture is not acceptable. In Figure 12.5 the possible hand positions are divided into 3 areas corresponding to these recommendations (area 1: 2 minutes; area 2: 1 minute; area 3: 0 minutes). As stated before, we estimate that for a mean discomfort of 2, 95% of the population will have less than "strong discomfort" (score 5 on the 10-point scale). The body part(s) in which discomfort is felt, depends on the posture. All healthy subjects who adopt the same posture (independent of the study) perceive discomfort in approximately same body part(s). Postures with hand positions at or below 50% SH are terminated by discomfort in the lower back and legs. In postures with hand positions at or above 100% SH, the shoulders and arms are the critical body parts. Also, it appears that a larger work distance results in a higher discomfort in shoulders and arms.

Applications and Limitations of the Method

The evaluation method for hand positions relates to:

- Standing postures
- Postures without external force
- Postures that are symmetric in the sagittal plane
- Postures that are maintained without rest pauses (static work)

   • Healthy, young adults.

The classification can be a guidance in practical situations for occupational health officers, designers, labor inspectors, and ergonomists to match the working time to the working posture.

   The recommendations are based on data with a large variation, caused by differences in intra- and interindividual characteristics and study design (including the task). One should be cautious when putting these recommendations into practice.

   The method has been developed for pure static postures without body motions. In most working postures minor changes in posture and loading may occur. This may result in partial recovery due to changing the critical muscle group or variations of the muscle effort. Another point of attention is that in many work situations body parts are supported by a table, an armrest, or a machine. This support unloads the muscles and the joint. It can be assumed that the MHT of a "dynamic" or supported posture is longer than the MHT of a static posture. For these kinds of working situations, the recommendations are expected to be relatively safe.

   In this study, posture was defined by the position of the hands with respect to the feet. This definition can influence the variation in MHT. The subjects were free to choose body angles of the knees, hips, back, shoulders, and elbows. Variation in these body angles within the same posture may have caused variation in MHT.

   Using the data of Meijst et al. (1995), the relation between hand position and posture was studied (Miedema, 1992; Miedema et al., 1996). Also, we calculated the effect of the interindividual variation in posture on the muscle effort. We used the 2-Dimensional Static Strength Prediction Program (2DSSPP; Chaffin and Andersson, 1984) for calculation of the muscle effort. It appears that the body angles of the upper extremity show the largest variation; i.e., up to 30° variation in the shoulder. For postures with the shoulders as the critical body part, this implicates a range of muscle effort of 8 to 31% MVC. The angles of the lower extremity and trunk show a variation between 7° and 15°. The variation in body angles increases for lower working heights. In the postures with the hands on 50% shoulder height the interindividual variation in muscle effort (of the back) increases up to 50% MVC between subjects. Thus, the classification of standing working postures defined by hand position is mainly usable for the moderate and comfortable postures. For uncomfortable postures with a low working height (50%SH), one should be cautious about using the method.

Relationship with Other Methods

The evaluation method for hand positions was compared with biomechanical calculations. For all postures, the muscle effort of the critical muscle group was calculated by using the 2-Dimensional Static Strength Prediction Program (2DSSPP; Chaffin and Andersson, 1984). Both classifications are comparable. The muscle effort (posture and force) increases when the work distance increases and/or when the working height is very low or very high. In comfortable postures, the biomechanical load is relatively low compared with moderate and uncomfortable postures.

   Figure 12.5 coincides also with anthropometric data. Recommendations concerning the optimal work area for the hands for upright postures result in lines similar to those in Figure 12.5 (Burandt, 1978). The evaluation method for hand positions was also compared with the classification of the Ovako Working Postures Analyzing System (OWAS; Karhu et al., 1977). The OWAS postures have been classified into four categories by experts, including physicians, work analysts, workers, and ergonomists. The postures of OWAS action category 1 (no improvement needed) are classified in our holding time classification as comfortable postures. Postures from OWAS action category 2 (improvements may be necessary in the future) are classified in our holding time classification as moderate postures. Postures from OWAS action categories 3 and 4 (improvement is needed as soon as possible, and immediately, respectively) are classified as uncomfortable postures. In this comparison, 10 postures (of the 19 postures) do not correspond with each other. It appeared that our classification of hand positions is more strict than the OWAS classification. This can be explained by the fact that OWAS is based on male workers, on more dynamic postures, and on more heavy work with external loads (steel industry).

TABLE 12.3   Calculated absolute hand positions (horizontal and vertical distance) in meters. It is assumed that shoulder height is 83% of the total body height.

	North Europe		Central Europe		East Europe	
	♂	♀	♂	♀	♂	♀
25% shoulder height	0.38	0.35	0.37	0.35	0.36	0.34
50% shoulder height	0.75	0.70	0.74	0.69	0.73	0.68
75% shoulder height	1.13	1.05	1.11	1.04	1.09	1.01
100 shoulder height	1.50	1.40	1.47	1.38	1.45	1.35
125% shoulder height	1.88	1.75	1.84	1.73	1.81	1.69
150% shoulder height	2.25	2.10	2.21	2.07	2.18	2.03
25% arm reach	0.22	0.20	0.21	0.20	0.21	0.20
50% arm reach	0.44	0.41	0.43	0.40	0.42	0.39
75% arm reach	0.65	0.61	0.64	0.60	0.63	0.59
100% arm reach	0.87	0.81	0.85	0.80	0.84	0.78

Example of an Application of the Hand Position Method

When a painter is painting a ceiling with a brush while standing on scaffolding, he lifts the right elbow to shoulder height. With flexion and extension movements in the elbow and wrist, he moves the brush. The shoulder is held in the same static position. This hand position varies between 125%SH/25%AR and 125%SH/75%AR. The posture 125%SH/25%AR is classified as a comfortable posture with a maximum acceptable holding time of 2 minutes. The posture 125%SH/75%AR is defined as a moderate posture with a maximum acceptable holding time of 1 minute. The final recommendation has to be based on the worst occurring posture. In this situation, the maximum acceptable holding time is 1 minute.

Absolute Hand Positions

For application of the recommendations, it may sometimes be easier to indicate the absolute hand position rather than the hand position as a percentage of shoulder height and arm reach. The absolute hand position was calculated from the relative hand position and anthropometric data. We assume that for the European population the shoulder height is 83% of the total body height (Molenbroek, personal communication). For the shoulder height and arm reach of the population, the international anthropometric data of Jürgens et al. (1989) and the Dutch data of Molenbroek and Dirken (1987) were used. The absolute hand positions are listed in Table 12.3.

12.3   The Work–Rest Model for Static Postures

Description of the Method

A mathematical work–rest model (WR model) for static postures has been developed to help designers and occupational health officers in selecting the most effective measures for reducing physical load by comparing different combinations of work–rest schemes and muscle load. The model can also be useful to develop standards and guidelines and to evaluate the acceptability of specific working conditions by comparing them with these standards or guidelines. Figure 12.6 shows a schematic representation of the WR-model.

The WR model predicts the course of muscular fatigue and recovery during work with static postures and rest. The prediction is based on four input variables. The first is the *muscle effort*, which is expressed as a percentage of the muscle strength (% MVC) of the critical muscle group. The critical muscle group determines the MHT of a given posture and force exertion from the Sjøgaard curve (see Figure 12.2). It is assumed that the muscle effort remains constant during all work periods. The muscle effort can be

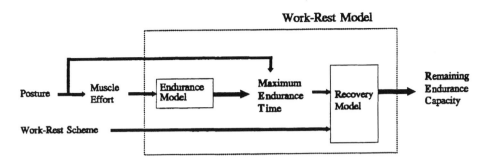

FIGURE 12.6 Schematic representation of the work rest model.

estimated with different techniques, such as EMG, biomechanical calculation, and possibly psychophysical methods.

The total work time and rest time are usually divided into a *number of work and rest periods* (*n*) with either constant or variable work and rest times. The *work time* (t_w) of a work–rest period is the duration of the muscle effort of the critical muscle group. The *rest time* (t_r) of a work–rest period is the length of time the critical muscle group can relax.

Based on the inputs, the WR model calculates the *maximum holding time* of the muscle effort. Also, the WR model calculates the course and minimum value of the *remaining endurance capacity (REC)*, which is the fraction of the MHT that a muscle effort can still be maintained continuously after a period of loading (and resting). The REC can be considered as the opposite of muscle fatigue and is considered to be the most important output variable of the model, because it indicates the (maximum) discomfort during that work–rest schedule (see Table 12.1 about the linear relationship between discomfort and time). For example, "no discomfort" (score 0 on the Borg scale) is felt at 100% REC, "strong discomfort" (score 5 on the Borg scale) is felt at 50% REC, and "extremely strong discomfort" is felt at 0% REC.

The model combines empirical studies on muscle fatigue and on muscle recovery during static contractions. The model equations were selected from the literature. The regression equation given by Sjøgaard (1986) was used for the relationship between muscle effort and MHT.

General Guidelines from the WR Model

From computer simulations and mathematical derivations, some general model predictions were formulated that can be considered general ergonomic guidelines for static postures. The first general guideline that can be given from the WR model predictions is that for a given total work and rest time and constant work and rest times, many short work–rest periods are better (i.e., generate less discomfort) than a few long work and rest periods.

For example, for a muscle effort of 20% MVC and total work time of 16 minutes and total rest time of 16 minutes, if the number of work and rest periods increases from 2 to 4, 8 and 16 minutes, the minimum REC increases from about 0 to about 40, 60, and 70%, respectively. This is illustrated by Figure 12.7. Furthermore, the model predicts that for a given number of variable work–rest cycles it is better to start with the longest work cycles.

Computer Program

Based on the model, a user-friendly program for a personal computer is being developed. With this program, the minimum REC (output) can be computed for a given combination of muscle effort, and number, duration, and distribution of work and rest periods (input). Instead of the minimum REC, the program can also calculate the required rest time for a given desired minimum REC. For each calculation a graph can be displayed, which shows the course of the REC during work and rest.

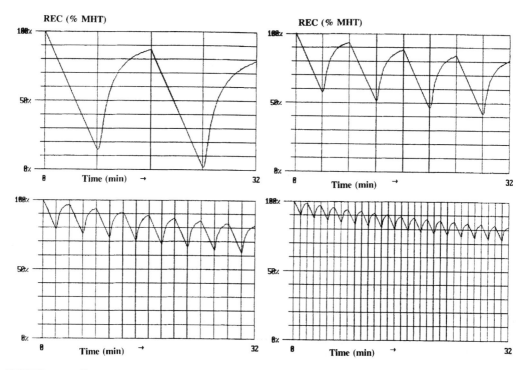

FIGURE 12.7 Illustration of the increasing REC with increasing number of work and rest periods (and constant total work and rest times).

The practical use of the program was tested by 13 occupational health officers. It appears that the program can be used by these practitioners to develop and support recommendations for changes of work and rest times during monotonous work with static postures. Before the program is released it will be tested more extensively.

Validity of the Method

The validity of the Work–Rest-model for static postures was tested in an experimental study with 10 subjects (Douwes and Dul, 1993). To validate the Work–Rest model predictions of REC, these predictions were compared with REC predictions from discomfort measurements during four different work–rest schedules, with the shoulder muscle effort varying from 20% to 36% MVC. During the work periods, the subjects were holding a weight in a seated posture with one arm elevated. Both the shoulder and elbow of that arm were 90° flexed forward.

Discomfort was measured using a method called *local musculoskeletal discomfort* (LMD), which has been developed and described by Van der Grinten and Smitt (1992). To record the level of discomfort, the LMD method uses the 10-point category ratio scale developed by Borg (1982). The location of discomfort is indicated on a modified body diagram of Corlett and Bishop (1976).

The results of the study (shown in Figure 12.8) indicate that the overall time pattern of REC during work–rest schedules is predicted well by the model. The maximum differences between model predictions and experimental REC values vary from 5% to 30% MHT. Both over- and underestimations occur. These results correspond well with the results of Serratos-Perez and Haslegrave (1992) in their study of the validity of the model of Milner. In this study ($n = 5$), eight different schedules of one work period and one rest period were performed. The muscle effort during work was 20%. Immediately after the rest periods, REC was measured. WR model predictions of REC appeared to overestimate the measured REC by 5% to 20% MHT.

Thus, it seems that WR model predictions are reasonably good on average but can differ up to 30% MHT in specific situations. These specific predictions could be improved by taking into account other

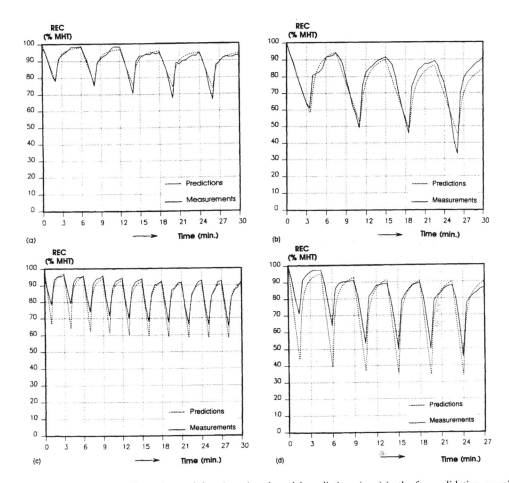

FIGURE 12.8   The course of experimental data (———) and model predictions (----) in the four validation experiments (n = 10): (a) muscle effort = 20% MVC; 2 min work, 4 min rest; (b) muscle effort = 21% MVC; 3.5 min work, 4 min rest; (c) muscle effort = 34% MVC; 1 min work, 2 min rest; (d) muscle effort = 36% MVC; 1.5 min work, 3 min rest (the total duration can be seen in the graphs).

factors that influence the REC. Therefore, the role of factors that may influence REC should be studied more thoroughly. In the meantime, it is advised to use the WR model only for comparing work–rest schedules and developing guidelines but not to evaluate specific situations in an absolute sense.

Possibilities of the WR Model

After further tests on the validity of the WR model, the model and its computer program may be a useful tool for evaluating static working postures. The model can be used to develop general ergonomic guidelines, such as those presented above. For specific working situations, the model can be used to select optimum combinations of muscle effort, work time(s), and rest time(s). Also, specific working situations may be assessed by comparing the minimum REC with a given standard, for example a limit value of 80% minimum REC such that "strong discomfort" is prevented.

Limitations of the WR Model

The WR model uses muscle fatigue as the only criterion to analyze static working postures. The load on the passive structures (i.e., ligaments, tendons) is not considered in the model. This load may be very important for the development of musculoskeletal disorders, in particular for extreme postures.

The model is developed for pure static postures without body motions. In most working postures minor changes in posture and loading may occur. This may result in partial recovery due to changing the critical muscle group or variations of the muscle effort. In that case, the model presumably under-estimates the REC and therefore gives a safe prediction of the REC.

The accuracy of the model depends mainly on the accuracy of the estimation of the muscle effort. Presently, no simple techniques are available to estimate this variable accurately. If the estimation of the muscle effort is not accurate, the model may only be useful for a comparison of working situations, and not for an absolute assessment.

It is assumed that one muscle group is the critical one that determines the maximum holding time of the posture. It is also assumed that the relationship between muscle effort and MHT is independent of several factors, such as the critical muscle group, the task that is being performed, and the role of muscles and passive structures in maintaining a posture.

The validity of these assumptions is not known. It is known that certain muscle groups are relatively more fatigue resistant than others because of differences in fiber type composition. In the future, more detailed modeling may be necessary. It is therefore important to assess the influence of these factors in future studies:

In the model, the muscle effort during work remains constant. In reality, however, it is expected that during static load the MVC will decrease, and because of a constant load, the muscle effort increases.

The present model can only be used for groups of people and not for individuals. The empirical data on muscle fatigue and recovery which were used in the model are average data for a group of people.

To extend WR model applications to prediction of risks of musculoskeletal disorders, information is needed on the relationship between REC (or discomfort) and the incidence of musculoskeletal disorders.

Example of an Application of the WR Model

Imagine a painter who paints a ceiling with a paint sprayer and works above shoulder height for many hours a day. To evaluate his physical load and to compare possible measures to reduce his load, we use the WR model. Suppose that the sprayer weights 5 kg. We measure the posture and use a biomechanical model to calculate the muscle effort at the different joints. It appears that the shoulder has the highest muscle effort, namely 26% MVC. When we put in 26% muscle effort the WR model tells us that the MHT of this load is 5.35 minutes. After 4 minutes of painting in this posture, the REC is 25.27%.

Suppose that the painter needs 40 minutes to paint one ceiling. What would be the course of the REC if we divide the 40 minutes into 10 periods of 4 minutes' work and add 4 minutes of rest in between the working periods?

In Figure 12.9 we can see that in this case the REC is 0% before the end of the task. So we need to add more rest time. To find out how much rest time is needed for the painter to finish his job without the REC exceeding 10%, the output parameter of the WR model can be changed into "needed rest time." We fill in 10% for the minimal REC value, and the WR model calculates that the rest periods should be at least 6.61 minutes.

However, according to our guidelines of the discomfort not exceeding 2 on the Borg scale, we would prefer that the REC not exceed 80%. With 10 work periods of 4 minutes, the model tells us that there is no solution in adding rest time. To solve this problem we have to reduce the work times or reduce the muscle load, by improving the posture or reducing the weight of the paint sprayer.

12.4   Standards for Working Postures Based on MHT Data

The MHT can be used to compare static loads of different postures (with or without force exertion) and is also playing an important role in the development of standards and guidelines for acceptable working postures and duration of static load and rest periods by national standardization committees, as well as in Europe (CEN; Comité Européen de Normalisation) and worldwide (ISO; the International Standard-ization Organization) committees. ISO/CD 11226 (1995; *Ergonomics — Evaluation of Working Postures*)

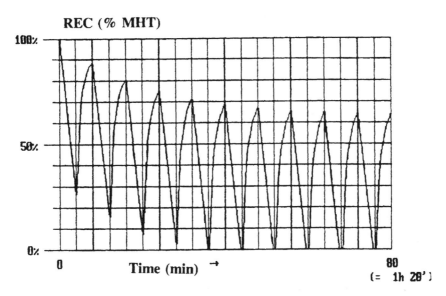

FIGURE 12.9 The course of the REC during 10 work and rest periods of 4 minutes, with a muscle effort of 26%.

contains an international standard to determine the acceptability of working postures. A comparable standard for working postures is being developed by the CEN under the machinery directive (prEN 1005-4).

Scope

The standard provides information for designers, employers, employees, and others involved in work, job, and product design. It specifies recommended limits for working postures with minimal external force exertion, while taking into account body angles and durations. The standard is meant to give reasonable protection to nearly the total healthy adult working population.

The Standard

The ISO standard for trunk inclination, head inclination, and upper arm elevation uses MHT data for judgment of acceptability. We will not present the complete ISO standard, but merely some examples of its use of MHT.

In the standard, trunk inclination is defined as the deviation angle from a neutral trunk position when viewed from the side. Head inclination is defined as the deviation angle from a neutral head position when viewed from the side. According to the standard, a trunk inclination of less than 20° is acceptable, a trunk inclination of between 20° and 60° needs to be evaluated with Table 12.4, and a trunk inclination of more than 60° is "not recommended." In Figure 12.10, the relationship between muscle effort and MHT is transferred to a relationship between trunk inclination and the maximum acceptable holding time, which is 20% of the MHT. When the trunk is fully supported, angles between 20° and 60° are acceptable without a time limit.

The inclination of the head is evaluated as acceptable when it is less than 25° and not recommended when it is more than 85°. A head inclination of between 25° and 85° needs to be evaluated with Figure 12.11. When the head is fully supported, these angles are acceptable without a time limit.

Depending on the purpose and situation different methods can be used for determining the working postures. These methods are observation, photography, video recordings, three-dimensional optoelectronic or ultrasound measuring systems, and body-mounted measuring devices such as inclinometers or goniometers are mentioned. Descriptions of these methods can be found in other parts of this book. For determining the MHT, the Sjøgaard curve or the WR model can be used.

TABLE 12.4   ISO-standard for trunk inclination
(with respect to the neutral posture when viewed from the side of the trunk).

	Acceptable	Go to Figure 12.10	Not Recommended
trunk inclination			
>60°	X	X	X
20°–60° without full trunk support	X		X
20°–60° with full trunk support	X		
0°–20°			
<0° without full trunk support			
<0° with full trunk support			

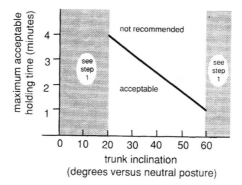

FIGURE 12.10   The relationship between trunk inclination and the maximum acceptable holding time.

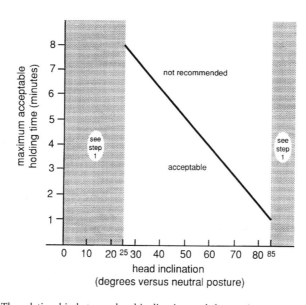

FIGURE 12.11   The relationship between head inclination and the maximum acceptable holding time.

Defining Terms

Critical muscle group: Muscle group with the largest muscle effort in a given posture.

Maximum Holding Time (MHT):   Maximum duration of a muscle effort without rest until maximum discomfort, from a rested state.

Maximum Voluntary Contraction (MVC):   Maximum isometric force exertion.

Muscle effort: Force (or moment) needed to maintain a posture (with or without external force exertion), as a percentage of the maximum force (MVC) or moment that can be exerted in the same posture.

Maximum acceptable holding time: Maximum acceptable time that a static posture can be held continuously without strong discomfort.

Rest time (t_r):   Duration of one relaxation period, or period in which the muscle is not loaded.

Remaining Endurance Capacity (REC):   Time that a muscle effort can still be maintained continuously after a period of loading (and resting), expressed as a percentage of the MHT of the muscle effort during loading periods.

Work time (t_w):   Duration of one work period, or period in which the muscle is loaded.

References

Björkstén, M. and Jonsson, B. 1977. Endurance limit of force in long-term intermittent static contractions. *Scand. J. Work Environ. and Health* 3: 23-27.

Borg, G.A.V. 1982. A category scale with ratio properties for intermodal and interindividual comparisons, in *Psychophysical Judgment and the Process of Perception*, eds. H.G. Geissler and P. Petzold, VEB Deutscher Verlag der Wissenschaften, Berlin.

Borg, G.A.V. 1990. Psychophysical scaling with applications in physical work and perception of exertion. *Scand. J. Work, Environ. and Health* 16(suppl): 55-58.

Boussenna, M., Corlett, E.N., and Pheasant, S.T. 1982. The relation between discomfort and postural loading at the joints. *Ergonomics* 25: 315-322.

Burandt, U. 1978. *Ergonomie für Design und Entwicklung*, pp. 34-37, Schmidt, Köln.

Caldwell, L.S. 1974. The load-endurance relationship for a static manual response. *Human Factors* 6(1): 71-79.

Chaffin, D.B. and Andersson, G.B.J. 1984. *Occupational Biomechanics*. John Wiley & Sons, New York.

Corlett, E.N. and Bishop, R.P. 1976. A technique for assessing postural discomfort. *Ergonomics* 19: 175-182.

Corlett, E.N. and Manenica, I. 1980. The effects and measurement of working postures. *Applied Ergonomics* 11: 7-16.

Douwes, M. and Dul, J. 1993. Studies on the validity of a work–rest model for static postures. *Proceedings of the International Association World Congress 1993. Ergonomics of Materials Handling*, Warsaw, Poland.

Dul J., Douwes M., and Miedema M.C. 1993. A guideline for the prevention of discomfort of static postures, in *Advances in Industrial Ergonomics and Safety V.* eds. R. Nielsen and K. Jorgensen, pp. 3-5. Taylor & Francis.

Dul, J., Douwes, M., and Smitt, P. 1994. Ergonomic guidelines for the prevention of discomfort of static postures can be based on endurance data. *Ergonomics* 37: 807-815.

Genaidy, A.M. and Karwowski, W. 1993. The effects of neutral posture deviation on perceived joint discomfort ratings in sitting and standing postures. *Ergonomics* 36: 785-792.

Grinten, M. Van der and Smitt, P. 1992. Development of a practical method for measuring body part discomfort, in *Advances in Industrial Ergonomics and Safety*, ed. S. Kumar, Taylor & Francis, Denver.

Hagberg, M. 1981. Electromyographic signs of shoulder muscular fatigue in two elevated arm positions. *Am. J. of Phys. Med.*, 60: 111-121.

Hagerup, A.B. and Time, K. 1992. Felt load on shoulder in the handling of 3 milking units with one- and two-handgrips in various heights. *Proceedings of International Scientific Conference on Prevention of Work-Related Musculoskeletal Disorders (PREMUS)*, pp. 105-107, Sweden.

ISO Document N 62. 1995. Final version of ISO/CD 11226 — Ergonomics — Evaluation of working postures, International Organization for Standardization, Delft, The Netherlands (not published).

Jürgens, H.W., Aune, I.A., and Pieper, U. 1989. *Internationale anthropometrischer Datenatlas. Bundesanstalt für Arbeidsschutz*, pp. 36-38, Dortmund.

Karhu, O., Kansi, P., and Kuorinka, I. 1977. Correcting working postures in industry: a practical method for analysis. *Appl. Ergonomics* 8: 199-201.

Keyserling, W.M., Punnett, L., and Fine, L.J. 1988. Trunk posture and back pain: identification and control of occupational risk factors. *Applied Ind. Hyg.* 3: 87-92.

Kilbom, Å. 1988. Intervention programmes for work related neck and upper limb disorders — strategies and evaluations. *Proceedings of the 10th Congress of the IEA*, pp. 33-47.

Manenica, I. 1986. *The Ergonomics of Working Postures: A Technique for Postural Load Assessment*, pp. 270-277, Taylor & Francis, London.

Meijst, W., Dul, J., and Haslegrave, C. 1995. Maximum holding times of static standing postures. Thesis of extended essay. TNO Institute of Preventive Health Care, Leiden, The Netherlands.

Miedema, M.C. 1992. Static working postures. *Part 1: Classification of static working postures on the basis of maximum holding time. Part 2: Secondary analysis on endurance data.* Thesis of extended essay. TNO Institute of Preventive Health Care, Leiden, The Netherlands.

Miedema, M.C., Douwes, M., and Dul, J. 1996. Recommended holding times for prevention of discomfort of static standing postures. *Industrial Ergonomics* 19: p 9-18.

Milner, N. 1985. *Modelling Fatigue and Recovery in Static Postural Exercise*, University of Nottingham, Nottingham. Ph.D. Thesis.

Molenbroek, J.F.M. and Dirken, J.M. 1987. Nederlandse lichaamsmaten voor ontwerpen, DINED-tabel (3e herziene versie). *Nederlands Tijdschrift voor Ergonomie* 12: 23-24.

Nag, P.K. 1991. Endurance limits in different modes of load holding. *Appl. Ergonomics* 22: 185-188.

Paoli, P. 1992. *First European Survey on the Work Environment 1991-1992.* European Foundation for the Improvement of Living and Working Conditions, Dublin.

Putz-Anderson, V. and Galinsky, T.L. 1993. Psychophysically determined work durations for limiting shoulder girdle fatigue from elevated manual work. *Int. J. of Industrial Ergonomics* 11 19-28.

Rohmert, W. 1960, Ermittlung von Erhohlungspausen für statische Arbeit des Menschen. *International Zeitschrift für angewandte Physiologie einschliesslich Arbeitsphysiologie* 18: 123-164.

Rose, L., Ericson, M., Glimskär, B., Nordgren, B., and Örtengren, R. 1992. Ergo-Index: Development of a model to determine pause needs after fatigue and pain reactions during work, in *Computer Applications in Ergonomics Occupational Safety and Health*, eds. M. Mattila and W. Karwowski pp. 461-468. Elsevier Science Publishers B.V., Amsterdam.

Serratos-Perez, J.N. and Haslegrave, C.M. 1992. In *Contemporary Ergonomics* ed. E.J. Lovesey, pp. 66-71. Taylor & Francis, London.

Sjøgaard, G. 1986. Intramuscular changes during long-term contraction, in *The Ergonomics of Working Postures*, eds. N. Corlett et al. Chapter 14. Taylor & Francis, London.

Taksic, V. 1986. *The Ergonomics of Working Postures: Comparison of Some Indices of Postural Load Assessment.* Taylor & Francis, London, pp. 278-282.

13

Low-Level Static Exertions

Gisela Sjøgaard
National Institute of Occupational Health
Denmark

Bente Rona Jensen
National Institute of Occupational Health
Denmark

13.1 Low-Level Static Exertions in the Workplace

Low-level static exertions have been identified as a risk factor for the development of cumulative trauma disorders or repetitive strain injuries from epidemiological studies. The exposure in terms of static exertions in the workplace has been assessed for different jobs and/or tasks based on electromyographic recordings from specific muscle groups (Table 13.1). Jobs characterized by relatively high static levels in neck and shoulder showed health outcomes in terms of musculoskeletal disorders in these body regions (Table 13.2). In the 1970s, static contractions of 15% MVC (maximum voluntary contraction) were considered to be tolerated for an "unlimited" period of time for a muscle.[1] However, later studies showed that if a contraction is to be maintained for just one hour, it may have to be as low as 8% MVC.[2] A permissible level of static muscle load of 2 to 5% MVC was then suggested.[3] However, it was observed that musculoskeletal disorders were frequent even in jobs with static levels of this magnitude, and it was suggested to reduce the acceptable static level, e.g., by job rotation.[4] Static levels as low as 0.5 to 1% MVC may relate to troubles in the shoulder region,[5,6] and most recently, statements have been brought forward that static loads are not acceptable at all if sustained frequently or over a long period of time. Actually, "working hours as a risk factor in the development of musculoskeletal complaints" has been proposed.[7] Such continuous revision of recommendations can be foreseen if we do not understand why low-level static exertions cause disorders. The acceptable limits or interventions recommended in the workplace will only reduce cumulative trauma disorders if the true risk factors that elicit adverse health outcome, are eliminated or minimized. Therefore, it is important to identify which aspect of these so-called low-level static exertions may be the risk factors. In this context, plausibility also plays an important role in risk identification, that is, possible physiological mechanisms of tissue degradation which may be causally related to the identified risk aspect. The term *low-level static exertions* will be discussed, followed by a presentation of possible short- and long-term physiological responses. Based on this, preventive strategies are presented.

TABLE 13.1 Electromyographic Data on Static (P = 0.1), Mean (P = 0.5), and Peak (P = 0.9) Muscle Load in the Shoulder Region During Different Work Tasks Expressed in Percentage of Maximal Electromyographic Activity or Percentage of Maximal Voluntary Force Development (%MAX).

Job	Muscles	P = 0.1 %MAX	P = 0.5 %MAX	P = 0.9 %MAX	References
Typewriting	m. trapezius	4	7	10	(8)
	m. deltoideus				
Office work	m. trapezius	1	4	—	(6)
CAD-work	m. trapezius	2	5	9	(9)
Industrial sewing	m. trapezius (r)	9	14	21	(10)
	m. trapezius (l)	9	16	25	
	m. infraspinatus	4	9	20	
Floor cleaning	m. trapezius	10	25	54	(11)
Assembly plant	m. trapezius	8	16	27	(12)
electronic work	m. deltoideus	7	13	28	
	m. infraspinatus	13	20	33	
Meat cutting	m. trapezius	6	10	17	(13)
Dental work	m. trapezius	9	13	18	(14)
Flight loading/unloading	m. trapezius	5	14	45	(15)
Letter sorting	m. trapezius	5	10	27	(16)
	m. deltoideus	5	14	19	
	m. infraspinatus	5	10	16	
Post office work	m. trapezius	6	14	33	(17)
(stamping)	m. infraspinatus	7	17	28	
Chocolate manufacturing work	m. trapezius	2	5	—	(6)

13.2   What Are "Static Exertions"?

Within the area of mechanics "static," in the strict sense means "no motion." In the workplace, truly static work postures are quite rare because most jobs include a number of movements to be performed often by the upper limbs. Even in supervision jobs, a number of objects have to be handled now and then.

According to the strict definition of static, one might suggest to use observation techniques to quantify how long a time a certain posture is maintained without any movement. But most likely this variable would fail to show a relationship to musculoskeletal disorders. For instance, lying in bed or sitting relaxed in a well-supporting chair is hardly considered a risk, although highly static. The reason is, of course, that no muscle exertions or contractions need to be performed in these conditions. Therefore, quantifying the true variable "static" when trying to identify risk factors is not sufficient. What we are looking for is the static muscle contraction that may induce an overload on the musculoskeletal system.

A profile of the muscular load during a period of work may be obtained by analyzing the amplitude probability distribution function (APDF) of the electromyographic signal (EMG). Such measurements for analysis of static muscle activity have been widely used in workplace studies, where the static level is defined as the probability level P = 0.1. For instance, a static level of 5% MVC means that the contraction level of the muscle is 5% MVC or above for 90% of the time, or in other words, only for 10% of the time is the muscle contraction below 5% MVC. This implies that muscular rest may occur for 10% of the recording time or less. The interpretation of a static contraction according to the APDF analysis has caused some confusion because the static level is actually defined in the time domain. Also, this variable does not give the information that the muscle is really performing a 5% MVC throughout the recording period; indeed, larger contraction forces may occur. Finally, this variable does not control for length changes of the muscle which means that the muscle contractions may well be dynamic. Nevertheless, redefining "static" in occupational settings has been a great "success." This is probably due to the time variable in essence being the real risk. But this was not intentional and no awareness has been paid to this fact by practitioners. Of note is that the risk factor probably is the *sustained* contraction.

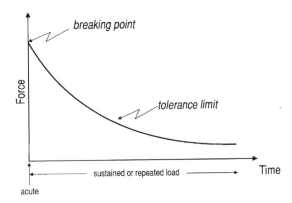

FIGURE 13.1 Tolerance limit for maximum force at breaking point depends on the type of muscle, e.g., cross-sectional area, state of training, and age. Further, the contraction mode, i.e., static or dynamic, is significant. For submaximal forces, the tolerance limit decreases with time, the rate of decrease depending on force magnitude and repetition frequency.

13.3 What Is a "Low Level"?

It may be surprising that low rather than high-level contraction forces seem to imply a risk. Of course, high forces can cause ruptures as seen in accidents where bone, ligament, tendon, or muscle are exerted beyond their breaking point, and in this sense, high forces imply a risk. However, low forces constitute a corresponding risk if repeated or sustained for a prolonged time. All structures, inert materials as well as biological tissues, are able to withstand a force characteristic to their structure. At high forces, disruption will occur when the breaking point is exceeded, and lesser forces repeated over time will eventually cause fatigue fracture (Figure 13.1). Repetitive force exertions are accumulated and cause eventual disruption, possibly not of the tissue as a whole but in terms of micro ruptures.

First of all, when evaluating the force level, the maximum strength of the muscle must be taken into account. This relates to the muscle's cross-sectional area, age, and state of training; and different muscle groups and subjects show highly different muscle strength. Therefore, exposure assessment in terms of force recordings in absolute numbers in Newtons (N) will not give sufficient information regarding the level of exertion. The maximum voluntary contraction (MVC) force must be recorded as well, and data must be presented in percentage of MVC as mentioned above regarding the EMG data.

Second, endurance time for muscles plays a significant role in this context. The relationship between force level and the time for which it can be sustained is depicted in the endurance time curve (Figure 13.2). At low force levels relative to the maximum strength, the muscle is capable of developing such force for long periods of time before being exhausted. Different muscles show highly different endurance capacity depending on muscle fiber type, anatomy, and state of training. But for every muscle there is a limit.

Third, in industry many low-level exertions include repeated static exertions or movements at quite high speed but with little displacement. When observing such tasks, often little attention is payed to the displacement, which is the cause for such exertions to be assessed as static. Also for intermittent static as well as dynamic contractions, endurance time curves exist.[2,18] It is for the dynamic contractions that the contraction level cannot be described only by the force in N or % MVC. The force–velocity relationship must also be taken into account, the relative load being higher when a specific force is developed with increasing speed (Figure 13.2). For instance, keyboard operators may press 200,000 keys a day or 500 per minute and a piano player strikes the tangents with finger movements at very high, maybe sometimes maximum, speed. For the unloaded limb, the EMG activity increases linearly with the velocity of the movement.[19] If, in addition to the movement velocity, there is an external force to overcome during the work tasks, this could imply maximum effort at high velocities even if the force level is low. Thus, low level cannot be assessed only in terms of % MVC, but the mode of contractions must also be taken

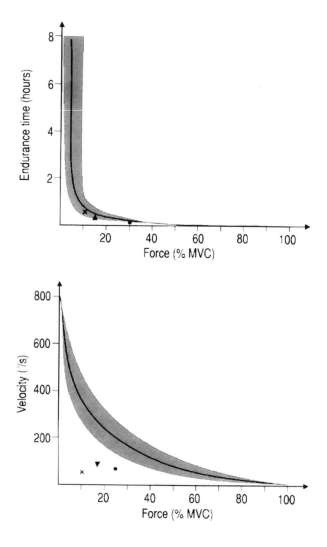

FIGURE 13.2 Upper part shows an endurance time curve for static contractions, which are defined as contractions at constant muscle length. Of note is the large range of endurance time at low forces, where the end point of exhaustion varies significantly, e.g., due to the level of motivation. Examples are given for handgrip (▲), shoulder abduction (x), and trunk extension (●). Lower part shows a force-velocity curve for dynamic contractions, only shown for concentric contractions, i.e., during muscle shortening. The maximum muscle force decreases with increasing velocity of shortening. Examples are given for shoulder movements during floor cleaning (●), shoulder movements during forking in agriculture (▼), and wrist movements during meat cutting (x). For the latter, it is seen that although the load is only 10% MVC, it is about 20% of the dynamic strength, since the maximal dynamic strength at this velocity is approximately 50% MVC.

into account. Ideally, the maximum dynamic voluntary contraction forces should be assessed and the work task evaluated in relation to the corresponding maximum force–velocity relationship.

In short, the term "low level" in the context of work-related static exertions refers to a working condition in which a muscle is activated at a level that can be maintained for a long period. This may be a true static contraction, sustaining a constant force and posture or varying in force within a limited range and without any movement. But even performing intermittent static or dynamic contractions (concentric/eccentric) at submaximal force velocities with small displacements and at intensities that can be maintained for a long time may be considered low-level static exertions in occupational

TABLE 13.2 One-Year Prevalence of Musculoskeletal Symptoms in Neck, Shoulder, Elbow/Forearm, and Hand/Wrist According to the Standardized Nordic Questionnaire[21]

Job	Sex	Number of workers	Neck (%)	Shoulder (%)	Elbow/forearm (%)	Hand/wrist (%)	References
Computer work	F	1285	49	48	13	25	(22)
	M	433	25	23	7	9	
Office work	F	643	48	48	12	22	(23)
	M	35	18	18	6	9	
CAD-work	F+M	149	70	—	41	52	(9)
Sewing machine operators	F	77	55	51	7	26	(24)
	F	303	57	53	7	28	
Cleaners	F	737	63	63	27	46	(25)
Assembly plant electronic work	F+M	25	64	56	—	—	(12)
Meat cutting	F	16	67	54	7	47	(23)
	M	114	39	62	15	47	
Meat cutting	(M)	2463	52	66	28	60	(13)
Dental work	F	43	49	40	19	40	(14)
	M	56	57	39	13	5	
Flight loading	M	808	30	31	14	22	(15)

From Kuorinka, I., Jonsson, B., Kilbom, Å., Vinterberg, H., Biering-Sørensen, F., Andersson, G., Jørgensen, K. Standardised Nordic questionnaires for the analysis of musculoskeletal symptoms. *Applied Ergonomics* 18(3):233-237, 1987.

settings.[20] Actually, when such exertions are measured by electromyography and analyzed by the above-mentioned APDF of the EMG, "static" levels of 5% MVC or more may be found. This means that a low-level static exertion is to be considered in the time domain and is characterized by workers being able to perform it for hours. The main feature is that the exertion is sufficiently low so that it can be sustained for a *prolonged* time, and the duration probably implies the risk.

13.4 Which Work Requirements Induce "Low-Level Static Exertions"?

Examples of jobs in which low-level static exertions are frequent are presented in Table 13.1 and 13.2. Additional job titles are numerous in the literature.[26] It is important in risk assessment to identify generic work requirements that induce these exertions. At random, requirements such as precision, speed, visual demand, and mental load can be mentioned. Also monotony or lack of variation is a characteristic that concerns working posture and movement as well as mental challenge. The same task is repeated over and over again most often by the hands. When operating with fast precise movements with the hands, there is a demand to stabilize the shoulder girdle. One reason is that the shoulders are the reference point for the upper limbs and if they move, the hands will be repositioned with respect to the motor control pattern for the upper limbs. Similarly, to control the position of the eyes, fixation of the neck is needed, and stable eye position is a prerequisite for most visual demands in industry. Interestingly, the fastest repositioning of the eyes can be performed when the neck and shoulder muscles are contracted up to 30% MVC.[27] When performing tasks at high speed, the stiffness of the musculoskeletal system must be increased. For this purpose, *co-contractions* are performed, which means that antagonistic muscles, i.e., muscles on each side of a joint, are contracting. This is especially common for the shoulder muscles. One reason is, as mentioned above, that the shoulder must be the stable fix point and "take-off" for arms and hands. Also the anatomy of the shoulder is such that is has the greatest mobility of all the joints in the body. It is a joint highly dependent on muscle stabilization, including *co-contractions*. These *co-contractions* have been shown to increase with increasing speed and precision demands.[19,28]

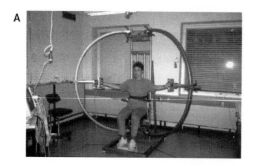

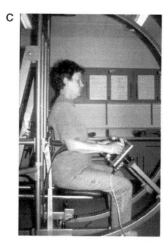

FIGURE 13.3   Experimental chair where the arm posture can be adjusted in any position of abduction (a) and flexion (b). The hands are grasping handles connected to 3-dimensional force transducers (c). Professor Bjørn Quistorff, University of Copenhagen, is acknowledged for the design.

13.5   Why Do "Low-Level Static Exertions" Imply a Risk?

As discussed above, it is not necessarily because exertions are static or at a low level that they imply a risk, but because such exertions are often sustained for prolonged periods of time. Additionally, often no sufficient recovery periods are allowed during such work tasks. A more informative term for the related risk factor would be *prolonged sustained or repeated muscle contractions*. According to the endurance curve, it is possible to sustain low-level exertions for a longer time than high-level exertions. It is likely, that this time factor is the risk. This hypothesis is supported by the physiological responses to such exertions, which constitute the plausibility. Standardized muscle contractions have been studied in combination with detailed physiological responses. In the following discussion, focus will be on mechanisms which may induce muscle damage.

    An example of a standardized setup for studying muscle contractions is shown in Figure 13.3. The test chair can be regulated for the subject to adopt any working posture and the force transducers connected to the hand grips allow for three-dimensional recordings. During specific work tasks, biomechanical calculations may then assess the relative load on various muscles or muscle parts/groups based on maximum contractions performed in identical postures and directions.[29]

    Intramuscular pressure and blood flow: With each muscle contraction, the tissue pressure (hydrostatic pressure) in the muscle increases in proportion to the force development. The absolute level in terms of mmHg varies widely between muscles and depends, among other things on the anatomy of the muscle itself as well as its surroundings. A bulky muscle attains higher pressures than a thin muscle, and a muscle

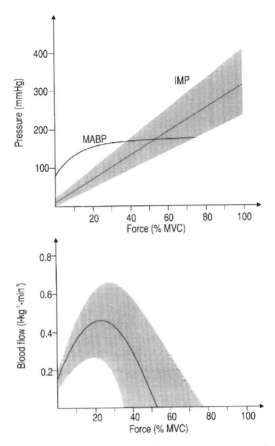

FIGURE 13.4 Upper part shows mean arterial blood pressure (MABP) and intramuscular pressure (IMP) with increasing contraction force. Lower part shows corresponding blood flow.

with bony surroundings or tight fascia shows relatively large increases because of the low compliance of these surroundings. At high contraction forces, the intramuscular pressure may attain values far above blood pressure (Figure 13.4) and obviously cause muscle blood flow to be occluded in areas where intramuscular pressure exceeds blood pressure, the highest pressures normally occurring deep in the muscles. However, even at low-level contractions, the complex microcirculatory regulation may become impeded. First of all, at low blood flow velocities it is not the mean blood pressure but the diastolic pressure that is decisive for maintenance of blood flow.[30] Further, with prolonged contractions, the muscle water content will increase[31] and correspondingly, the thickness of the muscle has been shown to increase.[32] Such a state of edematic tissue with increased volume will *per se* increase tissue pressure in a delimited closed muscle compartment with low compliance. At contraction levels in the order of 5 to 10% MVC, intramuscular pressures of 40 to 60 mmHg or more have been reported in muscles such as the *m. supraspinatus* in the shoulder.[33,34] Causal relationships between prolonged moderately increased tissue pressure and pathogenic changes have been studied extensively in relation to compartment syndromes.[35] Pressures above 30 mmHg maintained for eight hours have been shown to induce necrotic changes in the muscle even if no active contraction was performed and energy demand therefore was minimal.[36] One possible mechanism is that although initially blood flow is sufficient during low-level contractions, this may not be the case when the contraction is maintained for prolonged periods. Conditions with low flow and low perfusion pressure may provoke granulocyte plugging in the capillaries, which affects microcirculation, and may also facilitate formation of free radicals, which have a highly toxic effect.[37,38]

Metabolism: Adequate muscle blood flow is essential for muscle function because force development relies on the conversion of chemically bound energy to mechanical energy, a process also called energy turnover or metabolism. Some chemically bound energy or substrate is located in the muscle tissue (especially glycogen), but this may become depleted during prolonged activities. Therefore, the supply of substrates (including oxygen) to the muscle is crucial for such activities. The ultimate substrate in the conversion of chemical energy to mechanical energy is ATP, which is broken down in the myofibrils during the actin–myosin reaction. ATP is significant for the detachment of actin and myosin, and insufficiency of this process may cause rigor or contracture with massive pain. In normal muscle contractions, the actin–myosin reaction is initiated by the release of Ca^{2+} from the sarcoplasmic reticulum into the cytosol, and has been the focus in a number of studies on muscle fatigue. However, during the last decade, attention has been drawn also to the pathogenesis of Ca^{2+} -induced damage of muscle cells.[39] The reuptake of Ca^{2+} into the sarcoplasmic reticulum is an ATP-dependent process, which may be insufficient during prolonged activity because it accounts for up to 30% of the energy turnover during muscle activity. Further, energy crisis may result in an influx of Ca^{2+} from the extracellular space. Consequently, the cytosolic free Ca^{2+} is likely to be increased above normal for a prolonged time. This has serious implications for the phospholipids, including those in the muscle membrane. Ca^{2+} has a direct effect on phospholipase activity and, in addition, increases the susceptibility of the membrane lipids to free radicals, which have a highly toxic effect as mentioned above. Both these processes promote breakdown of the muscle membrane.[40] Finally, prolonged increased cytosolic Ca^{2+} concentration induces a Ca^{2+} load on the mitochondria and may eventually impair ATP formation, a sufficient concentration of which is a prerequisite for active force production. For more details see reference 41.

Motor control: Another important aspect during low force development is that although the muscle as a whole may not be metabolically exhausted, this may well be the case for single muscle fibers. The muscles are composed of different muscle fiber types and motor units with different recruitment thresholds. A stereotype recruitment order has been documented, which means that with increasing force, the low threshold motor units are always being recruited first.[42] Within a motor unit pool, various motor units may be alternating in activity pattern during a submaximal muscle contraction postponing fatigue to develop in each of the involved fibers.[30] However, in performing highly skilled movements and accurate manipulations, it is likely that the very same motor units are being recruited continuously. This holds true for pure static as well as slow force-varying and low-velocity dynamic contractions.[43,44] Additionally, contractions may be elicited due to reflexes, causing even more stereotype recruitment than during voluntary contractions. Mental load has been demonstrated to generate nonpostural muscle tension in shoulder muscles, and the same holds true for visual demands and neck muscles.[27,45-48] Also reflexes originating in the muscle itself from chemo- as well as mechanoreceptors may play a role, and recently the gamma-loop has been proposed to play a role in developing a potentially vicious circle.[49,50] The muscle fibers being continuously activated have been termed Cinderella fibers, because they are working from early to late.[51] A high energy turnover occurs in these fibers, and most likely they receive the least blood flow because tissue pressure increases in their vicinity due to the mechanical contraction impeding blood flow.[30] The pathogenic mechanisms described above regarding accumulation of Ca^{2+} and free radicals may be a concern, especially at the single muscle fiber level. Prolonged activity of specific motor units throughout an eight-hour working day may cause insufficient time for full recovery of these motor units due to a long-lasting element of fatigue,[52] which has been shown to occur in simulated occupational settings.[53] This may cause necrosis and, finally, cell destruction in these fibers. In line with this, fibers with marked degenerative characteristics have been found more frequently in muscle biopsies from patients with work-related chronic myalgia than in normal subjects in the trapezius muscle.[54] Interestingly, the degenerative fibers identified are slow twitch fibers, which connect with low threshold motor nerves.[55]

Perception of fatigue: When muscular work is performed over a prolonged period, fatigue develops. Fatigue may cause the work to be performed with less care or precision, and an accident can result. A fatigued worker is more likely to make a wrong movement, such as a slip and fall, leading to injury. However, even if an accident does not occur, prolonged fatigue without adequate time for recovery can lead to the development of musculoskeletal disorders. From the beginning of every muscle activity, the muscle is fatiguing, and muscle function is decreasing.[56] This condition is normally perceived as muscle fatigue. The perception of fatigue is a very useful mechanism for protecting the muscle against overload. Among other factors, the work-induced increase in potassium concentration in the interstitial space can help mediate the perception of fatigue in the central nervous system (CNS).[57] However, during very low level contractions, the accumulated increase in interstitial potassium may be subliminal to the threshold of the sensory afferents mediating the information to the CNS.[58] Also, in situations of machine-controlled work or heavy work pressure, the fatigue message is depressed, when it is not possible for the employee to take a rest. In other words, fatigue is ignored — consciously or subconsciously — which in the long term can have serious consequences.

The processes that take place in relation to fatigue are normally reversible for biological tissues, which is in contrast to inert tissues and a reason why we normally do not consider fatigue as dangerous. This means that muscles recover when resting after exertion. A rest period following muscular activity is therefore essential to enable the muscle to recover its full functional potential with regard to strength and endurance. Even an improvement or training effect of these variables may be obtained if optimal performance of activity and recovery periods is planned. There is no simple time equation for length of work and adequate length of a subsequent resting period. The process of recovery depends on the type of work that caused the muscle fatigue. For instance, so-called low-frequency fatigue and high-frequency fatigue are caused by fundamentally different biochemical changes in the muscle.[58] If fatigue is due to relatively high loads over a short time, the necessary recovery will be quicker than if fatigue is due to prolonged working at low load levels. Thus, if the same muscle group or group of fibers is activated continuously for a full working day of 7 or 8 hours, there is a risk that the muscle will not even be fully recovered by the next day. If such conditions persist for months or even years, they can ultimately inflict irreversible or chronic changes that may result in pain and impaired function.

13.6 How to Prevent Musculoskeletal Disorders from "Low-Level Static Exertions"

A prerequisite for effective prevention is knowledge of the cause of the disorder. The documentation so far of the time factor being essential gives the simple answer to this question: limit the time for each specific sustained muscle effort. Each single muscle cell and corresponding motor nerve and tendon demand recovery periods sufficiently long to attain full recovery. Time for recovery is not linearly related to time for activity. Rather, it increases exponentially. For example, if exhaustion is elicited by a high contraction force for 1 minute, then recovery is very fast, and after 2 to 3 minutes the same force can be performed again. But when a muscle is fatigued for 1 hour, it may take many hours for full recovery, and if the fatiguing process has lasted for an 8-hour working day, full recovery may not even have occurred the next morning when the next working day starts and the same tasks are to be managed. Interestingly, such sports activities as a marathon (lasting about 2 to 5 hours, depending on the state of training) are only performed a few times a year even by top athletes. Limits to prolonged activities are also seen in sport events such as the Tour de France or other endurance activities. Normally in the workplace, the worker is not totally exerted and often only part of the body is exerted. This means that somewhat shorter recovery will be acceptable, but still it is essential that the activity period is followed by a recovery period and that the duration of both is matched to the intensity and mode of contraction in the activity period (Figure 13.5).

In summary, it can be stated in concordance with an earlier discussion:[59] Human skeletal muscles are not adapted for continuous long-lasting activity. Indeed, no matter how low the exertion level is, rest periods are needed for the muscle to recover. Guidelines for low-level static exertions should therefore deal not just with the acceptable static level in percentage MVC. To recommend a reduction in exertion level from, for instance, 5 to 2% MVC will not help much physiologically; also it is not practical. Instead, we need guidelines for maximum acceptable *time limits* for prolonged sustained or repeated muscle

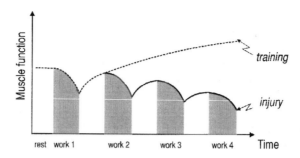

FIGURE 13.5   Muscle function in relation to work and recovery (duty circle).

contractions. This means that we must focus our attention on how long a muscle or group of muscles can tolerate maintaining or repeating the same low-level exertions. This is especially true when these same low-level exertions are imposed on a muscle day after day.

13.7 Recommendation for the Practitioner Regarding Job Profile and Workplace Design

- The workplace must allow for variation in working postures. This means, for example, that table and chair are easily adjustable. For instance, shifting frequently between sitting and standing is recommended, and instructions should be given to implement adjustments at frequent *time* intervals.

- The workplace must be designed based on principles of optimization and not minimization of mechanical workload. Therefore, it is recommended that work cycles include loads ranging from complete relaxation to moderately high contraction forces and velocities. Workers should be given the opportunity to optimize the phases of a duty cycle according to their capacity. It is important that the level or intensity of exertion changes over a wide range continuously over *time*.

- The job profile must allow for performing a variety of different tasks. The task variation should include variation regarding mental load as well as physical (mechanical) load on the musculo-skeletal system. If only specialized work tasks can be performed at each workstation, a job must include tasks at different workstations. For the use of tools, it is recommended that a variety of tools with different designs are used interchangeably. In combination, these variations should cause loading of different body regions and muscle groups regularly over *time*.

References

1. Rohmert, W. Problems of determination of rest allowances. *Applied Ergonomics* 4(3):158-162, 1973.
2. Björkstén, M. and Jonsson, B. Endurance limit of force in long-term intermittent static contractions. *Scand J Work Environ Health* 3:23-27, 1977.
3. Jonsson, B. Kinesiology. With special reference to electromyographic kinesiology, pp. 417-428, in *Contemporary Clinical Neurophysiology*, edited by Cobb, W.A., Van Duijn, H., Amsterdam, Elsevier Scientific Publishing Company, 1978.
4. Jonsson, B. The static load component in muscle work. *Eur J Appl Physiol* 57:305-310, 1988.
5. Veiersted K.B., Westgaard, R.H., and Andersen, P. Pattern of muscle activity during stereotyped work and its relation to muscle pain. *Int Arch Occup Environ Health* 62:31-41, 1990.
6. Jensen, C., Nilsen, K., Hansen, K., and Westgaard, R.H. Trapezius muscle load as a risk indicator for occupational shoulder-neck complaints. *Int Arch Occup Environ Health* 64:415-423, 1993.
7. Wærsted, M. and Westgaard, R.H. Working hours as a risk factor in the development of musculoskeletal complaints. *Ergonomics* 34(3):265-276, 1991.

8. Björkstén, M., Itani. T., Jonsson, B., and Yoshizawa, M. Evaluation of muscular load in shoulder and forearm muscles among medical secretaries during occupational typing and some nonoccupational activities, pp. 35-39, in *Biomechanics X-A*, edited by Jonsson, B., 1987, p. 35.

9. Jensen, C., Borg, V., Burr, H., Finsen, L., Olsen, H.B., Hansen, K., Juul-Kristensen, B., Vestergaard, K.B., Jensen, H.B., and Christensen, H. *Fysiske og psykosociale påvirkninger ved arbejde med computer-aided design (CAD). RAMBØLL. Rapport nr. 3.* København, Arbejdsmiljøinstituttet, 1996.

10. Jensen, B.R., Schibye, B., Søgaard, K., Simonsen, E.B., and Sjøgaard, G. Shoulder muscle load and muscle fatigue among industrial sewing-machine operators. *Eur J Appl Physiol* 67:467-475, 1993.

11. Søgaard, K., Fallentin, N., and Nielsen, J. Work load during floor cleaning. The effect of cleaning methods and work technique. *Eur J Appl Physiol* 73:73-81, 1996.

12. Christensen, H. Muscle activity and fatigue in the shoulder muscles of assembly-plant employees. *Scand J Work Environ Health* 12(6):582-587, 1986.

13. Christensen, H. (ed). *Udbeningsarbejde i svineslagterier.* København, Arbejdsmiljøinstituttet, 1996.

14. Finsen, L. *Biomechanical analyses of occupational work loads in the neck and shoulder. A study in dentistry* (Ph.D. thesis). Copenhagen, National Institute of Occupational Health, University of Copenhagen, 1995.

15. Jørgensen, K., Jensen, B, and Stokholm, J. Postural strain and discomfort during loading and unloading flight. An ergonomic intervention study, pp. 663-673. In *Trends in Ergonomics/Human Factors IV*, edited by Asfour, S.S., North-Holland, Elsevier Science Publishers B.V. 1987.

16. Jørgensen, K., Fallentin, N., and Sidenius, B. The strain on the shoulder and neck muscles during letter sorting. *Int J Ind Erg* 3:243-248, 1989.

17. Jørgensen, K. and Fallentin, N. *Lokal muskelbelastning og bevægeapparatssymptomer blandt ekspeditionspersonale på danske postkontorer.* København, August Krogh Institutet, University of Copenhagen, 1986.

18. Sjøgaard, G., Sejersted, O.M., Winkel, J., Smolander, J., Jørgensen, K., and Westgaard, R. Exposure assessment and mechanisms of pathogenesis in work-related musculoskeletal disorders: Significant aspects in the documentation of risk factors, in *Work and health. Scientific basis of progress in the working environment.* edited by Svane, O., Johansen, C., Luxembourg, European Commission, Directorate-General V, 1995.

19. Carpentier, A., Duchateau, J., and Hainaut, K. Velocity-dependent muscle strategy during plantar-flexion in humans. *J Electromyogr Kinesiol* 6:225-233, 1996.

20. Jørgensen, K., Fallentin, N., Krogh-Lund, C., and Jensen, B.R. Electromyography and fatigue during prolonged, low-level static contractions. *Eur J Appl Physiol* 57:316-321, 1988.

21. Kuorinka, I., Jonsson, B., Kilbom, Å., Vinterberg, H., Biering-Sørensen, F., Andersson, G., and Jørgensen, K. Standardised Nordic questionnaires for the analysis of musculoskeletal symptoms. *Applied Ergonomics* 18(3):233-237, 1987.

22. Aronsson, G., Åborg C., Örelius, M., Datoriseringens vinnare och förlorare. En studie av arbets-förhållanden inom statliga myndigheter och verk. *Arbete och Hälsa* 27:1-87, 1988.

23. Ydreborg, B., Bryngelsson, I., and Gustafsson, C. *Referensdata till Örebroformulären FHV 001 D (200 D), 002 D (202 D), 003 D, 004 D och 007 D. Data från 95 yrkesgrupper insamlade åren 1984-1989.* Örebro, Stiftelsen för yrkes- och miljömedicinsk forskning och utveckling i Örebro, 1989.

24. Schibye, B., Skov, T., Ekner, D., Christiansen, J.U., and Sjøgaard, G. Musculoskeletal symptoms among sewing machine operators. *Scand J Work Environ Health* 21:426-433, 1995.

25. Nielsen, J. *Occupational health among cleaners. (In Danish with English summary.) (Ph.D. thesis).* Copenhagen, National Institute of Occupational Health, University of Copenhagen, 1995.

26. Armstrong, T.J., Buckle, P., Fine, L.J., Hagberg, M., Jonsson, B., Kilbom, Å., Kuorinka, I.A.A., Silverstein, B.A., Sjøgaard, G., and Viikari-Juntura, E.R.A. A conceptual model for work-related neck and upper-limb musculoskeletal disorders. *Scand J Work Environ Health* 19(2):73-84, 1993.

27. Kunita, K. and Fujiwara, K. Relationship between reaction time of eye movement and activity of the neck extensors. *Eur J Appl Physiol* 74:553-557, 1996.

28. Sjøgaard, G., Laursen, B., Németh, G., and Jensen, B. High speed precision tasks increase muscle activity. *Book of Abstracts XVth Congress of the International Society of Biomechanics, Jyväskylä, Finland* 858-859, 1995(abstract).

29. Laursen, B. *Shoulder muscle forces during work. EMG-based biomechanical models (Ph.D. thesis)*. Copenhagen, National Institute of Occupational Health, Technical University of Denmark, 1996.

30. Sjøgaard, G., Kiens, B., Jørgensen, K., and Saltin, B. Intramuscular pressure, EMG and blood flow during low-level prolonged static contraction in man. *Acta Physiol Scand* 128:475-484, 1986.

31. Sjøgaard, G. Muscle energy metabolism and electrolyte shifts during low-level prolonged static contraction in man. *Acta Physiol Scand* 134:181-187, 1988.

32. Jensen, B.R., Jørgensen, K., and Sjøgaard, G. The effect of prolonged isometric contractions on muscle fluid balance. *Eur J Appl Physiol* 69:439-444, 1994.

33. Jensen, B.R., Jørgensen, K., Huijing, P.A., and Sjøgaard, G. Soft tissue architecture and intramuscular pressure in the shoulder region. *Eur J Morphol* 33(3):205-220, 1995.

34. Järvholm, U., Palmerud, G., Herberts, P., Högfors, C., and Kadefors, R. Intramuscular pressure and electromyography in the supraspinatus muscle at shoulder abduction. *Clin Orthop* 245:102-109, 1989.

35. Pedowitz, R.A., Hargens, A.R., Mubarak, S.J., and Gershuni, D.H. Modified criteria for the objective diagnosis of chronic compartment syndrome of the leg. *Am J Sports Med* 18(1):35-40, 1990.

36. Hargens, A.R., Schmidt, D.A., Evans, K.L., Gonsalves, M.R., Cologne, J.B., Garfin, S.R., Mubarak, S.J., Hagan, P.L., and Akeson, W.H. Quantitation of skeletal-muscle necrosis in a model compartment syndrome. *Bone Joint Surg (Am)* 63-A(4):631-636, 1981.

37. Schmid-Schönbein, G.W. Capillary plugging by granulocytes and the no-reflow phenomenon in the microcirculation. *Fed Proc* 46(7):2397-2401, 1987.

38. Jensen, B.R., Sjøgaard, G., Bornmyr, S., Arborelius, M., and Jørgensen, K. Intramuscular laser-Doppler flowmetry in the supraspinatus muscle during isometric contractions. *Eur J Appl Physiol* 71(4):373-378, 1995.

39. Jackson, M.J., Jones, D.A., and Edwards, R.H.T. Experimental skeletal muscle damage: The nature of the calcium-activated degenerative processes. *Eur J Clin Invest* 14:369-374, 1984.

40. Das, D.K. and Essman, W.B. *Oxygen Radicals: Systemic events and disease processes*. Karger, 1990.

41. Sjøgaard, G. and Jensen, B.R. Muscle pathology with overuse, in *Chronic upper limb musculoskeletal injuries in the workplace*, edited by Ranney, D., Philadelphia, U.S.A., W.B. Saunders Company, 1997.

42. Henneman, E. and Olson, C.B. Relations between structure and function in the design of skeletal muscles. *J Neurophysiol* 28:581-598, 1965.

43. Søgaard, K., Christensen, H., Jensen, B.R., Finsen, L., and Sjøgaard, G. Motor control and kinetics during low level concentric and eccentric contractions in man. *Electroenceph Clin Neurophysiol* 101:453-460, 1996.

44. Christensen, H., Søgaard, K., Jensen, B.R., Finsen, L., and Sjøgaard, G. Intramuscular and surface EMG power spectrum from dynamic and static contractions. *J Electromyogr Kinesiol* 5(1):27-36, 1995.

45. Westgaard, R.H. and Bjørklund, R. Generation of muscle tension additional to postural muscle load. *Ergonomics* 30(6):911-923, 1987.

46. Lie, I. and Watten, R.G. Oculomotor factors in the aetiology of occupational cervicobrachial diseases (OCD). *Eur J Appl Physiol* 56(2):151-156, 1987.

47. Wærsted, M. and Westgaard, R.H. Attention-related muscle activity in different body regions during VDU work with minimal physical activity. *Ergonomics* 39:661-676, 1996.

48. Wærsted, M., Eken, T., and Westgaard, R.H. Activity of single motor unit in attention-demanding tasks: firing pattern in the human trapezius muscle. *Eur J Appl Physiol* 72:323-329, 1996.

49. Johansson, H. and Sojka, P. Pathophysiological mechanisms involved in genesis and spread of muscular tension in occupational muscle pain and in chronic musculoskeletal pain syndromes: A hypothesis. *Med Hypotheses* 35:196-203, 1991.

50. Mense, S. Considerations concerning the neurobiological basis of muscle pain. *Can J Physiol Pharmacol* 69:610-616, 1991.

51. Hägg, G.M. Static work loads and occupational myalgia — a new explanation model, pp. 141-144. In *Electromyographical Kinesiology*, edited by Anderson, P.A., Hobart, D.J., Danoff, J.V., Elsevier Science Publishers B.V. 1991.

52. Edwards, R.H.T., Hill, D.K., Jones, D.A., and Merton, P.A. Fatigue of long duration in human skeletal muscle after exercise. *J Physiol (Lond)* 272:769-778, 1977.

53. Byström, S. Physiological response and acceptability of isometric intermittent handgrip contractions. *Arbete och Hälsa* 38:1-108, 1991.

54. Larsson, S., Bengtsson, A., Bodegård, L., Henriksson, K.G., and Larsson, J. Muscle changes in work-related chronic myalgia. *Acta Orthop Scand* 59(5):552-556, 1988.

55. Henriksson, K.G. Muscle pain in neuromuscular disorders and primary fibromyalgia. *Eur J Appl Physiol* 57(3):348-352, 1988.

56. Bigland-Ritchie, B., Cafarelli, E., and Vøllestad, N.K. Fatigue of submaximal static contractions. *Acta Physiol Scand* 128 (Suppl 556):137-148, 1986.

57. Sjøgaard, G. Exercise-induced muscle fatigue: The significance of potassium. *Acta Physiol Scand* 140 (Suppl. 593):1-64, 1990.

58. Sjøgaard, G. Potassium and fatigue: the pros and cons. *Acta Physiol Scand* 156:257-264, 1996.

59. Sjøgaard, G. Intramuscular changes during long-term contraction, pp. 136-143. In *The ergonomics of working postures. Models, methods and cases*, edited by Corlett, N., Wilson, J., Manenica, I., London and Philadelphia, Taylor & Francis, 1986.

14

Job Demands and Physical Fitness

Veikko Louhevaara
*Finnish Institute of Occupational
Health and University of Kuopio*

14.1 Introduction

Job demands include actual physical, mental, and social loading factors, all of which are needed for working in a productive and qualified manner. When job demands are considered in relation to physical fitness, the physical aspect of the work load is most important. Without the help of external power, the physical work load can only be handled by a worker's muscular performance in terms of dynamic and static muscle contractions.

In both industrialized and developing countries, there are numerous jobs requiring physical work in spite of rapid technological developments. Rutenfranz et al. (1990) estimated that about 20% of the workforce in industrialized countries are regularly exposed to heavy muscular work even though the proportion of conventional dynamic jobs with simple manual tools has decreased. On the other hand, static and repetitive tasks have increased in many jobs (Smolander and Louhevaara, 1996).

Various control measures affecting muscular work performance via a worker's physical fitness can be categorized as secondary preventive measures. These are the most relevant in physically heavy occupations in which primary preventive ergonomic measures involving technical and organizational arrangements at work prove to be insufficient. Usually the control measures on physical fitness include, broadly speaking, individual health promotion, a healthy and satisfying lifestyle, and the maintenance of work ability or productive aging (WHO, 1993; Ilmarinen and Louhevaara, 1994; Louhevaara and Ilmarinen,

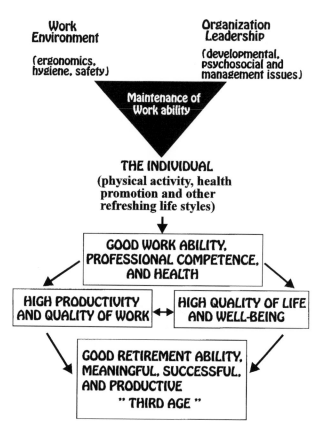

FIGURE 14.1   A concept for the maintenance of work ability and professional competence developed at the Finnish Institute of Occupational Health. The combination of various work-related and individual measures results in good work ability, professional competence, and health, as well as high productivity and quality of work. Simultaneously, the quality of life and the well-being of the workers improve. Thus, they also have better chances for a meaningful, satisfactory, and active "third age" after retirement.

1994). Of these measures, the most common ones are physical fitness training (exercise) and nutrition (Blair et al., 1996).

Individual physical fitness is one of the key elements of work ability which covers all capacities for coping with job demands without excessive over- or understrain. The concept of the maintenance of work ability involves measures in three areas, according to the triangle strategy developed at the Finnish Institute of Occupational Health. These areas are work and the environment (ergonomics, industrial hygiene, and safety on the job), organizational culture (psychosocial and management issues related to the job), and the individual worker (physical fitness training, health promotion, satisfying lifestyle) (Ilmarinen and Louhevaara, 1994; Ilmarinen et al., 1995) (Figure 14.1).

14.2   Physical Job Demands

According to the stress–strain concept introduced by Rutenfranz (1981), physical job demands i.e., muscular work load (stress, exposure, burden, exertion, effort) can be categorized as heavy dynamic work, manual materials handling (MMH), static postural work and repetitive work (Louhevaara, 1992) (Figure 14.2).

Heavy dynamic work with large muscle groups consists mainly of activities requiring the moving of a worker's own body mass, and his or her strain responses are mostly cardiorespiratory (overall) in nature. The load of heavy dynamic work increases in relation to moving speed, distance, the degree of ascent at the covered distance, and the amount of a worker's own body mass as well as the additional mass of

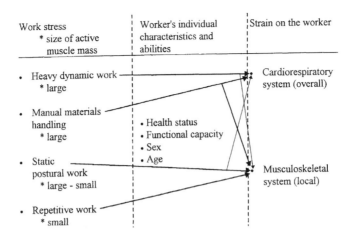

FIGURE 14.2 The effects of different types of physical (muscular) load on the cardiorespiratory and musculoskeletal system. The model has been adopted from the stress–strain concept, and individual characteristics and capacities are considered as intervening factors modifying strain responses due to physical work load.

personal protective equipment which must be worn in many heavy physical tasks. MMH involves mixed dynamic and static work with large muscle groups. The ordinary components of MMH are lifting, carrying, pushing, pulling, and holding external loads of various weights and sizes. Physical work load in MMH equally affects the cardiorespiratory and musculoskeletal system, whereas static postural and repetitive types of physical work loads predominantly produce musculoskeletal (local) strain responses (Louhevaara, 1992). Thus, the type of muscle contraction (dynamic versus static) and the amount of active muscle mass are very important factors as regards physical work load as well as the strain responses of a worker (Aminoff et al., 1996).

In addition to the above-mentioned aspects of muscular work, physical work load is greatly affected by the use of strength, the frequency of sudden peak load efforts, and work–rest regimens, as well as environmental factors such as basic thermal parameters (ambient temperature, relative humidity, and air velocity) (e.g., Kähkönen, 1993), and work rate or the intensity of work (e.g., Louhevaara et al., 1988).

Heavy dynamic work and MMH with large muscle groups are most often needed in jobs in forestry and agriculture, building, installation, transport, manual sorting, health care and home care, and cleaning, as well as in the work of fire fighters, police officers, and soldiers (Ilmarinen, 1984; Smolander et al., 1984; Ahonen et al., 1990; Louhevaara et al., 1990; Hopsu, 1993; Lusa, 1994; Soininen, 1995). Typical jobs with static postural and/or dynamic repetitive muscular work with small muscle masses are, for instance, electrical assembly and meat-processing (e.g., Jonsson et al., 1988; Viikari-Juntura et al., 1991).

The level of physical strain on a worker depends both on job demands and on his or her individual characteristics, capacities, skills, and motivation. Therefore, when considering optimal or acceptable physical job demands for different types of muscular work, one must base the criteria on cardiorespiratory, musculoskeletal (biomechanical), and subjective (psychophysical) strain responses. These may involve overall physiological changes, fatigue, symptoms and disorders in the whole body, or merely specific local changes, for instance, in a single small muscle group or joint.

The individual-based and multi-response-based evaluation of acceptable job demands arises from the principle that there are individual limits for overstrain and understrain as well as damage for each type of physical work load (Louhevaara, 1992). When physical job demands do not exceed the worker's individual capacities, his or her physiological organs adapt to the demands, and recovery is quick after the termination of work. If the job demands are too high, fatigue and various symptoms occur, work capacity and productivity diminish, and recovery is slow. Prolonged or repetitive overload or sudden peak loads may result in organic damage, i.e., injury, as well as work-related or occupational disease. On the other hand, muscular work at a specific intensity, frequency, and duration may also produce fitness training effects; similarly, muscular inactivity may cause detraining effects (Rohmert, 1983).

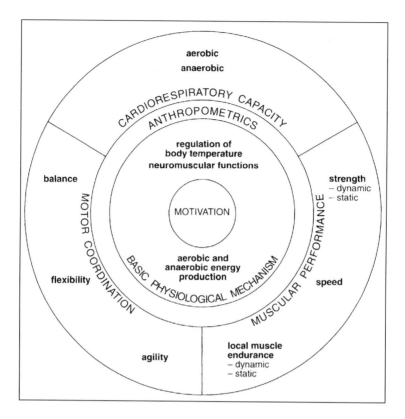

FIGURE 14.3 The main dimensions of physical fitness. The quality and quantity of output variables within cardiorespiratory capacity, muscular performance, and motor coordination depend on the basic physiological mechanisms and are affected by individual characteristics.

14.3 Physical Fitness

Physical fitness in terms of cardiorespiratory (aerobic) capacity, muscular performance (muscle strength and endurance), and motor coordination (body control) is based on basic physiological mechanisms: aerobic and anaerobic energy production, neuromuscular functions, and the regulation of body temperature. Anthropometric characteristics can be regarded as intervening factors in association with the output parameters of physical fitness. The utilization of different physical fitness capacities is done by means of voluntary muscle contractions which are impossible without an adequate level of motivation (Hollman and Hettinger, 1980, Åstrand and Rodahl, 1986) (Figure 14.3).

Physical work capacity is based on physical fitness. If the worker's profile of physical fitness and his or her anthropometrics are in harmony with the physical job demands, the situation is optimal or at least acceptable. The balanced situation is neither static nor permanent, but highly dynamic, and depends on changes both in job demands and individual capacities. Disturbances in the production process or the blackout of automatic data processing systems may change physical job demands completely within a few seconds. Similarly, sudden illness or injury may decrease a worker's level of physical fitness and work capacity dramatically. With the exception of these quick and often unexpected changes, the implementation of advanced technology and the unavoidable aging of workers will affect the sensitive balance between physical job demands and physical work capacity (WHO, 1993; de Zwart et al., 1995).

In this chapter, the physical job demands and fitness requirements are considered more thoroughly in the work of a fire fighter, a police officer, and a professional cleaner. The first two jobs can be classified as special occupations which occasionally encompass physically extreme and hazardous situations. These require near maximal or maximal physical exertion without any possibilities to apply even basic primary ergonomic controls. Professional cleaning belongs to basic service jobs and includes a lot of both heavy dynamic work and MMH activities. In many countries, and particularly in the Nordic countries, fire fighters taking part in operative tasks are almost without exception men. Also, the majority of the police officers on street duty are men. In professional cleaning, the situation is almost the reverse, as the workers are usually women. In the European Union, they are also often aging i.e., over 45 years old (WHO, 1993; Krueger et al., 1995).

14.4   Fire Fighters

Physically Demanding Fire-Fighting and Rescue Tasks

In the questionnaire study of Lusa et al. (1994), fire-fighting and rescue tasks were rated by 200 full-time fire fighters according to the job demands on physical fitness in terms of aerobic power (cardiorespiratory capacity), muscular performance (strength and endurance), and motor coordination (flexibility, agility, balance). Most of the respondents (79%) rated smoke-diving (entry into a smoke-filled space) with the use of the personal protective clothing system and a self-contained breathing apparatus (SCBA) as the task demanding most aerobic power. Almost half of the respondents (43%) felt that clearing passages with heavy manual tools required the highest demands on muscular performance. About three fourths (72%) of the respondents felt that the need for motor coordination was the greatest in fire-fighting and rescue operations on the roof. During the past five years, 54 to 71% of the respondents had carried out smoke-diving, clearing tasks, and roof work at least four times a year.

Demands on Aerobic Power

The oxygen consumption (energy expenditure) is linearly related to the amount and intensity of dynamic muscle work. The oxygen consumption, i.e., the oxygen supply to active muscles and tissues, is high in heavy dynamic tasks and many MMH jobs. This also sets high demands on the fitness or efficiency of the cardiorespiratory system.

In various fire-fighting and rescue operations with the protective clothing system and SCBA, such as smoke-diving, fire-suppression, ladder climbing, rescuing a victim, dragging a hose, and raising a ladder, mean oxygen consumption levels of 2.1 to 2.8 l/min have been reported by Lemon and Hermiston (1977), Louhevaara et al. (1985), Sothmann et al. (1990), and Lusa et al. (1993). Maximal exertion is frequent in peak loads and has been reported to correspond to the average oxygen consumption of 3.8 l/min and heart rate of 180 beats/min in young and healthy fire fighters (Lusa et al. 1993). In the heat, the effective regulation of body temperature is necessary for carrying out potential fire-fighting and rescue tasks. In prolonged or repetitive exposure to heat, heavy dynamic muscle work, and the wearing of personal protective equipment, the needed release of body heat is impossible, leading to the termination of work or to health hazards (White and Hodus, 1987; Ilmarinen et al., 1994).

In actual field operations with the protective clothing system and SCBA, the mean and particularly peak demands on the cardiorespiratory capacity have consistently been noted to be very high without any possibilities to alleviate the work load. Therefore, secondary preventive control measures are needed to guarantee the health and safety of fire fighters. In practice, this means that every fire fighter, regardless of age, exposed to physically demanding fire-fighting and rescue operations must be at the adequate level of cardiorespiratory fitness, i.e., maximal oxygen consumption.

Lemon and Hermiston (1977), Louhevaara et al. (1985), Sothmann et al. (1990), and Louhevaara et al. (1994) have recommended that a fire fighter should have a maximal oxygen consumption of 2.7 to 3.0 l/min and/or 34 to 45 ml/min per kilogram of body weight. For instance, if a fire fighter weights 75 kg and his maximal oxygen consumption is 3.0 l/min, his maximal oxygen consumption related to body weight is 40 ml/min/kg.

In Finland, the *Guide on Smoke-diving* (1991) gives a recommendation for a four-stage scale to classify fire fighters' maximal oxygen consumption, based on the research done at the Finnish Institute of Occupational Health (Table 14.1). According to the *Guide*, the cardiorespiratory fitness of male and female

TABLE 14.1   Tests and Their Classification for Assessing Male and Female Fire Fighters' Cardiorespiratory (Aerobic) Capacity and Muscular Performance (Strength and Endurance) (Maximal oxygen consumption = VO2max)

Test	Classification			
	Poor	Moderate	Good	Excellent
VO2max[a]				
(l/min)	≤2.4	2.5–2.9	3.0–3.9	≥4.0
(ml/min/kg)	≤29	30–35	36	≥50
Bench press (45 kg)				
(reps/60 s)	≤9	10–17	18–29	≥30
Sit-up				
(reps/60 s)	≤20	21–28	29–40	≥41
Squatting (45 kg)				
(reps/60 s)	≤9	10–17	18–26	≥37
Pull-up				
(max reps)	≤2	3–4	5–9	≥10

[a] Bicycle-ergometer or treadmill test

fire fighters in the age range of 20 to 63 years is considered sufficient for smoke-diving tasks when they attain a "good" result in the tests for maximal oxygen consumption (3.0 l/min and/or 36 ml/min/kg) (Lusa, 1994).

Demands on Muscular Performance

In the study of Lusa et al. (1991), the biomechanical features were evaluated in a simulated rescue–clearing task, in which a 9-kg power saw was lifted from the floor to ceiling level (211 cm) by seven young and six older fire fighters. The maximal muscular capacity of the subjects did not differ in regard to their maximal isokinetic muscle strength or to the results of the repetitive muscle performance tests.

In the task, the mean dynamic compression force at the L5/S1 disc was 6228 N, being equal for the young and older subjects. The peak torque for the back and knee extension was also similar for the subjects, and was, on average, 242 Nm and 120 Nm, respectively. The peak values corresponded to over 90% of their maximal isokinetic muscle strengths. The results showed that the lifting and handling of a heavy power saw produced a high load on the musculoskeletal system.

The tests on muscular performance in Table 14.1 are based on the study of Lusa et al. (1991) and the results of several pilot studies carried out at the Emergency Service Institute of Finland, Rescue Department of the City of Helsinki, Fire and Rescue Department of the City of Oulu, and the Fire and Rescue Department of the City of Jyväskylä (Louhevaara et al. unpublished results). The tests are mainly designed to follow up the fire fighters' physical fitness with special reference to their physical work capacity and muscular job demands. These tests have also been used since 1993 for the selection of applicants for the training courses of the Emergency Service Institute of Finland. The applicants should attain the minimum results determined by the Institute. In the actual selection of applicants for basic training courses, the required minimal level has been between "good" and "excellent" in each test presented in Table 14.1.

Conclusions

The fire fighters' job demands on physical fitness and work capacity are occasionally extremely high during their entire occupational career. Their cardiorespiratory and muscular fitness should be tested regularly in order to guarantee their work capacity for physically extreme operations. The use of fitness tests is necessary in planning and carrying out preventive measures for maintaining the health and work ability of fire fighters. The most effective and necessary measure is regular physical fitness training, which should be done either during working hours or in their leisure time.

14.5   Police Officers

Physical Fitness and Health

In operative police work, it is almost impossible to influence job demands or to change the work environment. Due to the strict selection criteria, a young police officer is healthy and physically fit when starting his or her career. However, in middle-age, the lifestyle of many police officers is sedentary and often associated with various health problems. These factors accelerate the unavoidable age-related reduction of physical fitness and, particularly after the age of 40 years, the physical fitness as well as work capacity of police officers often decline rapidly. Obesity, elevated blood pressure, high serum lipid levels, and musculoskeletal disorders and diseases are the most common health problems that also reduce the physical fitness needed to carry out job demands without excessive strain and increased hazards on health or safety (Soininen 1995).

Maintenance of Work Capacity

Soininen (1995) investigated in his intervention study the feasibility and effects of 8-month-long intensive and medium-intensive worksite fitness programs in 111 male police officers aged 40 to 54 years who were serving in three middle-sized cities. The intensive program (n = 48) included guided fitness training two hours per week during working hours and three training sessions per week during leisure time. The subjects (n = 24) in the medium-intensive program participated in three training sessions per week during their leisure time. Both interventions were separately carried out in two cities, whereas the subjects (n = 39) were the controls in the third city. The effects of the fitness programs were evaluated in regard to health, physical fitness, and subjective work ability, and were assessed with a questionnaire and laboratory and field measurements. The intensive fitness program was more feasible and efficient than the medium-intensive program. In the intensive group, the cardiovascular risk factors fell significantly, and both the maximal oxygen consumption and the index of the muscular fitness and flexibility improved 8%, on average. The corresponding improvements were smaller in the medium-intensive group. No positive changes were observed in the control group.

Physical Job Demands

In the field part of his intervention study, Soininen (1995) confirmed the previous observations of Smolander et al. (1985) on the physical job demands associated with police work. The mean physical work level was low, and the police officers spent, on average, about 60% of their working hours sitting. The most common workplaces were an office, a patrol car, and an alarm center. However, the questionnaire revealed that almost all police officers had a few physically extreme peak loads in their work every year. The peak loads usually occurred when the police officer had to catch or transport criminals, or intoxicated or mentally ill persons. The situations leading to physical peak load were usually very sudden and unexpected. They usually demanded a high level of muscular strength and endurance (carrying, lifting, pressing, wrestling, running, wrenching), often in unfavorable (dark, cold, and/or cramped space) environmental conditions.

Adequate Level of Physical Fitness

In order to manage the physical peak load situations successfully, police officers need to have a sufficient level of physical fitness. Smolander et al. (1984) suggested that the necessary level of the police officers operating in the street should be at least the same as the typical clients, who were 20- to 40-year-old men in Finland. Smolander et al. (1985) recommended that the test battery for evaluating the physical fitness of police officers should consist of items on anthropometrics, aerobic power (assessment of maximal oxygen consumption), muscular performance (assessment of muscle strength and endurance with repetitive tests), and flexibility. Later Soininen (1995) concluded that the sufficient level of each physical fitness dimension (aerobic power, muscle performance, and motor coordination) should be higher than the mean sex- and age-related reference values obtained from the population studies. The assessment of physical fitness should be done with reliable and feasible tests that could be completed in the occupational health services.

Based on the studies of Smolander et al. (1984, 1985) and Soininen (1995), the Finnish Ministry of Internal Affairs is preparing a guide for testing the work ability of police officers. Based on preliminary information, the guide will include tests on physical fitness and professional competence. Physical fitness will be determined by an indirect assessment of maximal oxygen consumption (submaximal bicycle-ergometer test or 2-km walking test), the evaluation of the strength and endurance of the trunk and arm muscles (sit-up test and the dynamic endurance and strength test for the upper arms with 5 or 10 kg barbells), as well as the determination of the flexibility of the hip and low back (sit and reach test) (Laukkanen, 1993; Pollock and Wilmore, 1994).

Conclusions

The selection criteria for police training seems to guarantee an acceptable level of physical fitness of police officers up to the age of 40 years. The physical demands of the activities of the police job are not sufficient to maintain the necessary level of physical fitness. Therefore, it must be obtained through regular and effective fitness training either during leisure time or working hours. Furthermore, after the age of 35 years the police officers' physical fitness should be followed and tested regularly by the occupational health services.

14.6   Professional Cleaners

Common Service Occupation

A professional cleaner is one of the basic service occupations, including e.g., hospital maids, home care workers, and kitchen and laundry workers. The professional cleaner is one of the most common occupations in the world. Krueger et al. (1995) estimated that in the European Union the number of full-time or part-time cleaners is about three million, and the great majority of them (95%) are women.

In many industrialized countries, the average age of the workforce is rapidly increasing due to the aging of the large post-war age cohort. For instance, the present Finnish professional cleaners are mostly over 45 years of age, i.e., aging workers. They have numerous problems in their work ability, resulting in a high incidence of sickness and a high frequency of early retirement (Hopsu, 1993; Hopsu et al., 1994). Moreover, during the past few years the serious economic depression has hampered the availability of adequate detergents, cleaning tools, and machinery and, on the other hand, the spaces to be cleaned have become substantially larger. Therefore, it is essential to maintain the work ability and to promote the professional competence and work motivation of the current professional cleaners with all possible means (Hopsu 1993).

Physical Fitness in Cleaning Work

Hopsu et al. (1994) reported the results of their intervention study on professional cleaners aimed at supporting their individual physical fitness and work capacity with physical fitness exercise and reducing their physical work load with ergonomic redesign measures. The intervention process was carried out using a participatory approach in the work units. Its main characteristic was to motivate workers and work units to participate in teamwork for developing their own physical fitness and work performance with individual health promotion and ergonomic measures.

The intervention and control group consisted of 30 and 32 female cleaners, respectively. The intervention group took part in individual physical fitness programs and also developed ergonomics of their work. The mean age of the subjects was 46 years. The length of the intervention period was one year. After the intervention, the subjects' cardiorespiratory and muscular performance had increased markedly and the work methods and tools had improved, resulting in a reduction of poor work postures. Due to the increased individual physical fitness level and work capacity, and the developments in ergonomics, the subjects' strain at work decreased according to the results of the heart rate. The improvements also positively affected subjective work ability. The intervention did not lower productivity.

Maintenance of Ability to Work and Physical Fitness

According to their intervention study, Hopsu et al. (1994) concluded that the maintenance of work ability, professional competence, and health of cleaners always needs to include a combination of measures which must affect work and the environment (ergonomics), work organization (teamwork and leadership), and the capacities of the individual workers. In this combination, the maintenance of physical fitness is an essential element. Furthermore, Hopsu et al. (1994) recommended that female professional cleaners should have at least average cardiorespiratory and muscular fitness without excessive weight in order to avoid overstrain. The fitness of cleaners should be evaluated regularly at intervals of five years by applying simple fitness tests in the occupational health services. The suggested test battery consisted of tests on anthropometrics (weight, height, and body mass index), aerobic power (2-km walking test), and the strength and endurance of the main muscle groups of the trunk (sit-up test, static endurance of the back muscles), legs (squatting test), arms (strength and endurance of upper arms with 5-kg barbells), and the flexibility of the trunk (bending forward and sideways in a standing position) (Laukkanen, 1993; Pollock and Wilmore, 1994).

Conclusions

Professional cleaning is a physically heavy job for women and requires physical fitness in terms of cardiorespiratory capacity, muscular performance, and motor coordination. The physical fitness level of professional cleaners should be higher than the average age-related values for women and should be tested in a pre-employment health examination, as well as followed regularly in periodic health examinations of the occupational health services. The participatory process based on the triangle strategy of the maintenance of work ability (Figure 14.1) proved to be efficient and feasible for improving the physical fitness and work capacity of cleaners; physical exercises and consideration of the ergonomic features of cleaning work by a teamwork-based developmental process were central elements.

14.7 Physical Exercise and Fitness

Several reports indicate that a relevant measure affecting individual health, functional capacity, and work ability is regular, moderate, and versatile physical exercise (fitness training) or activity. Together with moderate nutrition, physical exercise is effective in reducing body weight and lowering blood pressure and cholesterol levels, and it prevents ischemic heart disease (e.g., Fletcher, 1992; Lakka, 1994; Blair et al., 1996). Physical exercise is essential for sustaining cardiorespiratory, musculoskeletal, and nervous fitness and is, to a great extent, linked with good work ability regardless of the type of work (WHO, 1993). Physical exercise also has a positive effect on productivity, quality of work, absenteeism, and the turnover rate (Shephard, 1992). Therefore, physical exercise can be regarded as one of the basic elements in the maintenance of work ability (Figure 14.1).

Work ability is supported by physical fitness, and a variety of physical exercise modes is available for maintaining or improving different cardiorespiratory, musculoskeletal, and neuromuscular capacities involved in physical fitness (Figure 14.3). The quality and quantity of physical exercise, when aiming to improve work ability, depends on real job demands on physical fitness (see Figure 14.2) because training effects are highly specific in nature. It is obvious that jobs which encompass a lot of heavy dynamic work, i.e., high aerobic demands on the cardiorespiratory system, will not be benefited by a physical exercise program consisting of, for instance, near maximal weight lifting with a low number of repetitions. Instead, such jobs need physical exercise like brisk walking or jogging, which improve the capacity of the cardio-respiratory system. Moreover, both aerobic and strength exercise should be supplemented by training elements that aim for the promotion of sufficient motor coordination. Therefore, it is essential that physical exercise, targeting improved work ability, must be tailored to actual job demands. On the other hand, in jobs having low physical activity and high psychosocial demands, the main benefit of physical exercise may include its effects on mood, well-being, and mental work capacity.

Before starting physical exercise it is sometimes necessary to perform assessments of physical job demands (Louhevaara 1995). The assessments may help the prescription of appropriate training programs. However, in many cases just a careful observation of work and a brief worker interview may give the information needed for planning an adequate training program. If some standardized methods are necessary, the following basic work physiology and ergonomic methods can be used for this purpose:

- Job load and hazard analysis (Mattila 1985)
- Measurement of heart rate (Louhevaara et al., 1990)
- Basic Edholm Scale for estimation of energy expenditure (Ilmarinen et al., 1979)
- Rating of overall perceived exertion (Borg, 1970)
- Rating of local perceived exertion for the back, arms, and legs (Borg, 1970, Hultman et al., 1984).

Often, physical job demands are affected by environmental conditions and work rate. Thermal load increases muscular demands at work if the temperature, humidity, and air velocity substantially deviate from temperate ones, or if wearing heavy personal protective clothing and equipment is necessary. An unusually slow or rapid work rate may severely bias the reliable assessments of physical job demands.

The effects of work-site physical exercise interventions on physical fitness, work ability, health, and different work-related characteristics have been studied intensively at the Finnish Institute of Occupational Health during the past 10 years. The studies have focused on six occupational groups, including professional cleaners, nurses, home care workers, metal workers, fire fighters, and police officers.

The interventions lasted for 2 to 12 months, and musculoskeletal (dynamic and static muscle performance and motor coordination) and cardiorespiratory capacity (maximal oxygen consumption) were found to improve, on average, by 7 to 136% and 4 to 10%, respectively. Positive effects were also observed in subjective work ability and health, musculoskeletal symptoms, need for physiotherapy, absenteeism, strain at work, risk factors for ischemic heart disease (body fat, blood pressure, cholesterol, smoking), tolerance of shift work, and the mastering of work. The only drawback of the exercises was musculoskeletal injuries, which were usually minor, however (Louhevaara and Ilmarinen, 1995).

The results and experiences obtained from the interventions emphasize the following prerequisites for feasible physical exercise programs at the company level:

- Commitment and support of top management
- Commitment of the whole work unit
- Implementation entirely or partly during working hours
- Quick feedback on improvements in physical fitness
- Providing and strengthening of motivation
- Meaningful, versatile, and positive experiences from the exercise
- Skillful instruction and guidance (Louhevaara, 1995).

It is also very important that exercise programs are strictly confidential, voluntary, available for evaluation, and do not arouse feelings of guilt for anyone (WHO, 1988). Furthermore, all exercise programs should be based on close integration with other measures related to the triangle concept of the maintenance of work ability (Figure 14.1) (Louhevaara and Ilmarinen, 1995).

14.8   General Conclusions

In many jobs, heavy physical work requiring continuous or repetitive moving and MMH activities will remain indispensable in spite of rapid technological developments. In these types of jobs and particularly in a few specific occupations (e.g., fire fighters and police officers) in which primary preventive ergonomic controls on job demands are often difficult or impossible to apply, the workers need to have an adequate level of physical fitness in terms of cardiorespiratory capacity, muscular performance, and motor coordination. This is necessary in order to avoid over-strain and prevent health and safety hazards, as well as to attain good productivity and quality of work.

The required level of physical fitness depends on job demands and should be tested with reliable and feasible tests. In physically demanding and hazardous occupations like a fire fighter and a police officer, fitness tests should be done in the pre-employment situation and also regularly during the entire occupational career.

The only efficient way to improve physical fitness is regular fitness training. However, it is essential that physical training, targeting improved work ability, must be tailored to actual job demands. Both aerobic and strength exercise should be supplemented by training elements that aim for the promotion of sufficient motor coordination. On the other hand, in jobs having low physical activity and high psychosocial demands, the main benefit of physical exercise may include its effects on mood, well-being, and mental work capacity. Together with other individual health promotion measures physical training is an essential part of the concept for maintaining work ability, professional competence, and health. The best results are attained if individual measures (physical exercise) are carried out simultaneously with other developmental measures on work and the environment (ergonomics), as well as on organizational features of work (leadership).

References

Ahonen, E., Venäläinen, J.M., Könönen, U., and Klen, T. 1990. The physical strain of dairy farming. *Ergonomics* 33:1549-1555.

Aminoff, T., Smolander, J., Korhonen, O., and Louhevaara, V. 1996. Physical work capacity in dynamic exercise with differing muscle masses in healthy and older men. *Eur. J. Appl. Physiol.* 73:180-185.

Åstrand, P-O. and Rodahl, K. 1986. *Textbook of work physiology*, 3rd ed. McGraw-Hill, New York, NY.

Blair, S.N., Horton, E., Leon, A.S., et al. 1996. Physical activity, nutrition, and chronic disease. *Med. Sci. Sports Exerc.* 28:335-349.

Borg, G. 1970. Perceived exertion as an indicator of somatic stress. *Scand. J. Rehab. Med.* 2:92-98.

Fletcher, G.F. 1993. Statement of exercise: benefits and recommendations for physical activity programs for all Americans. *Circulation* 86:340-344.

Guide for Smoke-Diving. 1991. Finnish Ministry of Internal Affairs. Helsinki.

Hollmann, W. and Hettinger, Th. 1980. *Sportmedizin — Arbeits- und Trainingsgrundlagen.* 1. Lag. Schattauer Verlag, Stuttgart.

Hopsu, L. 1993. *Working conditions in jobs where the majority of workers are women: Cleaning work.* OECD panel group on women, work and health: National Report. Ministry of Social Affairs and Health. Helsinki.

Hopsu, L., Louhevaara, V., Korhonen, O., and Miettinen, M. 1994. Ergonomic and developmental intervention in cleaning work, in *Proceedings of the 12th Triennial Congress of the International Ergonomics Association.* Vol. 6: General Issues, pp. 159-160. University Press, Toronto, Canada.

Hultman, G., Nordin, M., and Örtengren, R. 1984. The influence of preventive educational programs on trunk flexion in janitors. *Appl. Ergonomics* 15:127-133.

Ilmarinen, J. 1984. Physical work load on cardiovascular system in different work tasks. *Scand. J. Work Environ. Health* 10:403-408.

Ilmarinen, J., Huuhtanen, P., Louhevaara, V., and Näsman, O. 1995. A new concept for maintaining work ability during aging, in *From Research to Prevention*, eds. Rantanen, J., Lehtinen, S., Hernberg, S., et al., pp. 123-127. Finnish Institute of Occupational Health, Helsinki, Finland.

Ilmarinen, J., Knauth, P., Klimmer, F., and Rutenfranz, J. 1979. The applicability of the Edholm scale for activity studies in industry. *Ergonomics* 22:369-376.

Ilmarinen, J. and Louhevaara, V. 1994. Preserving the capacity to work. *Ageing International* 21:34-36.

Ilmarinen, R., Griefahn, B., Mäkinen, H., and Kunemund, C. 1993. Physiological responses to wearing a fire fighter's turnout suit with and without microporous membrane in the heat, in *Proceedings of the Sixth International Conference on Environmental Ergonomics*, eds. Frim, J., Ducharne, M.B., and Tikuisis, P., pp. 78-79. Government of Canada, Montebello, Canada.

Jonsson, B., Persson, J., and Kilbom, Å. 1988. Disorders of the cerviobrachial region among female workers in the electronics industry. A two-year follow up. *Int. J. Ind. Erg.* 3:1-12.

Krueger, D. 1995. *Prevention of health and safety risks in professional cleaning and the work environment.* European Commission, Biomed 2 programme. Brussels, Belgium.

Kähkönen, E. 1993. *Comparison and error analysis of instrumentation and methods for assessment of neutral and hot environments on the basis of ISO standards.* Doctoral dissertation. University of Kuopio, Kuopio, Finland.

Lakka, T. 1994. *Leisure time physical activity, cardiorespiratory fitness, biological coronary risk factors and coronary heart disease: a population study in men in eastern Finland.* Doctoral dissertation. University of Kuopio, Kuopio, Finland.

Laukkanen, R. 1993. *Development and evaluation of a 2-km walking test for assessing maximal aerobic power of adults in field conditions.* Doctoral dissertation. University of Kuopio, Kuopio, Finland.

Lemon, P.W. and Hermiston, R.T. 1977. Physiological profile of professional fire fighters. *J. Occup. Med.* 19:337-340.

Louhevaara, V. 1992. Cardiorespiratory and muscle strain during manual sorting of postal parcels: A review. *J. Occup. Med.* (Singapore) 4:9-17.

Louhevaara, V. 1995. Assessment of physical work load at worksites: A Finnish-German concept. *Int. J. Occup. Safety Ergonomics* 1:144-152.

Louhevaara, V. 1995. *Feasibility of physical exercise in early rehabilitation of work ability.* 5th European Congress on Research in Rehabilitation, p. 105, Helsinki, Finland.

Louhevaara, V., Hakola, T., and Ollila H. 1990. Physical strain and work involved in manual sorting of postal parcels. *Ergonomics* 33:1115-1130.

Louhevaara, V. and Ilmarinen, J. 1995. Physical exercise as a measure for maintaining work ability during aging, in *The Paths to Productive Aging*, ed. Kumashiro, M., pp. 289-293. Taylor & Francis. Basingstoke.

Louhevaara, V., Teräslinna, P., Piirilä, P., et al. 1988. Physiological responses during and after intermittent sorting of postal parcels. *Ergonomics* 31:1165-1175.

Louhevaara, V., Tuomi, T., Smolander, J., et al. 1985. Cardiorespiratory strain in jobs that require respiratory protection, *Int. Arch. Occup. Health* 55:195-206.

Lusa, S. 1993. *Job demands and assessment of the physical work capacity of fire fighters.* Doctoral dissertation. University of Jyväskylä, Jyväskylä, Finland.

Lusa, S., Louhevaara, V., Smolander, J., et al. 1991. Biomechanical evaluation of heavy tool-handling in two age groups of firemen. *Ergonomics* 34:1429-1432.

Lusa, S., Louhevaara, V., and Kinnunen, K. 1994. Are the job demands on physical work capacity equal for young and aging firefighters? *J. Occup. Med.* 36:70-74.

Lusa, S., Louhevaara V., Smolander, J., et al. 1993. Physiological responses of firefighting students during simulated smoke-diving in the heat. *Am. Ind. Hyg. Assoc. J.* 54:228-231.

Pollock, M.L. and Wilmore, J.H. 1990. Exercise in health and disease. Evaluation and prescription for prevention and rehabilitation. 1st ed. W.B. Saunders Company, Philadelphia, NY.

Rohmert, W. 1983. Formen menslicher Arbeit, in *Praktische Arbeitsphysiologie*, eds. Rohmert, W. and Rutenfranz, J., 1. Lag. Georg Thieme Verlag, Stuttgart.

Rutenfranz, J. 1981. Arbeitsmedizinische Aspekts des Stressproblems, in *Stress, Theorien, Untersuchungen, Massnahmen*, ed. Nitsch, J.R. 1. Lag. Verlag Hans Huber, Bonn.

Rutenfranz, J., Ilmarinen, J., Klimmer, F., and Kylian, H. 1990. Work load and demanded physical performance capacity under different industrial conditions, in *Fitness for Aged, Disabled, and Industrial Worker*, ed. Kanenko, M. pp. 217-235. International Series on Sport Sciences, Vol. 20. Human Kinetics Books, Champaign.

Shephard, R.J. 1992. A critical analysis of work-site fitness programs and their postulated economic benefits. *Med. Sci. Sports and Exerc.* 24:354-370.

Smolander, J., Louhevaara, V., and Oja, P. 1984. Policemen's physical fitness in relation to the frequency of leisure time physical exercise. *Int. Arch. Occup. Environ. Health* 54:261-270.

Smolander, J., Louhevaara, V., Nygård, C-H., et al. 1985. Job demands and assessment of physical working capacity in policemen's occupational health service, in *Proceedings of the Ninth Congress of the International Ergonomics Association*, eds. Brown, I.D., Goldsmith, R., Coombes, K., and Sinclair, M.A., pp. 613-615, Taylor & Francis, Bournemouth, England.

Smolander, J. and Louhevaara V. 1998. Muscular work, in Stellman J.M., et al., eds. *Encyclopaedia of Occupational Health and Safety*. 4th ed. International Labour Office, pp. 1:29.28–29.31, Geneva, Switzerland.

Soininen, H. 1995. *The feasibility of worksite fitness programs and their effects on the health, physical capacity and work ability of aging police officers.* Doctoral dissertation. University of Kuopio, Kuopio, Finland.

Sothmann, M.S., Saupe, K.W., Jasenof, D., et al. 1990. Advancing age and the cardiorespiratory stress of fire suppression: Determining a minimum standard for aerobic fitness. *Human Performance* 3:217-236.

de Zwart, B.C.H., Frings-Dresen, M.H.W., and Dijk, F.J.H. 1995. Physical workload and the ageing worker: a review of the literature. *Int. Arch. Occup. Environ. Health* 68:1-12.

Viikari-Juntura, E., Kurppa, K., Kuosma, E., et al. 1991. Prevalence of epicondytis and elbow pain in the meat-processing industry. *Scand. J. Work Environ. Health* 17:38-45.

White, M.K. and Hodus, T.K. 1987. Reduced work tolerance associated with wearing protective clothing and respirators. *Am. Ind. Hyg. Assoc. J.* 48:304-310.

WHO (World Health Organization). 1988. *Health promotion for working populations.* Report of a WHO Expert Committee. WHO Technical Report Series 765, pp. 1-52, Geneva, Switzerland.

WHO (World Health Organization). 1993. *Ageing and work capacity.* Report of a WHO Expert Committee. WHO Technical Report Series 835, pp. 1-49, Geneva, Switzerland.

15

Psychosocial Work Factors

Pascale Carayon
Ecole des Mines de Nancy
France

Soo-Yee Lim
NIOSH

15.1   Introduction

This chapter examines the concept of psychosocial work factors and its relationship to occupational ergonomics. First, we provide a brief historical perspective of the development of theories and models of work organization and psychosocial work factors. Definitions and examples are then presented. Several explanations are given for the importance of psychosocial work factors in occupational ergonomics. Finally, measurement issues and methods for controlling and managing psychosocial work factors are discussed.

The role of "psychosocial work factors" in influencing individual and organizational health can be traced back to the early days of work mechanization and specialization, and the emergence of the concept of division of labor. Taylor (1911) expanded the principle of division of labor by designing efficient work systems accounting for proper job design, providing the right tools, motivating the individuals, and sharing of responsibilities between management and labor, and sharing of profits. This is known as the era of scientific management in which scientific methods are used to objectively measure work with the aim of improving its efficiency. These scientific methods involved breaking the tasks into small components or units, thus making work requirements and performance evaluations easy to define and monitor. Under these methods, work is simplified and standardized, therefore having a great impact on job and work processes. An analysis of psychosocial work factors in a job in this system would reveal that skill variety is minimal, workers have no control of the work processes, and the job is highly repetitive and monotonous. Such work system design can still be found in numerous workplaces.

As the workforce became more educated, individuals became more aware of their working conditions and environment, and began to seek avenues for improving their quality of working life. This is when the human relations movement emerged (Mayo, 1945), which raised the issue of the potential influence of the work environment on an individual's motivation, productivity, and well-being. Individual needs and wants were emphasized (Maslow, 1970). Thus, job design theorists incorporated worker behavior and work factors in their theories. The two theories of job enlargement and job enrichment formed the basis for many job design theories thereafter. These theories conceptualize the role of worker behavior and perception of the work environment in influencing personal and organizational outcomes. Job enlargement theory emphasized giving a larger variety of tasks or activities to the worker. While this was an improvement from the era of scientific management, the additional tasks or activities could be of a similar skill level and content: workers were performing multiple tasks of the same "kind." This has been called "horizontal loading" of the job, and is the opposite of job enrichment, which focused on the "vertical loading" of the job. Job enrichment aims at expanding the skills used by workers, while at the same time increasing their responsibility. Herzberg (1966), the father of the job enrichment theory, defined intrinsic and extrinsic factors (or motivation versus hygiene factors) that are important to worker motivation, thus leading to satisfaction or dissatisfaction, and psychological well-being. Intrinsic factors are related to the work (or job) conditions, such as having additional control over work schedules or resources, feedback, client relationships, skill use and development, better work content, direct communications, and personal accountability (Herzberg, 1974). Extrinsic factors are related to aspects of financial rewards and benefits and also to the physical environment. Herzberg indicated that extrinsic factors could lead to dissatisfaction with work, but not to satisfaction, while intrinsic factors could increase satisfaction with work. Herzberg's work demonstrated the complex relationships of job conditions, the individual's motivation, satisfaction, dissatisfaction, and psychological well-being. In a way similar to Herzberg's job enrichment theory, the Job Characteristics Theory (Hackman and Oldham, 1976) focused on the idea that specific characteristics of the job (i.e., skill variety, task identity, task significance, autonomy, and feedback) in combination with individual characteristics (growth need strength) would determine personal and work outcomes.

The Sociotechnical Systems Theory recognized two inter-related systems in an organization: the social system and the technical system. The main principle of the Sociotechnical Systems Theory is that the social and technical systems interact with each other, and that the joint optimization of both systems can lead to increased satisfaction and performance. The social system focused on the workers' perception of the work environment (i.e., job design factors) and the technical system emphasized the technology and the work processes used in the work (for example, automation, paced systems, and monitoring systems). In a study of coal mining (Trist and Bamforth, 1951), it was demonstrated that the technical system could impact the social system. In this study where semi-autonomous work groups were set up, workers were given opportunities to make decisions related to their work, and experienced better interactions with workers in their group, as well as task significance and completeness (see also Trist, 1981). Work by Trist and his colleagues showed that technological factors could influence both organizational and job factors. However, it was Davis (Davis 1980) who provided a conceptual framework and a set of principles that formulated the Sociotechnical theory. His framework called for a flattened management structure that would promote participation, interaction between and across groups of workers, enriched jobs, and most important, meeting individual needs. The Sociotechnical Systems Theory laid down the groundwork for the current understanding of how psychosocial work factors can be related to ergonomic factors by examining the interplay between the social and technical systems in organizations. Other recent theories and models of psychosocial work factors will be discussed later.

This rapid overview of the development of job design theories in the 20th century demonstrates the increasing role of psychological, social, and organizational factors in the design of work.

15.2 Definitions

Within the last decade, the role of psychosocial work factors on worker health has gained much popularity. However, the term of "psychosocial work factors" has been used loosely to define and represent many factors that are a part of, attached to or associated with the individuals. Some would consider what has been traditionally termed socioeconomic factors such as income, education level, and demographic or individual factors (e.g., age and marital status) as part of the psychosocial factors (Hogstedt, Vingard et al. 1995; Ong, Jeyaratnam et al. 1995). In order to understand psychosocial factors in the workplace, one needs to take into account the ability of an individual to make a psychological connection to his or her job, thus formulating the relationship between the person and the job. For instance, the International

Labour Office (ILO, 1986) defines psychosocial work factors as "interactions between and among work environment, job content, organizational conditions and workers' capacities, needs, culture, personal extra-job considerations that may, through perceptions and experience, influence health, work performance, and job satisfaction." Thus, the underlying premise in defining psychosocial work factors is the inclusion of the behavioral and psychological components of job factors. In the rest of the chapter, we will use the definitions proposed by Hagberg and his colleagues (Hagberg et al., 1995) because they are most highly relevant for occupational ergonomics.

Work organization is defined as the way work is structured, distributed, processed, and supervised (Hagberg et al., 1995). It is an "objective" characteristic of the work environment, and depends on many factors, including management style, type of product or service, characteristics of the workforce, level and type of technology, and market conditions. Psychosocial work factors are "perceived" characteristics of the work environment that have an emotional connotation for workers and managers, and that can result in stress and strain (Hagberg et al., 1995). Examples of psychosocial work factors include overload, lack of control, social support, and job future ambiguity. Other examples are described in the following section.

The concept of psychosocial work factors raises the issue of objectivity–subjectivity. Objectivity has multiple meanings and levels in the literature. According to Kasl (1978), objective data is not supplied by the self-same respondent who is also describing his distress, strain, or discomfort. At another level, Kasl (1987) feels that "psychosocial factor perception" can be less subjective when the main source of information is the employee but that this self-reported exposure is devoid of evaluation and reaction. Similarly, Frese and Zapf (1988) conceptualize and operationalize "objective stressors" (i.e., work organization) as not being influenced by an individual's cognitive and emotional processing. Based on this, it is more appropriate to conceptualize a continuum of objectivity and subjectivity. Work organization can be placed at one extreme of the continuum (that is the objective nature of work) whereas psychosocial work factors have some degree of subjectivity (see definitions above).

Psychosocial work factors result from the interplay between the work organization and the individual. Given our definitions, psychosocial work factors have a *subjective*, perceptual dimension, which is related to the *objective* dimension of work organization. Different work organizations will 'produce' different psychosocial work factors. The work organization determines to a large extent the type and degree of psychosocial work factors experienced by workers. For instance, electronic performance monitoring, or the on-line, continuous computer recording of employee performance-related activities, is a type of work organization that has been related to a range of negative psychosocial work factors, including lack of control, high work pressure, and low social support (Smith et al. 1992). In a study of office workers, information on psychosocial work factors was related to objective information on job title (Sainfort, 1990). Therefore, psychosocial work factors are very much anchored in the objective work situation, and are related to the work organization.

15.3 Examples of Psychosocial Work Factors

Psychosocial work factors are multiple and various, and are produced by different, interacting aspects of work. The Balance Theory of Job Design (Smith and Carayon-Sainfort, 1989) proposed a conceptualization of the work system with five elements interacting to produce a "stress load." The five elements of the work system are: (1) the individual, (2) tasks, (3) technology and tools, (4) environment, and (5) organizational factors. The interplay and interactions between these different factors can produce various stressors on the individual which then produce a "stress load" which has both physical and psychological components. The stress load, if sustained over time and depending on the individual resources, can produce adverse effects, such as health problems and lack of performance. The models and theories of job design reviewed at the beginning of the chapter tended to emphasize a small set of psychosocial work factors. For instance, the human relations movement (Mayo, 1945) focused on the social aspects of work, whereas the job characteristics theory (Hackman and Oldham, 1976) lists five job characteristics, i.e., skill variety, task identity, task significance, autonomy, and feedback. However, research and practice in

TABLE 15.1   Selected Psychosocial Work Factors and their Facets

1.	Job demands	Quantitative workload
		Variance in workload
		Work pressure
		Cognitive demands
2.	Job content	Repetitiveness
		Challenge
		Utilization and development of skills
3.	Job control	Task/instrumental control
		Decision/organizational control
		Control over physical environment
		Resource control
		Control over work pace: machine-pacing
4.	Social interactions	Social support from supervisor and colleagues
		Supervisor complaint, praise, monitoring
		Dealing with (difficult) clients/customers
5.	Role factors	Role ambiguity
		Role conflict
6.	Job future and career issues	Job future ambiguity
		Fear of job loss
7.	Technology issues	Computer-related problems
		Electronic performance monitoring
8.	Organizational and management issues	Participation
		Management style

the field of work organization has demonstrated that considering only a small number of work factors can be misleading and inefficient in solving job design problems. The balance theory proposes a systematic, global approach to the diagnosis and design or redesign of work systems that does not emphasize any one aspect of work. According to the balance theory, psychosocial work factors are multiple and of diverse nature.

Table 15.1 lists eight categories of psychosocial work factors and specific facets in each category. This list cannot be considered as exhaustive, but is representative of the most often studied psychosocial work factors.

The study of psychosocial work factors needs to be tuned in to the changes in society. Changes in the economic, social, technological, legal, and physical environment can produce new psychosocial work factors. For instance, in the context of office automation, four emerging issues are appearing (Carayon and Lim, 1994): (1) electronic monitoring of worker performance, (2) computer-supported work groups, (3) links between the physical and psychosocial aspects of work in automated offices, and (4) technological changes. The issue of technological changes applies nowadays to a large segment of the work population. Employees are asked to learn new technologies on a frequent, sometimes continuous, basis. Other trends in work organization include the development of teamwork and other work arrangements, such as telecommuting. These new trends may produce new psychosocial work factors, such as high dependency on technology, lack of socialization on the job and identity with the organization, and pressures from teamwork. Two APA publications review psychosocial stress issues related to changes in the workforce in terms of gender, diversity, and family issues (Keita and Hurrell, 1994), and some of the emergent psychosocial risk factors and selected occupations at risk of psychosocial stress (Sauter and Murphy, 1995).

15.4   Occupational Ergonomics and Psychosocial Work Factors

The emergence of macroergonomics has strongly contributed to the increasing interest in psychosocial work factors in the occupational ergonomics field (Hendrick, 1991; Hendrick, 1996). As shown above, the work factors can be categorized into the individual, task, tools and technologies, physical environment, and the organization (Smith and Carayon-Sainfort, 1989). They can also be described as either physical or psychosocial (Cox and Ferguson, 1994). Cox and Ferguson (1994) developed a model of the effects of physical and psychosocial factors on health. According to this model, the effects of work factors on health are mediated by two pathways: (1) a direct physicochemical pathway, and (2) an indirect psychophysiological pathway. These pathways are present at the same time, and interact in different ways to affect health. Physical work factors can have direct effects on health via the physicochemical pathway, and indirect effects on health via the psychophysiological pathway, but can also moderate the effect of psychosocial work factors on health via the psychophysiological pathway. This model demonstrates the close relationship between physical and psychosocial work factors in their influence on health and well-being.

The importance of psychosocial work factors in the field of occupational ergonomics emerges from several considerations.

1. Physical and psychosocial ergonomics are interested in the same job factors.
2. Physical and psychosocial work factors are related to each other.
3. Psychosocial work factors play an important role in physical ergonomics interventions.
4. Physical and psychosocial work factors are related to the same outcome, for instance, work-related musculoskeletal disorders.

First, some of the concepts examined in the physical ergonomics literature are similar to concepts examined in the psychosocial ergonomics literature. For instance, the degree of repetitiveness of a task is very important from both physical and psychosocial points of view. Physical ergonomists are more interested in the effect of the task repetitiveness on motions and force exerted on certain body parts, such as hands; whereas psychosocial ergonomists are concerned about the effect of task repetitiveness on monotony, boredom, and dissatisfaction with one's work (Cox 1985). In the physical ergonomics literature, an important job redesign strategy for dealing with repetitiveness is job rotation: workers are rotated between tasks which require effort from different body parts and muscles, therefore reducing the negative effects of repetitiveness of motions in a single task. From a psychosocial point of view, job rotation is one form of job enlargement (see above for a discussion of job enlargement). However, as discussed earlier, the psychosocial benefits of job rotation are limited because workers may be simply performing a range of similar, nonchallenging tasks. From a physical ergonomics point of view, job rotation is effective only if the physical variety of the tasks is increased; whereas from a psychosocial ergonomics point of view, job rotation is effective only to the extent of the content and meaningfulness of the tasks.

Second, physical and psychosocial work factors can be related to each other. For instance, the model proposed by Lim (1994) states that the psychosocial factor of work pressure can influence the physical factors of force and speed of motions. According to this model, workers may change their behaviors under the influence of work pressure, and, therefore, tend to exert more force or to speed up their work. Empirical evidence tends to confirm this relationship between work pressure (i.e., a psychosocial work factor) and physical work factors (Lim 1994). Another form of relationship between physical and psychosocial work factors is evident in the literature on control over one's physical environment. In this case, the psychosocial work factor of control is applied to one particular facet of the work, that is the physical environment. Control over one's physical environment can, therefore, have benefits from a physical point of view (i.e., being able to adapt one's physical environment to one's physical characteristics and task requirements), but also from a psychosocial point of view (i.e., having control is known to have many psychosocial benefits [Sauter et al., 1989]).

Third, psychosocial work factors are a crucial component of physical ergonomics interventions. In particular, the concept of participatory ergonomics uses the benefits of one psychosocial work factor, that is participation, in the process of implementing physical ergonomics changes (Noro and Imada, 1991). From a psychosocial point of view, using participation is important to improve the process and outcomes of ergonomic interventions. In addition, any type of organizational interventions, including ergonomic interventions, can be stressful because of the emergence of negative psychosocial work factors, such as uncertainty and increased workload (i.e., having more work during the intervention or the transitory period). Therefore, in any physical ergonomics intervention, attention should be paid to psychosocial work factors in order to improve the effectiveness of the intervention and to reduce or minimize its negative effects on workers.

Fourth, physical and psychosocial work factors can be related to the same outcome. One of these outcomes is work-related musculoskeletal disorders (WMSDs). There is increasing theoretical and empirical evidence that both physical and psychosocial work factors play a role in the experience and development of WMSDs (Hagberg et al. 1995; Moon and Sauter, 1996). Several mechanisms for the joint influence of physical and psychosocial work factors on WMSDs have been presented (Smith and Carayon, 1996). Therefore, in order to fully prevent or reduce WMSDs, both physical and psychosocial work factors need to be considered.

15.5   Measurement of Psychosocial Work Factors

From the occupational ergonomics point of view, the purpose of examining psychosocial work factors is to investigate their influence on and role in worker health and well-being. Thus, psychosocial work factors can be considered as predictors (i.e., independent variables), while worker health and well-being serve as the dependent variables or outcomes. The measurement or assessment of well-being can be classified into two levels of measures in terms of "context-free" (that is, life in general or general satisfaction) and "context-specific" (for example, job-related well-being) (Warr, 1994). It is the latter level of measure, "context-specific" that is relevant to the assessment of psychosocial work factors in the workplace. Table 15.1 shows a selected sample of the many different dimensions of jobs (for example, job demands, control, social support) that have been studied extensively. Furthermore, each dimension is made up of different facets that define and operationalize that particular dimension. For example, as shown in Table 15.1, the dimension of job demands consists of various facets, such as quantitative workload, variance in workload, work pressure, and cognitive demands; the dimension of job content includes repetitiveness, challenge on the job, and utilization of skills. It should be noted that Table 15.1 is not an exhaustive list of psychosocial work factors.

The most often used method for measuring psychosocial work factors in applied settings is the questionnaire survey. Difficulties with questionnaire data on psychosocial work factors are often due to the lack of clarity of the definitions of the measured factors or poorly designed questionnaire items that measure "overlapping" conceptual dimensions of the psychosocial work factor of interest. Measures of any one facet typically include several items that can be grouped in a "scale." Reliability of the scale is often being assessed by the Cronbach-alpha score method in which the intercorrelations among the scale items are examined for internal consistency. In general, it is recommended that existing, well-established scales be used in order to ensure the "quality" of the data (i.e., reliability and validity) and to be able to compare the newly collected data with other groups for which data has been collected with the same instrument (benchmarking).

The level of objectivity/subjectivity of the measures of psychosocial work factors will depend on the degree of influence of cognitive and emotional processing. For example, ratings of work factors by an observer cannot be considered as purely objective because of the potential influence of the observer's cognitive and emotional processing. However, ratings of work factors by an outside observer can be considered as more objective than an evaluative question answered by an employee about his/her work environment (e.g., "How stressful is your work environment?"). However, self-reported measures of psychosocial work factors can be more objective when devoid of evaluation and reaction (Kasl and Cooper, 1987). As discussed earlier, any kind of data can be placed somewhere on this objectivity/subjectivity continuum from "low in dependency on cognitive and emotional processing" (e.g., objective) to "high in dependency on cognitive and emotional processing" (e.g., subjective).

We discuss three different questionnaires which include numerous scales of psychosocial work factors. In addition, validity and reliability analyses have been performed on all three questionnaires. Two of these questionnaires have been developed and used to measure psychosocial work factors in various groups of workers or large samples of workers: (1) the NIOSH Job Stress questionnaire (Hurrell and

McLaney 1988), and (2) the Job Content Questionnaire (JCQ) (Karasek, 1979). The NIOSH Job Stress questionnaire is often used in the Health Hazard Evaluations performed by NIOSH. Translations of Karasek's JCQ exist in many different languages, including Dutch and French. The University of Wisconsin Office Worker Survey (OWS) is a questionnaire developed to measure psychosocial work factors in office/computer work (Carayon, 1991). This questionnaire covers a wide range of psychosocial work factors of importance in office and computer work. In addition to many of the psychosocial work factors measured by the NIOSH Job Stress Questionnaire or Karasek's JCQ, the OWS measures psychosocial work factors related to computer technology, such as computer-related problems (Carayon-Sainfort, 1992). The OWS questionnaire has been translated into Finnish, Swedish, and German. For all three questionnaires, data exist for various groups of workers in numerous organizations of multiple countries. This data can serve as a comparison to newly collected data and for benchmarking. Numerous other questionnaires for measuring psychosocial work factors exist, such as the Occupational Stress Questionnaire in Finland (Elo et al., 1994) and the Occupational Stress Indicator in England (Cooper et al., 1988). Other questionnaires are listed in Cook et al. (1981).

15.6 Managing and Controlling Psychosocial Work Factors

It is clear from the job design and occupational stress literature that jobs with negative psychosocial work factors, such as repetitiveness, no opportunity to develop skills, and low control, can have adverse effects on job performance and mental and physical health. Various approaches have been proposed to improve the design of jobs, such as job rotation and other forms of job enlargement, and job enrichment (see above). These strategies can be efficient to increase the variety in a job, to reduce the dependence on a particular technology or tool, and to increase worker control and responsibility. In particular, lack of job control is seen as a critical psychosocial work factor (Sauter et al., 1989). Providing a greater amount of control can be achieved by, for instance, allowing workers to determine their work schedules in accordance with organizational policies and production requirements, by allowing workers to give input into decisions that affect their jobs, by letting workers choose the best work procedures and task order, and by increasing worker participation in the production process. An experimental field study of a participation program showed the positive effects of participation on emotional distress and turnover (Jackson, 1983). According to the Sociotechnical Systems theory, autonomous work groups can be an effective strategy for increasing worker control and enriching jobs. Beyond increased control and improved job content, some forms of teamwork can have other positive psychosocial benefits, such as increased opportunity for socialization and learning.

Achieving the perfect job without any negative psychosocial work factors may not be feasible or realistic, given individual, organizational, or technological constraints and requirements. The balance theory (Smith and Carayon-Sainfort, 1989) proposes a job redesign strategy that aims to achieve an optimal job design. In this process, negative psychosocial work factors need to be eliminated or reduced as much as possible. However, when this is not possible, positive psychosocial work factors can be used to reduce the impact of negative psychosocial work factors. This balancing, or compensating, effect is based on the concept of the work system of the balance theory. The five elements of the work system (the individual, tasks, technology and tools, environment, and organizational factors) are interrelated: they can influence each other, and they can also influence the impact or effect of each other or their interactions. In this systems approach, negative psychosocial work factors can be balanced out or compensated by positive work factors.

Some trends in the field of organizational design and management may have positive characteristics from a psychosocial point of view. For instance, under certain conditions, the use of quality engineering and management methods can positively affect the psychosocial work environment, such as increased opportunity for participation, and learning and development of quality-related skills (Smith et al., 1989). However, other trends in the business world can have negative effects on the psychosocial work environment. For instance, downsizing and other organizational restructuring and reengineering may create highly stressful situations of uncertainty and loss of control (DOL 1995).

15.7   Conclusion

This chapter has demonstrated the importance of psychosocial work factors in the research and practice of occupational ergonomics. In order to clarify the issue at hand, we presented definitions of work organization and psychosocial work factors. It is important to understand the long research tradition on psychosocial work factors that has produced numerous models and theories, but also valid and reliable methods for measuring psychosocial work factors. At the end of the chapter, we presented examples of methods for managing and controlling psychosocial work factors.

Psychosocial work factors need to be taken into account in the research on and practice of occupational ergonomics. We have discussed the important role of psychosocial work factors with regard to physical ergonomics. In addition, given the constantly changing world of work and organizations, we need to pay even more attention to the multiple aspects of people at work, including psychosocial work factors.

References

Carayon, P. (1991). *The Office Worker Survey.* Madison, WI, Department of Industrial Engineering, University of Wisconsin-Madison.

Carayon, P. and Lim, S.-Y. (1994). Stress in automated offices. *The Encyclopedia of Library and Information Science.* A. Kent. New York, Marcel Dekker. Vol. 53, Supplement 16: 314-354.

Carayon-Sainfort, P. (1992). The use of computers in offices: impact on task characteristics and worker stress. *International Journal of Human Computer Interaction* 4(3): 245-261.

Cook, J. D., Hepworth, S. J. et al. (1981). *The Experience of Work.* London, Academic Press.

Cooper, C. L., Sloan, S. J. et al. (1988). *Occupational Stress Indicator.* Windsor, England, NFER-Nelson.

Cox, T. (1985). Repetitive work: Occupational stress and health. *Job Stress and Blue-Collar Work.* C. L. Cooper and M. J. Smith. New York, John Wiley & Sons: 85-112.

Cox, T. and Ferguson, E. (1994). Measurement of the subjective work environment. *Work and Stress* 8(2): 98-109.

Davis, L. E. (1980). Individuals and the organization. *California Management Review* 22(2): 5-14.

DOL (1995). *Guide to responsible restructuring.* Washington, D.C. 20210, U.S. Department of Labor, Office of the American Workplace.

Elo, A.-L., Leppanen, A., et al. (1994). The Occupational Stress Questionnaire. *Occupational Medicine.* C. Zenz, O. B. Dickerson and E. P. Horvarth. St. Louis, Mosby: 1234-1237.

Frese, M. and Zapf, D. (1988). Methodological issues in the study of work stress. *Causes, Coping and Consequences of Stress at Work.* C. L. Cooper and R. Payne. Chichester, John Wiley & Sons.

Hackman, J. R. and Oldham, G. R. (1976). Motivation through the design of work: test of a theory. *Organizational Behavior and Human Performance* 16: 250-279.

Hagberg, M., Silverstein, B., et al. (1995). *Work-Related Musculoskeletal Disorders (WMSDs): A Reference Book for Prevention.* London, Taylor & Francis.

Hendrick, H. W. (1991). Human Factors in organizational design and management. *Ergonomics* 34: 743-756.

Hendrick, H. W. (1996). Human factors in ODAM: an historical perspective. *Human Factors in Organizational Design and Management -V.* O. J. Brown and H. W. Hendrick. Amsterdam, The Netherlands, Elsevier Science Publishers: 429-434.

Herzberg, F. (1966). *Work and the Nature of Man.* New York, Thomas Y. Crowell Company.

Herzberg, F. (1974). The wise old turk. *Harvard Business Review*(September/October): 70-80.

Hogstedt, C., E. Vingard, et al. (1995). *The Norrtalje-MUSIC Study — An ongoing epidemiological study on risk and health factors for low back and neck-shoulder disorders.* PREMUS'95-Second International Scientific Conference and Prevention of Work-Related Musculoskeletal Disorders, Montreal, Canada.

Hurrell, J. J. J. and M. A. McLaney (1988). "Exposure to job stress — A new psychometric instrument." *Scandinavian Journal of Work Environment and Health* 14(suppl.1): 27-28.

ILO (1986). *Psychosocial Factors at Work: Recognition and Control.* Geneva, Switzerland, International Labour Office.

Jackson, W. E. (1983). Participation in decision-making as a strategy for reducing job-related strain. *Journal of Applied Psychology* 68: 3-19.

Karasek, R. A. (1979). Job demands, job decision latitude, and mental strain: implications for job redesign. *Administrative Science Quarterly* 24: 285-308.

Kasl, S. V. (1987). Methodologies in stress and health: past difficulties, present dilemmas, future directions. *Stress and Health: Issues in Research and Methodology.* S. V. Kasl and C. L. Cooper. Chichester, John Wiley & Sons: 307-318.

Kasl, S. V. and C. L. Cooper, Eds. (1987). *Stress and Health: Issues in Research and Methodology.* Chichester, John Wiley & Sons.

Keita, G. P. and Hurrell, J. J. J. (1994). *Job Stress in a Changing Workforce — Investigating Gender, Diversity, and Family Issues.* Washington, D.C., APA.

Lim, S. (1994). An integrated approach to cumulative trauma disorders in computerized offices: the role of psychosocial work factors, psychological stress and ergonomic risk factors. *IE.* Madison, WI, University of Wisconsin-Madison.

Maslow, A. H. (1970). *Motivation and Personality.* New York, Harper and Row.

Mayo, E. (1945). *The Social Problems of an Industrial Civilization.* Andover, MA, The Andover Press.

Moon, S. D. and Sauter, S. L. Eds. (1996). *Beyond Biomechanics — Psychosocial Aspects of Musculoskeletal Disorders in Office Work.* London, Taylor & Francis.

Noro, K. and Imada, A. (1991). *Participatory Ergonomics.* London, Taylor & Francis.

Ong, C. N., Jeyaratnam, J., et al. (1995). "Musculoskeletal disorders among operators of video display terminals." *Scandinavian Journal of Work Environment and Health* 21(1): 60-64.

Sainfort, P. C. (1990). *Perceptions of Work Environment and Psychological Strain Across Categories of Office Jobs.* The Human Factors Society 34th Annual Meeting.

Sauter, S. L., Hurrell, J. J. Jr., et al., Eds. (1989). *Job Control and Worker Health.* Chichester, John Wiley & Sons.

Sauter, S. L. and Murphy, L. R. (1995). *Organizational Risk Factors for Job Stress.* Washington, D.C., APA.

Smith, M. J. and Carayon, P. (1996). Work organization, stress, and cumulative trauma disorders. *Beyond Biomechanics — Psychosocial Aspects of Musculoskeletal Disorders in Office Work.* S. D. Moon and S. L. Sauter. London, Taylor & Francis: 23-41.

Smith, M. J., Carayon, P., et al. (1992). Employee stress and health complaints in jobs with and without electronic performance monitoring. *Applied Ergonomics* 23(1): 17-27.

Smith, M. J. and Carayon-Sainfort, P. (1989). A balance theory of job design for stress reduction. *International Journal of Industrial Ergonomics* 4: 67-79.

Smith, M. J., Sainfort, F., et al., Eds. (1989). *Efforts to Solve Quality Problems,* Secretary's Commission on Workforce Quality and Labor Market Efficiency, U.S. Department of Labor, Washington, D.C.

Taylor, F. (1911). *The Principles of Scientific Management.* New York, Norton and Company.

Trist, E. (1981). *The Evaluation of Sociotechnical Systems.* Toronto, Quality of Working Life Center.

Trist, E. L. and Bamforth, K. (1951). Some social and psychological consequences of the long-wall method of coal getting. *Human Relations* 4: 3-39.

16

Cognitive Factors

Philip J. Smith
The Ohio State University

The field of ergonomics can be divided into two broad categories: physical ergonomics and cognitive ergonomics (Kroemer et al., 1994; Van Der Verr, Bagnara, and Kempen, 1992). Given this division, the list of cognitive factors goes well beyond purely cognitive functions to encompass all mental activity. Thus, this list includes:

1. Psychomotor skills
2. Sensory and perceptual skills
3. Affective responses and motivation
4. Attention
5. Learning and memory
6. Language and communication
7. Problem solving and decision making
8. Group dynamics and teamwork

In terms of our understanding of these basic mental functions, cognitive ergonomics draws heavily upon the fields of psychology and linguistics (Anderson, 1993; Fleishman and Quaintance, 1989; Proctor and Dutta, 1995; Proctor and VanZandt, 1994; Rasmussen, 1988; Sheridan and Ferrell, 1974).

The cognitive issues within industrial ergonomics go beyond modeling these mental processes, however. The goal is to understand how the work environment interacts with the strengths and limitations of the worker to determine performance. For example, in considering the design of a set of controls and displays, it is critical to consider not only the characteristics of the worker, but also the impact of such factors as:

1. Intended functions of the controls and displays (supported tasks)
2. Relative and absolute positions of the controls and displays
3. The broader task environment
4. Consistency within the system
5. Consistency with other systems (population stereotypes or expectancies)

These task and environmental factors interact with characteristics of the operator, ranging from visual acuity and contrast sensitivity to the determinants of attention and the problem-solving strategies used to direct the access to and interpretation of information provided by displays.

It is, therefore, the interaction of human abilities with task and environmental factors that characterizes ergonomics. Based on this emphasis on the performance of the worker in some context, important design concepts and principles have been developed. For example, in the design of displays and controls, basic psychological findings such as the Gestalt "Laws of Good Form" (Gibson, 1986; Rock, 1995) have been incorporated in recommendations for achieving effective designs, such as the use of proximity or color coding to indicate functional groupings of displays and controls.

16.1   Example 1

To illustrate the role of cognitive factors in ergonomic evaluations, a series of examples are presented below. As a first simple example, consider the cognitive ergonomic concerns associated with designing a visual warning sign.

When designing a warning sign, one critical factor is its salience. The underlying cognitive factor is the selectiveness of human attention (which limits the ability of people to simultaneously attend to many different sources of information in the environment). Such *selective attention* is important in two senses. First, it is critical that the worker's attention be drawn to the warning sign. Second, it is equally important that attention be focused on the important contents of the warning sign.

A second cognitive factor (in the broad sense of the term) concerns legibility. Considerations like character size, stroke width and font, viewing distance, illumination and contrast are important, as well as characteristics of the worker, such as *visual acuity*.

A third factor is intelligibility: Will the worker correctly interpret the contents of the warning? And a fourth major factor is motivation: Even if the individual's attention is drawn to important contents of the warning and it is correctly read and interpreted, will behavior change? (Since the goal of ergonomics is to provide practical guidance, consideration of this last factor has lead to guidelines that recommend that, to avoid *design-induced error,* warning signs should be treated as a last resort and that, wherever feasible, designers should engineer out a predicted hazard rather than relying on a warning sign to prevent hazardous behavior.) Furthermore, such guidelines emphasize a major theme in cognitive ergonomics: **Designers need to play "psychologist" as part of the design process, predicting possible behaviors by prospective users of a system.**

Example 1, then, describes a range of cognitive factors that need to be considered in the design of a relatively simple artifact. Examples 2 through 4 further demonstrate the approach of cognitive ergonomics to design in more complex settings.

16.2   Example 2

Example 1 illustrated a fundamental concept in cognitive ergonomics: **Design is a prediction task.**

Application of this concept is further demonstrated in the design of documentation and help for computer software. Here, basic research has provided a model for how (many) software users approach documentation and help (Doheny-Farina, 1988; Kearsley, 1988).

According to this model, such assistance tends to be accessed when the user encounters difficulty, rather than through reading the software documentation and help prior to starting to use a new software system. This means the user accesses the help with a particular task or goal in mind. Then the user scans salient landmarks in the text (such as the bold subheadings on a help screen) looking for something that looks relevant. When the label for some section is judged relevant, the user then processes that localized portion of the text, using perceptual cues (such as spacing) to decide what constitutes that "chunk" of text, and reads only enough to develop a mental model of how to accomplish the desired task or goal (Janosky, Smith, and Hildreth, 1986).

From a designer's perspective, this qualitative model provides structure in predicting user performance:

1. Predict the alternative goals that users might have when accessing a given section of the documentation or help
2. Structure the text so easily identified headings can be scanned, using labels that will be judged as relevant to the appropriate user goals
3. Display the message so that all of its components appear to be visually related to the heading
4. Structure the contents of the text and graphics so that the reader is likely to process all of the critical contents. (For instance, a fact or instruction that is critical to forming a correct model of performance should not be placed last in a chunk of text, as it is less likely to be read in that position.)

Example 2, then illustrates how qualitative task-specific models can be developed to expand upon basic cognitive functions like perception and the focus of attention, supporting completion of detailed cognitive task analyses. It further illustrates how such models are used to identify specific steps or procedures to aid designers.

16.3 Example 3

The key importance of taking a broad systems view is even clearer when an industrial task such as visual inspection is considered (Drury and Prabhu, 1994). As an example, companies manufacturing glassware (such as glass bottles) traditionally have had workers view the bottles as they moved along a conveyor, trying to watch for and reject defective items. Defects can range from stones and bubbles in the glass to a crack in the lip of a jar.

The range of relevant factors includes:

1. Sensory and perceptual demands and skills (task characteristics such as *illuminance, contrast, effective size or visual angle, rate of movement,* and *time available for viewing,* as well as worker characteristics such as *acuity* and *contrast sensitivity*)
2. Attentional demands (particularly the impact of *vigilance decrements* over time on the task)
3. Memory and information processing rates (recalling and checking for different defects)
4. Decision making (deciding — given a very limited viewing time — whether or not to call a bottle defective)
5. Motivation

Physical and environmental factors are relevant as well, as discomfort can affect attentiveness.

In addition, the influence of a poorly designed inspection workstation can have an important impact not only on the inspector's attitude and motivation, but also on co-workers performing other types of jobs, as a result of group norms propagating throughout the plant. Thus, the designer of such a job needs to consider not only the direct impact of factors such as lighting and line speed, but also the impact of poor job design on worker motivation. In short, another important ergonomic perspective is that: **Design is a communication task.**

The message communicated by a poorly designed job is that the company isn't too concerned with the quality of the worker's performance. Such a message can induce low motivation on that job, which may then spread to workers elsewhere in the plant.

16.4 Example 4

Similarly, rich issues arise in considering the cognitive factors involved in designing decision support tools and systems to support training. As an illustration, consider the design of the Antibody IDentification Assistant (AIDA). This is a critiquing system (Fischer, 1991; Miller, 1986; Silverman, 1992) developed to assist with problem solving in blood banks and with embedded training both in the lab and as part of a college course on transfusion medicine.

AIDA as a Decision-Support System

AIDA was designed to provide assistance to blood bankers in performing a complex *abductive reasoning* task (Josephson and Josephson, 1994), the identification of antibodies in a patient's blood. This is an important step in determining compatible blood for a transfusion. As an abduction task, antibody identification exhibits a number of complexities, including a combinatorial explosion in the number of possible combinations of solutions (combinations of different antibodies), the potential for masking when multiple antibodies are present, and noise in the data.

Cognitive ergonomics plays a role in a number of steps when attempting to improve performance on such a task. These steps are illustrated below.

Problem Identification.   One application where an understanding of cognitive ergonomics is useful is in identifying the areas where performance needs to be improved. In particular, review of the literature on the causes of human error provides a sizable list of generic cognitive processes that characterize human performance of abduction or diagnosis tasks, and that produce errors in certain identifiable task environments (Bell, Raiffa, and Tversky, 1988; Chi, Glaser, and Farr, 1988; Fraser, Smith, and Smith, 1994; Hollnagel, 1993; Reason, 1990; Tversky and Kahneman, 1974). (For example, the use of a positive test strategy will produce a confirmation bias in certain situations [Klayman and Ha, 1987; Mynatt et al., 1978], leading to hypothesis fixation.) Such knowledge can be used to more efficiently and exhaustively identify important errors in a particular application by providing a top-down approach to problem identification.

In the case of antibody identification, empirical studies have demonstrated that a wide range of such predictable cognitive errors are made, including:

1. Slips
2. Perceptual distortions
3. Hypothesis fixation
4. Ignoring base rates
5. Biased assimilation

Some of these errors result from the use of simplifying (but fallible) domain-specific heuristics to reduce the cognitive load of the problem-solving task, but many are due to fundamental cognitive processes.

Cognitive Modeling.   In system design, cognitive ergonomics not only helps provide a top-down approach to identifying problems or sources of errors, it also provides guidance in modeling the cognitive processes and strategies involved in expert performance (Backland and McDermott, 1994; Clancey, 1986; Hoffman, 1987; Kolodner, 1993; Kuipers, 1994; Miller, 1984; Newell, 1990; Shortliffe, 1990; Wenger, 1987; Wyatt, Spiegel and Halter, 1992). A great deal is known about expert/novice differences in problem solving on tasks like abductive reasoning or diagnosis (Chi, Glaser, and Farr, 1988). This knowledge makes it possible to more efficiently and effectively model performance in a particular domain. For example, in performing antibody identification, experts use heuristic methods in order to keep the cognitive load manageable. However, since these heuristics are fallible, these experts also employ a more global strategy to reduce error. This metastrategy involves always collecting converging evidence using two or more independent strategies and data sources (Guerlain et al., 1994; Obradovich et al., 1996; Smith, Galdes, et al., 1991; Smith, Miller, et al., 1992).

Design.   Based on an understanding of needs of the user population and current problem-solving methods, cognitive ergonomics plays a role in answering a number of questions. First, there is the question of: **What general approach should be used to improve performance?**

This could range from training and education (Bailey, 1993; Fleming and Levie, 1993), to the design of an information system, to the design of an active decision support system, to full automation (eliminating the person from the job).

In the case of AIDA, the decision was to develop a system to provide both training and decision support. This decision was based on two ergonomic considerations. First, many of the errors observed in empirical studies of practitioners in the current environment were the result of slips (Norman, 1981).

While some of those might be reduced by better design of data displays, it is likely that a significant number would remain because of the nature of the task. This argues for use of the computer to avoid or catch such errors. Second, many other errors were the result of ignorance or misconceptions. This similarly supports the need for some type of intervention, such as a combination of training and active decision support. The third consideration, however, concerns the fallibility of the designer rather than the user. Evidence to date suggests that we do not know how to fully capture the expertise of people on such complex tasks, with the result that decision support systems (whether based on expert systems technology or optimization techniques) exhibit *brittleness* (Guerlain et al., 1994; Layton, Smith, and McCoy, 1994; Roth, Bennet, and Woods, 1987; Smith, McCoy, and Layton, 1997). This suggests the need to keep a well-trained person involved in the task.

This raises two additional questions if computers are to be used for decision support: **What technology should be used? What role should the computer play?**

Regarding the first question, human factors studies suggest that it is important that the user be able to develop an accurate mental model (Gentner and Stevens, 1983) of the functioning of the system, so that the user can work as an effective "partner" (Lehnert and Zirk, 1987). Such *cognitive compatibility* is important if the user is to compensate for the limitations of the designer (and the resultant brittleness of the technology). In the case of AIDA, based on such considerations, an expert systems approach was taken, modeling the computer's reasoning after the problem-solving strategies exhibited by human experts.

Regarding the second question, there are numerous variations, ranging from designing a system where the computer critiques the human user (which was the choice made for AIDA), to a system where the human critiques the computer's answer, to a system where the computer automatically completes subtasks at the request of the user. The best solution clearly depends upon the costs of different types of errors and the level of confidence that can be placed on the technology. Ergonomic studies clearly indicate, though, that systems based on either expert systems technology or optimization can induce errors in their users' reasoning when the user is asked to critique the computer's reasoning rather than vice versa. Such studies also show that the use of the computer as a critic can avoid such design-induced errors (as an example of such a *critiquing system* [Miller, 1986; Silverman, 1991], AIDA reduces errors by 33 to 62% on cases where it is fully competent and even reduces errors by 30% on cases where it is less than fully competent, because it prods the user to make more effective use of her own expertise [Obradovich, et al., 1996].) In short, consideration of the psychology of the user and of the designer, as well as the nature of the particular application, is necessary to make decisions about the appropriate roles for the user and for technology.

Ergonomics also has contributions to make at a more detailed level in developing technological support, specifically the design of the human–computer interface (Baecker, et al., 1995; Carroll, 1995; Gardiner and Christie, 1987; Norman, 1990; Treu, 1994; Tufte, 1990; Tufte, 1993). In the case of AIDA, this involved decisions about how to provide access to and display data more effectively, how to provide memory aids to reduce the memory load for the user, and about how to design an interface where the computer could unobtrusively monitor the user's performance and make meaningful, context sensitive inferences about when and where the user likely has made an error.

Evaluation. Finally, cognitive ergonomics has important contributions to make in system evaluation and usability testing. This includes the use of analytic methods, such as heuristic evaluation, as well as the design and analysis of empirical studies. For the latter approach, this includes guiding decisions about what tasks to use in testing, and the selection of what data to collect (such as concurrent verbal reports [Ericsson and Simon, 1993]).

AIDA as a Decision-Support System — Summary. In short, cognitive ergonomics provides a principled approach to the design of technological systems such as AIDA. It provides direction through the use of cognitive models of human problem solving and human error, as well as through the application of general design concepts and principles. It also prescribes the use of empirical methods in recognition of the fallibility of designers. In short, the role of cognitive ergonomics is to ensure that certain broad questions (such as the appropriate role for technology) are adequately addressed, as well as to provide guidance at a detailed level in the implementation of a design concept.

16.5   Summary

As illustrated above, one of the hallmarks of cognitive ergonomics is an emphasis on a broad systems approach. In evaluating cognitive factors, it is not enough to consider the strengths and limitations of the worker. The task and work environment must be considered as well as the broader organizational context. Furthermore, even when the focus of an analysis is on some cognitive task, physical ergonomics issues must be considered because the health and physical comfort of the worker can have a major impact on cognitive functions. Thus, cognitive ergonomics supports design by providing a link between our knowledge of basic psychological processes and their interaction with the tasks and tools within a work setting. The goal of this chapter has been to indicate the range of cognitive factors that need to be considered in designing different types of systems, and to illustrate how an understanding of these factors influences design. More details are found on specific topics in the following chapters of this book:

Chapter 8	Design of Information Devices and Controls
Chapter 34	How Complex Human–Machine Systems Fail: Putting "Human Error" in Context
Chapter 35	Human and System Reliability Analysis
Chapter 36	Some Developments in Human Reliability Assessment
Chapter 38	Managing the Speed–Accuracy Trade-off
Chapter 40	Design of Industrial Warnings

Defining Terms

Abductive reasoning: Reasoning or problem solving to determine the "best" explanation for a set of data.
Brittleness: A characteristic of computerized problem-solving systems (based on either expert systems technology or optimization techniques) where they provide inadequate or incorrect advice because they have incomplete or incorrect knowledge or an incomplete model of the domain.
Cognitive compatibility: In reference to computer systems, this term deals with the ability of the user to understand and effectively interact with the computer.
Contrast: A measure of the relative luminances of some visual target and its background.
Critiquing system: A problem-solving system (typically based on expert systems technologies) that plays the role of a critic, monitoring the performance of a person for potential errors and providing feedback when a potential problem is detected.
Design-induced errors: Errors made by a person using some product or system that could have been anticipated or predicted by the designer and could have been feasibly engineered out so that the resultant hazard no longer existed.
Illuminance: A measure of the amount of light falling on a surface.
Selective attention: A limitation in the ability of a person to attend to a large number of sources of information (such as multiple data displays) simultaneously.
Vigilance decrement: A decrease in performance over time on tasks such as inspection of a product for defects or monitoring of a sonar screen.
Visual acuity: A measure of the ability of a person to discriminate visual stimuli (such as characters in text strings).
Visual angle: A measure of the effective "size" of some visual target (such as a character in a text string). This effective size is a function of both the actual physical size of components of the target (such as the stroke width of the lines forming a character) and the distance of the viewer from that target.

References

Anderson, J.R. 1993. *Rules of the Mind*. Lawrence Erlbaum Associates: Hillsdale, NJ.
Bachant, J. and McDermott, J. 1994. R1 revisited: Four years in the trenches. *The AI Magazine*, 21-32.
Baecker, R.M., Grudin, J., Buxton, W.A., and Greenberg, S. 1995. *Readings in Human–Computer Interaction: Toward the Year 2000*. Morgan Kaufman: San Francisco, CA.

Bailey, G. D. (ed.). 1993. *Computer-Based Integrated Learning Systems.* Educational Technology: Englewood Cliffs, NJ.

Bell, D., Raiffa, H., and Tversky, A. 1988. Descriptive normative, and prescriptive interactions in decision making, in *Decision Making: Descriptive, Normative, and Prescriptive Interactions*, ed. D. Bell, H. Raiffa, and A. Tversky, pp. 9-30. Cambridge University Press: New York.

Carroll, J. (ed.) 1995. *Scenario-Based Design.* John Wiley & Sons, Inc.: New York.

Chi, M. T., Glaser, R., and Farr, M.J. (eds.). 1988. *The Nature of Expertise.* Lawrence Erlbaum: Hillsdale, NJ.

Clancey, W.J. 1986. From GUIDON to NEOMYCIN to HERACLES in twenty short lessons: ORN Final Report 1979-1985. *The AI Magazine*, 40-60.

Doheny-Farina, S., ed. 1988. *Effective Documentation: What We Have Learned from Research.* MIT Press: Cambridge, MA.

Drury, C. and Prabhu, P. 1994. Human factors in testing and inspection, in *Design of Work and Development of Personnel in Advanced Manufacturing,* ed. Salvendy, G. and Karwowski, pp. 331-354. W. Wiley: New York.

Ericsson, K.A. and Simon, H. 1993. *Protocol Analysis: Verbal Reports as Data.* MIT Press: Cambridge, MA.

Fischer, G., Lemke, A.C., Mastaglio, T., and Morch, A.I. (1991). The role of critiquing in cooperative problem solving. *ACM Transactions on Information Systems*, 9(3), 123- 151.

Fleishman, E. and Quaintance, M. (1984). *Taxonomies of Human Performance.* Academic Press: Orlando, FL.

Fleming, M. and Levie, H.H. (eds.). 1993. *Instructional Message Design: Principles from the Behavioral and Cognitive Sciences.* Educational Technology: Englewood Cliffs, NJ.

Gardiner, M. and Christie, B. (eds.). 1987. *Applying Cognitive Psychology to User Interface Design.* John Wiley & Sons: New York.

Gentner, D. and Stevens, A.C. (eds.). 1983. *Mental Models.* Lawrence Erlbaum: Hillsdale, NJ.

Gibson, J.J. 1986. *The Ecological Approach to Perception.* Lawrence Erlbaum: Hillsdale, NJ.

Guerlain, S., Smith, P.J., Gross, S.M., Miller, T.E., Smith, J.W., Svirbely, J.R., Rudmann, S., and Strohm, P. 1994. Critiquing versus partial automation: How the role of the computer affects human-computer cooperative problem solving, in *Human Performance in Automated Systems: Current Research and Trends*, M. Mouloua and R. Parasuraman, pp. 73-80. Lawrence Erlbaum Associates: Hillsdale, NJ.

Hoffman, R. 1987. The problem of extracting the knowledge of experts from experts from the perspective of experimental psychology. *The AI Magazine*, 8(2), 53-67.

Hollnagel, E. 1993. *Human Reliability Analysis Context and Control.* Academic Press: New York.

Janosky, B., Smith, P.J., and Hildreth, C. 1986. Online library catalog systems: An analysis of user errors. *Internatl. J. of Man-Machine Studies*, 25, 573-592.

Josephson, J. and Josephson, S. 1994. *Abductive Inference: Computation, Philosophy. Technology*, Cambridge University Press: New York.

Kearsley, G. 1988. *Online Help Systems: Design and Implementation.* Ablex: Norwood NJ.

Klayman, J. and Ha, Y.W. 1987. Confirmation, disconfirmation and information in hypothesis testing. *Psychological Review.* 94: 211-228.

Kolodner, J. 1993. *Case-Based Reasoning.* Morgan Kaufman: San Mateo, CA.

Kroemer, K., Kroemer, H., and Kroemer-Elbert, K. 1994. *Ergonomics: How to Design for Ease and Efficiency.* Prentice Hall: Englewood Cliffs, NJ.

Kuipers, B. 1994. *Qualitative Reasoning: Modeling and Simulation with Incomplete Knowledge.* MIT Press: Cambridge, MA.

Layton, C., Smith, P.J., and McCoy, E. 1994. Design of a cooperative problem-solving system for en-route flight planning: An empirical evaluation. *Human Factors*, 36, 94-119.

Lehner, P.E. and Zirk, D.A. 1987. Cognitive factors in user/expert-system interaction. *Human Factors*, 29, 97-109.

Miller, P. 1986. *Expert Critiquing Systems: Practice-Based Medical Consultation by Computer.* Springer-Verlag: New York.

Miller, R.A. 1984. INTERNIST-/CADUCEUS: Problems facing expert consultant programs. *Meth. Inform. Med.*, 23, 9-14.

Mynatt, C., Douherty, M., and Tweeney, R. 1978. Consequences of confirmation and disconfirmation in a simulated research environment. Q. J. *Expt. Psych.* 30:395-406.

Newell, A. 1990. *Unified Theories of Cognition.* Harvard University Press: Cambridge, MA.

Norman, D.A. 1981. Categorization of action slips. *Psychological Review.* 88, 1-15.

Norman, D.A. 1990. *The Design of Everyday Things.* Doubleday: New York.

Obradovich, J., Guerlain, S., Smith, P.J., Smith, J., Rudmann, S., Sachs, L., Svirbely, J., Kennedy, M., and Strohm, P. 1996. The Transfusion Medicine Tutor: The use of expert-systems technology to teach students and provide support to practitioners in antibody identification. *Proceedings of the 1996 International Conference on the Learning Sciences.* 249-255.

Proctor, R.W. and Dutta, A. 1995. *Skill Acquisition and Human Performance.* Sage: Thousand Oaks, CA.

Proctor, R.W. and VanZandt, T. 1994. *Human Factors in Simple and Complex Systems.* Allyn and Bacon: Boston.

Rasmussen, J. 1988. *Information Processing and Human–Machine Interaction: An Approach to Cognitive Engineering.* North-Holland: New York.

Reason, J. 1990. *Human Error.* Cambridge University Press: New York.

Rock, I. 1995. *Perception.* W.H. Freeman: New York.

Roth, E., Bennett, K., and Woods, D. 1987. Human interaction with an "intelligent" machine, *Internatl. J. of Man-Machine Studies*, 27, 479-525.

Sheridan, T.B. and Ferrell, W.R. 1974. *Man–Machine Studies: Information, Control, and Decision Models of Human Performance.* MIT Press: Cambridge, MA.

Shortliffe, E. 1990. Clinical decision-support systems, in *Medical Informatics: Computer Applications in Health Care,* ed. E. Shortliffe and L. Perreault, pp. 466-500. Addison-Wesley Publishing Company: New York.

Silverman, B.G. 1992. Building a better critic: Recent empirical results. *IEEE Expert*, 18-25.

Silverman, B.G. 1992. Survey of expert critiquing systems: Practical and theoretical frontiers. *Communications of the ACM*, 35(4), 106-128.

Smith, P.J., Galdes, D., Fraser, J., Miller, T., Smith, J.W., Svirbely, J.R. Blazina, J., Kennedy, M., Rudmann, S., and Thomas, D.L. 1991. Coping with the complexities of multiple-solution problems: a case study. *Internatl. J. of Man-Machine Studies*, 35, 429-453.

Smith, P.J., Giffin, Rockwell, T., and Thomas, M. 1986. Modeling fault diagnosis as the activation and use of a frame system. *Human Factors*, 28(6), 703-716.

Smith, P.J., McCoy, C.E., and Layton, C. 1997. Brittleness in the design of cooperative problem-solving systems: The effects on user performance. *IEEE Trans. on Syst., Man, Cybern.*, 27(3): 360-371.

Smith, P.J., Miller, T., Gross, S. Guerlain, S., Smith, J., Svirbely, J., Rudmann, S., and Strohm, P. 1992. The transfusion medicine tutor: A case study in the design of an intelligent tutoring system. *Proceedings of the 1992 Annual Meeting of the IEEE Society on Systems, Man and Cybernetics*, 515-520.

Treu, S. (1994). *User Interface Design: A Structured Approach.* New York: Plenum Press.

Tufte, E.R. 1993. *The Visual Display of Quantitative Information.* Graphics Press: Chesire, CT.

Tufte, E.R. 1990. *Envisioning Information.* Graphics Press: Chesire, CT.

Tversky, A. and Kahneman, D. 1974. Judgment under uncertainty: Heuristics and biases. *Science*, 185, 1124-1131.

Van der Veer, G.C., Bagnara, S., and Kempen, G.A.M. (eds.). 1992. *Cognitive Ergonomics: Contributions from Experimental Psychology.* North-Holland: Amsterdam.

Wenger, E. 1987. *Artificial Intelligence and Tutoring Systems: Computational and Cognitive Approaches to the Communication of Knowledge.* Kaufmann: Los Altos, CA.

Wyatt, J. and Spiegelhalter, D. 1992. Field trials of medical decision-aids: potential problems and solutions. *Proceedings of the 14th Annual Symposium on Computer Application in Medical Care*, 3-7.

For Further Information

Bogner, M.S., ed. 1994. *Human Error in Medicine*. Erlbaum: Hillsdale NJ.

Card, S., Moran, T., and Newell, A. 1983. *The Psychology of Human–Computer Interaction*. Erlbaum: Hillsdale, NJ.

Crane, J.G. 1992. *Field Projects in Anthropology: A Student Handbook* (3rd ed.). Waveland Press: Prospect Heights, IL.

Gaines, B. and Boose, J. 1989. *Knowledge Acquisition for Knowledge-Based Systems*. Academic Press: London.

Klein, G.A., Calderwood, R., and MacGregor, D. 1989. Critical decision method for eliciting knowledge. *IEEE Trans. Syst., Man, Cybern.*, 19(3).

Lajoie, S.P. and Lesgold, A. 1989. Apprenticeship training in the workplace: Computer- coached practice environment as a new form of apprenticeship. *Machine-Mediated Learning*, 3, 7-28.

Johnson-Laird, P. 1993. *Human and Machine Thinking*. Erlbaum: Hillsdale NJ.

Larkin, J. and Rainard, B. (1984). A research methodology for studying how people think. *Journal of Research in Science Teaching*, 21, 235-254.

Michie, D. 1986. *On Machine Intelligence*, 2nd ed. Ellis Horwood, Ltd.: Chichester.

Mitchell, C. and Saisi, D. 1987. Use of model-based qualitative icons and adaptive windows in workstations for supervisory control. *IEEE Trans. Syst., Man, Cybern.* 17, 573-593.

Moran, T. and Carroll, J., Eds. 1996. *Design Rationale: Concepts, Techniques, and Use*. Erlbaum: Hillsdale, NJ.

Moray N. 1986. Monitoring behavior and supervisory control, in *Handbook of Perception and Human Performance: Vol. 2, Cognitive Processes and Performance*, ed. K.R. Boff, L. Kaufman, and J.P. Thomas, pp. 40-51. John Wiley & Sons: New York.

Newell, A. and Simon, H.A. 1972. *Human Problem Solving*. Prentice-Hall: Englewood Cliffs, NJ.

Poulton, E. 1989. *Bias in Quantifying Judgments*. Lawrence Erlbaum Associates: Hillsdale NJ.

Rasmussen, J., Brehner, B., and Leplat, J. (eds.). 1991. *Distributed Decision Making: Cognitive Models for Cooperative Work*. John Wiley & Sons: New York.

Rasmussen, J., Pejtersen, A., and Goodstein, L. 1994. *Cognitive Systems Engineering*. John Wiley & Sons: New York.

Rouse, W. 1980. *Systems Engineering Models of Human–Machine Interaction*. North-Holland: New York.

Rubenstein, R. and Hersh, H. 1984. *The Human Factor: Designing Computer Systems for People*. Digital Press: Boston, MA.

Sanders, M. and McCormick, E. 1987. *Human Factors in Engineering and Design*. McGraw-Hill: New York.

Sanjek, R., ed. 1990. *Fieldnotes: The Making of Anthropology*. Cornell University Press: Ithaca, NY.

Shneiderman, B. 1987. *Designing the User Interface: Strategies for Effective Human–Computer Interaction*. Addison-Wesley: Reading MA.

Shafer, G. and Pearl, J. (Ed.). 1990. *Readings in Uncertain Reasoning*. Morgan Kaufman: San Mateo, CA.

Smith, J.B. 1994. *Collective Intelligence in Computer-Based Collaboration*. Lawrence Erlbaum: Hillsdale, NJ.

Tversky, A. 1982. *Judgment Under Uncertainty: Heuristics and Biases*. Cambridge University Press: Cambridge.

Wickens, C. 1992. *Engineering Psychology and Human Performance*, 2nd ed. Harper-Collins: New York.

Section II

Fundamentals of Work Analysis

17

Task Analysis

Anand Gramopadhye
Clemson University

Jatin Thaker
Clemson University

17.1 Background

Introduction

The need to improve the systems we work in has been the driving force behind human factors. Traditionally, studies in this area have been carried out by observing humans at work and by analyzing their work environments. The term "task analysis" refers to the formal approach of analyzing human performance in systems, and is comprised of the systematic recording and analysis of human work to identify human/system mismatches with an eye to ultimately designing superior systems. The goal, therefore, of any task analysis is to examine the existing human/machine systems in order to provide a basis for designing more efficient, effective systems that are based on known human capabilities.

This chapter contains a look at how task analysis has evolved to its current state, a discussion of the definitions of commonly used terms and descriptions of the task analysis procedure, an explanation of several different approaches to task analysis, and finally, a case study in which task analytic methodology is used to identify ergonomic interventions to minimize human error.

Historical Perspective

There are a number of different approaches to analyzing observable human behavior, each with its own name and purpose, but all sharing the broad goal of measuring human performance. For example, a time study is used primarily to determine the standard time needed to perform a job; an activity study is the more effective method for establishing an improved method of accomplishing a job; and link analysis is used to determine the best physical layout for a work area. Initially, the analysis of tasks focused on observable human behavior, but because automation has increased the number of cognitive and decision-making activities required of many operators, many task analyses now attempt to measure the cognitive activities which drive the observable behavior. Newer approaches, such as GOMS analysis, task knowledge structures, mental model development, and cognitive simulations, concentrate on describing and analyzing the operator's cognitive activities. Regardless of whether the objective of a particular analysis is to measure manual or cognitive tasks, it is undertaken to determine the performance requirements imposed by the task–hence the term task analysis.

Task analysis as we know it today did not develop in any predictable, or step-by-step fashion, but rather emerged from a variety of theories which developed somewhat independently from one another to address various contexts. This somewhat unstructured development has left us with a number of theoretical approaches to the study of human performance as well as with a number of terms, the definitions of which sometimes overlap. In a recent article summarizing the current status of task analysis, Stammers (1995) concluded that there is limited consensus on the terminology surrounding task analyses. It seems though, that the term "task analysis" is generally understood to refer to all activities involving the analyses of tasks and is, therefore, the term we will use throughout this chapter to mean just that (Drury et al., 1987; Kirwan and Ainsworth, 1992; Singleton, 1974; Stammers, 1995).

The growth of task analysis has closely followed that of human factors/ergonomics, because task analysis is central to the study of human factors and has become the expected means for beginning almost any human factors effort (Sheridan, 1997).

Task analysis as we know it today began early in this century with the work of Gilbreth (1911) and Taylor (1929). Taylor's work focused on describing a process and identifying ways to improve a job, but his interest was efficiency, not human wellness. Taylor's time-study approach, developed to determine time standards, did not address the ergonomic content or appropriateness of a particular task based on anthropometry or on human capabilities and limitations. Taylor's work was criticized in several quarters, but

"Taylor's greatest and lasting contribution to the science of industry is the approach he adopted. He approached problems which had been thought either not to exist or to be easily solved by common sense, in the spirit of scientific enquiry."

(Farmer, in Barnes, 1980)

Taylor used his scientific approach to describe, analyze, and improve the process of shoveling work at Bethlehem Steel Works. In the same vein, the Gilbreths expanded on this scientific approach, developing the motion study, a process that tried to determine the preferred method of doing work. As part of his work on motion study, Gilbreth identified 17 basic motions, called therbligs, which were common to all kinds of manual work, then used these basic motions to analyze the sequence of actions in a task. Today we use the combined term "motion and time study," as well as terms such as work measurement and work methods and design, to represent these initial techniques intended to analyze human work.

The U.S. Department of Labor also pursued job and task analysis in the 1930s. The intention of the Department of Labor was to formulate a consistent repertoire of personnel skills that could then be used in hiring, placement, and promotion.

The approaches developed by Taylor and Gilbreth were fine for measuring manual and repetitive tasks, but failed to address the cognitive components that made up those tasks. Realizing this limitation, Crossman (1956) proposed "mental therbligs," including planning and controlling activities, and developed the "sensori-motor" chart which drew links between planning/controlling and executing activities. Only much later were cognitive and information-processing tasks analyzed in a similar fashion (e.g., Card et al., 1983).

Work by Taylor, the Gilbreths, the Department of Labor, and Crossman, were important in lending credence to the potential for the systematic study of human work, but any look back at the development of task analysis would be remiss if it ignored the impact of the studies conducted by the U.S. military in the 1950s. Research by the military led to the development of a systematic task analytic process that defined performance requirements, training needs, and equipment design specifications. This process was especially helpful as personnel were becoming involved with increasingly sophisticated aircraft systems and complex weapon systems. In fact, the complexity of these systems led to changes in the very nature of the tasks humans would be expected to perform. The most commonly referenced work in this area is Miller's (1953) report "Method for Man–Machine Task Analysis," which used an approach that analyzed operators' tasks, then "linked" input to output. Later Miller developed a classification of tasks which used a codifying scheme based on the temporal patterns of a task. This classification system led to a taxonomic approach of classifying human performance. Since that time, a number of taxonomies of human performance have been developed (e.g., Fleishman and Quaintance, 1984). The advent of task analysis in the military provided a standardized approach to catalog skill requirements, and training and support needs. It also provided a basis for determining the requirements of new systems (e.g., Qualitative and Quantitative Personnel Requirements Information (QQPRI; Swain, 1962).

Outside of the United States though, specifically in the United Kingdom, the impetus for task analysis was driven by training needs. Early efforts in this direction were driven by the skills-based training movement (Crossman, 1956; Seymour, 1966). The focus here was to work with human operators to determine the knowledge an operator needed to perform subtasks, such as operating, inspecting, and disposing. Analysts tried to determine "how," "when," "why," etc., about a particular task, by observing a skilled operator over several cycles and using a questioning approach, or by performing the tasks themselves. The risk here, of course, was that analysts might assume that their own experience was, in all ways, a typical experience. Task information could, therefore, be misinterpreted using this technique. Another example where training has driven the growth of task analysis can be found in the works of Annett and Duncan (1967). They used a hierarchical task analytic (HTA) approach to establish training requirements. In an HTA approach, a task can be broken down to different levels, with each level covering more detail about the task. The number of levels to which a task was broken down was decided based on the degree of the analysis and various stopping rules. These rules are used to help the analyst determine the levels of hierarchy in a particular case (Stammers and Shepherd, 1995). The HTA approach has been used to define the training procedures and ergonomic requirements in complex process and nuclear industries (e.g., Duncan, 1974; Piso, 1981; Umbers and Reiersen, 1995). Examples of the use of HTA in other contexts can be found in the works of Shepherd (1989, 1993) and Carey, Stammers, and Astley (1989). Other approaches to task analysis within the human–computer interaction arena include Task Analysis for Knowledge Descriptions (TAKD) (Johnson et al., 1984), Task Action Grammars (TAG) (Payne and Green, 1986), and Task Knowledge Structure (TKS) (Johnson et al., 1988). Diaper (1989) provides a comprehensive coverage of various approaches to task analysis in the human–computer interaction arena.

Extensive use of task analysis outside of the military was first documented in Singleton's early work (1974). In the twenty-some years since these studies were conducted, we have seen an increase in the use of task analytic approaches to resolve ergonomic issues. In 1983, Drury et al. proposed a task analysis of aligning a lamp in the lamp holder of a copy machine using a column format. They found that it was very helpful to formally state the requirements for a good design between the task analysis and the ergonomic redesign. Stating the design requirements helped other members join the design phase, resulting in a participatory effort and greater ownership of the resulting design. In another effort, Armstrong et al. (1986) used task analysis based on traditional time and motion studies to identify risk factors in repetitive motion tasks. A similar approach had been used by Drury and Wick (1984) to analyze repetitive tasks in a shoe plant to identify potential ergonomic improvements. In their analyses, the tasks were observed and videotaped, then elementary job motions (based on therbligs) were determined for each body member. Their objective was to determine ergonomic interventions that would minimize musculoskeletal injuries. In this case, the investigators kept a detailed log of the body angles using the task description form. (Table 17.1 shows the task description form used for recording body angles data from Drury, 1987.) They administered both the body-part discomfort form developed by Corlett and Bishop (1976) and the general comfort rating scale developed by Shackel et al. (1979) several times during a shift to collect data on postural discomfort. They also recorded measures of productivity and quality. Their task analysis consisted of analyses of the raw descriptive data on body angles, forces, and discomfort. A major portion of the analysis involved categorizing the angles involved at each joint to determine just how close to the extreme value they came. For example, a measure that has proven to be effective in estimating exposure to Repeated Motion Injuries (RMI) problems of the wrist is simply the number of

TABLE 17.1   Task Description Form (Drury, 1987).

Job title:			Task 1	Task 2	Task 3	Task 4	Task 5	Task 6	Task 7	Task 8
FREQUENCY/BUNDLE										
BACK	Rotation									
	Lateral Bend									
	Flex/Ext									
NECK	Rotation									
	Lateral Bend									
	Flex/Ext									
SHOULDER	Rotation	R								
		L								
	Abd/Add	R								
		L								
	Flew/Ext	R								
		L								
ELBOW	Flexion	R								
		L								
FOREARM	Pron/Sup	R								
		L								
WRIST	Flex/Ext	R								
		L								
	Rad/Ulnar	R								
		L								
LEGS	Thigh to H	R								
		L								
	Shin to V	R								
		L								
	Foot to H	R								
		L								
	Rotation	R								
		L								
	Force	R								
		L								
POSTURE	Sit/Stand									
	Armrest									
	Foot Pedal									
	Backrest									
GRIP	Power	R								
		L								
	Tip pinch	R								
		L								
	Pulp pinch	R								
		L								
	Lat pinch	R								
		L								
FORCES	Push/Pull	R								
		L								
	Up/Down	R								
		L								
	In/Out	R								
		L								
	Fingers	R								
		L								
VIBRATION		R								
		L								
SHOCK		R								
		L								
LIGHTING	Luminance									
	Glare ?									

From Drury, C. G. (1987) A bio-mechanical evaluation of the repetitive motion injury potential of industrial jobs. *Seminars in Occupational Medicine*, Volume 2, No. 1, 41-49.

daily damaging wrist motions, i.e., any combination of a grip or external force with any nonzero wrist exposure. As part of the analysis, data from the body-part discomfort scale was summarized by determining the incidence of any nonzero body-part discomfort reading for each body part divided by the number of times the scale was administered. The data from the analysis was used to compare the demands placed on the operator with the operator's ability to meet those demands. Following this step, ergonomic interventions that would minimize ergonomic risks were identified. Finally, task analysis was applied as a recursive procedure to justify the choice of interventions and to measure the effect of those interventions.

Another method widely used for ergonomic analysis in Europe is the AET method (Landau and Rohmert, 1981). This method is based on the man-at-work system and the concept of stress and strain. The AET method initially analyzes an activity by determining the objects of work, the equipment, and the working environment. The objects of work are analyzed under material, energetic, and informational aspects, then analyzed by demands.

A more recent application of the use of a task analytic methodology for a specific industrial task is given by Gramopadhye et al. (1995). Gramopadhye et al. were faced with the task of improving inspection performance for a contact lens manufacturing company. The task description was designed to collect information on different aspects of the inspection task. Following this step, the different tasks within the inspection process were analyzed using Rasmussen's (1983) Skills-Rules-Knowledge (SRK) framework. The SRK framework was used to identify errors and to develop a taxonomy of errors. Following the development of the error taxonomy, a number of interventions, each expected to minimize inspection errors, were identified. Consequently, a computer-based inspection training program using simulated images of the contact lenses was developed to train inspectors to minimize inspection errors (Gramopadhye et al., 1998).

We have seen that a variety of approaches have been proposed to conduct task analysis. Unfortunately, there is no predetermined formula to ensure that investigators select the appropriate task analysis approach for any specific situation. The variety of approaches that we have, have emerged because of the variety of theoretical and methodological backgrounds and the range of contexts in which specific approaches have been developed (Stammers, 1995). Often the selection of any approach for task analysis depends upon the application domain, and more often (though not correctly), on the expertise the practitioner has in using a specific approach. In addition to the plethora of approaches available, another source of confusion about task analysis can be attributed to the various terms and related definitions that have appeared in the literature over the years. Despite the extensive use of task analysis in various domains, there is limited agreement among practitioners as to what constitutes a "task," or even what basic elements comprise a task analysis. Unless we resolve these theoretical issues, progress on task analysis, or at least our ability to communicate about it, will lag. The following section is intended to provide a better understanding of the theoretical issues underpinning task analysis.

17.2 Discussion of the Terms "Task" and "Task Analysis"

Task

A task can be defined as the unit element of analysis. In the context of human/machine systems, each task is considered to have an objective, and a specific input produces a system output. The definition of a "task" inherently defines the boundaries of a task. Based on the objectives and the level of analysis detail desired, the boundaries of a task can be expanded or collapsed. However, caution must be exercised by the practitioner when defining the boundaries of different tasks so that within any study, all tasks are comparable in terms of size (often defined by the scope, time, and level of effort required to perform them). Ensuring comparability between various tasks facilitates their comparison in terms of human performance requirements. The defining elements of a task are listed below.

1. Every task has a definite objective.
2. A task is a unit of activity conducted by one or more individuals.

3. An analyst must articulate each task's beginning and its end. The beginning is defined by an input to the system (e.g., information input, manual input, or a combination). The end of a task is defined by a system response (indication that the task objective has been completed).
4. Various tasks are associated with each other based on their objectives, functional relationship, and on space and time.

A task can be classified as continuous, discrete, or rule-based. Tracking tasks are examples of continuous tasks by which the operator has to control a system so that the system operates within predefined parameters. Most procedural tasks are discrete in nature. A typical example of such a discrete and procedural task is an assembly task. Rule-based tasks are a subset of the discrete task category. Rule-based tasks are found, for example, in diagnostic maintenance wherein the operator's next task is determined by the previous one.

Task Analysis

A task analysis is a study that aims to provide a systematic and comprehensive description of a task that is executed to achieve an overall system goal. Over the years, various definitions of task analysis have been proposed. McCormick (1976) considers task analysis to be the division of human work into component tasks and then the analysis of those components. Kirwan and Ainsworth (1992) define task analysis as the study of what an operator is required to do, in terms of actions and/or cognitive processes, to realize a system goal. A more general definition with a broader appeal in ergonomics is the following one, espoused by Drury (1983): "Task analysis is the comparison of the demands placed on the human operator (task demands) with the human's ability to meet those demands."

The task analytic approach has been used by system designers over a number of years (Meister, 1985; Johnson and Johnson, 1989). Figure 17.1 shows the role of task analysis in the overall ergonomic system design process (Pitkaar, Lenior, and Rijnsdorp, 1990). In the context of system design, task analysis has two main purposes, to design a new system, or to redesign an existing system (see Figure 17.2).

When designing a new system, the analyst must determine the system goals and the system functions that are used to achieve those goals. Following this step, the tasks needed to accomplish system functions are determined. Third, the analyst describes each task and determines the skills necessary to accomplish the task. Finally, task demands are compared with human capabilities and various function allocation alternatives (i.e., whether the task should be performed by the human alone, by the machine alone, or by a human who is assisted by a machine, etc.) are considered. The task analysis at this stage is a matter of articulating the task constraints and making them visible. These are the independent variables (inputs) that must be considered while performing the task, and the dependent variables which measure task performance (Sheridan, 1997).

In the redesign of an existing system, task analysis is used to identify problems and to suggest potential interventions that might remedy any problems. In these cases, task analysis is used to analyze complete systems or portions of existing systems in order to identify modifications to an existing system or to propose a completely new, better system. When viewed within the system design context, task analysis can be used for the following applications (adapted from Kirwan and Ainsworth, 1992):

1. System function allocation — decide on human–machine function allocation issues
2. Organizational issues — personnel selection, personnel qualification, skill requirements
3. Task design — identify the skills, procedures, and knowledge necessary to perform a task
4. Human–machine interface — workplace design, equipment/tool design
5. Human support requirements — design training and job aids
6. System reliability analysis — using data on human error to determine system reliability

The term "task analysis" has, unfortunately, been accepted to mean the simple description of a "task," in addition to the analysis of one. To resolve the resulting confusion, one needs to consider the various steps involved in conducting a task analysis. Referring to Figure 17.2, note that task analysis can be represented as consisting of the following steps:

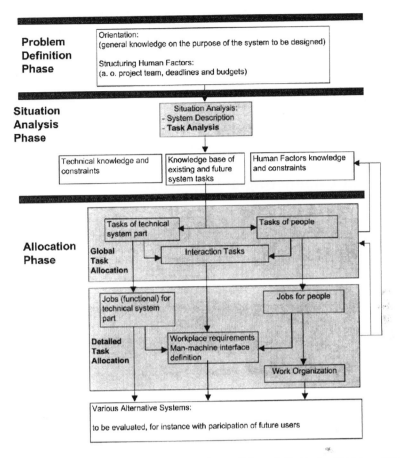

FIGURE 17.1 Role of task analysis within system design. (From Pitkaar, R. N., Lenior, T. M. J., and Rijnsdorp, J. E. (1990) Implementation of ergonomics in design practice: outline of an approach and some discussion points. *Ergonomics*, 1990, 33, 5, 583-587. With permission.)

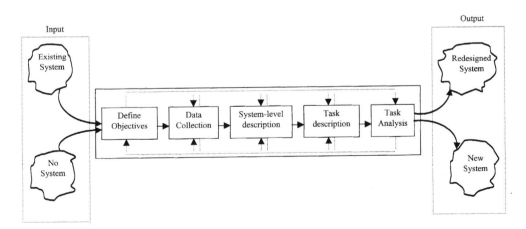

FIGURE 17.2 Steps in conducting task analysis.

1. Define objectives — establishing the scope of the task analysis
2. Data collection stage — collecting data on the task in question
3. System description stage — developing a system-level description

4. Task description — using the task description form to document or to describe the task
5. Analysis stage — this stage specifies human task requirements, synthesizes information, analyzes the task requirements with regard to human capabilities

Although, the various stages are detailed above, often analysts may not follow the above-mentioned steps exactly or use a variant of the above approach. For example, the data collection and task description stages might be done simultaneously. Or, the analyst may have to regress to an earlier stage or conduct several iterations at one or more stages. Finally, though, regardless of the approach used, it is the fourth stage that is most appropriately called the task analysis. Now that the various pertinent terms have been described, the following section discusses the steps involved in conducting a task analysis.

17.3   Steps in Conducting a Task Analysis

Define the Objectives and Scope of the Task Analysis

The first step in conducting a task analysis is to define the objectives/goals of the analysis to establish its scope. Doing this helps investigators identify the interactions between various personnel involved in all the phases of system design and development. At this stage, the analyst should also develop the detailed schedule required to successfully complete the task analysis and other system development activities. The following specific steps should be followed (in order):

- List the overall objectives of the task analysis
- Identify the personnel involved with the tasks to be studied
- Develop a detailed plan and schedule, identifying interactions with other personnel
- Obtain support for your study at all levels (operator level, supervisory level, management level)
- Identify the activities to be conducted by various members of the investigating team
- Develop a plan to accomplish these activities

Data Collection

Identifying the exact data to be collected is critical to conducting a task analysis. This phase involves defining the components and the subcomponents of the task. The type of analysis desired, the cost of the analysis, and the time available to conduct the analysis will often determine the level of detail required in the information to be collected.

System-Level Description

As a first step, an overall system level description should be developed to identify the goals of the system. Given the system-level goals, the inputs and outputs to the system should be determined. Depending upon the system under consideration, this process can be extensive. Various system description tools can be used to describe the system. The steps typically adopted to accomplish a system-level description (in order) are:

- Define the objective(s) of the system
- Develop a detailed verbal description of the system
- Use graphical approaches to clearly identify system components, the goals of each component, and the links among components; it is important to keep the stated objectives in mind while working on this step
- Identify inputs for each component of the system

Once the overall inputs, outputs, and system objectives are understood, it is critical to identify the various tasks required to achieve overall system objectives.

Task Description

After identifying the system components, the next step is to identify the different tasks within each system component. Detailed task information must be collected to ensure a comprehensive analysis. The first step in collecting task information is to develop a detailed verbal description of the task. Information is typically collected on the following items:

Task Name: A statement identifying the human performance requirement.

Task Objective: Outlines the goal of the task, so the analyst can develop links between this and other tasks and to identify relationships and commonalties with other tasks that have similar objectives and use similar methods, procedures, tools, or information.

Task Environment: Any specialized environment necessary to perform a particular task (e.g., specialized lighting to perform inspection, physical environment).

Time Required: The time required to perform the task (measured from when input is initiated to when output is obtained).

Tools/Equipment: The tools and type of equipment used to execute the task. The materials processed and products made.

Incidence/Frequency: Number of times per unit time (e.g., three times in an hour).

Control Actions/Input: Inputs or control actions necessary to execute a particular task. These could be manual, automated, or semi-automated.

Criticality of Task: Effect of task failure on system performance (often based on expert opinion).

Error Potential: Significant human errors that are likely to occur and have an impact on task objective(s) and system performance. Often based on observations and expert opinion.

Information: Information required by each task and the source of the information input (e.g., electronic, oral).

Knowledge: Cognitive skills (perception, attention, decision making) required to perform the task (useful in selection and training of appropriate personnel).

Outputs/Feedback: Response that tells the human that a particular task was completed. This feedback is often obtained by a change in the system state.

Rules: Specific procedures, rules, and guidelines needed to perform the task.

Sequence: The order in which the various subtasks are organized (e.g., serial or branching).

Skills: Human manual skills necessary to perform the task (useful in selecting and training of appropriate personnel).

Support Systems: Support systems to execute a task (e.g., decision-aiding tools used to perform diagnostic maintenance tasks or other personnel/equipment that supplies information or a service).

A variety of data collection techniques may be employed to collect task information. The cost of data collection, the type of task information, and the task analysis approach may often prescribe the choice of a specific data collection technique. A variety of these techniques are described in greater detail below.

Activity Sampling: Activity sampling was initially developed for time and work-study measurements. It can be conducted either by direct observation of the human or through video recordings. The method is used to collect data on the percentage of time spent on different tasks by a human in a system. By observing human activities at different time intervals, an analyst collects information on how humans spend their time. An important requirement in conducting activity sampling is that all activities are observable and discrete (distinguishable from the next activity), and occur for sufficient time periods so as to make measurement possible. Activity sampling is especially suitable to situations in which an operator has to do several different things but in no fixed order.

Critical Incident Technique: This technique is used to collect data on critical events that have the potential to significantly impact system performance. The critical incident technique is most useful in a system in which problems are suspected, but the source, nature, and severity of these problems are not known or completely understood. Data on critical incidents can be collected from operators because they work closely with the system and have first-hand knowledge of the system.

Observation: The direct observation of humans performing a task is a time-honored technique for collecting task information. The observer can watch an actual performance, a simulated performance, or a video recording of the performance of a task. This technique is useful for collecting initial information on the performance of a task and for verifying information collected through other data collection techniques. Another technique, closely linked to observation, is shadowing, wherein the analyst follows the operator around as the operator performs his daily work. The analyst is concerned not only with the tasks he/she is expecting the operator to perform, but also with any other task(s) the operator performs. Shadowing is particularly useful when an operator is required to perform a myriad of tasks and to interact with a number of personnel to perform these tasks. In these cases, the operator typically cannot recollect task-specific information when asked in an interview or in response to a questionnaire. It seems that the best way for an analyst to collect accurate information in these cases is by following the operators as they conduct their daily tasks. In collecting information by direct observation, though, analysts must be sensitive to the possible effect of their observing on the operator. This intrusion, caused by the operator's awareness of the analyst, might alter the operator's attitude or even the actual performance on the task.

Interviews: Interviewing domain experts is a popular technique for collecting information. Both structured and unstructured interviews can be used to collect task information. In a structured interview, the questions themselves and the order in which the questions are asked is predetermined. In an unstructured interview, the interviewer may have only a rough idea of the exact questions to ask, and may allow answers to a few initial questions to determine the direction of the interview. These unstructured or less structured interviews can be particularly helpful during the initial stages of data collection, whereas a structured interview may be more appropriate for collecting specific information. Interviews can also help analysts collect information that was missed during the direct observation of a person performing a task. When interviewing, analysts should be concerned with (1) asking probing questions to obtain more details about the task (e.g., Is the way the task is typically performed the "right," or "textbook" fashion, or is it a "quicker," way developed by the operator?) (2) identifying the interrelationships among various subtasks, and (3) identifying other ways the task might be performed. Interviews of both individuals and groups of personnel can be helpful.

Surveys/Questionnaires: Asking domain experts to complete questionnaires is another method of collecting task information. The advantage to using questionnaires is that information can be collected inexpensively from large numbers of participants in various geographical locations. The disadvantage is the often low response rate and the reduced amount of information obtained per respondent. Surveys and questionnaires are most popularly used for collecting opinion and attitude data and may be best used as a follow-up to an interview.

Verbal Protocols: The increase in automation has changed the nature of many jobs from manual to cognitive. Information on tasks with large cognitive components can often be best captured by asking the subject to "think aloud" while carrying out his actions. This technique, wherein subjects are asked to verbalize their actions (i.e., to explain what they are doing it, how they are doing and why they are doing it) while carrying out their task is called "verbal protocol." Here again, as with the shadowing technique, there is a risk that asking the subjects to verbalize their actions will interfere with task performance and/or change the way the task gets executed. It can be helpful, however, to compare the findings obtained from a verbal protocol with information gathered through other data collection techniques in an effort to corroborate the data.

Task Documentation: Information on the task can often be collected through task documentation, i.e., typical procedural steps are often clearly described in system documents. Drawings, specifications, log reports, company-wide procedures, operation manuals, and training and instructional material are a sampling of the variety of documents that can be used to collect information on a task.

It is important that a task description be comprehensive, have integrity and validity, and be in a format that is clear, useful, and easy for the analyst to understand. Alternate formats can be used to transcribe task descriptive information (e.g., flow charts, column format). Table 17.2. shows a sample of a column format used for describing a real-time shop floor control system (Anne and Greenstein, 1993).

TABLE 17.2 Task Description: Column Format

JOB DESIGN QUESTIONNAIRE

Task no.	Function	Task	Allocation	Information Required	Information Presented	Human Input	Comp. Input	Coordination	Cognitive Demand	Possible Errors	Consequences	Task Duration	Frequency	Who	Knowledge employed	Skill level	Task complexity	Task criticality	Comments
	System Management																		
1.1	Activate RTS																		
1.1.1		Place ON/OFF switch in "ON" position	Human	Position of ON/OFF switch	Switch label	Updated switch position		None	Low	Failure to locate switch	Inability to run the system	<1 min.	Once a day	Real time clerk or Supervisor	Basic operation of a PC	Low	Simple	High	
1.1.2		Access RTS directory	Human	Current directory	Prompt of current directory	Change directory command		None	Low	1. Specifying wrong directory 2. Issuing wrong command	Inability to run the system	<1 min.	Once a day	Real time clerk or Supervisor	Basic knowledge of DOS	Medium	Simple	High	Knowledge of when and how to perform this task must be currently memorized and is not displayed. Could display activation procedure above terminal and/or replace individual commands with one command that invokes a batch file.
1.1.3		Activate memory record manager	Human	Description of activation sequence	None currently	Btrieve command		None	Moderate	Issuing wrong command	Inability to run the system	<1 min.	Once a day	Real time clerk or Supervisor	Knowledge of RTS activation requirement and DOS	Low	Simple	High	
1.1.4		Load RTS and generate main menu	Human	Description of activation sequence	None currently	Btrieve command		None	Moderate	Issuing wrong command	Generation of no display or another display	<1 min.	Once a day	Real time clerk or Supervisor	Knowledge of RTS activation requirement and DOS	Medium	Simple	High	

Anne, M. and Greenstein, J. S. (1993) The design of a computer-based tool to aid the manager of a real-time manufacturing control system (*Tech. Report*). Clemson, SC: Clemson University, Department of Industrial Engineering.

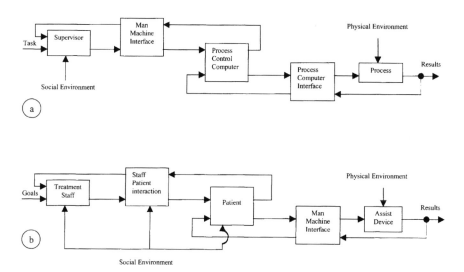

FIGURE 17.3   Flow chart showing the environment of a supervisor in process control and a patient in a rehabilitation center. (From Stassen, H. G., Steele, R. D., and Lymam (1989) A man-machine system approach in the field of rehabilitation: a challenge or necessity, in Baosheng Hu (ed.) *IFAC Analysis, Design and Evaluation of Man–Machine Systems*, PRC, 79-86. With permission.)

Task Analyses

Information collected during the task description stage is analyzed using one of the several approaches described below. The analysis stage applies various theories, models, and results of empirical ergonomic studies to the problem at hand. Although the list of approaches described below is not exhaustive, it does provide a representative list of some commonly used approaches. In addition, appropriate references have been included to lead the reader to opportunities for further study.

Flow Charting Approaches: Over the past several decades, various flow charting techniques have been used for task analysis. Flow charts or networks are graphical descriptions that can be used to describe and show links between the tasks and subtasks within a system. Figure 17.3 is a flow chart showing the environment of a supervisor in a process industry and a patient in a rehabilitation center (Stassen et al., 1989). A number of flow charting techniques are currently in use. The more popular of these are operation sequence diagrams, functional flow analysis, and decision diagrams. Figure 17.4 shows an operation sequence diagram (OSD) for baking soda crackers (Barnes, 1980). In this diagram, the various steps in the assembly process are outlined using operation sequence symbols. OSD can be depicted using alternate formats. Figure 17.5, an example of this, uses operation sequence symbols to depict the flow of information for the simple task of requesting a purchase order. The summary table shows the number of activities, times, and distance traveled, information which can be used to compare the method currently in use with the proposed method. Whatever the graphical representation, an operation sequence diagram assists by providing the analyst with a graphical representation of the tasks as they relate to both machine elements and other operators. This information is useful in helping analysts understand the links between various activities and then, consequently, to design more efficient systems. Similarly, functional analysis depicts the sequence of actions/functions that need to be performed by the system (Greer, 1981), and decision diagrams are often used to represent activities with decision components. Figure 17.6 shows a decision flow chart developed by Lock and Strutt (1985) to analyze error-likely situations in an inspection system.

Link Analysis: The objective of link analysis is to assist the analyst in identifying the relationships (links) between different system components (Chapanis, 1962). This approach can be used for arranging the physical layout of a screen display, instrument panel, workstations, and office, and help in understanding the communication links between individuals. Figure 17.7, adapted from Chapanis (1991), shows how

Baking Soda Crackers

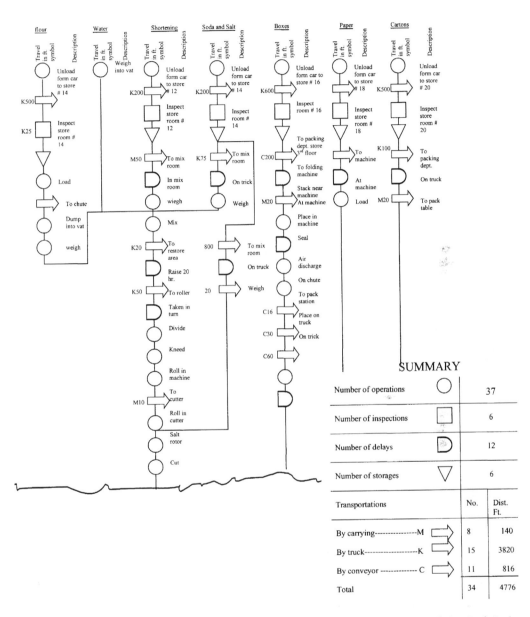

FIGURE 17.4 Assembly process chart: baking soda crackers. (From Barnes, R. (1980) *Motion and Time Study Design and Measurement of Work*. John Wiley & Sons Inc., New York. With permission.)

link analysis was used successfully to design an alternate physical layout for the *U.S.S. Louisville*. Initially a link table (see Table 17.3) was created by counting the number of times each system component interacted with the other. Then the associations between the different system components were identified. This information was also represented graphically using the schematic link diagram. In the link diagram, more frequently linked components are indicated by the greater number of lines shown connecting them. The information from Figure 17.7 and Table 17.3 was later used to devise an alternate schematic diagram which was sensitive to the links between the different system components (Figure 17.8).

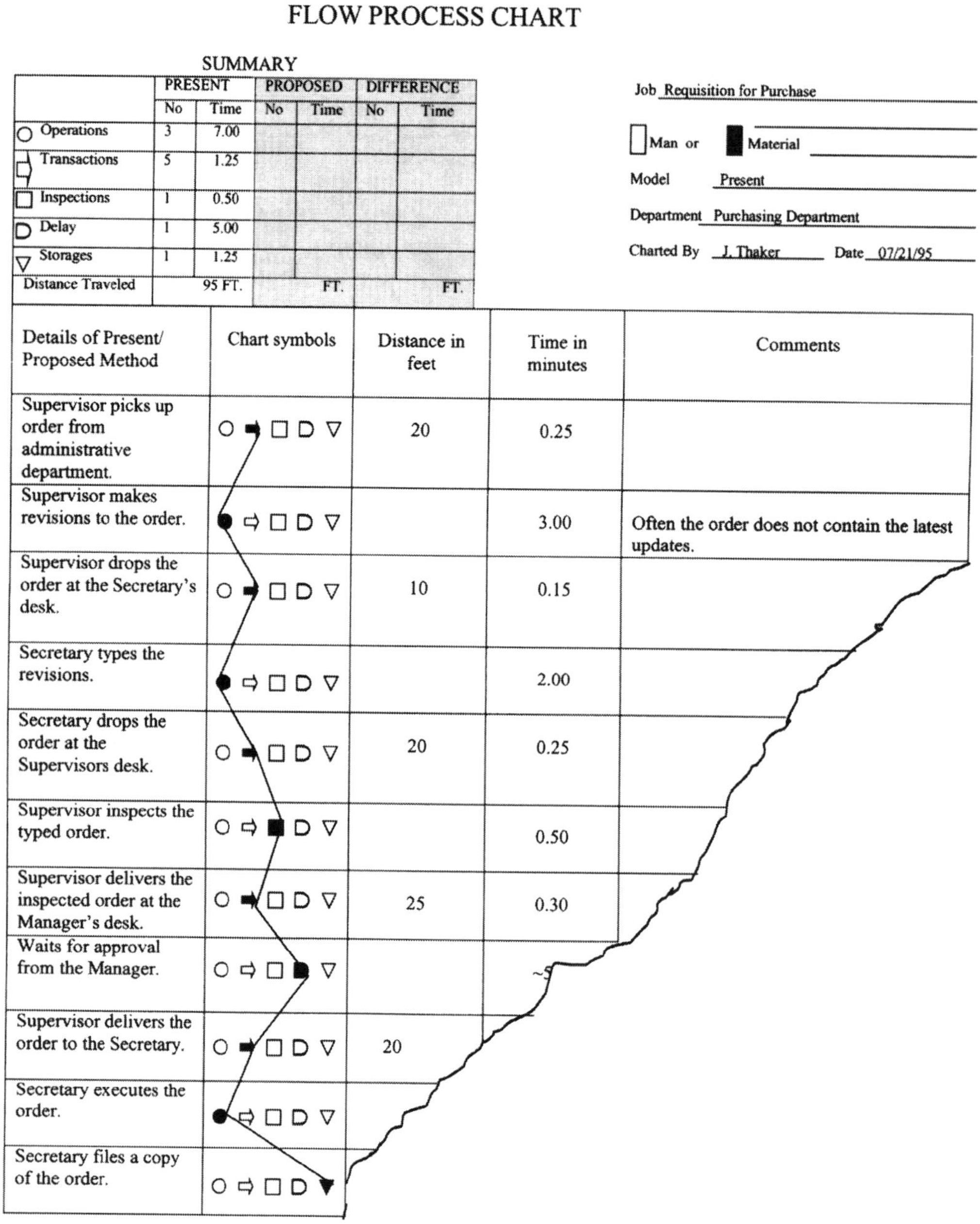

FLOW PROCESS CHART

SUMMARY

	PRESENT		PROPOSED		DIFFERENCE	
	No	Time	No	Time	No	Time
○ Operations	3	7.00				
Transactions	5	1.25				
☐ Inspections	1	0.50				
D Delay	1	5.00				
▽ Storages	1	1.25				
Distance Traveled	95 FT.		FT.		FT.	

Job Requisition for Purchase

☐ Man or ■ Material

Model   Present

Department Purchasing Department

Charted By   J. Thaker   Date 07/21/95

Details of Present/ Proposed Method	Chart symbols	Distance in feet	Time in minutes	Comments
Supervisor picks up order from administrative department.	○ ■ ☐ D ▽	20	0.25	
Supervisor makes revisions to the order.	● ⇨ ☐ D ▽		3.00	Often the order does not contain the latest updates.
Supervisor drops the order at the Secretary's desk.	○ ■ ☐ D ▽	10	0.15	
Secretary types the revisions.	● ⇨ ☐ D ▽		2.00	
Secretary drops the order at the Supervisors desk.	○ ■ ☐ D ▽	20	0.25	
Supervisor inspects the typed order.	○ ⇨ ■ D ▽		0.50	
Supervisor delivers the inspected order at the Manager's desk.	○ ■ ☐ D ▽	25	0.30	
Waits for approval from the Manager.	○ ⇨ ☐ ■ ▽		~5	
Supervisor delivers the order to the Secretary.	○ ■ ☐ D ▽	20		
Secretary executes the order.	● ⇨ ☐ D ▽			
Secretary files a copy of the order.	○ ⇨ ☐ D ▼			

FIGURE 17.5 Operation sequence diagram.

Critical Incident Analysis: The critical incident analysis technique is used widely with operational systems to study human error and has been described in great detail by Flanagan (1954). The technique has also found extensive use in the personnel and skills-related areas (Meister, 1985). This technique documents accidents, misses, and near misses using first-hand information from operators. Once such information is documented, human factors knowledge is applied to analyze accidents. The goal of this analysis is to lead analysts to hypothesize about the source of the errors, and to help them identify interventions to minimize errors in the future.

Workload Analysis: Workload analysis focuses on determining whether a human has been successful in completing assigned tasks within an allocated time. The technique focuses on measuring or estimating

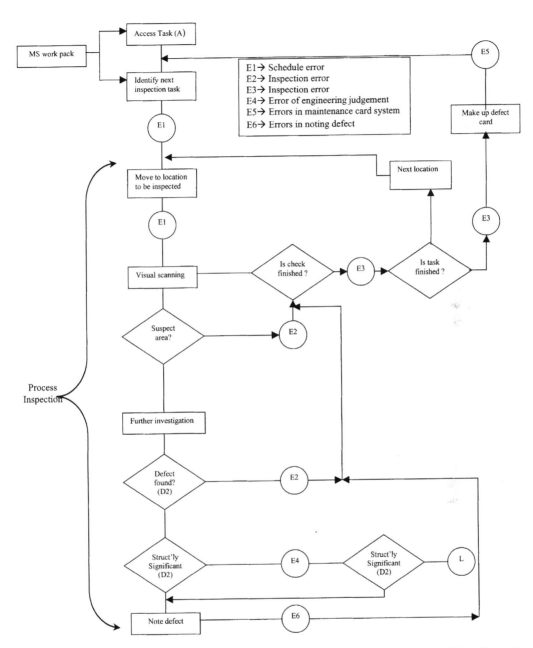

FIGURE 17.6 Inspection model flowchart. (From Lock, M. W. B. and Strutt, J. E. (1985) Reliability of in-service inspection of transport aircraft structures. *CAA Paper 85013*, London. With permission.)

the workload for different task segments as a function of time. Thus, knowing the workloads and human capabilities for different task segments, the analyst considers alternate function allocation strategies or provides additional resources for those tasks which have been identified as likely to overload the operator. In addition, the information on workloads and human capabilities can be used to specify or design hardware and software requirements. In a recent study, Wilson (1993) evaluated the workloads on aircraft pilots and the weapons systems officer during air-to-ground training missions using different workload measures. One such measure was the cardiac inter-beat intervals plotted as a function of different mission segment codes (see Figure 17.9).

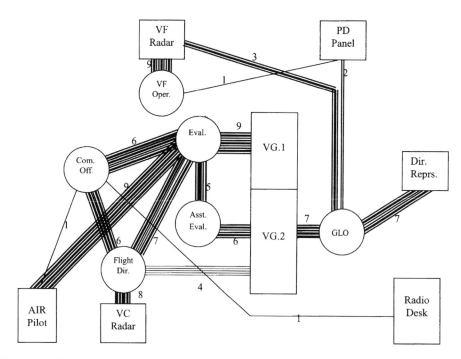

FIGURE 17.7   Schematic layout of the original arrangement of critical men and machines and the linkages between them. (From Chapanis, A. and Shafer, J. (1991) A workshop on human factors methods. *Annual Human Factors Society Meeting*, San Francisco, CA. With permission.)

TABLE 17.3   Linkages Between Men and Machine

		An Example of Link Analysis: Layout of the CIC Aboard the U.S.S. Louisville-2					
		MEN					
		Communications Officer	Evaluator	Assistant Evaluator	Gunnery Liaison Officer	Flight Director	VF Radar Operator
Men	Communications Officer		6			6	
	Evaluator	6		5		7	
	Assistant Evaluator		5				
	Gunnery Liaison Officer						
	Flight Director	6	7				
	VF Radar Operator						
	VC Radar					8	
	Air Pilot	1	9				
	Radio Desk	1					
	PD Panel				2		1
	VG Radar No. 1		9				
	VG Radar No. 2			8	7	4	
	VF Radar				3		9
Machines	Director Repeaters				7		

Chapanis, A. and Shafer, J. (1991) A workshop on human factors methods. *Annual Human Factors Society Meeting*, San Francisco, CA.

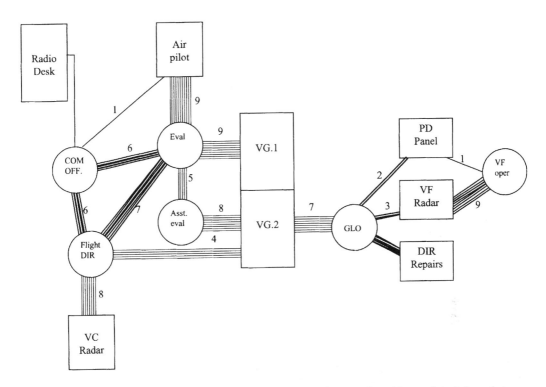

FIGURE 17.8 Schematic layout of the revised arrangement of critical men and machines and the linkages between them. (From Chapanis, A. and Shafer, J. (1991) A workshop on human factors methods. *Annual Human Factors Society Meeting*, San Francisco, CA. With permission.)

Hierarchical Task Analysis: The hierarchical task analysis (HTA) approach was originally an outgrowth of the work of Annett and Duncan (1967) and since then has been used extensively to study human error, human–machine interface design, human–computer interaction, training, job design, allocation of function, and assessment (Stammers and Shepherd, 1995). Simply stated, the HTA approach represents the task hierarchically.

To begin with, the HTA approach defines the overall system objectives, and later the tasks that are required to achieve system objectives. The tasks are redescribed in terms of a set of subtasks and subgoals and a plan that governs how the subtasks should be executed. Thus the final task analysis is hierarchical, with the level of detail increasing with each level. Each subtask can be examined to determine if it is defined to a sufficient level of detail; if not, the analyst can define it further. The level of detail is one of the critical issues in HTA, and various rules can be used to guide the analyst as to when the necessary level of detail has been attained. A commonly used rule is the $P \times C$ rule espoused by Annett and Duncan (1967). This rule states that further redefinition of the task is not necessary if the product of the probability (P) of inadequate performance and the cost (C) of inadequate performance is acceptable. The $P \times C$ rule can be applied to each task/subtask at a specific level. If the $P \times C$ value is unacceptable, then the task is broken down to greater levels of detail until an acceptable value of $P \times C$ is obtained for the lower level subtasks. Figure 17.10 shows an example of hierarchical task analysis applied to aircraft ramp maintenance activities (Mitchell, Bright, and Rickman, 1996). In this case, a human error classification scheme was developed and applied to a sample of operations in the HTA. This assisted the analysts in identifying and classifying errors and provided examples of potential human errors that might be expected to occur as operators perform generic maintenance tasks.

Fault Tree Analysis (FTA): The FTA is a quantitative top down approach that uses the and/or logic to estimate errors, representing tasks via a tree-like structure (Green, 1983; Stammers and Shepherd, 1995). The fault tree approach analyzes undesirable events often referred to as the "top event" by determining

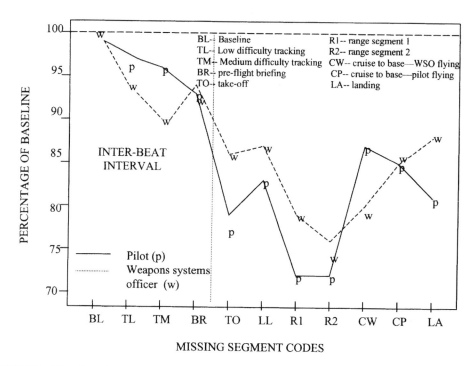

FIGURE 17.9   Mean Inter-beat Interval (IBI) percentage change from baseline for pilots and WSOs for each of the 11 segments. The vertical dotted line separates ground from flight segments. (From Wilson, G. F. (1993) Air-to-ground training missions: a psycho physiological workload analysis. *Ergonomics*, 36, 9, 1071-1087. With permission.)

what could cause it, either alone or in combination with other events. Herein "and" and "or" indicate the relationship between events. When two events are linked by "and," both events must occur in order for the parent event to occur, whereas when two events are linked by the term "or," the occurrence of one of the two events is sufficient to cause the parent event to occur. Probabilities can be assigned to individual events, enabling the analyst to estimate the probabilities of specific failures and those of the undesirable top event. The FTA has been used extensively to analyze the reliability and safety of complex systems (e.g., process industry, nuclear plant). Figure 17.11 shows the fault tree of an undesirable event in a nuclear power plant situation (Amendola et al., 1982). The FTA has also been used to a great degree to predict human error probabilities in complex systems — THERP approach (Swain and Guttman, 1980).

Failure Modes and Effects Analysis (FMEA): Unlike the FTA, which is a top-down approach, the FMEA is a bottom-up approach that has been successful in human reliability and error analysis studies (Hammer, 1985; Kirwan and Ainsworth, 1992). The FMEA approach is relatively straightforward. The analysis starts at the lowest level (i.e., task) and determines what effects a failure/error can have on system performance. The analyst typically starts with tasks and subtasks and identifies typical errors the human operator would be expected to make while executing the subtasks. Probability estimates or frequencies for each kind of error are estimated, following which the consequences of errors on system performance are deduced. If the consequences can be expected to be serious, further investigation is undertaken to identify interventions that can minimize errors or to design an error-tolerant system.

Behavior Taxonomy Approach: This approach focuses on developing behavioral taxonomies and classifying tasks based on different dimensions of performance. Over the years, various taxonomies have been developed to classify tasks (Miller, 1967; Fleishman and Quaintance, 1984). In addition, Gagne's (1977) taxonomy for learning tasks and the Position Analysis Questionnaire (McCormick, Jeanneret and Mecham, 1969) are commonly used in the areas of personnel selection and training.

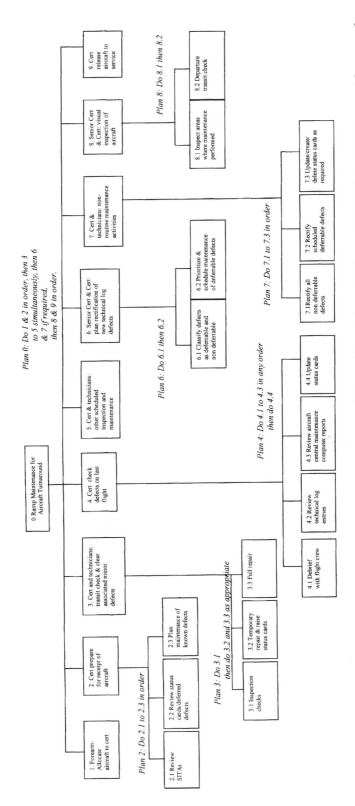

FIGURE 17.10 A hierarchical task analysis of aircraft ramp activities. (From Mitchell, K., Bright, C. K., and Rickman, J. K. (1996) A study into potential sources of human error in the maintenance of large civil transport aircraft. *CAA Paper 96004*, Lloyd's Register, Civil Aviation Authority, London. With permission.)

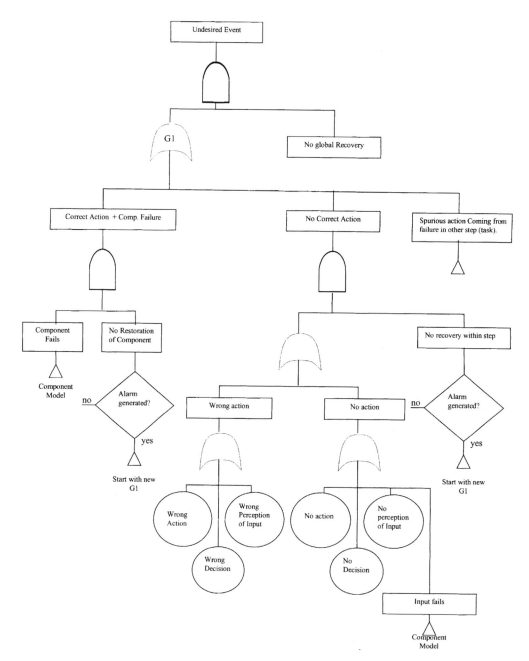

FIGURE 17.11   Fault tree analysis (FTA). (From Amendola, A., Mancini, G., Poucet, A. and Reina, G. (1982) Dynamic and static models for nuclear reactor operators — needs and examples. *IFAC Analysis, Design and Evaluation of Man-Machine Systems*, Germany, 103-110. With permission.)

Analysis of Cognitive Activities: Interest in human/computer interaction has resulted in several newer approaches to analyze the cognitive activities of humans as they interact with computers. Examples of such approaches include GOMS analysis (Card et al., 1983; Kieras, 1988), Task Analysis for Knowledge Structures (Johnson et al., 1984), Task Action Grammar (Payne and Green, 1986), mental model development, verbal protocol analysis (Bainbridge, 1991; Bainbridge and Sanderson, 1995), and cognitive simulation (Roth, Woods, and Pople, 1992).

17.4   Task Analysis: The Aircraft Inspection System

This section describes a study of the U.S. commercial aircraft inspection and maintenance system that was accomplished via a task analytic approach (Drury, Prabhu, and Gramopadhye, 1990; FAA, 1991; FAA, 1993a; Gramopadhye, Drury, and Prabhu, 1997). In this study, researchers analyzed the existing system, then identified potential ergonomic interventions. As is typical, following the identification of various interventions, the investigators undertook a detailed task analysis of specific subsystems with the intention that this analysis would lead to the implementation of specific interventions. A greater-than-usual level of detail is included here, since this document is intended to serve as a practical guide to others embarking on projects that require task analysis.

Introduction

The aircraft inspection/maintenance system is a complex one (Drury, Prabhu, and Gramopadhye, 1990; Drury, 1991; FAA, 1991) and is affected by a variety of geographically dispersed entities. These entities include large international carriers, regional and commuter airlines, repair and maintenance facilities, as well as the fixed-based operators associated with general aviation. An effective inspection is seen as a necessary prerequisite to public safety, so both inspection and maintenance procedures are regulated by the U.S. federal government via the Federal Aviation Administration (FAA). Investigators conducting this study found that while adherence to inspection procedures and protocols is relatively easy to monitor, tracking the efficacy of these procedures is not.

The maintenance process begins when a team that includes representatives from the FAA, aircraft manufacturers, and start-up operators schedule the maintenance for a particular aircraft. These schedules may be, and often are, later modified by individual carriers to suit their own scheduling requirements. These maintenance schedules are comprised of a variety of checks that must be conducted at various intervals. Such checks or inspections include flight line checks, overnight checks, and four different inspections of increasing thoroughness — the A, B, C, and most thorough and most time-consuming, D check. In each of these inspections, the inspector checks both the routine and non-routine maintenance of the aircraft. If a defect is discovered during one of these inspections, the necessary repairs are scheduled. Following these inspections, maintenance is scheduled to (1) repair known problems, (2) replace items because the prescribed amount of air time, number of cycles, or calendar time has elapsed, (3) repair previously documented defects (e.g., reports logged by pilot and crew, line inspection, items deferred from previous maintenance), and (4) perform the scheduled repairs (those scheduled by the team including the FAA representatives). In the context of an aging fleet, inspection takes an increasingly vital role. Scheduled repairs to an older fleet account for only 30% of all maintenance compared with the 60 to 80% in a newer fleet. This difference can be attributed to the increase in the number of age-related defects (FAA, 1991). In such an environment, the importance of inspection cannot be overemphasized. It is critical that these visual inspections be performed effectively, efficiently, and consistently over time. Moreover, 90% of all inspection in aircraft maintenance is visual in nature and is conducted by inspectors, so inspector reliability is fundamental to an effective inspection.

When the aircraft arrives at a maintenance site, the scheduled maintenance is translated into a set of job cards or work cards (i.e., instructions for inspection and maintenance). Initially, the aircraft is cleaned and access hatches opened so inspectors can view these various areas. This activity is followed by a thorough inspection — again, the inspection being primarily visual in nature. Since such a large part of the maintenance workload is dependent on the discovery of defects during inspection, it is imperative that this "incoming" inspection is completed as soon as possible after the aircraft arrives at the inspection maintenance site. At this point in the inspection process, inspectors are expected to discover those critical defects that will necessitate long follow-up maintenance times so this maintenance can be scheduled. Thus, there is a heavy inspection workload at the commencement of each inspection or check. It is only after the discovery of defects that the planning group can estimate the expected maintenance workload, order replacement parts, and schedule the maintenance. Frequently, maintenance facilities resort to overtime to accomplish these inspections. This overtime results in an increase in the total number of inspection hours, and often leads to prolonged work hours for the inspector. In addition, much of the inspection, including routine inspections on the flight line, are carried out during the night shift, because that is the time between the last flight of the day and first flight of the next.

During inspection, each defect is written up as a Non-Routine Card. This is translated into a set of work cards that identify the specific work needed to rectify the defect. Each of these defects is rectified by the maintenance crew. Rectifying each defect generates an additional inspection, typically called a "buy-back" inspection, conducted by the inspector to ensure that the work done by the maintenance crew meets the necessary standards.

The previous paragraph has pointed out that when an aircraft initially arrives at a maintenance site the inspection workload is quite large. As the service on the aircraft progresses, i.e., as maintenance crews begin work on the repairs, the inspection workload decreases. The inspection load increases again as maintenance tasks are completed. However, at this time the rhythm of the inspector's work is different; inspectors at this time are frequently interrupted as aircraft maintenance technicians request that inspectors conduct the required "buy-back" inspections of the completed work.

As in any system that is highly dependent on human performance, efforts made to reduce human errors by identifying human/system mismatches can have an impact on the overall effectiveness and the efficiency of the system. Given the backdrop of the inspection system, the objective of this particular study was to identify human/system mismatches and to design interventions that would improve human inspection performance in the aircraft maintenance system.

Objectives of the Study

The objectives of the study were fourfold:

1. To describe and analyze the existing aircraft inspection maintenance system
2. To identify human errors and develop a taxonomy of errors
3. To outline ergonomic interventions
4. To implement the specific ergonomic interventions expected to have the greatest impact on the system

Data Collection

The first stage of the assessment was to collect detailed information about the inspection process. A variety of data collection techniques were used to collect not only the process and procedural information and the idealized way of completing inspection tasks, but also information about the way tasks actually get accomplished. The main information sources for collection of task-descriptive data were the following:

1. *Observation and Shadowing:* Data was collected by observing aircraft inspection and maintenance operations at various sites, ranging from large international carriers to startup and regional operators of general aviation. This involved watching various inspectors and maintenance technicians accomplishing various tasks over several shifts.
2. *Interviewing:* Personal interviews conducted with inspectors, supervisors, aircraft maintenance technicians, lead mechanics and foremen, managers, planners, and other personnel associated with aircraft maintenance, were used to collect data on aircraft maintenance tasks. Interviews with system participants at all levels helped investigators to collect data on the structure and functioning of the system as well as to collect data on rare events such as system errors. This process was used to identify not only the prescribed way of working on the task but also the "quick and dirty" way that those tasks often really get completed.
3. *Documentation:* Information on aircraft inspection and maintenance procedures was obtained through company-wide procedures, Federal Aviation Authority mandated procedures (Federal Aviation Regulations — FARs), airworthiness directives, aircraft manufacturers manuals, and other documents.

Information from the above sources was used to obtain a basic understanding of the task(s) involved and to identify problem areas.

Task Description

The information on the inspection process was represented in various formats: (1) an inspection flow chart, (2) a task description based on generic inspection, and (3) a task description form. Following are more detailed descriptions of each of these formats.

Inspection Flow Chart: As a first step, a flow chart of inspection and maintenance activities was developed to illustrate the relationship between inspection and maintenance activities and the relationships among the personnel involved. A modified inspection/maintenance diagram is shown in Figure 17.12 (Kraus and Gramopadhye, in press)

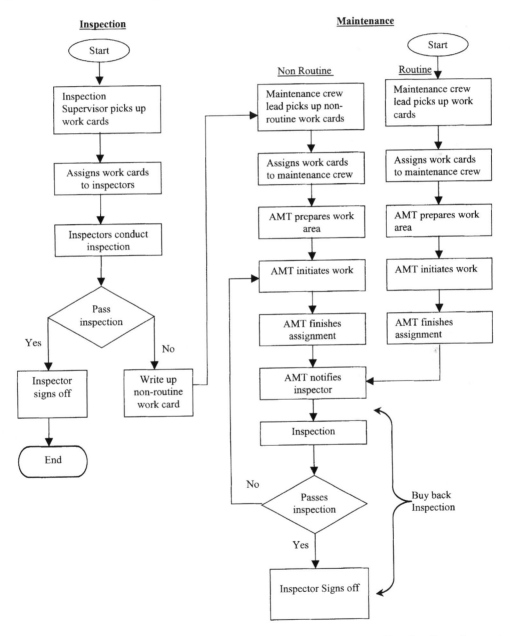

FIGURE 17.12 Inspection flow diagram. (From Kraus, D. and Gramopadhye, A. K. Transfer effects of computer-based team training. [To appear in the special issue on Human Factors in Aviation Maintenance in the International Journal of Industrial Ergonomics.]

Task Description Based on Generic Inspection Activities: Although general task description approaches are widely available (see Drury, 1983; Kirwan and Ainsworth, 1992), investigators in this case decided it would be advantageous to use an approach directly related to inspection. Literature describing human factors in inspections has produced the following generic list of inspection activities (e.g., Sinclair, 1984):

1. Present item to inspector
2. Search for flaws
3. Decide on rejection/acceptance of each flaw
4. Respond by taking necessary action.

It should be noted that not all steps may be required for all inspections. For example, some inspections may not necessitate search (e.g., color matching) and some may not necessitate decision making (e.g., absence of a rivet head on a lap splice). The description of various tasks within the aircraft inspection context is as shown in Table 17.4. The unit of task description was the work card (instructions outlining the specific inspection activity). A task was seen as continuing until a repair was completed and had passed airworthiness requirements. The work card was the unit of work assigned to one particular inspector on one physical assignment, but the actual time required to complete work shown on a work card is not consistent from work card to work card. Typically, inspectors were expected to complete the work outlined by the work card within one shift. However, work might have to be continued across shifts. The work listed on each work card could, potentially, require several inspections. Detailed task descriptions were obtained through site visits to various aircraft maintenance facilities.

Task Description Form: Data collected through interviews and site visits were transcribed using a standard working form (see Table 17.5), with a separate form for each of the five steps (initiate, access, search, decision making, and respond) in the generic task description. During the data collection phase, the human factors analysts remained with the inspectors, asking probing questions while these inspectors were actually at work. Following this step, knowledge of human factors models of inspection performance and the functioning of individual subsystems were used to identify specific subsystems and potential human/system mismatches under the Observations column in Table 17.5.

Task Analysis

Following this step, a detailed taxonomy of errors was developed from the failure modes of each task in aircraft inspection (Table 17.6.). This taxonomy, based on the failure modes and effects analysis (FMEA) approach, was developed due to the realization that a proactive approach to error control is necessary to identify potential errors. Thus, the taxonomy was aimed at the phenotypes of error (Hollnagel, 1989), that is, the observed errors. Using the generic task description of the inspection system, the goal or outcome of each task was postulated as shown in Table 17.6 (third column). These outcomes then formed the basis for identifying the failure modes of each task and included operational error data gained from the observations of inspectors and from discussions with various aircraft maintenance personnel, which were collected over a period of two years. Later, the frequencies of each kind of error were estimated, following which the consequence of errors on system performance was deduced. The error taxonomy provided the analysts a systematic framework to suggest appropriate intervention strategies. Following the classification of errors, the investigators, using their knowledge of human factors, identified interventions expected to minimize the errors. The role of the human factors effort was to change the human/machine system, thereby reducing the error incidence and making the system more reliable. Specifically, two types of interventions were considered: changing the system to fit the human, or changing the human inspector to fit the system. The specific interventions for five steps of the inspection process are listed in Table 17.7. Since then, various interventions for improving inspection performance have been implemented (FAA 1993b; Gramopadhye, Drury, and Sharit, 1993; Gramopadhye et al., 1997; Patel, Drury, and Prabhu 1993).

TABLE 17.4 Detailed Breakdown of Aircraft Visual Inspection by Task Step

Task Description	Visual Example	Task Step
1. Initiate	Get work card. Read and understand area to be covered.	1.1 Correct instructions written. 1.2 Correct equipment procured. 1.3 Inspector gets instructions. 1.4 Inspector reads instructions. 1.5 Inspector understands instructions. 1.6 Correct instructions available. 1.7 Inspector gets equipment. 1.8 Inspector checks/calibrates equipment
2. Access	Locate area on aircraft. Get into correct position.	2.1 Locate area to inspect. 2.2 Area to inspect. 2.3 Access area to inspect.
3. Search	Move eyes across area systematically. Stop if any indication.	3.1 Move to next lobe. 3.2 Enhance lobe (e.g., illuminate, magnify for vision, use dye penetrant, tap for auditory inspection). 3.3 Examine lobe. 3.4 Sense indication in lobe. 3.5 Match indication against list. 3.6 Remember matched indication 3.7 Remember lobe location. 3.8 Remember access area location. 3.9 Move to next access area.
4. Decision making	Examine indication against remembered standards (e.g., for dishing or corrosion).	4.1 Interpret indication. 4.2 Access comparison standard. 4.3 Access measuring equipment. 4.4 Decide on if it is a fault. 4.5 Decide on action. 4.6 Remember decision/action.
5. Respond	Mark defect. Write up repair sheet or if no defect, return to search.	5.1 Mark fault on aircraft. 5.2 Record fault. 5.3 Write repair action.
6. Repair	Repair area (drill out and repair rivet).	6.1 Repair fault
7. Buy back inspect	Visually inspect marked area.	7.1 Initiate. 7.2 Access. 7.3 Search. 7.4 Decision. 7.5 Respond.

From FAA (1991) *Human Factors in Aviation Maintenance — Phase One Progress Report*. DOT/FAA/AM-91/16, Office of Aviation Medicine, Washington, D.C.

17.5 Conclusion

This review has shown that task analysis has evolved over the years, with its growth linked closely to advances in technology and the growth of new application domains. A common theme that resonates throughout the chapter is that task analysis needs to be the basis for any human factors design effort. Ergonomists and human factors engineers need to bear this in mind as they embark upon any system design efforts. Furthermore, as practitioners continue to use different task analytic approaches, it is critical that they communicate the success and pitfalls in using a specific approach for a particular application. This will ultimately lead to the development of detailed principles or guidelines in using various task analytic approaches based on inputs available and outputs desired.

TABLE 17.5   Task Description Form

Task: Wing and Leading Edge Inspection   **Location:** Right Wing

Task Description	Task Analysis									
	Subsystems									Observations
	A	S	P	D	M	C	F	P	O	
Search										
1.0 Inspect slat structure, wiring and installation.										
1.1 Check wear on male duct of telescopic shaft duct.										
1.1.1 Check wear by moving it and seeing if it is loose.			X		X					No prescribed force.
1.2 Inspect slatwell area for corrosion and cracks.								X		No standards for wear.
1.2.1 Hold flashlight such that light falls perpendicular to the surface.								X		Holding flashlight for a long period at odd positions – strenuous.
1.2.2 Visually look for cracks and corrosion.			X	X	X					Lack of information on type of cracks and figures.
1.2.3 The visual indication confirmed by tactile and moves scrutinized inspection.			X	X	X					
1.3 Look for play in slat actuator nut.										

Attention: Number of time-shared tasks   Perception:

Memory: SSTS, Working, Long-term   Feedback: Quality, Amount, Timing

Senses: Visual, Tactile, Auditory   Decision: Sensitivity, Criterion, Timing

Control: Continuous, Discrete   Posture: Reading, Forces, Balances, Extreme Angles

From FAA (1991) *Human Factors in Aviation Maintenance — Phase One Progress Report.* DOT/FAA/AM-91/16, Office of Aviation Medicine, Washington, D.C.

TABLE 17.6   Task and Error Taxonomy for Visual Inspection (Tasks: Initiate and Access)

Task	Errors	Outcome
	1. INITIATE	
1.1 Correct instructions.	1.1.1 Incorrect instructions.	Inspector has correct and correctly working equipment, and understands instructions.
	1.1.2 Incomplete instructions.	
	1.1.3 No instructions available.	
1.2 Correct equipment procured.	1.2.1 Incorrect equipment.	
	1.2.2 Equipment not procured.	
1.3 Inspector gets instructions.	1.3.1 Fails to get instructions.	
1.4 Inspector reads instructions.	1.4.1 Fails to read instructions.	
	1.4.2 Partially reads instructions.	
	2. ACCESS	
1.5 Inspector understands instructions.	1.5.1 Fails to understand instructions.	
	1.5.2 Misinterprets instructions.	
	1.5.3 Does not act on instructions.	
1.6 Correct equipment available.	1.6.1 Correct equipment not available.	
	1.6.2 Equipment not available.	
	1.6.3 Equipment not working.	
1.7 Inspector gets equipment.	1.7.1 Gets wrong equipment.	
	1.7.2 Gets incomplete equipment.	
	1.7.3 Gets non-working equipment.	
1.8 Inspector checks/calibrates equipment.	1.8.1 Fails to check/calibrate.	
	1.8.2 Checks/calibrates incorrectly.	
2.1 Locate area to inspect.	2.1.1 Locate wrong aircraft.	Inspector with correct equipment at correct inspection site, is ready to begin inspection.
	2.1.2 Locate wrong area on aircraft.	
	2.1.3 Mis-locate boundaries of area.	
2.2 Area ready to inspect.	2.2.1 Cleaning work not completed.	
	2.2.2 Cleaning work incorrect.	
	2.2.3 Mtc. access tasks not completed.	
	2.2.4 Mtc. access tasks incorrect.	
	2.2.5 Parallel work prevents access.	
	2.2.6 Parallel work impedes inspection.	
2.3 Access area to inspect.	2.3.1 Access equipment not available.	
	2.3.2 Incorrect access equipment.	
	2.3.3 Access equipment poorly designed.	
	2.3.4 Access not physically possible.	
	2.3.5 Access discouragingly difficult.	
	2.3.6 Access dangerous to inspection.	

TABLE 17.6 (continued)   Task and Error Taxonomy for Visual Inspection (Tasks: Initiate and Access)

Task	Errors	Outcome
	3 SEARCH	
3.1 Move to next lobe.	3.1.1 Misses parts of access area. 3.1.2 Multiple searches of parts. 3.1.3 Too close/far between lobes. 3.1.4 Move to non-required area.	All indications located in all access areas.
3.2 Enhance lobe (e.g., illuminate, magnify for vision, use dye penetrant, tap for auditory inspection).	3.2.1 Enhance wrong area. 3.2.2 Enhance area inadequately. 3.2.3 Fail to use enhancing equipment.	
3.3 Examine lobe.	3.3.1 Fail to examine lobe. 3.3.2 Examine too short/long time. 3.3.3 Incorrect depth of examination. 3.3.4 Incomplete examination of lobe. 3.3.5 Fatigue from fixed posture.	
3.4 Sense indication in lobe.	3.4.1 Fail to attend lobe. 3.4.2 Fail to use cues present. 3.4.3 Fail to sense indication. 3.4.4 Sense wrong indication.	
3.5 Match indication against list.	3.5.1 Match against faults not listed. 3.5.2 Fail to match against full list. 3.5.3 Incorrect match.	
3.6 Remember matched indication.	3.6.1 Fail to record matched indication. 3.6.2 Forget matched indication.	
3.7 Remember lobe location.	3.7.1 Fail to record lobe location. 3.7.2 Forget lobe location.	
3.8 Remember access area location.	3.8.1 Fail to record access area location. 3.8.2 Forget access area location.	
3.9 Move to next access area.	3.9.1 Miss parts of area. 3.9.2 Multiple searches of parts. 3.9.3 Move to non-required area.	

Note: Search proceeds by successively examining each small area, called here a lobe, within a single area accessible without performing a new access, called here an access area. When all lobes have been examined in that access area, a new access is performed followed by a new search. The concept of a lobe comes from visual search where it is called a visual lobe.

TABLE 17.6 (continued)   Task and Error Taxonomy for Visual Inspection (Tasks: Initiate and Access)

Task	Errors	Outcome
	4 DECISION	
4.1 Interpret indication.	4.1.1 Classify as wrong fault type.	All indications located are correctly classified, correctly labeled as fault or no fault, and actions correctly planned for each indication.
4.2 Access measuring equipment	4.2.1 Choose wrong measuring equipment.	
	4.2.2 Measuring equipment not available.	
	4.2.3 Measuring equipment not working.	
	4.2.4 Measuring equipment not calibrated.	
	4.2.5 Measuring equipment wrongly calibrated.	
	4.2.6 Does not use measuring equipment.	
4.3 Access comparison standard.	4.3.1 Choose wrong comparison standards.	
	4.3.2 Comparison standard not available.	
	4.3.3 Comparison standard not correct.	
	4.3.4 Comparison incomplete.	
	4.3.5 Does not use comparison standard.	
4.4 Decide on if fault.	4.4.1 Type I error, false alarm.	
	4.4.2 Type II error, missed fault.	
4.5 Decide on action.	4.5.1 Choose wrong action.	
	4.5.2 Second opinion if not needed.	
	4.5.3 No second opinion if needed.	
	4.5.4 Call for buy-back when not required.	
	4.5.5 Fail to call for required buy-back.	
4.6 Remember decision/action.	4.6.1 Forget decision/action.	
	4.6.2 Fail to record decision/action.	
	5 RESPOND	
5.1 Mark fault on aircraft.	5.1.1 Fail to mark fault.	All faults and repair items are correctly recorded.
	5.1.2 Mark non-fault.	
	5.1.3 Mark fault in wrong place.	
	5.1.4 Mark fault with wrong tag.	
	5.1.5 Mark fault with wrong marker.	
5.2 Record fault.	5.2.1 Fail to record fault.	
	5.2.2 Record non-fault.	
	5.2.3 Record fault in wrong place.	
	5.2.4 Record fault incorrectly.	
5.3 Write repair action.	5.3.1 Fail to write repair action.	
	5.3.2 Write repair action for non-fault.	
	5.3.3 Write repair action for wrong place.	
	5.3.4 Mis-write repair action.	
	5.3.5 Specify buy-back if not needed.	
	5.3.6 Fail to specify needed buy-back.	
	6. REPAIR	
6.1 Repair fault.	6.1.1 Fail to repair fault.	All recorded faults correctly repaired and accessible for buy-back inspection.
	6.1.2 Repair non-fault.	
	6.1.3 Mis-repair fault.	
	6.1.4 Prevent access for buy-back.	

From FAA (1991) *Human Factors in Aviation Maintenance — Phase One Progress Report*. DOT/FAA/AM-91/16, Office of Aviation Medicine, Washington, D.C.

TABLE 17.7   Potential Strategies for Improving Inspection (FAA, 1991).

Task Step	Strategy	
	Changing Inspector	Changing System
Initiate	Training in visual inspection procedures (procedural training).	Redesign of job cards Feedforward of expected flaws
Access	Training in area location (knowledge and recognition training).	Better support stands Better area location systems Location for visual inspection equipment
Search	Training in visual search (cueing, progressive-part).	Task lighting Optical aids Improved visual inspection templates
Decision	Decision training (cueing feedback, understanding of standards).	Standards at the work point Pattern recognition off job aids Improved feedback to inspection
Response	Training in writing skills.	Improved fault marking Hands-free fault recording

From FAA (1991) *Human Factors in Aviation Maintenance — Phase One Progress Report.* DOT/FAA/AM-91/16, Office of Aviation Medicine, Washington, D.C.

References

Amendola, A., Mancini, G., Poucet, A. and Reina, G. (1982) Dynamic and static models for nuclear reactor operators — needs and examples. *IFAC Analysis, Design and Evaluation of Man-Machine Systems*, Germany, 103-110.

Anne, M. and Greenstein, J. S. (1993) The design of a computer-based tool to aid the manager of a real-time manufacturing control system (*Tech. Report*). Clemson, SC: Clemson University, Department of Industrial Engineering.

Armstrong, T. J., Radwin, R. Hansen, D. J., and Kennedy, K. W. (1986) Repetitive trauma disorders: job evaluation and design. *Human Factors*, 28(3), 325-336.

Annett, J. and Duncan, K. D. (1967) Task analysis and training design. *Occupational Psychology*, 41, 211-212.

Bainbridge, L. and Sanderson, P. (1995) Verbal protocol analysis, in J. R. Wilson and E. N. Corlett (eds.) *Evaluation of Human Work: A Practical Ergonomic Methodology.* Taylor & Francis: London, 169-201.

Bainbridge, L. (1991) Mental models in cognitive skill: the example of industrial process operation, in P. Bibby et al. (eds.) *Models in the Mind.* Academic Press: London.

Barnes, R. (1980) *Motion and Time Study Design and Measurement of Work.* John Wiley & Sons Inc., New York.

Card, S. K., Moran, T. P., and Newell, A. L. (1983) *The Psychology of Human Computer Interaction.* Hillsdale, NJ: Erlbaum.

Carey, M. S., Stammers, R. B., and Astley, J. A. (1989) Human computer interaction design: the potential and pitfalls of hierarchical task analysis, in D. Diaper (ed.) *Task Analysis for Human–Computer Interaction.* Ellis Horwood: Chichester, 56-70.

Chapanis, A. (1962) *Research Techniques in Human Engineering.* John Hopkins University Press: Baltimore.

Chapanis, A. and Shafer, J. (1991) A workshop on human factors methods. *Annual Human Factors Society Meeting*, San Francisco, CA.

Corlett, E. N. and Bishop, R. P. (1976) A technique for assessing postural discomfort. *Ergonomics*, 19, 175-182.

Crossman, E. R. F. W. (1956) Perceptual activity in manual work, *Research*, 9, 42-49.

Diaper, D. (1989) *Task Analysis for Human–Computer Interaction.* Ellis Horwood Limited: Chichester, England.

Drury, C. G., Paramore, B., Van Cott, H. P., Grey, S., and Corlett, E. N. (1987) Task analysis, Chapter 3.4, In G. Salvendy (ed.), *Handbook of Human Factors*. John Wiley & Sons Inc.

Drury, C. G. (1991) The maintenance technician in inspection. Chapter 3, in FAA (1991) *Human Factors in Aviation Maintenance — Phase One Progress Report*. DOT/FAA/AM-91/16, Office of Aviation Medicine, Washington, D.C., 45-103.

Drury, C. G., Prabhu, P., and Gramopadhye, A. K. (1990) Task analysis of aircraft inspection activities: methods and findings. *Proceedings of the Human Factors Society 34th Annual Meeting*, 1181-1184.

Drury, C. G. (1987) A bio-mechanical evaluation of the repetitive motion injury potential of industrial jobs. *Seminars in Occupational Medicine*, Volume 2, No. 1, 41-49.

Drury, C. G. and Wick, J. (1984) Ergonomic applications in the shoe industry. *Proceedings of the International Conference on Occupational Ergonomics*, Toronto, Canada, 489-493.

Drury, C. G. (1983) Task analysis methods in industry. *Applied Ergonomics*, 14(1), 19-28.

Duncan, K. D. (1974) Analysis techniques in training design, in E. Edwards and F. P. Lees (eds.) *The Human Operator in Process Control*. Taylor & Francis: London.

FAA (1991) *Human Factors in Aviation Maintenance — Phase One Progress Report*. DOT/FAA/AM-91/16, Office of Aviation Medicine, Washington, D.C.

FAA (1993a) *Human Factors in Aviation Maintenance — Phase Two Progress Report*. DOT/FAA/AM-93/5, Office of Aviation Medicine, Washington, D.C.

FAA (1993b) *Human Factors in Aviation Maintenance — Phase Two Progress Report*. DOT/FAA/AM-93/15, Office of Aviation Medicine, Washington, D.C.

Flanagan, J. C. (1954) The critical incident technique. *Psychological Bulletin*, 51, 327-358.

Fleishman, E. A. and Quaintance, M. K. (1984) *Taxonomies of Human Performance*. Academic Press: New York.

Gilbreth, F. B. (1911) *Motion Study*. D. Van Nostrand Co. Princeton.

Gagne, R. M. (1977) *The Conditions of Learning*. 3rd edition. Holt, Rinehart and Winston: New York.

Gramopadhye, A. K., Drury, C. G., and Prabhu, P. V. (1997) Training strategies for visual inspection. *International Journal of Human Factors in Manufacturing*, 7(3), 171-196.

Gramopadhye, A. K., Kimbler, D., Kimbler, E., Bhagwat, S, and Rao, P. (1995) Application of advanced technology to training for visual inspection. *Proceedings of the Human Factors and Ergonomics Society 39th Annual Meeting*, 1299-1304.

Gramopadhye, A. K., Drury, C. G., and Sharit, J. (1993) Training for decision making in aircraft inspection. *Proceedings of the Human Factors and Ergonomics Society 37th Annual Meeting*, 1267-1271.

Gramopadhye, A. K., Bhagwat, S., Kimbler, D., and Greenstein, J. (1998) The use of advanced technology for visual inspection training. *Applied Ergonomics*, Vol. 29, No. 3.

Green, A. E. (1983) *Safety Systems Reliability*. John Wiley: Chichester.

Greer, C. W. (1981) Human engineering procedure guide. *Report AFAMRL-TR-81-35*. Wright-Patterson Air Force Base, Ohio, USA,.

Hammer, W. (1985) *Occupational Safety Management and Engineering*. Prentice Hall, New Jersey.

Hollnagel, E. (1989) The phenotypes of erroneous actions: implications for HCI design, in G. R. S. Weir and J. L. Alty (eds.) *Human–Computer Interaction and Complex Systems*. Academic Press: London.

Johnson, P., Diaper, D., and Long, J. B. (1984) Task, skills and knowledge: task analysis for knowledge based descriptions, in Shackel, B. (ed.) *Interact'84 — Proceedings of the First IFIP Conference on Human Computer Interaction*. Amsterdam: North Holland, 23-27.

Johnson, H. and Johnson, P. (1989) Integrating task analysis into system design: surveying designers needs. *Ergonomics*, Vol. 32, No. 11, 1451-1467.

Johnson, P., Johnson, H., Waddington, R., and Shouls, A. (1988) Task related knowledge structures: analysis, modeling and application, in Jones, D. M. and Winder, R. (eds.) *People and Computers: From Research to Implementation*. Cambridge University Press: Cambridge, 35-62.

Kieras, D. (1988) Towards a practical goms model methodology for user interface design, in M. Helander (ed.), *Handbook of Human–Computer Interaction*. Elsevier Science Publishers B. V., North Holland, 135-157.

Kirwan, B. and Ainsworth, L. K. (1992) *A Guide to Task Analysis.* Taylor & Francis: London.

Kraus, D. and Gramopadhye, A. K. Transfer effects of computer-based team training. (To appear in the special issue on Human Factors in Aviation Maintenance in the International Journal of Industrial Ergonomics).

Landau, K. and Rohmert, W. (1981) AET — A new job analysis method, *Spring Annual Conference Proceedings.* Bern — Suttgart: Hans Huber.

Lock, M. W. B. and Strutt, J. E. (1985) Reliability of in-service inspection of transport aircraft structures. *CAA Paper 85013*, London.

McCormick, E. J., Jeanneret, P. R., and Machan, R. C. (1969) A study of job characteristics and job dimensions as based on the position analysis questionnaire. *Report No. 6*, Occupational Research Center, Purdue University, West Lafayette, Indiana, USA.

McCormick, E. J. (1976) Job and task analysis, in M. D. Dunette (ed.) *Handbook of Organizational and Industrial Psychology.* Rand McNally: Chicago.

Meister, D. (1985) *Behavioral Analysis and Measurement Methods.* Wiley: New York.

Miller, R. B. (1953) *A Method for Man–Machine Task Analysis.* Wright-Patterson Air Force Base, OH: Wright Air Development Center (DTIC AD-15721).

Miller, R. B. (1967) Task taxonomy: science or technology? *Ergonomics*, 10, 167-176.

Mitchell, K., Bright, C. K., and Rickman, J. K. (1996) A study into potential sources of human error in the maintenance of large civil transport aircraft. *CAA Paper 96004*, Lloyd's Register, Civil Aviation Authority, London.

Patel, S. C., Drury, C. G., and Prabhu, P. (1993) Design and usability evaluation of work control documentation. *Proceedings of the Human Factors and Ergonomics Society 37th Annual Meeting*, 1156-1160.

Payne, S. J. and Green, T. R. G. (1986) Task-action grammars: a model of the mental representation of task languages. *Human–Computer Interaction*, 2, 93-133.

Piso, E. (1981) Task analysis for process control tasks: The method of Annett et al., applied. *Journal of Occupational Psychology*, 54, 247-254.

Pitkaar, R. N., Lenior, T. M. J., and Rijnsdorp, J. E. (1990) Implementation of ergonomics in design practice: outline of an approach and some discussion points. *Ergonomics*, 1990, 33, 5, 583-587.

Rasmussen, J. (1983) Skills, rules, knowledge: signals, signs and symbols and other distinctions in human performance models. *IEEE Transactions: Systems, Man and Cybernetics*, Vol. SMC-13, 257-267.

Roth, E. M., Woods, D. D., and Pople, Jr., H. E. (1992) Cognitive simulation as a tool for cognitive task analysis. *Ergonomics*, 35, 10, 1163-1198.

Seymour, W. D. (1966) *Industrial Skills.* Isaac Pitman: London.

Shackel, B., Chidsey, K. S., and Shipley, P. (1979) The assessment of chair comfort. *Ergonomics*, 12, 269-306.

Shepherd, A. (1989) Analysis and training in information technology tasks, in D. Diaper (ed.) *Task Analysis for Human–Computer Interaction.* Ellis Horwood: Chichester, 15-55.

Shepherd, A. (1993) An approach to information requirements specification for process control tasks. *Ergonomics*, 36, 1425-1437.

Sheridan, T. (1997) Task analysis, task allocation and supervisory control, in M. Helander, T. K. Landauer, and P. Prabhu (eds.) *Handbook of Human–Computer Interaction.* Elsevier Science B. V., 87-105.

Sinclair, M. (1984) Ergonomics of quality control. Workshop document, *International Conference on Occupational Ergonomics*, Toronto.

Singleton, W. T. (1974) *Man–Machine Systems.* London: Penguin.

Stassen, H. G., Steele, R. D., and Lyman, L. (1989) A man-machine system approach in the field of rehabilitation: a challenge or necessity, in Baosheng Hu (ed.) *IFAC Analysis, Design and Evaluation of Man–Machine Systems*, PRC, 79-86.

Stammers, R. B. and Shepherd, A. (1995) Task analysis, in J. R. Wilson and E. N. Corlett (eds.) *Evaluation of Human Work: A Practical Ergonomic Methodology.* Taylor & Francis: London, 144-169.

Stammers, R. B. (1995) Factors limiting the development of task analysis. *Ergonomics*, Vol. 38, No. 3, 588-594.

Swain, A. D. (1962) System and task analysis, a major tool for designing the personnel subsystem. (*Report SCR-457*), Sandia Corp.: Albuquerque, New Mexico.

Swain, A. D. (1964) THERP (*Report SC-R-64-1338*) Sandia Corp.: Albuquerque, New Mexico.

Swain, A. D. and Guttman, H. E. (1980) Handbook of Human Reliability Analysis and Emphasis on nuclear power plant application (*NUREG/CR-1278*). Washington, D.C.: U.S.A.

Taylor, F. W. (1929) *The Principles of Scientific Management*. Harper and Bros. New York.

Umbers, I. G. and Reiersen, C. S. (1995) Task analysis in support of the design and development of a nuclear power plant safety system. *Ergonomics*, 38, 3, 443-454.

Wilson, G. F. (1993) Air-to-ground training missions: a psycho physiological workload analysis. *Ergonomics*, 36, 9, 1071-1087.

18

A Computer-Based Tool for Practical Ergonomic Job Analysis

Steven L. Johnson
University of Arkansas

18.1   Introduction

One of the simplest definitions of ergonomics is matching the physical, physiological, and psychological requirements of the job with the capabilities of the human operator. This goal is obviously not new and has long been an important component of ensuring the operational effectiveness of both military and commercial operations from the perspectives of productivity and product quality. Traditional industrial engineering efforts in methods and work measurement have also addressed job analysis and documentation from the time of Taylor and the Gilbreths. The adequacy of the match between operator capability and job requirements also affects a company's indirect costs related to absenteeism, turn-over rates, and training costs. In addition to reducing the quality of working life, a mismatch contributes to lost time, accidents and injuries, restricted work assignments, as well as workers' compensation costs. A factor that has increased industry's attention to job requirements is the need to be in regulatory compliance.

Management, labor, and governmental agencies have all increased their attention to ensuring that job characteristics are compatible with the abilities of operators because of two federal initiatives. The Americans with Disabilities Act (ADA) addresses the analysis of job requirements in the contexts of both "essential job functions" and "reasonable accommodations." The Occupational Health and Safety Administration's recent attention to musculoskeletal disorders to the upper extremities has also increased the focus on the characteristics of tasks that increase the risk of injuries and illnesses (i.e., risk factors), particularly in manufacturing, assembly, and processing facilities.

The direct and indirect financial costs to both the company and the employee, as well as the regulatory compliance implications of a mismatch between the job requirements and operator capabilities, necessitate an accurate method of analyzing and documenting the job characteristics. The focus of this chapter is on the various methods of performing an ergonomic analysis and documenting the job characteristics that can affect the occupational safety and health of the operators.

18.2   Documenting the Risk of Work-Related Musculoskeletal Disorders

There are three general approaches to appraising the characteristics of tasks that have been associated with increased risk of work-related injuries and illnesses. These are passive surveillance, active surveillance, and job-site analysis. *Passive surveillance* involves a review of the archival occupational safety and health records (i.e., OSHA logs, workers' compensation records, etc.). Additional data that can contribute to an assessment of potential discomfort that does not result in a reportable event, or even a visit to the medical department, are the requests for a job transfer and the absenteeism records for particular operations. Although this information can illuminate the extreme cases in a facility, the data are very noisy (i.e., incorrect names for operations, an injury that actually occurred as a result of a previous operation, etc.). In addition, for most organizations, it can be anticipated that the number of occurrences is small enough to render any statistical trend analysis erroneous. In fact, the potential for misinterpretation of such data can severely hinder an effective ergonomics effort.

Active surveillance generally involves soliciting information from the current employees pertaining to any discomfort or disorder they have experienced because of their work. The term "symptom survey" is often used in this context; although it is very likely that the term itself predisposes the employee to symptoms. Instead, labeling the instrument that requests information from the operator as to potential improvements as a "job improvement survey," prior to their indicating discomfort or pain, can have a very positive effect on the quality and usability of the information collected. It is the change in the responses over subsequent surveys that generally provides more useful information than that from a single application of an active surveillance instrument.

For both the active and passive surveillance methods, the obvious difficulty is differentiating between soreness that can occur naturally as someone begins or returns to a task and persistent discomfort that could indicate the onset of a disorder or injury. Another issue that relates to the effectiveness of active surveillance methods is the confidentiality of the data. That is, it is uncomfortable for many companies to collect data on discomfort or pain and not be able to identify the person in order to follow up on the reported problem. However, to obtain unrestrained information, and because the data could be interpreted as a form of medical record, confidentiality can be critical. One method of addressing this issue is to number the form and a corresponding tear-off signature area. The health care provider (i.e., usually the company nurse) is the only person who would have access to the cross reference of names with surveys. The health care provider can then use this information to follow up on medical problems that are reported.

A third general method of assessing the adequacy of the workplace design, work methods, tools, and equipment from an ergonomic perspective is the *job-site analysis*. There are two general approaches to job-site analysis: checklists and narrative reports. The narrative approach generally involves a qualified ergonomist with extensive training and experience in job-site analysis, evaluating each production operation in a facility. The product of the ergonomic analysis is a written report that discusses each operation, often accompanied with photographs and/or video recordings to illustrate the observations. It is often as important for operational personnel to understand the reasons some tasks do not experience problems as it is to understand the risks posed by problematic tasks. The major disadvantages of this narrative method of job-site analysis are the access to a qualified analyst, the time required to conduct the analysis and develop the report, and the resulting high cost. The primary benefit of this type of approach is that recommendations are provided that have the potential of reducing or eliminating the risk of fatigue, discomfort, and injury.

There have been some methodologies developed to assist in the analysis of injuries, specifically related to cumulative trauma disorders to the upper extremities. Drury and Wick (1984) developed a task analysis method to specifically document the posture, force, and frequency of tasks that could lead to injury. Jobs were videotaped and broken down into task steps (i.e., elements). The characteristics of the task elements were recorded on a task analysis form that addressed both the upper extremities and overall body posture.

The frequency of task characteristics that are considered to increase risk (i.e., the combination of radial deviation and pinch grip) were calculated. Each task element was considered "damaging" if it required a grip with deviation, flexion, or extension of the wrist. The compressive forces on the spinal discs were also calculated from the body angles and the amount of weight moved.

Drury (1987) developed a job analysis method that was broken down into two sections: task description and task analysis. The task description section involved analyzing jobs by using video recordings of the worker from five angles to develop an element breakdown documenting the body angles, posture, grip, forces, vibration, shock, and lighting. The frequency of each task was estimated by dividing the task working time over the day by the cycle time. The task analysis also included documenting the angles at each joint with respect to the maximum range of motion. The joint range-of-motion data were divided into zones: (1) no exposure (neutral to ± 10% of range), (2) low exposure (± 10% to ± 25% of range), (3) moderate exposure (± 25% to ± 50% of range), and (4) severe exposure (more than ± 50% of range). From the task description, each angle was replaced with a corresponding zone number. Drury used the term "Daily Damaging Wrist Motions (DDWM)" as a metric of exposure where a damaging wrist motion was defined as the frequency of any combination of a grip or external force with any nonzero wrist exposure.

Armstrong, Radwin, Hansen, and Kennedy (1986) developed a methodology for analysis, identification, and elimination of cumulative trauma disorders that involved fundamental industrial engineering work-methods procedures. The work content of each task element was evaluated with respect to risk factors: (1) repetitive or forceful sustained exertions, (2) shoulder posture (elbow above mid-torso reaching down and behind), forearm (inward or outward rotation with a bent wrist), wrist (palmar flexion, full extension, ulnar or radial deviation), hand (pinching), (3) mechanical stress concentrations, (4) vibration, (5) cold, and (6) gloves. If any risk factor was identified, it was recommended that modifications be made.

A structured job analysis procedure was developed by Keyserling, Armstrong, and Punnett (1991) to assist safety professionals in recognizing and evaluating exposures to risk factors. The method consisted of: basic job documentation, identification and evaluation of exposure to risk factors, and methods of controlling exposures to work-related risk, eliminating or reducing the risk factors to acceptable levels.

These various methods of job-site analysis span a continuum from recommendations based on expert judgment with little or no objective or quantitative documentation to very detailed measurement and analysis methods that characterize the biomechanical considerations of the job in great detail. However, in each case, there can be a significant amount of time and money devoted to the analysis procedure.

An alternative approach, that is generally quicker to implement, involves the use of a *checklist* that helps in the identification of the task characteristics that are associated with increased fatigue, discomfort, or injuries. Armstrong and Lifshitz (1986) developed a checklist that included physical stress, force, posture, workstation hardware, repetitiveness, and tool design. The final score was calculated as the fraction or percentage of the responses scored by a "yes." Subscores for each category of risk factors were calculated to indicate where the attention should be focused to control the problem.

The Occupational Safety and Health Administration (OSHA) has distributed a draft checklist associated with a proposed ergonomic standard. The objective of the checklist is to assist employers in the evaluation of the jobs in their facilities. The checklist includes a set of risk factors related to three categories: (1) the upper extremities, (2) back and lower extremities, and (3) manual handling of materials. Within each category, points are assigned, depending upon the importance of the factor (i.e., pinch grip more than two pounds) and the daily exposure time (2 to 4 hours, 4 to 8 hours, or more than 8 hours). The points are combined within each category, and the total score is compared to a trigger value. If the score is greater than the trigger value, the job was designated as a "problem job" that deserves a more detailed analysis.

Keyserling, Stetson, Silverstein, and Brower (1993) developed a checklist that included the following categories: environment, posture, metabolic rate, manual lifting, and the use of the upper extremities. Within each category, questions were asked that characterized the job. Each question was designed to evaluate the presence and/or duration of the risk factors. Each response resulted in a stress rating as follows: (1) zero (exposures were insignificant), (2) check (moderate exposures were present), and (3) star (substantial exposures were present). The number of elements within which it occurred was also recorded. An overall score from ergonomic stresses was calculated by summing the total number of checks and stars.

Both the narrative and the checklist methods of job-site analysis have advantages and disadvantages. The checklist is relatively quick and easily performed; however, it is generally quite incomplete, particularly with respect to the temporal aspects of the job, across task elements, tasks and recovery periods. In addition, even the checklist procedure generally requires some training on the administration of the checklist and the interpretation of the results. Most important, it is descriptive in that it indicates when a problem exists rather than being prescriptive in terms of providing recommended modifications to reduce the risks.

The more narrative approach to job analysis and documentation is generally much more complete, and the analyst generally includes recommended modifications, in addition to a description of the risks. However, this method requires a significant amount of training and experience in ergonomics, which results in the time required and the cost being relatively high. Another serious disadvantage of utilizing either an external or internal ergonomic expert to conduct the analysis and document the job characteristic is that the analyst often does not have a very complete knowledge of the job itself. That is, the impact of extraneous variables (different incoming material, alternative process conditions, etc.) can make an apparently accurate ergonomic analysis totally invalid because the conditions are not representative. In general, a somewhat incomplete knowledge of ergonomics is less dangerous than an incomplete knowledge of the operational process.

An alternative method of conducting a job-site analysis is to use a *computer-based system* to document and evaluate the task requirements. This approach has the advantage of being interactive and easily understood by operational personnel in the context of an active surveillance process or independently. The computer-based approach also provides a structured method of evaluating the postures and motions across task elements and tasks to determine the total amount of exposure and the time provided for recovery. The system can be used to simulate operations that have not yet been implemented, and it can be used to evaluate alternative configurations for the workplace design, work methods, tools, and equipment. As changes occur in the production process, the documentation can easily be updated by addressing only the individual tasks or task elements that have changed. Possibly the most important feature of the computer-based system is that it provides recommendations that can be used to reduce the characteristics of the tasks that increase postural and biomechanical stress, along with the potential for work-related musculoskeletal disorders.

Many of the modifications that have the potential of reducing or eliminating cumulative trauma disorders are not difficult to recognize or implement. The personnel already in place often have both the skills and knowledge to properly design the systems, given the proper analysis tools and resources.It is not sufficient to simply provide operational personnel with only a method of evaluating jobs using a rating scale that indicates that a hazard may exist. The computer-based job analysis system provides the resource necessary for individuals with little or no training in ergonomics to both evaluate jobs and suggest modifications that can reduce or eliminate the incidence of cumulative trauma disorders. In particular, the system is designed for production supervisors who have a full understanding of the job, but no training in ergonomics.

18.3   Format Used in the Job Analysis and Documentation System

The computer-based ergonomic job analysis system has two primary functions. First, it is an evaluation tool to analyze and document the characteristics of the job that are associated with musculoskeletal disorders (i.e., risk factors). Second, it is used by operational personnel (i.e., first-line supervisors and engineers) to prescriptively evaluate the workplace design, work methods, tools and equipment by suggesting effective modifications that can lead to more efficient operations, in addition to controlling work-related occupational injuries and illnesses. The completeness of the job description is of primary importance to the validity of an analysis system. One of the reasons that occupational safety records (i.e., OSHA 200 logs) are of less value than might be expected is that job titles are not applied consistently by all people. For example, the terminology used by production supervisors is often different from that used by the human resources department. Operators often do not know the name of the job that they were performing when they visit the company medical facility. In addition to ensuring that the terminology used to label the job is consistent, it is also important to consider the individual being analyzed. Obviously, the posture that occurs at a particular workplace can be different for a very short individual as opposed to a very tall person. The height and gender of the person being analyzed is entered into the system so that subsequent graphical presentations are adjusted to accurately represent the anthropometric considerations. This allows the workplace geometry to be entered relative to body landmarks (i.e., knee or waist height) rather than requiring absolute measurements.

Entering Job Characteristics

To ensure that the documentation is accurate and complete, it is often beneficial to have a complete list of all jobs for specific workstations. These are not necessarily the same as the job categories defined by the human resources department. The supervisor can subsequently identify and describe the tasks and task elements that constitute the job. In addition, information about the total work time, frequency and duration of the breaks, and job rotation are required to determine a valid evaluation of the exposure. If the time allocated to the various task elements and tasks do not correspond relatively closely to the total work time, the analysis will be invalid and the conclusions can be very misleading.

Figure 18.1 illustrates a task and task element manager that is used to enter the frequency and duration of the various task elements. The task duration is subsequently derived by the system based on the element times and frequencies. If the cumulative time differs from the total work time by more than 5%, a message is provided that informs the analyst of the magnitude of the discrepancy. If the total time differs by more than 15%, the resulting errors are considered excessive, and the analyst is not allowed to proceed until corrections are made.

The next portion of the data entry process involves the work environment and organizational characteristics of the task. In particular, the room and product temperature, task pacing (i.e., machine paced, incentive paced, etc.), and personal protective equipment (including glove types) can affect the postural and biomechanical stress experienced during a workday.

For a production supervisor to be able to characterize the postures and motions that occur during a task element, it is true that a picture is worth a thousand words. The graphical representation shown in Figure 18.2 illustrates the method of entering information related to the position and movement of the upper arm and shoulder. The beginning of each element is, by default, the ending position of the previous element. The term *repetitive* in this context does not refer to whether the task element occurs repetitively, but whether the motion occurs repetitively within the task element (i.e., multiple rotations when driving a single screw). Often, the left and right arms are performing the same activities, and this can be indicated without reentering the information.

Figures 18.3 and 18.4 illustrate the screens used to enter the information for the position and rotation of the forearm and the position and action of the wrist and hand. For each of the alternative choices provided to the analyst (i.e., production supervisor), a help screen can be accessed to define the terms and give examples that illustrate the concepts (i.e., pinch grip).

Frequently, much of the fatigue and discomfort experienced in industrial operations is due to the posture required. Figure 18.5 illustrates the screen that is used to indicate bending and/or twisting of the torso. Other postural considerations relating to sitting and standing are entered on the screen shown in Figure 18.6.

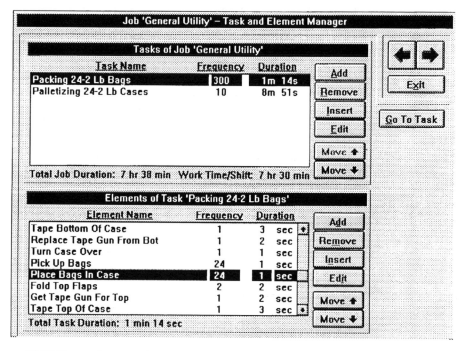

FIGURE 18.1   Screen used to enter task and element sequence, frequencies, and durations.

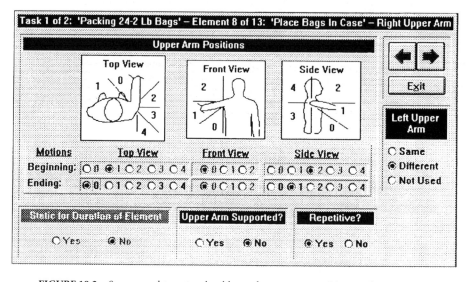

FIGURE 18.2   Screen used to enter shoulder and upper arm positions and movements.

Manual Material Handling Task Characteristics

Two approaches are included in the system to evaluate manual material handling tasks. First, the 1991 revision of the NIOSH Lifting Guidelines (Waters, Putz-Anderson, Garg and Fine, 1993) are used to develop the *recommended weight limit* (RWL) and the *lifting index* (LI). The second approach utilizes the tables developed by Liberty Mutual Insurance Company (Snook, 1978) that provide guidelines for lifting, lowering, pushing, pulling, and carrying. In general, the Liberty Mutual tables are more useful than the NIOSH guidelines as an engineering tool to evaluate and modify workplaces. The percent of the population capable of performing the task is more understandable and usable by operational personnel than is the

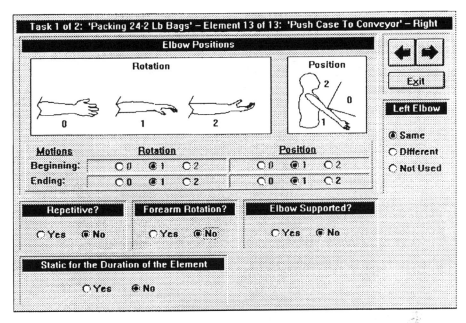

FIGURE 18.3   Screen used to enter elbow and forearm positions and movements.

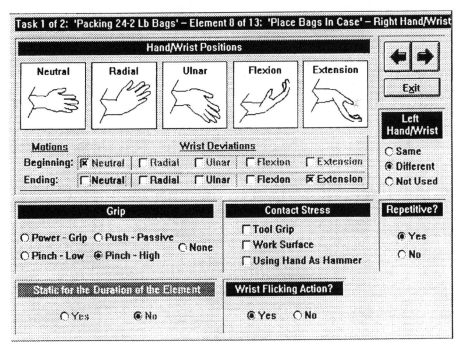

FIGURE 18.4   Screen used to enter wrist and hand positions and movements.

concept of a lifting index. Figures 18.7 and 18.8 illustrate the screens used to enter the data for the manual material handling aspects of each task element. Because it is generally easier for the production supervisor to document the geometry of the task using anatomical landmarks (i.e., knee or waist height) than to make measurements, a graphical method of entering the information is used. To eliminate the fact that knee heights are different for different people, the graphic is adjusted for the height of the person being analyzed, based on information entered previously.

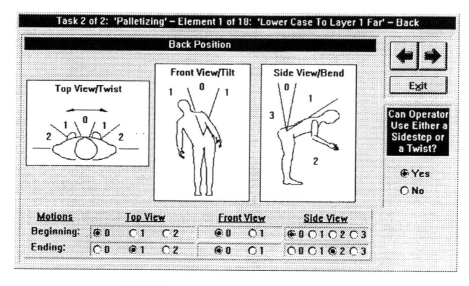

FIGURE 18.5   Screen used to enter back posture.

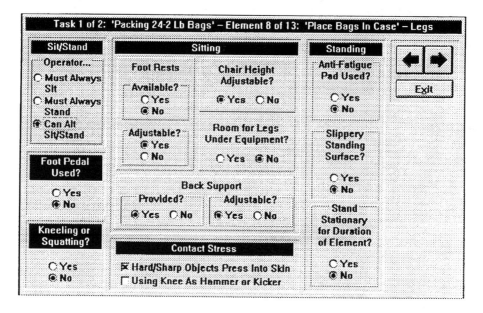

FIGURE 18.6   Screen used to enter lower extremity task characteristics.

Presentation of the Results and Recommendations of the Analysis

The common motions and postures are combined across task elements and tasks to represent the total exposure. The first set of screens present the motions per day and the exposure time for the various task characteristics (i.e., forearm rotated with arm extended). Figure 18.9 shows an example of a results screen for the arm. There are similar screens that report the results for the shoulder and the wrist/hand. Both the motions per day and the total exposure time are important in evaluating exposure to risk factors. The individual task elements that contribute to the particular activity are indicated by number.

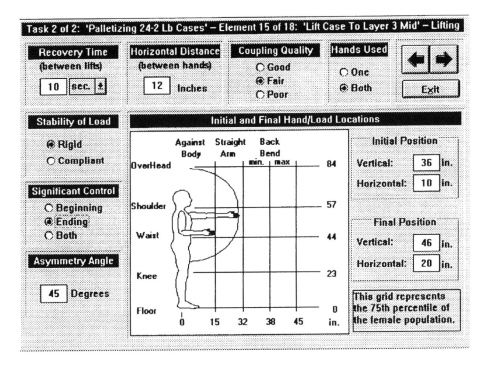

FIGURE 18.7   Screen used to enter lifting task characteristics.

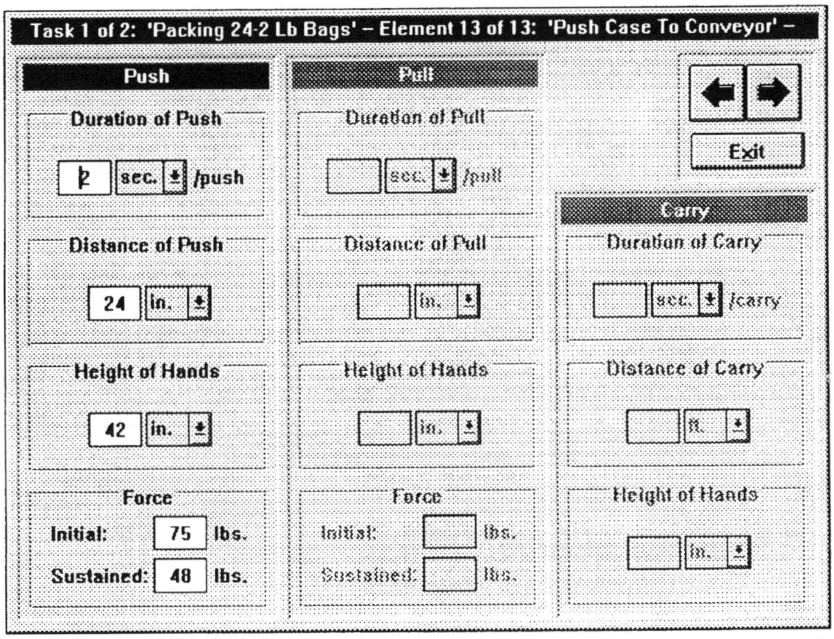

FIGURE 18.8   Screen used to enter push, pull and carry task characteristics.

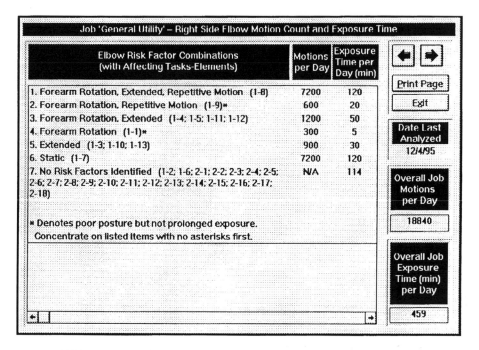

FIGURE 18.9 Results screen for the right-side elbow and forearm risk factors.

Although an indication of the task characteristics that surpass a threshold of risk (i.e., "problem" jobs) would obviously be helpful, the information necessary to provide such a conclusion is not currently available. It is the philosophy of the job analysis system that any and all awkward postures and motions should be addressed with the intention of reducing unnecessary fatigue and improving efficiency, as well as reducing work-related occupational injuries and illnesses. Therefore, the task of the analyst (production supervisor) is to continuously improve the workplace design, work methods, tools, and equipment to improve the effectiveness of the operations. The results screens can be used to identify the task elements contributing to risk and quantify the magnitude of the problem and the potential for improvement through modifications. As more information becomes available as to the dose–response relationship between the risk factors and musculoskeletal disorders, these can easily be incorporated into the system.

The screens that document the job characteristics that affect the operator's back and legs present the resulting exposure time in the particular postures (Figure 18.10). Whereas the previously discussed screens address the exposure of the body part, across task elements and tasks, the screen shown in Figure 18.11 documents the risk factors within a task element for the various parts of the body.

The results and recommendations for the manual material handling portions of the task are given together. Figures 18.12 and 18.13 illustrate a lifting and pushing task, respectively. The NIOSH lifting index and the recommended weight limit are included for the lifting tasks. In addition, the percent of males and females capable of performing the task element as determined by the Liberty Mutual Insurance data are included for lifting, lowering, pushing, pulling, and carrying. Figure 18.14 illustrates a screen that can be used by the analyst to evaluate the effects of task modifications. By changing the characteristics of the task and noting the changes in the results, alternative modifications can be evaluated, recommended, and supported.

The recommendations related to posture and biomechanical stress associated with non-material handling tasks also indicate both the number of motions and the total exposure time. Figure 18.15 illustrates an example of a recommendations screen for the hand and wrist.

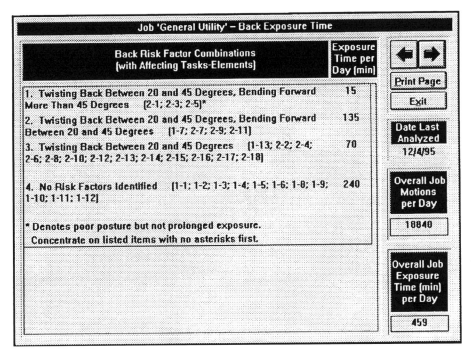

FIGURE 18.10   Results screen for the back posture factors.

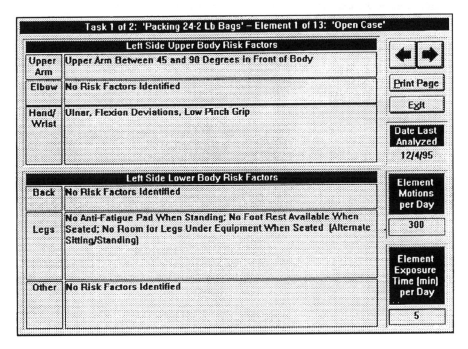

FIGURE 18.11   Results screen for an individual task element.

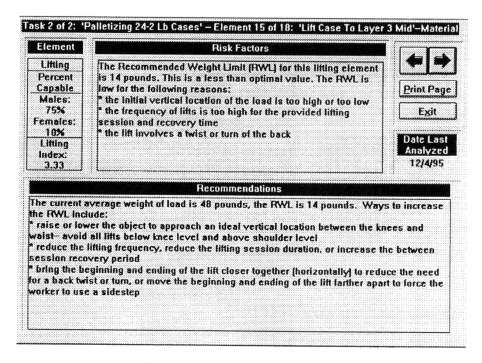

FIGURE 18.12   Results screen for a lifting task.

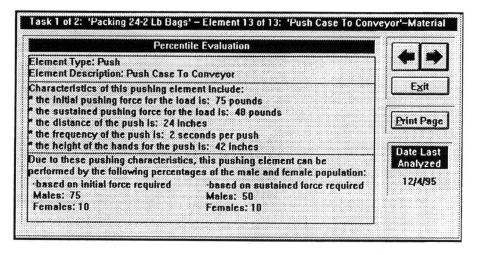

FIGURE 18.13   Results screen for a pushing task.

Many organizations prefer to have an assessment of how their jobs would be assessed using the most current OSHA checklist. Therefore, the computer-based system attempted to translate the job characteristics into the score format used for the checklist. The checklist is divided into the upper extremities, lower extremities and manual material handling. Figure 18.16 illustrates part of the checklist.

18.4   Discussion

As with any analysis process, the quality of the results and recommendations are highly dependent on the completeness and validity of the information entered by the analyst. That is, if the descriptions of

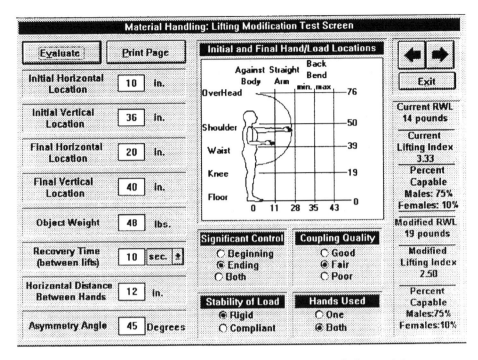

FIGURE 18.14   Lifting test screen to evaluate alternative task characteristics.

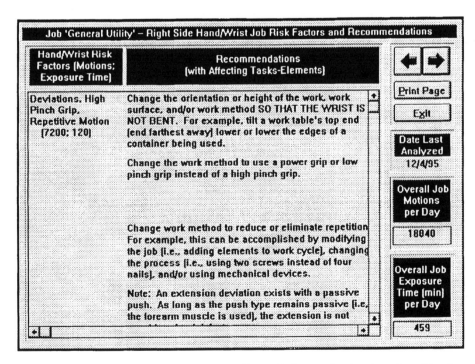

FIGURE 18.15   Recommendations screen to address risk factors related to the wrist and hand.

the task and task elements do not accurately represent the true job characteristics (physically or temporally), the results of the analysis will not be valid. Subsequently, the recommendations could be very inappropriate. It is also important to note that the results of the analysis provided by this system constitute

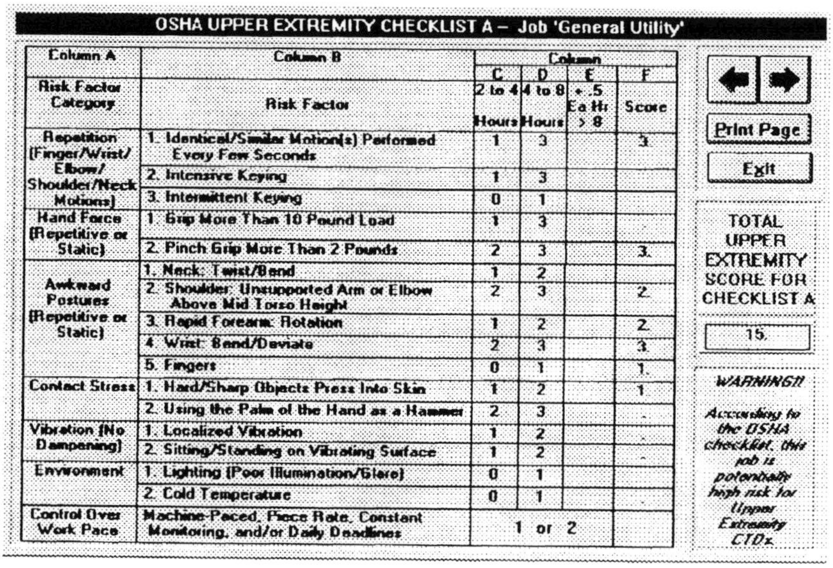

OSHA UPPER EXTREMITY CHECKLIST A – Job 'General Utility'						
Column A	**Column B**	**Column**				
Risk Factor Category	**Risk Factor**	**C** 2 to 4 Hours	**D** 4 to 8 Hours	**E** + .5 Ea Hr > 8	**F** Score	
Repetition (Finger/Wrist/ Elbow/ Shoulder/Neck Motions)	1. Identical/Similar Motion(s) Performed Every Few Seconds	1	3		3.	
	2. Intensive Keying	1	3			
	3. Intermittent Keying	0	1			
Hand Force (Repetitive or Static)	1. Grip More Than 10 Pound Load	1	3			
	2. Pinch Grip More Than 2 Pounds	2	3		3.	
Awkward Postures (Repetitive or Static)	1. Neck: Twist/Bend	1	2			
	2. Shoulder: Unsupported Arm or Elbow Above Mid Torso Height	2	3		2.	
	3. Rapid Forearm: Rotation	1	2		2.	
	4. Wrist: Bend/Deviate	2	3		3.	
	5. Fingers	0	1		1.	
Contact Stress	1. Hard/Sharp Objects Press Into Skin	1	2		1	
	2. Using the Palm of the Hand as a Hammer	2	3		-	
Vibration (No Dampening)	1. Localized Vibration	1	2			
	2. Sitting/Standing on Vibrating Surface	1	2		-	
Environment	1. Lighting (Poor Illumination/Glare)	0	1		-	
	2. Cold Temperature	0	1		-	
Control Over Work Pace	Machine-Paced, Piece Rate, Constant Monitoring, and/or Daily Deadlines	1 or 2				

Print Page
Exit

TOTAL UPPER EXTREMITY SCORE FOR CHECKLIST A

15.

WARNING!!

According to the OSHA checklist, this job is potentially high risk for Upper Extremity CTDx

FIGURE 18.16   Example of an OSHA checklist screen.

the beginning of the problem-solving process, not its completion. That is, there is no substitute for the knowledge and experience of operational personnel in the process of establishing effective and efficient methods of modifying the workplace design, work methods, tools, and equipment. In general, the operational personnel can usually develop applicable solutions to the problems, given that the problems are recognized in the first place. That is the goal of this analysis system. In addition, this system can be very useful in evaluating alternative designs before they are installed, thus reducing both the time and cost involved with retrofitting. By using the system to evaluate alternative configurations, the relative effects can both be documented and communicated in the decision-making process.

References

Armstrong, T. and Lifshitz, Y. (1986). Evaluation and design of jobs for control of cumulative trauma disorders. Presented at *Symposium on Ergonomic Interventions to Prevent Musculo-skeletal Injuries in Industry*, Denver, CO, October 9-10, 1986.

Armstrong, T., Radwin, R., and Hansen, D. (1986). Repetitive trauma disorders: job evaluation and design. *Human Factors*, 28(3), 325-336.

Drury, C. (1987). A biomechanical evaluation of the repetitive motion injury potential of industrial jobs. *Seminars in Occupational Medicine*, 2(1), 41-49.

Drury, C. and Wick, J. (1984). Ergonomic applications in the shoe industry. *Proceedings of the 1984 International Conference on Occupational Ergonomics*, 489-493.

Keyserling, W., Armstrong, T., and Punnett, L. (1991). Ergonomic job analysis: A structured approach for identifying risk factors associated with overexertion injuries and disorders. *Applied Occupational Environmental Hygiene*, 6(5), 353-363.

Keyserling, W., Stetson, D., Silverstein, B., and Brower, M. (1993). A checklist for evaluating ergonomic risk factors associated with upper extremity cumulative trauma disorders. *Ergonomics*, 36(7), 807-831.

Snook, S. (1978). The design of manual handling tasks. *Ergonomics*, 21(12), 963-985.

Waters, T.R., Putz-Anderson, V., Garg, A., and Fine, L. (1993). Revised NIOSH equation for the design and evaluation of manual lifting tasks. *Ergonomics*, 36(7), 749-776.

19

Job or Task Analysis for Risk Factors of Musculoskeletal Disorders

David J. Cochran
University of Nebraska-Lincoln

Terry L. Stentz
University of Nebraska-Lincoln

Brian L. Stonecipher
University of Nebraska-Lincoln

M. Susan Hallbeck
University of Nebraska-Lincoln

19.1 Introduction

There are numerous methods of analyzing jobs for ergonomic risk factors or ergonomic stressors (Keyserling et al., 1991; Ulin et al., 1992). Many companies, consultants, and others have devised job analysis methods. Most of these are proprietary or protected under copyright and as such are not available to everyone. There are some that are available in book form (Burke, 1992). These seem to be regimented and somewhat simplistic. Job analyses generally take one of three forms — checklist, interactive form-based, and narrative or open-ended methods. The checklist method is the simplest to conduct, but it has some very real drawbacks. Some will believe it can be used in place of training and experience, but in the hands of inadequately trained people, checklists can be and have been grossly misused. They do not typically allow for flexibility and may not ask the appropriate questions for a particular job unless they are extremely thorough and therefore long. The interactive form-based method takes considerable time and money to develop and to verify its efficacy. These sometimes have the same drawbacks that checklists have. Namely, they may not allow for flexibility and may not ask the appropriate questions for a particular job unless they are extremely thorough and therefore long. The narrative method is thorough, straightforward, relatively easy to learn, versatile, easy to use, and is generally accepted. On top of all this, it is an intuitively pleasing method.

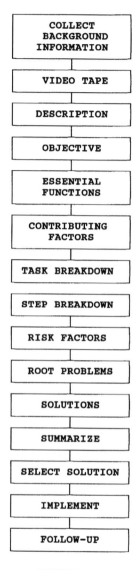

FIGURE 19.1

To conduct a narrative job analysis, background information is collected, the job in question is observed (live or preferably on videotape), a narrative description is written, the job objective is written, the essential functions are identified and written, and the contributing factors are identified and written. The job is then broken into tasks and then into steps. The steps are described; risk factors are identified for each step; root problems that cause these risk factors to be present are identified for each step; solutions are created and evaluated and a follow-up evaluation is conducted. This is nothing more than a general and logical procedure to thoroughly examine a job, and it is pictured in the flow chart in Figure 19.1. It is not tremendously difficult, but it does require some training, practice, and knowledge and awareness of the risk factors for musculoskeletal disorders. Experience and practice will improve the results.

19.2   Preparation

Prior to actually analyzing the job, collect background information and videotape the job. This should include all of the information necessary to be knowledgeable enough of the job to analyze it and propose feasible solutions. This is covered in some detail in Cochran et al., 1999. As a bare minimum, include the job title, work objective, work standard, major and minor tasks, the product, tools, equipment, materials, personal protective equipment, workstation layout, environmental conditions, any rotation scheme, preceding job, succeeding job, worker attributes, and required maintenance and repair. Also become knowledgeable in the physical, psychological, and medical problems experienced by workers on this job.

19.3   Job Analysis

Job Description

This section should inform the reader as to what the job consists of and how it is done. Describe the job in a narrative form. Cover each component in the sequence in which it occurs.

Job Objective

A short statement of what is to be accomplished by the job is sufficient here. This statement has the effect of setting a tone for the job analysis and can be referred to when questions arise about the necessity of particular parts of the job.

Job Essential Functions

Essential functions are basic and fundamental to the performance of a specific job. Essential functions are not marginal to the job or its performance. Enumeration of the essential functions is very useful for getting a job down to its essence. It can also be useful in complying with the Americans with Disabilities Act (see U.S. Equal Opportunity Commission, 1992).

Describe or list the functions of the job that are absolutely essential to its successful performance. The first consideration is whether employees in the position are actually required to perform the function. Reasons a function could be considered essential include that the position exists to perform the function; there are a limited number of other employees available to perform the function or among whom the function can be distributed; the function is highly specialized and the person in the position is hired for special expertise or ability to perform it. Examples of types of essential functions identified are activity tasks which have a process objective, physical movements and/or force exertions, body postures or positioning, cognitive operations and/or judgment-making, use of special knowledge, training, abilities or skills, or forms of communication and/or interaction with others or other programmed operational units.

Contributing Factors

At this point in the analysis, those things that are present in the job that might have an impact on the ergonomic problems of that job but are not identified in specific steps are identified. These are things such as incentive or piece work systems, overtime or unusual workdays, strictly controlled pace of work, intimidating management style, fear of job loss, and many others. Environmental factors that are present and may have an effect on the worker may also be included here.

Task Breakdown

Many jobs have more than one task. When these tasks are separate or dissimilar it is advisable to break the job into its separate tasks and treat each as a separate job.

Step Breakdown

Breaking a job into steps or subtasks is the first action to be taken. The whole job is observed (live or on tape). If it is an assembly line job, it will generally be very routine and easily described in steps. If it is a more complicated job and has been subdivided into tasks that are each analyzed separately, an overall summarization of the job, including all tasks, is required.

Another possibility exists when the job is complicated and has many different activities that do not occur in a set order or pattern. In this case, it is helpful to discern similar job activities and analyze each separately. Once again, an overall summarization of the job, including all activities, is required.

Step Description

Next, describe the job steps in ordinary language. This should describe what the person does — which hands are used, what tools are used, and what activities are conducted. The postures assumed and an estimate of the forces involved should be included. If the description of the step is longer than several sentences, the breakdown into steps may not be adequate or the description may be unnecessarily detailed.

Risk Factors

After the step description has been created, the risk factors present are noted and listed. It is useful for the body part involved to be identified and the risk factor described. It is not the purpose of this paper to present the risk factors, any quantification scheme, or their merits and demerits for predicting musculoskeletal disorders. That has been done in numerous other papers and will continue to be done as the science of ergonomics advances. Some recommended references are Armstrong (1983), Armstrong et al. (1986a), Rogers (1992), Marras and Schoenmarklin (1993), Marras and Lavender (1995), Putz-Anderson (1988), and Sommerich et al., (1993).

Root Problems

The root problems of the job that cause the risk factor or factors to be present are identified. There is an almost irresistible tendency to go from the risk factors to the solutions. This can be counterproductive. As an example, a risk factor may be extreme wrist flexion. The root problem may be improper working height, product orientation, lack of a jig or fixture, poorly designed tool, or a combination of these. The root problem is the cause of the risk factor being present, and it is important to specifically identify it. This step better points to the appropriate solutions than just identifying the risk factors that are present.

Solutions

Solutions are developed to reduce or eliminate as many of the problems as possible. Modifications of the workstation, tools, product, work methods, or the work organization are proposed. Creativity is emphasized here. The input and opinions of the workers doing the job are invaluable in creating and evaluating solutions. This also creates a "buy-in" situation, in which the workers are part of the solution and will try to make it work.

In order to facilitate the step breakdown, to enforce the idea of identifying the root problem, and to associate the solutions with the risk factors and root problems that they address, the authors have created a very simple form that forces the juxtaposition of these components of the job analysis. These are illustrated in Figures 19.2 through 19.5. The form asks for the rated importance of each risk factor with a scale from 0 to 4. Zero indicates that the rated importance to musculoskeletal disorders is none or negligible, and 4 indicates that the rated importance of the risk factor is high. The form also asks for the rated potential of each solution for a particular risk factor with a scale from 0 to 4. Zero indicates that the rated potential for reducing the root problem or risk factor, and therefore the musculoskeletal disorders, is none or negligible, and 4 indicates that the rated potential is high.

Job Analysis Summary

Risk Factors and Root Problems

This is a narrative section that summarizes the findings. It can be broken into two parts — step related and whole job related.

JOB: Loaf Mold Extractor Operator.	STEP NUMBER: 1	
DESCRIPTION: Operator pushes the cart of 10 molds into position to unload the molds. Approximate weight of the loaded cart is 800 pounds.		
RISK FACTOR (*)	**ROOT PROBLEM(S)**	**SOLUTION (**)**
Ulnar deviation, extreme extension, pronation, of both wrists. (3)	No handle.	Provide well designed handle(s). (2)
High pushing forces.(3)	Small wheels on cart	Larger wheels on cart(2)
	Large weight of cart	Larger wheels on cart(2)
	Uneven floors.	Floor maintenance.(2)
Slippery floor.(2)	Wet floor.	Better drainage.(2)
		Housekeeping.(1)
	Wrong shoe sole.	Appropriate shoes.(1)

* Include the rated importance.
** Include the rated potential.
Use a 0 to 4 scale. Zero indicates none and 4 the highest importance or potential.

FIGURE 19.2

JOB: Loaf Mold Extractor Operator.	STEP NUMBER: 2	
DESCRIPTION: The operator reaches into the storage rack, removes the mold lid, throws the lid about 4 feet to a bin. Lid weighs about 1 pound.		
RISK FACTOR (*)	**ROOT PROBLEM(S)**	**SOLUTION (**)**
Shoulder flexion, 45 degrees (0)	Height of mold in the cart.	Lower the mold height in the cart. (0)
Possible high wrist acceleration.	Location of the bin for lids.	Relocate the bin for a drop rather than a throw. (3)

* Include the rated importance.
** Include the rated potential.
Use a 0 to 4 scale. Zero indicates none and 4 the highest importance or potential.

FIGURE 19.3

The risk factors and the associated root problems found in the job steps are summarized in one or more paragraphs. Additionally, at this stage of the job analysis it may be useful to summarize overall levels of some factors for the entire job. In particular, the factors of repetition and static loading sometimes make more sense for the job and for the workday as a frequency or percentage of the time, respectively. Also, when the job involves different tasks, it is useful to determine the amount and percent of time spent on each. A summarization of the average cycle time, the range of cycle times, the active or working time, and the non-working time are often useful. From this, daily statistics can be developed and compared with production specifications and requirements.

JOB: Loaf Mold Extractor Operator.	STEP NUMBER: 3	
DESCRIPTION: The operator reaches into the storage rack and removes the double loaf mold, and carries mold to the tilt/extraction machine approximately 5 feet away. Approximate weight of the loaded mold is 70 to 75 pounds.		
RISK FACTOR (*)	**ROOT PROBLEM(S)**	**SOLUTION (**)**
High grasp force (3)	Mold weight	Eliminate lifting by slide transfer designed into the cart and T/E table (3)
	Very poor transfer design	
Ulnar deviation (3)	Mold design - bad handles	"
	Mold height/cart design	"
	Very poor transfer design	"
Shoulder flexion with high force, 45 degrees. (3)	Very poor transfer design	"
Lifting force 70 - 75 lb (4)	Mold weight	"
	Very poor transfer design	"

*   Include the rated importance.
**   Include the rated potential.
Use a 0 to 4 scale. Zero indicates none and 4 the highest importance or potential.

FIGURE 19.4

JOB: Loaf Mold Extractor Operator.	STEP NUMBER: 3 (continued)	
DESCRIPTION:		
RISK FACTOR (*)	**ROOT PROBLEM(S)**	**SOLUTION (**)**
Compression on calves from mold hitting while walking	Very poor transfer design	"
Slippery floor (3)	Same as Step 1 above.	Same as Step 1 above.

*   Include the rated importance.
**   Include the rated potential.
Use a 0 to 4 scale. Zero indicates none and 4 the highest importance or potential.

FIGURE 19.5

Finally, it is important to relate the risk factors found in the job and the injuries and illnesses found on the job. If these do not correspond, the analysis has been inadequate and further analysis is required.

Solutions

This is a narrative section that summarizes the possible solutions. It can also be broken into two parts — step related and whole job related. It states the possible solutions and which of the risk factors and root problems they will address. It can also give an appraisal of the sufficiency of change that these solutions will bring about, separately or as groups.

FOLLOW-UP EVALUATION			
JOB NAME: Loaf Mold Extractor Operator.			**DATE:**
CHANGE IMPLEMENTED	**(#)RISK FACTORS REDUCED (**)**	**RISK FACTORS CREATED (*)**	**RISK FACTORS UNCHANGED (*)**
Cart handles	(1)Ulnar deviation, extreme extension, and pronation (3)		
New floor surface	(1,2)High pushing forces (2)		
	(1)Slippery floor (1)		
Larger wheels	(1)High pushing forces (2)		
Shoe change	(1,3)Slippery floor (1)		
Drop chute for lids	(2)High wrist acceleration (3)		

\#   Step numbers in which the risk factor occurred.
*   Include the rated importance. Use a 0 to 4 scale. Zero indicates none and 4 the highest importance or potential.
**   Include the rated change. Use a -1 to 4 scale. Minus one indicates worse and 4 complete elimination.

FIGURE 19.6

19.4 Implementation and Follow-Up

The solution or solutions selected are tested and then implemented if they are found to be feasible, if they address the problems, and if they have the potential to improve the workers' safety and health. Follow-up is absolutely necessary to fine-tune the solution and to ensure that the problems have been addressed and that significant new ones have not been created. In order to facilitate the evaluation of changes that have been implemented, the authors have created another simple form that gives the change implemented, the risk factors reduced, the risk factors created, and the risk factors that remain unchanged. This form is illustrated in Figures 19.6 and 19.7. The form asks for the rated importance of each risk factor created with a scale from 0 to 4. Zero indicates that the rated importance to musculoskeletal disorders is none or negligible, and 4 indicates that the rated importance of the risk factor is very high. The form also asks for a rating of each changed risk factor with a scale from –1 to 4. Minus one indicates that the change made the root problem or risk factor worse, and 4 indicates that the root problem and its associated risk factor is completely eliminated. Implied in all of this is that a new job analysis is likely. These ratings are based on the analyst's judgment and experience. They can be improved upon by getting input from the people doing the job.

Worker input is critical in the follow-up. No matter how good the job may look to someone else, the person doing the job truly knows that job and the problems encountered. One example is that the job is fine as long as periodic maintenance is performed, but it becomes a real problem job when the maintenance is not performed. Also, worker input allows for fine-tuning and new suggestions.

19.5 Example Job — Loaf Mold Extractor Operator (100–500)

Job Description

In this job, the operator pushes a cart weighing hundreds of pounds into position (approximately 10 feet). Next, he removes a one-pound mold top and throws it into a bin about 5 feet away. He then lifts a heavy (70 to 75 lbs.) mold full of cooked product (lunch meat) from the cart, carries it to a table about 5 feet away, places it into a tilt/extraction table (T/E table), activates the T/E table, removes the endplates and throws them into a bin, removes the plastic coating and places it in a receptacle under the table, lifts

FOLLOW-UP EVALUATION			
JOB NAME: Loaf Mold Extractor Operator.		**DATE:**	
CHANGE IMPLEMENTED	**RISK FACTORS REDUCED (**)**	**RISK FACTORS CREATED (*)**	**RISK FACTORS UNCHANGED (*)**
Slide transfer designed into the T/E table	(3)High grasp force (4)		
	(3)Ulnar deviation (4)		
	(3)Shoulder flexion, high force 45 degrees (4)		
	(3)Lifting force (4)		
	(3)Compression on the calves from mold (4)		
		(4)Shoulder flexion, 45 degrees (0)	
		(4)Push force (1)	

\#   Step numbers in which the risk factor occurred.
*   Include the rated importance. Use a 0 to 4 scale. Zero indicates none and 4 the highest importance or potential.
** Include the rated change. Use a -1 to 4 scale. Minus one indicates worse and 4 complete elimination.

FIGURE 19.7

and carries one of the two loaves 4 feet to the slice/package cart, returns to the tilt/extraction table, carries the other of the two loaves 4 feet to the slice/package cart, returns to the T/E table, removes the mold, carries the mold back to the original cart, and places the mold in the cart. Each mold contains two logs or columns of lunch meat approximately 6" by 6" by 36". When the mold cart is empty, he pushes it (150 lbs.) 15 feet to a staging area. When the slice/package cart is full, he pushes it 5 feet to a staging area.

Job Objective

The objective of this job is to get a mold cart with full molds, remove the cooked product from the molds, place the product on the slice/package cart, move the empty mold cart with empty molds, and move the full slice/packaging cart to a staging area.

Job Essential Functions

The essential functions involve:

Pushing a very heavy cart — 800 pounds
Removing and disposing of mold lids
Transferring full molds to a tilt/extraction table
Removing plastic and an endplate from the product
Transferring cooked product to a cart
Transferring empty molds to a cart
Pushing a low-to-moderately heavy cart — 100 pounds
Pushing a heavy cart — 500 pounds

Contributing Factors

This job rarely involves overtime or high pressure. It is a secure job, and the worker doing it is usually left alone to do his job. The temperature is cool, but no drafts seem to be present. The worker does wear a rubber-type glove. Looking at all of this, there appear to be no significant contributing factors.

Task Breakdown

There is only one task in this job. It involves getting a mold cart with full molds, removing the cooked product from the molds, placing the product on the slice/package cart, returning the mold cart with empty molds, and moving the slice/packaging cart to a staging area. It is repeated throughout the workday.

Step Breakdown

The first three steps of the job analysis for this job are contained in Figures 19.2 through 19.5.

Risk Factor and Root Problem Summary

The stressors found in this job are ulnar deviation, extreme wrist extension, pronation of both wrists, high pushing forces, high grip forces, ulnar deviation, wrist flexion and extension, shoulder flexion and abduction, high arm force, inefficient side pull, trunk flexion, and trunk torsion, carrying heavy loads, toss or flip actions probably involving high wrist accelerations, molds that hit upper legs and knees as the worker walks, and wet floors.

These risk factors relate closely with the problems suffered by the workers doing this job. They have had tendinitis of the wrist, pain in the neck and shoulder, low back pain, and one incidence of a slip and fall resulting in a sprained ankle and bruises.

Solutions

Several solutions have been developed and are considered feasible. They are handles for the mold cart, larger wheels for the cart, different shoe soles, resurfaced floor, redesigned mold cart, redesigned T/E table, have the plastic stripped off mechanically, and a redesigned transfer to the slice/packaging cart. Well-designed handles for pushing or pulling the mold cart, larger wheels for the cart, better shoe sole, and a better floor surface will significantly reduce the risk factors associated with moving the mold cart. A well-designed cart and T/E table will significantly reduce the risk factors associated with mold transfer. A well-designed ramp, chute, or conveyor from the T/E table to the slice/packaging cart will significantly reduce the risk factors associated with product transfer.

Implementation and Follow-Up

Numerous solutions were implemented. Well-designed handles were installed on the cart for pushing. Larger wheels were installed on the cart. The floor was repaired such that it is less slippery and does not cause problems in pushing the cart. Appropriate shoe sole material was recommended to the employee for his next pair of boots. The cart carrying the molds and the T/E table were redesigned to allow pushing the molds from one to the other, thereby eliminating the lifting of molds. A convenient drop bin was provided to dispose of the mold lids. A hole was provided in the counter top to dispose of the endplates. An adjustable chute was installed so that the extracted loaf could be pushed onto the slice/package cart. Follow-up evaluation forms provided in Figures 19.6 and 19.7 show changes implemented, the risk factors reduced, the risk factors created, and the risk factors that are unchanged. Overall evaluation determined that this job was changed from one with considerable problems to one with very few problems. Even though the potential for musculoskeletal disorders is greatly reduced, periodic monitoring is recommended.

References

A Technical Assistance Manual on the Employment Provisions (Title I) of the Americans With Disabilities Act. (January 1992.) U.S. Equal Opportunity Commission, U.S. Government Printing Office, Superintendent of Documents, Washington, D.C.

Armstrong, T. J. (1983) *An Ergonomics Guide to Carpal Syndrome;* American Industrial Hygiene Association, Akron, OH.

Armstrong, T. J. (1986a) Ergonomics and cumulative trauma disorders, *Hand Clinics,* 1, 3.

Armstrong, T. J. (1986b) Repetitive trauma disorders: job evaluation and design, *Human Factors,* 28(3), 325-336.

Burke, M. (1992) *Applied Ergonomics Handbook,* Lewis Publishers, Inc., Boca Raton.

Cochran, D., Stentz, T., Stonecipher, B., and Hallbeck, S. Guide for videotaping and gathering data on jobs for analysis for risks of musculoskeletal disorders, in *Handbook of Industrial Ergonomics,* 1999.

Keyserling, W. M., Armstrong, T. J., and Punnett, L. (May 1991) Ergonomic job analysis: a structured approach for identifying risk factors associated with overexertion injuries and disorders, *Applied Occupational Environmental Hygiene,* 353-363, 6(5).

Marras, W. S. and Lavender, S. L. (1995) Biomechanical risk factors for occupationally related low back disorders, *Ergonomics,* 377, 38(2).

Marras, W. S. and Schoenmarklin, R. W. (1993) Wrist motions in industry, *Ergonomics,* 341, 36(4).

Putz-Anderson, V. (1988) *Cumulative Trauma Disorders, A Manual for Musculoskeletal Diseases of the Upper Limbs,* Taylor & Francis.

Rogers, S. H., A functional job analysis technique, *Occupational Medicine: State of the Art Reviews,* 7, 4.

Sommerich, C. M., McGlothlin, J. D., and Marras, W. S. (1993) Occupational risk factors associated with soft tissue disorders of the shoulder: a review of recent investigations in the literature; *Ergonomics,* 36(6), 697-718.

Ulin, S. S. and Armstrong, T. J. (1992) A strategy for evaluating occupational risk factors of musculoskeletal disorders, *Journal of Occupational Rehabilitation,* 2, 1, 35-49.

20

The AET Method of Job Evaluation

Kurt Landau
*University of
Technology–Darmstadt*

Regina Brauchler
University of Stuttgart–Hohenheim

Walter Rohmert
*University of
Technology–Darmstadt*

20.1 Objectives of Job Analysis

Planning, design, and evaluation of work should be preceded by an analysis of the job, the work tasks, and the resulting demands that are placed on the worker. This type of systematic analysis following a standard pattern is performed in only very few cases, mainly manual jobs in industry. Instead, *ad hoc* procedures relating to the individual case are used. No further use is made of the data after the immediate problem has been solved. This would, in any case, be impossible because the analytical instrument is either totally or, at least, partially inapplicable outside the confines of the company that used it. This means that companies regularly "reinvent the wheel." Analytical data that could be further evaluated for general purposes is not passed on, and the opportunity to further develop the discipline of ergonomics is lost. No taxonomies of jobs and tasks can be compiled, and questions relating to occupational research are left unanswered.

This raises the question of whether it would be possible to develop job analysis procedures that are universally applicable. Such procedures should cover the whole spectrum from heavy physical work to mental work, and they should be equally suitable for use in large and small operations and in different branches of industry.

For the evaluation of job analysis procedures, the following criteria, most of which have been developed by Frieling and Graf Hoyos, can be used (Frieling, 1975; Graf Hoyos, 1974):

The procedure should:

- Be based on a theoretical model that allows a practical interpretation of the results obtained with the job analysis
- Offer complete coverage of all demands that are present within a specific person-at-work system
- Offer maximum cost-effectiveness with regard to application, data processing, and data evaluation; the application of the procedure should allow standardization
- Go beyond a merely verbal work description and allow quantitative statements, at least at ordinal numbers

When applying the procedures, it should be possible to make statements as to:

- The reliability with which several raters analyze a person-at-work system at the same time (inter-rater-reliability)
- The reliability with which all items of the job analysis procedure can be rated (item reliability)

If job analysis is seen as an analysis of stress determinants, it can be assumed that it will be possible to make a quantitative evaluation of stress factors (generally rated on an ordinal scale) by duration, intensity, sequence, overlap, and time of occurrence within a job. If it is also claimed that the analysis procedure will provide information on the strains resulting from the stress patterns, this implies that the procedure is capable of:

- Producing repeatable qualitative and quantitative analyses of the strains arising (with the exception of emotional strains) (Luczak, 1975)
- Allocating psychological or physical strain to selected items qualitatively
- Rating specific items for the psychological or physical strains produced by them
- Helping to make quantitative evaluations of strains (ratings on a set scale) based on the results of physical or psychological examinations of the workers involved

This distribution describes the AET, a job analysis procedure which seeks to fulfill the requirements listed above.

The origins of AET (Arbeitswissenschaftliches Erhebungsverfahren zur Tätigkeitsanalyse; ergonomic job analysis procedure) date back to a study (Rohmert and Rutenfranz, 1975) ordered by the German government to investigate discrimination against women at work with respect to pay. A job analysis procedure was required that allowed a detailed investigation of workload and strain within a given person-at-work system. At that time, there was no job analysis procedure that could readily be used, although the PAQ (McCormick et al., 1969) seemed to provide a basis for the psychological items.

AET (Arbeitswissenschaftliches Erhebungsverfahren zur Tätigkeitsanalyse) job analysis (job evaluation) procedure was developed in 1978 (Landau et al., 1975; Landau, 1978; Rohmert and Landau, 1979); it has been continuously improved since then. An important characteristic is that when a firm applies the AET procedure to a specific job, the data are recorded in a central data bank at Darmstadt. The data bank now holds the analyses of over 7,000 jobs. Another important characteristic is that the AET procedure is applicable over a very wide range of jobs, blue collar as well as white collar, manual as well as engineering and executive, manufacturing as well as retail, office as well as factory.

For an analyst, the general procedure is to evaluate a job using 216 characteristics. The individual doing the job is not evaluated. The points for each characteristic are obtained and totaled. The pay for a worker is a function of the number of points, but the shape of the function varies with the firm.

20.2   AET — The Ergonomic Job Description Questionnaire

Theoretical Basis

The AET procedure is based, on the one hand, on the model of the person-at-work system and, on the other hand, on the concept of stress and strain. Concerning its criteria of classification of activities, in order to facilitate logical and deductive testing, AET is oriented toward the elements and flows of the person-at-work system.

The stress and strain concept of human work (see Rohmert et al., 1975; Luczak, 1975) assigns the work task to the "object area" (see Figure 20.1). Both the work task and the conditions of the work environment are included in the object area, from which job-specific and situation-specific demands result. These demands characterize the energetic-effective heaviness of work and the informative-mental difficulty of work. Partial stresses related to the work content are the result of duration and time distribution of work heaviness and difficulty.

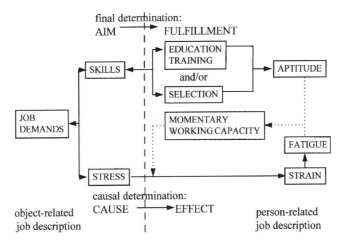

FIGURE 20.1 Stress-strain concept. (From Landau, K. and Rohmert, W. 1989. Introduction of the problems of job analysis, in Landau, K. and Rohmert, W. (eds.) *Recent Developments in Job Analysis, Proceedings of an International Symposium in Job Analysis.* University of Hohenheim, March 1989. London, New York, Philadelphia: Taylor & Francis, pp. 1-24. With permission.)

The partial stresses are expressed in factors and quantities describing the external effects of the person-at-work system on the working person (Luczak, 1975; Rohmert, 1983). Together with the situation-specific partial stresses that result from the physical and social working conditions, the work content-specific partial stresses determine the subjective activities. Activities are also influenced by the motivations and disposition of the working person; this can be partly derived from the required skill. These relationships form the "object area" of the enlarged stress–strain concept and the starting point of the job analysis with AET. The simultaneous consideration of the categories "tasks," "demands," "environmental conditions," and "required skills" as determinants of stress is the cornerstone of AET (Landau and Rohmert, 1989).

AET Structure with Regard to the Contents

To begin with, the person-at-work system is described. The description and scaling refers to the objects of work and to the equipment and the working environment. The objects of work are analyzed under material, energy, and information aspects. When a person is an object of work, the characteristics of the group of people he came from must be investigated. Finally, the qualities of material objects of work (like raw materials) are analyzed. The equipment, the work instruments (like tools, implements), and other operating materials (like hard- and software) are of interest. In this case, the ergonomic system of classification is completed by technical aspects. The analysis of the environment is related to the physicochemical conditions in the workroom, the organizational and social working conditions and to principles and methods of pay. There are 36 items for working and operating materials, 50 for the physical, organizational, and social working conditions, and 24 for the economic working conditions (see Table 20.1).

The object of investigation is the human activity in the person-at-work system; this is composed of the tasks which have to be carried out. The tasks ensue from the purpose of the person-at-work system. In this context part B of the analysis procedure represents a link between the tasks to be fulfilled and the resulting demands exerted on the working person. So, the analysis of the person-at-work system is followed by the task analysis, an evaluation by means of 31 items, subdivided according to the objects of work (see Table 20.2).

The task analysis is finally followed by the analysis of demands (see Table 20.3). Within this part the items were chosen in such a way as to take as many functions of the body as possible into account. This

TABLE 20.1   AET-Structure — Part A

A: ANALYSIS OF THE PERSON-AT-WORK SYSTEM	Number of characteristics	
1. Object of Work		33
Kind	(4)	
Characteristics	(28)	
Person as object of work	(1)	
2. Equipment		36
Means of production	(17)	
Changes in the state of the objects of work		
Change in the location of the objects of work		
Other means of production		
Other equipment	(19)	
for controlling the state		
for supporting human sense		
seat, worktable, workroom		
3. Working Environment		74
Physiochemical working environment	(12)	
Organizational and social conditions	(38)	
Temporal work organization		
Position on the work within the:		
operation process organization		
organization structure		
communication system		
Principals and methods of pay	(24)	
		143

TABLE 20.2   AET-Structure — Part B

B: ANALYSIS OF TASKS	Number of Characteristics
1. Related to Material Objects of Work	13
(Preparing, assembling, equipping, inserting, transporting, measuring, operating, checking, supervising …)	
2. Related to Abstract Objects of Work	8
(Planning, organizing, coding information, transcribing information, combining information, analyzing information …)	
3. Person-Related	8
(Representing, instructing, serving, attending …)	
4. Number and Frequency of Task Repetitions	2
	31

is true for the function of power and energy generation in the different organs (stress during postural work and static work, heavy dynamic work and active light work), as well as for the functions of perception, decision, and action in different mechanisms of information processing. The classification depends on the question of which area is required concerning the reception of information, decision, and action. In the case of the analysis of demands, one distinguishes between the demands in connection with the reception of information (17 items), with information processing (8 items), and with information output or activity (17 items). Besides the usual test–statistical requirements of an analysis procedure, a fundamental problem arises as a result of the possibility of theoretical justification. That is namely asking for the selection of characteristics within the outlined classification of the procedure and for the scale level of the selected characteristics.

TABLE 20.3   AET-Structure — Part C

C: ANALYSIS OF DEMANDS		Number of Characteristics
1. Range of Demands: Reception of Information		17
Organs of Sense for the Reception of Information	(6)	
Dimensions of Identification	(7)	
Forms of Identification	(2)	
Accuracy Necessary for the Reception of Information	(1)	
Vigilance	(1)	
2. Range of Demans: Decision		8
Complexity of decision	(1)	
Temporal scope of decision	(1)	
Necessary knowledge	(6)	
3. Range of Demands: Activity		17
Organs of activity and accuracy of activity for stress while acting:		
Caused by postural work	(6)	
Caused by static work	(4)	
Caused by heavy dynamic muscular work	(3)	
Caused by active light work	(4)	
		42

Particularly when selecting the items, the following basic facts have to be recognized and considered:

1. It is not possible that a fully completed catalogue of items can be expected as part of the selected theoretical concept, especially in view of the economy of the procedure.
2. Only those items that are important for numerous people-at-work systems should be included.
3. The selection of items implies a certain judgment; hence, "user-specific traits" — such as experience, standards of values, opinions — influence the system of analysis.

Thus, it must be noted that as the selection of the theoretical model is at the basis of the procedure, the selection, scaling, and description of items within this model entail a series of subjective assumptions. However, these assumptions are acceptable as long as several analysts achieve reliable results by applying the procedure.

Parallel to the development of AET, an attempt was made to limit the subjective influences on the construction of items by means of the concept of an iterative procedure development with respect to the experience of both company specialists and ergonomists.

If one acknowledges the subjective, author-specific influence of the selection of a theoretical model concept substantiating AET on the selection of items and on the determination of scale levels, codes, and aids of classification, then this means also recognizing the need to include technical criticism to achieve a progressive development of AET. The goals of the iterative AET development consisted of increasing its accessibility to users while safeguarding or improving the test–statistical quality criteria.

The relatively vague notion of "ease of use of AET" can be operationalized by the following desired qualities:

- A clear analysis
- Clear and comprehensible formulas
- Conformity of the definitions of AET with the meaning of terms in ergonomic literature (as far as possible)
- A time-saving and economical analysis and evaluation

The development of AET was oriented toward certain groups of users. In addition to ergonomically trained groups of users in research and practice, AET addresses ergonomists. Another category at which AET is aimed, and which is more a group of persons interested in AET rather than a group of users, includes people such as industrial psychologists, representatives of labor and management, and so forth.

TABLE 20.4   AET Versions and Reliability

AET Version	Number of Examined Jobs	Item Reliabilities	Position Reliabilities
AET (A) - Draft	26	0.57	0.64
AET (A)	42	0.65	0.87
AET (B)	4	0.71	0.71
	17 Analysts (Seminar classification)		
AET (C)	62	0.71	0.89
AET (D)	2	0.79	0.74
	22 Analysts (Seminar classification)		

Incorporating the practical experience and the extent of knowledge of these target groups into the iterative development of AET was accomplished by means of three AET seminars and several meetings of ergonomists with representatives of labor and management. These seminars were designed to take account of the broad area of AET's validity and application. The aims of the seminars were:

- To discuss basic questions on matters of the form and contents of AET
- To determine how long it takes to train experienced analysts in using AET
- To ascertain the reliability of use of AET in the corresponding version
- To state difficulties of application and to initiate suggestions for improvement

The iterative elaboration of AET, based on AET version (A) and carried out both on this basis and on the grounds of the results of the reliability studies, altogether resulted in three further versions (B,C,D) with corresponding drafts (see Table 20.4). The number of items was able to be reduced considerably from 390 to 216.

AET Structure with Regard to Form

The analysis of the job is done in the form of an observation interview, which means that necessary analytical data are collected first by observation of the job and working environment and second by interviewing the incumbent and the incumbent's superior.

Each AET item consists of an (underlined) question outlining the state of affairs to be grasped and indicates the code for classifying this item. In certain circumstances, examples are given as classification aids. The explanations clarify the questioning of the AET characteristic in view of extent, delineation, and classification, but they cannot be taken as a complete and binding instruction for the rating.

The items of the analysis of demands — which might be difficult to answer for somebody who is not sufficiently trained in ergonomics — contain additional classification aids in the form of "activity scales." The activity scale, based on previously investigated data, contains a series of grades of mainly illustrative activities. Equivalent to an ascending rating scale, we may suppose, at least approximately, an intensified demand (see Figure 20.2).

However, the selected job examples only represent an optional part of reality; an equidistant reproduction of the job examples an ordinal scales of the item concerned is not possible.

The classification of characteristics can only be done by means of the corresponding code, which follows the suggestions of McCormick et al. (1969). The different codes are as follows:

- Significance Code (S)

    The importance or significance of this aspect for the task should be estimated in relation to other tasks or activities. Use a range from 0 to 5.
- Duration Code (D)

    The code "duration" is based on a shift lasting eight hours. This is assumed even if the incumbent is a part-time worker. This is done in order to be able to compare the work content of jobs with different shift hours. Use a range from under 1/3 of shift time to whole shift time.

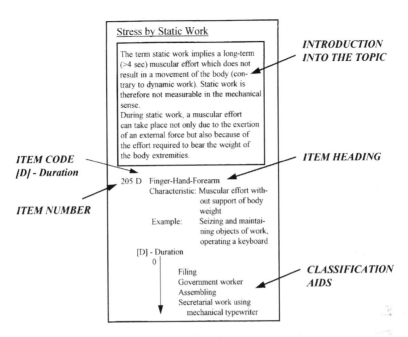

FIGURE 20.2   Example of an AET item.

- Frequency Code (F)

    This code characterizes the temporal distribution and position of stress sections. Use a range from 0 to 5.

- Alternative Code (A)

    The alternative code only asks about the presence of a characteristic: Does the work characteristic in question apply 1, or does it not apply 0?

- Exclusive Code (E)

    The exclusive code is always related to only one specific question, e.g., multiple properties of a working instrument. Use a range from 0 to 5.

An AET analysis in the field is directly followed by the coding of the AET items. This coding has to be done on standardized paper or in direct dialogue with the computer. The time required for an observation interview is usually 1 to 3 hours, depending on the analyst's practical experience, the type of job and the repetition rate of work processes.

AET Method of Evaluation

As a result of the various opportunities that exist to use AET data, the data can be used to answer fundamental questions arising in nearly all potential fields of application:

1. The AET codings can be used to characterize a person-at-work system.
2. The AET codings can be used to describe people-at-work systems or groups of people-at-work systems.
3. The characteristics revealed by all or most of the AET codings can be used to classify a person-at-work system.
4. Conversely, people-at-work systems can be grouped together according to common AET characteristics.

For these interrelated and basic requests, the best solution is through using univariate evaluation methods that are supplemented by a series of multivariate methods. Concerning univariate methods, well-known procedures of descriptive statistics and of profile analysis can be used. Procedures of multivariate analysis, such as cluster analysis, discrimination analysis, factor analysis, multidimensional scaling, etc. can be applied.

Examples of results obtained in various studies are given below. These are intended to give the user guidance in the evaluation and interpretation of AET codings, especially for the types of evaluation listed under the fundamental questions 2 and 3 above.

20.3   AET Applications and Example Evaluations

The evaluation of AET codings has to provide answers to the following questions:

1. What are the differences in the jobs regarding work content, objects of work, work instruments, workplace, working environment, and work organization in:
   different branches of industry
   different enterprises
   different departments
   different wage groups?
2. What are the differences in the job characteristics of jobs requiring different education and job-related training?
3. To what extent do the job characteristics of native and foreign workers differ?
4. To what extent do the job characteristics of industrial employees, office workers, executive personnel, and government workers differ?
5. Which jobs are particularly similar or differ markedly in view of different strain-relevant stress components?

Profile Analysis

A representation of the scores obtained by the groups of items in the form of profiles is suitable to give a graphic survey of the extent or the duration of stress experienced during the execution of jobs or groups of activities. This type of evaluation of the data derived from job analysis is designated "profile analysis," regardless of the fact that this term may possibly be used in a different sense by other disciplines. Figure 20.3 shows an example of a job profile obtained from AET analyses. It shows the characterization of the individual person-at-work system in laying bricks on the basis of one actual coding of the 216 AET-items. The grouping of the characteristics necessary for the execution of a profile analysis used by AET is derived from the structural items of the job analysis procedure. The method used for computing the profiles is explained in Landau and Rohmert (1981) and Rohmert and Landau (1983). The vertical plane shows the characteristics of the workplace and the types of demand, while the horizontal scale shows the maximum AET classification in percent.

Figure 20.4 uses the AET procedure to analyze the incidence of the various types of demand in 2,838 jobs usually occupied by males and 866 usually occupied by females in German industry. The upper bar represents the female jobs, the lower bar the male jobs. The analysis shows that the most important tasks for males involve operating, controlling, supervising, planning, organizing, and analyzing. The main tasks performed by females are checking, and also a variety of general, people-oriented service tasks. The tasks are divided into the stereotyped patterns of "typically male" and "typically female." Males perform more (complex) operating, controlling, and assembly tasks in which they are required to plan and organize their own work, while females, in addition to their additional job as mother or housewife, are employed in industry mainly for simple checking activities and also for people-oriented services. The jobs occupied by males were exposed to far higher levels of physical or chemical stress from the working environment. This applies both to factors like illumination, climatic conditions, vibration, and noise as well as to other environmental influences like noxious materials. The work hazards, including the frequency or probability of a work accident or an occupational disease, are rated higher for the male jobs.

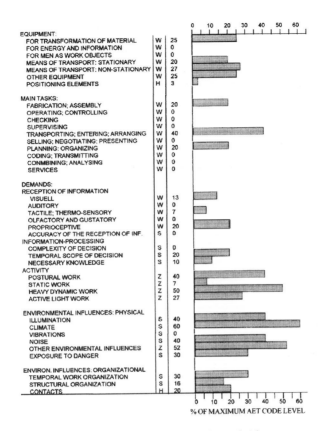

FIGURE 20.3 AET job profile: laying bricks.

The male jobs also show a higher level of demands in both the organization of working time (shift work), the sequence of operations, and overall planning. As night work by females was severely restricted by German law, it was and is largely a male preserve in industry, and the demands in this respect are therefore much higher for men.

Closer investigation shows that information reception and information processing place substantially higher degrees of certain types of demand on the male than on the female jobs. This led to a higher classification against the AET criteria for demands involving the reception of visual information, of auditory information, and proprioception. Similar levels for both male and female jobs were registered only for information reception via the senses of smell, taste, touch, and temperature sensitivity of the skin. There are only very slight differences between the sexes in the demands for accuracy of information reception. The male jobs involve tasks of greater complexity and, in some cases, greater critical stress. The level of knowledge required in male jobs is rated higher than in female jobs.

The proportion of physical work demands is very similar for both sexes. They are identical for static handling work and only insignificantly higher for males in the case of static holding work. In male jobs, longer periods of the shift are devoted to heavy dynamic work and, in female jobs, to active light work. This corresponds to the German role expectations in the division of work between men and women in industry, i.e., heavy dynamic work for men and active light work involving monotonous procedures for women.

Frequency Distributions

Tables 20.5 and 20.6 explain how company-, branch-, and sex-related position groups that were analyzed by means of the AET procedure can be evaluated in a simple way by analyzing the frequency distributions of levels. Table 20.5 shows percentages of the item levels "high" and "extreme" for selected characteristics of the AET procedure.

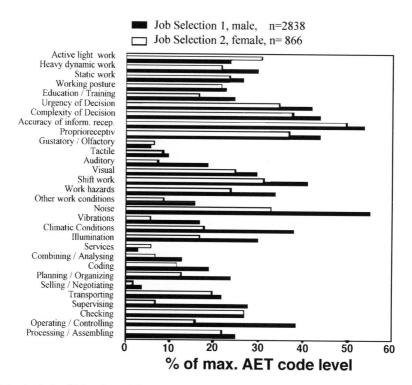

FIGURE 20.4 Analysis of job tasks and demands for 2,838 males and 866 females in German industry (216 AET items). (From Landau, K. and Rohmert, W. 1992. Evaluation of worker workload in flexible manufacturing industry. *The International Journal of Human Factors in Manufacturing*, 2 (4), pp. 369-388. With permission.)

TABLE 20.5 Frequency Distributions of High [4] and Extreme [5] AET Codings of Selected AET Items for In-Company Applications

| | % Level for AET Characteristics "HIGH" (4) and "EXTREME" (5) | | | | | |
| | Fabrication | | Preassembly | | Assembly | |
AET Item	Male	Female	Male	Female	Male	Female
Climate (71)	7	7	13	1	0	0
Vibrations (78)	87	25	2	0	18	17
Accuracy of Perception (17)	61	57	0	0	0	0
Decision Making Under Pressure of Time (19)	0	0	39	39	91	89

From Rohmert, W. and Landau, K. 1983. *A New Technique for Job Analysis*. London, Taylor & Francis. With permission.

For 7% of the workplaces of male workers, the climatic stress is rated "high" or "extreme." In the field of preassembly, 13% of the workplaces of male workers shows a high or extreme climatic stress. In the assembly department, an extraordinary climatic stress is to be found. Effects of mechanical vibrations in the fabrication department are rated "high" or "extreme" at 87% of the workplaces occupied by men and at 28% of the workplaces occupied by women. This method of evaluation can be continued in the same way for all characteristics of the procedure. If large-scale data collections related to branches are available, then it is possible to carry out an analysis of frequency distributions of levels of AET data due to conditions in the particular trade. This is shown in Table 20.6.

TABLE 20.6 Sex- and Branch-Related List of the Classifications High [4] and Extreme [5] for Selected AET Items as an Example of Frequency Distribution of AET Codings

	% Level for AET Characteristics "High" (4) and "Extreme" (5)			
	Occupied by Women	Occupied by Men	Metal Working Industry	Chemistry
Active Light Work (211)	32	6	24	22
Visual Identif. of Surface Structures (175)	34	25	34	39
Combining (161)	2	20	3	0
Noise (73)	95	40	35	43
Responsibility (109–112)	1	51	22	21

From Rohmert, W. and Landau, K. 1983. *A New Technique for Job Analysis*. London, Taylor & Francis. With permission.

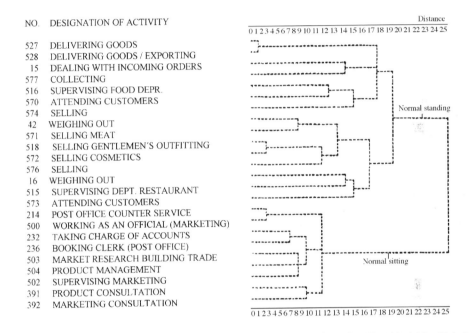

FIGURE 20.5 Example of an AET cluster analysis. (From Rohmert, W. and Landau, K. 1983. *A New Technique for Job Analysis*. London, Taylor & Francis. With permission.)

Cluster Analysis

The main aim of applying cluster-analytic methods to job analysis data is a reduction in the quantity of data. Apart from the generation of job groups, numerical relationships between the job groups must be established and interpreted. When carrying out the classification of activities, one must examine which conclusions can be drawn from the numerical relationship for the elaboration of a taxonomy of human activities. As far as ergonomics is concerned, the effects of specific work content on group composition and on numerical relationships between the groups are of particular interest. With the classification of large ergonomic data tools into smaller and comprehensive groups arises the possibility of investigating group scores further, i.e., the positions that are most representative for the group as well as for the respective workers, using (rather expensive) methods of occupational physiology and psychology.

Figure 20.5 shows one application of cluster analysis methods to AET data. A hierarchic cluster analysis has been carried out for 24 buying, selling, or trading jobs and is represented in the form of an activity dendrogram. In this analysis it is possible to distinguish that, with respect to working posture, two groups are clearly separated from the others.

1. Prevailing working posture is normal standing, with a small quota of the postures "bent standing," "crouching" and "kneeling." This group covers the working postures typical of selling activities in department stores and retail shops.
2. Prevailing body posture is sitting, with a small quota of standing. This posture is typical of commercial activities not to be classed with the retail trade. Significant activities in this group are the fields of marketing consultation and sales promotion.

The "dissection" of the results of a cluster analysis makes it possible to identify the job clusters at a given hierarchical level. In cases where jobs are grouped by sectors of industry, the mean duration of the observed postures per shift is shown graphically. As this involves the calculation of mean arithmetical values from data rated on an ordinal scale, it is advisable to interpret the results with caution.

Comparing different industries, Landau and Rohmert (1992) estimated, for example, the percentage of shift time involving static work (their Figure 6.1). The jobs in the iron and steel industry received the highest rating for static work, followed by the chemical industry, the automotive industry, and the services sector in that order. Static work mainly involves the use of the finger/hand/forearm region or the arm/shoulder/back region. Static work using the leg or foot region is of only minor importance in all the industries covered by this study.

The chemical industry has the highest percentages of heavy dynamic work and active light work. This is followed by the iron and steel industry in the case of heavy dynamic work, and by the automotive industry in the case of active light work. The percentages of heavy dynamic work and active light work are lowest in the services sector (Landau and Rohmert [1992] Figure 6.2).

Heavy dynamic work can involve either the arms and upper body muscles or the legs and pelvic muscles. Heavy dynamic stress on the legs and pelvic muscles was caused by walking, climbing, etc., in some cases with loads. Walking and climbing are still important work factors in the chemical industry, followed by the iron and steel industry, the automotive industry, and the service sector.

Active light work in the chemical industry involves mainly the finger/hand system. In the automotive industry, there are more gross motor activities using the hand/arm system. The foot/leg system is not used to any significant degree in any of the industries investigated (Landau and Rohmert [1992] Figure 6.3).

Shifting of Demands

The AET is capable of analyzing exceptionally high stress levels affecting specific body organs in people-at-work systems and also of quantifying tasks and demands at the workplace and, by extension, identifying shifts in demands such as:

- Cessation or addition of specific types of demand
- Changes in intensity of one or more types of demand
- Changes in duration of specific types of demand
- Changes in time spread of specific types of demand
- Changes in association between different types of demand.

Using the AET data, Landau and Rohmert (1992) compared eight jobs in mechanized assembly with 67 traditional assembly jobs in the automotive industry (Figure 20.7). The small size of the sample populations makes it necessary to interpret the present results with caution.

Visual information reception increases from 24% to 37% of the maximum AET score. There is an even higher increase, from 39% to 60% of the maximum score, in the requirement for accuracy of information reception. Proprioceptive information reception remains almost unchanged, while demands involving information reception by touch and via the thermosensors of the skin have been eliminated, because gripping actions are now carried out by the mechanized assembly. Demands involving auditory information reception, especially in cases where problems are starting to develop, increase because of the high noise level in the work environment.

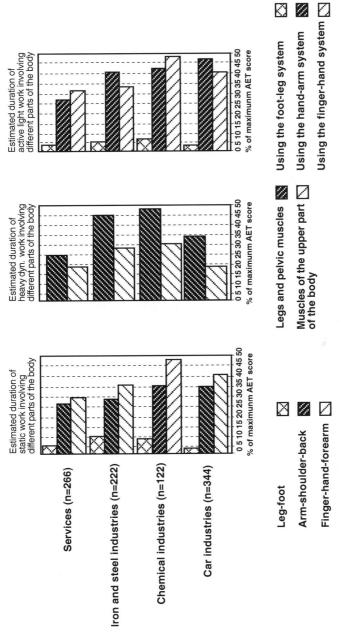

FIGURE 20.6 Estimated duration of physical work forms. An example of task analysis results obtained with AET (From Landau, K. and Rohmert, W. 1992. Evaluation of Worker Workload in Flexible Manufacturing Industry. *The International Journal of Human Factors in Manufacturing*, 2 (4), pp. 369-388. With permission.)

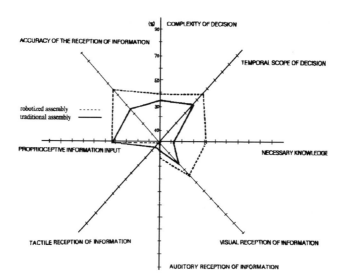

FIGURE 20.7   Shifts in demand involving information reception and information processing following robotization. An example of job analysis results obtained with AET. n1: solid line: 67 traditional assembly jobs; n2: dashed line: 8 jobs in mechanized assembly. (From Landau, K. and Rohmert, W. 1992. Evaluation of Worker Workload in Flexible Manufacturing Industry. *The International Journal of Human Factors in Manufacturing*, 2 (4), pp. 369-388. With permission.)

Demands relating to information processing produce an increase in qualification requirements. The amount of knowledge required increases from 12% to 41% of the maximum AET score. Decision complexity shows a small increase of 5%. However, the urgency of decisions rises from 43% to 55% because of the highly integrated assembly processes in which defects in the equipment have to be rectified without delay.

In conclusion, it can be stated that technological change *per se* does not necessarily bring an improvement in working conditions or a more humane work pattern. Industrial robots actually have both positive and negative effects for the workers. Their advantage is merely that they offer the option of a new production process better adapted to the needs of the workers. Whether this option and the option of improving the qualifications of the workers is exercised, depends very much on the attitudes of management and engineering departments.

20.4   Conclusion

Job analysis procedures are often criticized because of their unsuitability for a wide range of applications. The problem is that the development of universally applicable job analysis procedures is a costly business. The optimal solution is to design the procedure to be as universal as possible within the financial limitations imposed. Universality is defined in this case as suitability for the interpretation and generalization of data obtained from either different investigations, different sectors of industry or different companies.

Appropriation of the contributed AET procedure was done in several fields of applications in the last 20 years. The data obtained from 7,000 AET analyses checked and filed by us forms an available tool that allows detailed examinations, in particular the comparison of work tasks in different sectors of industry. The comparison of tasks and demands of work before and after the introduction of robotized, mechanized assembly is also suitable. Moreover, the use of statistical data analysis procedures for AET data provides significant aid in interpretation and data reduction. The cluster analysis in particular allows an economical application of job analysis in the fields of work design, demand analysis, personnel management, work and occupational research, detection of accident causes, and so forth.

To provide answers to ergonomic, technical, and epidemiological questions, it is necessary to collect all available data on potential deficiency and inadequate conditions at the workplace. Consequently, the main aim of further developments of job analysis procedures should be to maintain the number of items examined and to achieve maximum cost-effectiveness by applying the database obtained from successive investigations as a forecasting and indicative aid in the widest possible range of work situations and to achieve maximum cost-effectiveness by evaluating and interpreting the data.

References

Frieling, E. 1975. *Psychologische Arbeitsanalyse*, Stuttgart: Verlag W. Kohlhammer.

Frieling, E. and Hoyos, C. Graf. 1978. *Fragebogen zur Arbeitsanalyse–FAA*. Bern: Hans Huber Verlag.

Hoyos, Graf 1974. *Arbeitspsychologie*. Stuttgart: Verlag W. Kohlhammer GmbH.

Jeanneret, P.R. 1969. *A study of the job dimensions of "worker oriented" job variables and of their attribute profiles*. Lafayette: Unpublished doctoral dissertation, Purdue University.

Landau, K. 1978. *Das Arbeitswissenschaftliche Erhebungsverfahren zur Tätigkeitsanalyse — AET*, Dissertation, TH Darmstadt.

Landau, K. 1978. Das Arbeitswissenschaftliche Erhebungsverfahren zur Tätigkeitsanalyse — AET im Vergleich zu Verfahren der analytischen Arbeitsbewertung, *Fortschrittliche Betriebsführung*, 27, 1, 33-38.

Landau, K., Luczak, H., and Rohmert, W. 1975. Arbeitswissenschftliche Erhebungsbogen zur Tätigkeitsanalyse, in Rohmert, W. and Rutenfranz, J. (eds.) *Arbeitswissenschaftliche Beurteilung der Belastung und Beanspruchung an unterschiedlichen Arbeitsplätzen*. Der Bundesminister für Arbeit und Sozialordnung, Bonn.

Landau, K. and Rohmert, W. (eds.) 1981. *Fallbeispiele zur Arbeitsanalyse — Ergebnisse zum AET-Einsatz*. Bern: Huber.

Landau, K. and Rohmert, W. 1989. Introduction to the problems of job analysis, in Landau, K. and Rohmert, W., (eds.) *Recent Developments in Job Analysis*, Proceedings of an International Symposium on Job Analysis. University of Hohenheim, March, 14-15, 1989. London, New York, Philadelphia: Taylor & Francis, pp. 1-24.

Landau, K. and Rohmert, W. 1992. Evaluation of worker workload in flexible manufacturing industry. In: *The International Journal of Human Factors in Manufacturing*, 2 (4), 369-388.

Luczak, H. 1975. *Untersuchungen informatorischer Belastung und Beanspruchung des Menschen*. Fortschritts-Berichte der VDI-Zeitschriften, Reihe 10, 2, Düsseldorf, VDI-Verlag.

Luczak, H., Landau, K., and Rohmert, W. 1976. Faktoranalytische Untersuchungen zum Arbeitswissenschaftlichen Erhebungsbogen zur Tätigkeitsanalyse — AET. *Zeitschrift für Arbeitswissenschaft*, 30, 1976, 1, 22-30.

McCormick, E.J., Jeanneret, P.R., and Mecham, R.C. 1969. *The Development and Background of the Position Analysis Questionnaire (PQA)*. Occupational Research Center, Purdue University, Report No. 5.

McCormick, E.J., Mecham, R.C., and Jeanneret, P.R. 1972. *Technical Manual for the Position Analysis Questionnaire (PAQ)*. Lafayette: PAQ Services, Inc.

Rohmert, W., Rutenfranz, J., Luczak, H., Landau, K., and Wucherpfennig, D. 1975. Arbeitswissenschaftliche Beurteilung der Belastung und Beanspruchung an unterschiedlichen industriellen Arbeitsplätzen, in Rohmert, W. and Rutenfranz, J. (eds.) *Arbeitswissenschaftliche Beurteilung der Belastung und Beanspruchung an unterschiedlichen industriellen Arbeitsplätzen*. Der Bundesminister für Arbeit und Sozialordnung, Bonn, 15-250.

Rohmert, W. and Rutenfranz, J. 1975. *Arbeitswissenschaftliche Beurteilung der Belastung und Beanspruchung an unterschiedlichen industriellen Arbeitsplätzen*. Der Bundesminister für Arbeit und Sozialordnung, Bonn.

Rohmert, W. and Landau, K. 1979. *Das Arbeitswissenschaftliche Erhebungsverfahren zur Tätigkeitsanalyse (AET). Handbuch und Merkmalheft*. Bern, Stuttgart, Wien: Hans Huber Verlag.

Rohmert, W. 1983. Determination of stress and strain at real work places: Methods and results of field studies with air traffic control officers, in Moray, N. (Hrsg.): *Mental Workload, Its Theory and Measurement.* NATO Conference Series, Series III: Human Factors Col. 8 New York: Plenum Press 1979, 423-443.

Rohmert, W. and Landau, K. 1983. *A New Technique for Job Analysis.* London, Taylor & Francis.

For Further Information

For definitions and terminology used in job and task analysis, please refer to Landau, K., Rohmert, W. and Brauchler, R. 1997, *Task analysis: Part I — guidelines for the practitioner,* in Mital, A. and Kumar, S. 1997. One special Issue of the *International Journal of Industrial Ergonomics.* Elsevier Science Publishers North Holland (about 15 pages). Also refer to Landau, K. and Brauchler, R. 1997. *Task analysis: Part II — the scientific basis (knowledge base) for the guide,* in Mital, A. and Kumar, S. 1997. The same special issue of the *International Journal of Industrial Ergonomics.* Elsevier Science Publishers North Holland (about 30 pages).

Recent developments in job analysis in particular applications of job analysis procedures were discussed within the scope of an international symposium at the University of Hohenheim in March 1989. Please refer to Landau, K. and Rohmert, W. (eds.) 1989. *Recent Developments in Job Analysis — Proceedings of the International Symposium on Job Analysis.* (ISBN 0-85066-790-9) London, New York, Philadelphia: Taylor & Francis.

Actual investigation out of the AET data collection (7000 AET analysis) assessing physical load and musculoskeletal disorders was published in 1996. Please refer to Landau, K., Rohmert, W., Imhof-Gildein, B., Mücke, St., and Brauchler, R. 1996, *Risikoindikatoren für Wirbelsäulenerkrankungen (Schlußbericht) — Auswertung des Arbeitswissenschaftlichen Erhebungsverfahrens zur Tätigkeitsanalyse (AET-Datenbank) und Validierung eines neuen Arbeitsanalyseverfahrens.* (ISBN 3-89429-725-5) Bremerhaven, Wirtschaftsverlag NW. An abstract of this contribution is published by the German Bundesanstalt für Arbeitsschutz (eds.) 1996, *Problems and Progress in Assessing Physical Load and Musculoskeletal Disorders — Workshop vom 6. Oktober 1995 in der BAfAM (Bundesanstalt für Arbeitsmedizin).* (ISBN 3-89429-713-1) Bremerhaven, Wirtschaftsverlag NW.

In accordance with the basic European Guideline on Safety and Protection of Workers (89/391/EWG), the new European Union work safety laws stipulate an analysis of dangers at the workplace. These require the employer to carry out systematic workplace analysis to identify and eliminate the stresses caused by the work and the risks resulting from them. Computerized processing and evaluation of the data help to make the ABBA (Inspection of Workplace and Analysis of Stresses) system extremely economical. The methods used in the ABBA system are a further development of the Ergonomic Job Analysis Procedure (AET). The K-AET checklist comprises 102 items covering all the tasks and demands arising in a real person-at-work system. Supplements are used to collect data on special types of stress. Please refer to Landau, K., Maas, C., Schaub, Kh., Mücke, St., and Fischer, T. 1996. New software tools for developing job stress registers, in BAfAM (eds): *Tagungsband Nr. 10, Schriftenreihe der Bundesanstalt für Arbeitsmedizin 1996,* pp. 31-48

The use of ergonomic job analysis procedures and, in particular, a further development of the AET, to predict work-related diseases has already been successfully implemented in an expert system shell. The developed job analysis procedure used to predict work-related diseases was tested in a study of driving and control activities in the mining industry. Please refer to Landau, K. and Brauchler, R. 1994, Disease prediction using ergonomic knowledge bases, in Proceedings of the 12th Triennial Congress of the International Ergonomics Association, Volume 2, *Ergonomics in Occupational Health and Safety,* Toronto 1994, p. 159ff.

Brauchler, R. and Landau, K. 1992. Implementation of an epidemiological early-warning system by using rule induction algorithms, in Karwowski, W. and Mattila, M. *Computer Applications in Ergonomics, Occupational Safety and Health — Proceedings of the International Conference on Computer Aided Ergonomics and Safety,* Tampere, Finland, S. 249-254.

Landau, K. and Brauchler, R. 1992. Theoretical background and applications of epidemiological expert systems, in Karwowski, W. and Mattila, M. *Computer Applications in Ergonomics, Occupational Safety and Health — Proceedings of the International Conference on Computer Aided Ergonomics and Safety,* Tampere, Finland, S. 469-476.

21

Worker Strength Evaluation: Job Design and Worker Selection

Sean Gallagher
*National Institute for Occupational
Safety and Health*

J. Steven Moore
University of Texas Health Center

21.1   Introduction

Many jobs in industry severely tax the worker's musculoskeletal system and may approach or exceed the worker's maximum voluntary strength capabilities. When this occurs, there is evidence that the worker is at higher risk of experiencing a musculoskeletal disorder (Chaffin, 1978; Keyserling et al., 1980). It is for this reason that efforts have been taken over the last couple of decades to provide a means of evaluating the muscular strength capabilities of workers, so that jobs can be designed to eliminate taxing exertions, and to ensure that workers performing physically demanding jobs have the strength to safely perform required tasks.

The effectiveness of worker strength evaluation in reducing work-related musculoskeletal disorders (WMSDs) depends in large part on the purpose of the evaluation. Assessment of physical strength has been used for two primary purposes in the field of ergonomics: job design and worker selection. Job design has been the focus of the psychophysical method of strength assessment and is the technique most likely to have a positive impact in reducing WMSDs. In this technique, the strength of a population of workers is used to design the job so that the majority of workers find the exertion to be acceptable. Studies have indicated that designing tasks by this approach may reduce back injuries by up to 33% (Snook et al., 1978). Strength testing has also been used for the purpose of worker selection, that is, making sure that workers have sufficient strength to perform physically demanding jobs. In a sense, this approach to controlling WMSDs is antithetical to one of the primary tenets of ergonomics: design the job to fit the worker. Instead, worker selection seeks to "fit" a strong worker into a physically demanding job. Predictably, this technique does not result in nearly the same magnitude of reduction in injuries compared to the job design approach. However, some studies have indicated a partial success using this technique. It should be noted that such an effect has only been evident when the procedure is employed in an environment known to place workers at very high risk of injury. Furthermore, it must be noted that most studies that have examined worker selection procedures have been short term (usually a follow-up period of one year or less). There can be no guarantee that this approach will be successful in protecting workers over the long term.

Muscular strength is a very complex function that can vary greatly depending upon the methods of assessment (Gallagher et al., 1998). As a result, there is often a great deal of confusion and misunderstanding with regard to the appropriate uses of strength testing in the context of ergonomics. It is not uncommon to see techniques misapplied by those unfamiliar with the caveats and limitations associated with various strength-testing procedures. The purposes of this chapter will be threefold: to provide the reader with a basic understanding of human strength, to characterize various methods of strength testing, and to describe ways that these techniques have been used in the attempt to control work-related musculoskeletal disorders (WMSDs).

21.2 Definition of Muscular Strength

Before describing the various strength-testing procedures available, one must first understand what is meant by the term *muscular strength*. For the purposes of this paper, muscular strength will be defined as *the capacity to produce a force or torque with a voluntary muscle contraction* (Gallagher et al., 1998). It is important to note that the strength or force output measured is that which the subject is willing to produce, and is probably somewhat lower than what the muscle is capable of producing in absolute terms (Chaffin and Andersson, 1991). It has been estimated that the maximal voluntary strength a subject is willing to put forth may be as much as 30% lower than the physiological tolerance of the muscle–tendon–bone system (Hettinger, 1961).

21.3 Measurement of Human Strength

We do not currently have the ability to directly measure the force or tension developed within the muscle of a living person (Kroemer et al., 1994). If this were possible, it might greatly simplify the analysis of worker strength. Lacking this ability, we must use indirect measurement techniques in which we measure (externally) the forces or torques generated at some interface between the person and a measurement device. This is important to realize because there are a multitude of ways that such an interface can be constructed, each of which can (and will) influence the resulting strength measure.

Consider the isometric elbow flexion measurement depicted in Figure 21.1 (Gallagher et al., 1998). Were we able to measure the muscle force directly, we would find that the muscle was developing a force of 1,000 Newtons (N). Being unable to do so, we must measure the forces external to the body using a force cuff. But where should we place the force cuff, close to the elbow joint or near the wrist? As will be demonstrated, the force reading will be dramatically affected depending on where the cuff is placed.

In this figure, the tension developed by the muscle acts through a lever arm of distance a. In so doing, it creates a torque about the elbow joint equal to $F_m \times a$. Assuming that the exertion is static (nothing moves), measured forces (on the gauge) will equal the elbow flexor torque divided by the distance that the gauge's associated force cuff is from the elbow joint. That is,

$$Q = (F_m \times a)/b \tag{21.1}$$

or

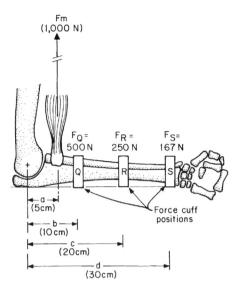

FIGURE 21.1   Given a constant muscle force (F_m), forces measured at various distances from the elbow will result in different force readings (F_Q, F_R, or F_S).

$$R = (F_m \times a)/c \qquad\qquad (21.2)$$

or

$$S = (F_m \times a)/d \qquad\qquad (21.3)$$

As we move the interface (a force cuff) from the elbow to the hand, the measured force will decrease. So, what can we say is the maximal force that can be generated in elbow flexion?

The answer is that it depends on how and where the forces are being measured.

This example highlights several important points. One central idea that should be understood is that "muscular strength is what is measured by an instrument" (Kroemer et al., 1994). One can see from the example given above that it would be entirely possible to have a case where two groups of subjects have (in actuality) identical muscle strength, but where differences in measurement techniques indicate wildly different strength capabilities. People using strength data must understand in detail how the measurements were done. Thus, a record of a person's strength describes what the instrumentation measured when the person voluntarily produced a muscle contraction in a specific set of circumstances with a specific interface and instrumentation (Gallagher et al., 1998).

21.4   Types of Muscular Strength

Muscular exertions can be divided into those which produce motion about a joint (*dynamic* exertions), and those which do not (*isometric* or *static* exertions). The vast majority of occupational tasks involve dynamic exertions. Unfortunately, the complexity of dynamic tasks (where one has to deal with factors such as velocity and acceleration) makes quantification of this type of strength more difficult (Chaffin and Andersson, 1991; Kroemer et al., 1994). For example, there may be great variability in speed of contraction with different people performing a given task. This, in turn, has a large bearing on the forces that can be produced by the muscles (Åstrand and Rodahl, 1977). Static exertions, on the other hand, are easier to quantify, but do not accurately represent muscle forces where the activity is very dynamic in nature (Kroemer et al., 1994). Neither of these types of strength testing is inherently better than the other — the important thing is to make sure that the test that is used is appropriate for the application being studied.

Isometric Strength

Tests of isometric strength involve application of a force against a stationary load-measuring device (Chaffin, 1996). Because of the relative simplicity of isometric strength tests, standardized procedures have been developed (Caldwell et al., 1974). The recommended protocol describes several control measures to standardize the execution and reporting of tests of static strength. For example, the recommended exertion duration is four to six seconds, with 30 seconds' to two minutes' rest provided between tests. Instructions are to be carefully stated to inform subjects of potential risk and use of the test results, and to prevent coercion or undue incentives to the subject during the exertions. The recommendations also detail methods of standardizing test postures, body supports, and restraint systems, as well as the control of environmental factors (temperature, humidity, noise, spectators, etc.). These procedures have been widely accepted, and have helped unify the techniques used to test isometric muscle strength by researchers around the world.

Dynamic Strength

In contrast to isometric strength testing, a number of different techniques exist to examine dynamic strength capabilities. One type of dynamic strength assessment is that involving measurement of *isoinertial* strength. Kroemer (1983) and Kroemer et al. (1990) define the isoinertial technique of strength assessment as one in which *mass properties of an object are held constant,* as in lifting a given weight over a predetermined distance. Several strength assessment procedures possess the attribute in this definition. Most commonly associated with the term is a specific test developed to provide a relatively quick assessment of a subject's maximal lifting capacity using a modified weight-lifting device. Another is a technique where the subject is asked to provide an estimate of an acceptable (submaximal) load, under set conditions (frequency and duration of lift, a specified lifting task, etc.). This technique is called the *psychophysical* methodology (Snook, 1978). Both will be discussed in greater detail later in the chapter.

Dynamic strength can also be evaluated using tests of *isokinetic* strength (Hislop and Perrine, 1969). This procedure evaluates dynamic strength *throughout a range of motion and at a constant velocity.* Such an exercise allows the muscle to contract at its maximum capability at all points throughout the range of motion. At the extremes of the range of motion of a joint, the muscle has the least mechanical advantage, and the resistance offered by the machine is correspondingly lower. Similarly, as the muscle reaches its optimal mechanical advantage, the resistance of the machine increases proportionally.

These are the most common tests of dynamic strength used in ergonomics; however, others are available. For example, there are devices that can measure force exerted during a constant acceleration exertion, those measuring strengths in an eccentric (muscle lengthening) mode, and several others (Kroemer et al., 1990). However, most of these have been used primarily for research purposes and not for worker strength evaluations. These devices are beyond the scope of the present chapter.

21.5   Factors Affecting Muscular Strength

Before discussing the use of physical strength assessment in job design and worker selection, it is important for the reader to understand some of the factors that can influence muscular strength. These may include personal factors, variables related to the task, or environmental factors (Ayoub and Mital, 1989). The following sections describe some of the major factors known to have a significant influence on strength test performance.

Personal Factors
Gender

There is a distinct difference between males and females in terms of muscular strength. On the average, the muscle strength of women is about two thirds that of men; however, the difference is variable according to which muscle group is examined. For example, for certain muscle groups women may have only 35% (on the average) the strength of the same muscle group in men. For other muscle groups, the difference

TABLE 21.1 Psychological Factors Affecting Maximal
Muscular Strength and Their Likely Effects

Factor	Likely effect
Feedback of results	Positive
Instructions on how to exert strength	Positive
Arousal of ego involvement, aspiration	Positive
Pharmaceutical agents	Positive
Startling noise, subject's outcry	Positive
Hypnosis	Positive
Setting of goals, incentives	Positive or negative
Competition, contest	Positive or negative
Verbal encouragement	Positive or negative
Spectators	?
Deception by researcher	?
Fear of injury	Negative
Deception by subject	Negative

From Kroemer, K.H.E. and Marras, W.S., 1981. Evaluation
of maximal and submaximal static muscle exertions. *Human
Factors*, 25: 643-653. With permission.

between genders may be as little as 15% (Chaffin and Andersson, 1991). Women tend to perform relatively better when the task involves lower extremity muscle groups, and relatively poorer when the exertion requires a great deal of upper body strength. Some of the difference in strength between men and women can be accounted for by the difference in body size between the genders. However, it seems that even when one accounts for the difference in size, there remains a 20% difference in strength between males and females (Åstrand and Rodahl, 1977).

Age

Muscle strength generally reaches a peak in an individual's late 20s or early 30s, and begins a gradual decline thereafter. In general, the strength of the 40-year-old is approximately 5% less than that achieved at its peak. By the time the individual is 65, strength is 20% below its peak. However, it should be duly noted that a regimen of strength training can significantly influence the rate of decline (Åstrand and Rodahl, 1977).

Anthropometry

Anthropometry is the study of the physical dimensions and composition of the human body (Stramler, 1993). Certain anthropometric measures appear to be related to the amount of strength of which a subject is capable. The measures that most highly relate to strength are lean body mass (body mass corrected for fat), and limb cross-sectional data (obtained from measurements of circumference) (Chaffin and Andersson, 1991). Stature (a subject's height) does not appear to be highly related to strength.

Motivation

Many psychological factors, especially subject motivation, can have a marked influence on measured strength. These factors can have both positive and negative effects. Table 21.1 illustrates certain factors that may influence subject motivation, and the expected effect (Kroemer and Marras, 1981).

Task Influences

Posture

The posture adopted by the body can have a major impact on the expression of human strength. For example, the angle of a joint during an exertion can profoundly affect measured muscle strength. As illustrated in Figure 21.2, elbow flexion strength is highest when the joint is at 90 degrees. As the joint deviates from that angle, less force can be developed. Whole-body posture has also been shown to have a large effect on strength. For example, lifting strength is much lower when a subject is kneeling or seated as compared with standing (Gallagher et al., 1988; Yates and Karwowski, 1987).

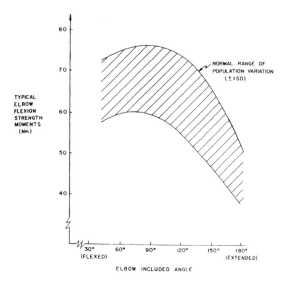

FIGURE 21.2   Change in elbow flexion strength as a result of changes in the angle of the elbow. (From *Occupational Biomechanics,* Chaffin, D.B. and G.B.J. Andersson, © 1991 by John Wiley & Sons, Inc. Reprinted by permission of John Wiley & Sons, Inc.)

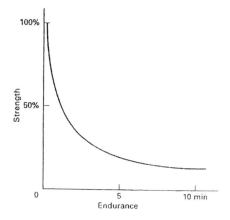

FIGURE 21.3   Endurance time as a function of required strength. (From Rohmert, W., 1966. *Maximal Forces of Men Within the Reach Envelope of the Arms and Legs* (in German), Research Report No. 1616, State of Northrhine — Westfalia, Westdeuscher Verlag Koeln-Opladen. With permission.)

Duration of Exertion

The amount of force that can be sustained during an exertion depends on the length of time of that exertion. Figure 21.3 illustrates this point. An exertion requiring 100% of a subject's maximal voluntary strength can only be sustained for a short period of time. However, as strength requirements are reduced, the exertion can be maintained for longer periods (Rohmert, 1966).

Velocity of Contraction

In work activities, muscular forces are usually applied through a range of motion and may be performed at a wide range of velocity of movement. As illustrated in Figure 21.4, the peak force (or torque) generated by a muscle decreases with increasing velocity of movement. In other words, higher forces can be produced at slow movement speeds. Muscles cannot generate as much force during high-velocity movements (Fox and Mathews, 1981).

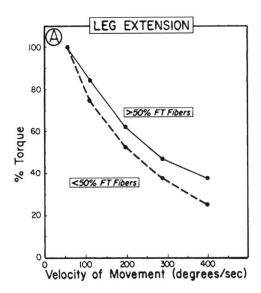

FIGURE 21.4 An example of the force–velocity relationship of muscle: forces generated by a muscle decrease with increasing velocity of movement. (From *The Physiological Basis of Physical Education and Athletics,* Fox, E.L. and D.K. Mathews, © 1981 By CBS College Publishing. With permission.)

Muscle Fatigue

Muscles that are highly stressed become fatigued, which correspondingly reduces the amount of strength that can be produced. Muscular fatigue appears to be dependent on the blood flow through the muscle. When a muscle is tightly contracted, blood flow is impeded, and the delivery of oxygen to the muscle is thereby reduced. Very low levels of exertion (less that 15% of maximal contraction) can be performed for long periods of time without excessive muscular fatigue (Åstrand and Rodahl, 1977).

Environmental Influences on Strength

Temperature and Humidity

Changes in temperature and humidity can affect strength capabilities, particularly at higher levels. Snook and Ciriello (1974) reported that an increase in Wet Bulb Globe Temperature (an index of heat stress) from 20 degrees C to 27 degrees C resulted in a 20% decrease in lifting capacity, a 16% reduction in pushing strength and an 11% reduction in carrying capacity. The effect of cold environments of strength capacity has not been well studied.

21.6 Purposes of Strength Measurement in Ergonomics

There are a number of reasons people may want to collect human strength data. Among the most common is collecting population strength data which can be used to build an anthropometric database; create design data for products, tasks, equipment, etc.; and for basic research into the strength phenomenon. This chapter will discuss two common uses of physical strength assessment in ergonomics: job design and worker selection and placement.

Strength Assessment for Job Design

Perhaps the most effective use of worker strength evaluations is in the area of job design (Snook, 1978). Job design has been a primary focus of the psychophysical method of determining acceptable weights and forces. In this technique, subjects are typically asked to adjust the weight or force associated with a

TABLE 21.2 Excerpt from the Liberty Mutual Tables for Maximum Acceptable Weight of Lift (kg) for Males and Females

Gender	Box Width (cm)	Distance of Lift (cm)	Percent Capable	Floor Level to Knuckle Height One Lift Every							
				5 sec	9 sec	14 sec	1 min	2 min	5 min	30 min	8 hr
Males	49	51	90	7	9	10	14	16	17	18	20
			75	10	13	15	20	23	25	25	30
			50	*14*	*17*	20	27	30	33	34	40
			25	*18*	*21*	25	34	38	42	43	50
			10	*21*	*25*	29	40	45	49	50	59
Females	49	51	90	*6*	7	8	9	10	10	11	15
			75	*7*	9	9	11	12	12	14	18
			50	*9*	10	11	13	15	15	16	22
			25	*10*	*12*	13	16	17	17	19	26
			10	*11*	*14*	15	18	19	20	22	30

Italicized values exceed 8-hour physiological criteria (energy expenditure).

Source: Snook, S.H. and Ciriello, V.M., 1991. The design of manual handling tasks: revised tables of maximum acceptable weights and forces. *Ergonomics* 34(9):1197-1213. With permission.

task in accordance with their own perception of what is an *acceptable workload* under specified test conditions (Snook, 1985). It can be seen from this description that this technique does not attempt to evaluate the maximum forces a subject is capable of producing. Instead, it evaluates a type of "submaximal," endurance-based estimate of acceptable weights or forces.

In the context of lifting tasks, the following procedure is usually used in psychophysical strength assessments. The subject is given control of one variable, typically the amount of weight contained in a lifting box. There will usually be two 20-minute periods of lifting for each specified task: one starting with a light box (to which the subject will add weight), the other starting with a heavy box (from which the subject will extract weight). The box will have a hidden compartment containing an unknown (to the subject) amount of weight, varied before each test, to prevent visual cues to the subject regarding how much weight is being lifted. The amount of weight selected during these two sessions is averaged and is taken as the maximum acceptable weight of lift for the specified conditions. In psychophysical assessments, the subject is instructed to work consistently according to the concept of "a fair day's pay for a fair day's work": working as hard as he or she can without straining or becoming unusually tired, weakened, overheated, or out of breath (Snook and Ciriello, 1974).

As psychophysical strength data are collected on large numbers of subjects, it becomes possible to design jobs so that they are well within the strength capabilities of the vast majority of workers. One criterion that is often used is to design the job so that 75% of workers rate the load as acceptable (Snook et al., 1978). Studies have indicated that if workers lift more than this amount, they may be three times more likely to experience a low back injury. On the other hand, designing jobs in accordance with this criterion has the potential to reduce the occurrence of low back injuries by up to 33% (Snook et al., 1978).

Several authors have published comprehensive tables on loads deemed acceptable by workers over a wide variety of industrial tasks. Of these, the most comprehensive are those developed by Snook and colleagues (Snook et al., 1978; Snook and Ciriello, 1991), which detail maximum acceptable loads for lifting, lowering, pushing, pulling, and carrying for both male and female workers. Table 21.2 presents an excerpt from one such tabulation, dealing with acceptable weights of lift for male and female workers for a floor level to knuckle height lift (Snook and Ciriello, 1991). If one wanted to design a lifting task requiring 4.1 lifts/min (one lift every 14 seconds) so that it was acceptable to 75% of females and 90% of males for the given box dimensions, one could go to these tables and determine that the acceptable weights of lift are 9 kg and 10 kg, respectively. Such a job should be designed so that the weight lifted is no more than 9 kg, or approximately 20 pounds.

As with all strength evaluation techniques, there are both advantages and disadvantages to the psychophysical approach. The major advantages of the psychophysical approach include the fact that it allows a realistic simulation of industrial tasks, and is capable of simulating intermittent as well as continuous types of lifting tasks. Furthermore, psychophysical results appear to be very reproducible, and are related to the occurrence of low back pain (Snook, 1985). However, disadvantages can also be identified. Perhaps the most important of these is that psychophysical results sometimes exceed what experts feel are safe according to other criteria, such as biomechanical stress or physiological cost (Snook, 1985). Furthermore, the technique is subjective. That is to say, it relies on self-reporting by the subject. Given the large number of overexertion injuries that occur every year, it is not clear that workers can always tell how much weight is safe for them to lift.

Despite the disadvantages that may be present with this technique, it is the only strength evaluation procedure that focuses on using the acquired data to develop a permanent engineering (job design) solution to the control of low back pain. It would be expected that this approach would afford a greater level of protection to workers than the worker selection techniques that follow.

Worker Selection and Placement

The purpose of worker selection and placement programs is to ensure that jobs which involve heavy physical demands are not performed by those lacking the necessary strength capabilities (Chaffin, 1996). It should be noted that this method is not the preferred strategy of the ergonomist, and there continues to be some controversy regarding the effectiveness of the approach, but there is some support for worker selection procedures in the literature (Chaffin et al., 1978; Keyserling et al., 1980). However, it is clear that specific conditions need to be met for this procedure to have a chance of success.

The process of strength evaluation for worker selection should be approached with caution on many fronts. In the first place, issues of unfair discrimination may be raised if appropriate testing procedures are not used (Chaffin and Andersson, 1991). If strength is to be used as a screening criterion, it is critical that the strength test employed is directly related to specific work requirements. Accurate representations of working postures are also important, as strength in one posture cannot accurately predict strength in another posture. Chaffin and Andersson (1991) suggested that the following criteria be used for methods of worker selection:

1. Is it safe to administer?
2. Does it give reliable, quantitative values?
3. Is it related to specific job requirements?
4. Is it practical?
5. Does it predict risk of future injury or illness?

Prior to testing, a history should be taken to ensure the worker does not have any cardiovascular or musculoskeletal problems that would increase the risk of taking the test.

Research has shown that worker selection cannot consist of general tests of strength (Battié et al., 1989; Troup et al., 1981; Mostardi et al., 1992). Such tests do not appear to be helpful in identifying those at risk of overexertion injury (strong workers seem to experience injury rates similar to those less strong). Instead, worker strength measures must be tied to a biomechanical analysis of workplace demands in order to predict those having increased risk of injury (Chaffin et al., 1978; Keyserling et al., 1980). There are two key principles that must be considered regarding the use of strength assessment for purposes of worker selection. These principles deal with the job relatedness of the test employed, and use of strength tests only under conditions where they have shown the ability to identify workers at high risk of injury.

The literature has shown that worker selection is only effective when a worker's strength capacity is equated with the demands of the job. All too often, emphasis is placed on collecting data on the former attribute, while the latter receives little or no attention (Chaffin, 1996). As will be illustrated, strength data in the absence of information regarding job demands is insufficient for purposes of worker selection. Consider the following scenario: an employer has an opening for a physically demanding job and wishes to hire an individual with strength sufficient for the task. This employer decides to base his employment decision on a strength test given to a group of applicants. Naturally, the employer selects the applicant with the highest strength score to perform the job. The employer may have hired the strongest job applicant; however, what this employer must understand is that he may not have decreased the risk of injury to his employee if, for example, the demands of the job still exceed this individual's capacity. This illustration should make it clear that only through knowing about the person's capabilities *and* the job demands might worker selection protect workers from WMSDs.

A second issue that must be considered when worker selection is to be implemented is that of the test's predictive value. The predictive value of a test is a measure of its ability to determine who is at risk of future WMSD. In the case of job-related strength testing, the predictive value appears to hold only when testing individuals for jobs where high risk is known (Chaffin, 1996). Strength testing does not appear to predict the risk of injury or disease to an individual when job demands are low or moderate. Furthermore, as noted previously, the effectiveness of worker selection techniques has not been demonstrated in long-term studies, only in relatively short-term investigations. It is unclear whether such tests will predict workers at risk of injury over the long term.

Finally, it should be noted that muscular strength is only one factor in a complicated and poorly understood mechanism of injury. A host of other tissues (such as the tendons, ligaments, and joint surfaces) may be deformed or injured by the stresses they experience, whether or not the muscles are able to develop sufficient strength for the job. Thus, it can be said that adequate muscular strength is necessary for safe performance of physical work but is not in itself sufficient for protection against injury.

21.7   Isometric Analysis

When a worker is called upon to perform a physically demanding lifting task, moments (or torques) are produced about various joints of the body by the external load (Chaffin and Andersson, 1991). Often these moments are augmented by the force of gravity acting on the mass of various body segments. For example, in a biceps curl exercise, the moment produced by the forearm flexors must counteract the moment of the weight held in the hands, as well as the moment caused by gravity acting on the center of mass of the forearm. In order to successfully perform the task, the muscles responsible for moving the joint must develop a greater moment than that imposed by the combined moment of the external load and body segment. It should be clear that for each joint of the body, there exists a limit to the strength that can be produced by the muscle to move ever-increasing external loads. This concept has formed the basis of isometric muscle strength prediction modeling (Chaffin and Andersson, 1991).

The following procedures are generally used in this biomechanical analysis technique. First, workers are observed (and usually photographed or videotaped) during the performance of physically demanding tasks. For each task, the posture of the torso and the extremities is documented at the time of peak exertion. The postures are then recreated using a computerized software package, which calculates the load moments produced at various joints of the body during the performance of the task (Chaffin and Andersson, 1991). The values obtained during this analysis are then compared to population norms for isometric strength obtained from a population of industrial workers. In this manner, the model can estimate the proportion of the population capable of performing the exertion, as well as the predicted compression forces acting on the lumbar discs resulting from the task.

Figure 21.5 shows an example of the workplace analysis necessary for this type of approach. Direct observations of the worker performing the task provide the necessary data. For example, the load magnitude and direction must be known (in this case a 200 N load acting downward), the size of the worker, the postural angles of the body (obtained from photographs or videotape), and whether the task requires one or two hands. Furthermore, the analysis requires accurate measurement of the load center relative to the ankles and the low back. A computer analysis program can be used to calculate the strength requirements for the task, and the percentage of workers who would be likely to have sufficient strength capabilities to perform it. Results of this particular analysis indicate that the muscles at the hip are most

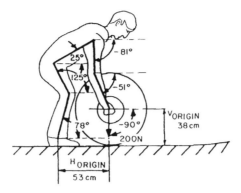

FIGURE 21.5   Postural data required for analysis of joint moment strengths using the isometric technique. (From *Occupational Biomechanics,* Chaffin, D.B. and G.B.J. Andersson, © 1991 by John Wiley & Sons, Inc. Reprinted by permission of John Wiley & Sons, Inc.)

stressed, with 83% of men having the necessary capabilities but only slightly more than half of women would have the necessary strength in this region. These results can then be used as the basis for determining those workers who have adequate strength for the job. However, such results can also serve as ammunition for recommending changes in job design (Chaffin and Andersson, 1991).

21.8   Isoinertial Testing

The Strength Aptitude Test

The Strength Aptitude Test (SAT) is a classification tool for matching the physical strength abilities of individuals with the physical requirements of jobs in the Air Force (McDaniel et al., 1983). The SAT is given to all Air Force recruits as part of their preinduction examinations. Results of the SAT are used to determine whether an individual has the minimum strength criterion which is a prerequisite for admission to various Air Force Specialties (AFSs). The physical demands of each AFS are objectively computed from an average physical demand weighted by the frequency of performance and the percent of the AFS members performing the task. Objects weighing less than 10 pounds are not considered physically demanding and are not considered in the job analysis. Prior to averaging the physical demands of the AFS, the actual weights of objects handled are converted into equivalent performance on the incremental weight lift test using statistical procedures developed over years of testing. These relationships consider the type of task (lifting, carrying, pushing, etc.), the size and weight of the object handled, and the type and height of the lift. Thus, the physical job demands are related to, but are not identical to, the ability to lift an object to a certain height. Job demands for various AFSs are reanalyzed periodically for purposes of updating the SAT (McDaniel, 1994).

In this technique, a preselected mass, constant in each test, is lifted by the subject (typically from knee height to knuckle height, elbow height, or to overhead reach height). The amount of weight to be lifted is relatively light at first, but the amount of mass is continually increased in succeeding tests until it reaches the maximal amount that the subject voluntarily indicates he/she can handle. Figure 21.6 shows an example of an isoinertial strength testing device. At the time of this writing, over 2 million Air Force personnel have been tested using this procedure.

A unique aspect of this technique is that it is the only strength measurement procedure discussed in this document where results are based on the success or failure to perform a prescribed criterion task (Kroemer, 1983). The criterion tasks studied have typically included lifting to shoulder height, elbow height, or knuckle height.

FIGURE 21.6   An isoinertial weight-lifting device. (From Kroemer, K.H.E., 1983. An isoinertial technique to assess individual lifting capacity. *Human Factors,* 25: 493–506. With permission.)

When developing the SAT, the Air Force examined more than 60 candidate tests in an extensive, four-year research program and found the incremental weight lift to 1.83 m to be the single best test of overall dynamic strength capability, which was both safe and reliable (McDaniel, 1994). This finding was confirmed by an independent study funded by the U.S. Army (Myers et al., 1984). This study compared the SAT to a battery of tests developed by the Army (including isometric and dynamic tests) and compared these with representative heavy demand tasks performed within the Army. Results showed the SAT to be superior to all others in predicting performance on the criterion tasks.

The Progressive Inertial Lifting Evaluation (PILE)

Another variety of isoinertial strength test is the Progressive Isoinertial Lifting Evaluation (PILE) (Mayer et al., 1988a, b). Instead of using a weight rack as shown in Figure 21.6, the Progressive Isoinertial Lifting Evaluation (PILE) is performed using a lifting box with handles and increasing weight in the box as it is lifted and lowered. Subjects perform two isoinertial lifting/lowering tests: one from floor to 30" (LUMBAR) and one from 30" to 54" (CERVICAL). Unlike the isoinertial procedures described above, there are three possible criteria for termination of the test: (1) voluntary termination due to fatigue, excessive discomfort, or inability to complete the specified lifting task; (2) achievement of a target heart rate (usually 85% of age predicted maximal heart rate); or (3) when the subject lifts a "safe limit" of 55 to 60% of his/her body weight. Thus, contrary to the tests described above, the PILE test may be terminated due to cardiovascular factors, rather than when an acceptable load limit is reached.

Since the PILE was developed as a means of evaluating the degree of restoration of functional capacity of individuals complaining of chronic low back pain (LBP), the initial weight lifted by subjects using this procedure is somewhat lower than the tests described above. The initial starting weight is 3.6 kg for women and 5.9 kg for men. Weight is incremented upward at a rate of 2.3 kg every 20 seconds for women, and 4.6 kg every 20 seconds for men. During each 20-second period, four lifting movements (box lift or box lower) are performed. The lifting sequence is repeated until one of the three endpoints is reached. The vast majority of subjects are stopped by the "psychophysical" endpoint, indicating the subject has a perception of fatigue or overexertion. The target heart rate endpoint is typically reached in older or large individuals. The "safe limit" endpoint is typically encountered only by very thin or small individuals.

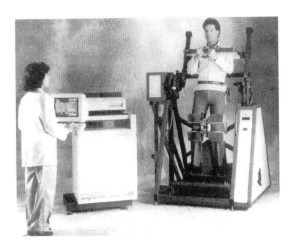

FIGURE 21.7   An isokinetic trunk flexion and extension device used to evaluate lumbar muscle strength. (Cybex Medical, Division of Henley Healthcare, Sugarland, TX.) (Photo courtesy of Henley Healthcare.)

Mayer et al. (1988b) developed a normative database for the PILE, consisting of 61 males and 31 females. Both total work (TW) and force in lbs. (F) were normalized according to age, gender, and a body weight variable. The body weight variable, the adjusted weight (AW), was taken as actual body weight in slim individuals, but was taken as the ideal weight in overweight individuals. This was done to prevent skewing the normalization in overweight individuals.

21.9   Isokinetic Tests

A technique of dynamic testing that has been growing in popularity is that dealing with the measurement of isokinetic strength (Hislop and Perrine, 1969). As defined previously, this technique evaluates muscular strength throughout a range of motion and at a constant velocity. It is important to realize that people do not normally move at a constant velocity (Kroemer et al., 1990). Instead, human movement is usually associated with significant acceleration and deceleration of body segments. Thus, there is a perceptible difference between isokinetic strength and free dynamic lifting. In the latter instance, subjects may use rapid acceleration to gain a weight lifting advantage, as in the Strength Aptitude Test described above. Acceleration is not permitted in isokinetic tests of strength.

The majority of isokinetic devices available on the market focus on quantifying strength about isolated joints or body segments, for example, trunk extension and flexion (see Figure 21.7). This may be useful for rehabilitation or clinical use, but isolated joint testing is generally not appropriate for evaluating an individual's ability to perform occupational lifting tasks. One should not make the mistake of assuming, for instance, that isolated trunk extension strength is representative of an individual's ability to perform a lift. In fact, lifting strength for a task may be almost entirely unrelated to trunk muscle strength (Himmelstein and Andersson, 1988). Strength of the arms or legs (and not the trunk) may be the limiting factor in an individual's lifting strength. For this reason, machines that measure isokinetic strengths of isolated joints or body segments should not be used as a method of evaluating worker capabilities related to job demands in most instances.

Many investigators have used dynamic isokinetic lifting devices specifically designed to measure whole-body lifting strength (Pytel and Kamon, 1981; Kishino et al., 1985) (see Figure 21.8). These devices typically have a handle connected by a rope to a winch which rotates at a specified isokinetic velocity when the handle is pulled. Studies using this type of device have demonstrated good correlations between isokinetic Dynamic Lift Strength (i.e., a lift from floor to chest height) and the maximum weights individuals were willing to lift for infrequent tasks using the psychophysical approach (Pytel and Kamon, 1981). Thus, under certain circumstances, this device appears to have some validity for assessment of job-related dynamic lifting strength capabilities of individuals. Some investigators have attempted to

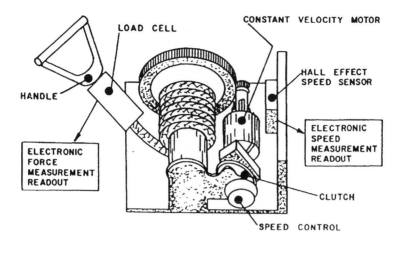

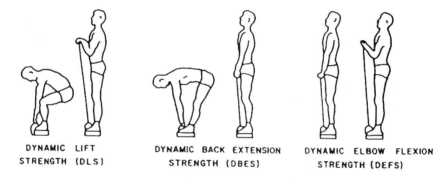

FIGURE 21.8   An isokinetic device used to evaluate whole-body lifting strengths. (From Pytel, J.L. and Kamon, E. Dynamic strength test as a predictor for maximal and acceptable lift, *Ergonomics*, 24: 663–672. With permission.)

modify this type of instrument by providing a means to mount it so that isokinetic strength can be measured in vertical, horizontal, and transverse planes (Mital and Vingararamoothy, 1984). However, while advances have been made in the use of isokinetic devices for worker strength evaluation, this procedure cannot be considered fully developed in the context of worker selection procedures.

21.10   Conclusions

In spite of advances in measurement techniques and an explosive increase in the volume of research, our understanding of human strength remains in its preliminary stages. It is clear that muscle strength is a highly complex and variable function dependent on a large number of factors. It is not surprising, therefore, that there are not only substantial differences in strength between individuals, or that strength measurements for a single individual can vary a great deal even during the course of a single day. Strength is not a fixed attribute — strength training regimens can increase an individual's capability by 30 to 40% or more. Disuse can lead to muscle atrophy (Åstrand and Rodahl, 1977).

The use of physical strength assessment in ergonomics has focused on both job design and worker selection techniques. Of these, the former has a much greater potential to significantly reduce WMSDs. Worker selection techniques must be considered a method of last resort — where engineering changes or administrative controls cannot be used to reduce worker exposure to WMSD risk factors. This technique has only shown a moderate effect in truly high risk environments, and only in short-term studies. It is not known whether worker selection procedures have a protective effect over the long term.

References

Åstrand, P.O. and Rodahl, K., 1977. *Textbook of Work Physiology*, McGraw-Hill Book Company, New York, 681 pp.

Ayoub, M.M. and Mital, A., 1989. *Manual Materials Handling*. London: Taylor & Francis, 324 pp.

Battié, M.C., Bigos, S.J., Fisher, L.D., Hansson, T.H., Jones, M.E., and Wortley, M.D., 1989. Isometric lifting strength as a predictor of low back pain. *Spine* 14:851-856.

Caldwell, L.S., Chaffin, D.B., Dukes-Dobos, F.N., Kroemer, K.H.E., Laubach, L.L., Snook, S.H., et al., 1974. A proposed standard procedure for static muscle strength testing. *American Industrial Hygiene Association Journal* 35:201-206.

Chaffin, D.B., 1996. Ergonomic basis for job-related strength testing, in *Disability Evaluation* (1st edition, Demeter, S.L., Andersson, G.B.J., Smith, G.M. eds.), Mosby, St. Louis, MO, Chapter 17, 159-167.

Chaffin, D.B. and Andersson, G.B.J., 1991. *Occupational Biomechanics* (2nd ed.), New York: John Wiley & Sons, 518 pp.

Chaffin, D.B., Herrin, G.D., and Keyserling, W.M., 1978. Pre-employment strength testing: an updated position. *Journal of Occupational Medicine* 20(6):403-408.

Fox, E.L. and Mathews, E.L., 1981. *The Physiological Basis of Physical Education and Athletics* (3rd ed.), Philadelphia: Saunders College Publishing, 677 pp.

Gallagher, S., Marras, W.S. and Bobick, T.G., 1988. Lifting in stooped and kneeling postures: Effects on lifting capacity, metabolic costs, and electromyography of eight trunk muscles. *Int. J. Ind. Erg.* 3:65-76.

Gallagher, S., Moore, J.S., and Stobbe, T.J., 1998. *Physical Strength Assessment in Ergonomics*, American Industrial Hygiene Association Ergonomics Guide, AIHA Press, Fairfax, VA, 60 pp.

Hettinger, T., 1961. *Physiology of Strength*, Springfield, IL: Charles C Thomas.

Himmelstein, J.S. and Andersson, G.B.J., 1988. Low back pain: risk evaluation and preplacement screening. *Occupational Medicine: State of the Art Reviews*. 3(2):255-269.

Hislop, H. and Perrine, J.J., 1967. The isokinetic concept of exercise. *Physical Therapy* 47:114-117.

Keyserling, W.M., Herrin, G.D., and Chaffin, D.B., 1980. Isometric strength testing as a means of controlling medical incidents on strenuous jobs. *Journal of Occupational Medicine* 22(5):332-336.

Kishino, N.D., Mayer, T.G., Gatchel, R.J., Parish, M.M., Anderson, C., Gustin, L., and Mooney, V., 1985. Quantification of lumbar function: Part 4: Isometric and isokinetic lifting simulation in normal subjects and low-back dysfunction patients, *Spine 10* (10): 921-927.

Kroemer, K.H.E., 1983. An isoinertial technique to assess individual lifting capability. *Human Factors* 25(5): 493-506.

Kroemer, K.H.E. and Marras, W.S., 1981. Evaluation of maximal and submaximal static muscle exertions. *Human Factors*, 25: 643-653.

Kroemer, K.H.E., Marras, W.S., McGlothlin, J.D., McIntyre, D.R., and Nordin, M., 1990. On the measurement of human strength. *International Journal of Industrial Ergonomics*, 6: 199-210.

Kroemer, K.H.E., Kroemer, H.B., and Kroemer-Elbert, K.E., 1994. *Ergonomics: How to Design for Ease and Efficiency*. Englewood Cliffs, NJ: Prentice-Hall.

Mayer, T.G., Barnes, D., Kishino, N.D., Nichols, G., Gatchell, R.J., Mayer, H., and Mooney, V. (1988a). Progressive isoinertial lifting evaluation — I. A standardized protocol and normative database. *Spine* 13(8): 993-997.

Mayer, T.G., Barnes, D., Nichols, G., Kishino, N.D., Coval, K., Piel, B., Hoshino, D., and Gatchell, R.J. (1988b): Progressive isoinertial lifting evaluation — II. A comparison with isokinetic lifting in a chronic low-back pain industrial population. *Spine* 13(8): 998-1002.

McDaniel, J.W., 1994. Personal communication.

McDaniel, J.W., Shandis, R.J., and Madole, S.W., 1983. *Weight Lifting Capabilities of Air Force Basic Trainees*. AFAMRL-TR-83-0001. Wright-Patterson AFBDH, Air Force Aerospace Medical Research Laboratory.

Mostardi, R.A., Noe, D.A., Kovacik, M.W., and Porterfield, J.A., 1992. Isokinetic lifting strength and occupational injury: A prospective study. *Spine* 17(2): 189-193.

Myers, D.O., Gebhardt, D.L., Crump, C.E., and Fleishman, E.A., 1984. *Validation of the Military Entrance Physical Strength Capacity Test (MEPSCAT).* U.S. Army Research Institute Technical Report 610, NTIS No. AD-A142 169.

Pytel, J.L and Kamon, E., 1981. Dynamic strength test as a predictor for maximal and acceptable lift. *Ergonomics 24(9):663-672.*

Rohmert, W., 1966. *Maximal Forces of Men Within the Reach Envelope of the Arms and Legs* (in German), Research Report No. 1616, State of Northrhine — Westfalia, Westdeuscher Verlag Koeln-Opladen.

Snook, S.H., 1978. The design of manual handling tasks. *Ergonomics 21:963-985.*

Snook, S.H., 1985. Psychophysical considerations in permissible loads. *Ergonomics 28(1):327- 330.*

Snook, S.H. and Ciriello, V.M., 1974. Maximum weights and work loads acceptable to female workers. *J. Occup. Med.,* 16: 527-534.

Snook, S.H. and Ciriello, V.M., 1991. The design of manual handling tasks: revised tables of maximum acceptable weights and forces. *Ergonomics 34(9):1197-1213.*

Snook, S.H., Campanelli, R.A., and Hart, J.W., 1978. A study of three preventive approaches to low back injury. *J. Occup. Med 20(7):478-481.*

Stramler, Jr., J. H., 1993. *The Dictionary for Human Factors/Ergonomics.* Boca Raton, FL, CRC Press, 413 pp.

Troup, J.D.G., Martin, J.W., and Lloyd, D.C.E.F., 1981. Back pain in industry. A prospective Study. *Spine,* 6: 61-69.

Yates, J.W., Karwowski, W., 1987. Maximum acceptable lifting loads during seated and standing work. *Applied Ergonomics* 18: 239-243.

22

Dynamic Workplace Factors in Manual Lifting

William S. Marras
The Ohio State University

22.1 Introduction

One need only observe a worker perform an industrial materials handling task to appreciate the fact that dynamics or motion is an integral part of any lifting task. The significance of this dynamic component of the task has been recognized since the days of Isaac Newton, who demonstrated that force is a product of mass and acceleration ($F = m \times a$). Until recently, the field of ergonomics has not had many tools that were capable of assessing the effects of lifting dynamics. The significance of lifting dynamics has been suggested in much of the literature, however, the importance of considering motion has been under-appreciated.

Several studies indicate that risk of a low back disorder may increase significantly when motion occurs during a lift. Punnet et al. (1991) have shown that the vast majority of low back disorders in automobile assembly plants cannot be explained simply by the weight of the object lifted or the instantaneous posture in many high-risk jobs that are dynamic. Bigos and associates (1986) also suggested that those workers who performed dynamics lifting tasks were at three times the risk of low back disorder compared to those who were exposed to static awkward postures. Marras and colleagues (1993) identified the levels of exposure to dynamic trunk motions that were associated with high and low risk of occupationally related low back disorders. Thus, this recent evidence suggests that dynamics can clearly influence the ability to safely lift objects.

There are few ergonomic tools currently available to evaluate the dynamic aspects of lifting tasks. For the most part, dynamic analyses and controls have been developed along the lines of: (1) biomechanical loading principles, (2) psychophysical and dynamic strength assessments, and (3) kinematic evaluations based upon epidemiologic trends. This chapter will review the development of dynamic analytic techniques in these three areas as well as demonstrate their application.

22.2 Biomechanical Loading of the Spine During Dynamic Lifting

Techniques to assess and control occupational low back disorder exposure risk have evolved over the past three decades. Early attempts at controlling low back disorders (LBDs) at work were based on biomechanical logic. This logic assumes LBD risk can be assessed by comparing the load imposed upon the spine to the tolerance limits for the vertebral endplates. (See Chapter 10 for a description of the LBD injury process.) Spinal loading is a function of both *external* forces and *internal* forces acting on the body. External forces are the result of load mass characteristics and the body's system dynamics imposing loads on the spine. For example, the object mass being lifted (or the mass of the arms and trunk) is acted upon by the forces of gravity in order to load the spine. Internal forces refer to forces generated internal to the musculoskeletal system in response to these external forces. For example, the muscles and passive tissues (e.g., ligaments) respond to an external load by generating force within the musculoskeletal system that counteract the external load. Since the internal force-generating structures vary in their geometric orientation, it is important that their vector orientation is realistic in biomechanical modeling efforts. This is essential so that the nature and magnitude of spine loading (compression versus shear) during a task can be accurately assessed. Since the tolerance limits of the spine vary dramatically in compression versus shear versus torsion, accurate assessments of spine loading are crucial to an understanding of occupational risk. This review will focus on the ability of the various assessment tools to accurately evaluate internal forces and the subsequent spine loading which will define the subsequent risk of LBD.

The early biomechanical models of occupational low back loading simplified the analyses of the musculoskeletal system in two ways. First, these models assumed that the internal force-generating system could be simplified by representing the body as a cantilever system. This assumption permitted the internal force to be simplified by assuming the musculoskeletal system generates force using one equivalent muscle that best counteracts the external load. Under realistic conditions, the internal forces are generated by a complex activation of many muscles. This complex activation pattern creates vectors of loading on the spine that can be quite complex. Assumptions regarding a single equivalent internal force-generation system greatly simplify this situation.

The second major assumption in these early models was that the lifting activity could be represented by a static system. Early models assumed that the loads imposed upon the musculoskeletal system during lifting can be represented by observing the body at one instant in time while it was in a static (frozen) posture. These models were thought to be appropriate for situations where no appreciable acceleration was present during the lift. In other words, lifts were assumed to be slow and smooth.

One of the early attempts to quantitatively assess the loads on the spine as a function of an occupational lifting task was developed by Chaffin and Baker (1969). This model assumed that the body could be considered a cantilever system with a single equivalent muscle counterbalancing the externally applied moment. It also assumed the lift could be represented by a static model of the lifting situation. Chaffin and Baker developed a two-dimensional (coplanar) model that predicted the load imposed upon the spine as well as the moments at the major joints that would be required to balance the external load. When the moments were compared to the working population's static strength, a strength "benchmark" was available which would describe the percent of the population who would be expected to have adequate strength to perform the task. Later, this model was expanded to a three-dimensional assessment (Chaffin and Muzaffer, 1991). This model used the same logic as the two-dimensional coplanar model but permitted the body to be represented in static asymmetric postures. This model logic was also extended to include the influence of dynamic motion of the object lifted. Frievalds and associates (1984) were able to account for the effects of body segment dynamics and made predictions about the activity of single equivalent muscles in the musculoskeletal system in response to dynamic motion. In this model, the assumption that the lift could be represented by a static system was relaxed, and motion during lifting was allowed. However, the assumption that the body behaved as a cantilever system with one equivalent muscle supplying the internal force was maintained.

The internal force-generation system of the trunk has been recognized as a multiple muscle system, especially during motion (Winter and Woo, 1990). In order to accurately understand the nature of the loading on the spine during dynamic lifting, the coactivation pattern of the muscles and other internal force-generation mechanisms must be understood. Studies have demonstrated that the muscles that are active and the timing of the muscle activation change dramatically as dynamic movement conditions and asymmetry conditions of the body change (Marras and Reilly, 1988; Marras et al., 1986; Marras and Mirka, 1993). Hence, the significance of muscle coactivation upon spine loading predictions has been duly recognized in the literature.

Several attempts have been made to predict the activity of the multiple muscle system so that this information can be used in lifting models. Schultz and Andersson (1981) described a multiple muscle system consisting of the 10 major muscle groups that would be identifiable when passing a cutting plane through the lumbar level of the trunk. Using classical engineering techniques, they were able to describe the three forces and three moments acting upon the spine due to the external load being supported. In order to predict internal muscle activity, they tried two approaches. First, they simply made assumptions about which muscles would be active in a given task. This typically involved limiting the muscle activity to agonist muscles with the greatest mechanical advantage. Second, they employed linear optimization techniques to deal with the indeterminate nature of the problem (6 force and moment equations and 10 unknowns [muscles]). This attempt provided a solution to the problem without having to make simplifying assumptions, however, the solutions often did not match observations, especially under dynamic lifting conditions. Other optimization techniques have been attempted (e.g., Bean et al., 1988; Rathske, 1995), however, most suffer from the basic problem that linear optimization is incapable of providing more non-boundary solutions than the number of functional constraints. In most optimization models, the force and moment equations provide 5 functional constraints (plus an optimization function), and there are at least 10 unknowns (muscles and internal forces). Thus, the solution to the problem is destined to be indeterminate (Marras, 1988). In addition, optimization has only been attempted and has had limited success for static trunk loading conditions. Other techniques have also attempted to predict the activity of the multiple muscle system. Neural network techniques (Nussbaum et al., 1997) have been employed to predict the activity of the muscles under static conditions. This technique appears more valuable for classifying muscle responses as typical or nontypical. In addition, once the lifting situation changes, networks must be retrained for specific conditions. Stochastic models of muscle activities have also been produced (Mirka and Marras, 1993). These models indicate the probability that a certain combination of muscles would be performed given the dynamic characteristics of the lift. The limitation with this approach is that the lifting situation must be simulated many times for each given lifting task in order to accurately assess the loading characteristics (and probabilities) of the spine.

As demonstrated by these modeling efforts, it is a very difficult task to predict the activity of the trunk musculature during manual lifting tasks. Muscle activity must be determined in order to reasonably predict spine loading. Under static loading conditions, significant variability in muscle recruitment has been documented, and this variability becomes even greater when dynamic motion is present in the lift (Marras and Mirka, 1993). Specifically, large amounts of trunk muscle coactivity occur during dynamic activity that must be assessed in order to reasonably assess spinal loading characteristics.

This complexity in muscle recruitment has prompted several researchers to develop biologically assisted models of the trunk. In biologically assisted models, the person performing the task of interest is biologically monitored, and this recorded activity is used as model input in order to assist in the assessment of internal force development in the trunk. The most commonly used biological signal for these purposes is electromyography (EMG) of the trunk musculature. McGill and Norman (1985) were the first to develop an EMG-assisted model of the trunk. They developed an elegant anatomically detailed EMG-assisted model of bending motions containing 50 muscles and 12 ligaments. Other models (Thelan et al., 1994) have demonstrated that it is possible to assess simultaneously spinal loads in three dimensions under static loading conditions. The degree of complexity in these models also becomes very large and necessitates many simplifying assumptions about how many of the nonmonitored muscles behave.

An EMG-assisted model as been developed that is specifically designed to assess spine loading under dynamic manual lifting conditions. This model was specifically designed as a tool to evaluate manual lifting under actual lifting conditions. The goal in model development was to make the model only as

FIGURE 22.1   Geometric representation of the trunk used in the EMG-assisted model. The trunk is modeled as two plates that move dynamically throughout an exertion.

complex and detailed as needed to accurately reflect spinal loading. The model has been developed and validated so that it can accurately assess dynamic trunk loading under symmetric and asymmetric lifting conditions (Marras and Sommerich, 1991a, b; Granata and Marras, 1993; 1995), twisting force generation conditions (Marras and Granata, 1995), and lateral bending lifting conditions (Marras and Granata, 1997). The general structure of the model is shown in Figure 22.1. Geometrically, the model assumes that the trunk can be represented by two plates, one passed through the thorax, the other passed through the pelvis. Trunk muscle forces are represented by 10 vectors connecting the two plates. EMG electrodes monitor the activity of 10 major muscle groups in the trunk and each EMG signal is adjusted for muscle velocity, length, cross-sectional area, and force capacity before the muscle force is calculated. The contributions of each force vector is summed in each cardinal direction within the trunk to predict spine compression and shears as well as the moments imposed upon the spine. The moment calculation is used as a validation check for model fidelity. A person performing a materials handling task while standing on a force plate can provide an independent measure of spinal moments. If the measured moment imposed about the spine correlates well with the model predicted moment, then it is likely that the model is performing well. This model has been used extensively to evaluate dynamic materials handling tasks in industry. It has the advantage of being able to simultaneously evaluate many aspects of the lifts such as muscle activities, trunk kinematics, muscle coactivation, trunk muscle internal moment contributions, spinal loading, etc. Since the model resides in a Windows environment it makes it possible to evaluate multiple trunk parameters (e.g., muscle activities, kinematics, loading, etc.) while the lifting task is observed. Figure 22.2 shows an example of such an evaluation.

   One should also consider the limitation associated with biomechanical assessments of dynamic lifting situations. Accurate biomechanical analyses can provide a measure of loading imposed upon the spine and when compared with spine load tolerance data (Brinckman, 1990; Evans and Lissner, 1965; Sonoda, 1962) can offer an objective measure of the risk associated with a one-time performance of a dynamic lift. However, in order to assess the risk of repeated lifts as would occur in the workplace, the loading models must be performed over multiple observations of the lifting situation throughout the workday. In addition, the loading must be compared to either spine tolerances observed during repetitive loading of the spine (Brinckmann, 1990) or a finite element model of the spine must be employed to predict when the spine structures would fail under repetitive loading conditions. Such tolerance data are still under development, and therefore, accurate risk predictions based upon biomechanical analyses may be premature.

   Current investigations on spine tolerance limits have changed to explore more of the cumulative trauma aspects of spine loading. The current trend in tolerance research involves investigations relative to the

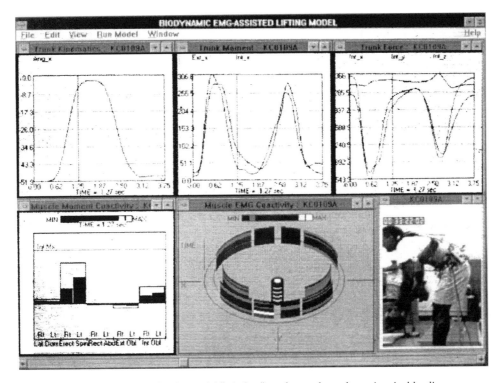

FIGURE 22.2   Example of a model "window" used to evaluate dynamic spinal loading.

biochemical changes that occur in spinal tissue as a function of age, spine load, and repetitive loading even at low level. This work is in stark contrast to the early investigations that explored strictly the mechanical limits of loading. However, most of this work is still in its early stages. In addition, such discussions are outside the scope of this chapter.

22.3   Psychophysical and Strength Assessments of Dynamic Lifts

Another approach used to assess dynamic lifting has been psychophysical analyses. In a psychophysical analysis, the weight of the object lifted is adjusted until the weight is judged as subjectively acceptable to the lifter for a given lifting situation. Using this logic, data sets have been collected and tabulated that describe the amount of weight that is acceptable to both males and females given lifting task variables. Task variables documented include the frequency of the lift, the duration of the lifting period, box size and dimensions, height of the lift, movement distance, number of people involved in the lift, the symmetric or asymmetric conditions of the lift, shape of the object lifted, load distribution (number of hands used for lifting), coupling conditions, load stability, and direction of applied force (pushing versus pulling). Typically, workplaces are designed according to a criteria that the workplace would be acceptable to 75% of the females in the population based upon these data.

An example of this type of data is shown in Table 22.1. These data were collected by Snook and Ciriello (1991) under very precisely controlled conditions. The effects of changing the task dynamics can be appreciated through an examination of this table. As shown here, for any given task conditions the weight judged acceptable decreases as the frequency of the lift increases. Greater frequency implies that workers are moving faster. Thus, these analyses indirectly account for dynamic workplace factors in manual lifting.

A related measure for the assessment of dynamic workplace factors is that of dynamic strength assessment. Dynamic strength has been evaluated for the whole body as well as for the back isolated from the rest of the body. In whole body dynamic strength testing, the dynamic characteristics of the object lifted are controlled or measured, and the force imposed on the object is the variable of interest. In

TABLE 22.1 An Example of Psychophysical Data Used to Evaluate the Acceptability of a Lifting Condition.

Width‡	Distance§	Percent†	Floor level to knuckle height One lift every								Knuckle height to shoulder height One lift every								Shoulder height to arm reach One lift every							
			5	9	14	1	2	5	30	8	5	9	14	1	2	5	30	8	5	9	14	1	2	5	30	8
			s	s		min	min	min		h	s	s		min	min	min		h	s	s		min	min	min		h
75	76	90	6	7	9	11	13	14	14	17	8	10	12	13	14	14	16	17	6	8	9	10	10	11	12	13
		75	9	11	13	16	19	20	21	24	10	14	16	18	18	19	21	23	8	10	12	14	14	14	16	17
		50	12	15	17	22	25	27	28	32	13	17	20	22	23	24	26	29	10	13	15	17	17	18	20	22
		25	15	18	21	28	31	34	35	41	16	21	24	27	27	28	32	35	11	16	18	21	21	22	24	27
		10	18	22	25	33	37	40	41	48	19	24	28	31	32	33	37	40	14	18	21	24	24	25	28	31
	51	90	6	8	9	12	13	15	15	17	8	11	13	15	15	16	18	19	6	8	9	12	12	12	14	15
		75	9	11	13	17	19	21	22	25	11	15	17	20	20	21	23	25	8	11	12	15	15	16	18	20
		50	13	15	18	23	26	28	29	34	14	19	21	25	25	26	29	32	10	14	16	19	20	20	23	25
		25	16	19	22	29	33	35	36	42	17	23	26	30	31	32	36	39	13	17	19	23	24	25	27	30
		10	19	22	26	34	38	42	43	50	20	26	30	35	36	37	41	45	15	19	22	27	27	29	32	35
	25	90	8	9	11	13	15	16	17	20	10	13	15	18	18	19	21	23	7	10	11	14	14	14	16	18
		75	11	13	15	19	22	24	24	28	13	17	20	23	24	25	27	30	10	13	15	18	18	19	21	23
		50	15	18	21	26	29	32	33	38	17	22	25	30	30	31	35	38	12	16	19	23	23	24	27	29
		25	18	22	26	33	37	40	41	48	20	27	30	36	36	38	42	46	15	20	22	28	28	29	32	35
		10	22	26	31	38	44	47	49	57	23	31	35	42	42	44	49	53	17	23	26	32	32	34	38	41
	76	90	7	8	10	13	15	16	17	20	8	10	12	13	14	14	16	17	7	9	10	12	12	13	14	16
		75	10	12	14	19	22	24	24	28	10	14	16	18	18	19	21	23	9	11	13	16	16	17	19	21
		50	14	16	19	26	29	32	33	38	13	17	20	22	23	24	26	29	11	15	17	20	21	21	24	26
		25	17	20	24	33	37	40	41	48	16	21	24	27	27	28	32	35	13	18	20	25	25	26	29	31
		10	20	24	28	38	43	47	48	57	19	24	28	31	32	33	37	40	15	21	23	28	29	30	33	36

Maximum acceptable weights and forces (data columns are frequency‑of‑lift conditions; column headings appear on the preceding page). Row labels are box width (‡) and vertical distance of lift (§); the two data groups per box width correspond to 51% and 25% of the industrial population (*).

Box width 49 cm — 51%

Vertical distance (cm)								
90	18	16	14	14	14	11	9	7
75	23	21	19	18	18	14	12	9
50	29	27	24	23	23	18	15	12
25	35	32	29	28	28	21	19	14
10	41	37	34	32	32	25	22	16

Box width 49 cm — 25%

Vertical distance (cm)								
90	20	18	17	18	19	14	11	8
75	30	25	25	26	28	18	15	11
50	40	34	33	35	38	23	20	14
25	50	43	42	44	48	27	24	17
10	59	50	49	52	57	32	28	20

Box width 76 cm — 51%

Vertical distance (cm)								
90	19	18	16	15	15	13	11	8
75	25	23	21	20	20	17	15	11
50	32	29	26	25	25	21	19	14
25	39	36	32	31	30	26	23	17
10	45	41	37	36	35	30	26	20

Box width 76 cm — 25%

Vertical distance (cm)								
90	23	20	19	20	22	16	13	10
75	33	29	28	30	32	21	17	14
50	45	38	37	40	43	27	22	18
25	56	48	47	50	54	32	28	21
10	67	57	56	59	64	37	33	25

Box width 34 cm — 51%

Vertical distance (cm)								
90	21	19	17	16	16	12	10	9
75	27	25	22	21	21	16	13	11
50	35	32	28	27	27	21	17	14
25	42	38	34	33	33	25	20	16
10	48	44	40	38	38	29	23	19

Box width 34 cm — 25%

Vertical distance (cm)								
90	24	22	21	23	26	18	15	11
75	34	31	30	33	34	23	20	14
50	46	40	39	44	43	29	25	18
25	57	50	49	55	52	35	30	21
10	68	59	58	66	60	41	35	25

‡Box width (the dimension away from he body) (cm).
§Vertical distance of lift (cm).
*Percentage of industrial population.
Italicized values exceed 8 h physiological criteria (see text).

Adapted from Snook, S.H. and Ciriello, V.M. (1991) The design of manual handling tasks: revised tables of maximum acceptable weights and forces, *Ergonomics*, 34(9), 1197–1213.

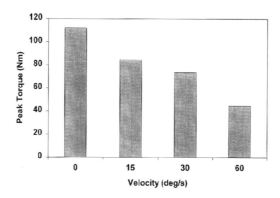

FIGURE 22.3   Trunk torque production as a function of increasing trunk velocity.

isolated back testing a dynamometer is aligned with a point of rotation on the back and the rotational motion characteristics are either controlled or documented while the torque-generation capacity of the trunk is monitored. The difference in the information derived between these two approaches is related to the specificity of the analysis desired. If one wants to assess dynamic lifting capacity without concern for the specific level of capacity of any given joint, then one might be interested in whole body strength. In this situation, strength is limited by whatever joint is weakest in the kinetic chain in the body. For example, if one is limited by shoulder strength in a lift, the strength of the shoulder will define the limit of the whole body lift. However, we will not have any idea about which joint limits the lift or how much capacity and relative level of protection is present in the other joints in the kinetic chain.

Whole body dynamic strength assessments have been assessed in several ways. Pytel and Kamon (1981) as well as Kamon et al. (1982) have used a dynamometer to document peak force capacity during dynamic whole body lifting. They limited the isokinetic speed of the object lifted to 0.73 m/s and observed the force that could be generated by a subject. In these studies, the peak force generated in the vertical direction was documented as both students and industrial workers lifted a handle from floor to chest height, from the floor to knuckle height, and when pulling a handle from knuckle height to chest height. These test positions were similar to some of the positions tested for psychophysical analyses. However, as expected, the force documented was much greater in these tests because the maximum force generation was of interest as compared to the acceptable weight of lift, as is documented in psychophysical studies. When compared to whole body isometric lifting strength, these dynamic forces were lower by about one third.

Isolated dynamic trunk strength testing has also been well documented in the literature. In this situation, the subject is placed in a dynamometer and the strength generated about the L5/S1 joint has been recorded through motions with different dynamic characteristics. Marras and associates (1987) have demonstrated the capacity to generate torque with the back changes as a function of the increasing velocity, increasing asymmetry, and as a function of forward trunk flexion angle. Figure 22.3 demonstrates how trunk strength changes with trunk angular velocity. This type of information can be compared to the strength required by a task (as assessed by dynamic models) to estimate the risk of trunk muscle overexertion given the motion characteristics of a dynamic lifting task. In this manner, the percent of the population that would exceed their trunk strength could be predicted and this, in turn, could provide a measure of back injury risk.

One needs to consider the advantages and disadvantages of psychophysical and dynamic strength information when evaluating and designing a manual lifting task. In particular, one must distinguish the objectives of evaluating whether one is capable of *performing* a task and whether one is at *risk of injury* when performing a materials handling task. The advantages of both psychophysical and dynamic strength data is that one has an objective measure of task performance and can therefore predict the percentage of the population that is capable of performing a task. When using psychophysical data to evaluate dynamic tasks, one must assume that the task will be performed at a speed and pace similar to that of the subjects used in the development of the database. However, in realistic work situations, it is a common observation that workers will work very rapidly to finish a materials handling task so that they can maximize their rest time. Thus, assumptions about the dynamic characteristics matching those of the database may not hold. When using psychophysical data, one also assumes that if lifting a load that is judged acceptable to a person they are less likely to suffer an injury. On one hand, there is evidence that the psychological perception of risk is a significant factor in the reporting of back injuries on the job (Bigos et al., 1986). On the other hand, there is also a belief that perceived acceptability may not be associated with risk over the long run. For example, many people choose to smoke and would call that risk acceptable. However, the literature is clear in that smoking dramatically increases the chances of physiological problems. Thus, one can make the argument that people are not good judges of risk perception. Therefore, psychophysical acceptance may not provide a level of protection for the working population. The argument can also be made that back injury is cumulative, thus, the load level that is acceptable at one point in time may not be acceptable as cumulative trauma occurs. Similar arguments can be made for the cautious application of dynamic strength data. Even though one has the capacity to perform a task, that does not mean they will not become injured, especially when the task is performed repeatedly. Thus, psychophysical and dynamic strength data must be used with caution and should be one of many tools used for the consideration of workplace design.

22.4 Assessments Based on Trunk Kinematics and Historical Observations of Risk

A unique approach to the assessment of occupationally related low back disorder risk associated with dynamic lifting at the worksite has recently been developed (Marras et al., 1993, 1995). This approach considers trunk motion characteristics in three-dimensional space along with traditionally documented characteristics of the workplace in a multiple logistic regression model of occupationally related LBD risk. This model requires documentation of trunk motions during a given occupational task. One means to facilitate these measurements is through the use of a trunk goniometer. A trunk goniometer, called the lumbar motion monitor or LMM, was developed to document the rotational position, velocity, and acceleration characteristics of the trunk during work tasks.

The LMM was used to document the trunk motion patterns and workplace characteristics of over 400 workers in jobs that have been associated historically with varying degrees of risk. Jobs with at least three years of historical LBD risk data (derived from medical records) were documented for trunk motion and workplace characteristics and used to form two risk groups (Marras et al. 1993). One group consisted of materials handling jobs that result in tasks associated with no recordable LBDs (0 incidences per 200,000 hours of exposure). The other group consisted of jobs that could be classified as having a high risk of LBD (average of 26.4 LBDs per 200,000 hours exposure and at least 12 incidences per 200,000 exposures). The analyses of the data indicated that high-risk jobs were more likely to involve rapid trunk motions that occurred between several planes of the body. Figure 22.4a and b show examples of low-risk and high-risk work cycles, respectively. These figures show the three-dimensional position of the thorax relative to the pelvis. The points in the figure are spaced one-sixtieth of a second apart. Thus, the farther apart the points, the faster the trunk is moving. Note how the low-risk motions shown in Figure 22.4a are confined to one or two planes of the body, and the speed of motion is low. On the other hand, Figure 22.4b indicates motions that are occurring in all three planes of the body and much more rapidly.

Multiple logistic regression models were developed that best discriminate between high-risk and low-risk groups of low back disorder based upon characteristics of the job and worker trunk motion patterns. Of the 114 variables associated and documented for each job, five factors were used to create a multiple logistic regression model that best discriminates between membership in the high- and low-risk groups. These five factors consist of: (1) frequency of lifting, (2) load moment (load weight multiplied by the distance of the load from the spine), (3) average twisting velocity (measured by the LMM), (4) maximum sagittal flexion angle through the job cycle (measured by the LMM), and (5) maximum lateral velocity

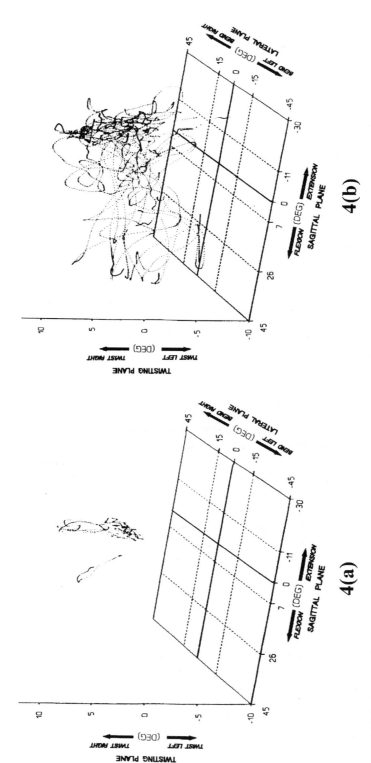

FIGURE 22.4 An example of trunk motion in a (a) low-risk and (b) high-risk job. Each point in the figure indicates the position of the thorax relative to the pelvis in three-dimensional space. The points are spaced one-sixtieth of a second apart. Thus, the more space between points the faster the motion.

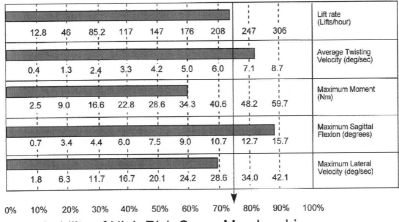

FIGURE 22.5   The LMM risk model. The five factors in the model are scaled according to the relative risk associated with each variable. A combination of the variables indicates overall probability of increased LBD risk that is shown at the bottom of the scale.

(measured by the LMM). These five variables have been scaled relative to their association with high-risk group membership. Figure 22.5 shows the risk model along with the appropriate scaling that can be used for workplace design and assessment purposes. The figure shows that by considering the combination of these five factors the probability of high-risk group membership can be predicted. Using this technique it is possible to address the issue of how much exposure to a risk factor is too much exposure, while considering the interactive effects of the other risk factors. This model has been shown to have a high degree of predictability (odds ratio = 10.7) compared to previous attempts to assess work-related low back disorder risk. Validity studies currently under way have indicated that the model is at least as sensitive as originally documented (Marras et al., 1993) and most likely more predictive than documented.

A similar assessment has been performed of the LMM database that indicates it is possible to assess the *severity* of the risk associated with dynamic lifting tasks (Marras et al. 1995). This assessment is very similar to the one just described with the exception that the scaling of the five risk variables (as shown in Figure 22.5) is different. Ongoing validity studies have also indicated that this is a very sensitive technique for dynamic manual lifting assessment.

This model can be used to assess current risk associated with the design of material handling tasks or it can be used to assess the expected risk associated with modifications or redesign of a job. In these cases it would be necessary to "mock up" the workplace and test a worker performing 5 or 6 repetitions of the job. The advantage of this assessment is that the evaluation provides information about risk that would take years to derive from historical accounts of incidence rates.

22.5   Applications of Dynamic Assessment Tools

This section will provide an example of how some of the dynamic assessment tools discussed in this chapter can be used for the evaluation of dynamic manual lifting assessments. The example shows how the three dynamic tools can be used to assess the job of an order selector in food distribution warehouses.

It is widely known that the order selection task in food distribution centers places the worker at risk of occupationally related low back disorders (LBDs). This job is associated with one of the greatest incidence rates of LBD in the United States. The National Association of Wholesale Grocers of America (NAWGA) and the International Foodservice Distribution Association (IFDA) disclosed that 30% of the injuries reported by food distribution warehouse workers were attributable to back sprains/strains

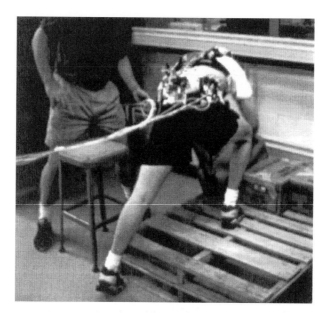

FIGURE 22.6   An order selector performing a stock picking task and wearing the experimental apparatus required to perform an LMM risk analysis and EMG-assisted model analysis.

(Waters, 1993). In addition, over a five-year period, it was found that back injuries could account for nearly 60% of lost workdays (NIOSH, 1992). Hence, grocery item selectors have an incidence of low back pain that is at least as severe as other manual materials handling jobs.

One approach to controlling this risk consists of manipulating the characteristics of the object or box to be handled in the food distribution center. A committee organized by the Food Marketing Institute (FMI) was interested in considering the various options available to them in order to help mediate the risk of work-related LBD in these food distribution centers. Among the options considered are: (1) reducing the weight of the boxes, (2) reducing the size of the boxes, or (3) incorporating handles into the boxes. However, it is currently unknown what effect these changes to the box characteristics would have on the loading of the spine and the subsequent risk of low back disorder. Such decisions would have a significant financial impact on the manufacturers of the items in the food distribution centers because it would require them to significantly change the packaging system for all products. Therefore, the FMI committee was interested in assessments that could *realistically* assess the significance of the contemplated changes to the box or case design.

The objective of this assessment was to determine the change in LBD risk and spine loading associated with selecting cases (in a warehouse environment) that varied as a function of: (1) weight (40, 50, and 60 lbs.), (2) size (2681 or 1584 cu. in.), and (3) the existence of handles or hand holds. In addition, in order to assess the problem in context, these variables were explored as a function of where the case was on a pallet. Ten experienced order selectors were evaluated as they selected cases from a slot (storage bin) on to a pallet jack. During the different experimental trials the case weight, case size, and case coupling (handles) conditions were varied. Workers were instructed to pick the entire complement of cases from the pallet so that they could be observed picking from all locations on a pallet. While the workers were lifting cases, they were being continuously monitored so that LBD risk could be assessed. Figure 22.6 shows the workplace environment of an order selector performing an order selection task.

Risk of LBD was evaluated through three assessment tools. First, a lumbar motion monitor (LMM) risk model was used that compared work conditions to those identified historically as being associated with a high risk of LBD injury. Second, the EMG-assisted biodynamic model described earlier was used to more specifically evaluate spine loading characteristics associated with the various potential changes in the case design. Third, the psychophysical acceptance was assessed as a function of each condition.

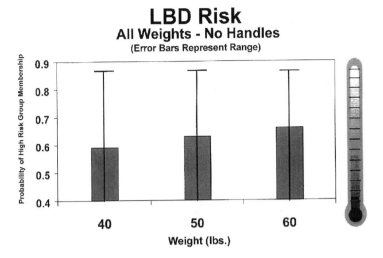

FIGURE 22.7 LMM risk as a function of case weight lifted.

Collectively, this information provided a rich understanding of the cost-benefits associated with the various case parameters under consideration.

As discussed above, the assessment tools assess risk by very different means. The LMM risk model *represents risk* in terms of the probability (based upon historical trends in industry) that the conditions of the task (job profile) resemble a high-risk situation. The EMG-assisted biodynamic model assesses *spine loading* that is assumed to be associated with risk of developing LBD. Finally, the psychophysical tool determines the percentage of the population that would be expected to find a load acceptable. These methods were used to assess the risk of LBD as a function of each case characteristic considered. In order to facilitate interpretation of these results a risk "thermometer" concept was used along with the graphical representation of the results. This risk thermometer can be used to assist in risk assessment regardless of the evaluation system used to determine risk. When the values on the graph are within the upper regions of the thermometer a highly risky (dangerous) condition is indicated, whereas when the values on the figure are aligned with the lower region of the thermometer, the conditions are safe. A continuous spectrum between these extremes permits one to make relative judgments between conditions.

Case Weights

The effect of case weight upon risk of LBD is indicated in Figure 22.7. This figure shows the results of the LMM risk model, however, similar trends were indicated when spine loading was assessed. Generally, these results indicate that as the weight of the case increases the risk increases in a rather linear fashion. The *average* risk associated with a 40-lb. case is close to the acceptable range, whereas, the 60-lb. case approaches the danger zone on the risk thermometer. Of particular significance is the fact that under each condition the range of the data (indicated by the error bars) is extremely large. Thus, for any given case weight, the risk associated with the condition can be either extremely safe or extremely risky. Further analyses indicated that the position of the case on the pallet more specifically defined risk than did case weight.

Position on the Pallet

These analyses have shown that the case position on the pallet is instrumental in defining LBD risk. Figure 22.8 indicates spine loading as a function of pallet position for the 40-lb. cases. Note that for this condition most of the spine loadings associated with the top and middle layers of the risk are within the safe region on the risk thermometer. However, the data indicate that the vast majority of the observations

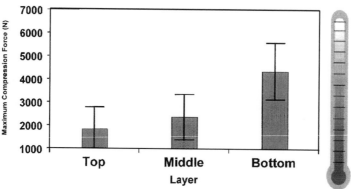

FIGURE 22.8   Spine loading (compression) associated with the location of a 40-lb. case on a pallet.

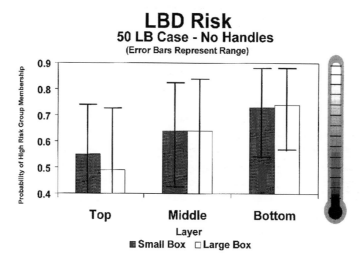

FIGURE 22.9   LMM risk associated with two different case sizes while lifting a 50-lb. case from different regions of a pallet.

associated with lifting from the bottom layer of the pallet imposed a significant risk of LBD to the workers. Similar trends were noted when the LMM risk model was used as a measure. The risk increased substantially for the bottom layer of the pallet when the 50- and 60-lb. cases were lifted. Thus, this analysis shows that the risk was significant only for lifts from the bottom layer of the pallet.

Case Size

The LMM risk analyses did indicate that case size was a significant factor in determining risk. However, the practical significance of this finding was questionable. Figure 22.9 shows the effects of case size for the 50-lb. case as a function of location on the pallet. This indicates that the only benefit of varying case size occurs at the top layer of the pallet and the evaluation of risk as a function of location has shown that little risk exists at the top layer of the pallet. The differences noted at the bottom (problematic) layer of the pallet were not significant enough to justify the control of case size. Therefore, there is no practical benefit of controlling case size among the sizes considered in this study.

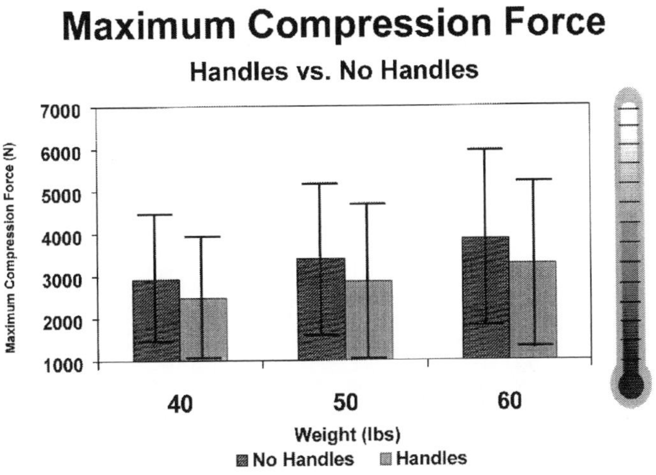

Maximum Compression Force
Handles vs. No Handles

FIGURE 22.10   Spine compression as a function of handles and case weights.

Handles

The effects of case handles on spine loading compared to no handle conditions is illustrated in Figure 22.10. This figure indicates that *on the average* the effect upon spine loading of incorporating handles into a box is approximately equivalent to reducing the case weight by 10 pounds. However, as mentioned previously, serious risk only occurs at the bottom layer of the pallet. Therefore, if one considers the effects of handles upon spine loading as a function of the different levels of the pallet, the situation shown in Figure 22.11 becomes apparent. This figure shows that none of the conditions yield all of the data within the safe potion of the risk thermometer. However, at the bottom layer the 40-lb case with handles yields the lowest risk situation, with approximately 40% of the observations within the safe range of the risk thermometer and only 3% of the observations within the danger range. By contrast, the 40-lb box without handles and the 50-lb. box with handles increase the percent of the observation within the dangerous risk range to about 7 to 10% of the observations. Furthermore, a 50-lb. box without handles further increases risk to where lifting from the bottom layer would yield less than about 12% of the observations in the acceptable zone and over 20% of the observations within the danger zone. A summary of the percentage of data within the biomechanical benchmark zones of the risk thermometer is shown in Table 22.2. This table provides a means to quantitatively consider the trade-offs associated with the various workplace variables under consideration. For comparison purposes, similar information was provided in tabular form in Tables 22.3 and 22.4 for the LMM risk assessment and psychophysical acceptance, respectively. Table 22.3 indicates that subtle differences can be noted in risk between the different box conditions. Table 22.4 indicates that from a psychophysical perspective none of the conditions would be acceptable because none of the conditions would be judged acceptable to 75% of the females. Note the different sensitivities of information that can be derived given the different approaches. One should also note the dramatically different levels of effort required to perform the various analyses. EMG-assisted biomechanical models require significant time, effort, and resources, whereas LMM risk assessments and psychophysical assessments are quite quick and relatively easy to perform.

   This analysis has demonstrated how one can pinpoint which case parameter variables or features are worthy of consideration for inclusion in the food distribution environment for the purpose of reducing the risk of work-related LBD. In general, the following conclusions were drawn.

- Risk of LBD increases linearly as case weight increases.
- The greatest risk and loading of the spine occur during lifts from the bottom layers of the pallet. The other layers of the pallet pose acceptable lifts regardless of the case weight.
- Case size has a significant effect on risk, but the difference has no practical meaning. There is no reason to control case size based upon the range of sizes explored in this study.

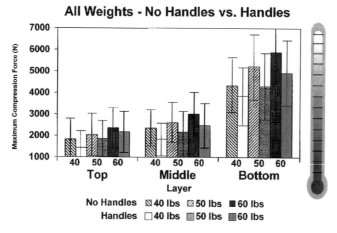

FIGURE 22.11   The effects of handles and the location of the case on a pallet on spine loading.

TABLE 22.2   Summary of Spine Compression Loading Compared to Spine Tolerance "Benchmarks" as a Function of Case Weight, Handles, and Location on a Pallet.

		Case Weight					
		40 lbs		50 lbs		60 lbs	
		Handles	No Handles	Handles	No Handles	Handles	No Handles
Top Layer	≤3400 N	99.1	94.6	92.5	88.2	91.7	84.3
	3400 to 6400 N	0.9	5.4	7.5	11.8	8.3	15.7
	>6400 N	0.0	0.0	0.0	0.0	0.0	0.0
Middle Layer	<3400 N	93.5	86.7	87.4	79.0	84.6	70.2
	3400 to 6400 N	6.5	13.3	12.6	21.0	15.4	29.1
	>6400 N	0.0	0.0	0.0	0.0	0.0	0.7
Bottom Layer	<3400 N	40.3	27.4	29.7	12.3	12.7	2.7
	3400 to 6400 N	56.3	65.5	59.7	65.3	71.7	64.4
	>6400 N	3.3	7.1	10.7	22.3	15.7	32.9

Safe   (<3400 N)
Caution   (3400–6400 N)
Danger   (>6400)

- Handles have a significant effect upon spine loading. The effect is particularly significant when lifting from the lowest levels of the pallet. The 40-lb box, when combined with handles, represents the condition with the lowest level of risk even when lifting from the lowest level of the pallet. Handles have the effect upon spine loading of lowering the case weight by 10 lbs. Thus, a 50-lb case with handles can reduce spine loading to the level of a 40-lb case without handles.

These findings provided some practical solutions to FMI for the design of the distribution center, and these recommendations were incorporated into Food Industry Guidelines. This example demonstrates the range of information that is obtainable from these dynamic assessment tools. Psychophysical methods are easiest to apply. However, as this example shows, the specific conditions of a work situation may not match the psychophysical database well as was the case for the pallet middle level order selecting in this example. The LMM risk model is straightforward and easy to use and provides a realistic assessment of the risk of LBD risk probability. This risk model can relate to specific work conditions since it interprets the kinematics associated with a particular work condition. Finally, in order to derive more precise information about the task, such as actual spinal loading information, a much more sophisticated

TABLE 22.3 Summary of LMM Risk Assessment of an Order Selecting Task as a Function of Case Weight, Handles, and Location on a Pallet.

	Probability of High Risk Group Membership	Case Weight					
		40 lbs		50 lbs		60 lbs	
		Handles	No Handles	Handles	No Handles	Handles	No Handles
Top Layer	<30%	0.0%	5.0%	0.0%	0.0%	0.0%	0.0%
	30% to 70%	100.0%	95.0%	90.0%	90.0%	90.0%	85.0%
	>70%	0.0%	0.0%	10.0%	10.0%	10.0%	15.0%
Middle Layer	<30%	5.0%	0.0%	0.0%	0.0%	0.0%	0.0%
	30% to 70%	85.0%	85.0%	85.0%	70.0%	60.0%	65.0%
	>70%	10.0%	15.0%	15.0%	30.0%	40.0%	35.0%
Bottom Layer	<30%	0.0%	0.0%	0.0%	0.0%	0.0%	0.0%
	30% to 70%	50.0%	50.0%	55.0%	45.0%	35.0%	45.0%
	>70%	50.0%	50.0%	45.0%	55.0%	65.0%	55.0%

Safe   (<30%)
Caution (30% to 70%)
Danger   (>70%)

TABLE 22.4 Summary of Psychophysical Assessment of Order Selection Task. The Values in the Table Indicate the Percentage of the Industrial Population that Would Find the Specific Conditions Acceptable.

	Case Weight											
	40 lbs				50 lbs				60 lbs			
	Handles		No Handles		Handles		No Handles		Handles		No Handles	
Gender	M	F	M	F	M	F	M	F	M	F	M	F
Top Layer	75%	0%	75%	0%	50%	0%	50%	0%	25%	0%	25%	0%
Middle Layer	n/a	n/a	n/a	n/a	n/a	n/a	n/a	n/a	n/a	n/a	n/a	n/a
Bottom Layer	50%	0%	25%	0%	25%	0%	10%	0%	10%	0%	0%	0%

Safe   (<30%)
Caution (30% to 70%)
Danger   (>70%)

assessment is necessary (EMG-assisted model). However, this requires a great deal of effort and may not be practical for "on the job" workplace assessments. The correct level of assessment depends upon the consequences of information. In this example, the FMI wanted to make sure that the answer to their problem was realistic because it would require many manufacturers to make significant, expensive changes to their production process and would affect national guidelines. Therefore, they elected to seek the most complete and richest information possible. Had the magnitude of the problem not been so great and the consequences so costly, less complex analyses (the LMM risk model or psychophysical assessments) may have been adequate.

References

Bean, J.C., Chaffin, D.B., and Schultz, A.B. (1988) Biomechanical model calculation of muscle forces: A double linear programming method. *J. Biomechanics*, 21 (1) 59-66.

Bigos, S.J., Spengler, D.M., Martin, N.A., Zeh, J., Fisher, L., Nachemson, A., and Wang, M.H. (1986) Back injuries in industry: A retrospective study. II. Injury factors. *Spine*, 11(3), 246-251.

Brinckmann, P., Biggemann, M., and Hilweg, D. (1988) Fatigue fracture of human lumbar vertebrae. *Clinical Biomechanics*, Supplement No. 1, 1988.

Chaffin, D.B. (1969) A computerized biomechanical model: development of and use in studying gross body actions. *J. Biomechanics*, 2, 429-441.

Chaffin, D.B. and Muzaffer, E. (1991) Three-dimensional biomechanical static strength prediction model sensitivity to postural and anthropometric inaccuracies. *IIE Transactions*, 23(3), 215-227.

Evans, F.G. and Lissner, H.R. (1965) Studies on the energy absorbing capacity of human lumber intervertebral discs. *Proceedings of the Seventh Strapp Car Crash Conference*, Springfield, IL.

Frievalds, A., Chaffin, D.B., Garg, A., and Lee, K. (1984) A dynamic evaluation of lifting maximum acceptable loads. *J. Biomechanics*, 17, 251-262.

Granata, K.P. and Marras, W.S. (1993) An EMG-assisted model of loads on the lumbar spine during asymmetric trunk extensions, *J. Biomechanics*, 26(12), 1429-1438.

Granata, K.P. and Marras, W.S. (1995), An EMG-assisted model of trunk loading during free-dynamic lifting. *J. Biomechanics*, 28(11), 1309-1317.

Kaymon, E., Kiser, D., and Pytel, J. (1982) Dynamic and static lifting capacity and muscular strength of steelmill workers. *Am. Ind. Hygiene Assoc. J.*, 43, 853-857.

Marras, W.S. and Granata, K.P. (1995) A biomechanical assessment and model of axial twisting in the thoraco-lumbar spine. *Spine*, 20(13), 1440-1451.

Marras, W.S. and Granata, K.P. (1996) An assessment of spine loading as a function of lateral trunk velocity. *J. Biomechanics*, (30)7, 697-703.

Marras, W.S., Lavender, S.A, Leurgans, S., Rajulu, S., Allread, W.G., Fathallah F., and Ferguson, S.A., (1993) The role of dynamic three dimensional trunk motion in occupationally-related low back disorders: the effects of workplace factors, trunk position and trunk motion characteristics on injury. *Spine*, 18(5), 617-628.

Marras, W.S., Lavender, S.A., Leurgans, S., Fathallah F., Allread, W.G., Ferguson, S.A., and Rajulu, S. (1995) Biomechanical risk factors for occupationally related low back disorder risk. *Ergonomics*, 38(2), 377-410.

Marras, W.S. and Mirka, G.A.(1992) A comprehensive evaluation of asymmetric trunk motions. *Spine*, 17(3), 318-326.

Marras, W.S. and Reilly, C.H.(1988) Networks of internal trunk loading activities under controlled trunk motion conditions. *Spine*, 13(6), 661-667.

Marras, W.S. (1988) Predictions of forces acting upon the lumbar spine under isometric and isokinetic conditions: A model-experiment comparison. *International Journal of Industrial Ergonomics*, 3(1), 19-27.

Marras, W.S., Rangarajulu, S.L., and Wongsam, P.E. (1987) Trunk force development during static and dynamic lifts. *Human Factors*, 29(1), 19-29.

Marras, W.S. and Sommerich, C.M., (1991) A three dimensional motion model of loads on the lumbar spine, Part I: Model structure. *Human Factors*, 33(2), 123- 137.

Marras, W.S. and Sommerich, C.M. (1991) A three dimensional motion model of loads on the lumbar spine, Part II: Model validation. *Human Factors*, 33(2), 139-149.

Marras, W.S., Wongsam, P.E., and Rangarajulu, S.L.(1986) Trunk motion during lifting: The relative cost. *International Journal of Industrial Ergonomics*, 1(2), 103-113.

McGill, S.M. and Norman, R.W. (1985) Dynamically and statically determined low back moments during lifting. *J. Biomechanics*, 18, 877-885.

NIOSH 1992 Interim Report, HETA 91-405, March 1992.

Nussbaum, M.A., Martin, B.J., and Chaffin, D.B. (1997) A neural network model for simulation of torso muscle coordination. *J. Biomechanics*, 30(3), 251-258.

Punnett, L., Fine, L.J., Kyserling, W.M., Herrin, G.O., and Chaffin, D.B. (1991) Back disorders and nonneutral trunk postures of automotive assembly workers. *Scand. J. Work Environ. Health*, 17:337-346.

Pytel, J.L. and Kamon, E. (1981) Dynamic strength test as a predictor for maximal and acceptable lifting. *Ergonomics*, 24, 663-672.

Raschke, U., Martin, B.J., and Chaffin, D.B. (1996) Distributed moment histogram: A neurophysiology based method of agonist and antagonist trunk muscle activity prediction. *J. Biomechanics*, 29(12), 1587-1596.

Schultz, A.B. and Andersson, G.B.J. (1981) Analysis of loads on the lumbar spine. *Spine*, 6:76-82.

Snook, S.H. and Ciriello, V.M. (1991) The design of manual handling tasks: revised tables of maximum acceptable weights and forces, *Ergonomics*, 34(9), 1197-1213.

Sonoda, T. (1962) Studies of the strength for compression, tension, and torsion of the human vertebral column. *J. Kyoto Perfect. Med. Univ.*, 71, 659-702.

Thelan, D.G. and Schultz, A.B. (1994) Identification of dynamic myoelectric signal-to-force models during isometric lumbar muscle contractions. *J. Biomechanics*, 27(7), 907-919.

Waters, T.R., Putz-Anderson, V., Garg, A., and Fine, L.J. (1993) Revised NIOSH equation for the design and evaluation of manual lifting tasks. *Ergonomics*, 36(7), 749-776.

Winter, J.M. and Woo, S.L.-Y. (1990) *Multiple Muscle Systems*, Springer-Verlag, New York.

23

Push–Pull Force Limits

Sheik N. Imrhan
University of Texas at Arlington

23.1 Introduction

Pushing and pulling (push–pull) activities involves static or dynamic muscular force exertions for moving or stabilizing objects. Neither the direction of the force nor the motion of the object need be perfectly linear or in the horizontal plane so long as deviation is small. In cases where the deviation is great, the exertion can be considered a hybrid of two types, such as lift–push, lift–pull, etc. as used by Pheasant et al. (1982). Push–pull activities occur in many types of work environments — shipping and receiving, moving, warehousing, agriculture and farming, retailing, etc. — and are becoming more common as a result of efforts to minimize lifting, lowering, holding, and carrying, the most debilitating and costly manual materials handling (MMH) activities. Baril-Gingras and Lortie (1990), in studying over 900 tasks, estimated that nearly half of all materials handling activities were push–pull ones. However, push–pull activities also account for a significant amount of overexertion musculoskeletal disorders (NIOSH, 1981; Klein, Jensen, and Sanderson, 1984; Troup and Edwards, 1985), accounting for approximately 20% of all back injuries from MMH activities.

Most push–pull activities can be distinguished by the general body posture, the number of hands involved, and whether the exertion is static or dynamic. The choice depends on a number of factors: standing posture and two-handed exertions for strong forces and significant control, and sitting posture and one-handed exertions for moderate and weak forces; dynamic activity for moving loads (e.g., industrial carts) from one point to another and for activating electromechanical mechanisms (e.g., in lawn mower and outboard motor engines), and static activity for supporting objects and operating certain mechanisms, such as levers in transportation vehicles.

Pulling and pushing can be achieved with both the arms and legs, but this chapter is concerned only with the arm activities because they constitute almost all of the pulling and pushing in the workplace and are they ones associated with health concerns. In the workplace, both push and pull forces are applied for moving objects out of their positions to create space for other objects or to allow passage of other objects, though this may imply inefficiency in workplace design; moving objects from one position to another where they can be stored or further manipulated; and stabilizing an object, as in preventing a box from sliding along an incline. In addition, pull forces only are applied for activating mechanical devices, such as engines on small boats and lawn mowers, and opening packages, as in pinching and tearing plastic wrappers on consumer products; and push forces only are applied for shutting lids on consumer products, and activating electrical and electronic switches with the fingers.

The choice of using a push or pull may also depend on the designs of the workplace and object being handled. For example, it is difficult and often impossible to pull a large box that has no handles, especially if it is heavy. Instead, a worker may simply change position and push it, assuming the workplace can accommodate such change of position. Similarly, a cart with defective steering in the front wheels may be pulled on a handle instead of being pushed.

The choice of a hand–object contact depends on the size and shape of handle, the intended activity, or the amount of force desired. Power grips should be the best choice when great push–pull forces are desired. However, such forces may not be achieved without a handle of adequate size and shape to accommodate a power grip. As mentioned by Kumar (1995), handle orientation may also be important. For some objects, such as such as a large television box or a panel with no handles, a flat-palm contact may be the only recourse. When contact area is small — not more than about a few square centimeters — only pinch grips may be possible (Imrhan and Sundararajan, 1992), but in these cases the expected force application is usually relatively small. Hook grips are applicable where handles are not large enough for power grips and the expected forces are only of moderate or small magnitudes. Unlike pulling, great push forces can be achieved without a firm grip on a handle and even without a handle. In the latter case, a flat palm contact may be appropriate.

23.2   Factors Influencing Push–Pull Strengths

Studies in pushing or pulling have shown that wide differences in strength capacities are obtained according to the type of exertion (static or dynamic), number of hands performing the exertion (one or two), and general body posture (standing, sitting, kneeling, or lying down). Therefore, in discussing push–pull force limits and applying data from the literature to assess task, equipment, or workplace design, one must identify the exertions according to these or similar classifications. However, many other independent variables have been shown to affect people's capacity to exert pull or push forces. They are discussed below.

Handles

The type of handle influences the maximal attainable force. In general, for maximal pulling the person should have a comfortable power grip on the handle, with maximum contact along the width of the hands (i.e., no overlapping of the hands), and with adequate clearance at the sides of the objects. For cylindrical handles, a diameter of about 1.5 to 2 in. (3.8 to 5.1 cm) (Ayoub and LoPreisti, 1971) and length of 5 in. (13 cm), which are recommended for other manual materials tasks, are also suitable for pushing and pulling. A poor hand–handle or hand–object interface acts like a weak link in the chain of force transmission and can limit push–pull force capacity severely. Fothergill, Grieve, and Pheasant (1991), for example, found up to 65% decrease in strength when poor hand–handle interfaces were used for pulling and pushing. A poor hand–handle interface also fails to capture advantageous handle-handle locations (Fothergill et al., 1991). A handle is always necessary for effective pulling but not always for pushing. When, instead of a handle, a flat surface is available for pushing, great forces can still be achieved, as demonstrated by Kroemer (1969 and 1974). Push–pull tasks and work equipment should also be designed, when possible, to position the forearm in mid-supination/pronation where the muscles in the forearm are in least tension.

One-Handed versus Two-Handed Exertions

The difference between one- and two-handed maximal efforts is much smaller than is generally believed and is dependent on the height of force application and body posture. Data from Fothergill et al. (1991) and Chaffin, Andres, and Garg (1983) indicate that the ratio of one- to two-handed pushes and pulls varies from 0.64 to 1.04, depending on the conditions of force exertion. The ratio is high (well above 0.50) because (1) push–pull strengths come not only from the arms but also the trunk, which is active in one-handed exertions, and (2) one-handed exertions afford greater freedom in a person's posture and, hence, more effective use of the body's weight and center of gravity (Fothergill et al., 1991). The ratio is greater when arm strength is the limiting factor in the exertions, as when pushing or pulling at high positions, such as above the shoulder, and standing erect.

Body Posture

The strength of a muscular exertion depends, to a large extent, on body posture, which defines the magnitude of the biomechanical advantages of the various lever systems contributing to muscular torques about the body's joints. Standing, sitting, and kneeling are probably the three most common of these postures. Davis and Stubbs (1980) investigated two-handed horizontal push–pull forces but their data are not definitive — standing forces were stronger than kneeling ones only in some situations. For one-handed exertions, Mital and Fard (1990) showed that standing one-handed dynamic pull strength is about 37% stronger than sitting strength.

Strength in any of these three whole-body postures is dependent on the geometrical configuration of other body segments, such as foot position, hand–handle height (height of force application), and the angles in the arms and legs. These factors are discussed in later sections in this chapter. It must be noted also that the best posture for static exertions is not necessarily the best for dynamic exertions (Lee, Chaffin, Herrin, and Waikar, 1991), and dynamic postures are difficult to monitor because they change continuously as the body moves (Resnick and Chaffin, 1995).

Foot Position

In standing exertions, foot and hand placements determine the effective posture. The placement of the feet relative to the hand–handle contact influences the stability of the body and, hence, the leverage for pulling and pushing efforts. In general, static standing exertions are enhanced when the feet are separated, one foot in front the other (Ayoub and McDaniel, 1974). For pushing, one should lean forward with the rear (pivoting) foot positioned behind the body's center of mass; and for pulling, with the front (pivoting) foot in front of the center of mass. The leaning postures allow people to use their body weight more effectively in counteracting the push or pull force at the hands and enhance strength (Kroemer and Robinson, 1971; Ayoub and McDaniel, 1974; Warwick, Novack, Schultz, and Berkeson, 1980; Pheasant et al., 1982; Chaffin et al., 1983). Even when not leaning, stronger forces are realized when the feet are apart, one in front of the other, than when placed side by side (Chaffin et al., 1993; Daams, 1993), and pushing or pulling in free-style posture has been shown to yield considerably greater forces than in standardized postures with the feet together or one in front of the other. The amount of leaning and shoe–floor traction influence the distance of the foot from the hand–handle contact. In general, static MVC push force is greater than pull force when the feet are separated in the fore–aft plane and the body is leaning, but are of about the same strength when the feet are close together, side by side (Chaffin et al., 1983). In addition, one can achieve an optimal push–pull posture over a wider range of angles of force application when the feet are separated compared to when they are together (Fothergill et al., 1991).

Standing, Sitting, and Kneeling

Given the stability of the body when standing and the added strength from using the legs, one would expect that standing push–pull strengths are greater than kneeling ones. However, the data from Davis and Stubbs (1980) indicate that this depends on the arm position when pulling or pushing horizontally.

In some cases, and for some age groups, kneeling generated MVC forces of about the same magnitudes as when standing. There is no clear pattern from their study, however, to warrant a simple generalization. Moreover, Gallagher (1989) found that kneeling generated greater MVC forces than standing. Sitting and standing push–pull strengths are difficult to compare directly because of the great differences in the geometry in the arms and legs and in other factors. However, it is worth mentioning that Mital and Faard (1990) found isokinetic pulls while sitting unrestrained to be 73% as strong as while standing.

Height of Application of Forces

The optimal height for application of push or pull forces depends on general body posture (especially the angles in the arms and knees), and the degree of leaning forward (for pushing) or backward (for pulling). There seems to be wide agreement that best height for static pushing or pulling in the horizontal direction while standing is between chest and knee, with pulling height being lower than pushing height (Martin and Chaffin, 1972; Ayoub and McDaniel, 1974; Chaffin et al., 1983; Kumar, Narayan, and Baccus, 1995). Recommending an exact height would be unwise since the optimal height has been shown to depend on such conditions as type of exertion (static or dynamic), angle at the elbow, frictional characteristics at the shoe–floor contact, etc. However, various recommendations seem to indicate that two-handed pushing while standing is strongest at about elbow to hip height, and pulling at about hip to knee height. The direction of the applied force modifies this relationship. Pheasant and Grieve (1981) showed that the optimal height for pulls gets lower (to 25 cm from 63 cm), below the knee, and for pushes gets higher (to 175 cm from 100 cm), above the shoulder, when the forces are exerted at an angle upward (lift–pull and lift–push).

Horizontal Foot Distance and Reach Distance

The horizontal foot distance (HFD) is the perpendicular distance between the ankle (of the rear foot, for separated feet) and hand–handle contact, when standing. For pushing, the pivoting foot is the rear one, and for pulling it is the front one. The reach distance (RD) is the distance from the shoulder to the hand–handle contact, when sitting. Both HFD and RD define the configuration of the upper body, and HFD also defines the configuration of the lower body. They influence push–pull strengths significantly, but in different ways. In standing, strength increases with RD (Martin and Chaffin, 1972), but reach distance is itself influenced by a number of factors: amount of space available, foot- or shoe–floor contact, height of the handle, etc. (Kroemer, 1974; Chaffin et al., 1983). Ayoub and McDaniel (1974) have shown that the best HFD for static exertions is about 90 to 100% of shoulder height behind the hand for pushing and 10% in front of the hand for pulling. There is evidence that, when sitting, peak static push strength for a given starting position of the arm follows an increasing–decreasing trend with reach distance (Lower, et al., 1977) and both dynamic and static pull strength increases with inrcreasing reach distance (Mital and Faard; 1990). An analysis of data from VanCott (1972) also indicates that static pull strength increases with reach distance. Dynamic (isokinetic) pull strength in a single pull follows an increasing–decreasing trend over the range of pull (Imrhan and Ayoub 1985; Garg and Beller, 1990; Imrhan and Ramakrishnan, 1992), and there is some evidence that, for any given velocity of dynamic pull, there is an optimal arm configuration for maximal pull strength (Imrhan and Ayoub, 1990).

Direction of Push/Pull Force Exertion

Direction of pushing and pulling can be described with respect to (1) the transverse plane and (2) the sagittal or coronal plane. The strongest direction of force exertion should be that in which body's reaction force is maximized at the contact of the body with the floor, seat, etc. When standing and pulling from below the horizontal or pushing from above the horizontal (that is, at an angle to the horizontal), the reaction at the shoe–floor contact can be increased and, therefore, push–pull force increased, as found by Garg and Beller (1990) and Imrhan and Ramakrishnan (1992). When sitting, however, and especially if restrictions are placed on motion of the upper body (e.g., chest harness), this relationship may be modified and some other direction may be the strongest, as found by Imrhan and Ayoub (1988). In their study, horizontal pulls at shoulder level were stronger that angled pulls (toward the shoulder) above and below the horizontal. Pushing on a high handle upward at an angle to the horizontal or pulling on a low handle upward at an angle can produce a horizontal force component that is stronger than if these exertions were directed along the horizontal (Fothergill et al., 1991).

Pushing or pulling in the horizontal is strongest with the hand directly in front of the shoulder, for one-handed exertion, (Lower et al., 1977) or with both hands directly in front of the body, for two-handed exertions (Kumar, 1994). However, when pulling or pushing at an angle to the sagittal plane (sideways), reacton force and, hence, muscular force is less. One- or two-handed push–pull strength decreases rapidly as the arm moves across the body to the left or right (Mital and Fard, 1990; Kumar and Garand, 1992; Kumar et al., 1995). Mital and Faard (1990) showed that the decrease is more rapid to the left of the body than to the right for right-handed exertions. Pulling with the hand directly toward the body has been found to be about 10% stronger than across the body, (Mital and Fard 1990; Imrhan and Ramakrishnan, 1992); and pulling with the fingers (pinch–pulling) across the body horizontally is not significantly different in strength from pulling across the body obliquely (e.g., from left waist to right shoulder) (Imrhan and Sundararajan, 1992).

Body Support

Research has shown that a body support can enhance push–pull strengths by as much as 50% (Kroemer, 1974). The support helps to enhance the reaction force to the muscular exertions and is best positioned perpendicular to the exerted force. Supports are suitable only for static exertions, however. When standing, a panel, wall, or fixed footrest is effective, especially for pushing (Kroemer, 1974). When sitting, a stable chair backrest or footrest (Caldwell, 1964) is effective for pushing, and a harness is effective for pulling (Imrhan and Ayoub, 1985). Pheasant et al. (1982) also found that a low ceiling improved push–press strength because people were able to brace themselves on it, but it weakened lift strength because it constituted an obstacle.

Gender Differences

A female's absolute push–pull strengths are significantly weaker than a man's, and the female/male strength ratio varies considerably depending on posture and type of strength. The ratios fall mostly in the range 0.5 to 0.9 (Fothergill et al., 1991; Kumar et al., 1995), which is similar to the ratios found in other MMH activities. The higher ratios (less inequality) occur when the legs are more influential in the exertion, such as in pulling at low to medium height (Fothergill et al., 1991) and when strengths are weaker in general, such as exertions in planes away from the sagittal plane (Kumar et al., 1995).

Back Muscle Forces and L5/S1 Disc Compression Forces

MMH tasks, including pulling and pushing, that produce compression at the L5/S1 spinal disc in excess of 3400 N are considered hazardous (NIOSH, 1981). The results of certain studies (Lee et al., 1989; Chaffin et al., 1983; Resnick and Chaffin, 1995) show that maximal dynamic pushing and pulling create hazardous L5/S1 compression, especially when the exertions are performed at low heights (hip to knee), and that pulling creates greater compression than pushing. Disc compression also increases as push–pull forces or speed of walking increases (Lee et al., 1989).

Body-Support Traction

The shoe–floor or seat–buttocks coefficient of friction (COF) when standing or sitting is an important determinant of MVC or maximal acceptable force. (Fox, 1967; Kroemer and Robinson, 1971; Kroemer, 1974). Kroemer (1974) demonstrated that push force can increase by as much as 50% when the COF of shoe–floor contact increases from 0.3 (poor traction) to 0.6 (good traction). One can compensate for low traction by using foot and back supports.

Coefficient of friction can influence posture and self-selected speed of movement dramatically and is also influenced by the height of the push–pull exertions (Resnick and Chaffin, 1995; Lee et al., 1991). Differences in the effects of posture on strength are more pronounced when the coefficient of friction is low. Designers can avoid this situation by ensuring high traction on walking and sitting surfaces where pushing and pulling are common.

Distance of Movement of Body

The maximal acceptable dynamic push–pull loads (limits) that can be attained depends on the distance the load is moved. Clearly, the greater the distance the lighter the load. Maximal acceptable dynamic push–pull force limits over various distances have been established by Snook and Ciriello (1991).

Speed of Push/Pull

Resnick and Chaffin (1995) found that speeds of push selected voluntarily by subjects for short walking distances (1.5 m or 5 ft.) were much slower (0.2 to 1.1 m/s) than those proposed by the methods and time measurements (MTM) system for longer distances (1.8 m/s). These authors concluded that slower speeds than the MTM ones should be used for pushing heavy loads, especially over short distances.

Maximum walking speed for pulling and pushing loads depends on a number of factors including the shoe–floor COF, person's strength, load, type of handle, and height of handle. Stronger two-handed pushes while walking are executed at greater speeds than weaker ones, and heavier loads are pushed at slower speeds than lighter loads, regardless of people's strengths. Resnick and Chaffin (1995) found the speeds for stronger pushes on a light load (45 kg) to average 0.75 m/s for a voluntarily selected "hard push" and 0.4 m/s for "easy push." With a 450-kg cart load, subjects pushed at 0.44 m/s and 0.3 m/s, respectively. These authors also found that push speed increased as handle height decreased from shoulder to elbow to knee. Peak dynamic pull strength is achieved at a later position in the pull range as velocity of pull increases (Imrhan and Ayoub, 1985 and 1990).

Anthropometry

Body weight is the most influential anthropometric variable on push–pull strength, especially when it is used to enhance force directly, as when leaning forward in pushing and leaning backward in pulling (Kroemer, 1969; Ayoub and MacDaniel, 1974). Imrhan and Sundararajan (1992) also found body weight to be the most highly correlated anthropometric variable, even for strengths (finger pull) that are not affected by posture. As for almost all types of working strengths, no single anthropometric variable or combination of variables is suitable for predicting push or pull force accurately (Imrhan, 1983; Kumar et al., 1995). However, anthropometric variables of the upper body, especially the arm, can be used in combination with task variables, such as speed, arm position, etc., to significantly improve the predictive power of models of push or pull strength (Imrhan, 1983 and 1988).

Environmental Stress

One study (Snook and Cirello, 1974b) has found that workload for push decreased by 16% as environmental temperature changed from WBGT temperature of 17.2°C (63.0°F) to 27.0°C (80.6°F). Until more specific data becomes available, one may apply these findings to the design of jobs and administrative controls for protecting workers subjected to heat stress.

Endurance

The body posture and other conditions associated with enhanced maximal strength seem to be the same conditions that are associated with enhanced endurance. Caldwell (1964) showed that the sitting body posture that generated the greatest pull strength was the one in which endurance at submaximal strength (at 80% MVC) was also greatest. Presumably, in this posture, the body's levers are at the best overall mechanical advantage and require the least effort for eliciting and maintaining a specified level of strength. However, there is no other push–pull study that substantiates this relationship.

Frequency

Maximal push or pull forces are expected to decrease as frequency of task performance increases. Snook and Ciriello (1991) provide data from psychophysical push–pull experiments over several frequencies (from once per 6 seconds to once per 8 hours), showing that maximal acceptable forces decrease with increasing frequency. As for other MMH activities, this kind of decrease is nonlinear.

23.3 Push–Pull Magnitudes and Safe Force Limits

Safe push–pull force exertion limits may be interpreted as the maximum force magnitudes that people can exert without injuries (for single strong muscular exertions) or cumulative trauma disorders (for repeated exertions) of the upper extremities under specified conditions. Safe limits for static exertions should not be the same as for dynamic exertions.

Static Standing Forces

Many factors influence the magnitude of a static MVC force (single exertion) and, therefore, it would be unwise to recommend a single value for either push or pull force limits for task or workplace design. In addition, even for a given set of conditions, such as handle height, arm and leg posture, etc., MVC values differ considerably across published studies. Average static two-handed MVC push forces have ranged from about 400 to 620 N in males and from about 180 to 335 N in females when there is no bracing of the body, and pull forces from about 310 to 370 N in males and 180 to 270 N in females. Bracing, as mentioned before, can enhance force by as much as 50%.

Dynamic Standing Forces

Dynamic two-handed push–pull forces are not as strong as static ones. Dynamic push forces (mostly in moving industrial carts) have ranged from 170 to 430 N in males and 200 to 290 in females, and push forces from 225 to 500 N in males and 160 to 180 N in females. Initial forces (required to set a stationary object in motion) are generally lower than sustained forces (required to keep an object moving) in pushing and pulling tasks. Snook and Ciriello (1991) observed that maximal acceptable initial pulling force was 13% lower than pushing force, and maximal acceptable sustained force, 20% lower.

The most useful guidelines on dynamic push–pull force limits have been published by Snook and Ciriello (1991). These authors have proposed maximal acceptable force limits (MAFs) for males and females (four tables) in comfortable work conditions by combining two sets of data: the first from several early studies (Snook, Irvine, and Bass, 1970; Snook and Ciriello, 1974a, b; Snook, 1978; Ciriello and Snook, 1978), and the second from four subsequent studies found in Ciriello and Snook (1983), Ciriello, Snook, Blick, and Wilkinson (1990), and Ciriello, Snook, and Hughes (1991). Partial reproductions of the final four tables for pushing and pulling in males and females are given in Tables 22.1 through 22.4 in this chapter. Given the large number of ways in which a person can move an object from one location to another, and that many are not embodied in these tables, it is necessary that the tables be interpreted carefully to apply their values to task and workplace design.

The tables are the psychophysically determined maximal forces that people are willing to accept if they were to perform the push or pull activities as a normal eight-hour job. The forces are stated as a function of other work-related independent variables for both males and females. These are:

1. Distance of push/pull: 2.1, 7.6, 15.2, 30.5, 45.7, and 61.0 m (or 7, 25, 50, 75 and 100 ft.)
2. Frequency of push/pull: The same frequencies are not given for all distances, but each distance has force limits for one exertion per 8 hr., 30 min., 5 min., and 2 min. Greater frequencies are quoted for the smaller distances.

TABLE 23.1   Maximum Acceptable Forces of Pull for Females (kg)

Height	Percent	2.1 m pull							45.7 m pull				
		6	12	1	2	5	30	8	1	2	5	30	8
		s		min				h	min				h
						Initial forces							
135	90	13	16	17	18	20	21	22	12	13	14	15	17
	75	16	19	20	21	24	25	26	14	16	17	18	20
	50	19	22	24	25	28	29	31	17	18	20	21	24
	25	21	25	28	29	32	33	35	19	21	23	24	27
	10	24	28	31	32	36	37	39	22	24	25	27	31
57	90	15	17	19	20	22	23	24	13	14	15	17	19
	75	17	20	22	23	26	27	28	16	17	18	20	22
	50	20	24	26	27	30	32	33	18	20	22	23	26
	25	23	27	30	31	35	36	38	21	23	25	27	30
	10	26	31	34	35	39	40	43	24	26	28	30	34
						Sustained forces							
135	90	6	9	10	10	11	12	15	6	6	7	7	9
	75	8	12	13	14	15	16	20	8	9	9	9	12
	50	10	16	17	18	19	21	25	10	11	11	12	16
	25	13	19	21	21	23	25	31	12	13	14	14	19
	10	15	22	24	25	27	29	36	14	15	16	17	23
57	90	5	8	9	9	10	11	13	5	6	6	6	8
	75	7	11	12	12	13	14	18	7	8	8	8	11
	50	9	14	15	16	17	18	23	9	10	10	11	14
	25	11	17	18	19	21	22	27	11	12	12	13	19
	10	13	20	21	22	24	26	32	12	14	14	15	20

Height = vertical distance from floor to hand–object (handle) contact.

Percent = percentage of industrial workers capable of exerting the stated forces in work situations.

From Snook, S.H. and Ciriello, V.M. 1991. The design of manual handling tasks: revised tables of maximum acceptable weights and forces. *Ergonomics*, 34(9): 1197–1213. With permission.

3. Height (vertical distance from floor to hands) at which push/pull was exerted. Different heights are given for males and females — 144, 95, and 64 cm (or 57, 37.4, 25.2 in.) for males; and 135, 89, and 57 (53.1, 35.0, and 22.4 in.) for females.

4. The percentage of workers (10, 25, 50, 75, and 90%) who are capable of sustaining the particular force during a typical eight-hour job.

The data are also given for both initial force (force required to get object in motion) and sustained force (force required to keep object in motion). Each number in a table, therefore, corresponds to *the maximum initial or sustained push or pull force which a given percent percentage of the population can exert for a specified distance, at a specified frequency, and at a specified height without a significant chance of being injured or developing cumulative disorders.* Profound judgment must be used when applying the results of these tables in the workplace because the tables do not represent all possible push–pull conditions, and because the tabled conditions are represented by discrete states. It may be necessary to interpolate force limits for variable values not stated in the tables but which are within the stated ranges, for example, pulling or pushing over a distance of 9.1 m (30 ft.); but, this must be done with caution since the relationships are not linear.

Note that body posture is not a variable in these tables, even though it is one of the most influential variables in force exertion. This is because, as stated earlier, posture changes continuously during dynamic activities. The best description of posture relevant to these tables and the appropriate assumptions for work conditions are:

TABLE 23.2   Maximum Acceptable Forces of Push for Females (kg)

Height	Percent	2.1 m push							45.7 m push				
		6	12	1	2	5	30	8	1	2	5	30	8
		s		min				h	min				h
		Initial forces											
	90	14	15	17	18	20	21	22	12	13	14	15	17
	75	17	18	21	22	24	25	27	15	16	17	19	21
135	50	20	22	25	26	29	30	32	18	19	21	22	25
	25	24	25	29	30	33	35	37	20	22	24	26	29
	10	26	28	33	34	38	39	41	23	25	27	29	33
	90	11	12	14	14	16	17	18	11	12	12	13	15
	75	14	15	17	17	19	20	21	13	14	15	16	18
57	50	16	17	20	21	23	24	25	15	17	18	19	22
	25	79	20	23	24	27	28	30	18	19	21	22	25
	10	21	23	26	27	30	31	33	20	22	23	25	28
		Sustained forces											
	90	6	8	10	10	11	12	14	5	5	5	6	8
	75	9	12	14	14	16	17	21	7	8	8	8	11
135	50	12	16	19	20	21	23	28	9	10	11	11	15
	25	16	20	24	25	27	29	36	11	13	13	14	19
	10	18	23	28	29	32	34	42	14	15	16	17	22
	90	5	6	8	8	9	9	12	5	5	5	6	7
	75	7	9	11	12	13	14	17	7	7	8	8	11
57	50	10	13	15	16	17	18	23	9	10	10	11	15
	25	12	16	19	20	22	23	29	11	13	13	14	19
	10	15	19	23	23	26	28	34	13	15	16	16	22

Height = vertical distance from floor to hand–object (handle) contact.
Percent = percentage of industrial workers capable of exerting the stated forces in work situations.
From Snook, S.H. and Ciriello, V.M. 1991. The design of manual handling tasks: revised tables of maximum acceptable weights and forces. *Ergonomics*, 34(9): 1197–1213. With permission.

1. The force limits apply to subjects walking in unrestricted postures and exerting two-handed symmetrical forces on handles (with comfortable grips) with the hands in approximately the same horizontal plane.
2. The walking surface is flat and level, and provides enough traction to prevent slipping of the feet. In the experiments in which the tabled data were developed, subjects wore shoes that provided high friction with their contact surface (a treadmill walkway).
3. The forces in the tables are the horizontal components of the applied muscular forces. In the experiments, subjects directed their exertions in the horizontal.
4. Environmental conditions are considered comfortable. The experiments were conducted in a climate controlled chamber at 21°C (80.6°F) and 45% humidity.
5. The people pushing and pulling are physically fit and accustomed to manual labor.
6. There is no obstruction to body movement due to clothing or obstacles along the path of movement.

If we wished to evaluate an existing MMH system, we can compare the load being pushed or pulled with the value in the table corresponding most closely to the actual task conditions. Tasks with high significant risks for injuries can then be identified. If we wished to design a task to prevent injuries we can use the tabled values as force limits for the tasks. Where there are differences in task conditions from the experimental conditions on which the tables are based, as listed above, we should use our judgment and make appropriate adjustments, if necessary, to the table values.

TABLE 23.3 Maximum Acceptable Forces of Pull for Males (kg)

Height	Percent	2.1 m pull							45.7 m pull				
		6	12	1	2	5	30	8	1	2	5	30	8
		s		min				h	min				h
						Initial forces							
	90	14	16	18	18	19	19	23	10	11	13	13	16
	75	17	19	22	22	23	24	28	12	14	16	16	20
144	50	20	23	26	26	28	28	33	15	16	19	19	24
	25	24	27	31	31	32	33	39	17	19	22	22	28
	10	26	30	34	34	36	37	44	20	22	25	25	31
	90	22	25	28	28	30	30	36	16	18	21	21	26
	75	27	30	34	34	37	37	44	19	22	25	25	31
64	50	32	36	41	41	44	44	53	23	26	30	30	37
	25	37	42	48	48	51	51	61	27	30	35	35	43
	10	42	48	54	54	57	58	69	30	34	39	39	49
						Sustained forces							
	90	8	10	12	13	15	15	18	6	7	8	9	10
	75	10	13	16	17	19	20	23	7	9	10	11	14
144	50	13	16	20	21	23	24	28	9	11	12	14	17
	25	15	20	24	25	28	29	34	11	13	15	17	20
	10	17	22	27	28	32	33	39	12	14	17	19	23
	90	11	14	17	18	20	21	25	8	9	11	12	15
	75	14	19	23	23	26	27	32	10	12	14	16	19
64	50	17	23	28	29	32	34	40	13	15	17	20	23
	25	20	27	33	35	39	40	48	15	18	21	24	28
	10	23	31	38	40	45	46	54	17	20	24	27	32

Height = vertical distance from floor to hand–object (handle) contact.

Percent = percentage of industrial workers capable of exerting the stated forces in work situations.

From Snook, S.H. and Ciriello, V.M. 1991. The design of manual handling tasks: revised tables of maximum acceptable weights and forces. *Ergonomics*, 34(9): 1197–1213. With permission.

One-Handed Force Magnitudes

All sitting push–pull forces described in this chapter are one-handed ones. Daams (1993) provides data for standing one-handed exertions. As for two-handed standing forces, one-handed forces vary considerably among studies with similar variables, and within individual studies depending on the test conditions or variables. An examination of published studies of one-handed MVC forces achievable by people range widely among studies and even among test conditions within individual studies. Thus, generalizations on recommended forces are not easy to promote. Two are mentioned below. It would be more appropriate to state the ranges of these forces. Average static standing push–pull forces have ranged from 70 to 134 N and sitting forces from 350 to 540 N. Dynamic pull forces have ranged from 170 to 380 N in females and from 335 to 673 N in males when sitting, for almost all studies. Average pull forces in males, while lying down prone, have ranged from 270 to 383 N and push forces, 285 to 330 N (Hunsicker and Greey, 1957). Davis and Stubbs (1980) and Mital, Nicholson, and Ayoub (1993) have published guidelines for one-handed push–pull forces, but they are not specific enough and must be interpreted with caution. They do not cover the wide range of conditions (posture, reach, static or dynamic contraction, handles, etc.) that can occur in a typical push-pull task. David and Stubbs (1980) recommended one-handed occasional standing static push forces for three age groups (under 40 yrs., 41 to 50 yrs., and 51 to 60 yrs.) for different reach distances in the range 5 to 70 cm. The values from their graphs range from 30 kg at 5 cm to 15 kg at 70 cm for under 40-yrs. males, with the values decreasing by 1 to 2 kg

TABLE 23.4 Maximum Acceptable Forces of Push for Males (kg)

Height	Percent	2.1 m pull							45.7 m pull				
		6	12	1	2	5	30	8	1	2	5	30	8
		s		min				h	min				h
						Initial forces							
	90	20	22	25	25	26	26	31	13	14	16	16	20
	75	26	29	32	32	34	34	41	16	18	21	21	26
144	50	32	36	40	40	42	42	51	20	23	26	26	33
	25	38	43	47	47	50	51	61	24	27	32	32	39
	10	44	49	55	55	58	58	70	28	31	36	36	45
	90	19	22	24	24	25	26	31	12	14	16	16	20
	75	25	28	31	31	33	33	40	16	18	21	21	26
64	50	31	35	39	39	41	41	50	20	22	26	26	32
	25	38	42	46	46	49	50	59	24	27	31	31	39
	10	43	48	53	53	57	57	68	27	31	36	36	44
						Sustained forces							
	90	10	13	15	16	18	18	22	7	8	10	11	13
	75	13	17	21	22	24	25	30	10	11	13	15	18
144	50	17	225	27	28	31	32	38	12	14	17	19	23
	25	21	27	33	34	38	40	47	15	18	21	24	28
	10	25	31	38	40	45	46	54	18	21	24	28	33
	90	10	13	16	16	18	19	23	7	8	9	11	13
	75	14	18	21	22	25	26	31	9	11	12	14	17
64	50	18	23	28	29	32	33	39	12	14	16	18	22
	25	22	28	34	35	39	41	48	14	17	20	23	27
	10	26	32	39	41	46	48	56	17	20	23	26	31

Height = vertical distance from floor to hand–object (handle) contact.

Percent = percentage of industrial workers capable of exerting the stated forces in work situations.

From Snook, S.H. and Ciriello, V.M. 1991. The design of manual handling tasks: revised tables of maximum acceptable weights and forces. *Ergonomics*, 34(9): 1197–1213. With permission.

for 41 to 50-yrs. males and by 4 to 8 kg for 51 to 60-yrs. males. They also recommended a 30% decrease in these values for pushing–pulling at frequencies greater than once per minute. Mital, Nicholson, and Ayoub (1993) recommend the following guidelines for a typical workday for standing work: A push force of 107 N (24 lbs.) for an exertion of less than once per minute, and 73 N (16.5 lbs.) for greater frequencies; a pull force of 98 N (22 lbs.) for less than once per minute and 67 N (15 lbs.) for greater frequencies.

Pinch–Pull Force Magnitudes

Pinching and pulling with one hand while stabilizing the object with the other hand has been observed in male adults to yield forces of 100, 68, and 50 N when using the lateral, chuck, and pulp pinches, respectively (Imrhan and Sundararajan, 1992; Imrhan and Alhaery, 1994).

23.4 Conclusions

Push–pull strengths depend on numerous task-related factors. The strength variations have been examined from controlled laboratory experiments and have yielded profound insights into the characteristics of these strengths. However, much is still not known about push–pull strengths or acceptable loads (forces). The use of these strengths for establishing force limits for the design of tasks, equipment, and workplace still depend strongly on data gathered from simulation of typical occupational activities. The best compromise, at present, is to use the results of simulated experiments, as published by Snook and

Ciriello (1991), and modify the data according to the information available on the nature of push–pull strengths.

Defining Terms

Body support: An object which a person can brace on to enhance muscular force exertion.

Coefficient of (static) friction (μ): The ratio of the maximum force (F) acting along the area of contact between an object in contact with another body to the weight (N) of the object (or the force pressing the two bodies together); that is, $\mu = F/N$. Its value depends on the types of materials in contact with each other. Coefficient of friction for wet (slippery) surfaces may be below 0.2 and for dry surfaces with very good traction, above 0.8.

Horizontal foot distance: The perpendicular distance between the ankle of the pivoting foot (for separated feet) and hand–handle contact, when standing and pushing or pulling.

Manual materials handling: Moving or stabilizing loads with the hand(s) mostly by pulling, pushing, lifting, lowering, or carrying.

Maximum acceptable force: The maximal force (in pushing, pulling, or any other activity) a person is willing to exert voluntarily under work conditions during a workday with the opinion that the force will not cause undue discomfort or strain. It is usually determined experimentally by simulating work conditions in a laboratory.

Maximal voluntary contraction (MVC): A muscular contraction in which a person applies the strongest effort (for lifting, pushing, pulling, etc.) to the point where he or she does not suffer from significant muscular discomfort or pain. The resulting force or torque, measured with an appropriate instrument, is called the person's MVC strength for that particular task.

Push: A muscular effort applied to an object such that the object moves or tends to move away from the body. Neither the direction of movement nor the applied force need be linear or directly toward the body, as long as the deviation is not sharp.

Pull: A muscular effort applied to an object such that the object moves or tends to move closer to the body. The same conditions for direction as in "push" apply.

Push–pull strength (force): The maximal force (peak or 3-second mean) exerted on an object at the hand–object interface in a single MVC push or pull exertion, under specified conditions.

Reach distance: The perpendicular distance from the shoulder to the hand at the hand–object contact.

Safe force limits: Safe push–pull force exertion limits are the maximum force magnitudes that people can exert without injuries (for single strong muscular exertions) or cumulative trauma disorders (for repeated exertions) of the upper extremities under specified conditions.

Wet bulb globe temperature: A temperature value (in degrees Celsius or Fahrenheit) that represents the combined effect of air temperature, natural wet bulb temperature (reflecting humidity), and radiant temperature (solar load). It is used for evaluating environmental heat stress conditions.

References

Ayoub, M.M. and McDaniel, J.W. 1974. Effect of operator stance and pushing and pulling tasks. *Transactions of American Institute of Industrial Engineers*, 6, 185-195.

Ayoub, M.M. and LoPreisti, P. 1971. The determination of an optimum size handle by use of EMG. *Ergonomics*, 14: 509-518.

Caldwell, L.S. 1964. Body position and strength and endurance of manual pull. *Human Factors*, 6: 479-484.

Baril-Gingras, G. and Lortie, M. 1990. Les modes operatoires et leur determinants: Etudes des activites de manutention dans une grand entreprise de transport, in *Proceedings of the 23rd Annual Conference of HFAC*, Ottawa, pp. 137-142.

Chaffin, D.B., Andres, R.O., and Garg, A. 1983. Volitional postures during maximal push/pull exertions in the sagittal plane. *Human Factors*, 25: 541-550.

Ciriello, V.M. and Snook, S.H. 1978. The effects of size distance, height, and frequency on manual handling performance, in *Proceedings of the Human Facttors Society 22nd Annual Meeting*, Detroit, MI, pp. 318-322.

Ciriello, V.M. and Snook, S.H. 1983. A study of size, distance, height, and frequency effects on manual handling tasks. *Human Factors*, 25: 473-483.

Ciriello, V.M., Snook, S.H., Blick, A.C., and Wilkinson, P.L. 1990. The effect of task duration on psychophysically determined maximum acceptable weights and forces. *Ergonomics*, 33: 187-200.

Daams, B.J. 1993. Static force exertion in postures with different degrees of freedom. *Ergonomics*, 36: 397-406.

Davis, P.R. and Stubbs, D.A. 1980. *Force Limits in Manual Work*. Guilford: IPC Science and Technology Press.

Fothergill, D.M., Grieve, D.W., and Pheasant, S.T. 1991. Human strength capabilities during one handed maximum voluntary exertions in the fore and aft plane. *Ergonomics*, 35: 203-212.

Gallagher, S. 1989. Isometric pushing, pulling and lifting strengths in three postures, in *Proceeding of the HFS 33rd Annual Meeting*, Santa Monica, CA, pp. 637-640.

Garg, A. and Beller, D. 1990. One handed dynamic pulling strength with special reference to speed, handle height and angles of pulling. *Intl. J. Ind. Ergon.*, 6: 231-240.

Hunsicker, P.A. and Greey, G. 1957. Studies in human strength. *Research Quarterly*, 28:109

Imrhan, S.N. 1983. *Modeling Isokinetic Strength of the Upper Extremity*. Ph.D. Dissertation, Texas Tech University Lubbock, TX.

Imrhan, S.N. and Ayoub, M.M., 1985, An analysis of rotary and pull strength of the upper extremity, in *Trends in Ergonomics/Human Factors II*, ed. R.E. Eberts and C.G. Eberts, Elsevier Science Publishers, B.V., North Holland.

Imrhan, S.N. and Ayoub, M.M. 1988. Predictive models of upper extremity rotary and linear pull strength. *Human Factors*, 30(1): 83-94.

Imrhan, S.N. and Ayoub, M.M. 1990. The arm configuration at the point of peak dynamic pull strength. *Intl. J. Ind. Ergon.*, 6: 9-15.

Imrhan, S.N. and Alhaery, M. 1994. Finger pinch-pull strengths: large sample statistics, in *Advances in Industrial Ergonomics and Safety VI*, ed. F. Aghazadeh. Taylor & Francis, London, pp. 595-597.

Imrhan, S.N. and Ramakrishnan, U. 1992. The effects of arm elevation, direction of pull and speed of pull on isokinetic pull strength. *Intl. J. Ind. Ergon.*, 9: 265-273.

Imrhan, S.N. and Sundararajan, K. 1992. An investigation of finger pull strengths. *Ergonomics*, 35(3): 289-299.

Klein, B.P., Jensen, R.C., and Sanderson, L.M. 1984. Assessment of workers' compensation claims for back strains/sprains. *J. Occup. Med.*, 26(6): 443-448.

Kroemer, K.H.E. 1969. *Push Forces Exerted in Sixty-Five Common Working Positions*. ARML-TR, WPAFB, Ohio, 68-143.

Kroemer, K.H.E. 1974. Horizontal push and pull forces. *Applied Ergonomics*, 5(2): 94-102.

Kroemer, K.H.E. and Robinson, D.E. 1971. *Horizontal Static Forces Exerted by Men Standing in Common Working Postures on Surfaces of Various Tractions*. Aerospace AMRL-TR, WPAFB, Ohio, 70-114.

Kumar, S. and Garand, D. 1992. Static and dynamic strength at different reach distances in symmetrical and asymmetrical planes. *Ergonomics*, 35: 861-880.

Kumar, S. 1994. The back compressive forces during maximal push–pull activities in the sagittal plane. *J. Human Ergol.*, 23: 133-150.

Kumar, S., Narayan, Y., and Bacchus, C. 1995. Symmetric and assymetric two-handed pull-push strength of young adults. *Human Factors*, 37(4): 854-865.

Lee, K.S., Chaffin, D.B., Herrin, G.D. and Waikar, A.M. 1991. Effect on handle height on lower back loading in cart pushing and pulling. *Applied Ergonomics*, 22(2):117-123.

Lee, K.S., Chaffin, D.B., Waikar, A.M., and Chung, M.K. 1989. Lower back muscle forces in pushing and pulling. *Ergonomics*, 32: 1551-1563.

Lower, R.S., Schutz, R.K., and Sadosky, T.S. 1977. A prediction model of arm push strength in the transverse plane, in *Proceedings of the Human Factors Society 21st Annual Meeting*, Human Factors Society, Santa Monica, CA, pp. 132-136.

Martin, J.B. and Chaffin, D.B. 1972. Biomechanical computerized simulation of human strength in sagittal-plane activities. *AIIE Trans.* 4(1): 19-28.

Mital, A., Nicholson, A.S., and Ayoub, M.M. 1993. *A Guide to Manual Materials Handling*. Taylor & Francis, Washington, D.C.

Mital, A. and Faard, H.F. 1990. Effects of sitting and standing, reach distance, and arm orientation on isokinetic pull strengths in the horizontal plane. *Intl. J. Ind. Ergon.* 6(3): 241-248.

NIOSH 1981. *Work Practices Guide for Manual Lifting*. U.S., D.H.S.S., Pub. No. 81-122.

Pheasant, S.T. and Grieve, D.W. 1981. The principal features of maximal exertions in the sagittal plane. *Ergonomics*, 24(5): 327-338.

Pheasant, S.T., Grieve, D.W., Rubin, T. and Thompson, S.J. 1982. Vector representations of human strength in whole body exertion. *Applied Ergonomics*, 13(2), 139-144.

Resnick, M.L. and Chaffin, D.B. 1995. An ergonomic evaluation of handle height and load in maximal and submaximal cart pushing. *Applied Ergonomics*, 26(3): 173-178.

Snook, S.H. and Ciriello, V.M. 1974a. Maximum weights and workloads acceptable to female workers. *J. Occ. Med.*, 16: 527-534.

Snook, S.H. and Ciriello, V.M. 1974b. The effects on heat stress on manual handling tasks. *Am. Ind. Hyg. Assn. J.*, 31: 681-685.

Snook, S.H., Irvine, C.H., and Bass, S.F. 1970. Maximum weights and workloads acceptable to male industrial workers. *Am. Ind. Hyg. Assn. J.*, 31: 579-586.

Snook, S.H. 1978. The design of manual tasks. *Ergonomics*, 21: 963-985.

Snook, S.H. and Ciriello, V.M. 1991. The design of manual handling tasks: revised tables of maximum acceptable weights and forces. *Ergonomics*, 34(9): 1197-1213.

Troup, J.D.G. and Edwards, F.C. 1985. *Manual Handling and Lifting. An Information and Literature Review with Special Reference to the Back*, Her Majesty's Stationery Office, London.

Van Cott, H.P. and Kinkade, R.G. (eds.) 1972. *Human Engineering Guide to Equipment Design*, McGraw-Hill Book Company, New York.

Warwick, D., Novack, G., Schultz, A., and Berkeson, M. 1980. Maximal voluntary strengths of male adults in some lifting pushing and pulling activities. *Ergonomics*, 23, 49-54.

For Further Information

The design of manual handling tasks: revised tables of maximum acceptable weights and forces. by S.H. Snook and V.M.Ciriello, 1991, *Ergonomics*, 34(9) 1197-1213. This paper gives detailed tables of maximum acceptable force limits (for eight hours of work) for many manual materials handling tasks, including pushing and pulling.

Human Engineering Guide to Equipment Design edited by H.P Van Cott and R.G. Kinkade, R.G., 1972, Washington, D.C., U.S. Govt. Printing Office. This book gives several tables of data on push–pull static strengths in standing, sitting, and lying down postures.

Manual Materials Handling by M.M. Ayoub and A. Mital, 1989, Taylor & Francis, NY. This book gives tables of some recent data on push–pull activities.

24
Force Exertion in User–Product Interaction

Brechtje J. Daams
Delft University of Technology

24.1   Introduction

How to crack a nut? How to design a nutcracker with which to crack a nut? What force needs to be applied to a nutcracker designed to crack that nut? What force can be applied by the intended user of that nutcracker designed to crack that nut?

Questions like these arise in the process of product design. The forces users can and will exert in the use of a product are important criteria for the design of that product. It is usual to distinguish consumer products from professional products. This is a useful distinction in so far as it relates to the differences between the composition of a group of professional users and a group of normal consumers, and to the differences in the way both groups use the products. Consumer products are usually bought by their prospective users. These users are usually nonspecialists with diverse backgrounds, different levels of education, and of varying strength. Consumers include children, the elderly, the physically disabled, and the world's largest minority group, women.

Professional products are, as a rule, used frequently, possibly up to 8 hours a day, five days a week, all year round. A group of professional users is more homogeneous than a group of consumers. Users of professional products are aged between about 18 and 65 and are generally healthy. Even so, their strength may vary substantially. Both types of products and users are considered in this chapter.

In general the design limits for a product are dictated by the comfort, efficiency, and safety of its operation. Products should be designed so that even the weakest of the intended users are able to operate, use, or handle them. Maximal forces may be required of users for a short time in cases like emergencies, but in general a comfortable level of force is preferred.

Products should also be designed to withstand the largest forces strong users may exert. Accidental damage or breakage must be avoided, as this may lead to anything from minor annoyance to serious injury.

Consumer products that cause discomfort will be difficult to sell. Professional products that cause health damages, either in the short term or in the long term, will be banned from the workplace. Both from an ethical and an economical point of view it is therefore important for designers to adapt products to the force capabilities of their prospective users.

The use product designers may make of the results from research on force exertion is the central issue of this chapter. First, a guideline is given on how to consider force exertion in the design process. A short survey of results pertaining to relevant aspects of force exertion is summarized in a list of rules of thumb. Finally, recommendations are given on how to determine the relevance of results of literature and research to a particular design.

24.2 Considering Force Exertion in Product Design

The Design Process

Design is an iterative process. This implies that there is no clear-cut path from a problem posed to a product designed to solve that problem. Nevertheless, specific activities can be discerned in the design process and some order in these activities is advisable (Roozenburg and Eekels, 1991). The starting point of a design is the definition of the problem to be solved, an assessment of its setting, and the collection and analysis of information. On the basis of this, a program of requirements for the future product is formulated. Ideas are generated, sketches are made, good ones are worked out, and of these the best one is selected through comparison with the requirements. Details are then dealt with and technical drawings and/or a model of the design is made. If possible, a test version of the product is evaluated through user trials. The results of these trials may be incorporated in an improved version of the design.

In the next section an attempt is made to outline how considerations of the force capabilities of users may be incorporated in the design process as it has been described above.

Considering Force Exertion in the Design Process

Problem definition. A product fulfills a need and solves a problem. Nuts need to be cracked because people wish to eat nuts. Grass should not be left to grow too long. These are both problems for which a number of solutions already exist. A good and accurate definition of the problem is essential to an appropriate solution. The problem definition should refer to the problem to which the product should be the solution. Defining the problem by stating a solution is not correct. It remains, however, common practice, in particular with principals. They may tell the designer, for example, that the problem is that she has to design a nutcracker, or that the problem is that no lawn mower suitable for elderly people exists. Such a problem definition is unnecessarily restrictive. It will result in yet another nutcracker and the umpteenth lawn mower. To define a problem as a solution (or a product) limits creative thought. Due consideration of users' force capabilities at the problem definition stage may lead to new, unexpected, and unconventional solutions that would be better in the given situation. These solutions do not necessarily involve designed products. Thus the sale of peeled and cracked nuts is an alternative to the sale of nutcrackers. For the elderly, perhaps a walking frame with lawn-mowing option would be a good solution to the problem of long grass. More generally, keeping a sheep or goat, or even changing the lawn into maintenance-free paving, may be a good solution. Although these are extreme examples, they illustrate the point.

Assessment of the setting. The function(s) of the product, the target user group, the circumstances of use, and the relevant behavior of the users should be established or estimated. For example, it makes a great difference whether the future users will be children or adults, male or female, and whether they will wear gloves or not. Will they be able to brace themselves to exert force? On a product that is hand-held and can be manipulated in the best position (like a jam jar), the maximal force that can be exerted is different from that on a fixed object where obstructions can hinder the adoption of the optimal posture. If the product is to be used frequently, this aspect should certainly be taken into account when establishing the maximal force. A product used by a person in a state of panic may demand a different force (and less precision!) of the user than one used for leisure purposes.

For example, after assessment it may be clear that the target group consists of the general public aged ten and upward (all minorities who are able to walk included); that the function of the product is to mow the lawn; that the situation of use is outdoors on a lawn, often in the sun and sometimes in wind and rain; that the users may be wearing a minimum of clothing and footwear, that the product will be used for a period of a few minutes to an hour once every week; that the exerted force will be dynamic; that there is no possibility for the users to brace themselves while exerting force; and that, if the force that has to be exerted is too great, people will buy a motorized lawn mower right away or next time.

Choosing design principles. The proposed function of the product must be translated into an action which can be performed optimally by the user. In this view, the product is the concrete intermediate between the two. At this stage, a designer should not be concerned with concrete products, but rather with selecting the best working principle for force exertion. Knowledge of the principles of physics and human force exertion and essential logical thinking are indispensable. The question a designer should ask is not "How much force can be exerted this way?" but "Is this the optimal way to exert force, or is there a better one?" A thorough evaluation of principles will make the selection of relevant information on force exertion in the subsequent stages of the design process more effective. It will help to avoid the risk of being bogged down by an excess of information, resulting in the dreaded "designer's block."

Assessment of information needed. Once the design principles have been established, it can be inferred which information on force exertion is needed. The users, their postures, the direction of the exerted force, and bracing possibilities are known within certain limits. The force capabilities are distributed over a wide range. The average or the median of this distribution is generally not a very useful measure for product design. Products designed for the average force cannot be operated by the weaker end of the population distribution, and may be damaged or broken by the strongest users. For design purposes, the weakest users are often more relevant than the strongest, and beginners more so than experienced users. The same applies to the postures and the directions of the exerted forces. It depends on the requirements of the product whether the lower maximally exertable forces, or the higher maximally exertable forces, or maybe both, should indicate the design limits. In addition to deciding whether the weakest or the strongest users are the most relevant for specific design purposes, the designer ought to decide and clearly define which percentile of the chosen population he or she is designing for: P_5, P_1, or P_{99}? Especially with extreme percentiles this can make an enormous difference for the values involved, and consequently for the design.

Here a comment should be made on the widespread habit of quoting the P_5 as a "normal" design maximum. To the annoyance of millions of people, many designers think it is accepted, standard, or even good practice, to exclude the upper or lower 5% (or both) of a population. This "P_5-P_{95} syndrome" results in products that cannot be used effectively, or comfortably, by 10% of the users.

It is important to note that products that are designed for the lower percentiles of the population (where forces are concerned, these are the weaker persons) can be easier to use or operate by the average user. Those products may sell even better to "normal" users than standard products, because they too appreciate clear, simple features and light operation. Unless, of course, these products get stigmatized as being specially made for the disabled and the elderly. Neither strong nor weak users want to be seen with a product that visibly classifies them as weak. If this negative image can be avoided, products designed for the weak can be a (commercial) success with everyone.

Taking again the example of the lawn mower, one can now look up which force can be exerted, for instance, comfortably for a few minutes up to an hour by 99% of the general public aged 10 years and older (all minorities that are able to walk included), pushing horizontally with two hands on a handle with a diameter of, for example, 3 cm, at elbow height while walking (dynamic force), without bracing themselves. External factors which should be taken into account include clothing, footwear, and the weather.

Gathering of information. The most relevant and exact information is obtained by custom research using a representative sample of the target user group in the required postures. This will best serve to reduce design uncertainty.

One should never take one's own strength as a measure "to get an indication." The large dispersion of data on force exertion (Sanchez and Grieve, 1992; Sanders and McCormick, 1993) goes unnoticed when measurements are limited to one or a few persons. We are none of us the average person, but still there are many designers who presume that if they can exert a certain force, anybody else must be able to do so too.

If custom research, even on a small scale, is impossible, literature is the alternative, though at best it will give only an indication of the range of forces involved. Caution is required with the application of data from literature. Generally the subjects and the experimental setting will not be similar, or even relevant, to those of the product. Consequently, such figures will be difficult to use. They should only be applied with the necessary comments and safety margins.

Definition of requirements. When it is known within which limits the product is to function, the program of requirements can now be extended with requirements concerning force exertion. As with all requirements, those on force exertion should preferably be operational. Clearly defined limits for compliance (preferably in hard numbers) must be included in the requirements. Only thus can a design later be tested for compliance with these requirements.

Design. In the design stage, the requirements are translated into a design for a product, in which they are assimilated as much as possible. Conflicting requirements sometimes lead to the inevitable compromise, a process inherent to design. Obviously, some requirements, especially those pertaining to safety, should not be compromised. It is the art of the designer to come up with a good product, despite any conflicts that may have arisen.

Follow-up. The first prototypes should be tested with subjects and evaluated, and the product should, if possible, be improved accordingly. This is important, especially if the information forming the basis for the program of requirements is obtained from literature.

Examples

The outline of the way considerations of force exertion are incorporated in the design process is best illuminated with some examples from practical experience. They illustrate the point that questions on force exertion in user–product interaction can never receive a standard, ready-to-use answer.

A first example of a product to be (re)designed is a large professional cheese slicer, as used in supermarkets (see Figure 24.1). Present cheese slicers are fitted with a handgrip at the end of a blade that rotates around a pivot. Enlargement of the blade (or the arm of the moment) reduces the force needed to slice, but increases the movement of arm and hand. The handgrip is positioned in the same direction as the blade, so that the wrist is in an uncomfortable position when exerting force. The slicer is usually positioned on a table or bench, so that the force is exerted on the handle from about shoulder height to about elbow height. The users are professional women and men between 18 and 65. Instead of trying to find out how much force can be exerted in such a situation, thought is given to a more comfortable way to slice. Suggested improvements include lower positioning of the equipment, so that the force is exerted with the hand at elbow height and lower, and body weight can be utilized. A further suggestion is to change the handgrip so that the wrist need not be flexed to extreme angles. If possible, the movement should be a translation instead of a rotation, so that force needs to be exerted in one direction only. The optimal length of the arm, weighing the length of the stroke against the force needed to operate the cheese slicer, may be determined experimentally.

A second product is a wheelchair for children in Sri Lanka who suffer from the consequences of polio. Their legs are paralyzed, so the vehicle has to be moved by using the arms. For some mechanical reason, force is to be exerted in one direction only: either pushing or pulling. Deciding on pulling horizontally, based solely on the fact that in this direction the greatest force can be exerted, would be wrong. When pulling in a horizontal direction, the child will tend to pull himself/herself out of the seat, which is uncomfortable and prevents maximal force exertion, even if the child were to be strapped to the back of the seat. When pushing, however, the child will be able to brace himself/herself against the back of the seat, which is more comfortable, and allows more force to be exerted. Therefore, pushing against the

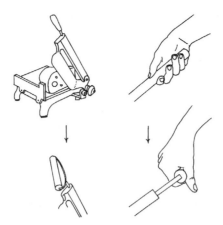

FIGURE 24.1 Suggestion for improvement of a professional cheese slicer. (From Daams, B.J. 1994. *Human Force Exertion in User–Product Interaction.* Delft University Press, Delft. With permission.)

back of the seat is preferred. It should be noted, however, that to exert force in both directions in a cyclic movement is better for the development of the muscles. Uneven development of muscle groups may lead to incorrect loading of the joints and related problems in the future. The use of two arms instead of one will, of course, allow more force to be exerted, and will also stimulate more symmetrical physical development in the child. To know the maximal forces, these should be measured for the actual children concerned. Although there are some data on maximal pushing and pulling abilities of European children in literature (e.g., Steenbekkers, 1993), this information cannot be used, because it cannot be assumed that children who are disabled and from a different ethnic group exert maximal forces equal to those exerted by able-bodied European children.

For the design of a portable or rolling easel, information on maximal exertable push, pull, and carry forces were asked for. The easel is intended to be used for outdoor painting and should be easily transportable by a person on foot. The target group consists of elderly people, so the required force forms an important aspect of the design. Carrying is no option, because it will certainly require more force and energy than simply rolling the easel along. Rather than pushing it, a wheeled object is preferably pulled along, because this makes it easier to negotiate ramps and curbs, as a straightforward analysis of the physics involved shows. To decrease rolling resistance, few and large wheels are advised. The force necessary to stabilize and maneuver the easel should be as small as possible, because energy and attention should not be diverted from the main activity, pulling the easel along. Therefore, two wheels are preferred over one.

The original question thus reduces to how much force elderly people can exert when pulling something along. If the lowest part of the distribution (say the P_1) of elderly women is included in the target group, the force that can be exerted is practically zero, for these people have barely sufficient strength to walk about unsupported, and will not have much force left to pull easels around. These considerations lead to the conclusion that exact information on the strength of the users is no longer relevant. The easel should be designed to be pulled with the minimal possible force to allow the largest possible proportion of the target user group to use it.

24.3 Summary of Literature

Literature is the alternative if custom research tailored to a specific product design, even on a small scale, is impossible. Finding the relevant information can be difficult though, because literature on force exertion is sparse and far between, many variables play a role, and research is generally aimed at working conditions, rehabilitation, and sports rather than the use of products.

Results quoted in literature have been obtained mainly for standardized and static postures, and concern maximal forces, which are usually measured for about four seconds. In practice, however, only the position of a handle or control is known and subjects will exert static or dynamic submaximal force for any time and in any posture they feel like, and to some extent with different muscle groups.

Nevertheless, some results from literature can be generalized and may serve as rules of thumb. These rules can be useful to find a good technical principle for force exertion in the early stages of a design process. Their validity should be checked at a later stage of the design process when product ideas are tested in more detail. The summary of these rules of thumb we give here has been arranged according to the variables person, posture, product, environment, and force.

Person

Gender. The physical strength of women is roughly half to two thirds that of men. The ratio of mean female to mean male maximal force varies for various adult subject groups between 35% (Sing and Karpovich, 1968) and 88% (Fothergill et al., 1992); for elderly people over 75 years of age between 19% (Page, 1981) and 68% (Mathiowetz et al., 1985); and for children between 74% and 103% (Mathiowetz et al., 1986 and Steenbekkers, 1993). Although males and females differ significantly in strength, the effect of gender is small after allowing for body size and composition.

Age. Maximal strength is attained between 20 and 30 years, it is relatively stable from 20 up to 59 years and starts to decline between approximately 35 to 60 years, but there is no close agreement on the exact ages in literature. The large dispersion in the quoted age limits reflects the disagreement on the exact ages found in literature. There is also little agreement on how rapidly strength declines with age. This may be related to the increase of inter-individual variance with age.

Anthropometric variables. The correlation between maximal force exertion and body height and weight is in general significant but not very high, except for children.

Laterality. The difference in strength between the preferred and the nonpreferred side of the body ranges between "not significant" for finger pull strength of right-handed subjects (Imrhan and Sundararajan, 1992) and the nonpreferred side exerting 87% of the hand grip strength of the preferred side (Mathiowetz et al., 1985).

Motivation. With auditory encouragement and visual feedback, maximal force exertion increases about 10% (Peacock et al., 1981).

Gloves. When gloves are worn by users, the maximally exertable force may decrease. Gloves do not enhance grip strength and torque, and in general reduce both up to 79% (Swain et al., 1970; Chen et al., 1989; Vincent and Tipton, 1988; McMullin and Hallbeck, 1991; Wang, 1991).

Product

Torque. Maximal torque, but not maximal force, increases with increasing diameter of knobs and lids. The optimal diameter is determined by the hand size of the subjects and roughness of the surface. Imrhan and Loo (1988) found an optimal diameter of 83 mm for a screw top container with a smooth surface. More torque can be exerted on a control, handle, or lid which is not round but offers an opportunity for a good grip (like a paddle, Bordett et al., 1988). The longer the lever, the larger the resulting moment.

Pull, lift, and carry. A good handle for lifting and carrying by adults should be at least 115 mm long, about 25 to 50 mm in diameter, with a hand clearance of 30 to 50 mm (Drury, 1980 and Hsia and Drury, 1986). For various force directions, various cross-sectional shapes of handle are recommended (Cochran and Riley, 1986). Sharp corners, edges, ridges, finger grooves, and curvatures should be avoided (Drury, 1980). However, if a handle is curved, it should be convex, following the natural curve of the gripping hand. Soft, smooth, nonslip surfaces are preferred. Cold and hard surfaces and vibration are to be avoided. The handle should be oriented such that it can be used without undue deviation of the wrist. Depending on the task, handle height can influence force exertion.

Grip force. Grip force appears to be maximal with a handle separation of about 55 to 60 mm. This separation is related to hand size (Fransson and Winkel, 1991; Radwin and Oh, 1991).

Environment

Support. More push and pull force is exerted maximally when the subjects can brace themselves, for example against a backrest, armrest, or footrest (Rohmert et al., 1987; Caldwell, 1962). For exerting large foot forces, a lumbar support is recommended. For subjects who are standing and reaching far forward, a support at the level of the pelvis can reduce the moment at the lower back with 30%, which results in a more comfortable posture (Frankel et al., 1984).

Temperature. The optimum ambient air temperature for sustained contractions of grip force appears to be 18°C. At other temperatures, endurance is shorter (Clarke et al., 1954).

Acceleration. Accelerations up to 5 *g* affect endurance but do not affect strength. Arm movements are effective up to 6 *g*, wrist and finger movements up to 12 *g* (Morgan et al., 1963; Woodson, 1981).

Posture

Free posture. Force can be exerted optimally in a freely adopted posture (Daams, 1993). Even small constraints in posture are found to have a considerable effect on the measured strength (Haslegrave, 1992).

Statics. Nature may assist in finding the most efficient way to exert force through the laws of statics. Moment is the product of lever and force. Adapting the size of the lever will change the resulting moment or change the required force.

Exerting force with use of body weight can make a task more pleasant.

The maximal force in a certain posture allowed by the laws of statics can be calculated from the height of the handle, the position of the center of mass, the pivot around which rotation would start, and sometimes the coefficient of friction with the environment. If this calculated force exceeds the force that physiologically can be exerted by the muscles of the subject, his or her maximal force is not limited by the posture. If, however, the calculated maximal force is less than the physiologically exertable force, it is clear that the posture will limit the force that the subject can maximally exert.

One/two handed. Two-handed strength commonly exceeds one-handed strength but is found to be less than twice the value. The ratio of one- versus two-handed force exertion in the sagittal plane (pushing, pulling, lifting, pressing, and all their combinations), ranges from 0.61 to 1.04, as measured by Fothergill et al. (1991) and Sanchez and Grieve (1992).

Joint angles. Extreme joint angles should be avoided during force exertion, especially if repeated often or maintained for sustained periods of time. Maximal force can be limited at extreme joint angles. Grip force and pulp, chuck, lateral, and three-jaw chuck pinch force are maximal with the wrist in neutral position (Imrhan, 1991; Hallbeck and McMullin, 1991; McMullin and Hallbeck, 1991; Fernandez et al., 1991).

Maximum leg force occurs when the knee is slightly bent.

Task distance. Hand grip and torque forces generally are greater if the task is close to the individual's body rather than at arm's length (Woodson, 1981). When pushing with one hand and lifting with one or both hands, the closer to the body, the more force can be exerted (The Materials Handling Research Unit, 1980). Static and dynamic lifting strengths are inversely related to reach distance (Kumar, 1990). For pulling, it seems to be the reverse: one-armed isokinetic pulling strength increases with reach distance (Mital and Faard, 1990). For comfortable force exertion, the arms should stay below shoulder height (Wiker et al., 1990).

Finger position. Maximal push and slide forces of the thumb are larger than those of the other fingers, and all types of force exerted with the pad of the fingertip are larger than those exerted with the tip of the finger. Maximal forces exerted with the flat of the hand exceed all finger forces (Bandera et al., 1985).

Asymmetric postures. Maximal lifting strength decreases with increasing asymmetry of posture (Kumar, 1990).

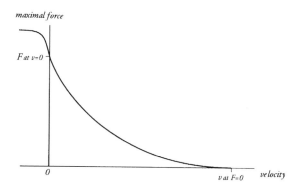

FIGURE 24.2 Force-velocity characteristics of skeletal muscle, showing a decrease of maximal tension (maximal force) as the muscle shortens and an increase as it lengthens (Adapted from Hof, A.L. 1987. Spiermechanica, in *Biomechanica*. Edited by R. Huiskes. Samson Stafleu, Alphen aan de Rijn. [in Dutch])

Endurance. The weight of limbs should be taken into account when considering the exertion of smaller forces over extended periods of time; hence sometimes the support of limbs is favored.

Force

Rest. A long and heavy task can be relieved by including periods of rest between start and finish. Frequent and short rest periods are much more effective than a few of long duration, the total time of rest being equal.

Precision. Maintained isometric muscle work reduces the precision of manual performance (Hammar-skjöld et al., 1989).

Dynamic force. Maximal dynamic force decreases with increasing speed. Figure 24.2 shows the theoretical relation between the exerted force and velocity of contraction for a separate muscle (Hof, 1987). Maximal static force and maximal eccentric force in general exceed maximal concentric force. The smaller the range of motion, the smaller the maximal exerted force (Ayoub et al., 1981 and 1982; Hafez et al., 1982). With cycling, peak force is exerted at 90° past the top dead center in each revolution (Hoes et al., 1968; Sargeant et al., 1981). Optimal power output is not attained at maximum velocity nor at zero velocity (static force), but somewhere in between. With cycling, maximal power is generated at a velocity of 110 rpm (Sargeant et al., 1981).

Endurance time. Maximal endurance time increases consistently with lower force levels (see Figure 24.3). There is no agreement on the influence of age, sex, muscle group, and subject on maximal endurance time (Elbel, 1949; Rohmert, 1960 and 1965; Caldwell, 1963 and 1964; Byrd and Jeness, 1982; Sato et al., 1984; Bishu et al., 1990; Deeb et al., 1992; Deeb and Drury, 1992; Daams, 1994). When significant differences are found, they are small.

Although long endurance of low force levels is possible, it is not desirable nor advisable. The maximal endurance time at very low force levels should not be considered infinite. Force is easier to maintain if the relative force level is low, if the subject is not subject to boredom, and may change his posture from time to time. These last two conditions apply even more to the lower force levels. Consequently, dynamic and/or cyclic force exertion should be considered preferable. This corresponds with old physiological insights and with the findings of Sjøgaard (1986) and Ulmer et al. (1989).

(Dis)comfort. Discomfort increases with increasing force, work/rest ratio, and task duration (Wiker, 1991).

For dynamic one-handed pulling, it is suggested that exertion of maximal force at high speed is more comfortable than at low speed (Garg and Beller, 1990).

The ratio of "comfortable" to maximal endurance time of force exertion may range from 0.50 at 80% and 0.19 at 15% of maximal force (see Figure 24.3 and next paragraph).

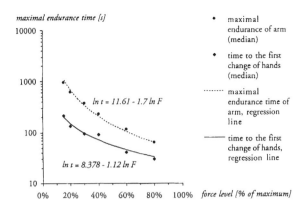

FIGURE 24.3   Medians and interpolations of maximal endurance time and time to the first change of hand, for arms only. (From Daams, B.J. 1994. *Human Force Exertion in User–Product Interaction.* Delft University Press, Delft. With permission.)

24.4   Recommendations for Research and the Use of Literature

Using Literature

General guidelines as outlined in the previous paragraph are useful tools in the early stages of the design process. They help to define the principal directions of design solutions. As the designs are worked out in more detail in subsequent stages of the design process, the need for more exact information arises. This information is best obtained from custom research tailored to the design at hand. Failing that, a designer will have to resort to literature. In this paragraph requirements are formulated that will guide the design of custom-tailored research and will help to determine the relevance, representativeness, and validity of data obtained from literature.

Physical strength and force exertion are important research topics in industrial ergonomics, military ergonomics, rehabilitation, and sports. Results of this research are reported in a vast and growing body of literature. Research in the different areas of ergonomics is diverse with respect to purpose, method, and subject group. The literature in the list of references with this chapter reflects this diversity. Relatively little of this research is directed toward force exertion in user–product interaction, but what there is tends to be too specific, aimed at one particular product or a select group of products. Designers would profit from research according to a more general product-oriented principle with a large variety of subjects. Such an approach ideally would lead to an atlas of human force exertion as outlined by Daams (1994).

For force exertion data from literature to be useful in product design, two essential requirements must be fulfilled at least. First, the experiment must be specified in all relevant detail. Journals and handbooks can make an important contribution here by requiring, and allowing for, extensive descriptions of the experiments. A second important requirement is that the experiment is similar to the real-life situation considered. In other words, the outcome of the experiment must be relevant to and valid for this intended use of the data. This is often neglected by both designers and researchers. For example, standardized postures inflicted by researchers on their subjects tend to be artificial and reduce the relevance of experiments for real-life situations (Daams, 1993).

Regarding the subjects, it is important to know their number, sex, age, and other characteristics that may be of importance, such as level of training or type of employment. For consumer products, it is imperative to include children, elderly, and disabled people in the sample. In any case, both females and males should be investigated.

A minimum number of subjects is needed to ensure a reproducible and sometimes representative result. To establish whether forces exerted in different situations are significantly different, a dozen or more subjects are needed. To establish a force exertion which is representative for a whole population, the main requirement is that the sample should match the population regarding the distribution of sex, age, build, profession, health, level of training, level of ability, etc. In total, several hundred subjects may be necessary. In both cases, the actual minimum sample size depends on the size of the variation coefficient (the standard deviation divided by the average force) which affects the power of the test, and on the accuracy with which the actual distribution of forces in the population needs to be known. The calculation of the sample size depends on the sort of test (paired or unpaired). For more information, we refer to standard textbooks on statistics.

It is important to know the posture, the direction of force exertion, the type of force, the exertion time, the laterality, the shape and size of the handle, the position of the handle, and other factors that possibly influence the result.

At present, most literature on force exertion concerns static, maximal force exertion in standardized postures. Characteristic of forces exerted on products, however, is that they are most often dynamic, they last for any time between a second and a few minutes, that usually they are not maximal, and that the users adopt a "natural" posture while exerting force.

Thus, there is a need for more information on dynamic, submaximal (comfortable) force, on endurance time and on force exerted in a "natural" (or "free") posture.

Conducting Research

Accounting for the factors "posture," "endurance time of sub-maximal force," and "(dis)comfort" in an appropriate manner will benefit research for product design and increase its relevance. In the following paragraphs, these aspects along with measuring methods are discussed.

Posture. In assessing human force exertion, the use of standardized postures can lead to inaccurate prediction of the forces and postures that occur in everyday life. A standardized posture is generally considered to yield more reliable data (implicit in Caldwell et al., 1974). Even if this were so, force data acquired from measurements in standardized postures may not have predictive validity.

For pushing, pulling, and exerting torque in various postures, research shows that the forces measured in free postures are highly reproducible, even with extreme handle heights and when the place to which the force is to be applied is not fixed (Daams, 1993). The exerted forces are equally reproducible in free and standardized postures. The difference in average force, though, is considerable and significant: much more force can be exerted in free posture. Furthermore, postures of subjects during force exertion in free posture show a remarkable intra-individual reproducibility. This measurement method thus not only yields information on force, but also on the standing room needed by the subjects, which is also relevant to design.

Research on force exertion in free posture, therefore, may be considered most suitable for product design research.

Endurance. Several investigations have been conducted into the endurance of submaximal force. There is, however, no standardized way to measure endurance. In general, endurance time is measured as a function of a percentage of maximal force. To enable the use of the results in product design and allow a comparison with literature or a reproduction of the experiment, several additional aspects of the measuring method should be known. In particular the limitation of the movements subjects are allowed to make, the limits between which the force has to be maintained, the definition of the point at which the measurement ends, and the way maximal force is determined should be registered. This last aspect not only influences maximal endurance time, but even affects the influence of other variables on the results (Daams, 1994).

In general, these aspects are not mentioned in present literature, which makes the results unsuitable for application to product design.

For a reproducible measurement of endurance time during force exertion, the following method was found to yield good results:

- First, maximal force is measured according to the method of Caldwell et al. (1974): 2 seconds buildup and 4 seconds maximal force. No feedback or encouragement is given.
- The subject is then asked to exert the required force (unaware of the level) and maintain it for as long as possible. The limb which exerts force is supported. During measurement of endurance time, again no encouragement is given and no conversation is allowed, as it appears that talking and the presence of other people both influence the endurance time positively.
- The range within which the force is allowed to vary during a measurement should be fairly wide to prevent unintentional, premature ending of the measurement, e.g., by muscle tremors, sneezing, flagging attention, and other minor actions by the subject. Such an unintentional end of a measurement proves to be very frustrating to subjects. X-t recordings of measurements show that, with a narrow limit at which warning beeps sound (10% above and below the required force level) and a wide limit at which the measurement ends (50% above and below the required force level), forces are maintained sufficiently close to the required level.

There is no need to limit the movement of subjects, because other than slight movements cause the exerted force to exceed one of the warning or even end limits. This motivates the subject to return to the previous posture, or it ends the measurement altogether.

Reproducible results obtained with this method are shown in Figure 24.3.

Comfort. It is hard to define what constitutes a comfortable force. In ergonomics, comfort is generally defined as the "absence of discomfort." In literature, various methods are used to measure (un)comfortable force exertion, resulting in different values (Arnold, 1991; Berns, 1981; Bordett et al., 1988; Kanis, 1993; Schoorlemmer and Kanis, 1992; Garg and Beller, 1990; Wiker, 1991; Schutz, 1972; Sato et al., 1984). Apparently, there is not one single discomfort level, and which discomfort level is measured depends on the measuring method. In all experiments, however, there is one common factor: the subjects are asked to think about and indicate their feelings of (dis)comfort. This will inevitably make them concentrate on registering and comparing any feelings of discomfort, thereby disturbing the measurement.

Product designers are interested in discomfort as the "absence of comfort." They wish to prevent a situation in which a user operating a product starts to note discomfort and gets annoyed. The following measurement is intended to catch the moment at which the endurance of force exertion is no longer comfortable, without involving subjects consciously in the judging process. The underlying assumption is that the moment at which the user starts to note discomfort is indicated by a spontaneous change of posture. Here this method is illustrated for horizontal push, but it is assumed to be applicable to other postures and forces, keeping in mind both the influence of posture on the results, especially at lower force levels, and the reproducibility of the results.

Subjects are asked to exert a certain percentage of maximal push force with one hand (see Figure 24.4), for a time which is three times the expected maximal endurance time at that force level, according to the formula by Rohmert (1960 and 1965). They are told they can change hands whenever they feel like it. The time to the first change of hands is taken as an indication of the moment at which discomfort is noted. Figure 24.3 shows the results of an actual experiment. The ratio to the maximal endurance time ranges from 0.50 at 80% to 0.19 at 15% of maximal force. This kind of "comfortable endurance time" has proven to be reproducible. It may serve product designers as a useful and valid measure for the onset of discomfort. Apart from probably being more relevant to product design than maximal endurance time it is also a good deal more comfortable to measure both for subject and experimenter.

24.5 Conclusion

Comfort, efficiency and safety of operation are important properties of a product. To incorporate these properties into products, designers have to be aware of the forces the intended users can and will exert on these products. This awareness should permeate the whole design process, from the conceptual stage

FIGURE 24.4   Posture during research on endurance time and (dis)comfort: pushing with one hand. (From Daams, B.J. 1994. *Human Force Exertion in User–Product Interaction.* Delft University Press, Delft. With permission.)

where rules of thumb may assist in finding the right direction for a design, to the detailing of the final design, where exact information on the expected forces determines the design limits. This information can be derived from custom-tailored research or from literature. The information from both these sources should not be used blindly, but it should be judged time and again on its relevance to and validity for a particular design. Such an intelligent and critical attitude is the key to successful integration of force exertion data in product design.

Defining Terms

Comfort: Comfort is generally defined as "absence of discomfort."

Comfortable endurance (time): The time during which a force can be maintained without discomfort by a subject. There is no standard procedure to measure comfortable endurance. A usable and reproducible method is proposed in this chapter. In general, comfortable endurance is measured as a function of a percentage of maximal force.

Concentric force exertion: Dynamic force exertion during which the muscles shorten.

Dynamic force exertion: Force exertion during which the length of the muscles changes. Consequently, the segments of the body that are involved rotate relative to each other. There is no standard procedure to measure dynamic force. Dynamic force exertion is expressed in Newtons (N).

Eccentric force exertion: Dynamic force exertion during which the muscles lengthen.

Endurance (time): See *Comfortable endurance (time)* and *maximal endurance (time).*

Laterality: The difference between the preferred and the nonpreferred side of the body.

Maximal force: The maximal force which subjects can exert in an experiment. Static maximal force is generally measured according to the method of Caldwell et al. (1974). Static force exertion is built up for a second, without jerk, and the maximum is maintained for a few (3 to 5) seconds. No feedback or encouragement is given.

Maximal endurance (time): The maximal time during which a force can be maintained by a subject. There is no standard procedure to measure maximal endurance. A usable and reproducible method is proposed in this chapter. In general, maximal endurance is measured as a function of a percentage of maximal force.

P_5-P_{95} syndrome: The habit of quoting the P_5 and/or the P_{95} as a "normal" design maximum, thus excluding 5 or even 10% of the population from using a product effectively or comfortably.

Power: A measure of work. It is expressed in watts, as the product of exerted force and velocity (Nm/s).

Static force exertion: Force exertion during which the length of the muscles does not change. Consequently, the speed of the movement during force exertion is zero. Force exertion is expressed in Newtons (N). When the force exertion is maximal, some authors refer to this measured variable as "Maximal Voluntary Contraction" or MVC.

References

Arnold, A.-K. 1991. An ergonomic approach to the design of consumer packaging, in *Interface '91, Proceedings of the 7th symposium on Human Factors and Industrial Design in Consumer Products* (pp. 138-143). Edited by D. Boyer and J. Pollack. The Human Factors Society, Santa Monica, CA.

Ayoub, M.M., Gidcumb, C.F., Beshir, M.Y., Hafez, H.A., Aghazadeh, F., and Bethea, N.J. 1981. *Development of an Atlas of Strengths and Establishment of an Appropriate Model Structure.* Institute for Ergonomics Research, Texas Tech University, Lubbock, Texas.

Ayoub, M.M., Gidcumb, C.F., Reeder, M.J., Beshir, M.Y., Hafez, H.A., Aghazadeh, F., and Bethea, N.J. 1982. *Development of a Female Atlas of Strengths.* Institute for Ergonomics Research, Texas Tech University, Lubbock, Texas.

Bandera, J.E., Kern, P., and Solf, J.J. 1985. *Ergonomische Kenngrößen für Kontaktgreifarten.* Bundesanstalt für Arbeitsschutz, Dortmund. [in German]

Berns, T. 1981. The handling of consumer packaging. *Applied Ergonomics,* 12: 153-161.

Bishu, R.R., Myung, R.H., and Deeb, J.M., 1990. Evaluation of handle positions using force/endurance relationship of an isometric holding task, in *Proceedings of the Human Factors Society 34th Annual Meeting* (pp. 684-687). The Human Factors Society, Santa Monica, CA.

Bordett, H.M., Koppa, R.J., and Congelton, J.J. 1988. Torque required from elderly females to operate faucet handles of various shapes. *Human Factors,* 20 (3): 339-346.

Byrd, R. and Jeness, M.E., 1982. Effect of maximal grip strength and initial grip on contraction time and on areas under force-time curves during isometric contractions. *Ergonomics,* 25 (5): 387-392.

Caldwell, L.S. 1962. Body stabilisation and the strength of arm extension. *Human Factors,* 4: 125-130.

Caldwell, L.S., 1963. Relative muscle loading and endurance. *Journal of Engineering Psychology,* 2: 155-161.

Caldwell, L.S., 1964. The load-endurance relationship for a static manual response. *Human Factors,* 6: 479-484.

Caldwell, L.S., Chaffin, D.B., Dukes-Dobos, F.N., Kroemer, K.H.E., Laubach, L.L., Snook, S.H., and Wasserman, D.E. 1974. A proposed standard procedure for static muscle strength testing. *American Industrial Hygiene Association Journal,* 35: 201-206.

Chen, Y., Cochran, D.J., Bishu, R.R., and Riley, M.W. 1989. Glove size and material effects on task performance, in *Proceedings of the Human Factors Society 33rd Annual Meeting* (pp. 708-712). The Human Factors Society, Santa Monica, CA.

Clarke, R.S.J., Hellon, R.F., and Lind, A.R. 1954. The duration of sustained contractions of the human forearm at different muscle temperatures. *J. Physiol.,* 143: 454-473.

Cochran, D.J. and Riley, M.W. 1986. The effects of handle shape and size on exerted forces. *Human Factors,* 28 (3): 253-265.

Daams, B.J. 1993. Static force exertion in postures with different degrees of freedom. *Ergonomics,* 36 (4): 397-406.

Daams, B.J. 1994. *Human Force Exertion in User–Product Interaction.* Delft University Press, Delft.

Deeb, J.M. and Drury, C.G., 1992. Perceived exertion in isometric uscular contractions related to age, muscle, force level and duration in *Proceedings of the Human Factors Society 36th Annual Meeting* (pp. 712-716). The Human Factors Society, Santa Monica, CA.

Deeb, J.M., Drury, C.G., and Pendergast, D.R., 1992. An exponential model of isometric muscular fatigue as a function of age and muscle groups. *Ergonomics,* 35 (7/8): 899-918.

Drury, C.G. 1980. Handles for manual materials handling. *Applied Ergonomics,* 11 (1): 35-42.

Elbel, E.R., 1949. Relationship between leg strength, leg endurance and other body measurements. *Journal of Applied Physiology,* 2 (4): 197-207.

Fernandez, J.E., Dahalan, J.B., Halpern, C.A. and Viswanath, V. 1991. The effect of wrist posture on pinch strength, in *Proceedings of the Human Factors Society 35th Annual Meeting* (pp. 748-752). The Human Factors Society, Santa Monica, CA.

Fothergill, D.M., Grieve, D.W., and Pheasant, S.T. 1991. Human strength capabilities during one-handed maximum voluntary exertions in the fore and aft plane. *Ergonomics,* 34 (5): 563-573.

Fothergill, D.M., Grieve, D.W., and Pheasant, S.T. 1992. The influence of some handle designs and handle height on the strength of the horizontal pulling action. *Ergonomics*, 35 (2): 203-212.

Frankel, V.H., Nordin, M., and Snijders, C.J. 1984. *Biomechanica van het skeletsysteem*. De Tijdstroom, Lochem. [in Dutch]

Fransson, C. and Winkel, J. 1991. Hand strength: the influence of grip span and grip type. *Ergonomics*, 34 (7): 881-892.

Garg, A. and Beller, D. 1990. One-handed dynamic pulling strength with special reference to speed, handle heights and angles of pulling. *International Journal of Industrial Ergonomics*, 6: 231-240.

Hafez, H.A., Gidcumb, C.F., Reeder, M.J., Beshir, M.Y. and Ayoub, M.M. 1982. Development of a human atlas of strengths, in *Proceedings of the Human Factors Society 26th Annual Meeting* (pp. 575-579). The Human Factors Society, Santa Monica, CA.

Hallbeck, M.S. and McMullin, D.L. 1991. The effect of gloves, wrist position, and age on peak three-jaw chuck pinch force: a pilot study, in *Proceedings of the Human Factors Society 35th Annual Meeting* (pp. 753-757). The Human Factors Society, Santa Monica, CA.

Hammarskjöld, E., Ekholm, J., and Harms-Ringdahl, K. 1989. Reproducibility of work movements with carpenters' hand tools. *Ergonomics*, 32 (8): 1005-1018.

Haslegrave, C.M. 1992. Predicting postures adopted for force exertion: thesis summary. *Clin. Biomech.*, 7 (4): 249-250.

Hoes, M.J.A.J.M., Binkhorst, R.A., Smeekes-Kuyl, A.E.M.C., and Vissers, A.C.A. 1968. Measurement of forces exerted on pedal and crank during work on a bicycle ergometer at different loads. *Int. Z. angew. Physiol. einschl. Arbeitsphysiol.*, 26: 33-42.

Hof, A.L. 1987. Spiermechanica, in *Biomechanica*. Edited by R. Huiskes. Samson Stafleu, Alphen aan de Rijn. [in Dutch]

Hsia, P.T. and Drury, C.G. 1986. A simple method of evaluating handle design. *Applied Ergonomics*, 17 (3): 209-213.

Imrhan, S.N., 1991. The influence of wrist position on different types of pinch strength. *Applied Ergonomics*, 22 (6): 379-384.

Imrhan, S.N. and Loo, C.H. 1988. Modelling wrist-twisting strength of the elderly. *Ergonomics*, 31 (12): 1807-1819.

Imrhan, S.N. and Sundararajan, K. 1992. An investigation of finger pull strengths. *Ergonomics*, 35 (3): 289-299.

Kanis, H. 1993. Operation of controls on consumer products by physically impaired users. *Human Factors*, 35 (2): 305-328.

Kumar, S. 1990. Symmetric and asymmetric stoop-lifting strength, in *Proceedings of the Human Factors Society 34th Annual Meeting* (pp. 762-766). The Human Factors Society, Santa Monica, CA.

Materials Handling Research Unit, the, 1980. *Grenzen van de voor handenarbeid benodigde lichaamskracht*. IPC Science & Technology Press Ltd., Luxembourg. [in Dutch]

Mathiowetz, V., Kashman, N., Volland, G., Weber, K., Dowe, M., and Rogers, S. 1985. Grip and pinch strength: normative data for adults. *Arch. Phys. Med. Rehabil.*, 66: 68-74.

Mathiowetz, V., Wiemer, D.M., and Federman, S.M. 1986. Grip and pinch strength: norms for 6- to 19-year-olds. *The American Journal of Occupational Therapy*, 40 (10): 705-711.

McMullin, D.L. and Hallbeck, M.S. 1991. Maximal power grasp force as a function of wrist position, age and glove type: a pilot study, in *Proceedings of the Human Factors Society 35th Annual Meeting* (pp. 733-737). The Human Factors Society, Santa Monica, CA.

Mital, A. and Faard, H.F. 1990. Effects of sitting and standing, reach distance, and arm orientation on isokinetic pull strengths in the horizontal plane. *International Journal of Industrial Ergonomics*, 6: 241-248.

Morgan, C.T., Cook, J.S., Chapanis, A., and Lund, M.W. (eds.) 1963. *Human Engineering Guide to Equipment Design*. McGraw-Hill, New York.

Page, M. 1981. An ergonomics evaluation of a reclosable pharmaceutical container with special reference to the elderly. *Ergonomics*, 24 (11): 847-862.

Peacock, B., Westers, T., Walsh, S., and Nicholson, K. 1981. Feedback and maximum voluntary contraction, *Ergonomics*, 24 (3): 223-228.

Radwin, R.G. and Oh, S. 1991. Handle and trigger size effects on power tool operation, in *Proceedings of the Human Factors Society 35th Annual Meeting* (pp. 843-847). The Human Factors Society, Santa Monica, CA.

Rohmert, W. 1960. Ermittlung von Erholungspausen für statische Arbeit des Menschen. *Int. Z. angew. Physiol. enschl. Arbeitsphysiol.*, 18: 123-164. [in German]

Rohmert, W. 1965. Physiologische Grundlagen der Erholungszeitbestimmung. *Arbeit und Leistung*, 19 (1): 1-28. [in German]

Rohmert, W., Mainzer, J., and Kahabka, G. 1987. Analyse biomechanischer und physiologischer Engpässer beim Ausüben von Stellungskräften. *Zeitschrift für Arbeitswissenschaft*, 41 (2): 114-130. [in German]

Roozenburg, N.F.M. and Eekels, J. 1991. *Produktontwerpen, struktuur en methoden*. Lemma, Utrecht. [in Dutch]

Sanchez, D. and Grieve, D.W. 1992. The measurement and prediction of isometric lifting strength in symmetrical and asymmetrical postures. *Ergonomics*, 35 (1): 49-64.

Sanders, M.S. and McCormick, E.J. 1993. *Human factors in engineering and design*, 7th edition. McGraw-Hill, New York.

Sargeant, A.J., Hoinville, E., and Young, A. 1981. Maximum leg force and power output during short-term dynamic exercise. *Journal of Applied Physiology: Respirat. Environ. Exercise Physiol.*, 51 (5): 1175-1182.

Sato, H., Ohashi, J., Iwanaga, K., Yoshitake, R., and Shimada, K. 1984. Endurance time and fatigue in static contractions. *Journal of Human Ergology*, 13: 147-154.

Schoorlemmer, W. and Kanis, H. 1992. Operation of controls on everyday products, in *Proceedings of the Human Factors Society 36th Annual Meeting* (pp. 509-513). The Human Factors Society, Santa Monica, CA.

Schutz, R.K. 1972. *Cyclic Work–Rest Exercise's Effect on Continuous Hold Endurance Capacity*. Unpublished Ph.D.-dissertation. University of Michigan, Ann Arbor.

Singh, M. and Karpovich, P. 1968. Strength of forearm flexors and extensors in men and women. *Journal of Applied Physiology*, 25 (2), 177-180.

Sjøgaard, G. 1986. Intramuscular changes during long-term contraction, in *The Ergonomics of Working Postures*. (pp. 136-143). Edited by E.J. Corlett, I. Manenica and J. Wilson. Taylor & Francis, London.

Steenbekkers, L.P.A. 1993. *Child Development, Design Implications and Accident Prevention*. Physical Ergonomics Series. Delft University Press, Delft.

Swain, A.D., Shelton, G.C., and Rigby, L.V. 1970. Maximum torque for small knobs operated with and without gloves. *Ergonomics*, 13 (2): 201-208.

Ulmer, H.-V., Knieriemen, W., Warlo, T., and Zech, B. 1989. Interindividual variability of isometric endurance with regard to the endurance performance limit for static work. *Biomed. Biochem. Acta*, 48 (5/6), 504-508.

Vincent, M.J. and Tipton, M.J. 1988. The effect of hand protection and cold immersion on grip strength, in *Contemporary Ergonomics 1988, Proceedings of the Ergonomics Society Conference* (pp. 323-327). Edited by E.J. Lovesey. Taylor & Francis, London.

Wang, M.-J.J. 1991. The effect of six different kinds of gloves on grip strength, in *Towards Human Work: Solutions to Problems in Occupational Health and Human Work*. (pp. 164-169). Edited by M. Kumashiro and E.D. Megaw. Taylor & Francis, London.

Wiker, S.E. 1991. Fatigue, discomfort and changes in the psychometric function with repetitive pinch grasps, in *Designing for Everyone. Proceedings of the 11th Congress of the International Ergonomics Association* (pp. 368-370). Edited by Y. Queinnec and F. Daniellou. Taylor & Francis, London.

Wiker, S.E., Chaffin, D.B., and Langolf, G.D. 1990. Shoulder postural fatigue and discomfort. *International Journal of Industrial Ergonomics*, 5 (2):133-146.

Woodson, W.E. 1981. *Human Factors Design Handbook*. McGraw-Hill, New York.

25

Rapid Upper Limb Assessment (RULA)

E. N. Corlett
University of Nottingham

25.1 Introduction

During the last two decades there has been an increasing awareness of the problems arising in the upper limb, wrist, elbow, or shoulder, as a consequence of the continuous use of the muscles and joints during work. First signs were evident in the users of computers. Due to a poor understanding of the tasks involved and inadequate task analyses, these were ascribed to users being work shy, bored, or in one famous phrase in Britain, "egg shell personalities."

Detailed studies in, for example, Australia and the United States revealed that although psychosocial factors could contribute to the problems being experienced it was the mechanical component that predominated. This was compounded from the frequency of use of the joint involved, the posture adopted by the limb, and the forces exerted during the tasks. Where these factors were appropriately analyzed and changes introduced, the incidence of injuries could be significantly reduced or even eliminated. Changes typically included the reorganization of the task.

The urgent requirement for methods to investigate what had become a major problem across the world gave rise to many reports and papers. Health and safety bodies demanded measures and in the European Union the international requirement was that employers were required to assess the risks of the problems arising, and to make the necessary changes. These requirements were embodied in a "daughter" directive of the main European Union Health and Safety Directive No. 89/391/EEC, and it is a requirement of these directives that they are incorporated into the laws of the nation states that make up the European Union. In Britain the "daughter" directive concerning work with display screen equipment was published by the government as *"Work-Related Upper Limb Disorders — A Guide to Prevention,"* (*Health and Safety Executive*, 1990).

For organizations unused to risk assessment, the demand to undertake such a process was formidable, although national health and safety bodies in the various European countries issued guidance notes to assist with the activity. However, many of these just discussed some possible sources of upper limb problems but gave no methodology for assessment, use of the results, or indications of the most desirable directions for change. The RULA procedure was developed as a consequence of this lack.

25.2   Requirements

What was needed was a procedure by which an individual, with relatively little training, could assess a workplace, recognize from the assessment the major points that implied a risk to the user, and know from the results of the analysis what actions to take to reduce the risk. The system developed was not a system for diagnosing the presence or otherwise of upper limb disorder in the worker, it was to assess whether the workplace could present a hazard to that worker whereby he or she may be at risk of suffering upper limb problems. Thus the technique was devised as a proactive risk assessment tool, rather than a response to already reported injuries.

The procedure uses workers themselves, from whom the measurements are taken. Since it is workers, not machines, who suffer the problems, it is necessary to do more than pursue a relatively impersonal observation. The experience of these workers is relevant to probe the problem. Hence, apart from the measurements arising from the use of the RULA technique, some additional information is needed if the risks are to be more reliably assessed.

The aim was to produce a system that used observation of the workers doing their jobs, and a minimum of other questions and measurements, to identify the most probable points presenting the risks. These procedures had to give their results in such a form that they would demonstrate to the analyst where changes had to be made, and in what directions. The same procedures had to be usable after the changes had been made, so as to provide measures of improvement in risk more rapidly than retrospective counts of injuries.

25.3   The RULA Procedure

The model on which the RULA procedure was based was the OWAS method, developed by the Finnish Institute of Occupational Health and the OVAKO Steel Company to investigate problems of lifting and of back injuries. This process required that segments of the body be judged on a simple scale, producing a sequence of numbers which were matched against a grid. The value of the numbers and their position on the grid told the analyst the severity of the posture, while taking actions that reduced the magnitude of the numbers improved the situation. RULA extends this assessment process to the control of upper limb disorders.

As indicated earlier, the factors contributing to upper limb disorders are more than posture. They include frequency of use, forces, and the availability of changes in the work cycle. Hence, each of these must be part of the assessment if a reasonable judgment of the likely risk is to be obtained. Furthermore, posture is not just that of the wrist and arm; the whole body posture can tell us something about the adequacy, or otherwise, of the task. (See Figure 25.1.)

To perform a RULA assessment, observations are made of the limb and body postures for those parts of the work cycle the analyst considers to give the most frequent use of the joint, or the most extreme joint angles. For each of the chosen parts of the work cycle, the position of the upper and lower arm and the wrist are assessed, using the diagrams of Group A, and entered on the score sheet in the boxes marked A. The positions of the neck, trunk, and legs are then assessed, using the information in the diagram marked Group B, and entered on the score sheet under B. (See Figure 25.2) Note that for posture score A, the left and the right arms are assessed separately, and both a score A and a Grand Score derived for each. Of course, both arms may not require assessing if, in the judgment of the assessor, one of them is not at risk.

From each of these sets of observations a score can be obtained, using Table A or Table B in Figure 25.3 for the observation sets A and B, respectively. These tables combine the effects of the postures to give an initial estimate of risk, but as yet we have not taken into account the possible contributions of force and frequency.

To include these we now turn to the boxes dealing with the Muscle Use score and the Forces or Load scores (Figure 25.4). The former allows us to modify the initial estimate, which is based on posture, to account for aspects of muscle use known to influence upper limb problems, while the latter takes into account the effects of loadings imposed on the limb, which also affect whether or not some injury will be suffered.

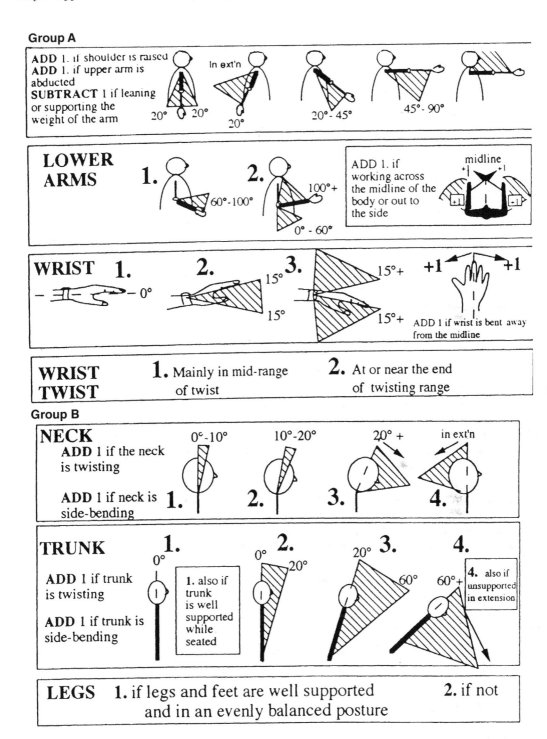

FIGURE 25.1

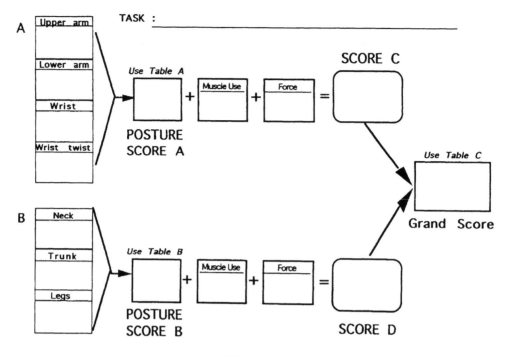

FIGURE 25.2

Having incorporated these two additional factors, we now sum the three values for each of sections A and B of the score sheet, arriving at score C and score D (Figure 25.5). It is with these two scores (C and D) that we now enter the Grand Score table, to find the final score (Figure 25.6).

It is this final score which gives us an estimate of the risk potential for the task. This Grand Score table was developed in a research program that involved experiments in which the various risk factors were manipulated, and an extensive study of the literature, resulting in a minimum set of numbers which would give a usable indication of severity without implying a greater precision than the system would achieve (McAtamney 1994).

As we move across the Grand Score table from the top left to the bottom right, the numbers get larger. This is emphasized by the shading, which darkens as we take this direction. The higher the number, the greater the risk of musculoskeletal symptoms arising. To put some numbers on this, we have proposed a number of action levels, as follows:

Action level 1: A score of one or two indicates that the posture is acceptable if it is not maintained or repeated for long periods.

Action level 2: A score of three or four indicates that further investigation is needed and changes may be required.

Action level 3: A score of five or six indicates that investigation and changes are required soon.

Action level 4: A score of seven or more indicates that investigation and change are required immediately.

Having found the Action level, it is then necessary to decide the most appropriate changes to reduce the risk. It will be evident, from an inspection of the score sheets, that as the situation worsens the numbers get higher. Hence, a first approach to improvement is to identify changes that will reduce the value of the scores, aiming to reduce the high numbers first. This will improve the total score, although a change in one of the values is unlikely to reduce the total scores by enough; several changes are usually necessary.

TABLE A   Upper Limb Posture Score

UPPER ARM	LOWER ARM	WRIST POSTURE SCORE 1		2		3		4	
		TWIST 1	2	TWIST 1	2	TWIST 1	2	TWIST 1	2
1	1	1	2	2	2	2	3	3	3
	2	2	2	2	2	3	3	3	3
	3	2	3	3	3	3	3	4	4
2	1	2	3	3	3	3	4	4	4
	2	3	3	3	3	3	4	4	4
	3	3	4	4	4	4	4	5	5
3	1	3	3	4	4	4	4	5	5
	2	3	4	4	4	4	4	5	5
	3	4	4	4	4	4	5	5	5
4	1	4	4	4	4	4	5	5	5
	2	4	4	4	4	4	5	5	5
	3	4	4	4	5	5	5	6	6
5	1	5	5	5	5	5	6	6	7
	2	5	6	6	6	6	6	7	7
	3	6	6	6	7	7	7	7	8
6	1	7	7	7	7	7	8	8	9
	2	8	8	8	8	8	9	9	9
	3	9	9	9	9	9	9	9	9

TABLE B   Neck, Trunk, Legs Posture Score

NECK POSTURE SCORE	TRUNK POSTURE SCORE 1		2		3		4		5		6	
	LEGS 1	2	LEGS 1	2	LEGS 1	2	LEGS 1	2	LEGS 1	2	LEGS 1	2
1	1	3	2	3	3	4	5	5	6	6	7	7
2	2	3	2	3	4	5	5	5	6	7	7	7
3	3	3	3	4	4	5	5	6	6	7	7	7
4	5	5	5	6	6	7	7	7	7	7	8	8
5	7	7	7	7	7	8	8	8	8	8	8	8
6	8	8	8	8	8	8	8	9	9	9	9	9

FIGURE 25.3

25.4   Some Complementary Measures

All possible actions of the hands are not represented in the RULA tables, so it is necessary to record certain other possible actions. Investigators can add to the ones listed here from their experiences of the particular requirements of the businesses they are engaged in. Other aspects of bodily constraints are also listed here, as many of them influence the ability of operators to change positions and·to rest various parts of their bodies during the working day.

Hands

Are the hands or fingers doing pounding, gripping, exerting a force, or performing stretching or twisting actions? Is a pinch grip required? If a handle is used, is it too large, or too small, to be gripped easily? Is

MUSCLE USE SCORE

Give a score of 1 if the posture is;

mainly static, e.g. held for longer than 1 minute
repeated more than 4 times/minute

Static posture, held longer than 1 minute: score = 1
Moderate posture, not static, not highly repetitive: score = 0
Highly repetitive posture, repeated more than 6 times/min.: score = 1

FORCES OR LOAD SCORE

0.	1.	2.	3.
• No resistance or less than 2kg intermittent load or force	• 2-10kg intermittent load or force	• 2-10kg static load • 2-10kg repeated load or force • 10kg or more intermittent load or force	• 10kg or more static load • 10kg or more repeated loads or forces. • Shock or forces with a rapid buildup.

FIGURE 25.4

it too short or does it have pressure points that press into the palm? Is it difficult to grasp due to gloves, or a smooth surface?

Workpoint

Useful approximations for suitable work heights are:

Precision work: 100 to 200 mm above elbow height
Assembly work, with hand support: 50 to 70 mm above elbow height
Work requiring free hand movement: at, or just below, elbow height
Heavy work, requiring force: 100 to 300 mm below elbow height

Possible sources of obstructions, which may force inadequate postures, are awkwardly placed controls or visual requirements requiring reaching or bending, obstructions to the legs and feet, structures that prevent the operators from sitting, such as inadequate knee room.

Common problems with seating, which increase the difficulties of workers achieving good postures and changes of postures, are seats which are not adjustable, which have no padding, which dig into the users' legs, which prevent the users from swinging round to reach to the side, with a poor backrest which does not allow good lumbar support when in use.

Discomfort

Apart from the physical dimensions of the workpoint, as mentioned above, the responses of the users can give good guidance to assist with locating hazards. For full details of Body Part Discomfort recording, reference should be made to an appropriate textbook, such as that listed in the bibliography to this chapter. But even brief inquiries among workers concerning body discomforts can reveal the sources of

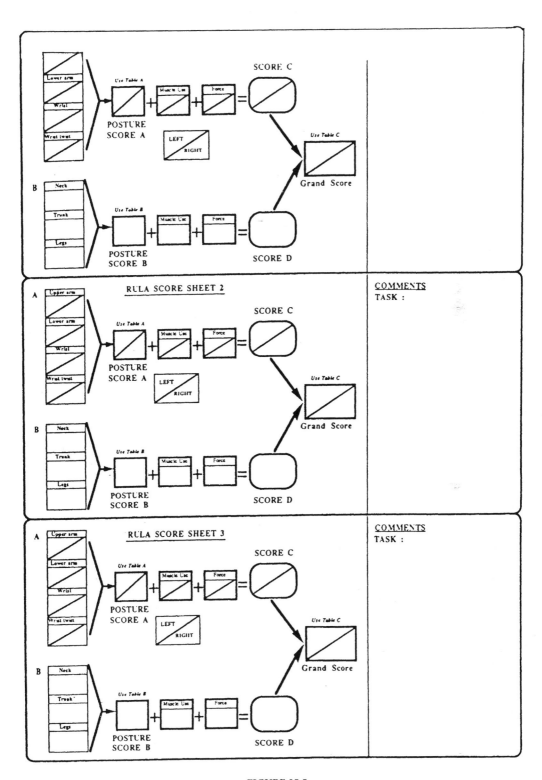

FIGURE 25.5

SCORE D (NECK, TRUNK, LEGS)

		1	2	3	4	5	6	7+
SCORE C (UPPER LIMB)	1	1	2	3	3	4	5	5
	2	2	2	3	4	4	5	5
	3	3	3	3	4	4	5	6
	4	3	3	3	4	5	6	6
	5	4	4	4	5	6	7	7
	6	4	4	5	6	6	7	7
	7	5	5	6	6	7	7	7
	8	5	5	6	7	7	7	7

TABLE C Grand Score Table

ACTION LEVEL 1	A score of one or two indicates that posture is acceptable if it is not maintained or repeated for long periods.
ACTION LEVEL 2	A score of three or four indicates further investigation is needed and changes may be required.
ACTION LEVEL 3	A score of five or six indicates investigation and changes are required soon.
ACTION LEVEL 4	A score of seven or more indicates investigation and changes are required immediately.

FIGURE 25.6

difficulties. Most musculoskeletal injuries give early warning by the growth of discomfort at the site, or in neighboring muscles. It is thus useful to ask workers, perhaps at intervals during the day, to point out those sites where discomfort has appeared during work, starting with the most uncomfortable first.

The results of such inquiries should complement the RULA analyses. Least discomfort will arise when the limbs are in positions represented by a value of one in each of the diagrams in the illustrations of Groups A and B. An increase in the value of the score will probably be accompanied by an increase in the discomfort levels reported, although this is not invariably true. If discomforts are found, they should be taken seriously as representing early warning of potential injuries, and action for change should be taken early. Sometimes management will not be too comfortable with inquiries of this sort, but it is very rare to find other than accurate responses to these inquiries about discomfort. What is more, they provide specific information relating to each individual's working condition, which will give valuable guidance for taking effective action.

25.5   Final Comments

The use of RULA is primarily as a survey tool. Although many, if not most, of the changes which will be required as a result of its use will be straightforward and obvious, there will be cases where more detailed investigations will be required. One user found that, in more than 8,000 workplaces, something like 80% of the changes could be done immediately and for little cost, while the rest could include some changes which were difficult and required time and money. Hence, it should not be assumed that after RULA all is well! An open mind should be maintained concerning further investigations, and the technique should also be used after changes to see if there is evidence of improvement in the scores, as well as in the experiences of work.

What is very useful about the technique is that relatively unskilled personnel can use the method after modest training; indeed, it has been taught to workers, who have then gone on to improve their own workplaces. One development has been the computerization of the procedure so that it can be available to all the computer users in a company, who can assess themselves using the display on their own screens. On completing their assessment, the scores are shown and suggestions for improvement are offered (Lueder,1996).

Methods engineers can find it a valuable tool that can be combined with their other techniques without any serious increase in the time needed to investigate working methods. It can be used, too, when new workplaces are being proposed, to see that they do not create unsuitable work situations. It will be evident that it can be used in mock-ups for new workplaces, giving assurance that the various features of the workplace will suit all users.

References

For information on the OWAS method, consult The Centre for Occupational Safety, Lonnrotinkatu 4B, SF-00120 Helsinki, FINLAND.

Health and Safety Executive. (1990) *Work-Related Upper Limb Disorders — A Guide to Prevention.* HSE Books, Sudbury, UK.

Lueder, R. K. (1996) A proposed RULA for computer users. *Proceedings of the Ergonomics Summer Workshop,* UC Berkeley Center for Occupational and Environmental Health, San Francisco.

McAtamney, L. (1994) *Interrelationship of Risk Factors Associated with Upper Limb Disorders in VDU Users.* PhD thesis, University of Nottingham, UK.

McAtamney, L. and Corlett, E. N. (1993) RULA: a survey method for the investigation of work-related upper limb disorders. *Applied Ergonomics,* 24,(2) 91-99.

McAtamney, L. and Corlett, E. N. (1992) *Reducing the Risks of Work-Related Upper Limb Disorders: A Guide and Methods.* The Institute for Occupational Ergonomics, The University of Nottingham, UK.

Wilson, J. R. and Corlett, E. N. (1995) *Evaluation of Human Work: A Practical Ergonomics Methodology.* (2nd ed.) Taylor & Francis, London.

26

OWAS Methods

Markku Mattila
*Tampere University of Technology
Finland*

Mika Vilkki
*Tampere University of Technology
Finland*

26.1 Introduction

Musculoskeletal disorders are one of the biggest occupational health problems in industrialized countries. According to national statistics, the proportion of musculoskeletal diseases of all occupational diseases in Finland was 31% (Kauppinen et al., 1994) and in the United States, 44% (Bureau of Labor Statistics, 1996). It was also noted that approximately 10% of occupational accidents resulted from sudden movements, lifting, repetitive motions, or overuse (Federation of Accident Insurance Institutions, 1995). Physical workload has been recognized as a factor affecting workers' health at several jobs. For example, about 33% of occupational diseases attributed to construction sites in Finland were linked to ergonomic factors associated with manual tasks (Federation of Accident Insurance Institutions, 1995).

Poor working postures constitute one of the main risk factors for musculoskeletal disorders (Burdorf et al., 1991), ranging from minor back problems to severe handicapping (Keyserling, 1986; Åaras et al., 1988). The effects of poor postures will continue unless proactive steps are taken to evaluate and reduce the problem. Therefore, it is essential to recognize early the patterns of work-related musculoskeletal symptoms and disorders and their risk factors in the workplace (Kuorinka and Forcier, 1995). More suitable working postures may have a positive effect on workers' musculoskeletal systems, and may allow for more effective control of work performance and reduction in the number of occupational accidents (Corlett et al., 1979; Wangenheim et al., 1986; Genaidy et al., 1990).

One practical method for analyzing and controlling poor working postures in industry is OWAS, the Ovako Working Posture Analysis System (Karhu et al., 1977).

26.2   Theoretical Background For OWAS

Development of the OWAS Method

The Ovako Working Posture Analysis System (OWAS) was first developed in the Finnish steel industry in the 1970s. The need to identify and assess poor working posture arose from the fact that many jobs at the steel mill included physically stressful tasks. These problems had led to an increasing number of sick leaves and early retirements due to musculoskeletal disorders (Heinsalmi, 1985).

The project for working posture improvement was started at the plant. The jobs at the steel mill were studied and 680 photographs of different working postures were taken. This material represented nearly all existing typical work situations at the steel mill (Karhu et al., 1977). This material was analyzed and sorted by the researchers in order to create a classification system for the postures. The researchers were able to identify 84 typical posture combinations of back, arms, and legs. In practical experiments these proved to cover the most common postures in the steel industry. The typical working postures were combinations of four back postures, three arm postures, and seven leg postures.

After the typical postures were chosen, the usefulness of the system was tested. Twelve work design engineers were taught the system, and they then analyzed 28 tasks in the steel plant. The results were promising and led to the precise reliability testing of the method. During this assignment, 52 tasks were analyzed and over 36,000 postures were recorded. This analysis showed high inter-observer reliability but slightly lower inter-worker reliability (Karhu et al., 1977).

In order to evaluate the discomfort and health effects of the different postures, 32 experienced steel workers evaluated each posture in a four-point rating scale ranging from "normal posture with no discomfort and health effects" to "extremely bad posture, short exposure leads to discomfort, ill effects on health possible." The postures were then evaluated and their risks to the musculoskeletal system were assessed by a group of specialists consisting of international ergonomists. Based on these evaluations, the final classification of postures into different action categories for preventive measures was made.

Working Posture Classification in the OWAS Method

The 84 working postures classified in the OWAS system cover the most common and easily identifiable work postures for the back, arms, and legs. An estimate for load handled by the person observed is also made in connection with the posture.

Each classified posture of the OWAS is determined by a four-digit code in which the numbers indicate the postures of the back, the arms, and the legs, as well as the load/effort needed.

Back

In the OWAS system the first digit in the posture code indicates the posture of the back. There are four choices for the different back postures: (1) back straight, (2) back bent, (3) back twisted, and (4) back bent and twisted (Table 26.1).

Arms

The second digit in the observation code indicates the posture of the arms. There are three choices for the arm postures in the OWAS system: (1) both arms below shoulder level, (2) one arm at or above shoulder level, and (3) both arms at or above shoulder level (Table 26.2).

Legs

The third digit in the posture code indicates the posture of legs. There are seven choices for the postures of the legs in the OWAS system: (1) sitting, (2) standing on two straight legs, (3) standing on one straight leg, (4) standing or squatting on two bent legs, (5) standing or squatting on one bent leg, (6) kneeling, and (7) walking (Table 26.3).

TABLE 26.1 Definition of Four Codes for the Back Postures in the OWAS System

1 BACK STRAIGHT

"Back straight" means that worker's back is less than 20° (the angle of the lines which go between head–hips and legs) bent forward or sideways or less than 20° twisted (the angle between shoulders and hips).

2 BACK BENT

"Back bent" means that worker is in a posture in which the upper body is bent forward or backward 20° (the angle of the lines which go between head–hips and legs) or more.

3 BACK TWISTED (OR BENT SIDEWAYS)

"Back twisted" means that the back is twisted 20° or more (as defined above) or bent sideways 20° or more.

4 BACK BENT AND TWISTED

"Back bent and twisted" means a situation where back is bent (like in case 2) and simultaneously twisted (like in case 3).

TABLE 26.2 Definition of Three Codes for the Arm Postures in the OWAS Method

1 BOTH ARMS BELOW SHOULDER LEVEL

"Both arms below shoulder level" means a situation in which both arms are completely below shoulder level.

2 ONE ARM AT OR ABOVE SHOULDER LEVEL

"One arm at or above shoulder level" means that one arm or part of it is at or above shoulder level.

3 BOTH ARMS AT OR ABOVE SHOULDER LEVEL

In "Both arms at or above shoulder level" both arms are fully or partly at or above shoulder level.

TABLE 26.3 Definition of the Seven Postures for Legs in the OWAS Method

1 SITTING

"Sitting" means that the weight of the body is supported on the buttocks. In this posture the legs are also below the buttocks.

2 STANDING ON BOTH STRAIGHT LEGS

"Standing on both straight legs" means that the weight of the body is supported on two straight legs. The knee angle is more than 150°.

3 STANDING ON ONE STRAIGHT LEG

"Standing on one straight leg" is a situation in which one leg is straight and the weight of the body is completely supported by that leg. The knee angle is more than 150°.

4 STANDING OR SQUATTING ON BOTH FEET, KNEES BENT

In this posture the weight of the body is on both legs and both knees are bent on a 150° or smaller angle.

5 STANDING OR SQUATTING ON ONE FOOT, KNEE BENT

In this posture the weight of the body is on one leg, and it is also bent from the knee. The knee angle is 150° or smaller.

6 KNEELING ON ONE OR BOTH KNEES

In this posture the person is kneeling either on both knees or one knee.

7 WALKING OR MOVING

In this posture the person is walking or moving around at the workplace.

TABLE 26.4   Definition of Three Codes for the Load Handled in the OWAS Method

---

1   LOAD/USE OF FORCE ≤ 10 KG

Weight handled or force needed is 10 kg or less.

2   LOAD/USE OF FORCE >10 KG ≤ 20 KG

Weight handled or force needed is exceeds 10 kg but is less than 20 kg.

3   LOAD/USE OF FORCE > 20 KG

Weight handled or force needed exceeds 20 kg.

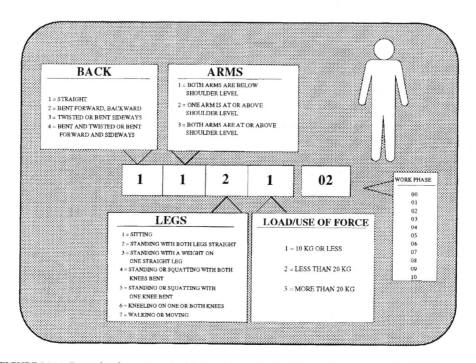

FIGURE 26.1   Example of a posture classification and codes for different body parts in the OWAS system.

Load/Use of Force

The fourth digit in the OWAS code indicates how big a load the person is handling or how much force must be used in the operation. Load/use of force has three alternatives: (1) less than 10 kilograms, (2) between 10 and 20 kilograms, and (3) more than 20 kilograms (Table 26.4).

The OWAS posture classification is presented in Figure 26.1. In addition to the posture code, it is possible to use the fifth digit to indicate the work phase or task the person was doing when the posture was observed. An example of the numerical code 1121 shows the usage of the OWAS system (Figure 26.1). Code 1121 indicates that the worker's back is straight (back: 1). He works with both arms below shoulder level (arms: 1), has his weight on two straight legs (legs: 2), and handles a load less than 10 kg (load/use of force: 1). The fifth digit indicates the work phase in which the posture occurs, for instance, carrying. The use of the work phase code and the analysis of different tasks among the observed material became possible through the computerized OWAS applications.

TABLE 26.5 The OWAS Action Categories for Prevention

Action Category	Explanation
1	Normal and natural postures with no harmful effect on the musculoskeletal system — **No actions required**
2	Postures with some harmful effect on the musculoskeletal system — **Corrective actions required in the near future**
3	Postures have a harmful effect on the musculoskeletal system — **Corrective actions should be done as soon as possible**
4	The load caused by these postures has a very harmful effect on the musculoskeletal system — **Corrective actions for improvement required immediately**

Making Observations for OWAS Data Gathering

The frequency of different postures and their proportional share of the working time are determined by observations. The basic idea in the observational technique is to collect material through observations made at set intervals over a given period of time. This provides an overall picture of the job studied. In the OWAS analysis, the observations are made using visual, split-second observations at the moment when the observer glances at the worker. The observations are collected in the original OWAS method using special forms made for this purpose. When the computer program for OWAS data gathering (OWASCO) is used, a portable laptop computer is utilized for the data input.

The observations should be done in the actual work situation, field conditions. In some cases, video-tapes can be used. The advantage of using videotapes is that the observer has as much time as he/she wants to look at the observed posture. The videotapes can also easily and effectively be used in recalling the actual work situations when providing feedback from the posture study or when teaching new work methods (Mattila et al., 1992).

The observations are made using an equal interval system where the interval between observations normally is either 30 or 60 seconds. The reason for this is that in field conditions it is often too hard for the observer to use shorter intervals. Shorter intervals are suggested in cases when it is possible to make the analysis using the videotapes, or the nature of the task requires it, e.g., short work cycles. The observation period shouldn't exceed 40 minutes without 10 minutes for resting.

The frequencies of work postures and their relative proportion (%) of the working time are calculated from the observation results. The error limits associated with the mean relative proportions of work posture are calculated for 95% probability, using a random system formula. The error limits become smaller as the total number of observations increases. The error limits in mean values based on 100 observations are ±10%, and with 400 observations, ±5%. Mean values obtained through observation can be considered sufficiently reliable when the error limits are under 10% (Louhevaara and Suurnäkki, 1991).

The Analysis of Recognized Work Postures in the OWAS System

The posture combinations and the relative proportions of certain postures are classified into four action categories for improvement needs. The classification of the postures is based on the risk assessment of musculoskeletal disorders and the physical load on the subjects musculoskeletal system. The action category indicates the urgency and priority for corrective measures. The action categories for prevention range from 1, no actions required, to 4, corrective measures needed immediately (Table 26.5). This categorization based on risk assessment was originally constructed by physicians, work analysts, and workers and then revised and validated by an international group of experts (Karhu et al., 1977).

The classification for individual posture combinations indicates the level of risk of injury or harmful effects caused by that classified posture (combination of the postures of back, arms, and legs and the load handled) for the musculoskeletal system. If the risk for musculoskeletal disorder is high, then the action category indicates the need and urgency for corrective actions. The action categories for each individual postures are presented in Figure 26.2.

back	arms	1			2			3			4			5			6			7			use of force
		1	2	3	1	2	3	1	2	3	1	2	3	1	2	3	1	2	3	1	2	3	
1	1	1	1	1	1	1	1	1	1	1	2	2	2	2	2	2	1	1	1	1	1	1	
	2	1	1	1	1	1	1	1	1	1	2	2	2	2	2	2	1	1	1	1	1	1	
	3	1	1	1	1	1	1	1	1	1	2	2	3	2	2	3	1	1	1	1	1	2	
2	1	2	2	3	2	2	3	2	2	3	3	3	3	3	3	3	2	2	2	2	3	3	
	2	2	2	3	2	2	3	2	3	3	3	3	4	4	3	4	3	3	4	2	3	4	
	3	3	3	4	2	2	3	3	3	3	3	4	4	4	4	4	4	4	4	2	3	4	
3	1	1	1	1	1	1	1	1	1	1	2	3	3	3	4	4	4	1	1	1	1	1	
	2	2	2	3	1	1	1	1	1	1	2	4	4	4	4	4	4	3	3	3	1	1	
	3	2	2	3	1	1	1	2	3	3	4	4	4	4	4	4	4	4	4	1	1	1	
4	1	2	3	3	2	2	3	2	2	3	4	4	4	4	4	4	4	4	4	2	3	4	
	2	3	3	4	2	3	4	3	3	4	4	4	4	4	4	4	4	4	4	2	3	4	
	3	4	4	4	2	3	4	3	3	4	4	4	4	4	4	4	4	4	4	2	3	4	

FIGURE 26.2 Action category for each individual OWAS classified posture combination.

The second classification is based on the time spent in different postures for each body part. This classification examines the relative proportion postures of the back, the arms, and the legs during the observation period (Figure 26.3). The same four action categories used in the classification of individual posture combinations are used here. The postures for each body part are counted together and when the relative proportion of certain posture during the observation period exceeds fixed limits, the action category changes from lower to higher. This indicates that the urgency of corrective actions is increasing. The OWAS system doesn't have a classification for the relative proportion of the use of force/load handled. In cases when heavy materials handling occurs, the situation must be evaluated separately in each case. For such evaluations, a biomechanical analysis is useful.

The Data Analysis and Reporting

When an adequate amount of postural data has been gathered, the results of the OWAS study are analyzed and reported. The use of a computerized OWAS application makes the analysis fast and more versatile than the traditional pen-and-paper method. The use of a computerized application is strongly recommended, but the traditional method is functional if the OWAS software is not available.

In the data analysis, all the collected postures are counted together and processed on the summary form. The basic analysis of the collected data includes two items: (1) the calculation of the percentages of posture combinations and (2) the proportional shares of postures of different body parts, falling into different action categories.

If a more detailed analysis is needed, the data may be analyzed according to different work phases. This helps in finding the most difficult tasks and operations in the work analyzed. It is also possible to analyze the most difficult postures (categories 3 and 4) separately and study these situations more carefully. This analysis makes it possible to quickly examine which tasks need corrective actions most urgently.

In the reporting of the OWAS study, the results are printed out and graphs are drawn from the distributions. If videotapes are available they can be very helpful in reporting the results of the postural study.

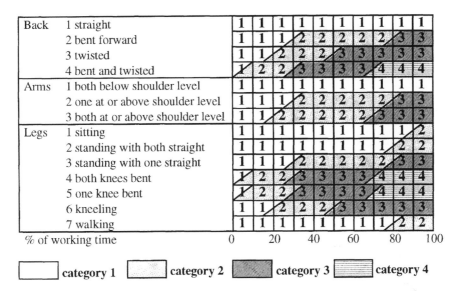

% of working time   0   20   40   60   80   100

☐ **category 1** ☐ **category 2** ▨ **category 3** ▤ **category 4**

FIGURE 26.3 Action categories for the relative proportions of the postures of the different body parts.

Procedure to Handle OWAS Results

It is highly recommended to establish a cooperative team to handle the results of OWAS analysis. This group could consist of: for example, the analyst, representatives of the workers, management, occupational health care specialist, and other persons who may be involved in job redesign and development. When the team has the results and identifies the problematic task associated with the poor work postures from the videotapes, it is easier and more effective to provide corrective measures and redesign the working methods (Kivi and Mattila, 1991; Mattila et al., 1992).

Reliability and Validity

The inter-observer reliability of the OWAS method has been tested in many jobs in different industries. In all reported cases the inter-observer reliability was high, averaging over 90% (Karhu et al., 1977; Louhevaara and Suurnäkki, 1991; Mattila et al., 1993). The posture of the back is most difficult to for observers to distinguish.

Leskinen and Tönnes (1993) studied the validity of the OWAS system by comparing the postures collected using the OWAS system with the accurate work postures recorded using the electronic Selspot II camera system. The OWAS system proved to give nearly the right picture about the postural load. Kuusela (1994) investigated the biomechanical load caused by typical OWAS postures, using a special biomechanical software. This study showed that the biomechanical basis of the OWAS system is correct with the exception of a few situations.

26.3 Computer-Aided Applications

The first application was introduced in the late 1980s by Tampere University of Technology in Finland. This system was based on a portable programmable HP-71B calculator used for data collection. The data was then transferred to a PC computer through a special interface linkage. The analysis of the collected data was then done by the analysis program in the PC computer (Kivi and Mattila, 1991). The next step in the software development was introduced in 1991 (Mattila et al., 1992). This system used a portable Canon X-07 computer for data collection. The main improvement compared to the previous system was that the data collection unit was already capable of performing a limited amount of data analysis. In order to achieve a comprehensive data analysis the data had to be transferred to the PC (Figure 26.4).

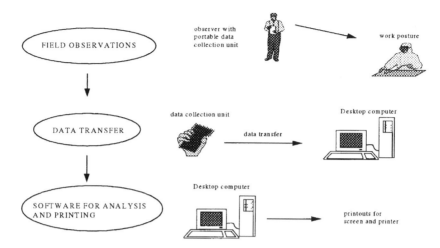

FIGURE 26.4   The OWAS data processing system used in the early computerized versions.

```
OWASCO 2.3  (C) Tampere University of Technology, 1993.

                 Give the OWAS code: 11110

BACK              ARMS                 LEGS                    LOAD
1 Straight        1 2 below shoulder   1 Sitting               1 <10 kg
2 Bent            2 1 below shoulder   2 Standing on 2 legs    2 <20 kg
3 Twisted         3 2 above shoulder   3 Standing on 1 leg     3 >20 kg
4 Bent and Twisted                     4 Standing on 2 bent knees
                                       5 Standing on 1 bent knee
                                       6 Kneeling
WORKPHASE                              7 Walking
0 Lifting
1 Carrying
2 Packing
3 Adjusting
4 Other                                      <-      MOVE     ->
5 Not defined
6 Not defined                            RETURN = ACCEPT
7 Not defined
8 Not defined                               N       = MISSED CASE
9 Not defined

                                                        ESC = QUIT
              10:43:51   N:o observations: 0
```

FIGURE 26.5   Data input interface of the OWASCO computer program.

The third-generation OWAS software was developed in 1992 (Vilkki et al., 1993). The light weight and reduced prices of PC notebook computers made it possible to use the same computer for data collection and analysis. This allowed faster analysis of data and immediate feedback at the worksite. The software program package is called OWASCO & OWASAN. In this system the OWASCO program is used for data collection and the OWASAN program is used for the analysis of data.

In the data analysis the background information for the work to be observed and the chosen analysis interval are given before starting the observation. In this stage it is also necessary to divide the observed job into work phases. If the observations are made properly using the work phase coding, it makes it easier in data analysis to identify the problematic tasks and to create measures for job redesign. The data input screen of the OWASCO program allows the choice of body postures from the menus to be always visible on the screen (Figure 26.5). A built-in clock with a "beep" sound indicates the moment when the observation should be done.

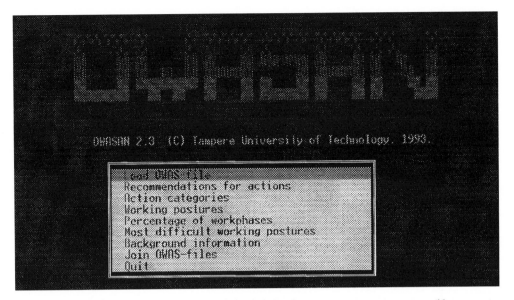

FIGURE 26.6   The main menu of the OWASAN computer program.

The OWASAN data analysis program calculates different types of analysis from the collected data. The functions of this program include: (1) recommendations for actions, (2) action categories, (3) listing of working postures, (4) percentage of work phases, (5) most difficult working postures, (6) printing the background information sheet, and (7) joining the OWAS files together (Figure 26.6). It is possible get the analysis from one particular work phase or the whole analyzed OWAS data.

The results can be printed on the screen, on a file for export to other software, e.g., word processing, or for a printer. An example of a printout of the "Recommendations for Actions" sheet is presented in Figure 26.7.

Other OWAS applications have also been introduced. Kant et al. (1990) introduced a computerized version of the OWAS system and tested it in garages. Pinzke (1992) introduced a software analysis method more precise than the OWAS method but using the same action categories.

26.4   Experiences from OWAS Usage in Different Fields

The OWAS method was originally developed in the steel industry. The jobs at the steel mill included a large variety of working postures combined with heavy materials handling. Due to the nature of the work for which the system was developed, it is most suitable in the analysis of physical work with clearly observable postures of different body segments.

The OWAS method has been applied over the years in various industries in many countries. Most of the reported cases are analyses of jobs including heavy physical work and dynamic working postures.

Steel Mill

Karhu et al. (1981) reported an OWAS study of two jobs: bricklaying and installation of a roll unit in a steel mill. In this study, poor working postures were indicated and corrective measures were suggested. The main result of this study was decreased number of poor working postures and considerably higher productivity.

Mining Industry

Heinsalmi (1985) reported a working posture study in the Finnish mining industry. In this study, the postures of a drilling unit were examined. The proportion of working postures in the OWAS categories

```
RECOMMENDATIONS FOR ACTIONS

COMPANY: Factory Inc.
DEPARTMENT: Assembly
THE JOB OBSERVED: Welding
OBSERVER: Ilkka
INTERVAL: 20

Whole material.   Total: 20

                 0%  10% 20% 30% 40%  50%  60% 70%  80%  90% 100%
----------------------------------- BACK -----------------------------------
Straight         1111111111111111111111111111111111■11111111111   75.00 %

Bent             111111111■11/222222222222222222222222/333333333   20.00 %

Twisted          111■1111/2222222222222/33333333333333333333333   5.00 %

Bent&Twisted     ■1/2222222222/333333333333333333/44444444444444   0.00 %
----------------------------------- ARMS -----------------------------------
Free             11111■11111111111111111111111111111111111111111   10.00 %

One up           1111111111111/222222222222222222222222/33■333333   85.00 %

Both up          111■11111/2222222222222222222222/33333333333333   5.00 %
----------------------------------- LEGS -----------------------------------
Sitting          111■1111111111111111111111111111111111111111/2222   5.00 %

Stand.with 2     1111111111111111111111111111111■11111/2222222222   60.00 %

Stand.with 1     1111111111111■2222222222222222222222/333333333   30.00 %

Both bent        11/■2222222222/3333333333333333/44444444444444   5.00 %

One bent         ■1/2222222222/33333333333333333/4444444444444   0.00 %

On knees         ■11111111/2222222222222/33333333333333333333333   0.00 %

Walking          ■1111111111111111111111111111111111111/222222222   0.00 %
----------------------------------- LOAD -----------------------------------
< 10 kg          Find out the reasons for differe■t loads . . .   65.00 %

< 20 kg          Find out the reason■ for different loads . . .   35.00 %

> 20 kg          ■ind out the reasons for different loads . . .   0.00 %
----------------------------------------------------------------------------
```

FIGURE 26.7 An example of the "Recommendations for Actions" sheet printed with the OWASAN program.

2 through 4 was 5.4%. Changes in the machinery layout in the workplace decreased the share of poor postures to 1.0%.

Cleaning

Hopsu and Louhevaara (1991) used the OWAS method to measure postural load in cleaners work during an intervention study. The intervention included educational training and ergonomic job redesign. The results show a decrease from 39% to 25% in the amount of postures in categories 2 through 4. The constancy of the improvement was evaluated in a follow-up study a year after the intervention.

Garage

Kant et al. (1990) used the OWAS system in the analysis of the jobs of 84 mechanics in 42 garages. In this study, the use of the computerized OWAS system made it possible to easily identify the most difficult work phases. Some tasks, e.g., working at the side of the car, had 18% of posture combinations in categories 3 and 4. But while using the vehicle lift in the same job, the number of category 3 and 4 postures was only 4%.

Construction

Kivi and Mattila (1991) used the OWAS method to analyze twelve jobs at construction sites. This study showed that many jobs at construction sites include a high number of poor working postures. The most difficult jobs found in this study were cement worker and repair worker. The proportion of OWAS category 3 and 4 postures in these jobs was 28% and 18%, respectively. In this study, the computerized OWASCO & OWASAN program package was found useful in pinpointing the needs for improvement.

Mattila et al. (1993) analyzed the working postures of carpenters doing hammering tasks. The OWAS method showed a high incidence of poor postures in many tasks, but it was not possible to find differences in working postures caused by different types of hammers.

Rohmert et al. (1993) did an extensive study of tilers work. In this study, about 12,000 working postures were analyzed. The results showed that in tiling work more than 60% of the postures require corrective actions.

Paper Mill

Mattila et al. (1992) analyzed jobs in the paper mill industry. In the maintenance and adjustment tasks performed while the machine was stopped, the proportion of the OWAS category 3 and 4 postures was highest, in some cases up to 50%. The OWAS method proved to be suitable for analysis of this type of job also.

Railways

Peereboom (1993) presented a strategy for how to use the OWAS method. He used this approach to study the postures of train mechanics and railway maintenance workers.

Manufacturing

Vilkki et al. (1993) studied manual materials handling tasks in manufacturing. The percentages of postures in categories 3 and 4 were relatively low, but heavy lifting occurred frequently. Therefore, a biomechanical analysis was combined with the OWAS analysis. The computerized OWAS method proved to be an effective way to analyze poor postures during the redesign process of a production line.

Nursing

Nurses' working postures were studied by Engels et al. (1994). The nurses jobs were analyzed in two wards. This study showed that working postures of nurses were slightly harmful. Differences in postural load between wards were found.

Farming

Nevala-Puranen (1995) reported a study of farmers' postural load. In this study, poor postures were identified by the OWAS method and new techniques were taught. For example, in milking the proportion of category 3 and 4 postures decreased from about 50% to 20%. The study showed positive and permanent changes in farmers' working postures.

26.5 Conclusions and Future Prospects

The OWAS method has shown its functionality in many studies in different types of industries over the years. It is a useful way to get an overall picture of the postural load caused by different postures in different jobs and to direct the improvements in the working methods used at work. The method is feasible and the computer-aided applications have made the analysis more effective and more versatile.

The OWAS method effectively supports the modern safety management at the plant and the preventive occupational health care especially, while it includes a risk assessment procedure to show the most problematic postures and to pinpoint the priorities for corrective actions.

The OWAS analysis has proved to give good background information for participatory ergonomics. It is possible to use the results at the workplace immediately after the analysis in discussions to improve the working postures and to redesign the working methods. The utilization of the results is most effective in a cooperative team which consists of, e.g., the occupational health care professional, management and worker representatives, the person who did the analysis, and others who are involved with the analyzed work. The analysis of the OWAS results by a cooperative team gives a good basis for open discussion at the workplace. In this way, the OWAS method is a successful tool to support continuous improvement at the workplace. Such an approach allows the improvement of tasks, job redesign, and new working methods development. The videotapes and photographs taken during the study help the team to visualize the actual situations.

In the future there seems to be the possibility for automatic posture recording which will improve the accuracy and the speed needed for the analysis. In Manual materials handling tasks and jobs where loads handled play an essential role, it would be beneficial if biomechanical analysis could be integrated in the OWAS method. These are some challenges for future development for the OWAS method.

References

Åaras, A., Westgaard, R.H., and Strandew, E. 1988. Postural angles as an indicator of postural load and muscular injury in occupational work situations, *Ergonomics* 31, 915-933.

Burdorf, A., Govaert, G., and Elders, L. 1991. Postural load and back pain of workers in the manufacturing of prefabricated concrete elements et al., *Ergonomics* 34, 909-918

Bureau of Labor Statistics, 1996. *"Characteristics of injuries and illnesses resulting in absences for work, 1994."* ftp://stats.bls.gov/pub/news.release/osh2.txt/, U.S. Department of Labor.

Corlett, E.N., Madeley, S.J., and Manenica, J. 1979. Posture targeting: a technique for recording work postures, *Ergonomics* 24, 795-806.

Engels, J.A., Landeweerd, J.A., and Kant, Y. 1994, An OWAS-based analysis of nurses' working postures. *Ergonomics*, 37, 5, 909-919.

Federation of Accident Insurance Institutions, 1995. *Työtapaturma- ja ammattitautitilasto 1993 ("Occupational accidents and diseases 1993")*. Tapaturmavakuutuslaitosten liitto, Helsinki, p. 96.

Genaidy, A., Karwowski, W., and Musavinezhad, S.H. 1990. Computer-aided ergonomics: a tool for control of musculoskeletal injury, in Karwowski, W., Genaidy, A., and Asfour, S.S. (eds.), *Computer-Aided Ergonomics*, Taylor & Francis, 8-28.

Heinsalmi, P. 1985. Method to measure working posture loads at working sites (OWAS), in Corlett, N., Wilson, J., and Manenica, I. (ed.), *The Ergonomics of Working Postures*. Taylor & Francis, London, pp. 100-104.

Hopsu, L. and Louhevaara, V. 1991. The influence of educational training and ergonomic job redesign intervention on the cleaners' work: a follow up study, in Quéinnec, Y. and Daniellou, F. (eds.), *Designing for Everyone*. Taylor & Francis, pp. 534-536.

Kant, I., Notermans, J.H.V., Borm, P.J.A. 1990, Observation of working postures in garages using the Ovako Working Posture Analyzing System (OWAS) and consequent workload reduction requirements. *Ergonomics* 33, 209-220.

Karhu, O., Kansi, P., and Kuorinka, I. 1977. Correcting working postures in industry. A practical method for analysis. *Applied Ergonomics* 8, 199-201.

Karhu, O., Härkönen, R., Sorvali, P., and Vepsäläinen, P. 1981. Observing working postures in industry: Examples of OWAS application. *Applied Ergonomics* 12, 13 — 17.

Kauppinen, T., Vaaranen, V., Vasama, M., Tokkanen, J., and Jolanki, R. 1994. *Occupational Diseases in Finland in 1993*, The Finnish Institute of Occupational Health, Helsinki, p. 102 (in Finnish).

Kivi, P. and Mattila, M. 1991. Analysis and improvement of work postures in the building industry: application of the computerized OWAS method. *Applied Ergonomics* 22 (1), 43-48.

Kuorinka, I. and Forcier L. (eds.). 1995. *Work Related Musculoskeletal Disorders (WMSDs)*, Taylor & Francis, London.

Kuusela, J. 1994. *Working Postures and Their Biomechanical Loading*, Tampere University of Technology, Finland, Unpublished technical report (in Finnish).

Leskinen, T. and Tönnes, M. 1993. Validity of observation methods used for the evaluation of working postures, *Työ ja Ihminen*, 7(1993):4, 299-314. (in Finnish, with English summary).

Long, A.F. 1992. A computerized system for OWAS field collection and analysis, in Mattila, M. and Karwowski, W. (eds.), *Computer Applications in Ergonomics, Occupational Safety and Health*. Elsevier, Amsterdam, pp. 353-358.

Louhevaara, V. and Suurnäkki, T. 1992. *OWAS: A Method for the Evaluation of Postural Load During Work*. Institute of Occupational Health and Centre for Occupational Safety, Helsinki, p. 23.

Mattila, M. and Kivi, P. 1991. Analysis of problematic working postures and manual lifting in building tasks, in Quéinnec, Y. and Daniellou, F. (eds.), *Designing for Everyone*. Taylor & Francis, London.

Mattila, M., Vilkki, M. and Tiilikainen, I. 1992. A Computerized OWAS analysis of work postures in the papermill industry, in Mattila, M. and Karwowski, W. (eds.), *Computer Applications in Ergonomics, Occupational Safety and Health*. Elsevier, Amsterdam, pp. 365-372.

Mattila, M., Karwowski, W., and Vilkki, M. 1993. Analysis of working postures in hammering tasks on building construction sites using the computerized OWAS method. *Applied Ergonomics* 24(6), 405-412.

Nevala-Puranen, N. 1995. Reduction of farmers' postural load during occupationally oriented medical rehabilitation. *Applied Ergonomics* 26, 6, 411-415.

Peereboom, K.J. 1993. A strategy for using the OVAKO working posture analyzing system (OWAS) to determine the physical load of actions, in Marras, W.S., Karwowski, W., Smith, J.L., and Pacholski, L. (eds.), *The Ergonomics of Manual Work*. Taylor & Francis, London, pp. 245-248.

Pintzke, S. 1992. A computerized method of observation used to demonstrate injurious work operations, in Mattila, M. and Karwowski, W. (eds.), *Computer Applications in Ergonomics, Occupational Safety and Health*. Elsevier, Amsterdam, pp. 359-364.

Rohmert, W., Wakula, J., and Schildge, B. 1993. Analysis of working postures of tilers, in Marras, W.S., Karwowski, W., Smith, J.L., and Pacholski, L. (eds.), *The Ergonomics of Manual Work*. Taylor & Francis, London, pp. 33-40.

Wangenheim, M., Samuelson, B., and Wos, H. 1986. ARBAN — a force ergonomic analysis method, in Corlett, E.N., Wilson J.R., and Manenica, I. (eds.) *The Ergonomics of Working Postures*, Taylor and Francis, 243-255.

Vilkki, M., Mattila, M., and Siuko, M. 1993. Improving work postures and manual materials handling tasks in manufacturing: A case study, in Marras, W.S., Karwowski, W., Smith, J.L. and Pacholski, L. (eds.), *The Ergonomics of Manual Work*. Taylor & Francis, London, pp. 273-276.

27

Ergonomics of Hand Tools

Andris Freivalds
The Pennsylvania State University

27.1 Introduction

Tools are as old as the human race itself. The hands and feet could be considered tools given to the human by nature. However, tools as we know them were developed as extensions of the hands and feet to amplify the range, strength, and effectiveness of these limbs. Thus, the early human, by picking up a stone, could make the fist heavier and harder, producing a more effective blow. Similarly, by using a stick, a longer and stronger arm was created.

The exact time when humans began to use and to make tools is not known. Leakey (1960), during his excavations in Africa, uncovered evidence that more than a million years ago the prehistoric human was already a tool-maker, using stones for chipping and bones for leather work. Similarly, Napier (1962) indicated that with changing tasks, such as converting from the power to precision grip, there was similar change in the anatomy of the hand as well as development of tools. An important milestone occurred when the stone tools were provided with handles some 35,000 years ago. The addition of the handle increased the range and speed of action and increased the kinetic energy for striking tasks. A still later change in tool development occurred with the change in tasks from food gathering to food production. New tools were required and developed accordingly. Surprisingly, many of these tools, with minor improvements and refinements, are still in use today. The reasons for such stagnation could be twofold: either the tool reached an optimal form very quickly with no room for improvement or there was no impetus for further improvement. The latter is the resigned view that since a tool has been used by so many people for so many years, no further improvement is possible. The former view is obviously not true since Lehmann (1953) noted the existence of over 12,000 different styles of shovels in Germany in the 1930s, all essentially used for the same task. Indeed, the last great change in tool development occurred with the start of the Industrial Revolution, with a change in task, from food production to manufacturing of goods.

The parallel development of tools with changing technology has given rise to another problem. The current technology explosion has proceeded too quickly to permit the gradual development of tools appropriate for the new industrial tasks. The instant demands for new and specialized tools to match the needs of technology has, in many cases, bypassed the testing needed to fit these tools to human users. This has resulted in a variety of hand-tool-generated work stressors and an increased incidence of cumulative trauma of the hand, wrist, and forearm, reducing productivity, disabling individuals, and increasing the medical costs for industry.

Cumulative trauma disorders (CTDs) are injuries to the musculoskeletal system that develop gradually as a result of repeated microtrauma. Because of the slow onset and relatively mild nature of the trauma, the condition is often ignored until the symptoms become chronic and more severe injury occurs. These problems are a collection of a variety of problems including repetitive motion disorders, carpal tunnel syndrome, tendinitis, ganglionitis, tenosynovitis, and bursitis, with these terms sometimes being used interchangeably. There are four major work-related factors that seem to lead to the development of CTD: (1) use of excessive force during normal motions, (2) awkward or extreme joint motions, (3) high amounts of repetition of the same movement, and (4) the lack of sufficient rest to allow the traumatized joint to recover. The most common symptoms associated with CTD include: pain, restriction of joint movement, and soft tissue swelling. In the early stages there may be few visible signs, however, if the nerves are affected, sensory responses and motor control may be impaired. If left untreated, CTD can result in permanent disability (Putz-Anderson, 1988).

The cost of CTD in U.S. industry, although not all due to improper tool design, is quite high. Data from the National Safety Council (1993) suggest that 15 to 20% of workers in key industries (meatpacking, poultry processing, auto assembly, and garment manufacturing) are at potential risk for CTD and that in 1991 some 223,600 cases or 61% of all occupational injuries were associated with repetitive actions. The worst industry was manufacturing, while the worst occupational title was butchering with 222 CTD claims per 100,000 workers (Putz-Anderson, 1988). With such high rates and average costs of $30,000 per case, NIOSH, in its *Year 2000 Objectives,* has targeted the reduction of CTD incidence from 82 to 60 cases per 100,000 overall workers and from 285 to 150 in certain manufacturing industries (NIOSH, 1989a).

The proper selection, evaluation, and use of hand tools is a major ergonomic concern. The following review will discuss the basic principles involved in tool design and the desirable attributes for specific tools.

27.2   Principles and Problems of Tool Design

General Principles

An efficient tool has to fulfill some basic requirements (Drillis, 1963):

1. It must effectively perform the function for which it is intended. Thus, an axe should convert a maximum amount of its kinetic energy into useful chopping work, cleanly separate wood fibers, and be easily withdrawn.
2. It must be properly proportioned to the body dimensions of the operator to maximize efficiency of human involvement.
3. It must be designed to match the strength and work capacity of the operator. Thus, allowances have to be made for the gender, age, training, and physical fitness of the operator.
4. It should not cause undue fatigue, i.e., it should not demand unusual postures or practices that will require more energy expenditure than necessary.
5. It must provide sensory feedback in the form of pressure, some shock, texture, temperature, etc., to the user.
6. The capital and maintenance costs should be reasonable.

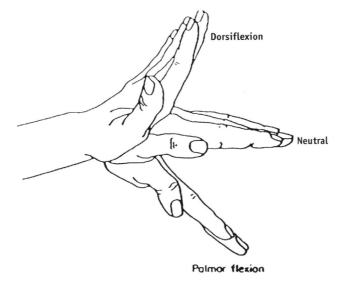

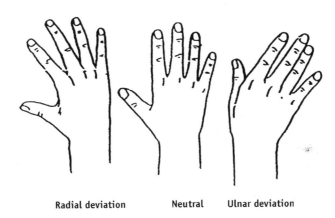

Radial deviation   Neutral   Ulnar deviation

FIGURE 27.1   Types of wrist movement.

Types of Grip

To better understand the design principles of hand tools, it is necessary to have a brief description of the anatomy and functioning of the human hand as well as some of the diseases that can result from its misuse. The human hand is a complex structure of bones, arteries, nerves, ligaments, and tendons. The fingers are controlled by the extensor carpi and flexor carpi muscles in the forearm. The muscles are connected to the fingers by tendons which pass through a channel in the wrist, formed by the bones of the back of the hand on one side and the transverse carpal ligament on the other. Through this channel, called the carpal tunnel, pass also various arteries and nerves. The bones of the wrist connect to two long bones in the forearm, the ulna and the radius. The radius connects to the thumb side of the wrist and the ulna connects to the little finger side of the wrist. The orientation of the wrist joint allows movement in only two planes, each at 90° to the other (Figure 27.1). The first gives rise to *palmar flexion* and *dorsiflexion* (or *extension*). The second movement plane gives *ulnar* and *radial deviation*.

The manual dexterity produced by the hand can be defined in terms of a *power grip* and a *precision grip*. In a power grip, the tool, whose axis is more or less perpendicular to the forearm, is held in a clamp formed by the partly flexed fingers and the palm, with opposing pressure being applied by the thumb (Figure 27.2). There are three subcategories of the power grip differentiated by the line of action of force:

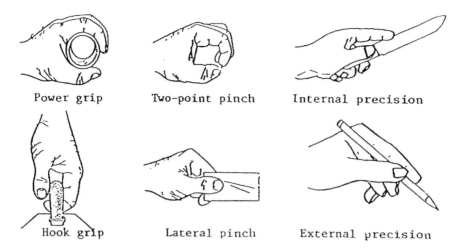

FIGURE 27.2   Types of grip.

(1) force parallel to the forearm, as in sawing; (2) force at an angle to the forearm, as in hammering; and (3) torque about the forearm, as when using a screwdriver. As the name implies, the power grip is used for power or for holding heavy objects.

In a precision grip, the tool is pinched between the flexor aspects of the finger and the opposing thumb. The relative position of the thumb and fingers determines how much force is to be applied and provides a sensory surface for receiving feedback necessary to give the precision needed. There are two types of precision grip: (1) internal, in which the shaft of the tool (e.g., knife) passes under the thumb and is thus internal to the hand; and (2) external, in which the shaft (e.g., pencil) passes over the thumb and is thus external to the hand. The precision grip is used for control. Other grips are just variations of the power or precision grip and include the hook grip, for holding a box or handle, a two-point pinch, and a lateral pinch (Figure 27.2).

Static Muscle Loading

When tools are used in situations in which the arms must be elevated or tools have to be held for extended periods, muscles of the shoulders, arms, and hands may be loaded statically, resulting in fatigue, reduced work capacity, and soreness. Abduction of the shoulder with corresponding elevation of the elbow will occur if work has to be done with a pistol-grip tool on a horizontal workplace. An in-line or straight tool reduces the need to raise the arm and also allows for a neutral wrist posture.

Prolonged work with arms extended can produce soreness in the forearm for assembly tasks done with force. By rearranging the workplace so as to keep the elbows at 90°, most of the problem can be eliminated (Figure 27.3). Of course, the orientation of the tool as related to the work surface can modify this posture. This will be discussed further in the next section. Similarly, continuous holding of an activation switch can result in fatigue of the fingers and reduced flexibility.

Awkward Wrist Position

As the wrist is moved from its neutral position, there is loss of grip strength. Starting from a neutral wrist position, full pronation decreases grip strength by 12%, full flexion/extension by 25%, and full radial/ulnar deviation by 15% (Terrell and Pursewell, 1970). The percent of maximum grip strength available can be quantified by ($r^2 = .854$, $p < .001$):

$$\%Grip = 98.6 - 8.8 \times PS - 25.2 \times FE - 16.3 \times RU \qquad (27.1)$$

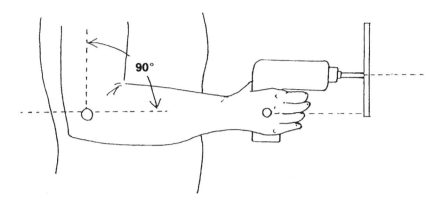

FIGURE 27.3 Optimum working posture with elbow bent at 90°.

where

PS = 1 if the wrist is fully pronated and 0 if in a neutral position or supinated
FE = 1 if the wrist is fully flexed or extended and 0 if in a neutral position
RU = 1 if the wrist is fully in radial or ulnar deviate and 0 if in a neutral position

Furthermore, awkward hand positions may result in soreness of the wrist, loss of grip, and, if sustained for extended periods of time, occurrence of *carpal tunnel syndrome*. To reduce this problem, the workplace or tools should be redesigned to allow for a straight wrist, i.e., lowering work surface and edges of containers, tilting jigs toward the user (Figure 27.4), using a pistol handle on power tools for vertical surfaces and in-line handles for horizontal surfaces (Figure 27.5), etc. Similarly, the tool handle should reflect the axis of grasp, such as a pistol grip on knives (Figure 27.6) (Armstrong et al., 1982).

Tissue Compression

Often, in the operation of hand tools, considerable force is applied by the hand. Such actions can concentrate considerable compressive force on the palm of the hand or the fingers, resulting in *ischemia*, obstruction of blood flow to the tissues, and eventual numbness and tingling of the fingers. Handles should be designed to have large contact surfaces to distribute the force over a larger area (Figure 27.7) or to direct it to less sensitive areas such as the tissue between the thumb and index finger. Similarly, finger grooves or recesses in tool handles should be avoided. Since hands vary considerably in size, the grooves will accommodate only a fraction of the population.

Gender

Female grip strength typically ranges from 50 to 67% of male strength (Konz, 1990; Chaffin and Andersson, 1984), i.e., the average male can be expected to exert approximately 500 N, while the average female can be expected to exert approximately 250 N. An interesting survey by Ducharme (1975) examined how tools and equipment that were physically inadequate for female workers hampered their performance. The worst offenders were crimpers, wire strippers and soldering irons. Females have a twofold disadvantage — an average lower strength and an average smaller grip span. Ducharme (1975) concluded that women could be integrated more quickly and safely into the work force if tools were designed to accommodate their smaller hand dimensions.

Handedness

Alternating hands permits reduction of local muscle fatigue. However, in many situations this is not possible because the tool use is one-handed. Furthermore, if the tool is designated for the user's preferred

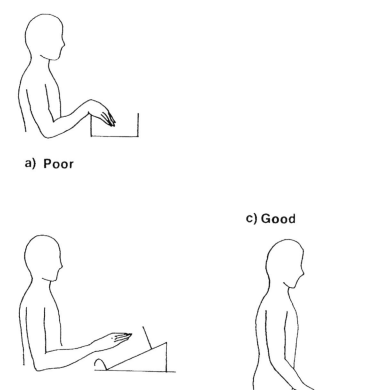

FIGURE 27.4 Proper orientation of jigs and containers.

hand — which for 90% of the population is the right hand — then 10% are left out (Konz, 1974). Laveson and Meyer (1976) gave several good examples of right-handed tools that cannot be used by a left-handed person, e.g., a power drill with side handle on the left side only, a circular saw, and a serrated knife beveled on one side only. Miller and Freivalds (1987) found right-handed males to show a 12% strength decrement in the left hand, while right-handed females showed a 7% strength decrement. Both left-handed males and females had nearly equal strengths in both hands. They concluded that left-handed subjects were forced to adapt to a right-handed world. Using time study, Konz and Warraich (1985) found decrements, ranging from 9% for an electric drill to 48% for manual scissors, for using the nonpreferred hand as opposed to the preferred hand.

Posture

In general, unless the posture is extreme, i.e., standing versus lying, torque exertion capability is not affected substantially by posture (Mital, 1986). The height at which torque was applied had no influence on peak torque exertion capability.

Repetitive Finger Action

If the index finger is used excessively for operating triggers, symptoms of *trigger finger* develop. Thus, trigger forces should be kept low, preferable below 10 N, to reduce the load on the index finger. Two- or three-finger operated controls are preferable (Figure 27.8); finger strip controls or a power grip bar are even better. For a two-handled tool, a spring-loaded return saves the fingers from having to return the

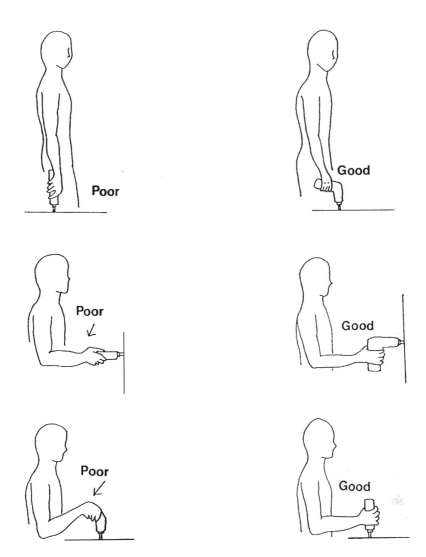

FIGURE 27.5 Proper orientation of power tools in the workplace.

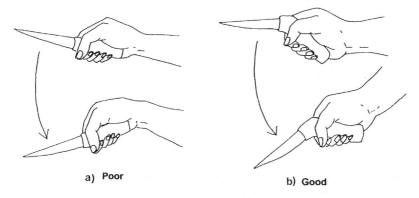

FIGURE 27.6 Pistol-grip knife.

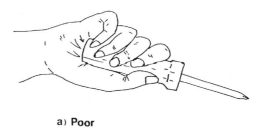

a) **Poor**

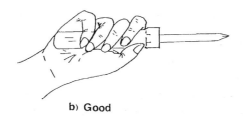

b) **Good**

FIGURE 27.7   Avoiding tissue compression in tool design.

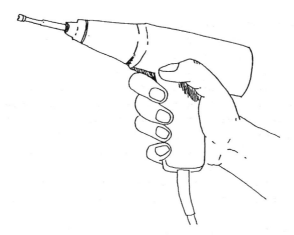

FIGURE 27.8   Three-finger trigger for power tools.

tool to its starting position (Eastman Kodak, 1983). In addition, the high number of repetitions must be reduced. Although critical levels of repetitions are not known, attempts have been made to identify the maximal number of exertions per hour or shift that can be tolerated. Most of these have been regarding wrist movements, but could reasonably apply to the fingers as well. Luopajarvi et al. (1979) found high rates of muscle–tendon disorders in assembly line packers with over 25,000 movements per day, while NIOSH (1989b) found similar problems in workers exceeding 10,000 motions per day.

Handle Diameter

Power grips around a cylindrical object should surround the circumference of the cylinder, with the fingers and thumb barely touching. However, the actual diameter may vary due to the task as well as the hand size of the operator. Thus, for a power grip on screwdrivers, Rubarth (1928) recommended a diameter of 40 mm. For minimum *EMG* activity, Ayoub and LoPresti (1971) found a 51-mm handle diameter to be best. However, based on the maximum number of work cycles completed before fatigue and on the ratio of grip force to EMG activity, they suggested a 38-mm diameter. For handles on boxes, Drury (1980) found diameters of 31 to 38 mm to be best in terms of the least reduction in grip strength. Using various handles of noncircular cross section, Cochran and Riley (1986) found the largest thrust forces in handles of 41-mm equivalent circular diameter (based on a total 130 mm circumference) for both males and females. For manipulation, however, smaller handles of 22 mm were found to be the best. Eastman Kodak (1983), based on company experience, recommended 30 to 40 mm with an optimum of 40 mm for power grips. Thus, one can summarize that handle diameters should be in the range of 30 to 50 mm, with the upper end best for maximum torque and the lower end for dexterity and speed.

Handle Length

The length of the handle has been studied to a lesser extent. For cut-out as well as normal handles, there should be enough space to admit all four fingers. Hand breadth across the metacarpals ranges from 71 mm for a 5th percentile female to 97 mm for a 95th percentile male (Garrett, 1971). Thus, 100 mm may be a reasonable minimum, but 120 mm may be more comfortable. Eastman Kodak (1983) recommended 120 mm. If the grip is enclosed or gloves are used, even larger openings are recommended. For an external precision grip, the tool shaft must be long enough to be supported at the base of the first finger or thumb. A minimum value of 100 mm is suggested (Konz,1990). For an internal precision grip, the tool should extend past the palm, but not so far as to hit the wrist (Figure 27.7) (Konz, 1990).

Handle Shape

Rubarth (1928) investigated handle shape and concluded that, for a power grip, one should design for maximum surface contact so as to minimize unit pressure of the hand. A tool with a circular cross section was found to give largest torque. Pheasant and O'Neill (1975) concluded that the precise shape of the handle was irrelevant and recommended simple knurled cylinders. Similarly, Cochran and Riley (1986) found that no one shape may be perfect, and that shape may be more dependent on the type of task and motions involved than initially thought. The circular cross section was found to be worst and a triangular one best for thrusting forces; triangular shape was slowest for a rolling type of manipulation, while a rectangular shape of height/width ratios from .67 to .8 appeared to be a good compromise for many tasks. A further advantage of a noncircular cross section is that the tool does not roll when placed on a table. It should also be noted that handles may not always have the shape of a true cylinder except for a hook grip. For screwdriver-type tools, the handle end is rounded to prevent undue pressure at the palm. For hammer type tools, the handle may have some flattening curving to indicate the end of the handle.

A final note on shape is that T-handles yield much better performance than straight screwdriver handles. Pheasant and O'Neill (1975) reported as much as a 50% increase in torque. Optimum handle diameter was found to be 25 mm and optimum angle was 60°, i.e., a slanted T (Saran, 1973). The slant allows the wrist to remain straight and, thus, generate larger forces.

Grip Surface, Texture, and Materials

For centuries wood was the material of choice for tool handles. Wood was readily available and easily worked. It has good resistance to shock and thermal and electrical conductivity, and has good frictional qualities even when wet. Since wooden handles can break and stain with grease and oil, there has been a shift to plastic and even metal. However, metal should be covered with rubber or leather to reduce shock and electrical conductivity and increase friction. Such compressible materials also dampen vibration and allow a better distribution of pressure, reducing the feeling of fatigue and hand tenderness (Fellows and Freivalds, 1991). The grip material, however, should not be too soft, otherwise sharp objects, such as metal chips, will get embedded in the grip and make it difficult to use. Grip surface area should be maximized to ensure a pressure distribution over as large an area as possible. Excessive localized pressure sometimes causes pain that forces workers to interrupt their work. Pressure/pain thresholds of around 500 kPa for females and 700 kPa for males have been found, with the thenar and os pisiforme areas being most sensitive (Fransson-Hall and Kilbom, 1993). During maximal power grips these values are greatly exceeded.

The frictional characteristics of the tool surface vary with the pressure exerted by the hand, the smoothness and porosity of the surface, and the type of contamination (Buchholz et al., 1988; Bobjer et al., 1993). Sweat increases the coefficient of friction, while oil and fat reduce it. Adhesive tape and suede provide good friction when moisture is present (Buchholz et al., 1988). The type of surface pattern as defined by the ratio of ridge area to groove area shows some interesting characteristics. When the hand is clean or sweaty, the maximum frictions were obtained with high ratios (i.e., maximizing the hand–surface contact area), while when the hand is contaminated, maximum frictions were obtained with low ratios (i.e., maximizing the capacity to channel away contaminants) (Bobjer et al., 1993).

Angulation of Handle

As discussed previously, deviations of the wrist from the neutral position under repetitive load can lead to a variety of cumulative trauma disorders as well as decreased performance. Therefore, angulation of tool handles, e.g., power tools, may be necessary so as to maintain a straight wrist. The handle should reflect the axis of grasp, i.e., about 78° from the horizontal, and should be oriented so the eventual tool axis is in line with the index finger (Fraser, 1980). This principle has been applied to various tools such as pliers and soldering irons, as mentioned previously.

An interesting extension of this concept has been promoted as Bennett's handle (Emanual et al., 1980). Bennett developed this concept based on the angle formed by the index finger and the life line under the thumb. This angle of 19°, used for his handles, is claimed to maintain a straight wrist, generate increased strength and control, and reduce stress, shock, and fatigue. Konz (1986) conducted a variety of tests to evaluate the effectiveness of Bennett's handle on a hammer in comparison with a standard hammer. Subjects preferred a 10° bend but performed no better than with a standard hammer. Knowlton and Gilbert (1983) used cinematography to evaluate a curved and conventional claw hammer. Bilateral grip strength was measured before and after a task, nail driving. The curved hammer produced a smaller strength decrement and caused less ulnar deviation than the conventional hammer. Thus, a bent handle may give some benefits.

Grip Span for Two-Handled Tools

Grip strength and the resulting stress on finger flexor tendons vary with the size of the object being grasped. A maximum grip strength is achieved at about 45 to 50 mm on a dynamometer with parallel sides, or about 75 to 80 mm on a dynamometer with handles angled inward (Chaffin and Andersson, 1984). At distances different from the optimum, percent grip strength decreases (Figure 27.9) as defined by ($r^2 = .99$, p<.001):

$$\%\text{Grip} = 100 - .11 \times S - 10.2 \times S^2 \tag{27.2}$$

where S = given grip span minus optimum grip span in cm.

Because of the large variation in individual strength capacities, and to accommodate 95% of the population, maximal grip requirements should be limited to less than 90 N.

Weight

For non-striking applications, the weight of the hand tool will determine how long it can be held or used and how precisely it can be manipulated. For tools held in one hand with the elbow at 90° for extended periods of time, Eastman Kodak (1983) recommended a load of no more than 2.3 kg. For precision operations, tool weights greater than .4 kg are not recommended unless a counter-balanced system is

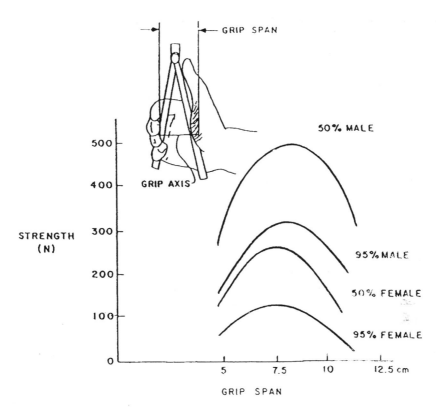

FIGURE 27.9   Grip strength as a function of grip span. (From Chaffin, D.B. and Andersson, G. 1984. *Occupational Biomechanics.* John Wiley & Sons, New York. With permission.)

used. Heavy tools, used to absorb impact or vibration, should be mounted on a truck to reduce effort for the operator. In addition, the tool should be well balanced, with its center of gravity as close as possible to the center of gravity of the hand (unless the purpose of the tool is to transfer force, as in a hammer). Thus, the hand or arm muscles do not need to oppose any torque development by an unbalanced tool.

Gloves

Gloves are often used with hand tools for safety and comfort. Safety gloves are seldom bulky, but gloves worn in subfreezing climates can be very heavy and interfere with grasping ability. Wearing woolen or leather gloves may add 5 mm to the hand thickness and 8 mm to the hand breadth at the thumb, while heavy mittens add 25 mm and 40 mm, respectively (Damon et al., 1966). More important, gloves reduce grip strength 10 to 20% (Cochran et al., 1986) and manual dexterity performance times 12 to 64% (Weidman, 1970). Neoprene gloves slowed performance times by 13% over barehanded performance, over terry cloth by 36%, over leather by 45%, and over PVC by 64%. Thus, there is a trade-off to be considered between increased injury and reduced performance without gloves and reduced performance with gloves. Perhaps the tool can be redesigned even more to reduce the need for gloves.

Vibration

Vibration is a separate and very complex problem with powered hand tools. Vibration can induce *white finger* syndrome, the primary symptom of which is a reduction in blood flow to the fingers and hand due to *vasoconstriction* of the blood vessels. As a result, there is a loss of sensory feedback and decreased performance. In addition, vibration may contribute to the development of carpal tunnel syndrome, especially in jobs with a combination of forceful and repetitive exertions (Silverstein et al., 1987). It is generally recommended that vibrations in the critical range of 2 to 200 Hz (Lundstrom and Johansson, 1986) be avoided. The exposure to vibration can be reduced through a reduction in the driving force and the use of vibration-damping materials on either the tool or the hand (gloves) (Andersson, 1990).

27.3   Attributes of Common Industrial Nonpowered Tools

Shovels

Shovels are used to lift, move, and toss loose dirt or sand. The blade is fastened to the shaft through a socket which, if stamped from a flat sheet, is generally rolled over to form a crimp known as a frog. The shaft may either taper to an end or have a handle. The handle traditionally has been of a T form, but more lately is of a D form. Long shafts are generally 1.2 to 1.7 m long, while the short D handle is about .7 m long. However, for an unconstrained posture and task, long-handled shovels are 18% more efficient (in terms of energy expenditure per sand shoveled) than short-handled shovels (Freivalds, 1986b). The external angle of the shaft with respect to the blade is called the lift and provides the tool with added leverage. Based on a trade-off between smaller low-back compressive forces and lower energy cost, Freivalds (1986b) found the optimum lift angle to be approximately 32°. Shovel weight should be as small as possible, especially if one considers that this is unproductive weight that has to be lifted along with the load moved. Weights should be below 1.5 kg (Freivalds and Kim, 1990). Blade size depends very much on the density of the material being handled; the less dense the material the larger the blade size. For a given material (foundry sand), Freivalds and Kim (1990) found that a blade-size/weight ratio of .068 m^2/kg is optimum, i.e., for a 1.5 kg shovel the blade size is .1 m^2 with a resulting load weighing 4.4 kg.

In terms of maximum amount of sand shoveled per given time period, the optimum shoveling rate is in the range of 18 to 21 scoops/min. The optimum shovel load ranges from 5 to 11 kg, depending on the decision criterion to be used, i.e., for high rates of shoveling (18 to 20 scoops/min), the lower end of the load range (5 to 7 kg) may be more appropriate (which follows the principle of reducing static loading), while for lower rates (6 to 8 scoops/min), the higher end of the load range (8 to 11 kg) may be acceptable (which follows the principles of increasing efficiency with larger loads).

This was determined by F. W. Taylor in the first scientific study on shoveling at the Bethlehem Steel Works in 1898, in which he was able to reduce the manpower requirements for material handling from 600 to 140 men (Taylor, 1913).

Shoveling throw height is a trade-off between increased efficiency for higher heights at a cost of an increased energy expenditure. Since shoveling performance stays reasonably constant up to a height of 1.3 m, an acceptable throw height may be as high as 1 to 1.3 m. Similarly, shoveling performance remains fairly constant up to a distance of 1.2 m (Freivalds, 1986a).

Hammers

Hammers are striking tools designed to transmit a force to an object by direct contact and thereby change its shape or drive it forward. The tool's efficiency in doing this may be defined as the ratio of the energy utilized in striking to the energy available in the stroke. This efficiency is maximized by placing the tool mass center as close as possible to the center of action, i.e., increasing the mass of the tool head relative to the handle. Another aim is to transform as much of the kinetic energy of the hammer into deforming an object's shape as possible. Thus, the mass of the hammer should be small relative to the mass of the forging and anvil. On the other hand, in driving a nail, the intent is to transform the kinetic energy of the hammer into the kinetic energy of the nail. Then the mass of the hammer should be great in relation to the mass of the nail. The overall mechanical efficiency for hammering a 15-cm nail into a wooden block can be as high as 57% (Drillis, 1963). However, there is a limit to the weight that can be placed in the head of the hammer. Increasing the head weight decreases angular velocity and ultimately the total kinetic energy, in addition to increasing physiological energy costs (Widule et al., 1978; Corrigan et al., 1981).

Saws

The action of heavy sawing requires a power grip with repetitive flexion and extension at the elbow, while the action of light sawing involves a precision grip with manipulation of the wrist. For the former, pistol grips are utilized, while for the latter a cylindrical screwdriver-type handle provides the best precision grip. Gläser (1933) found that for forestry work, a two-handed action provided more force and better performance but at a higher energy cost. Most efficient was the kneeling posture in which less torso support was needed and less energy was expended. Typically, Western saws cut as they are pushed through the wood, while Oriental saws cut as they are pulled through the wood. Although sawing times were not significantly different, energy expenditure was significantly lower (26%) for the Oriental saws (Bleed et al., 1982).

Pliers

Pliers and related tools — wire strippers, pincers, and nippers — are tools with a head in the form of jaws; jaws can have a variety of configurations, i.e., joints which may be simple or complex. Sometimes the handles are straight, but more typically they are curved outward to conform roughly to the shape of the grasp. The grasp, depending on use, can be of the precision or power type. In their simple form, pliers are a very common tool and, if used casually for short periods of time, will give reasonable performance with little fatigue. However, the relationship of the handles to the head forces the wrist into ulnar deviation, a posture which cannot be held repeatedly or for prolonged periods of time without fatigue or occurrence of cumulative trauma disorders. A further problem is that ulnar deviation reduces the range of wrist rotation by 50%, possibly reducing productivity. By bending the handles of the pliers instead of the wrist, Tichauer (1976) was able to reduce the stress on the operator's wrists and overall injury rates by a factor of six at Western Electric Co.

Other factors to be considered in the design of pliers were detailed by Lindstrom (1973): a working grip (subjectively chosen on an adjustable grip) width of 90 mm for men and 80 mm for women and a handle length of 110 mm for men and 100 m for women. For repeated or continuous operation, the required working strength should not exceed 33 to 50% of the individual's maximum strength. To minimize the applied pressure to the soft tissue of the palm (keep less than 200 to 400 kPa), the handles should be enlarged and flattened. Thus, indentation of the handles for the fingers is undesirable. Encasing the basic metal handles in a rubber or plastic sheath provides insulation and improves the tactile feel. Also, spring-loaded handles eliminate the need to manually open the handles.

Screwdrivers

The handles of screwdrivers (and similar tools: files, chisels, etc.) can either be used with a precision grip for stabilization or a power grip for torque. Crucial factors are also the size, shape, and texture of the handle. Applied torque increases with an increase in the diameter of the handle (Pheasant and O'Neill, 1975). Differences in the precise shape of handles appears to be less significant, as long as the hand does not slip around the handle (Cochran and Riley, 1982). Thus, knurled cylinders allow for significantly greater torque production than smooth cylinders. Further details can be found in the previous section on handle shape.

Knives

Although a very old tool, the knife has recently appeared in the literature as a possible cause for the increase in cumulative trauma disorders in food processing due to the long hours of static loading on the forearm flexors and the slippery handles requiring high grip forces (Armstrong et al., 1982; Karlqvist, 1984). For poultry processing, Armstrong et al. (1982) suggested a pistol-type grip (Figure 27.6) to allow the operator to hold the blade and the forearm horizontal so as to eliminate ulnar deviation and wrist flexion. A circular or elliptical handle with a circumference of 99 mm, as well as a strap, were recommended to allow the hand to relax between exertions without losing the grip on the knife. Similarly, for the fish canning industry, Karlqvist (1984) fitted some knives with a pistol-grip handle and others with larger diameter handles for better balance and movement. Note that one knife design will not be best for all jobs.

27.4   Attributes of Common Industrial Power Tools

Power Drills

In a power drill, or other power tools, the major function of the operator is to hold, stabilize, and monitor the tool against a workpiece, while the tool performs the main effort of the job. Although the operator may at times need to shift or orient the tool, the main function for the operator is to effectively grasp and hold the tool. A drill is comprised of a head, body, and handle, with all three, ideally, being in line. The line of action is from the line of the extended index finger so that in the ideal drill, the head is off-center with respect to the central axis of the body. Handle configuration is important, with the choices being pistol-grip, in-line, or right-angle. As a rule of thumb, in-line and right-angle are best for tightening downward on a horizontal surface, while pistol-grips are best for tightening on a vertical surface (Figure 27.5) with the aim being to obtain a standing posture with a straight back, upper arms hanging down, and a straight wrist (Figure 27.3). For the pistol grip, this results in the handle being at an angle of approximately 78° with the horizontal (Fraser, 1980).

Another important factor is the center of gravity. If it is too far forward in the body of the tool, a turning moment is created, which must be overcome by the muscle of the hand and forearm, creating muscular effort additional to that required for holding, positioning, and pushing the drill into the workpiece. The primary handle should be placed directly under the center of gravity, such that the body juts out behind the handle as well as in front. For heavy drills, a secondary supportive handle may be needed, either to the side or preferably below the tool, such that the supporting arm can be tucked in against the body rather than being abducted. Also tool balancers should be utilized for heavy tools (Fraser, 1980).

Nutrunners

Nutrunners, especially common in the automobile industry, are used to tighten nuts, screws, and other fasteners. They come in a variety of handle configurations, torque outputs, shut-off mechanisms, speeds, weights, and spindle diameters. Torque levels range from .1 to 5000 Nm and, for pneumatic tools, are generally lumped into approximately 22 power levels (M1.6 to M45), depending on motor size and gearing required to drive the tool. The torque is transferred from the motor to the spindle through a variety of mechanisms such that the power (often air) can be quickly shut off once the nut or other fastener is tight. The simplest and cheapest mechanism is a direct-drive, which is under the operator's control, but, because of the relatively long time it takes to release the trigger once the nut is tightened, transfers a very large *reaction torque* to the operator's arm. Mechanical friction clutches will allow the spindle to slip, reducing some of this reaction torque. A better mechanism for reducing the reaction torque is the air-flow shut-off which automatically senses when to cut off the air supply as the nut is tightened. A still faster mechanism is an automatic mechanical clutch shut-off. The most recent mechanisms include the hydraulic pulse system where the rotational energy from the motor is transferred over a pulse unit containing an oil cushion (filtering off the high-frequency pulses as well as noise) and a similar electrical pulse system, both of which, to a large extent, reduce the reaction torque (Freivalds and Eklund, 1993).

Variation of torque delivered to the nut depends on a variety of conditions including properties of the tool, the operator of the tool, properties of the joint, i.e., the combination of the fastener and material being fastened (ranging from soft, with the materials having elastic properties such as body panels, to hard, when two stiff surfaces, such as pulleys on a crankshaft, are brought together), stability of the air supply, etc. The torque experienced by the user (the reaction torque) depends on the above factors plus the torque shut-off system and may contribute to the development of cumulative trauma disorders. In general, using electrical tools at lower than normal rpm levels or underpowering pneumatic tools resulted in larger reaction torques and more stressful ratings. Pulse-type tools produced the lowest reaction torques and were rated as less stressful. It was hypothesized that the short pulses "chop up" or allow the inertia of the tool to resist the reaction torque. Another possibility is to provide reaction torque bars (Freivalds and Eklund, 1993).

27.5 Discussion

Currently, the most important issue regarding tool design is the reduction in the potential for the development of cumulative trauma disorders. Until about 10 years ago, tool usage was little changed from the days of the Industrial Revolution. The operator used tools manually in the manufacturing of goods. The operations required considerable forces, which was somewhat leveraged by the appropriate tool. Because of the manual nature of the tasks and the forces involved, the operations were fairly slow. With the advent of automation, the excessive force levels were eliminated, and many task elements relegated to the human could be eliminated. The operator performs a smaller part of the original task, which now can be speeded up because the machine handles most of the work. Unfortunately, the elements still left to the human operator become more limited in scope and thus more repetitious in nature. This incomplete and unergonomic automation has led to an upsurge in CTD cases, especially if the repetition is combined with excessive wrist deviations and forceful exertions. The threshold for injury for any single risk factor is not known, let alone for a combination of factors. There is also, probably, a trade-off between each of those factors that need to be quantified for exposure and threshold levels.

Tied in with frequency is the trade-off with productivity. Any reduction in frequency of tool usage may have a direct result in decreasing productivity. One alternative to maintaining constant productivity is to rotate operators for a critically repetitive task. But then again it is necessary to know threshold levels of frequency and how much rest, at which intervals, must be provided to recover from the trauma induced from repetitive tool usage. Also, it is important to know whether performing a greater variety of movements from those in the injurious task will allow the body to recover or will only delay recovery. These are all issues that haven't been fully addressed.

Another issue not fully resolved is the trade-off between manual and power tool usage. Most researchers, based on the limited human force capacity and greater fatiguability as compared to machines, have advocated the use of power tools. Unfortunately, power tools, whether powered electrically or pneumatically, produce some vibration. Vibration damping typically requires either an increase in the inertial mass (at the cost of increasing the weight of the tool and increasing the fatigue of the user) or vibration-absorbing systems which introduce a bit of "slop" in the hand–handle interface absorbing the vibrations (again at the cost of reducing control of the tool). Power tools, also, have a tendency to produce reaction torques, which can be reduced by using pulse-type tools (at the cost of increasing vibration) or by using reaction bars (again at the cost of limiting the control or maneuverability of the tool). These are issues that need to be clarified further.

Recent research indicates that power grip capabilities can be increased by a better understanding of the pressure distribution of the hand while using a tool or by improving the frictional characteristics of the tool handle surface. Perhaps the development or application of new polymers to tool handles can improve the efficiency of tool usage. Also new ways of measuring the hand–handle interface, such as the "data glove" of Yun et al. (1996), can provide more accurate information on this topic.

Most current work has addressed the power grip for tools. However, most power requirements are being fulfilled by machines, leaving the human operator to perform more precise tasks that currently cannot be easily replicated by machines. Unfortunately, there is very little information on precision or pinch grips and the precision aspects of tools. Questions on grip design and force exertion capabilities for precision grips, as well as occupational injury risk during work with high demands on precision, need to be studied further.

Epidemiological considerations are also important in substantiating proper ergonomic designs. Unfortunately, at present, there are few good studies that support good ergonomic tool design or clearly indicate the deficiencies in such designs. More injury data for both hand and powered tools are needed.

A final but very important consideration is the adaptation of tools for a more diverse population. With the aging of the worker population and the passage of the Americans with Disabilities Act, it is imperative that tools also be usable by individuals with a wide range of capabilities. This is both a challenge and an opportunity for ergonomists and tool designers to put their skills to effective use.

Defining Terms

Carpal tunnel syndrome: Compression of the median nerve in the carpal tunnel of the wrist resulting in pain, tingling, and/or paralysis in the fingers.

Cumulative trauma disorders: The collection of problems occurring in the upper extremities due to repetitive motion.

Dorsiflexion: *See Extension.*

EMG (electromyography): Electrical activity of the muscle.

Extension: Movement that decreases the angle between two adjacent bones.

Flexion: Movement that decreases the angle between two adjacent bones.

Ischemia: Occlusion of blood flow in an artery.

Palmar flexion: *See Flexion.*

Power grip: Hand grip such that the thumb opposes partly flexed fingers, barely overlapping. Provides maximum power.

Precision grip: Hand grip such that the thumb opposes only the first or second fingers, resulting in a much lower gripping force, but greater precision.

Radial deviation: Bending the wrist in the direction of the thumb.

Reaction torque: The transfer of torque to the operator's arms as a nut tightens and before the tool cuts power.

Trigger finger: Swollen tendon sheath resulting in the tendon being locked, such that attempts to move the finger cause a snapping or jerking movement.

Ulnar deviation: Bending the wrist in the direction of the little finger.

Vasoconstriction: Ischemia of the peripheral blood flow.

White finger syndrome: Occupation vibration syndrome characterized by finger blanching due to ischemia of the digital arteries.

References

Andersson, E.R. 1990. Design and testing of a vibration attenuating handle, *Int. J. Indus. Erg.* 6:119-125.

Armstrong, T.J., Foulke, J.A., Joseph, B.S., and Goldstein, S.A. 1982. Investigation of cumulative trauma disorders in a poultry processing plant. *Am. Indus. Hygiene Assoc. J.* 43:103-116.

Ayoub, M. and LoPresti, P. 1971. The determination of an optimum size cylindrical handle by use of electromyography. *Ergonomics* 14:509-518.

Bleed, A.S., Bleed, P., Cochran, D.J., and Riley, M.W. 1982. A performance comparison of Japanese and American hand saws, *Proc. Human Factors Soc.* Santa Monica, CA, 26:403-407.

Bobjer, O., Johansson, S.E., and Piguet, S. 1993. Friction between hand and handle. Effects of oil and lard on textured and non-textured surfaces; perception of discomfort. *Appl. Erg.* 24:190-202.

Buchholz, B., Frederick, L.J., and Armstrong, T.J. 1988. An investigation of human palmar skin friction and the effects of materials, pinch force and moisture. *Ergonomics* 31:317-325.

Chaffin, D.B. and Andersson, G. 1984. *Occupational Biomechanics.* John Wiley & Sons, New York.

Cochran, D.J., Albin, T.J., Riley, M.W., and Bishu, R.R. 1986. Analysis of grasp force degradation with commercially available gloves. *Proc. Human Factors Soc.* Santa Monica, CA. 30:852-855.

Cochran, D.J. and Riley, M.W. 1986. An evaluation of knife handle guarding. *Human Factors* 28:295-301.

Corrigan, D.L., Foley, V., and Widule, C.J. 1981. Axe use efficiency — a work theory explanation of an historical trend. *Ergonomics* 24:103-109.

Damon, A., Stoudt, H.W., and McFarland, R.A. 1966. *The Human Body in Equipment Design.* Harvard University Press, Cambridge, MA.

Drillis, R.J. 1963. Folk norms and biomechanics. *Human Factors.* 5:427-441.

Drury, C.G. 1980. Handles for manual materials handling. *Appl. Erg.,* 11:35-42.

Ducharme, R.E. 1975. Problem tools for women. *Indus. Eng.* Sep:46-50.

Eastman Kodak Co. 1983. *Ergonomic Design for People at Work.* Lifetime Learning Pub., Belmont, CA.

Emanual, J.T., Mills, S.J., and Bennett, J.F. 1980. In search of a better handle. *Proc. Symp.* Human Factors Indus. Design Consumer Products. Tufts University, Medford, MA, pp. 34-40.

Fellows, G.L. and Freivalds, A. 1991. Ergonomics evaluation of a foam rubber grip for tool handles. *Appl. Erg.* 22:225-230.

Fraser, T.M. 1980. *Ergonomic Principles in the Design of Hand Tools.* International Labour Office. Geneva, Switzerland.

Fransson-Hall, C. and Kilbom, Å. 1993. Sensitivity of the hand to surface pressure. *Appl. Erg.* 24:181-189.

Freivalds, A. 1986a. The ergonomics of shovelling and shovel design — a review of the literature. *Ergonomics* 29:3-18.

Freivalds, A. 1986b. The ergonomics of shovelling and shovel design — an experimental study, *Ergonomics* 29:19-30.

Freivalds, A. and Eklund, J. 1993. Reaction torques and operator stress while using powered nutrunners, *Appl. Erg.* 24:158-164.

Freivalds, A. and Kim, Y.J. 1990. Blade size and weight effects in shovel design, *Appl. Erg.* 21:39-42.

Garrett, J. 1971. The adult human hand: some anthropometric and biomechanical considerations. *Human Factors.* 13:117-131.

Gläser, H. 1933. *Beiträge zur Form der Waldsäge und zur Technik des Sägens,* Ph.D. Dissertation, Everswalde, Germany.

Karlqvist, L. 1984. Cutting operations at canning bench — a case study of handtool design, *Proc. 1984 Inter. Conf. Occup. Erg.* Human Factors Association of Canada, Rexdale, Ont. pp. 452-456.

Knowlton, R.G. and Gilbert, J.C. 1983. Ulnar deviation and short term strength reductions as affected by a curve handled ripping hammer and a conventional claw hammer, *Ergonomics* 26:173-179.

Konz, S. 1974. Design of handtools. *Proceedings Human Factors Soc.* Santa Monica, CA, 18:292-300.

Konz, S. 1986. Bent hammer handles. *Human Factors* 27:317-323.

Konz, S. 1995. *Work Design.* 4th ed. Publishing Horizons. Worthington, OH.

Konz, S. and Warraich, M. 1985. Performance differences between the preferred and non-preferred hand when using various tools. *Ergonomics International '85* (I.D. Brown, R. Goldsmith, K. Coombes and M.A. Sinclair, eds.), Taylor & Francis, London, pp. 451-453.

Laveson, J.K. and Meyer, R.P. 1976. Left out "lefties" in design. *Proc. Human Factors Soc.* 20:122-125.

Leakey, L.S.B. 1960. Finding the world's earliest man. *Natl. Geog.* 118:420-435.

Lehmann, G. 1953. *Praktische Arbeitsphysiologie.* Thieme Verlag, Stuttgart, Germany.

Lindstrom, F.E. 1973. *Modern Pliers.* Bahco Verktyg, Enköping, Sweden.

Lundstrom, R. and Johansson, R.S. 1986. Acute impairment of the sensitivity of skin mechanoreceptive units caused by vibration exposure of the hand. *Ergonomics* 29:687-698.

Luopajarvi, T., Kuorinka, I., Virolainen, M., and Holmberg, M. 1979. Prevalence of tenosynovitis and other injuries of the upper extremities in repetitive work. *Scand. J. Work Env. Health.* 5, Sup. 3:48-55.

Miller, G. and Freivalds, A. 1987. Gender and handedness in grip strength. *Proc. Human Factors Soc.* Santa Monica, CA, 31:906-909.

Mital, A. 1986. Effects of body posture and common hand tools on peak torque exertion capabilities, *Appl. Erg.* 17:87-96.

Napier, J. 1962. The evolution of the hand. *Sci. Am.* 207:56-62.

National Safety Council. 1993. *Accident Facts.* Chicago, IL.

NIOSH. 1989a. *Occupational Safety and Health, Year 2000 Objectives.* National Institute for Occupational Safety and Health, Center for Disease Control, Atlanta, GA.

NIOSH. 1989b. *Health Hazard Evaluation — Eagle Convex Glass, Co.,* HETA-89-137-2005, National Institute for Occupational Safety and Health, Cincinnati, OH.

Pheasant. S.T. and O'Neill, D. 1975. Performance in gripping and turning — a study in hand\handle effectiveness. *Appl. Erg.* 6:205-208.

Putz-Anderson, V. 1988. *Cumulative Trauma Disorders.* Taylor & Francis, London.

Rubarth, B. 1928. Untersuchung zur Festgestaltung von Handheften für Schraubenzieher und ähnliche Werkzeuge. *Industrielle Psychotechnik* 5:129-142.

Saran, C. 1973. Biomechanical evaluation of T-handles for a pronation supination task. *J. Occup. Med.* 15:712-716.

Silverstein, B.A., Fine, L.J., and Armstrong, T.J. 1987. Occupational factors and carpal tunnel syndrome. *Am. J. Indus. Med.* 11:343-358.

Taylor, F.W. 1913. *The Principles of Scientific Management.* Harper & Bros. New York.

Terrell, R. and Purswell, J. 1976. The influence of forearm and wrist orientation on static grip strength as a design criterion for hand tools. *Proc. Human Factors Soc.* Santa Monica, CA, 20:28-32.

Tichauer, E.R. 1976. Biomechanics sustains occupational safety and health. *Indust. Eng.* pp. 46-56 (Feb).

Weidman, B. 1970. *Effect of Safety Gloves on Simulated Work Tasks.* AD 738981, National Technical Information Service, Springfield, VA.

Widule, C.J., Foley, V., and Demo, F. 1978. Dynamics of the axe swing, *Ergonomics* 21:925-930.

Yun, M.H., Cannon, D., Freivalds, A., and Thomas, G. 1997. An instrumented glove for grasp specification in virtual reality based point and direct telerobotics, *IEEE Trans. Sys. Man Cyber.* 27: 835-847.

For Further Information

The article by R. J. Drillis, *Folk norms and biomechanics* is a good introduction for the historical evolution of tools.

Ergonomic Principles in the Design of Hand Tools by T.M. Fraser is a very good overall reference for tool design.

V. Putz Anderson's *Cumulative Trauma Disorders* is a good introduction into cumulative trauma disorders that may result from poor job practices and poor tool design.

Ergonomic Design for People at Work by the Human Factors Section at Eastman Kodak Co. has many very good practical applications of tool design as related to industrial practice.

(See References for complete citations.)

The Human Factors and Ergonomics Society (P.O. Box 1369, Santa Monica, CA 90406, USA) and The Ergonomics Society (Devonshire House, Devonshire Sq., Loughborough, Leic. LE11 3DW, UK) are good sources of information, various publications, newsletters, annual conferences, etc. that relate to ergonomics in general.

28

Computer-Aided Design and Human Models

J. Mark Porter
Loughborough University

Keith Case
Loughborough University

Martin T. Freer
Loughborough University

28.1   Introduction

With the CAD/CAM systems available today it is quite possible for some products to progress from concept design through production without requiring full-size physical models or mock-ups to perform any necessary evaluations. This helps to reduce the time scale of product design considerably. However, as most CAD/CAM systems provide little or no information concerning the needs of the end user, there is considerable danger that design decisions may only consider the engineering, styling, legislative, and financial constraints for the product, with the ergonomics issues comparatively unassessed until the design is completed.

It is essential that the ergonomics input take place throughout the design process, but nowhere is it more important than at the concept and early development stages. Basic ergonomics criteria, such as the adoption of healthy and efficient postures for the range of future users, need to be satisfied very early because there is usually only limited scope for modification later on without considerable financial and time penalties. Traditionally, the various ergonomics criteria were assessed during a product's development by conducting user trials with hand-built prototypes. These prototypes were not constructed just in order to perform ergonomics trials, they were often made first and foremost to visualize the product in 3D and to determine how to efficiently deal with the legal and manufacturing issues. As the visualization and engineering functions are now increasingly being performed with digital prototypes or mock-ups displayed on computer graphics terminals, it is clear that there is an urgent need to provide computer-based ergonomics assessment functions as well. The computer modeling of people (known as *man-modeling* since its inception in the 1960s but now increasingly termed *human modeling* in the 1990s) provides the ability to construct 2D or 3D models from anthropometric data which can be articulated between the body segments to simulate a wide variety of postures. These human models can then be used, in conjunction with the CAD model of the product being designed, to conduct computer-based user trials to assess criteria such as fit, reach, vision, and the resulting constraints upon posture. Such predictions enable the ergonomist to be more proactive in the design process and to be able to work closely with the other design team members to achieve ergonomic solutions to the design within the various financial, legal, engineering, and aesthetic constraints.

28.2   Human Models, Past and Present

Kinematic modeling enables the spatial evaluation of workplaces where either the human or parts of the physical environment are placed in different positions over time. This type of modeling is the focus of this chapter, while kinetic (or dynamic) modeling, usually associated with assessing the body's response to large external forces such as those experienced in car crash simulations, is not covered.

Kinematic human modeling and global systems, past and present, include ADAPS (Delft University, The Netherlands, see Post and Smeets, 1981), ANYBODY and ANTHROPOS (IST GmbH, Germany), APOLIN (Grobelny et al., 1992), BOEMAN (Boeing Co., USA), BUFORD (Rockwell International, USA), CAR (Naval Air Development Centre, USA), COMBIMAN and CREW CHIEF (Armstrong Aerospace Medical Research Laboratory, USA, see McDaniel, 1990), CYBERMAN (Chrysler Co., USA), Envision/ERGO (Deneb Robotics Inc., USA), ERGODATA (Laboratoire d'Anthropologie Appliquée. France), ERGOMAN (see Coblenz et al., 1991), ergoSHAPE (Institute of Occupational Health, Finland, see Launis and Lehtela, 1992, 2D only), ergoSPACE (Institute of Occupational Health, Finland, see Launis and Lehtela, 1990), FRANKY (G.I.T., Germany, see Elias and Lux, 1986), JACK (University of Pennsylvania, USA, see Badler et al., 1993), MDHMS (McDonnell Douglas, USA), MANNEQUIN (Biomechanics Corporation of America), MINTAC (Kuopio Regional Institute of Occupational Health and the University of Oulu, Finland, see Kuusisto and Mattila, 1990), RAMSIS (BMW and other car manufacturers, Germany), SAFEWORK (Genicom Consultants, Canada, see Fortin et al., 1990), SAMMIE (SAMMIE CAD Ltd. and Loughborough University, UK, see Porter et al., 1995), TADAPS (University of Twente, The Netherlands, see Westerink et al., 1990) and WERNER (Institute of Occupational Health, University of Dortmund, Germany, see Kloke, 1990).

Comparisons between some of these systems can be found in Dooley (1982), Rothwell (1985), Porter et al. (1993, 1995), and Das and Sengupta (1995), and it is not the intent of this chapter to present a detailed description of each human modeling system. It is important to appreciate that the quality of a product's ergonomics has more to do with the design team's judgment and ability to incorporate sound ergonomics principles in the design than to the use of any specific human modeling system (Das and Sengupta, 1995). Such systems do not automate the design process by creating ergonomics solutions to a set of specified inputs, rather they should be regarded as tools to be used by the design team. An increasing number of system developers and users have created information pages on the world wide web. Figures 28.1 through 28.5 show plots of the human models from JACK (as used by MIDAS, a U.S. Army–NASA product), MDHMS, and SAMMIE. These plots were taken from their respective web sites listed at the end of this chapter. Figures 28.6 and 28.7 show plots from SAFEWORK and Envision/ERGO.

The differences between the systems can be examined in terms of a number of features.

- The complexity of the human model (e.g., 2D or 3D, number of body segments, surface details for visualization or specified only by anthropometry)
- Whether the joint angles are constrained to possible angles or whether impossible postures could be inadvertently set
- The anthropometric databases available
- The extent of control over the size of individual body segments (fixed, linear scaling possible or direct control with data
- Whether there are extensive CAD facilities integrated fully with the human model or whether the design model has to be ported to and from another CAD system
- The ability of the workplace modeler to provide functional modeling (e.g., several components can be modeled and collectively called "driver's door", which can be rotated as one unit around the hinge point), hidden line views (lines behind solids removed), color surface shading, reflections, shadows and textures

FIGURE 28.1 MDHMS model of an aircraft maintenance operation. (Source: MDHMS web page.)

FIGURE 28.2 JACK models showing the use of textures and shadowing for imparting added realism. (Source: University of Pennsylvania web page.)

- How the system can assess the human model's reach, fit, or vision (the flexibility of these assessments varies considerably from simple reach to a point to automated volumetric reach, from visually inspected clearance to automated intersecting solid detection routines, and from the display of eye point location to the display of perspective views and mirror views from either eye)
- Whether the system provides strength data or calculates torque loads on selected joints.

The various systems cost from as little as a few hundred U.S. dollars up to $60,000 for a software license. Some systems can run on a PC, but many require a Sun or Silicon Graphics workstation or equivalent. Usability is a key requirement of such systems, and a fast response time often requires a high specification computer.

FIGURE 28.3   JACK performing a simple reach task in a cockpit. (Source: MIDAS (US Army-NASA) web page.)

FIGURE 28.4   JACK wearing protective clothing and equipment for aircrew. (Source: MIDAS (US Army-NASA) web page.)

Basic Functionality of the Systems

These systems are intended to be used as a predictive tool for the assessment of the capabilities of people when interacting with the designed physical environment. The basic functionality that is required is listed below in bold, in each case followed by a brief discussion of the relevant issues:

- **3D modeling of people** of the selected sex, age, nationality, and occupational groups. This is achieved using published anthropometric data, if indeed it exists for the population being examined.

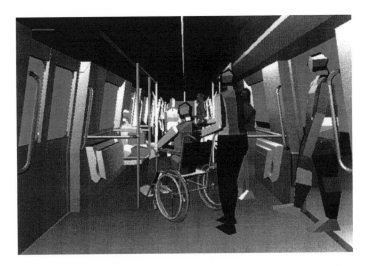

FIGURE 28.5 SAMMIE models used to evaluate accommodation in a railway carriage. (Source: Loughborough University web page.)

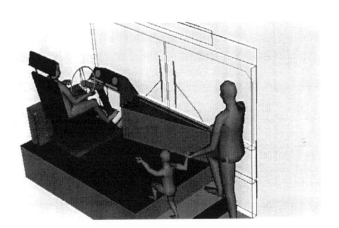

FIGURE 28.6 Design of a bus driver's cabin using SAFEWORK. (Courtesy of Safework Incorporated.)

FIGURE 28.7 Simulation of fuel tank assembly using Envision/ERGO. (Courtesy of Deneb Robotics Incorporated.)

FIGURE 28.8   SAMMIE model of an individual person recorded by the LASS bodyscanner system.

The current databases have several shortcomings, basically because they were established with little consideration for the needs of 3D human modeling systems. For example, surveys record external body dimensions, whereas computer models need joint-to-joint dimensions in addition. The limited number of anthropometric dimensions recorded in surveys leave many gaps when having to fully define a 3D computer model. Should the human model remain as true as possible to the real data or should artistic license be granted to model more "realistic" models? The danger with the latter approach is that the designer (be it a stylist, industrial designer, engineer, or ergonomist) may come to believe the "added" data and, for example, feel confident that it is possible to design seat profiles based upon such highly detailed, but fictitious, models.

The relatively recent technique of body scanning, whereby thousands of data points can be recorded from the surface of the body, makes it possible to model individual people with considerable accuracy. Figure 28.8 shows a SAMMIE model constructed using scanned data from the LASS system (e.g., Jones et al., 1989).

- **knowledge base of comfort angles** for the major joints of the body

Human models come with various numbers of joints. Those with relatively few (e.g., fewer than 20) do not have detailed models of the hands or spine. With such details the number of joints can be well over 100. This large number of degrees of freedom in the human model's posture poses problems for the user who has to decide how to position the model realistically. The problem is made easier in some systems with the provision of automated reach tests, inverse kinematics, and grasping behaviors such that the model's hand can reach, grasp, and operate specified handles. This is done automatically, ensuring that the various joint angles do not exceed maximum or comfortable ranges as specified in the published literature. Such data on comfort angles are widely available for application areas such as computer workstations and cars. However, closer examination often reveals disagreement in the literature or the recognition that the recommended postural angles are based only on theoretical analysis. For example, when assessing a computer workstation design, should the human models be positioned sitting upright with a 90 degree trunk–thigh angle as generally recommended by many sources, or should the seat have a reclined backrest (Grandjean et al., 1984) or a forward tilting seat cushion (Mandal, 1984)?

The interrelationships between joints, such as the knee and hip, are typically not considered when using comfort angle recommendations. For example, the range of comfortable backrest angles is affected by any constraints to the knee angle, such as is experienced in a low sports car seat.

- ability to **model the proposed workstation in 3D, together with the simulation of ranges of adjustment** to be incorporated into the design.

FIGURE 28.9 SAMMIE evaluation of the Fiat Punto before full-size prototypes were available for road trials.

This "working" model of the product being developed is an essential part of a human modeling ergonomics design system because the human model needs to interact with the design in order to assess the physical characteristics of the interface. The requirements for an ergonomics model of a prototype design are, however, substantially different from the needs of other forms of CAD systems because the extremely detailed geometric information from an engineering CAD package is rarely required for ergonomics evaluations. Furthermore, human modeling systems should be used at the concept stages in design in order to help define the initial design specification, rather than just evaluating it at a later stage after the engineering criteria have been satisfied. Engineering CAD models rarely have the functionality of the various components under investigation embedded in their data structure (e.g., seat adjustment ranges, mirror rotation constraints), so these must be added to the ergonomics model. In many cases, it will be easier to create specific models for an ergonomics evaluation rather than simply transfer in detailed engineering models. Some human modeling systems have their own integrated CAD facilities which permit modifications to both the human models and the product being designed to be easily carried out. Other human modeling systems have only very crude CAD facilities, and these work in association with established commercial CAD systems, requiring the porting from system to system of either the human model or the product model. Software standards are continuing to be developed to improve the porting of geometric and functional information across different makes of computer hardware and software (Case et al., 1991).

- ability to **assess the kinematic interaction between the models of people and the workstation,** specifically in terms of the issues of user fit (e.g., headroom and legroom in a car), reach (e.g., to the steering wheel, gear selector, and pedals), and vision (e.g., of the road environment, both directly and in the mirrors, and the instrument binnacle).

The assessments focus on whether or not the people modeled can work efficiently at the workstation and can adopt a "comfortable" posture (i.e., within the ranges of joint angles considered acceptable). Figure 28.9 shows a SAMMIE model of the prototype Fiat Punto car in which a large male driver is simulating reversing the car, simultaneously assessing reach to the clutch pedal with one foot, reach to the steering wheel with one hand, and reach to the gear selector with the other, twisting in the seat and assessing vision around the head restraint, past the rear seat occupants, and through the rear window to the window environment. The same analysis can be conducted with a small female driver with the seat, steering wheel, and head restraint adjusted to suit her needs, within the ranges specified by the prototype design.

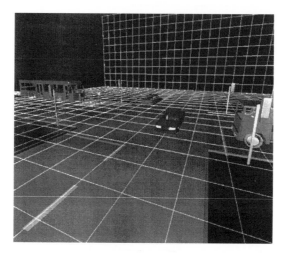

FIGURE 28.10   General view showing layout of SAMMIE models of vehicles, pedestrians, and road infrastructure.

FIGURE 28.11   Small male driver's view (5th percentile sitting eye height) from the coach shown on the left of Figure 28.10. Note that only two pedestrians are seen crossing in front of the cab.

Figures 28.10 through 28.12 show how simply and clearly the interactions between people, machines, and the physical environment can be presented. Figure 28.10 shows a general perspective view of a SAMMIE model of a road junction with various vehicles and pedestrians in position for the subsequent evaluation of driver vision. Figures 28.11 and 28.12 show a small and large male driver's view from one of the vehicles (which required the prior positioning of these drivers within the cab using the adjustment ranges as appropriate). This evaluation revealed that small drivers are more at risk of not being able to see children crossing directly in front of the cab.

  • ability to make **iterative modifications to the design** to achieve optimum compromises.

Figure 28.13 shows one possible solution to the problem of poor driver vision directly in front of the cab. This solution involved the specification of the size, orientation, and radius of curvature of a supplementary mirror so that the range of drivers could see a wide-angle view across the front of the cab. Iterative modifications to the workstation model can be easily made to improve the human models' posture. Some systems provide information on static strength or calculate torque loads on certain joints,

FIGURE 28.12 Large male coach driver's view (99th percentile sitting eye height) of the same scene as shown in Figure 28.12. Note that the large driver can see the child crossing in front of the cab, but the small driver cannot.

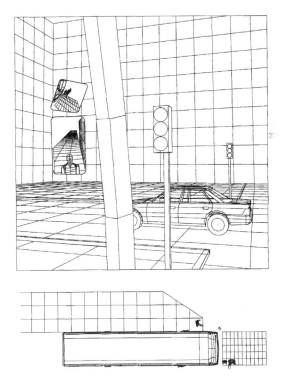

FIGURE 28.13 Coach driver's view in the exterior mirrors. The lower mirror shows the reflected field of view alongside the coach, including a legislative field of view at ground level and a pedestrian. The upper mirror reflects a wide-angle view across the front of the cab showing a child with an adult.

providing information to help identify more efficient designs in this respect. Human modeling systems have most to offer at the concept stage of design when they can be used to explore possible options for a design. Design is all about working within constraints, and sometimes challenging these constraints, to achieve the best compromises.

28.3 Benefits of Using Human Modeling CAD Systems

SAMMIE CAD has operated an ergonomics design consultancy service since 1978 and has now completed more than 150 commercial projects for more than 40 national and international clients. Some of these projects are discussed to illustrate the benefits derived from the use of a 3D human modeling CAD system.

The Formal Specification of the Future Users

It is crucial to the success of a design project to determine exactly who the intended users of a design will be. While seemingly an obvious starting point, it is often not at all clear in the client's mind. Any SAMMIE evaluation requires the investigation of how well human variability, in terms of size and shape, is accommodated by that design. This forces the client to make important decisions about acceptable accommodation range (e.g., 5th to 95th percentile or wider for a particular dimension) and the user population in terms of nationality, sex, and age groups at the earliest stage of design. For example, in an evaluation of a helicopter redesign we were able to demonstrate to the client that the existing aircraft chosen as a starting point, initially without particular regard to the users, was not capable of accommodating the population extremes (97.5th percentile Dutch male pilots and 25th percentile female pilots of other European nationalities) without structural changes so great as to warrant an almost completely new airframe. As a consequence, the project was aborted at an early stage, well before any full-size mock-ups were constructed or other major development costs incurred.

The Formal Specification of the Tasks

The next step is to help the client to establish a clear definition of all the tasks the users will be required to perform so they can be simulated in the evaluation. Often, this process identifies conflicts between various task functions. For example, SAMMIE was used in the design of the Brussels Tram 2000 (Figures 28.14 through 28.17). It was established that the driver had two equally important but conflicting tasks, namely driving the vehicle and selling tickets to passengers. A cab designed to allow ease of operation, optimum visibility and comfortable postures while driving was found to be severely compromised by the requirement to have the driver swivel around and sell tickets while remaining seated (given insufficient space for the driver to stand during ticketing operations). Since SAMMIE is a visual medium, it was possible to clearly demonstrate the problem to the rest of the design team and together look for solutions by quickly developing and investigating a variety of alternative seat swivel mechanisms and rotation points in the SAMMIE model. By group effort, a mechanism was developed that allowed the seat to move and swivel so that both tasks could be easily accomplished and that was feasible, cost effective, and did not require major changes to the cab or console structure.

The Formal Consideration of Other Factors

Because human models are used as predictive tools, it is important to have many other factors specified while conducting any evaluations. Traditionally, these other factors were often identified at the working prototype testing stage. This creates the need for an early dialogue with all stakeholders in the design process. These other factors are wide ranging, covering all aspects of the people involved, their job design, as well as the organizational and psychosocial issues. Another important consideration often overlooked by the client is the physical environment and its possible effects upon user task performance. A recent project examined control design for a new European fighter aircraft in which the control would only be used when the aircraft was "out of control". This posed several issues which the engineers had not considered because the pilot had always expected to be "in control" when considering the design of other controls. The motion conditions under which this particular control might be used are so severe that the "normal" usability criteria for acceptable reach and vision identified were totally inappropriate.

STIB

FIGURE 28.14 Stylist's concept sketches for the Brussels Tram 2000. (Courtesy of Design Triangle, UK).

STIB

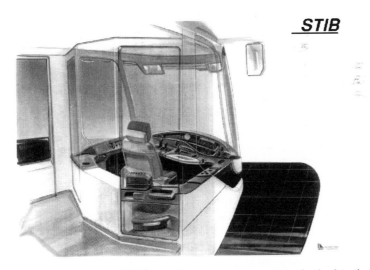

FIGURE 28.15 Stylist's rendering of a SAMMIE model showing a prototype design for the driver's cab in the Tram 2000. Note the ticketing desk behind the seat, which posed postural problems for the driver. The identified solution was to provide a seat which swivelled with offset centers of rotation so that it swung closer to the desk.

Proactive Ergonomics

Having specified who the future users will be, the tasks they will perform, and under what circumstances, it is then possible to construct appropriate human models and position them in "working" postures with only the simplest of workstation models. Areas of common reach and vision for various sized human models can be identified for the placement of the primary controls and displays before these items have

FIGURE 28.16   Advertising poster showing the new Tram 2000 in operation.

FIGURE 28.17   SAMMIE model used in the development of the driver's cab and the passenger areas.

been actually designed in detail. The minimum volume of space required by the human models to adopt their various task postures is easily observed at the earliest stages of design, and this provides the possibility of ensuring that sufficient clearances are provided as the design progresses. Such information and feedback can be provided throughout the development process, allowing the ergonomics issues to be considered proactively. This simultaneous consideration of people issues and engineering issues promotes the identification of optimum compromises, which are essential for a successful design. Such an approach was used in the design of the Lightweight Sports Car, a prototype car exhibited at the 1996 U.K. Motor Show. SAMMIE models of the defined future drivers and passengers (ranging from 99th percentile U.S.

FIGURE 28.18 SAMMIE model ported into Alias for developing the styling and engineering around the defined future occupants. (Photograph courtesy of Dr. C.S. Porter, Coventry University, UK.)

dimensions for both occupants down to 5th percentile Japanese male dimensions) were positioned in comfortable and appropriate postures before being ported into Alias, a 3D computer-aided styling system (Figure 28.18). This enabled the styling and engineering to develop around the human models (Porter and Porter, 1998), ensuring that the final design was both functional and aesthetic.

Reduction in Project Time-Scale

Another recent project involved the development of the driver's cab for the new Amsterdam tram, which provides a good example of the time savings achievable. The SAMMIE model was based on the bare minimum of engineering hard points (including the floor plan, the external body work, the crash resistant structure, and the rear wall) as soon as they were established. With a detailed ergonomics specification of the users and their vision and posture requirements, it was possible to quickly determine the required seat movement envelope and begin to develop a set of surfaces for controls and displays based upon the reach and vision capabilities of the user population (Figure 28.19). The engineers were provided with 3D coordinate and modeling data for an ergonomically designed workstation from which they could build their own CAD model within a matter of days. A mock-up was in fact built directly from the SAMMIE model of the driver's cab (Figure 28.20) and evaluated by a sample of Dutch tram drivers. The design was fully accepted without any changes being made to the SAMMIE design.

Iterative Design and Evaluation

SAMMIE CAD was involved in the development of the driver's cab for the new Lantau express train for Hong Kong's new airport. The designers (Design Triangle, U.K.) sketched a number of exterior forms for initial development (Figure 28.21). The SAMMIE model of the cab structure was built within a single day, and work was started to develop a suitable driver's workstation well in advance of any engineering drawing or engineering CAD work (Figures 28.22 and 28.23). The client subsequently had several changes of mind regarding the external form, which required major changes to the cab body and reduced the space available inside. In addition, the cab space was further reduced as the electrical equipment requirement for space grew, and the carriages were also shortened to enable them to pass through tunnels with a tighter radius of turn than usual. These changes were made to the model as they arose, allowing the assessment of their effect on the workstation ergonomics immediately. Indeed the client later decided that the passenger emergency evacuation route had to be through the front of the train, which effectively cut the cab into three parts. SAMMIE was used to explore how this requirement might be accommodated

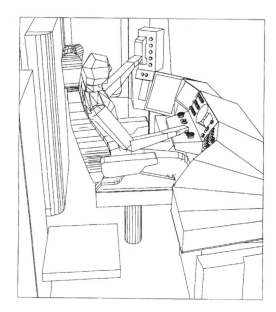

FIGURE 28.19   SAMMIE model of the driver's cab for the Amsterdam Tram was developed before any detailed engineering took place.

FIGURE 28.20   This mock-up of the Amsterdam Tram was made exactly to the SAMMIE specification and successful trials with Dutch drivers were conducted.

with the minimum number of compromises, mostly by reducing the amount and size of equipment required by the driver in order to fit a usable workstation into the smaller space. One novel solution that arose from this was the provision of a chair that can be slid away from the workstation and into a recess in the rear wall (Figure 28.24) to improve cross-cab access and allow sit/stand operation, while still providing a high-quality seat system. (There was insufficient space to swivel a seat, and commercially available flip-up seating would not satisfy the stated comfort criteria.) A full-size mock-up was built from the SAMMIE design (see Figure 28.25) which was used to confirm the easy evacuation of passengers in an emergency.

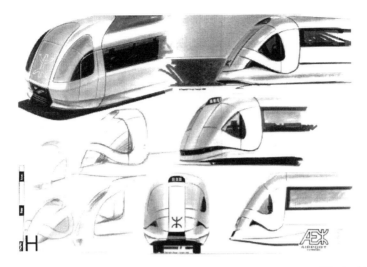

FIGURE 28.21   Stylist's concept sketches for the Lantau Line Express Train. (Courtesy of Design Triangle Ltd., UK.)

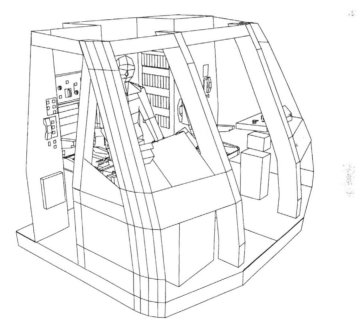

FIGURE 28.22   SAMMIE model of the Express Train cab showing the effects of the structural design and electrical requirements upon the accommodation space.

Cost-Effective Ergonomics

Any reduction in development times should be advantageous from a financial point of view. In addition, the cost of a sophisticated human modeling system is often less than the cost of making just one full-size mock-up using simple materials such as wood and glass fiber. Fitting trials or similar studies also incur extra costs for space rental, staff costs, subject payment, and so on. Because human modeling CAD systems enable the ergonomics input to be provided much earlier in the design process, this reduces the likelihood of expensive modifications being necessary at later stages.

FIGURE 28.23   Stylist's rendering of a SAMMIE model of the driver's workstation for the Express Train. (Courtesy of Design Triangle, UK.)

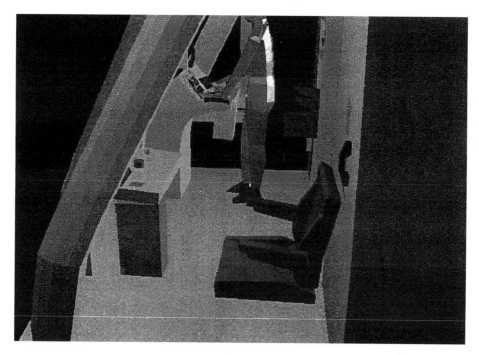

FIGURE 28.24   SAMMIE model of the Express Train cab showing the sit/stand workstation with the seat recessed to improve cross cab access.

Improved Communication

The ergonomist can work most effectively when in collaboration with other members rather than acting as a critic assessing the efforts of others. Collaboration is encouraged by human modeling CAD systems as the system can act as a focal point for the design team even before detailed engineering

FIGURE 28.25   Full size mock-up of the Express Train cab constructed for user trials in Hong Kong.

development work commences. The ergonomics problems with a proposed design can be presented visually and accurately, both in perspective with color surface rendering or as engineering drawings. This ensures efficient communication within the design team and leads naturally to solution-oriented action. The use of a human modeling CAD system is systematic and objective in its approach, which enables all stakeholders in a project (such as the designers, manufacturers, installers, operators, maintainers, recyclers) to examine any assumptions and constraints and to question the conclusions drawn. They can easily visualize any design problems identified and also have a direct involvement with the investigation of alternatives. This gives the ergonomist the opportunity to be proactive and to support the other design team members using communication methods (i.e., computer graphics) that are completely natural for them.

SAMMIE uses CAD techniques both as a tool for ergonomics analyses and as a medium for communication. The work is conducted "on screen" and it requires a high degree of interaction between the members of the design team. The working computer models are often very simple, and this helps to focus on the important ergonomics issues. However, at the completion of this work, we often spend nearly as much time again constructing a more detailed and aesthetically pleasing model, and this helps to impart a greater sense of validity. It is much more persuasive to the client to embody a good ergonomics specification into a CAD model than it is to present a written report listing ergonomics recommendations.

28.4   Issues of Validity

We have always advocated that human modeling systems should not replace user trials with full-size mock-ups, unless the design or the design modifications are so simple as to not warrant concern. In-depth user trials can reveal problems with so many more issues including long-term discomfort, effects of fatigue, negative transfer of training, error rate, performance, and the acceptance of the product. Many designers, engineers and ergonomists are expectantly waiting for the all-singing, all-dancing human modeling system to appear. The likelihood of such a system being developed in the near future seems remote. However, the advantage of using human modeling systems is that it is possible to build full-size mock-ups with the confidence that few, if any, modifications will be necessary to physically accommodate the users. The detailed evaluation of criteria such as those above can proceed without delay and without the extra costs of getting the basics right.

Human-modeling CAD systems have the potential to offer considerably more validity as a simulation of people than the traditional 2D manikins that are overlaid on drawings. This is because they can be made to simulate specific user groups and tasks in three dimensions. Also, 2D manikins are often used in a very simplistic way without a knowledge of their origin in terms of user type (e.g., military/civilian data, age range, nationality, size, date of survey) and application (e.g., erect or slumped sitting height, percentile values for individual body segments). For example, designers may have a 5th percentile adult female manikin and 50th and 95th percentile adult male manikins. The nature of these manikins gives support to the notion that people come either short and short limbed, tall and long limbed, or somewhere in between. It has been repeatedly demonstrated that this is not true and that the intercorrelation between body dimensions is rather poor (e.g., Haslegrave, 1980). Statistically, it is not possible for an individual to be 5th or 95th percentile in all vertical body dimensions and still be 5th or 95th percentile in stature. The manikin designer can resort to other techniques to ensure that the manikins are statistically correct, for example by calculating median values or using regression equations to describe component body dimensions for groups of men or women of a given stature and weight. Whichever method is chosen to define a variety of statistically correct manikins, there is still the problem of estimating the percentage of people accommodated by a particular design. A common mistake made by many designers is to use the 5th percentile female stature and 95th percentile male stature manikins to assess a workstation, assuming that if both of these manikins can be accommodated then so can 95% of the adult population. This is an incorrect assumption because it implies that those people designed out, because either their sitting height, hip breadth, or leg length, for example, are greater than 95th percentile male values, are all the same people. Similarly, all those with sitting eye height, arm length, or leg length smaller than 5th percentile female values are assumed to be the same individuals. Because these dimensions are not strongly correlated, these assumptions are incorrect. A study of air crew selection standards and design criteria analysis reported by Roebuck, Kroemer, and Thomson (1975, p. 268) illustrates the problem perfectly. It was shown that nearly half of the air crew were designed out when the 5th to 95th percentile range was used on a large number of body dimensions (in this case 15). Even limiting the number of dimensions to just 7 (sitting height, eye height-sitting, shoulder height-sitting, elbow rest height, knee height, forearm-hand length, and buttock-leg length) designed out over 30% of the available air crew.

This brief overview of the inherent problems with the use of manikins indicates the benefits that can arise from using 3D human modeling CAD systems with variable anthropometry. For example, the CAD operator has complete control of the dimensions or percentile values of each segment of the human model (assuming the system offers this facility) and can interactively change them in a matter of seconds. For example, a model of 99th percentile male stature with short arms and long legs for such a stature can be used to determine the rearmost position required of an adjustable steering wheel. Unfortunately data at this level of detail are not commonly presented in surveys, one of the exceptions being the anthropometric survey of Royal Air Force Aircrew by Simpson and Hartley (1981).

There is an important distinction between the evaluation of which percentile values are accommodated for a particular design dimension and the evaluation of what percentage of the population will be accommodated in all respects. Three-dimensional human modeling systems offer significant advantages in both respects and, in particular, the latter. Roebuck (1995) discusses two statistical methods by which some human modeling systems predict the percentage of the target population that will be accommodated by a particular workstation design. CAR and MDHMS use Monte Carlo methods to generate a large number of theoretical human models, each one of which represents a possible case that could occur in a population of people without violating any of the underlying anthropometric statistics for that population. Another statistical approach is Principal Component Analysis, a version of which is used in the SAFEWORK system.

SAMMIE is currently developing a dataset of body scans and anthropometric dimensions, including joint centers and joint mobility, from a carefully selected sample of people representative of the British and European population. Each person can then be modeled individually within SAMMIE and automatically positioned in a prototype workstation model according to a range of predefined criteria. Evaluations of fit, reach, vision, and the required postures will then be conducted, automatically resulting in the identification of those individuals who failed to successfully complete any of these tests.

Whenever possible we aim to combine the use of SAMMIE with the more traditional ergonomics methods. For example, our involvement with a major supermarket chain commenced with a survey of the musculoskeletal discomfort reported by staff in every work area (Porter et al., 1991). This allowed us to expose the high-risk workstations, namely the delicatessen and the cashier's workstation, and to subsequently model these using SAMMIE. This computer analysis simulated the users, both staff and customers, and their tasks, and the detailed postural analysis revealed several casual factors for the reported discomfort. Modifications were then made to the computer models of these workstations in order to improve the working postures. These designs were subsequently mocked-up and fine-tuned in terms of other more subjective attributes, such as the aesthetic issues.

In a similar way, SAMMIE was more recently used in the development of the Fiat Punto. The system was used to model the prototype Punto from engineering drawings and then to investigate driver accommodation for a variety of nationalities (Porter, 1995; Figure 28.9). It was considered to be essential that the Punto would accommodate drivers of all sizes. Two prototype Puntos were subsequently made available for the development of the seat with detailed assessments of the prototypes and competitor cars from 20 carefully selected members of the public driving over a 60-mile test route. This attention to detail by the manufacturer was rewarded when the Punto was voted European Car of the Year in 1995.

Human modeling systems are increasingly being used by designers and engineers who may not all have a thorough training in ergonomics issues relevant to the design of equipment and workplaces. This can result in the human models being regarded as just another CAD model, the size and shape of which can be specified and positioned within set constraints. It is all too easy to forget the important differences between designing and manufacturing the product, where variables can be specified, and evaluating the design using human models. These human models are also specified by the designer, but it is essential that they cover the wide range of sizes, shapes, functional mobility, and postural preferences that are exhibited by the population of future users of the product. Ergonomists learn the hard way that people are all different. Their experiences are based on trials, interviews, and accident reports. There is a concern that the use of human modeling systems by operators without this appreciation will lead to standardized procedures being developed taking little account of such differences. It does not necessarily follow that people will hold dangerous pieces of equipment by the handle as intended. Computer people may do as they are instructed, but real people, particularly when poorly trained, fatigued, under stress, working to a tight schedule, and so on, must not be expected to be so disciplined.

References

Badler, N.I., Phillips, C.B., and Webber, B.L., (1993). *Simulating Humans: Computer Graphics Animation and Control*, Oxford University Press, N.Y.

Case, K., Bonney, M.C., and Porter, J.M., 1991. Computer graphics standards for man modelling, *Computer Aided Design*, 23, 4, 257-268.

Coblenz, A., Mollard, R., and Renaud, C., 1991. Ergoman: 3-D representations of human operator and man-machine systems. *International Journal of Human Factors in Manufacturing*, 167-178.

Das, B. and Sengupta, A.K., 1995. Computer-aided human modeling programs for workstation design, *Ergonomics*, 38, 9, 1958-1972.

Dooley, M., 1982. Anthropometric modelling programme — a survey. *IEEE Computer Graphics and Applications*, 2, 17-25.

Elias, H.J. and Lux, C., 1986. Gestatung ergonomisch optimierter Arbeitsplatze und Produkte mit Franky und CAD (The design of ergonomically optimized workstations and products using Franky and CAD). *REFA Nachrichten*, 3, 5-12.

Fortin, C., Gilbert, R., Beuter, A., Laurent, F., Schiettekatte, J., Carrier, R., and Dechamplain, B., 1990. SAFEWORK: A micro-computer aided workstation design and analysis, new advances and future developments. Genicom Inc., Montreal Quebec.

Grandjean, E., Hunting, W., and Nishiyama, K., 1984. Preferred VDT workstation settings, body posture and physical impairments, *Applied Ergonomics*, 15, 99-104.

Grobelny, J., Cyewski, P., Karwowski, W., and Zurada, J., 1992. APOLIN: a 3-dimensional ergonomic design and analysis system, in *Computer Applications in Ergonomics, Occupational Safety and Health*, eds. M. Mattila and W. Karwowski, pp. 129-135. Elsevier Science Publishers BV, Amsterdam.

Haslegrave, C.M., 1980. Anthropometric profile of the British car driver, *Ergonomics*, 23, 436-67.

Jones, P.R.M., West, G.M., Harris, D.H., and Read, J.B., 1989. The Loughborough anthropometric shadow scanner, *Endeavour, New Series*, 13, 4, 162-168.

Kloke, W.B., 1990. WERNER: a personal computer implementation of an extensive anthropometric workplace design tool, in *Computer-Aided Ergonomics*, eds. W. Karwowski, A.M. Genaidy, and S.S. Asfour, pp 57-67. Taylor & Francis, London.

Kuusisto, A. and Mattila, M., 1990. Anthropometric and biomechanical man models in computer-aided ergonomic design structure and experiences of some programs, in *Computer-Aided Ergonomics*, eds. W. Karwowski, A.M. Genaidy, and S.S. Asfour, pp 104-114. Taylor & Francis, London.

Launis, M. and Lehtelä, J., 1990. Man models in the ergonomic design of workplaces with the micro-computer, in *Computer-Aided Ergonomics*, eds. W. Karwowski, A.M. Genaidy and S.S. Asfour, pp 68-79. Taylor & Francis, London.

Mandal, A.C., 1984. What is the correct height of furniture?, in *Ergonomics and Health in Modern Offices*, ed. E. Grandjean, pp. 471-476, Taylor & Francis, London.

McDaniel, J.W., 1990. Models for ergonomic analysis and design: COMBIMAN & CREW CHIEF, in *Computer-Aided Ergonomics*, eds. W. Karwowski, A.M. Genaidy, and S.S. Asfour, pp. 138-156, Taylor & Francis, London.

Porter, J.M., 1995. The ergonomics development of the Fiat Punto — European Car of the Year 1995, *Proceedings of the IEA World Conference 1995*, Rio de Janeiro, Brazil, eds. A. de Moraes and S. Marino, pp. 73-76. Associacao Brasileira de Ergonomia.

Porter, J.M., Almeida, G.M., Freer, M.T., and Case, K., 1991. The design of supermarket workstations to reduce the incidence of musculo-skeletal discomfort, in *Designing for Everyone and Everybody*, eds. Y. Queinnec and F. Daniellou pp. 1122-1124. Taylor & Francis, London.

Porter, J.M., Case, K., Freer, M.T., and Bonney, M.C., 1993. Computer-aided ergonomics design of automobiles, in *Automotive Ergonomics*, eds. B. Peacock and W. Karwowski, pp. 43-78, Taylor & Francis, London.

Porter, J.M., Freer, M., Case, K., and Bonney, M.C., 1995. Computer aided ergonomics and workspace design, in *Evaluation of Human Work: A Practical Ergonomics Methodology*, 2nd edition, eds. J.A. Wilson and E.N. Corlett, pp. 574-620. Taylor & Francis, London.

Porter, J.M. and Porter, C.S., 1998. Turning automotive design "inside-out". *International Journal of Vehicle Design*, Vol. 19, No. 4, 385-401.

Post, F.H. and Smeets, J.W., 1981. ADAPS: Computer aided anthropomerical design. *Tijdschrift voor Ergonomic*, 6, (4), 11-18 (in Dutch).

Roebuck, J.A., 1995. *Anthropometric Methods: Designing to Fit the Human Body*, Human Factors and Ergonomics Society, USA.

Roebuck, J.A., Kroemer, K.H.E., and Thomson, W.G., 1975. *Engineering Anthropometry Methods*, John Wiley & Sons, New York.

Rothwell, P.L., 1985. Use of man-modelling CAD systems by the ergonomist, in *People and Computers: Designing the Interface*, eds. P. Johnson and S. Cook, pp. 199-208. Cambridge University Press, U.K.

Simpson, R.E. and Hartley, E.V., 1981. Scatter diagrams based on the anthropometric survey of 2000 Royal Air Force Aircrew (1970/71), *Royal Aircraft Establishment Technical Report 81017*, Farnborough, Hampshire, England.

Westerink, J., Tragter, H., Van Der Star, A., and Rookmaaker, D.P., 1990. TADAPS: a three-dimensional CAD man model, in *Computer-Aided Ergonomics*, Taylor & Francis, London, pp 90-103.

For Further Information

Books

A wide selection of human modeling systems is individually presented in the following two books: *Computer-Aided Ergonomics*, eds. Karwowski, W., Genaidy, A. and Asfour, S.S., 1990. Taylor & Francis Ltd., London; and *Computer Applications in Ergonomics, Occupational Safety and Health*, eds. Mattila, M. and Karwowski, W., 1992. Elsevier Science Publishers B.V., The Netherlands.

A detailed description of the development of the Jack system is presented in: *Simulating Humans: Computer Graphics Animation and Control*, Badler, N.I., Phillips, C.B., and Webber, B.L., 1993. Oxford University Press, N.Y.

Database

An up-to-data database concerning the technical specifications of the major commercially available human modeling systems is available from the CSERIAC Program Office, AL/CFH/CSERIAC Bldg. #248, 2255 H Street, Wright-Patterson AFB, OH 45433-7022, USA. Tel: +513 255-4842, Fax: +513 255-4823. Acknowledgment is given to Aaron Gayman and Chris Sharbaugh for kindly supplying information for this chapter.

Web Sites

Several system developers and users have created web pages providing graphic images and current information. These include:

Deneb/ERGO, Auburn Hills, Michigan, USA: http://www.deneb.com/ergo.html

JACK at the University of Pennsylvania, USA: http://www.cis.upenn.edu/~hms/jack.html

McDonnell Douglas Human modeling system (MDHMS), Long Beach, California, USA: http://pat.mdc.com/LB/LB.html

MIDAS application of Jack at NASA Ames Research Center, Moffett Field, California, USA: http://ccf.arc.nasa.gov/af/aff/midas/MIDAS_home_page.html

SAFEWORK at Genicom Consultants Inc., Montreal (Quebec): http://www.safework.com

SAMMIE at Department of Design and Technology, Loughborough University, Leicestershire, UK: http://www.lboro.ac.uk/departments/cd/docs_dandt/staff/Porter/sammie.html

29

A Guide to Computer Software for Ergonomics

Jari Järvinen
Motorola

Hongzheng Lu
Lucent Technologies

29.1 Introduction

Usually ergonomics or human factors specialists attempt to optimize work with computers and software. The purpose of this chapter is to review the computer software that can help ergonomists, researchers, and practitioners at their work.

A vast amount of ergonomics knowledge has been generated by researchers, but it is not always effectively communicated to the practitioners who design, redesign, and evaluate products, tools, equipment, facilities, and environments. Tight schedules of the design projects, constantly changing design specifications, and the increasing complexity of work systems and equipment create situations in which the ergonomic considerations may be ignored by the designers. Moreover, the diversity of the available ergonomic literature may confuse a designer who does not have a comprehensive understanding of the field.

Recent developments in ergonomics software and other computer-aided tools make specialized knowledge accessible and present it in a suitable form to practitioners. The designers can now more easily obtain feedback needed to design products for human use and to optimize manufacturing and maintenance processes already at the design phase, when the cost of making changes is still relatively low. The incorporation of ergonomic knowledge through the use of ergonomics software at the early stages of the design process also reduces the number of prototypes needed to bring a product to market, and enables the manufacturer to create products that are cheaper to make and provide a better fit for the user.

New software applications in the field of ergonomics are being developed at an accelerating pace. This guide cannot be considered a complete catalog of ergonomics software. Selected, and most applicable, commercially available software from various application areas are described. Although ergonomics software can provide an access to applicable ergonomics knowledge and data, the human is still the main component in the design and analysis processes. The computer excels in computation, data processing, and storage, while creativity is the human's strength.

29.2 Fundamentals

Since the 1960s several computer programs have been created to assist ergonomics practitioners. The first ergonomics computer programs were based on geometric human models. More recently, a variety of analysis modules and databases have been incorporated into these human modeling systems. Examples of other currently available software include ergonomics analysis programs (without human modeling capabilities), computerized database programs, and programs that provide a computer operator with guidance in ergonomic work techniques and reminders about rest breaks. A general description of different types of ergonomics software is presented in this section.

Human Modeling Systems

Human modeling is not a new design tool. For decades, wooden and plastic dimensionally accurate templates have been used in the design of automobiles and aircraft. The increase in the computing power of the workstations and personal computers has made it possible to use computerized human modeling tools at the design stage of workplaces and equipment.

The human modeling programs use anthropometrically accurate human models to simulate human size and motions as they do their work or use a product or tool. The product or environment model is usually created using a CAD modeling software. The design can then be evaluated using a human modeling system that is either a modular component of the CAD software or an independent, stand-alone human modeling system. Some of the stand-alone systems incorporate basic CAD modeling capabilities, and/or the CAD models can be imported to the human modeling system, or in some cases, the human models can be exported to the CAD system.

Different human models have been developed for different needs. The simple models are more or less static, easy to use, do not require much computer memory, and thus, can be used in almost any computer for routine design tasks. The more sophisticated human models are more versatile and can animate human motions in real time. However, these systems are complex and require special knowledge, skills, and training to be used effectively. Some of the sophisticated human modeling programs require powerful graphic workstations, or minicomputers, and are used mainly in the automotive and aerospace industries, as well as in academia and the military.

With the increasing computing power of personal computers, the sophistication of the PC-based human modeling systems can also be increased. Recently, relatively sophisticated human modeling systems have been developed that do not require investments in high-end hardware and software, and thus, are easily accessible for small and midsize companies. Some organizations, such as NASA, have adopted a two-tier approach that uses PC-based human modeling systems (such as custom version of MQPro) for routine design tasks, and high-end workstation programs (such as Jack) for modeling complex human–machine systems and for *virtual reality* applications.

Some of the ergonomics software use *virtual reality* (*VR*) to simulate human responses and fit to a proposed design. VR is a high-end user interface that involves real-time simulation and interactions through multiple sensorial channels (Burdea and Coiffet, 1994). The VR user is immersed in a computer-generated world via a boom or head-mounted display, and may "fly" or "walk" through the virtual world and interact with it (Mourant, 1994). In so-called desktop VR, the user wears stereo glasses and/or a head tracking system. This is sometimes called nonimmersion or "fishbowl" VR.

Other Ergonomics Software

Ergonomics analysis programs usually consist of computerized algorithms or equations used in ergonomics analyses, such as biomechanical analysis or energy expenditure prediction. Some established analysis methods, such as OWAS, have also been computerized. These programs do not usually include graphical human or environment modeling capabilities.

Other types of commercially available ergonomics software designed for ergonomics practitioners include computerized databases consisting of anthropometric information, design guidelines, and other information. The databases usually combine data from several sources, and the sophisticated user interfaces and database management procedures of the modern programs can efficiently provide the required design information for the practitioner.

Recently, due to the rapidly increased use of computers and the recognition of the problems related to excessive computer use, new types of ergonomics programs have emerged. Instead of assisting a designer or an ergonomics practitioner, these programs provide the computer user with guidance on ergonomic work techniques and stretching exercises, and reminds him/her about rest breaks.

29.3   Applications and Examples

Selected examples of ergonomics software from various application areas are described in this section. The examples were selected from the following areas: human modeling systems for personal computers (MQPro, ergoSHAPE) and workstations (SAMMIE, Deneb/ERGO, Jack, Safework, MDHMS), ergonomics analysis programs (ErgoEASER, ErgoMOST, NIOSH Lifting Equation, OWASwin, 3D SSPP, Energy Expenditure Prediction Program), database programs (PeopleSize), and training/stretching software (WorkSmart).

Human Modeling Systems for Personal Computers

MQPro

MQPro is a PC-based human modeling system and ergonomic design software. With a few mouse clicks, the program creates anthropometrically accurate, three-dimensional human models representing several ethnic groups, *percentiles,* and *somatotypes.* These models can be manipulated to any human-compatible position and viewed from any angle, distance, or perspective. The views can be printed, plotted, or exported to other graphics software for further enhancement of the image. The human models can walk, bend, reach, and grasp objects. Although equipped with a set of basic three-dimensional modeling and editing tools, MQPro is designed with import/export capabilities so it can be used with other graphics software, such as AutoCAD and 3D Studio.

MQPro has evolved from the human modeling package Mannequin. It has been developed by BCAM International and used for ergonomics consulting services in product and workplace evaluation and design projects. Custom versions of the software have been provided to select organizations, such as NASA and university research centers. The custom applications have ranged from specific interface needs to the integration of motion capture hardware (Flock of Birds).

Main features: anthropometric database of eleven populations, including 1988 U.S. Army (Natick) and NASA-STD-3000; adult male, adult female, and child human models with three somatotypes; 2.5th, 5th, 50th, 95th, 97.5th percentile human models or customize option for each body part; normal, customized, or free joint range of motion; human model representation in five levels of detail ranging from stick-figure to skeleton and high-resolution humanoid; physically challenged human models with wheelchairs, canes, and crutches; field of vision cones; reach envelopes for hands and feet; reach analysis; 2D and 3D drawing and editing tools; wire frame, shading, and hidden line removal capabilities; ability to "see" through selected human model's eyes; placement of camera, field of view, and light source for any three-dimensional perspective view; *Revised NIOSH 1991 lifting equation;* simulation of lifting, pushing, pulling by adding external forces and torques in any direction on any body part; calculation of reaction joint forces and torques due to external load and body posture; presentation of anthropometric information, joint angles, joint forces, and torques in tabular form; frame-by-frame animation; and animated walking on specified path. Some of the MQPro's capabilities are illustrated in Figures 29.1 and 29.2.

System requirements: 486 PC with co-processor or Pentium, Windows 3.1 or higher, Windows 95, or Windows NT, 4MB RAM, 4MB of available hard disk space, VGA display. Price range: less than $1,000.

FIGURE 29.1   Two human models and three-dimensional models of a backhoe and jackhammer created using MQPro. (Courtesy of HumanCAD Systems)

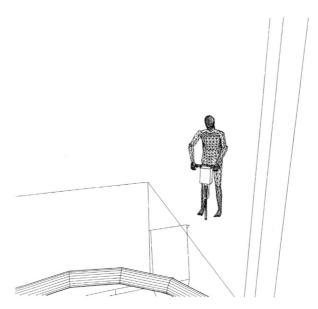

FIGURE 29.2   A field of view as seen by the MQPro human model sitting in the backhoe seat. (Courtesy of HumanCAD Systems)

ErgoSHAPE

ErgoSHAPE is a two-dimensional human modeling system developed by the Institute of Occupational Health in Finland to supplement a widely used PC-based AutoCAD design program. The system is used as a module of AutoCAD and consists of drawing files, menu files, and AutoLISP program files. Recently, a version of ergoSHAPE for the Designer computer-aided design program was developed.

The ergoSHAPE system consists of the following parts: (1) two-dimensional 50th percentile male and female human models based on the anthropometric dimensions of the northern European and North American populations, (2) biomechanical calculations in two dimensions; the stress is calculated as a percentage of the maximal static muscle strength, (3) recommendation charts providing design guidelines, such as dimensions for seated and standing workstations, and (4) ErgoTEXT text files, that include ergonomic design guidelines covering several areas of ergonomics.

ErgoSHAPE human models can also illustrate the curves indicating the viewing angles and distances and the reach zones in two dimensions. A more detailed description can be found in Launis and Lehtelä (1990).

System requirements: IBM-compatible PC, AutoCAD or Designer software. Price range: less than $1,000.

Human Modeling Systems for Workstation Computers

SAMMIE

SAMMIE (System for Aiding Man–Machine Interaction Evaluation), developed by the University of Nottingham, is one of the first commercially available human modeling systems. A detailed description of SAMMIE can be found in Case et al. (1990).

The human model of SAMMIE was originally based on data from Dreyfus's anthropometric measurements (Dreyfus, 1966). The limb and body segment lengths of the model can be varied by percentiles or by explicitly defining the dimensions for each body segment. Anthropometry data files can also be created and modified by the designer. In addition, the somatotypes, which define the flesh outline of the model, can be varied.

The user can build a three-dimensional environment around the human model by solid modeling techniques. The user can specify primitive geometric shapes and assemble these to models of equipment or environments.

The human model can interact with the created environment model in several ways. Logical relationships between the body segments are included. The limbs can only be moved within normal human ranges of motion. The joint constraints can be defined by the user. The following ergonomic evaluations can be performed: reach assessment, fit assessment, visual field assessment, and posture evaluation against "normal" constraints.

SAMMIE has been used for the following applications, among others: power station control consoles, automobile design, visualization of car driver views in the English Channel Tunnel, financial trading facilities, helicopter concept design, and underground design.

System requirements: mainframe or workstation computer (Apollo or Sun), UNIX operating system.

Deneb/ERGO

Deneb/ERGO human-factors analysis software has been developed to be used in a workstation platform, such as Silicon Graphics, with a UNIX operating system. It is used in conjunction with Deneb's IGRIP modeling software.

Main features: 5th, 50th, and 95th percentile anthropometric male and female models that move realistically; enables creation of generic assembly motion routines and sequences; human models have strength capabilities, i.e., they can "lift" as much as average humans; NIOSH 1991 lifting equation; energy expenditure prediction of the simulated tasks (Garg's model); time measurements of the simulated tasks (MTM-UAS); evaluation of the injury potential of the simulated tasks; and immersive VR capability.

System requirements: workstation computer (Silicon Graphics), UNIX operating system, Deneb's IGRIP modeling software package.

Jack

Jack is a human modeling software package developed at the Center for Human Modeling and Simulation at the University of Pennsylvania.

Jack provides a three-dimensional interactive environment for controlling articulated figures. It features a detailed human model and includes realistic behavioral controls, anthropometric scaling, task animation and evaluation systems, strength-guided motion, view analysis including viewing the environment model through the human model's eyes, automatic reach and grasp, collision detection and avoidance, and other tools for a wide range of applications. Dynamic torques can be computed throughout animated motions and compared against strength data to assess the validity of motions and postures.

Jack also provides an interface to other hardware such as Cyberglove and Ascension Flock of Birds, which enable virtual reality applications.

System Requirements: Silicon Graphics IRIS 4D, INDIGO, or INDY workstations, Z-buffer, three-button mouse, SGI Operating System 4.0.1 or higher, 16 MB RAM. Price range: $10,000 to 20,000.

Safework

Safework is a modular human modeling system that runs in Silicon Graphics workstations. Safework is available in four versions with increasing functionality. The following features are included in the most advanced package, which contains all available functions and options: six basic anthropometric three-dimensional human models (5th, 50th, and 95th percentile male and female); user access to anthropometric variables and data (1988 U.S. Army Natick laboratories); seven different somatotypes; manipulation through inverse and direct kinematics; fully articulated hand and spine models; normal and restricted joint mobility; coupled range of motion; vision analysis, including binocular, ambinocular, and monocular vision, and various vision cones; animation; postural, comfort angle and reach analysis modules; collision detection; files can be imported from most CAD programs; basic geometric modeling capabilities.

System requirements: Silicon Graphics workstation Indy or Indigo2 R4400 (recommended), 32 MB RAM or higher (recommended: 64 MB or higher), Graphics Engine XZ (recommended: Graphics Engine Impact), IRIX 5.3. Price range: $30,000 to $60,000 depending on the version.

McDonnell Douglas Human Modeling System

In addition to three-dimensional biomechanical human models and virtual models of equipment and environments, the McDonnell Douglas Human Modeling System (MDHMS) offers detailed eye articulation, hand and finger articulation, and shoulder articulation. The model can simulate rotating eyeballs, which enables the illustration of what a human can see more realistically than most virtual reality applications can.

This software package has been used to study human-machine interactions and other human activities in the crew stations of a fighter/attack aircraft, crew-return vehicle, commercial aircraft, and extravehicular mobility unit. McDonnell Douglas' suppliers are also using MDHMS to evaluate the assembly and maintenance of their aircraft components before they are sent to McDonnell Douglas. This software requires a powerful workstation computer with a UNIX-based operation system.

Ergonomics Analysis Programs

ErgoEASER

ErgoEASER (Ergonomics Education, Awareness, System Evaluation & Recording) is a set of PC-based interactive tools for evaluating computer workstations and lifting tasks, and recommending appropriate controls.

ErgoEASER has been developed by Pacific Northwest National Laboratory (PNNL), the U.S. Department of Energy (DoE), and the U.S. Department of Defense (DoD), in consultation with the U.S. Department of Labor/Occupational Safety and Health Administration (DOL/OSHA). ErgoEASER is a work in progress, and is being shared with federal agencies and made available to the public.

ErgoEASER allows the user to evaluate workplaces and tasks to identify risk factors that should be addressed. Users may then modify variables affecting the risk factors and determine which variables will eliminate or reduce the hazards when adjusted.

Currently, ErgoEASER consists of three components: (1) getting started, (2) awareness and reporting, and (3) analysis modules for lifting and VDT workstations.

The "Getting Started" module includes background information on occupational ergonomics. The "Awareness and Reporting" module consists of examples and photographs of hazardous postures. A risk factor checklist is included to help the user assess his/her own company's work situations. Based on entered information, the software may recommend additional analyses. The analysis modules assist the user in more detailed evaluations to support the design of VDT workstations and lifting tasks. The user interactively enters variables that describe specific work situations, such as key VDT workstation and operator dimensions. The user can then view the simulated postures and observe the ergonomic effects resulting from workstation configurations. The software highlights the body parts with an increased stress level. The user can make adjustments by changing the values of the variables. Finally, the software generates a report that documents task specifications and recommended solutions. The Lifting Analysis module incorporates the 1991 NIOSH Lifting Equation.

ErgoMOST

ErgoMOST software was developed to analyze physical risk factors associated with repetitive motion. It is a module of the MOST (Maynard Operation Sequence Technique) Work Measurement Systems software package. ErgoMOST focuses on methods for improvements in the workplace to minimize ergonomic stress and maximize motion efficiency, and it can be used as a stand-alone application, or it can interact with other MOST software products. Relative Ergonomic Stress Indices are assigned based on used force, posture, repetition, grip, and vibration. Twelve body joints are evaluated by combining ergonomic analysis with defined methods using MOST. ErgoMOST also provides a range of reports and parameters for displaying results. Users can select a step- or job-level report, a report by body parts or ergonomic areas, and text or graphic formats.

  System requirements: IBM PC-compatible 486 33 MHz or higher, 8MB RAM, Windows 3.1 or higher.

NIOSH Lifting Equation

The revised NIOSH Lifting Equation can be used to evaluate simple and well-defined lifting tasks. Specific lifting parameters are entered as inputs, and the equation provides a Recommended Weight Limit (RWL) for the lift. The revised NIOSH Lifting Equation is discussed in detail elsewhere in this book.

  The NIOSH Lifting Equation is included as an analysis module in some of the human modeling systems described above, such as MQPro and Deneb/Ergo. In addition, numerous software sources that include the equation in computerized form are listed in the NIOSH web page (http://www.cdc.gov/niosh/home-page.html).

OWASwin

OWAS (Ovako Working posture Analyzing System) is a method for the evaluation of postural load during work based on a systematic observation and classification of working postures (Karhu et al., 1977). It is an observational sampling technique in which a four-digit code is used to describe postures of the body parts and the required force. Postures are observed at a set time interval. The coded posture combinations are classified into four action categories which allow the most stressful activities to be identified and indicate the priorities for corrective measures.

  The action categories along with recommended actions are defined as follows: (1) normal posture: *no actions required*, (2) the posture is slightly harmful: actions to change the posture should be taken *in the near future*, (3) the posture is distinctly harmful: actions to change the posture should be taken *as soon as possible*, and (4) the posture is extremely harmful: actions to change the posture should be taken *immediately*. More information about the OWAS method is presented in Chapter 26.

  The computerized version of OWAS consists of data collection and data analysis modules. The data collection module is used either in field conditions, using a portable computer and observing "live" work, or in the laboratory, where the videotaped work is observed. The postural data are collected using visual, split-second observations. The postures of the back, arms, and legs, as well as weight of load or used force are identified and recorded using a predefined coding system. During the posture coding the subtask or activity can also be recorded along with the posture codes. The subtask information can be used to focus interventions on the most demanding subtasks.

  The data analysis module allows the postural data to be analyzed in two ways. One way is to examine the combined posture of the back, arms, legs, and exerted effort, and determine its effect on the musculoskeletal system. The other way is to examine the relative time spent in a particular posture for each body part and determine the time effect on the musculoskeletal system.

  The analysis will result in the following output: number of observations, percentage of individual postures in each action category, relative time spent in each posture, relative time spent in each subtask, number of observations of each posture, and remedial action recommendations.

  System requirements: Windows 3.1 or higher. Price range: less than $1,000.

3D Static Strength Prediction Program

The 3D Static Strength Prediction Program (3D SSPP) can be used to evaluate the physical demands of a prescribed job. Both DOS and Windows versions of the program are available. The program is based on years of research at the Center for Ergonomics at the University of Michigan concerning work-related human biomechanical and static strength capabilities. The program can be applied to a worker's motions in three-dimensional space. However, the effects of acceleration and momentum must be negligible. The program is most useful in the analysis of the slow movements used in heavy materials handling tasks that can be described as a sequence of static postures.

The first estimation of the working posture to be evaluated can be created by entering the position of the hands and the load on the hands. The posture can be modified by changing or entering joint angles for selected joints. The posture and force direction are presented using a stick figure in the analysis screen. A three-dimensional illustration of the posture is also available. The modeling of the work environment is not possible.

The analysis results are presented using the following output screens: (1) analysis summary; (2) anthropometric data, including link lengths and weights, joint angles, and balancing status of the analyzed posture; (3) moments at different joints; (4) moments about joint movement hinges including percentage of population with sufficient strength capability; (5) low back muscle and disc forces (L5/S1 disc); (6) spinal analysis summary including resultant moments and forces at L2/L3 and L4/L5 spinal segments; and (7) low back compression optimization summary including resultant forces of all muscles involved in the lumbar area.

Energy Expenditure Prediction Program

The Energy Expenditure Prediction Program (EEPP) software is used to predict the total energy expenditure and energy expenditure rate in performing a job. It is based on the assumption that a job can be divided into simple tasks, or activity elements, and that the average metabolic energy rate of the job can be predicted by knowing the energy expenditure of the simple tasks and the time duration of the job.

Other Ergonomics Software

PeopleSize

PeopleSize is a computerized database with a graphical mouse-driven interface consisting of most human dimensions that are relevant to design. The database is a comprehensive and validated anthropometry reference compiled from over 70 sources. Each body dimension in the database is based on the average of several available sources for each major racial group, weighted by sample size. The data sources are listed in the Help file of the software.

Main features: dimension-specific data for any percentile; adjustments for clothing (gloves, shoes, head gear, winter clothes, summer clothes, etc.) and sitting posture (slumped or erect); calculates percentiles for given measurements; program output can be stored to the "log" to be used later; displays detailed descriptions of measurement methods and terminology.

System requirements: Windows 3.1 or higher, 2 MB RAM, 1.5 MB free disk space, EGA or better graphics, mouse. Mac System 7 or higher, 2 MB RAM + 2 MB virtual memory, 2 MB free disk space.

WorkSmart Stretch Software

WorkSmart stretch software reminds the user about a stretch break every 50 minutes during the workday. WorkSmart offers four groups of stretches designed to benefit different areas of the body affected by daily computer-related muscle stress: Neck and shoulders, upper extremities, lower body, and back. The software displays instructions and graphic demonstrations for the stretch exercises. The user can also reschedule the break by postponing it by 10 to 50 minutes.

System requirements: Windows 3.1 or above, 3.5" Floppy drive, mouse, 386 IBM compatible or above, 16 color VGA with palletized VGA display driver, or 256 color VGA.

Defining Terms

Percentile: The frequency distributions for each measurement of population size are expressed in percentiles. A percentile indicates the percentage of population at or below certain measure.

Revised NIOSH Lifting Equation: An equation, developed by NIOSH, for calculating a recommended weight for specified two-handed lifting tasks.

Somatotype: The morphological type of a human body.

Virtual reality (VR): A high-end user interface that involves real-time simulation and interactions through multiple sensorial channels. A VR user is immersed into a computer-generated world via boom or head-mounted display, and may "fly" or "walk" through the virtual world and interact with it.

References

Burdea, G. and Coiffet, P. 1994. *Virtual Reality Technology*, John Wiley & Sons, New York.

Dreyfus, H. 1966. *The Measure of Man — Human Factors in Design*, Whitney Library of Design, New York.

Case, K., Porter, J.M., and Bonney, M.C. 1990. SAMMIE: a man and workplace modelling system. In *Computer-Aided Ergonomics*, eds. W. Karwowski, A.M. Genaidy and S.S. Asfour, p. 31-56. Taylor & Francis, New York.

Karhu, O., Kansi, P., and Kuorinka, I. 1977. Correcting working postures in industry: A practical method for analysis. *Applied Ergonomics*. 8:199-201.

Launis, M. and Lehtelä, J. 1990. Man models in the ergonomic design of workplaces with the microcomputer, in *Computer-Aided Ergonomics*, eds. W. Karwowski, A.M. Genaidy, and S.S. Asfour, p. 68-79. Taylor & Francis, New York.

Mourant, R.R. 1994. *Virtual Reality and Ergonomics*. Workshop presented at IEA 12th Triennial Congress.

T.R. Waters, V. Putz-Anderson, and A. Garg, 1994. *Application Manual for the Revised NIOSH Lifting Equation*, DHHS (NIOSH) Publication No. 94-110, NTIS, Springfield, VA.

For Further Information

Deneb/ERGO: Deneb Robotics, Inc., 3285 Lapeer Road West, P.O. Box 214687, Auburn Hills, MI 48321-4687.

Energy Expenditure Prediction Program: University of Michigan Software, 3003 South State Street, Suite 2071, Ann Arbor, Michigan 48109-1280, USA.

ErgoEASER®: Pacific Northwest National Laboratory, P.O. Box 999, Richland, WA 99352.

ErgoMOST: H.B. Maynard and Company, Inc., Eight Parkway Center, Pittsburgh, PA 15220, Internet: www.hbmaynard.com.

ErgoSHAPE: Institute of Occupational Health, Ergonomics Unit, Topeliuksenkatu 41 a A, 00250 Helsinki, Finland.

Jack®: Transom Technologies, Inc., 201 South Main St., Suite 1000, Ann Arbor, MI 48104. Internet: http://www.cis.upenn.edu/~hms/jack.html.

Karwowski, W., Genaidy, A.M., and Asfour, S.S. (eds.) 1990. *Computer-Aided Ergonomics*, Taylor & Francis, New York.

MQPro: HumanCAD Systems, 3100 Steeles Avenue West, Concord, Ontario L4K 3R1, Canada. Internet: http://www.mqpro.com.

PeopleSize: Friendly Systems Ltd., 443 Walton Lane, Loughborough, LE12 8JX, England.

Safework: Les Consultants Genicom, Inc., 3400 de Maisonneuve West, Suite 1430, Montreal (PQ), H3Z 3B8, Canada. Internet: http://www.safework.com/index.html.

SAMMIE: SAMMIE CAD Ltd., Quorn, Loughborough, Leicestershire, UK.

WorkSmart Stretch Software: Ergodyne Corporation, 1410 Energy Park Dr., STE 1, ST Paul, MN 55108-9950.

30

Guide for Videotaping and Gathering Data on Jobs for Analysis for Risks of Musculoskeletal Disorders

David J. Cochran
University of Nebraska-Lincoln

Terry L. Stentz
University of Nebraska-Lincoln

Brian L. Stonecipher
University of Nebraska-Lincoln

M. Susan Hallbeck
University of Nebraska-Lincoln

30.1 Introduction

Videotaping jobs for a detailed ergonomic analysis at a later time is common practice. This is an especially effective method for documenting and evaluating the risk factors for musculoskeletal disorders (MSD). The quality of the analysis cannot be any better than the quality of the videotape and related data gathered. Therefore, it is critical that the videotaping be done correctly. The authors have videotaped and/or analyzed hundreds of jobs in industry. In that process we have learned what works and what does not. The following is a compilation of that knowledge. This guide used unpublished information developed by NIOSH (V. Putz-Anderson and D. Habes) circa 1991 but is primarily based on the experience of the authors. The methodology applies to a very wide variety of industry and job types. The flow or sequence of activities is presented in Figure 30.1.

30.2 Preparation and Equipment

Prior to actually taping the job, it is advisable to collect background information. This will normally include job title, work objective, work standard, major and minor tasks, the product, tools, equipment, workstation layout, environmental conditions, any rotation scheme, worker attributes, work flow, and the jobs just preceding and just following this job. It is also important to know the injury and illness history of the job so that the videotape will feature the appropriate views of the workstation and worker. This information is all included in the job analysis sheets included in this article. Schedule the taping with appropriate managers

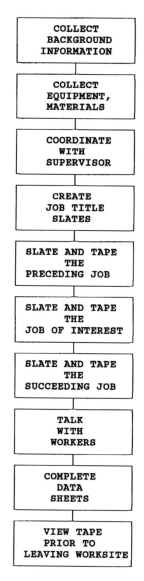

FIGURE 30.1   Flow chart of the activities involved in videotaping a job.

and supervisors and ask them to inform their workers that videotaping will take place. Make certain that the jobs to be taped will be performed during the scheduled taping session and will be typical of all of the work normally done. It is often constructive to have a meeting with the supervisor prior to taping the job to get as much of the information gathering done as possible and to iron out any potential difficulties. Prior contact with workers to be taped is advised to make them comfortable with the process and to assure them of the objectives. In some cases, the workers' permission is required before taping can begin.

Prior to entering the work area, it is advisable to create job title information slates or visual starters to introduce each job to be taped. This includes making a job title information slate for the job immediately before the job of interest and one for the job immediately after. With a dark marker in very large print (computer printing in large print also works very well) write the company, date, location, job name or title, shift, and any other pertinent information desired (see Figure 30.2). It is important that the job title information slate contain all of the pertinent information and be easily readable in the video.

XYZ COMPANY
DATE 6/6/66
TIME - 16:30
LOCATION - Z LINE
JOB NAME - HACKER
SHIFT - DAY

FIGURE 30.2   Job Title Information Slate.

In videotaping, two records are being created. The first is the videotape. The second is a log that is on paper or a voice recording. Maintain the log in the same sequence as the jobs on the videotape. This log is to contain information and notes about the job that will not be on the videotape. The log should have data sheets (discussed later) associated with each job attached to it. The person analyzing the tape must be able to follow the videotape and the log easily. Make certain that the job name is consistent on the videotape and the log.

Equipment that is necessary for taping includes:

Video camera with at least two charged batteries good for 2 hours each
    Date and time, including seconds, on the tape is highly recommended
An adequate supply of videotape
Objects of known length
    Dollar bills (6 inches)
    Yard stick with obviously alternating colors every 6 inches
Tape measure
Clipboard or tablet

Equipment that is useful but may not be absolutely required:

Dictating recorder
Force-measuring equipment
Stop watch
Light meter
Sound pressure meter
Thermometer
Air movement measuring equipment
Laptop computer
Reflective tape or dots to put on worker joints
Tripod or monopod

30.3   Videotaping Procedure

Be sure the light is adequate. If you are taping a worker wearing dark clothing against a light background the picture quality will generally be inadequate. To remedy this you may select another worker wearing lighter clothing, have the worker wear lighter clothing, or provide additional lighting.

Videotape the job title information slate for the job. If possible, verbally record all of the slate information on the audio portion of the tape while taping the slate. On-screen date, time, job name, or other useful information on the tape should be used if available. If this capability is limited, always put the time, including seconds, in the picture because the date is already on the job title information slate and these times can be useful when the tape is analyzed. Note on the log the job name and the first individual to be taped. If more than one individual is taped, list them in the order of taping with a very short description for future identification.

The objective is to get a good representation of all aspects of the job. Tape all tasks and subtasks of the job. Tape each task for 5 to 10 minutes and try to include at least 10 repetitions of all of its parts.

This is especially true for those angles that you consider most revealing of the task or the most important for future analysis. Sometimes when the task is long, there will be repeated cycles within it. In these cases it is important to get videotape of approximately 10 repetitions or cycles of everything included in each task and/or the entire job. Record all the idle time or other time between the cycles and/or the task activities. It is important that the camera be held as steady as possible. Where possible, use a tripod or at least a monopod. Another possibility is to lean or brace yourself or the camera against a solid, nonvibrating, object. Walking with the camera while recording is very detrimental to the quality of the videotape produced. Therefore, avoid walking with the camera while recording unless absolutely necessary. When walking and taping, slowness and steadiness are critical.

The sequence of activities in the actual taping is presented in Figure 30.3. Begin taping each task with a view of the entire workstation and then go to a whole-body view of the worker (Figure 30.4a). The seat or chair along with the standing surface should be included in this view. Two or more cycles are recommended before zooming in on the body area of interest. Resist the temptation to zoom in and out as the task progresses. A single view is almost always better. If the hands, arms, shoulders, neck, back, and/or hips are important, it is usually advisable to keep the whole upper body, including the buttocks, in view (Figure 30.4b). If the feet, legs, and/or hips are important, it is usually advisable to keep the whole lower body, including the buttocks, in view (Figure 30.4c). A more restricted viewing area is rarely useful. If it is useful, it should be supplemental to the larger body views. Videotape from angles which will allow determination of hand, wrist, arm, back, or other important postures. Tape from both sides and the front if possible. A view from above, as in Figure 30.4c, is helpful but often not possible.

Videotape the hand tools, jigs, fixtures, materials, parts, and personal protective equipment involved in the job. Videotape one or two cycles of the tasks immediately preceding and following the task in question. Where more than one worker is doing a job, record several of them. Try to tape workers who exhibit different experience or skill levels and those with a variety of body sizes and types.

In our experience, the same problems with the quality of videotapes occur repeatedly. The most frequently encountered bad characteristics are:

Too close
No complete workstation view
Not steady
Insufficient time or cycles
Move from one view to another too quickly or frequently
Frequent zoom in and out
No slate
No job before and after
Poor light or contrast
Audio narrative that is unintelligible
Back view
Walking or moving with the line
Lack of dimensional cues

30.4 Additional Data to Be Gathered

Even though the videotape lets the analyst see the job, it does not have all of the information needed to properly analyze a job and to propose viable solutions. Therefore, before, after, or during (if another person is available) the taping, it is useful to collect as much information about the job as possible. For this purpose six data sheets, contained in Figures 30.5 through 30.10, are included in this article. These data sheets attempt to structure all of the information the analyst might need. For any specific job, there may be blanks and whole data sheets that are not necessary or are not appropriate. If there is a certainty

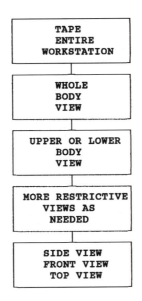

FIGURE 30.3   Flow chart of videotaping a job.

that the information is not going to be used or useful then don't collect it. Also there will be information that is pertinent to that job that is not addressed in the data sheets. These sheets are as complete as we can make them at this time. We recommend modifying or improving them for your specific needs.

The data sheets are, for the most part, self explanatory, but some of the items may need a little explanation. The Job/Task Information Data Sheet contains blanks for the basic information needed in evaluating the job. The "Work Typical" asks for a yes or no response as to whether the work being taped is what is normally done for that job.

For the Workstation Dimensions Data Sheet the forces can sometimes be measured directly. When this is not possible, they can be arrived at indirectly. This is done by having the workers exert a force on force-measuring equipment that they feel is equivalent to that exerted on the job or by having them estimate the forces and give this information verbally. When getting the force estimates indirectly it is best to get them from as many workers as possible and sometimes to do the job yourself and estimate the forces. It is also worthwhile to estimate what the percent of the maximum force is for those workers doing the job. This can be done by having the workers exert their maximum force on force-measuring equipment or by having them estimate the percent of their maximum force this job requires.

The Tool Data Sheet; Materials, Parts, Bins, Pallets, etc. Data Sheet; and Personal Protective Equipment Data Sheet are self explanatory. When possible, collect examples of tools, materials, parts, bins, and personal protective equipment to accompany the videotape. Having examples of the items used in a job accompany the videotape is preferable to completing the data sheet. However, listing them on the appropriate data sheets is still advisable. The sketches called for in these data sheets should have the dimensions on them.

The Maintenance Data Sheet is also self explanatory. The information gathered on this data sheet is very important. This information is rarely gathered and may have a tremendous impact on the job analysis and redesign. In our experience, lack of or poor maintenance has been the major cause of job or workstation problems, and dealing with that has been the only remedy necessary.

During or just after the videotaping process, it is useful to get input from the supervisor and employees. This should include their thoughts as to what parts of the job are causing difficulty, what the real problems are, and their ideas as to what the possible solution might be. Additionally, the Rated Perceived Exertion (RPE) (Borg, 1982) and a body part discomfort survey (Corlett 1976), can be used to determine how strenuous the job is and which parts of the workers bodies are affected.

FIGURE 30.4　Video frames of an individual at work. a) Seated view of the whole body that includes the workstation. b) Seated view of the upper body. c) Standing lower body view. d) Seated overhead and forward view.

30.5 Expected Results

The expected results of this process should be a videotape that is easy for a competent person to analyze. In addition, all of the details and information that cannot be put on a videotape will be available for the analyst. With this complete package the analysis will go smoothly with a minimum of delay and confusion. The analysis produced will not be deficient because of the documentation of the job. Additionally, the documentation created will be useful in follow-up and for demonstrating the original job. It is also useful to videotape and collect data about the job after it is changed. These materials can be used in training and demonstration.

FIGURE 30.4 (continued)

30.6 Summary

Video taping a job for ergonomic analysis is not a hit or miss operation. Care and planning are required and a systematic approach is recommended. The steps recommended in this paper are stated in the flow chart of Figure 30.1. Proper taping and data gathering will make job analysis for ergonomic purposes much easier and more valid. The care and time spent properly videotaping and gathering data about the job will pay off in the long run.

DATA SHEET #1 - JOB/TASK INFORMATION

JOB NAME(S): _____.

JOB LOCATION:　　Bldg._____.　Floor:_____.　　Department:_____.

　　　Line:_____.　　Work typical?_____.　　Exceptions _____

Brief Job Description:_____

_____

_____.

Preceding job:_____　Succeeding job:_____

Number of employees on this job:_____.　　Shifts:_____.

Control of work pace:　Worker:_____　　Line_____

Line speed :_____.

Effective line speed:_____ (pieces/minute or hour per worker).

Incentive or piece work? _____.

Work Time: (hours and or minutes)　Per Day:_____.

　　Per Week:_____　Schedule:_____.

Jobs rotated with_____ at _____ intervals.　If the rotation scheme is more complicated

describe it more completely here or on a separate page.

Temperature:　Area:_____　　　Product:_____　　Drafts:_____

PPE (gloves etc.):_____.

Tools:_____.

Materials:_____.

Training? _____

LIST THE TAPED WORKER'S NAMES OR IDENTIFICATION HERE, ON THE BACK, OR ANOTHER

PAGE.

FIGURE 30.5　Data Sheet #1 — Job/Task Information.

DATA SHEET #2 - WORKSTATION DIMENSIONS

JOB NAME(S): _____.

WORK SURFACE AND/OR WORKING AREA

1. Height:_____.   2. Depth:_____.   3. Width:_____.

4. Angle:_____.   5. Product Height:_____.   6. Hand Height:_____.

SEATING

7. Chair make and model:_____.

8. Adjustable?_____.   Height:_____   Range:_____.

                 Seat Pan:_____.   Range:_____.

                 Back:_____.   Range:_____.

                 Arm Rests:_____.   Range:_____.

9. Foot Rest: Needed?_____. Available?_____.   Adequate:_____.

10. Thickness of work surface:_____.   11. Leg room adequate?_____.

REACHES - Repeat 12 through 18 for all major reaches.

12. Description:_____.

13. Distance:_____.   14. Height:_____.

15. Angle (away from straight ahead):_____.

16. Angle (+ up and -down):_____.

17. Load Carried (actual weight when possible):_____.

18. Frequency:_____.

THROWS:

19. Distance_____ Height_____ Weight_____ Angle (away from straight ahead)_____.

FORCES (It may be necessary to use another sheet or the back of this one):

20. Push:_____   21. Pull:_____

22. Pinch:_____   23. Grasp:_____
23. Lifting or carrying:_____.

SKETCH THE WORKSTATION WITH DIMENSIONS ON ANOTHER PAGE:

FIGURE 30.6   Data Sheet #2 — Workstation Dimensions.

DATA SHEET #3 - TOOL

1. Name of tool:_____.

2. Job(s) used on:_____.

3. Tool weight?_____lbs.   4.   Balance?_____.

5. Handle(s):   Span:_____inches.   Length:_____inches.

                   Material:_____.   Texture:_____.

                   Pressure points?_____.

6. Counterbalanced?_____.   Needed?_____.   Appropriate tension?_____.

7. Place for the tool in the workplace (e.g., holster, fixture)?_____.

8. Is the tool powered?_____.

        Type?   Torque:_____.   Reciprocating/Vibrating:_____.

                   Other (describe):_____.

        Power source? Air?_____.   Exhaust away from the hand?_____.

                   Electric?_____.   Hydraulic?_____.

                   Other (describe):_____.

9. Vibration?   Present:_____.   Measured:_____.

10. Heat Source:_____ Temperature:_____.

11. Cold Source:_____ Temperature:_____.

12. Forces:_____

SKETCH TOOL WITH DIMENSIONS HERE:

FIGURE 30.7   Data Sheet #3 — Tool.

DATA SHEET #4 - MATERIALS, PARTS, BINS, PALLETS, ETC.

If the material or part described here is of a reasonable size and inexpensive ask for an example piece.

JOB NAME(S):_____.

1.   Name of object or material other than tools handled?_____

_____

2.   Describe:

               Name:_____

               Material:_____

               Size:_____

               Sharp Edges:_____

               Weight:_____lbs.

3.   Hot?_____. Temperature:_____.

4.   Cold?_____. Temperature:_____.

5.   Comments or additional description:

SKETCH THE PART WITH DIMENSIONS HERE.

FIGURE 30.8   Data Sheet #4 — Materials, Parts, Bins, Pallets, Etc.

DATA SHEET #5 - PERSONAL PROTECTIVE EQUIPMENT

1.   Name of equipment:_____.

2.   Purpose:_____.

3.   Job(s) used on:_____.

4.   Description:_____.

5.   Available Sizes:_____.

6.   Fit:_____.

7.   Material:_____.

8.   Texture:_____.

9.   Interference?_____.

Comments:_____

_____

_____.

SKETCH ITEM WITH DIMENSIONS HERE.

FIGURE 30.9   Data Sheet #5 — Personal Protective Equipment.

DATA SHEET #6 - MAINTENANCE

WORKSTATION

Nature: _____

_____

Frequency (now/required): _____

_____

Impact on the work/worker: _____

_____

TOOLS

Nature: _____

_____

Frequency (now/required): _____

_____

Impact on the work/worker: _____

_____

_____

PPE

Nature: _____

_____

Frequency (now/required): _____

_____

Impact on the work/worker: _____

_____

_____

FIGURE 30.10 Data Sheet #6 — Maintenance.

References

Borg, G. A. V. 1982. Psychological bases of perceived exertion. *Medical Science Sports Exercise* 14(5): 377
Corlett, E. N. and Bishop, R. P. 1976. A technique for assessing postural discomfort. *Ergonomics* 19(2): 175-182.

31

Physiological Instrumentation

Danuta Koradecka
Central Institute for Labour Protection
Warsaw

J. Bugajska
Central Institute for Labour Protection
Warsaw

31.1   Introduction

Physiological load of occupational physical work can be assessed on the basis of a rate of energy expenditure or oxygen consumption by the body. In many industrial tasks, it is also important to determine

0-8493-1802-5/03/$0.00+$1.50
© 2003 by CRC Press LLC

how much force must be generated by muscles to lift or support the weight of objects or to maintain the steady body position. However, these data are unsatisfactory when individual work tolerance has to be predicted. For this purpose, it is necessary to establish a relationship between the oxygen requirement for a given job to the individual maximal oxygen uptake or relation of force required to perform any task to the individual muscle strength (maximal voluntary contraction force). Moreover, many additional factors in the working environment, such as high or low ambient temperature, noise, etc., or psychological stress can influence work tolerance. Thus, the assessment of strain imposed by work and accompanying environmental factors should also include direct evaluation of the body responses to the work load or exposure to specific factors. Among them are indices of cardiovascular and respiratory system function, changes in body temperature, rate of sweating, muscle strength, muscle electrical activity (EMG), or indices of fatigue on the central nervous system such as disturbances in movement coordination, lowered level of arousal, etc.

31.2   Heart Rate (HR)

Description of the Parameter

Heart beat frequency (heart rate) is the most important and the simplest parameter of cardiovascular system function. Physical work, as well as many environmental stimuli, evoke an immediate acceleration of HR and, subsequently, an increase in the volume of blood ejected by the heart per unit of time (cardiac output). This response is usually followed by an increment in the amount of blood ejected during a single contraction of cardiac ventricles and changes in the distribution of blood flow among various organs due to the local constriction or dilatation of blood vessels. The changes in the cardiovascular system function allow the delivery of more oxygen and nutrients to active tissues, e.g., skeletal muscles, and the removal of metabolic waste products from the cells. The system also plays an important role in hormone transport from endocrine glands to target organs and in thermoregulation.

Specialized cells in the heart play a role of internal pacemakers due to their unique ability to produce electrical impulses initiating rhythmic cardiac muscle contractions. An acceleration of HR may be simply caused by increased blood flow to the heart, e.g., due to the dynamic contraction of skeletal muscles acting as pumps promoting venous return. However, the changes in HR may also be evoked by neural impulses from the autonomic nervous system and adrenal hormones, i.e., adrenaline and noradrenaline. There are two branches of the autonomic nervous system affecting the internal pacemaker of the heart: one of them — called sympathetic — increases HR, while the other — parasympathetic — exerts an opposite effect. Both branches of the autonomic system are under the control of the central nervous system and receive information from other brain centers and peripheral tissues. Circulating in blood, noradrenaline and adrenaline act in concert with the sympathetic nerves enhancing HR. These mechanisms adjust HR and, subsequently, cardiac output to the current needs. It is worth noting, however, that during psychological stress the increases in HR may exceed the real needs considerably.

At rest, under comfortable conditions, HR is usually 60 to 80 beats/min. It shows, however, spontaneous fluctuations of various frequency, known as physiological HR variability (HRV). Fast fluctuations of a frequency corresponding to the respiratory rhythm are caused mainly by the changes in the parasympathetic nervous system activity, while the slow fluctuations depend on both sympathetic and parasympathetic activities.

During dynamic exercise, HR is directly proportional to the intensity of the exercise. The increases in HR are greater during small muscle group exercise, e.g., arms, than during large muscle group involvement, e.g., legs, in spite of the same load. The maximal HR attained during exhausting efforts in young, healthy people usually exceeds resting values three times. The ability to reach the highest heart rate during maximal exercise decreases with age (Table 31.1). As a result, the same heart rate for a younger and older person implies different physiological load, imposed on the circulatory system in these people.

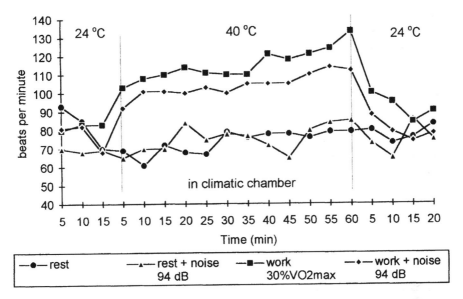

FIGURE 31.1 Comparison of heart rate at rest and during physical work (30% VO_{2max}) performed at 40°C (climatic chamber) accompanied by noise of 94 dB. (From Kurkus-Rozowska B. et al., Changes in blood pressure observed as a combined effect of heat and noise during physical load, *Arch. Complex Environmental Studies*, 8, 3-4, 11-16, 1996. With permission.)

Heart rate during exercise shows a high correlation with relative physical load expressed as the percentage of an individual maximal oxygen uptake (VO_{2max}). It has been shown that at 50% of maximal load (50% VO_{2max}), the heart rate of young, healthy men is 130 beats per minute, and at 30%, it is approximately 110 beats per minute. In women, a load of 50% VO_{2max} results in a heart rate increase up to 140 beats per minute. Stress and environmental stimuli, such as high or low temperature or noise, increase HR at rest and may exaggerate its response to exercise (Figure 31.1).

The increase in heart rate in response to acute stimuli is a result of an increase in sympathetic activity, a decrease in parasympathetic activity, or a combination of the two. The deceleration of heart rate occurring, e.g., during face cooling (diving reflex), results from an increase in parasympathetic activity, a decrease in sympathetic activity, or both (Papillo and Shapiro, 1990). Recently, spectral analysis of heart rate variability has come into use as a useful and sensitive index of autonomic function, particularly for the evaluation of mental stress (Pagani et al., 1991).

Description of the Parameter Measurement

Measurement of HR under various field conditions is quite easy. It can be made palpable, i.e., by putting a finger tip on an artery (e.g., carotid or radial artery) and counting the number of beats per minute. The method is based on the artery wall pulsation phenomenon. This is the most common method. However, it is not very precise and is not always possible to administer, e.g., during work requiring arm movements.

It is also possible to count heart rate by listening to the sound of the working heart with a stethoscope or using a microphone registering the acoustic signal, processing it, and displaying HR. This method can be used during physical effort.

Automatic manometers (of the so-called capacitive type) use the phenomenon of capacity changes connected with the filling of vessels according to the rhythm of the pulse rate. This information is electronically processed and HR is shown on a screen.

The electrocardiograph detects, amplifies, and prints out changes in the electrical potential between electrodes attached to the skin surface. These changes are caused by electrical impulses generated by

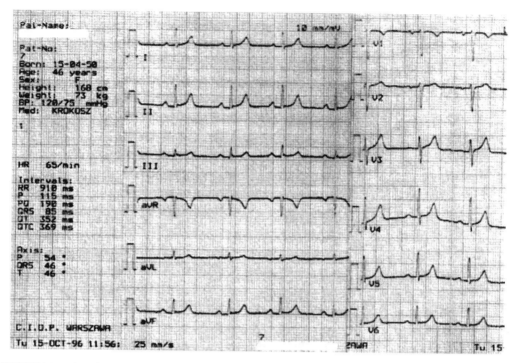

FIGURE 31.2   Sample ECG record with automatic measurement of heart rate calculated on the basis of R-R waves (in milliseconds).

cardiac pacemakers and spread through the heart. They show a regular pattern with the peak potential, called R waves, occurring as the impulse spreads through the heart ventricles. The time interval(s) between the two successive peak R waves (R-R interval) corresponds to the period of beat cycle and is the reciprocal of HR (60/R-R interval = beats/min). The electrocardiograph can be easily recorded during various activities using electrodes attached to the chest.

In the case of older ECG equipment, the measurement of the R-R intervals has to be performed manually (putting a special ruler to the ECG curve). In newer ECG equipment, this function is taken over by a processor and heart rate is displayed on a screen (Figure 31.2). In the ECG equipment which works in the Holter system, ECG signals are monitored and "remembered" for 24 hours. Statistical data can be processed by a computer, pathological changes can be assessed, and the course of this parameter over time can be shown (Figure 31.3).

Another possibility is to register heart rate in a given time with equipment in the form of a watch, which — via radio waves — registers signals from an electrode placed on the chest. This information can be sent through an appropriate interface to a computer. A computer program processes the data statistically and graphically (Figure 31.4).

Interpretation of the Parameter

Mean heart rate at rest is 60 to 80 beats per minute. If daytime mean heart rate falls below 50 beats per minute, it is a case of so-called bradycardia, which is considered a pathological symptom in people of average physical capacity. The phenomenon of functional bradycardia takes place in the case of people with high physical capacity, especially in endurance-trained athletes. It is connected with a great volume of blood ejected from the heart during each contraction, which — even with lower heart rate — allows the delivery of an adequate amount of blood to the tissues (the so-called athlete's heart). On the other hand, heart rate at rest exceeding 100 beats per minute should be considered a symptom of tachycardia. It is often connected with myocardial ischemia or neurohormonal hyperexcitability related to stress or some illnesses (e.g., hyperthyreosis).

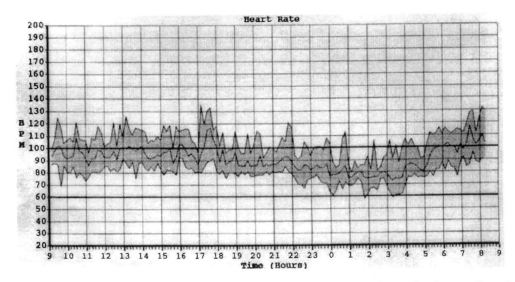

FIGURE 31.3 Sample record of the measurement of heart rate over 24 hours with the use of an electrocardiograph.

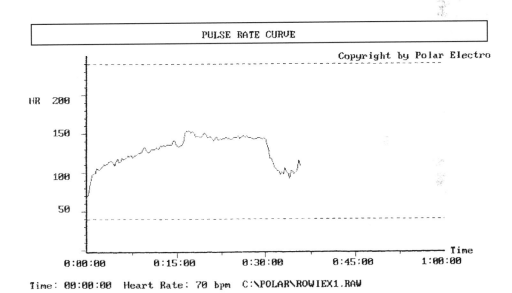

FIGURE 31.4 Sample record of a heart rate measurement performed during exercise with a watch-like device collecting the signal via radio waves from an electrode placed on the chest.

Maximal heart rate during exercise for healthy people can be estimated by subtracting the age of the examined person from the number 220.

31.3 Minute Pulmonary Ventilation (VE) and Respiratory Rate (RR)

Description of the Parameter

The respiratory and cardiovascular systems act in concert to provide a system of delivering oxygen to tissues and remove carbon dioxide from them. Respiratory rate (number of breaths during 1 min), tidal

TABLE 31.1   Correlation Between Maximal Heart Rate and Age

age (years)	Maximal Heart Rate per minute (HR_{max}/min)	
	mean	range
10	210	190–215
15	203	185–218
20–29	193	173–213
30–39	185	165–205
40–49	176	156–196
50–59	168	148–188
60–69	162	141–181
70–79	153	133–173
80–89	145	125–165

From Kozłowski St., Nazar K., *Introduction to Clinical Physiology,* PZWL 1995 (in Polish).

volume (TV), that is the volume of air inspired during a single breath (depth of breathing), and the product of these two variables — minute pulmonary ventilation — are the main parameters of respiratory system function. The diffusion of gases occurs in the pulmonary alveoli. It depends on the differences in partial pressures of oxygen and carbon dioxide between the alveoli and blood. The maintenance of balance in partial pressures of blood gases and blood pH requires coordination of the respiratory and cardiovascular systems. This is accomplished by involuntary regulation of rate and depth of breathing by the respiratory centers in the brain. These centers establish the breathing pattern by sending out impulses to the respiratory muscles. The respiratory centers receive information on the chemical environment in the body, from the areas of the brain sensitive to carbon dioxide and pH, and from the special receptors located in the aorta and carotid artery that are sensitive to partial pressure of oxygen and carbon dioxide in blood and its pH. Accumulation of carbon dioxide and the subsequent increase in body fluid acidity (drop of pH) appear to be the strongest stimulators of respiration. However, the respiratory system function may also be influenced by other neural factors, among them sensory impulses from working muscles, emotional distress, etc. There is also a possibility of voluntary control of breathing through the cerebral cortex. This can be, however, overridden by the involuntary controlling mechanisms.

Minute ventilation (V_E) increases in direct proportion to the effort intensity and oxygen uptake (VO_2) (Figure 31.5): up to 70% of maximal oxygen consumption (VO_{2max}) in people with low physical capacity, and up to 85% in people with high physical capacity. Exceeding this level results in hyperventilation, which is an increase in the proportion of ventilation to oxygen consumption. The beginning of hyperventilation coincides with an increase in carbon dioxide percentage in the expired air, an abrupt rise of the level of lactic acid in blood, and a decrease of blood pH. The above-mentioned phenomenon is illustrated in Figure the 31.6.

Description of the Parameter Measurement

Respiratory rate can be determined approximately by watching chest movements. However, this method is not applicable if there is little chest mobility during breathing. In order to precisely determine respiratory rate, spirometers — which measure changes of mechanical parameters on the basis of the air movement during inspiration or expiration — can be used.

Recently, electronic spirometers, in which air flow sensors measure breathing, have been introduced. Devices that measure man's metabolism by analyzing the expiratory gases can also measure respiratory rate.

In order to measure minute ventilation of the lungs, it is necessary to put on a face mask connected to a gas meter and to measure the volume of gas expired over a minute.

The measurement of ventilation in newer types of devices can be made by monitoring the respiration rate (RR) and tidal volume (TV) with an air flow sensor.

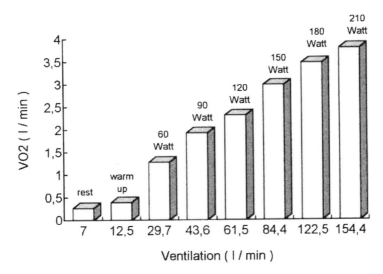

FIGURE 31.5 Sample of relationship between oxygen uptake (VO_2) and minute ventilation of the lungs (V_E) in a male subject during different physical load (in Watts). (From Konarska, M., Kurkus-Rozowska, B., Krokosz, A., and Furmanik, M., Application of pulmonary ventilation measurements to assess energy expenditure during normal and massive muscular work, *Proceedings of the 12th Triennial Congress of the IEA*, 3, Human Factors Association of Canada, 1994. With permission.)

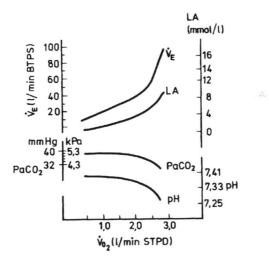

FIGURE 31.6 Changes in pulmonary ventilation, blood lactic acid, blood pH, and blood $PaCO_2$ in relation to oxygen uptake, during exercise of increasing intensity. (Adapted from Kozłowski St., Nazar K., *Introduction to Clinical Physiology*, PZWL 1995 [in Polish]. With permission.)

Interpretation of the Parameter

Mean respiratory rate at rest in healthy people is 10 to 20 breaths per minute. In small children, respiratory rate is higher. In adults, respiratory rate decreases with an increase of tidal volume.

During exercise, respiratory rate increases to 30 to 60 breaths per minute. The number of breaths is directly proportional to the intensity of the performed exercise, and it stabilizes at a fixed level after a few minutes of effort.

Respiratory rate exceeding 60 per minute has low efficiency from the point of view of pulmonary ventilation, while causing rapid fatigue of the respiratory muscles.

There is often an increase of respiratory rate per minute at rest in people with changes in the respiratory system related to a lower tidal volume, which can be caused by constriction of the respiratory tract, (e.g., asthma). This can also take place while breathing hot air and during illnesses that are accompanied by higher internal temperature.

The mean minute ventilation of the lungs at rest is 6 to 12 l/min. During exercise, minute ventilation can increase 25-fold in relation to ventilation at rest.

Maximal minute ventilation of the lungs (VE_{max}) during exercise is different for people of different physical capacity. In people of low physical capacity it is 70 to 90 l/min; in people of average physical capacity it is 110 to 130 l/min, in people of high physical capacity it is 15 to 160 l/min. In athletes maximal minute ventilation of the lungs can reach 200 l/min. In women, the maximal minute ventilation is lower than in men of the same age because of smaller respiratory volume. Maximal minute ventilation of the lungs decreases with age.

Minute ventilation of the lungs below the suggested standards can be the first signal of respiratory obduration related to the pathological changes in the respiratory system (e.g., among cigarette smokers).

31.4   Maximal Oxygen Uptake (VO_{2max})

Description of the Parameter

Maximal oxygen uptake is the ability to take oxygen during maximal physical exercise. It is the best indicator of the efficiency of the oxygen transport system in the body and the ability of tissues to utilize oxygen in metabolic processes. Oxygen consumption depends mainly on maximal cardiac output, the mass of skeletal muscles, and the capacity of aerobic biochemical processes yielding energy for muscular work. Maximal pulmonary ventilation, blood volume, the number of red blood cells, and hemoglobin content also play an important role in VO_{2max} limitation.

Maximal oxygen uptake (VO_{2max}) is a basic parameter for evaluating physical capacity because:

- It determines the amount of oxygen per minute which man is able to consume in order to satisfy the demand for oxygen during physical load. Exercise during which oxygen requirement exceeds VO_{2max} can be maintained only for a few minutes.
- It makes it possible to calculate the relative load expressed as % VO_{2max} (i.e., it makes it possible to calculate the load, which permits work for a long time without developing muscle fatigue and the accumulation of lactic acid in the blood).

Maximal oxygen uptake of a given person, expressed per kg of body mass per minute, allows evaluation of individual physical work capacity and classification of it as low, average, high, or very high (Table 31.2). The ability of oxygen uptake decreases with age (Figure 31.7).

The amount of oxygen uptake during submaximal effort correlates with relative load and, as a result, also with physiological parameters, e.g., heart rate (Figure 31.8). Physiological responses to exercise such as an increase of minute pulmonary ventilation, increase of heart rate, blood pressure and changes in peripheral blood flow are similar in various persons if the relative work load expressed as percentage of VO_{2max} is the same.

Description of the Parameter Measurement

The amount of maximal oxygen consumption is defined during maximal or supramaximal exercise as a result of a difference in oxygen concentrations in inspired air (oxygen concentration in the surrounding air) and in expired air. However, most often the value of VO_{2max} is predicted on the basis of the correlation of oxygen consumption and heart rate during submaximal exercise, according to Åstrand's nomogram (Åstrand and Rodahl, 1986).

Interpretation of the Parameter

The evaluation of maximal oxygen consumption according to sex, age, and physical capacity expressed as milliliters of O_2 per minute per kilogram is shown in Table 31.2.

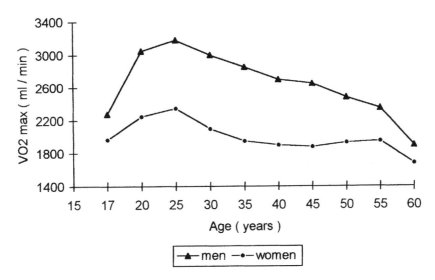

FIGURE 31.7 Maximal oxygen uptake (VO$_{2max}$) calculated on the basis of heart rate during exercise (submaximal) in male (a) and female (b) subjects. (From Kozłowski St., Nazar K., *Introduction to Clinical Physiology*, PZWL 1995 [in Polish]. With permission.)

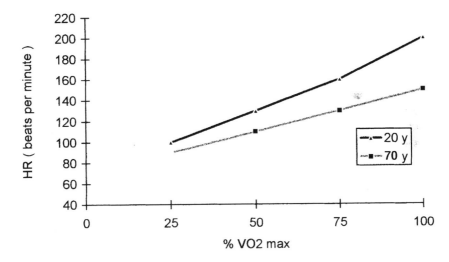

FIGURE 31.8 Relationship between heart rate (HR) and relative physical load (given as percent of maximum oxygen uptake by the body, %VO$_{2max}$) in subjects 20 and 70 years old. (From Kozłowski St., Nazar K., *Introduction to Clinical Physiology*, PZWL 1995 [in Polish]. With permission.)

31.5 Energy Expenditure

Description of the Parameter

Energy expenditure is the total energy liberated by the body in order to maintain life functions (basic or rest metabolism), body temperature at rest, and support muscle work during exercise. The amount of the energy used for basic metabolism depends on body mass, sex, age, and health. The amount of energy required to perform physical work depends on the mass of the involved muscles, body position during work, and work intensity.

TABLE 31.2 Maximal Oxygen Uptake for People of Various Ages in ml/min/kg

Sex/Age	Capacity				
	Very Low	Low	Average	High	Very High
Women					
20–29	28	29–34	35–43	44–48	>49
30–39	27	28–33	34–41	42–47	>48
40–49	25	26–31	32–40	41–45	>46
50–65	21	22–28	29–36	37–41	>42
Men					
20–29	38	39–43	44–51	52–56	>57
30–39	34	35–39	40–47	48–51	>52
40–49	30	31–35	36–43	44–47	>48
50–59	25	26–31	32–39	40–43	>44
60–69	22	22–26	27–35	36–39	>40

From Åstrand P-O., Rodahl K.: Textbook of Work Physiology, McGraw-Hill, N.Y. 1986.

Energy expenditure can be used to evaluate the physical load of work and to determine balance, i.e., an equilibrium between intake in the form of food and energy expenditure, e.g., when treating obesity.

Energy expenditure is expressed in W/m^2 per body surface of the subject (previously in kcal/min and kJ/min).

Description of the Parameter Measurement

Energy expenditure can be measured directly in calorimetric chambers or indirectly on the basis of oxygen consumption during exercise. The amount of energy expended can also be estimated on the basis of the measurement of minute ventilation of the lungs, which is in direct proportion to oxygen consumption (up to 75% VO_{2max} of a given person).

At present, oxygen consumption is measured with electronic oximeters. Energy expenditure calculated on the basis of oxygen consumption is the result of multiplying oxygen consumption (in liters per minute) by the oxygen energy equivalent, taking into account the respiratory quotient RQ (the ratio of carbon dioxide to oxygen in expired air). The result is expressed in kJ/min.

$$Energy\ expenditure\ (kJ/min) = VO_2(l/min) \times energy\ equivalent\ of\ O_2\ (for\ a\ given\ RQ)$$

By using equipment for measuring minute ventilation of the lungs, it is possible to determine energy expenditure on the basis of the simple relation between ventilation and oxygen consumption (Figure 31.9). Datta-Ramanathan's equation is the simplest and most commonly used:

$$energy\ expenditure\ (kcal/min) = VE\ (STPD) \times 0.21$$

where VE = minute ventilation of the lungs in l/min, STPD = standard conditions of dry gas volume for 0°C and for atmospheric pressure of 760 mm Hg.

Interpretation of the Parameter

Energy expenditure is an absolute measure of work load. However, the physiological load and, subsequently, work tolerance depend on the individual working capacity. The relation of oxygen uptake or the known oxygen requirement for a given task to VO_{2max} of the working person is often used for the assessment of physical strain in occupational work. The suggested physical work load is 30% VO_{2max} during a shift. Load which exceeds 50% VO_{2max} per shift can adversely affect the worker's health in the long run. Energy expenditure was measured and evaluated in industry, in a large number of subjects

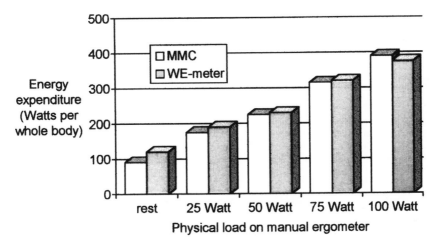

FIGURE 31.9 Comparison of the energy expenditure evaluation using MMC (Metabolic Measurement Card, on the basis of oxygen uptake) and WE-meter (electronic ventilation meter, on the basis of minute ventilation) during exercise on a manual ergometer. (From Konarska, M., Kurkus-Rozowska, B., Krokosz, A., and Furmanik, M., Application of pulmonary ventilation measurements to assess energy expenditure during normal and massive muscular work, *Proceedings of the 12th Triennial Congress of the IEA*, 3, Human Factors Association of Canada, 1994. With permission.)

TABLE 31.3 Work Load Classification on the Basis of Net Energy Expenditure in kJ/min

	Energy Expenditure (kJ/min)	
Physical Work	Men	Women
Light	8.4–20.5	6.3–14.2
Moderately hard	20.9–30.0	14.7–22.6
Hard	31.4–41.4	23.0–30.0
Very hard	41.9–51.9	31.4–39.3
Extremely hard	>52.0	>39.8

Christensen E. H.: Physiological evaluation of work in the Nykroppa iron works, in *Ergonomics Society Symposium on Fatique*. Floyd W.F. ed. Lewis, London 1953.

performing various tasks. The data can be found in the available literature (Passmore and Durnin, 1967). Because of that, energy expenditure evaluation and physical work load classification mostly apply to a "standard person." Net energy or oxygen requirement can be calculated for a shift (after subtracting the value of basic metabolism) and compared with the data in the work load classification table (Table 31.3).

31.6   Skin Temperature and Core Temperature

Description of the Parameter

Skin temperature and core temperature allow us to evaluate the efficiency of the thermoregulatory system in relation to changes in ambient temperature, exercise, etc. (Figure 31.10). This efficiency is important because of the necessity to maintain constant core temperature within a narrow range of temperatures for particular organs (e.g., approximately 36°C for muscles and approximately 37.2°C for the brain). The average body core temperature is 37°C. Core temperature varies according to the circadian cycle (approximately 37°C during the day and approximately 36°C at night). In small children, the temperature is slightly higher because of the smaller number of sweat glands. The average core temperature changes as a result of metabolism changes in some diseases (e.g., thyropathy).

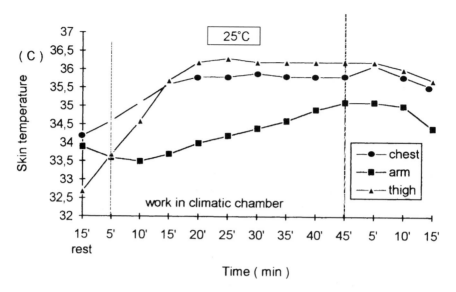

FIGURE 31.10 Sample of the changes of skin surface temperature during work (30% VO_{2max}) performed in oil protective clothing, at 25°C. (From Marszałek A., Sawicka A., Necessity to limit the duration of work performed in oil protective clothing, *Proceedings of 8th Congress of Polish Association of Occupational Medicine,* 1996 (in Polish).

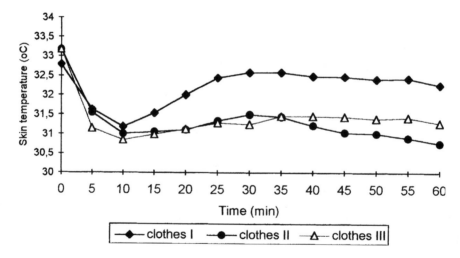

FIGURE 31.11 Time course of mean weighted skin temperature during exercise of 30% VO_{2max}, performed in three kinds of protective clothing. Clothes I = cold weather clothing for working in cold store; clothes II = cold weather and rain clothing for airport mechanics; clothes III = cold weather and rain clothing for construction workers. (From Marszałek A., Sołtyński K., Sawicka A.: Physiological evaluation of cold protective clothing, *Int. J. Occup. Safety & Ergonomics,* 1995, 1, 3, 235. With permission.)

Skin temperature is measured on the body surface. Local differences in skin temperature result from the skin blood flow difference. Because of that, skin temperature is measured in several places (from 3 to 10 points). The result in every spot is multiplied by the ratio characteristic for blood supply of a given body part. In this way, mean weighted skin temperature is calculated in accordance with the 8-point formula (ISO 9886, 1992). (Figure 31.11).

$$T_{sk} = 0.07T_{s1} + 0.175T_{s3} + 0.175T_{s4} + 0.07T_{s5} + 0.07T_{s6} + 0.05T_{s7} + 0.19T_{s10} + 0.2T_{s13}$$

where

T_{s1} = temperature of the forehead
T_{s3} = temperature of the right scapula
T_{s4} = temperature of upper left front part of the chest
T_{s5} = temperature of the upper right arm
T_{s6} = temperature of lower left arm
T_{s7} = temperature of the dorsum of the left palm
T_{s10} = temperature of the front part of the right thigh
T_{s13} = temperature of the back part of the left calf

Core temperature is influenced by ambient temperature and physical exercise. During physical exercise a considerable amount of thermal energy is produced. Out of 5 kcal produced during metabolism, only about 1 kcal is converted into work, and the rest into heat that has to be dispersed. Otherwise, core temperature would rise, and that would make further work impossible. It could also pose a life hazard (heat stroke).

The thermoregulatory mechanism utilizes the following physical phenomena:

- Redistribution of heat produced inside the body to skin through conduction and convection
- Redistribution of body heat to the environment or inversely through radiation, convection, and conduction
- Evaporation of sweat from skin and evaporation of water from pulmonary alveoli and mucous membranes

Description of the Parameter Measurement

Core temperature of sick adults is usually measured with a mercury thermometer in the armpit or in the mouth under the tongue. In the case of small children, a thermometer is usually put in the rectum.

Physiologists measure core temperature in the rectum, 5 to 8 cm deep, or in the esophagus or near the tympanic membrane using appropriate thermistor sensors or thermocouples. Skin temperature is measured with special thermistor sensors or thermocouples placed in specific places of the body.

Interpretation of the Parameter

Average core temperature measured with a mercury thermometer in the mouth is 37°C, in the armpit, 36.6°C.

Man's core temperatures below 27°C and over 42°C most often result in death. The rise of core temperature by more than 3°C, resulting from the impairment of the sweating mechanism or heat exchange problems, can be hazardous because of the disturbance of the functions of the circulatory system.

Average skin temperature measured in thermoneutral conditions is 34 to 36°C.

31.7 Sweating

Description of the Parameter

Sweating and sweat evaporation is the most effective way to remove heat surplus from the body. The efficiency of thermoregulation depends on the number of sweat glands in skin, the pace of sweat secretion, environmental conditions influencing sweat evaporation, and on the efficiency of the circulatory system responsible for transporting heat from working muscles to the skin. Sweat secretion increases in the 3rd to 5th second of exercise. First, in a reflex way, sweat glands placed on the trunk are activated. After 5 to 10 minutes of work, the number of active glands increases due to the rise of core temperature. After exceeding 60 to 70% maximal oxygen uptake (VO_{2max}), sweat secretion decreases as a result of lower flow of blood to the skin, which is related to the increase of blood flow in the working muscles.

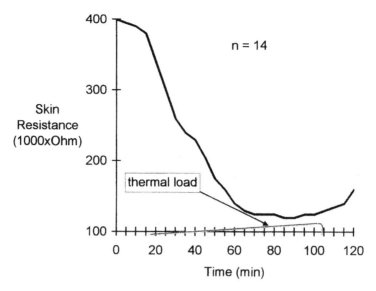

FIGURE 31.12   Changes of chest skin resistance R_s in subjects with exogenous thermal load (Adapted from Grucza R.: *Thermoregulation Model in Case of Endogenic and Exogenic Thermal Load of the Organism*, CMDiK PAN Warszawa, 1979, [in Polish].)

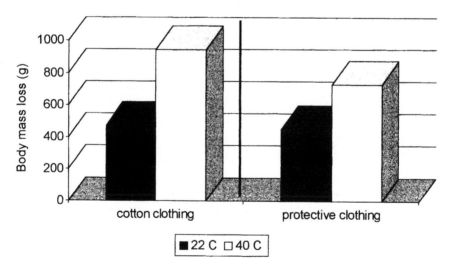

FIGURE 31.13   Sample of body mass loss (g) in subjects as an indicator of water loss when sweating during work at 30% VO_{2max}, in two kinds of protective clothing was compared with cotton clothing (control). (From Bugajska J. et al. Development of methods for evaluation of protective clothing for compliance with occupational physiology and hygiene, *International Journal of Occupational Safety and Ergonomics*, submitted for publication.

Description of the Measurement

Sweat can be measured with a device registering changes of skin resistance when sweating is low (Figure 31.12) or with hygrometers in the case of heavy sweating or high ambient temperature. Sweat rate can also be calculated indirectly on the basis of body mass loss measurement (Figure 31.13).

Interpretation of the Measurement

The ability to sweat depends on acclimatization to high temperature, age, race, etc. During maximal exercise the acclimatized person can secrete up to 1 liter of sweat per hour and 3 to 4 liters of sweat per working shift.

31.8 Tremometry

Description of the Parameter

Tremometry is a method used for measuring motor efficiency as far as hand tremor (trembling) and visual–motor coordination are concerned.

Method of Measurement

Testing is performed with a tremometer consisting of a desk in which holes of various shapes and sizes are cut out and a metal pen. The subject is to contour all cut-out shapes with a pen in such a way that side and bottom areas of the holes are not touched. The examination is made in the same position (standing or sitting) without resting the hand on the device's desk.

The result of the examination: time of performing the task, number of errors, time of errors.

Interpretation of the Measurement

Interpretation consists in comparing the results before and after a particular load factor is applied.

31.9 Critical Flicker Frequency Threshold

Description of the Parameter

Critical flicker frequency (CFF) threshold is considered to be an indicator of the level of stimulation of cortical centers, tiredness caused by work, and physiological changes resulting from the influence of alcohol or some pharmacological agents.

Method of Measurement

CFF is measured with the Flicker Test. The test is conducted with a device consisting of a measuring instrument and a black tube allowing both-eyed observation of the source of flickering light without any inflow of light from the outside. In the tube there will be a central flickering red light of 1 cm^2, shining with the intensity of 7.8 milliamperes. This light is surrounded by eight diodes, also red, shining with continuous light.

The range of flicker frequency changes is 0 to 99.9 Hz. The change of flicker frequency takes place automatically. The accuracy of the result is ±0.1 Hz.

The subject is asked to react (by pressing a button) when he/she notices that the flickering light becomes continuous (in the case of ascending threshold) or when the person notices that the continuous light has started to flicker (in the case of descending threshold). Ascending threshold of light flickering is measured during the increase of light flicker frequency from 30 Hz upward; descending threshold, on the other hand, is measured when flicker frequency decreases from 99.9 Hz downward.

The number of measurements depends on the examination; most frequently a six-fold measurement is used.

The result of the examination: light flicker frequency when the subject notices the beginning of flickering (descending threshold) or cessation of flickering (ascending threshold) measured in Hz/s.

Interpretation of the Measurement

Interpretation consists in comparing the results before and after a particular load factor is applied.

31.10 Muscle Fatigue

Description of the Parameter

The term fatigue is used to describe sensations of general tiredness and accompanying decrement in the muscle ability to generate force or power. There are many causes of fatigue, among them a failure of the muscle fiber contractile mechanism, exhaustion of muscle energy substrates, such as high-energy phosphates and glycogen, the accumulation of metabolic waste products (lactic acid, hydrogen ion, ammonia), and failure of the near-muscular impulse transmission are considered. The contribution of various mechanisms to the limitation of muscle working ability differ in various types of work and none of the above-mentioned causes alone can explain all aspects of fatigue. It is generally accepted that the central nervous system is also the site of fatigue. In fact, perceived discomfort, decrease of alertness, and coordination disturbances always accompany strenuous muscular work and even precede the physiological limitation in muscle performance.

Muscle fatigue can be estimated by methods which register the electrical activity of muscle (electromyographic signal — EMG) and the exerted force. A comparison of the measured parameters before, during, and after exercise can be used as an indicator of fatigue. EMG is used for the evaluation of fatigue of particular muscles, whereas the measurement of force is used for the evaluation of whole muscle groups, e.g., muscles of the upper or lower limb, or dorsal muscles.

Description of the Measurement

Electromyography

Muscle fatigue may be indicated on the EMG signal record as a shift of the signal power spectrum toward lower frequencies (Bugajska et al., 1993; Lindstrom et al., 1970). In such a case, the parameter determining the median power frequency (MPF) is used to evaluate muscle fatigue.

Moreover, muscle fatigue can be evaluated on the basis of the changes in the following EMG parameters: ZC (the number of signal crossing of zero, the so-called Zero Crossing) (Hagg, 1981) and the signal amplitude value (Lindstrom et al., 1970).

The values of the coefficients of slope of the regression line between parameters indicating muscle fatigue and the time of the measurement constitute an indicator of muscle fatigue (Roman-Liu, 1996; Sundelin et al., 1992). The cause of the changes of the value of the EMG signal parameters is not clearly determined. Production of lactic acid resulting in pH change in the studied muscles is indicated. The rise in potassium concentration in the intercellular fluid (Bigland-Ritchie et al., 1979) connected with fatigue is given as another cause.

Decrease of Maximal Force or Time of Its Maintenance

As a result of fatigue, the maximal exerted force or the time of maintenance of this force decreases. Dynamometric force measurements at certain time intervals can be used to obtain (as in the case of EMG signal) the coefficient of inclination of regression lines between the measured value and the time indicating muscle fatigue.

During dynamic activities such as a jump, a power indicator can also be used. The subject performs, with no breaks, several maximal vertical jumps. The reaction force of the base (measured on a dynamometric platform) is registered and on this basis the power of each bounce is calculated.

Power is defined by an equation of simple regression:

$$P = a + b \times t$$

where P = power of each cycle, t = time in which a given power is reached, and a, b = equation coefficients. Coefficient b, the angle of the slope of the line, is an indicator of fatigue of the lower limb muscles.

31.11   Muscle Strength

Definition of the Parameter

Muscle strength depends on the muscle size (the sum of the cross-sectional areas of every muscle fiber) and neural mechanisms that recruit fibers during muscle contraction. The more fibers are recruited simultaneously, the greater the force that can be generated. Moreover, the frequency of each fiber stimulation results in the force increment. The pattern of muscle fiber stimulation can be assessed by EMG. There is a close correlation between the force of contraction and integrated muscle electrical activity.

Muscle strength (Ważny, 1977; Zorski, 1985) indicates man's ability to generate force during muscle contractions. It can be also defined as the ability to overcome and oppose external resistance.

Muscle strength is characterized by structural and geometric as well as energy and information factors. All these factors also have an influence on muscle strength effect during performance. According to the equation of muscle contribution understood as the value of the moment of muscle strength (Fidelus, 1989):

$$M_z = \rho \sum_{i=1}^{n} p_i r_i(\alpha) \left\{ \frac{F_i}{F_{io}} \left[\frac{l_i}{l_{io}} (\alpha) \right] \right\} \frac{U_i}{U_{i\,max}}, \; [\mathrm{Nm}],$$

muscle strength effect is influenced by:

M_z = moment of external forces in relation to a human body
ρ = muscle tension identical for a given muscle group, $[\mathrm{Nm}^{-2}]$
p_i = physiological cross-section of the i-th muscle, $[\mathrm{m}^2]$
r_i = arm of the force of i-th muscle, $[\mathrm{m}]$
α = articular angle $[°]$

$\left\{ \dfrac{F_i}{F_{io}} \left[\dfrac{l_i}{l_{io}} (\alpha) \right] \right\}$ = dependence of the active force on the length of the i-th muscle, which is a function of the articular angle
F_o = value of the muscle active force when its length at rest is l_o
$U_i/U_{i\,max}$ = ratio of the current EMG value $[\mathrm{V}]$ of the i-th muscle to maximal value of EMG for a given muscle.

In order to solve the equation, all values, except for tension, must be known. The calculated value of muscular tension also contains a certain factor of proportionality between the mean values of the figures used in the equation and the actual figures of the subject.

In mechanics, force is understood as a vector of force acting on a given particle. According to Newton's second principle of dynamics, force:

$$m \times a = F$$

where m = mass (of the particle), a = acceleration of the body, F = force (Zorski, 1985).

In order to estimate muscular strength, it is best to measure the moments of strength developed by muscle groups in relation to a particular center of rotation; this results from the analysis of correlation between muscle strength and geometric parameters (Buśko et al., 1991).

Method of Measurement

Measurement can be performed in two ways: indirectly and directly. In the case of an indirect measurement, the force and its arm are measured, and then the following multiplication takes place:

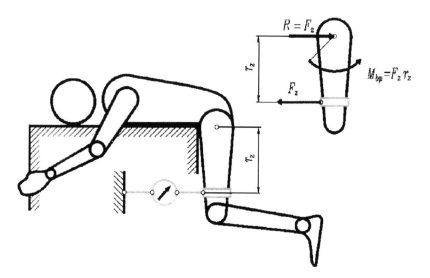

FIGURE 31.14   Measurement of the torque of extensors of the hip joint (M_{bp} — torque, F_z — external force, r_z — arm of external force, R — reaction force). (From Tokarski T. Dependence between muscle force and dynamic parameters developed on a dynamometric platform and bicycle, Doctoral thesis, Warsaw 1994. With permission.)

$$M = r \times F$$

where M = torque, r = arm of force (measured with a ruler), F = force (read from the dynamometer).

In the case of a direct measurement, the torque is measured with a torque meter.

During the measurements, the conditions in which the length of muscles allows the development of maximal torque are established. In most cases, muscles develop maximal torque in the range of 60 to 120° (Komi, 1979; Pieter et al., 1989). Not all authors, however, agree. This can result from using various methods of measuring the torque (Jensen et al., 1971; Kowalk et al., 1993). Most frequently, the measurement of moments of muscle forces is performed at 90° (Fidelus et al., 1984; Fugl-Meyer et al., 1980), in such a way that the center of gravity of the free rotating part (e.g., shank or forearm) is situated under or above the axis of rotation in a joint, so the torque resulting from weight of this part of the body is zero. It is easier to set the angle at 90° than, e.g., 60° or 120°; besides, when using the indirect method of calculating the torque, only one operation of multiplication of force and arm (sin 90° = 1) read from the dynamometer is made. (see Figure 31.14.)

Static conditions during the measurement are ensured by stabilizing the neighboring body parts.

Measuring Equipment

Indirect measurement of torque (Figure 31.14): stabilizing frame, stabilizing belts, dynamometer, ruler.

Direct measurement of torque (Figure 31.15): stabilizing armchair, torque meter, amplifier, digital voltmeter.

31.12   Body Position

Description of the Parameter

Body position is the position of individual body parts in relation to one another. Body position is defined by the angles of individual joints.

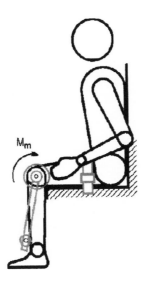

FIGURE 31.15 Measurement of torque of knee joint extensors (M_m). (From Tokarski T. Dependence between muscle force and dynamic parameters developed on a dynamometric platform and bicycle, Doctoral thesis, Warsaw 1994. With permission.)

There are several methods to evaluate the position at work. One is analysis of the posture at work with the use of audiovisual techniques. Such systems can analyze the movement of the human body recorded on a videotape. The system can consist of the following parts: video camera, videocassette recorder, computer, card for image processing, and a computer program for movement analysis (Macellari et al., 1983). Movement is recorded with a camera, which films the movement of individual body parts to which markers have been attached. Then, the image from the camera is converted into a digital form that can be read by a computer. As a result, it is possible to determine the position of all the markers, that is, individual body parts in space and time (Normand et al., 1983) (Figure 31.16). On the basis of the obtained values, computer analysis allows calculation of, among others, articular angles and the speed and angle acceleration in individual joints.

Another method is recording with position sensors placed on the body.

Description of the Measurement

Equipment with pendulum potentiometer sensors can be used for such measurements (Fig, 31.17). Sensors placed on the head, back, and arm make it possible to record and then analyze the deflection of the head, back, or arm from the position indicated as initial (Aaras and Stranden, 1988). The movement is analyzed in two planes, frontal and sagittal.

OWAS (Ovako Working Posture Analysis System)

OWAS (Karhu et al., 1986) can be used to make a direct and objective evaluation of load caused by the work posture. With the help of OWAS, a quantitative analysis of standard work postures can be performed, taking into consideration the external forces by classifying the position of the back, arms, and legs, and the external load. Combinations of positions of individual body parts (back, arms, legs), taking into consideration the external load, are grouped into four categories. This method can be used in examinations conducted directly at the workstation. By conducting an observation with a device for coding the work posture or with a video camera, it is possible to evaluate the time of holding and the frequency of changing a given work posture and to estimate the muscular load related to it.

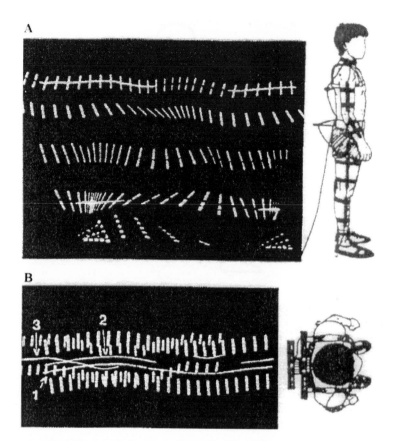

FIGURE 31.16   Position of individual body parts when moving, seen from the side (A) and from above (B) recorded with a light-emitted diodes and video cameras. (From Normand M. C., Richards C. L., Filion M., Dumas F., Tardif D.: A simplified method for tridimensional analysis of gait movements, in *Biomechanics IX-B,* Winter A. et al. (eds.), Illinois, 1983, 255-259.)

Maximum Time of Holding a Work Posture

For a full analysis of the working conditions, after defining and evaluating the work posture, holding time should be estimated for each of those postures. Maximum Holding Time (MHT) has been defined for several chosen positions (Karhu et al., 1986). MHT is the time a specific posture can be held (nonstop or with intervals) without the risk of fatigue or musculoskeletal system problems.

When there is a great variability of positions, MHT can be much higher than for a position held continuously. By comparing the holding time of a given position with MHT, it is possible to evaluate working conditions. Green, yellow, and red zones in a three-zone rating system of evaluation were marked in the following way to indicate safe holding time for each position:

> Green zone — holding time at 0 to 30% MHT
> Yellow zone — holding time at 30 to 50% MHT
> Red zone — holding time at over 50% MHT

UniTOR

UniTOR is a computer-aided method of recording and analyzing the work process. This is a computer program made for recording, statistically processing data, and documenting the work process in accordance with the TOR (Time — Object — Recording) method (Gedliczka et al., 1993). It is especially used in recording the process of current work.

FIGURE 31.17 Pendulum potentiometer sensors placed on the head, back, and arm.

References

Aaras A., Stranden E.: Measurement of postural angels during work. *Ergonomics*, 31, 5, 1988, 935-944.

Åstrand P-O., Rodahl K.: *Textbook of Work Physiology*, McGraw-Hill, N.Y. 1986.

Bigland-Ritchie B., Jones D., Woods J.: Excitation frequency and muscle fatigue, electrical responses during human voluntary and stimulated contractions, *Experimental Neurology*, 64, 1979, 414-427.

Bugajska J., Roman D., Koradecka D.: Analysis of fatique during repetitive manual work (hand-grip), in *The Ergonomics of Manual Work*, Marras W., Karwowski W., Smith J. L., Pacholski L., (eds.) Taylor & Francis, London, 1993.

Bugajska J., Szmauz-Dybko M., Sawicka A., Tokarski T.: Development of methodologies for evaluation of protective clothing for compliance with occupational physiology and hygiene. *International Journal of Occupational Safety and Ergonomics*, submitted for publication.

Buśko K., Musiał W., Wychowański M.: *Instructions to Exercises in Biomechanics*, Academy of Physical Education edition, Warszawa 1991.

Christensen E. H.: Physiological evaluation of work in the Nykroppa iron works, in *Ergonomics Society Symposium on Fatique*. Floyd W.F. (ed.) Lewis, London 1953.

Fidelus K., Urbanik Cz.: The influence of various types of muscle effort on the effect of strength and speed training, *Biol. Sport*, 1984, 1, 186-198.

Fidelus K.: *Biomechanic Outline of Physical Exercises*, part 1, Academy of Physical Education edition, Warszawa 1989.

Fugl-Meyer A. R., Gustafsson L., Burstedt Y.: Isokinetic and static plantar flexion characteristics, *Eur. J. Appl. Physiol.*, 1980, 44, 221-234.

Gedliczka A., Goralczyk A., Otręba R., Wolska A.: New method of timekeeping, the power of time-object-recording, in *The Ergonomics of Manual Work*, Marras W. et al. (eds.), Taylor & Francis, London 1993.

Greenleaf J. E.: Hyperthermia and exercise. *Internat. Rev. Physiol.: Environmental Physiology III T.*, 20, Robertson D. (ed.) University Park Press, Baltimore, 1979.

Grucza R. et al.: Dynamics of sweating in men and women during passive heating, *Europ. J. Appl. Physiol.*, 1987, 51, 309-314.

Grucza R.: *Thermoregulation Model in Case of Endogenic and Exogenetic Thermal Load of the Organism*, CMDiK PAN Warszawa, 1979, (in Polish).

Hagg G.: Electromyographic fatigue analysis based on the Number of Zero Crossings, *European Journal of Physiology*, 1981, 391, 78-80.

Hagg G., Suurkula J.: Zero crossing rate of electromyograms during occupational work and endurance tests as predictors for work related myalgia in the shoulder/neck region, *European Journal of Applied Physiology*, 62, 1991, 436-444.

Ikeda M., Sato K., Oshima M., Physiological workload of train drivers on a suburban commuter railway, *RTTI Raport*, 1988, Mae., 2, 3, 13-16.

Ishibashi Y., On the degree of fatique of workers in a sewing factory for three different ages, *Jap. J. of Science of Clothing*, 1982, Oct., 26, 1, 19-26.

Iwasaki T., Kurimoto S., Noro K., The change in colour critical flicker fusion (CFF) values and accommodation times during experimental repetitive tasks with CRT display screens, *Ergonomics*, 1989, 32 (3), 293-305.

Jansen A. A., de Gier J. J., Slanger J. L., Alcohol effects on signal detection performance, *Neuropsychobiology*, 1985, 14/2/, 83-87.

Jansen A. A., de Gier J. J., Slanger J. L., Diazepam-induced changes in signal detection performance: a comparison with the effects on the critical flicker-fusion frequency and the digit symbol substitution test, *Neuropsychobiology*, 1986, 16/4/, 193-197.

Jensen R. H., Smith G. L., Johnston R. C.: A technique for obtaining measurements of force generated by the hip muscle, *Arch. Phys. Med. Rehabil.*, 1971, 52, 201-215.

Karhu U., Kansi P., Kuorinka I.: Correcting working postures in industry: a practical method for analysis, *Applied Ergonomics*, 1986, 8, 199-201.

Komi P V.: Neuromuscular performance factors influencing force and speed production, *Scand. J. Sports Sci.*, 1979, 12, 417-466.

Komi P.: Electromyographic, mechanical and metabolic changes during static dynamic fatigue, in Knuttgen G. H., Vogel J. A. and Poortmans J., *Biochemistry of Exercise*, Vol. 13, pp. 197-215. International Series on Sport Science. Human Kinetics Publishers Inc., Champaign, IL, 1983.

Konarska M. et al. Application of pulmonary ventilation measurements to assess energy expenditure during manual and massive muscular work, *Proceedings of the 12th Triennial Congress of the International Ergonomics Association*, Vol. 3, Human Factors Association of Canada, 1994.

Kowalk D. J., Besser M. P., Vaughan Ch. L., Bowsher K. F.: Abduction — adduction moments of the knee during stair ascent and descent, *Abstracts of International Society of Biomechanics XIVth Congress*, 1993, Vol. 1, pp. 716-717.

Kozłowski St., Nazar K., *Introduction to Clinical Physiology*, PZWL 1995 (in Polish).

Kurkus-Rozowska B. et al. Changes in blood pressure observed as a combined effect of heat and noise during physical load, *Archives of Complex Environmental Studies*, Vol. 8, No. 3-4, pp 11-16, 1996.

Lindstrom L., Magnusson R., Petersen J.: Muscular fatigue and action potential conduction velocity changes studied with frequency analysis of EMG signals, *Electromyography*, 1970, 4, 341-356.

Łuczak A., Sobolewski A., 1995, The results of long-term observations of critical flicker frequency threshold (CFF), *Ergonomia*, 18, 2, 179-187.

Macellari V., Rossi M., Bugarini M.: Human motion monitoring using the CoSTEL system with reflective markers, in *Biomechanics IX-B*, Winter A. et al. (eds.), Illinois 1983, 260-264.

Marek T., Pieczonka- Błaszczyk W., Method of critical flicker frequency measurement and its accuracy, *Zeszyty Naukowe U. J.*, 1979, 29.

Marszałek A., Sołtyński K., Sawicka A.: Physiological evaluation of cold protective clothing, *Int. J. Occup. Safety & Ergonomics*, 1995, 1, 3, 235.

Matsumoto K., Sasagawa N., Kawamori M., Studies of fatigue of hospital nurses due to shiftwork, *Jap. J. of Industrial Health*, 1978, 20, 81-93.

Misawa T., Shigeta S., An experimental study of work load on VDT performance, part 1, Effects of polarity of screen and colour of display, *Jap. J. of Industrial Health*, 1986, Nov., 28, 6, 420-427.

Misawa T., Shigeta S., An experimental study of work load on VDT performance, part 2, Effects of difference in input devices, *Jap. J. of Industrial Health*, Nov., 1986, 28, 6, 462-469.

Ogiński A., Koźlakowska-Swigoń L., Pokorski J., Iskra-Golec, I., CFF as the strain indicator of control desks working in continuous motion, *Przegląd Lekarski*, 1981, 38, 9, 695-700.

Nadel E. R.: *Problems with Temperature Regulation During Exercise.* Academic Press, London 1977.

Normand M. C., Richards C. L., Filion M., Dumas F., Tardif D.: A simplified method for tridimensional analysis of gait movements, in *Biomechanics IX-B,* Winter A. et al. (eds.), Illinois, 1983, 255-259.

Nygaard E.: Woman and exercise–with special reference to muscle morphology and metabolism, in *Biochemistry of Exercise IV B,* Poortmans J. R., Niset G. (eds.) Univ. Park Press, Baltimore 1981.

Pagani M., Mazzuero G., Ferrari A., Liberati D., et al.: Sympathovagal Interaction During Mental Stress, *Circulation,* 1991, 83, (suppl. II).

Papillo J. F., Shapiro D.: The cardiovascular system, in *Principles of Psychophysiology-Physical, Social and Inferential Elements,* Cacioppo J. T., Tassinary L. G., (eds.), CUP, Cambridge, 1990.

Passmore R., Durnin J. U. G. A., *Energy, Work and Leisure,* Heinemann Educational Books, London, 1967.

Pieter W., Hijmans J., Taffe D.: Isokinetic leg strength of taekwondo practitioners, *Asian J. Phys. Education,* 1989, 12, 3, 55-64.

Rewerski W., Kozłowski St., Wróblewski T., Korolkiewicz K.: *Thermoregulation,* PZWL, Warszawa 1973.

Roman-Liu, D., Wittek, A., Kędzior, K., Musculoskeletal Load assessment of the upper limb positions subjectively chosen as the most convenient, *International Journal of Occupational Safety and Ergonomics,* 1996, 2, 4.

Seki K., Hugon M., Fatigue subjective et degradations de performance en environnement hyperbarea saturation, *Ergonomics,* 1977, 20, 2, 103-119.

Shephard R. J.: *Human Physiological Work Capacity.* CUP, Cambridge 1978.

Smolander J., Kolari P., Korhonen O., Ilmarinen R.: Skin blood flow during incremental exercise in a thermoneutral and hot dry environmental, *Europ. J. Appl. Physiol.,* 1987, 56, 273-280.

Sołtyński K. et al.: A method for determining thermal load in a high intensity radiation field of short duration, *Proceedings of the 12 Triennial Congress of the International Ergonomics Association,* Toronto, 1994.

Sołtyński K., Konarska M.: Body heat balance a man with deficient sweat rate subjected to physical work in hot environment, *International Journal of Occupational Safety & Ergonomics,* 1996, submitted for publication.

Sundelin G., Hagberg M.: Electromyographic signs of shoulder muscle fatigue in repetitive arm work paced by the methods-time measurement system. *Scand. J. Work, Environ. Health,* 1992, 18, 262-268.

Ważny Z.: *Training of Muscle Strength,* Sport i Turystyka, Warszawa 1977.

Ważny Z.: *Glossary of Athletic Training,* Academy of Physical Education edition, Warszawa 1994.

Zorski H. red.: *Technical Mechanics,* Ed. PWN, Warszawa 1985.

prEN 1005-4, Safety of Machinery — Human physical performance — Part 4: Working postures during machinery operation.

32

Video-Based Measurements of Human Movement

Peter M. Quesada
University of Louisville

32.1   Introduction

Quantitative analysis of human movement, via video motion measurement, can be a powerful means for addressing important ergonomics issues (e.g., potential modifications of occupational tasks, directed at preventing acute and/or cumulative trauma injury) (Berguer, 1997; Boston, 1995; Garg, 1991; Gracovetsky, 1990; Kumar, 1990, 1994; Lee, 1994; Pascarelli, 1993; Rudy, 1995). Statistics and correlations pertaining to injury incidence can suggest potential hazards of certain occupational tasks; however, statistical inferences cannot explain mechanisms by which injuries occur due to task performance. Identifying such mechanisms generally requires quantitative understanding of the kinematics, and often the kinetics, of a task. Actual knowledge of injury mechanisms is typically vital for effecting preventative modifications.

Video-based motion measurement is an effective and advantageous method of obtaining quantitative human movement data for a variety of ergonomics and biomechanics applications (Asato, 1993; Boninger, 1997; Cooper, 1996; Gracovetsky, 1989, 1995; Peterson, 1996; Robinson, 1993; Roosmon, 1993). Video-based motion measurements, which can provide absolute positional data for individual body segments, are commonly preferable to electrogoniometric measurements, which quantify only relative orientations of adjacent segments. Video motion measurement can also be less cumbersome for human subjects, and can allow individuals to perform tasks more naturally.

FIGURE 32.1   Depiction of subject with markers placed at anatomic locations. Solid circles represent markers that are visible in the view shown, while empty circles represent markers that are hidden in the view shown.

32.2   Overview of Video Motion Measurement Equipment

Cameras

Video cameras are the most obvious, if not most essential, items required for video motion measurement. A wide range of cameras are, or can be made, suitable for collecting motion data. Suitable cameras include both analog and digital types, as well as cameras with short-range, medium-range, and long-range lenses. The types of cameras available can influence the range of movement tasks that can be measured; or, alternatively, the need to measure a specific movement task can influence the selection of cameras to be used.

Image Capture Devices

For each frame of video data, two-dimensional camera coordinates for markers, placed on the body surface are obtained from each camera used. Two-dimensional data from a single camera may be sufficient to draw meaningful conclusions for a limited number of applications. More commonly, however, two-dimensional data from multiple cameras are integrated to determine three-dimensional positional data, for each marker, with respect to a global coordinate system (GCS), fixed to a given point in the testing object space. As with the types of cameras, the number of cameras available also affects the range of movement tasks that can be measured.

Body Surface Markers

Preparation of subjects involves placement of markers on a subject's body surface (Figure 32.1). Such markers generally do not substantially restrict subject movement, thereby permitting measurement of a wide range of movements. Markers are either positioned on palpable bony landmarks (e.g., lateral maleoli, anterior superior iliac spines, acromion processes, etc.) or attached to rigid fixtures that are placed on body segments (often in the vicinity of segment mass centers). Specific sites of marker or fixture placement are generally related to the biomechanical model that will be used to compute kinematic variables. Conversely, biomechanical model development can be affected by ease or difficulty of identifying specific marker placement locations.

Markers can be characterized as being either passive or active. Passive markers reflect light from sources attached, or in close proximity, to each camera. Both infrared and incandescent light are commonly used to illuminate passive markers. Lights can be positioned slightly above and behind a camera, or around the periphery of a camera lens (Figure 32.2). In these or other light/camera arrangements, the light source

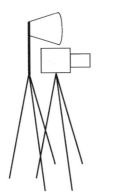

FIGURE 32.2 Depiction of light/camera arrangements for illuminating passive markers. On the left is a single light, positioned just above and behind a camera. The right side depicts an array of lights around the periphery of a camera lens.

must be out of the view of the camera in whose vicinity the source is positioned. Suitable arrangements do not necessarily require, however, that each light source be out of view for all cameras. Incandescent lighting is generally held constant, while infrared lighting is often strobed. Active markers generate their own light to be viewed by cameras. Active markers can be more cumbersome than passive markers if lead wires are needed to deliver power. With some video motion systems, active markers are strobed as a means of providing automated marker identification. Colored lights have also been used to assist or automate marker identification. In addition to colored lights, light emitting diodes and infrared lights have been used as active markers.

Load Measurement Devices

Computation of kinetic variables associated with human movement commonly involves use of various load measuring devices and application of inverse dynamic formulations. Force platforms are among the most common devices used for collecting raw kinetic data. These devices measure the forces imposed on them, generally through the feet, in three component directions, as well as the moments of force about three axes oriented along the measured force directions. Platform force and moment measurements can be used to determine the location of the resultant force's point of application (Figure 32.3). The location of this point is essential if resultant force data is to be integrated with video-generated kinematic data to estimate internal loading variables such as joint moments and contact forces. A variety of devices are available for collecting kinetic data associated with body segments other than the feet. Uniaxial load cells can measure simple tensile and compressive loads, such as the net load on the hands due to pulling on a rope. Pressure sensitive mats can be used to record the load distribution at a seating interface. Additionally, and somewhat more indirectly, accelerometers placed on an object of known mass can record object acceleration data that are suitable for determining loads placed on an object by an individual.

32.3   Overview of Video Motion Measurement Tasks

Collection of three-dimensional video motion data begins with a camera calibration procedure. The purpose of this process is to determine the position and orientation, with respect to the GCS (Figure 32.4), of each camera to be used. Placement of body surface markers generally follows camera calibration or proceeds simultaneously, if sufficient personnel are available. With cameras calibrated and markers placed, video data collection proceeds as the subject performs the task(s) to be studied. Initial processing then reduces raw video data (i.e., two-dimensional coordinates for each marker, in each camera's image plane)

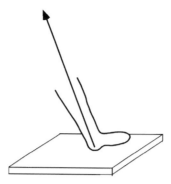

FIGURE 32.3   Depiction of a resultant force measurement with a force platform. The length of the vector indicates the magnitude of the force, the orientation indicates the force's direction, and the origin of the vector indicates the point of load application.

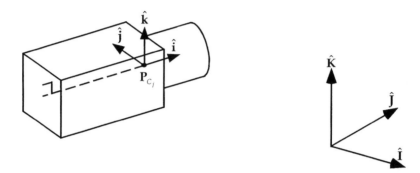

FIGURE 32.4   Depiction of a camera with its position, $\mathbf{P}_{C_j}$, and LCS axes directions indicated with respect to a GCS.

to generate three-dimensional coordinates, with respect to a GCS, for each marker at every sampling interval. Based on these three-dimensional marker coordinates, as well as estimated coordinates of body locations without markers, biomechanical variables are then computed. Computation of motion variables is generally the final step in the video measurement process. Variables can be formulated to quantify both temporal and spatial characteristics of a movement task.

32.4   Calibration of Object Space

Successful collection and reduction of video motion data requires a quantitative knowledge of the location and orientation of each camera to be used. Consequently, execution of a calibration procedure invariably accompanies sessions of video motion data collection, to determine camera location and orientation information (Figure 32.4). Camera location can be described as the position, in the GCS, of a characteristic point (e.g., the focal point). Similarly, camera orientation can be quantified by the orientation/direction of a local coordinate system (LCS), which typically has one axis perpendicular to the image plane. Traditionally, camera calibration has involved imaging of several markers, whose three-dimensional coordinates are known with respect to the GCS, and that are arranged to fill the object volume (Figure 32.5). Early variations of this technique require that each marker be identified in each camera view, while more current applications have markers identified automatically. In either case, camera locations and orientations are determined once marker identification has been completed.

   The mechanical process of calibrating video motion systems has changed little (until very recently) as technology for video motion measurement has progressed from using cine film and other similar photographic means, to using video cameras for acquiring raw movement data. Advancements in computational algorithms, however, have provided substantial improvements in both accuracy and data processing

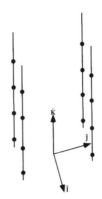

FIGURE 32.5   Depiction of a calibration, reference marker arrangement.

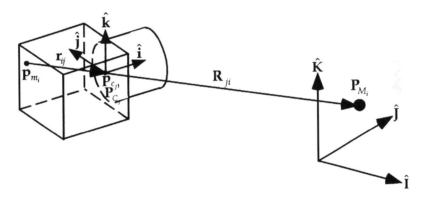

FIGURE 32.6   Depiction of LCS-based vector, $\mathbf{r}_{ij}$, from marker image to camera focal point, and GCS-based vector, $\mathbf{R}_{ji}$, from focal point to marker location.

time associated with camera calibration (Abdel-Aziz, 1971; Marzan, 1975; Woltring, 1976; Whittle, 1982). A substantial volume of software, that incorporates the more basic algorithms, is available in the public domain. Many of the more sophisticated algorithms, however, have been commercialized as the demand for software products to perform camera calibration as well as marker coordinate generation has grown. Consequently, details of current calibration and coordinate generation algorithms will not be discussed here.

Complete mastery of camera calibration details is unnecessary for most exercises in video motion measurement. An overview of the principles involved with the process, however, should assist in arranging appropriate camera configurations and calibration object volumes. To understand, fundamentally, the camera calibration process, one can visualize a pair of vectors, associated with marker, i, that is of interest (Figure 32.6). The first vector, $\mathbf{r}_{ij}$, originates from the projected image of the marker, $\mathbf{p}_{m_i}$, on the backplane of camera, j (i.e., image plane, j), and terminates at the camera focal point. The second vector, $\mathbf{R}_{ji}$, originates from the focal point and terminates at the marker, $\mathbf{P}_{M_i}$. Vector $\mathbf{r}_{ij}$ and point $\mathbf{p}_{m_i}$ are described with respect to an LCS, which is attached to the camera and defined by three mutually perpendicular unit vectors, two unit vectors lying in the image plane and the third being perpendicular to the image plane. Vector $\mathbf{R}_{ji}$ and point $\mathbf{P}_{M_i}$ are described for a GCS, fixed to the laboratory space. These vectors are thus expressed as:

$$\mathbf{r}_{ij} = \mathbf{p}_{c_j} - \mathbf{p}_{m_i} \quad (\text{in LCS}) \tag{32.1a}$$

$$\mathbf{R}_{ji} = \mathbf{P}_{M_i} - \mathbf{P}_{C_j} \quad (\text{in GCS}) \tag{32.1b}$$

where $\mathbf{p}_{c_j}$ and $\mathbf{P}_{C_j}$ designate the camera focal point in the LCS and GCS, respectively. For ideal flat camera lenses and backplanes, these vectors can be rendered colinear by rotating the LCS to be parallel with the GCS. This process is mathematically accomplished by multiplying $\mathbf{r}_{ij}$ by an appropriate rotation matrix, $[\mathbf{M}_{R_j}]$(that is a function of variables describing the relative orientation of the of the LCS with respect to the GCS):

$$\mathbf{r}'_{ij} = \left[\mathbf{M}_{R_j}\right]\mathbf{r}_{ij} \tag{32.2}$$

Subsequently, the GCS vector is directly proportional to the rotated LCS vector:

$$\mathbf{r}'_{ij} = k_{ij}\mathbf{R}_{ji} \tag{32.3a}$$

$$\left[\mathbf{M}_{R_j}\right]\mathbf{r}_{ij} = k_{ij}\mathbf{R}_{ji} \tag{32.3b}$$

$$\mathbf{r}_{ij} = k_{ij}\left[\mathbf{M}_{R_j}\right]^{-1}\mathbf{R}_{ji} \tag{32.3c}$$

$$\left[\mathbf{P}_{c_j} - \mathbf{P}_{m_i}\right] = k_{ij}\left[\mathbf{M}_{R_j}\right]^{-1}\left[\mathbf{P}_{M_i} - \mathbf{P}_{C_j}\right] \tag{32.3d}$$

where k_{ij} is a constant pertaining to the combination of marker, i, with camera, j. This vector relationship yields three component equations for a given marker, i, and camera, j. Marker position, $\mathbf{P}_{M_i}$ (in the GCS) is known, while location of the projected marker image, $\mathbf{p}_{m_i}$ (in the LCS), is measured for most traditional camera calibration protocols. Unknown variables include: global camera position components (constituting $\mathbf{P}_{C_j}$), the camera orientation parameters (contained within $[\mathbf{M}_{R_j}]$) and the proportionality constant, k_{ij}. One or more of the LCS components of $\mathbf{p}_{c_j}$ may also be unknown in some calibration algorithms. Consequently, the equations generated by a single marker are insufficient to determine the unknown variables associated with a camera's location and orientation. Fortunately, each additional reference marker will generate three similar vector component equations, while introducing an additional unknown, k_{ij}. Continued addition of reference markers will, ultimately, yield a sufficient number of equations to estimate the location and orientation of camera, j. In most applications the number of reference markers exceeds the minimum required, and, thus provides some redundancy to the process. The large number of equations, typically involved, generally dictate the use of numerical algorithms to effect a solution. The process is repeatable for each camera to be used in the desired video motion measurement task.

Estimations of GCS coordinates for unknown marker positions tend to be better when markers are near the positions of reference markers used during calibration (Figure 32.7). Consequently, reference markers are typically arranged to fill, as much as possible, the space to be occupied by markers during data collection.

The process of determining unknown GCS coordinates of markers, imaged during video motion measurement, utilizes the same vector relationship (Equation 3) described for camera calibration. Application of this relationship at each desired sampling interval will yield, as unknown variables, the marker GCS coordinates as well as a value of k_{ij} for a given camera, j, with a view of marker, i. The number of unknowns generated, thus exceeds the number of equations available for a single application of this relationship. Generation of sufficient equations to solve for all unknowns associated with a marker at a specific sampling interval, requires at least two applications of the vector relationship, with one occurring for each camera with a view of the marker. Consequently, estimation of a marker's GCS coordinates requires that the marker be visible to a minimum of two cameras.

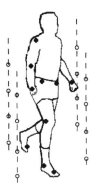

FIGURE 32.7   Depiction of subject with markers moving through a calibrated object space. Empty circles represent locations of reference markers that were present during calibration. Some body markers (e.g., left wrist) are close to reference marker locations, while others (e.g., right shoulder) are rather far from reference locations.

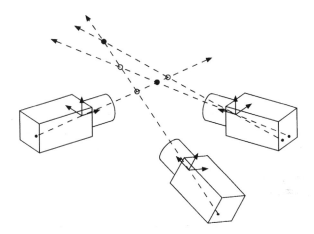

FIGURE 32.8   Depiction of markers (larger solid circles) imaged by multiple cameras. Vectors originating from images of markers (smaller solid circles), on camera image planes, cross at marker locations; however they can also cross at other "ghost" marker locations, indicated by empty circles.

The process of determining the GCS coordinates, for a given marker at a specified sampling interval, can be described graphically as identifying the intersection of two or more vectors in three-dimensional space (Figure 32.8). Each vector originates from a camera's backplane image of the marker, passes through the camera focal point, and continues indefinitely through the marker and into three-dimensional space.

This graphical representation visually demonstrates the process of determining GCS marker coordinates; however, it can also suggest the potential for identifying false, or "ghost" markers when vectors passing through different markers unintentionally intersect with each other (Figure 32.8). Generation of ghost markers tends to complicate further processing of marker coordinate data to compute kinematic variables associated with the movement tasks being measured. Ghost markers can occur when multiple cameras are positioned and directed in a common plane that multiple markers tend to occupy during measured movements. This undesirable effect can be largely avoided by arranging cameras to be positioned and directed in different planes.

Recently, a different calibration technique has been developed in which a rigid rod of known length and with two markers attached at the ends is moved throughout the desired object volume while camera data are collected. Considering some of the basic concepts developed previously, one can perhaps appreciate the potential strengths of this technique, such as the potential to generate a large number of calibration points and the capacity for those points to fill a large portion of the object volume. The computational algorithms for this calibration technique involve somewhat more sophisticated programming logic and will not be discussed further.

32.5   Placement of Body Surface Marker

Placement of body surface markers generally follows camera calibration or proceeds simultaneously if sufficient personnel are available. Two approaches to marker placement are commonly used. In one approach, markers are placed individually at specific anatomical locations (Kadaba, 1990). In the other approach, a cluster of markers (three or more) is attached to a rigid or semirigid fixture that is placed on a body segment (Antonsson, 1989). Selection of locations for marker placement is dictated by the ease with which such locations can be reproduced and/or the biomechanical modeling for which marker coordinate data will serve as input. When applying individual markers, reproducibility of marker placements is particularly essential if comparisons are to be made between data from different subjects or subject groups, or between data from a single subject acquired under different conditions or on separate occasions. Typically, individual marker placements are most repeatable when they are located directly over bony prominences, such as the lateral malleolus.

32.6   Collection and Initial Processing of Video Motion Data

With cameras calibrated and markers placed, video data collection proceeds as the subject performs the task(s) to be studied. Once motion data are collected, the process for reducing raw video data to generate marker coordinate information will depend on the type of system used to acquire the data. The data reduction process will typically include marker identification (either manual or automated identification, or a combination of the two) and reconstruction of image plane marker coordinates (performed via software). The order in which these two elements are performed is often the factor that determines the amount of time needed to complete initial data reduction. Earlier motion systems required that markers be identified from the perspective of each camera used, with three dimensional reconstruction following marker identification. For many of the more recent systems, however, three dimensional reconstruction is performed first, followed by marker identification from any arbitrary perspective. The computational basis for determining three-dimensional marker coordinates from two-dimensional image plane data is essentially the same whether marker identification precedes or follows reconstruction.

32.7   Computation of Biomechanical Variables

With three-dimensional marker coordinates determined for each frame of video data, biomechanical variables can be computed. For some applications computations are based upon surface movement (i.e., coordinate trajectories of the markers placed on the body surface), while for other purposes, calculations are based upon skeletal movement. This latter approach requires that three-dimensional coordinates be determined for skeletal locations (typically joint centers) at which markers cannot be placed. Surface based calculations may be sufficient when motion is desired in just a single place (e.g., sagittal plane). Quantification of movement in multiple planes, however, generally requires application of skeletal-based computational techniques (Kadaba, 1990; Vaughan, 1992).

Estimation of coordinate data for skeletal locations, often referred to as "virtual markers," is commonly based on mathematical models involving surface marker coordinates, at times in conjunction with previously estimated coordinates for other virtual markers. Such models may take the form of statistical regression equations that can be evaluated directly, or multiple constraint equations requiring sequential or simultaneous solution (Kadaba, 1990; Vaughan, 1992). Some of these approaches permit estimation of virtual marker coordinates for a given frame using other marker data for the same frame only. Other techniques exist, however, for which virtual marker coordinates are determined for a given frame using marker data from both preceding and succeeding frames.

A commonly employed technique for computing hip joint center coordinates is a typical example of using regression-based relationships to approximate virtual marker coordinates. This technique involves placement of a marker at each of the anterior superior iliac spines (ASISs) and placement of a third marker at the sacrum or at the end of a wand, such that the marker lies in the plane formed by sacrum and the ASIS markers. Three unit vectors are then determined as follows:

$$\hat{\mathbf{j}}_{pelvis} = \frac{\mathbf{P}_{M_{left\ ASIS}} - \mathbf{P}_{M_{right\ ASIS}}}{\left| \mathbf{P}_{M_{left\ ASIS}} - \mathbf{P}_{M_{right\ ASIS}} \right|} \tag{32.4a}$$

$$\hat{\mathbf{k}}_{pelvis} = \frac{\left(\mathbf{P}_{M_{left\ ASIS}} - \mathbf{P}_{M_{right\ ASIS}} \right) \times \left(\mathbf{P}_{M_{sacral}} - \mathbf{P}_{M_{right\ ASIS}} \right)}{\left| \left(\mathbf{P}_{M_{left\ ASIS}} - \mathbf{P}_{M_{right\ ASIS}} \right) \times \left(\mathbf{P}_{M_{right\ ASIS}} \right) \times \left(\mathbf{P}_{M_{sacral}} - \mathbf{P}_{M_{right\ ASIS}} \right) \right|} \tag{32.4b}$$

$$\hat{\mathbf{i}}_{pelvis} = \hat{\mathbf{j}}_{pelvis} \times \hat{\mathbf{k}}_{pelvis} \tag{32.4c}$$

where $\mathbf{P}_{M_{right\ ASIS}}$, $\mathbf{P}_{M_{left\ ASIS}}$, and $\mathbf{P}_{M_{sacral}}$ are vectors containing three-dimensional coordinates in the GCS of the right ASIS, left ASIS, and sacral markers, respectively. The vectors, $\hat{\mathbf{i}}_{pelvis}$, $\hat{\mathbf{j}}_{pelvis}$, and $\hat{\mathbf{k}}_{pelvis}$, are unit vectors (i.e., magnitude of each equals unity) that form a cartesian coordinate system describing the orientation of the pelvis with respect to the GCS. The positions of the hip joint centers, $\mathbf{P}_{M_{right\ hip\ center}}$ and $\mathbf{P}_{M_{left\ hip\ center}}$, can be described as

$$\mathbf{P}_{M_{right\ hip\ center}} = \mathbf{P}_{M_{right\ ASIS}} + \left| \mathbf{P}_{M_{right\ ASIS}} - \mathbf{P}_{M_{left\ ASIS}} \right| \left(-a_1 \hat{\mathbf{i}}_{pelvis} + a_2 \hat{\mathbf{j}}_{pelvis} - a_3 \hat{\mathbf{k}}_{pelvis} \right) \tag{32.5a}$$

$$\mathbf{P}_{M_{left\ hip\ center}} = \mathbf{P}_{M_{left\ ASIS}} + \left| \mathbf{P}_{M_{right\ ASIS}} - \mathbf{P}_{M_{left\ ASIS}} \right| \left(-a_1 \hat{\mathbf{i}}_{pelvis} - a_2 \hat{\mathbf{j}}_{pelvis} - a_3 \hat{\mathbf{k}}_{pelvis} \right) \tag{32.5b}$$

where a_1, a_2, and a_3 are constants equal to the fractions of the inter-ASIS distance that an ASIS marker position must be translated in the respective unit vector directions, in order to arrive at a hip joint center.

As indicated, determination of virtual marker locations via application of multiple constraints is also possible. Knee joint centers are often approximated in such a manner, following identification of hip joint centers as described previously. This approach typically involves coordinate data for two markers in addition to hip joint center coordinates. One marker (with coordinate vector, $\mathbf{p}_{M_{right/left\ knee\ marker}}$) is placed on the lateral knee joint such that it lies on the assumed joint axis extended. The other marker (with coordinate vector, $\mathbf{p}_{M_{right/left\ femoral\ marker}}$) is placed on the femur or at the end of a femoral wand such that it lies approximately in a plane defined by the assumed rotational axis and the hip joint center (recall that a vector and a point off the vector uniquely define a plane). Considering the right lower limb for this discussion, a unit vector, $\hat{\mathbf{u}}_I$, perpendicular to the plane is first computed as

$$\hat{\mathbf{u}}_I = \frac{\left(\mathbf{P}_{M_{right\ hip\ center}} - \mathbf{P}_{M_{right\ knee\ marker}} \right) \times \left(\mathbf{P}_{M_{right\ femoral\ marker}} - \mathbf{P}_{M_{right\ knee\ marker}} \right)}{\left| \left(\mathbf{P}_{M_{right\ hip\ center}} - \mathbf{P}_{M_{right\ knee\ marker}} \right) \times \left(\mathbf{P}_{M_{right\ femoral\ marker}} - \mathbf{P}_{M_{right\ knee\ marker}} \right) \right|} \tag{32.6}$$

Next, two intermediate unit vectors, $\mathbf{u}'_{II}$ and $\mathbf{u}'_{III}$ are determined as

$$\hat{\mathbf{u}}'_{II} = \frac{\left(\mathbf{P}_{M_{right\ hip\ center}} - \mathbf{P}_{M_{right\ knee\ marker}} \right)}{\left| \left(\mathbf{P}_{M_{right\ hip\ center}} - \mathbf{P}_{M_{right\ knee\ marker}} \right) \right|} \tag{32.7a}$$

$$\hat{\mathbf{u}}'_{III} = \hat{\mathbf{u}}'_{II} \times \hat{\mathbf{u}}_{I} \tag{32.7b}$$

Based upon an assumption that the knee joint axis is perpendicular to a vector from knee center to hip center, $\hat{\mathbf{u}}'_{III}$ can be rotated about $\hat{\mathbf{u}}_{I}$ by some angle, ϕ, to obtain vector, $\hat{\mathbf{u}}_{III}$, in the direction of the joint axis. The angle ϕ, vector $\hat{\mathbf{u}}_{III}$, and knee center location, $\mathbf{p}_{M_{right\,knee\,center}}$, are computed as

$$\phi = \sin^{-1}\left[\frac{\frac{1}{2}\left(knee\ width + marker\ diameter\right)}{\left|\mathbf{p}_{M_{right\,hip\,center}} - \mathbf{p}_{M_{right\,knee\,marker}}\right|}\right] \tag{32.8a}$$

$$\hat{\mathbf{u}}_{III} = \hat{\mathbf{u}}'_{II}\sin(\phi) + \hat{\mathbf{u}}'_{III}\cos(\phi) \tag{32.8b}$$

$$\mathbf{p}_{M_{right\,knee\,center}} = \mathbf{p}_{M_{right\,knee\,marker}} + \frac{1}{2}\left(knee\ width + marker\ diameter\right)\hat{\mathbf{u}}_{III} \tag{32.8c}$$

where knee width is measured from lateral to medial femoral condyle.

Computation of motion variables is generally the final step in the video measurement process. Variables can be formulated to quantify both temporal and spatial characteristics of a movement task. Temporal variables describe times or durations associated with the task being measured (e.g., time to initiate movement following a stimulus) and are typically single values. Angular variables that quantify absolute or relative orientation of body segments (e.g., elbow flexion angle) represent typical spatial variables. Rates of changes of relative segment angles are potential variables that incorporate both temporal and spatial attributes.

By identifying the video frames associated with initiation and completion of a task, the task duration, t_{task}, is readily computed as

$$t_{task} = \left(frame_{f} - frame_{i}\right)\Delta t \tag{32.9}$$

where $frame_{f}$ and $frame_{i}$ are the frame numbers for the initiation and finish of the task, respectively, and Δt is the video sampling rate. Similar expressions can quantify the time required to perform a specific portion of a task. Alternatively, it is often useful to quantify the duration of a task component as a percentage of the overall task duration:

$$\%_{task\,component} = \frac{frame_{b} - frame_{a}}{frame_{f} - frame_{i}} \times 100\% \tag{32.10}$$

where $frame_{a}$ and $frame_{b}$ are the frame numbers for the start and end of the task component. Such normalized time values are often useful for comparing task component durations when overall task durations differ.

Kinematic variables computed from marker coordinate data can quantify both translational and rotational movement of one or more body segments with respect to either a GCS or one another. Marker coordinate data are often filtered prior to calculating kinematic variables. Absolute translation of a body segment (e.g., forearm) between two sampling intervals, i and f (e.g., the beginning and end of a gear shift task) is computed as

$$\Delta \mathbf{P}_{A} = \mathbf{P}_{A,f} - \mathbf{P}_{A,i} \tag{32.11}$$

where $\mathbf{P}_{A,i}$ and $\mathbf{P}_{A,f}$ are the position vectors at frames i and f, respectively, for location A (e.g., the mass center) on the body segment of interest. Translational velocity, $\mathbf{v}_{A,i}$, and acceleration, $\mathbf{a}_{A,i}$, of point A at frame, i, can be discretely approximated as

$$\mathbf{v}_{A,i} = \frac{1}{2(\Delta t)}\left(\mathbf{P}_{A,i+1} - \mathbf{P}_{A,i-1}\right) \tag{32.12a}$$

$$\mathbf{a}_{A,i} = \frac{1}{(\Delta t)^2}\left(\mathbf{P}_{A,i+1} - 2\mathbf{P}_{A,i} + \mathbf{P}_{A,i-1}\right) \tag{32.12b}$$

Estimations for velocity and acceleration can also be obtained as the first and second time derivatives, respectively, of time functions for, i.e., $\mathbf{P}_A$:

$$\mathbf{v}_A(t) = \frac{d}{dt}\mathbf{P}_A \tag{32.13a}$$

$$\mathbf{a}_A(t) = \frac{d^2}{dt^2}\mathbf{P}_A(t) \tag{32.13b}$$

Often the functions fitted to the experimental data are piecewise polynomials.

Kinematic variables that quantify the orientation or rotation of body segments are often of greater interest than translational variables. Marker coordinate data can be used to compute either two- or three-dimensional body segment orientations or rotations, with respect to a GCS or with respect to one another. Most of these computations inherently assume that body segments behave as rigid bodies (i.e., that the distance between any two points on a segment remains constant).

Measurements of two marker (or virtual marker) positions ($\mathbf{P}_A$ and $\mathbf{P}_B$) per body segment (a marker need not be unique to a particular body segment) permit orientations and rotations of each segment to be computed based upon a single unit vector per segment (i.e., two-dimensional computation). The absolute orientation of a body segment (e.g., upper arm) can be quantified by a unit vector, $\hat{\mathbf{u}}_s$, directed from one marker to the other:

$$\hat{\mathbf{u}}_s = \frac{\mathbf{P}_B - \mathbf{P}_A}{\left|\mathbf{P}_B - \mathbf{P}_A\right|} \tag{32.14}$$

The relative orientation of two segments, i and j (e.g., upper arm and lower arm), can be described by an angle, ϕ, between the projection of the unit vectors, $\hat{\mathbf{u}}_{s,i}$ and $\hat{\mathbf{u}}_{s,j}$:

$$\phi = \cos^{-1}\left(\hat{\mathbf{u}}_{s,i} \bullet \hat{\mathbf{u}}_{s,j}\right) \tag{32.15}$$

These expressions are generally intended to be evaluated with three-dimensional marker coordinate data; however, they can be applied to two-dimensional coordinates as well. Two-dimensional applications assume the axis of rotation to be perpendicular to the coordinate plane. This requirement, unfortunately, limits the utility of two-dimensional data to movements occurring in a single plane. Occupational movements adhering to this restriction are rather limited.

Three-dimensional computation of absolute and relative segment orientations and rotations involves associating a three-dimensional local coordinate system (LCS) with each body segment of interest. Each LCS is described by three mutually orthogonal unit vectors, $\hat{\mathbf{u}}_1$, $\hat{\mathbf{u}}_2$, and $\hat{\mathbf{u}}_3$ (Figure 32.9). Calculation of these unit vectors requires measurement of at least three marker (or virtual marker) positions ($\mathbf{P}_A$, $\mathbf{P}_B$,

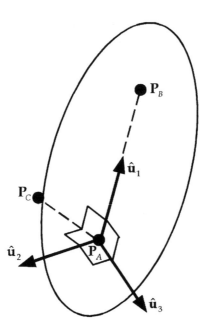

FIGURE 32.9 Depiction of segment LCS determined from segment marker positions.

and $\mathbf{P}_C$) per segment for each sampling interval. A given marker position can potentially be used in the computations of more than one segment LCS. For example, a virtual marker at the knee joint center could be used in determining LCSs for both thigh and lower leg segments. As with previously described two-dimensional computations, rigid body assumptions are also inherent to these three-dimensional kinematic calculations. The LCS unit vectors for a segment are computed as

$$\hat{\mathbf{u}}_1 = \frac{\mathbf{P}_B - \mathbf{P}_A}{\left|\mathbf{P}_B - \mathbf{P}_A\right|} \tag{32.16a}$$

$$\hat{\mathbf{u}}_2 = \frac{\left(\mathbf{P}_B - \mathbf{P}_A\right) \times \left(\mathbf{P}_C - \mathbf{P}_A\right)}{\left|\left(\mathbf{P}_B - \mathbf{P}_A\right) \times \left(\mathbf{P}_C - \mathbf{P}_A\right)\right|} \tag{32.16b}$$

$$\hat{\mathbf{u}}_3 = \hat{\mathbf{u}}_1 \times \hat{\mathbf{u}}_2 \tag{32.16c}$$

The unit vectors, $\hat{\mathbf{u}}_1$, $\hat{\mathbf{u}}_2$, and $\hat{\mathbf{u}}_3$, and, can be designated, for a segment, as $\hat{\mathbf{i}}_s$, $\hat{\mathbf{j}}_s$, and $\hat{\mathbf{k}}_s$, though not necessarily in that order. The order of designation, while somewhat arbitrary, generally follows the convention of forming a right-hand coordinate system (i.e., $\hat{\mathbf{i}}_s \times \hat{\mathbf{j}}_s = \hat{\mathbf{k}}_s$; $\hat{\mathbf{j}}_s \times \hat{\mathbf{i}}_s = -\hat{\mathbf{k}}_s$).

With an LCS determined for each body segment of interest, orientation of any given segment can be described with respect to any other coordinate system. Consequently, one can quantify a segment's absolute orientation (i.e., with respect to a GCS) or relative orientation (i.e., with respect to another LCS). One technique, for quantifying orientation of an LCS, involves rotating a reference coordinate system, $\hat{\mathbf{i}}_r$, $\hat{\mathbf{j}}_r$, and $\hat{\mathbf{k}}_r$, (either the GCS or another LCS) through angles ϕ, θ, and ψ about its three coordinate system axes, in succession so as to render it parallel to the LCS of interest ($\hat{\mathbf{i}}_s$, $\hat{\mathbf{j}}_s$, $\hat{\mathbf{k}}_s$). Many rotation sequences are possible, however, as an example one can consider rotation through angle, ϕ, about $\hat{\mathbf{i}}_r$ (which transforms $\hat{\mathbf{j}}_r$ and $\hat{\mathbf{k}}_r$ to $\hat{\mathbf{j}}_r'$ and $\hat{\mathbf{k}}_r'$); then rotation through angle, θ, about (which transforms $\hat{\mathbf{i}}_r$ and $\hat{\mathbf{k}}_r'$ to $\hat{\mathbf{i}}_r'$ and $\hat{\mathbf{k}}_r''$); and last, rotation through angle, ψ, about $\hat{\mathbf{k}}_r''$ (which transforms $\hat{\mathbf{i}}_r'$ and $\hat{\mathbf{j}}_r'$ to $\hat{\mathbf{i}}_r''$ and $\hat{\mathbf{j}}_r''$). Parallelism between the LCS of interest and the rotated reference coordinate system dictates that

$$\begin{bmatrix} \hat{\mathbf{i}}''_r \\ \hat{\mathbf{j}}''_r \\ \hat{\mathbf{k}}''_r \end{bmatrix} = \begin{bmatrix} \hat{\mathbf{i}}_s \\ \hat{\mathbf{j}}_s \\ \hat{\mathbf{k}}_s \end{bmatrix} \tag{32.17}$$

Transformation of the reference coordinate system via three successive rotations can be computed by multiplying a rotation matrix, $[_r\mathbf{M}_s]$, with the original LCS unit vectors:

$$[_r\mathbf{M}_s]\begin{bmatrix} \hat{\mathbf{i}}_r \\ \hat{\mathbf{j}}_r \\ \hat{\mathbf{k}}_r \end{bmatrix} = \begin{bmatrix} \hat{\mathbf{i}}_s \\ \hat{\mathbf{j}}_s \\ \hat{\mathbf{k}}_s \end{bmatrix} \tag{32.18}$$

For the rotation sequence described above, the rotation matrix takes the form

$$[_r\mathbf{M}_s] = \begin{bmatrix} \cos\psi\cos\theta & \sin\psi\cos\phi + \cos\psi\sin\theta\sin\phi & \sin\psi\sin\phi - \cos\psi\sin\theta\cos\phi \\ -\sin\psi\cos\theta & \cos\psi\cos\phi - \sin\psi\sin\theta\sin\phi & \cos\psi\sin\phi - \sin\psi\sin\theta\cos\phi \\ \sin\theta & -\cos\theta\sin\phi & \cos\theta\cos\phi \end{bmatrix} \tag{32.19}$$

From the indicated matrix multiplication, three of nine potential expressions can be chosen, from which the three angular rotations, ϕ, θ, and ψ, can be determined. One possible set of expressions is

$$\sin\theta = \hat{\mathbf{k}}_s \cdot \hat{\mathbf{i}}_r \tag{32.20a}$$

$$\cos\theta\cos\phi = \hat{\mathbf{k}}_s \cdot \hat{\mathbf{k}}_r \tag{32.20b}$$

$$\cos\psi\cos\theta = \hat{\mathbf{i}}_s \cdot \hat{\mathbf{i}}_r \tag{32.20c}$$

From these relationships, the rotations can be determined as

$$\phi = \cos^{-1}\left\{ \frac{\hat{\mathbf{k}}_s \cdot \hat{\mathbf{k}}_r}{\cos\left[\sin^{-1}\left(\hat{\mathbf{k}}_s \cdot \hat{\mathbf{i}}_r\right)\right]} \right\} \tag{32.21a}$$

$$\theta = \left[\sin^{-1}\left(\hat{\mathbf{k}}_s \cdot \hat{\mathbf{i}}_r\right)\right] \tag{32.21b}$$

$$\psi = \cos^{-1}\left\{ \frac{\hat{\mathbf{i}}_s \cdot \hat{\mathbf{i}}_r}{\cos\left[\sin^{-1}\left(\hat{\mathbf{k}}_s \cdot \hat{\mathbf{i}}_r\right)\right]} \right\} \tag{32.21c}$$

Angles computed for rotations about each of the LCS axes are often referred to as "Cardan" angles. Alternatively, rotations could be performed such that the first and third rotations are about the same axis, and the second rotation is about one of the remaining axes. Such a rotation scheme results in angles that are commonly termed "Euler" angles.

For some applications, particularly those involving rotational kinetics, calculation of segment angular velocities and acceleration may be desirable. Segment angular velocity and acceleration calculations are somewhat more complicated than translational velocity and acceleration computations. Angular velocity, $\mathbf{\Omega}$, and acceleration, $\dot{\mathbf{\Omega}}$, of a segment are determined as vectors with components Ω_x, Ω_y, Ω_z and $\dot{\Omega}_x$, $\dot{\Omega}_y$, $\dot{\Omega}_z$ in the $\hat{\mathbf{i}}_s$, $\hat{\mathbf{j}}_s$, and $\hat{\mathbf{k}}_s$ directions, respectively. For the rotation sequence described above the angular velocity and acceleration components can be expressed as:

$$\Omega_x = \dot{\phi}\cos\theta\cos\psi + \dot{\theta}\sin\psi \tag{32.22a}$$

$$\Omega_y = -\dot{\phi}\cos\theta\sin\psi + \dot{\theta}\cos\psi \tag{32.22b}$$

$$\Omega_z = \dot{\phi}\sin\theta + \dot{\psi} \tag{32.22c}$$

$$\dot{\Omega}_x = \ddot{\phi}\cos\theta\cos\psi - \dot{\phi}\dot{\theta}\sin\theta\cos\psi - \dot{\phi}\dot{\psi}\cos\theta\sin\psi + \ddot{\theta}\sin\psi + \dot{\theta}\dot{\psi}\cos\psi \tag{32.22d}$$

$$\dot{\Omega}_y = -\ddot{\phi}\cos\theta\sin\psi + \dot{\phi}\dot{\theta}\sin\theta\sin\psi - \dot{\phi}\dot{\psi}\cos\theta\cos\psi + \ddot{\theta}\cos\psi - \dot{\theta}\dot{\psi}\sin\psi \tag{32.22e}$$

$$\dot{\Omega}_z = \ddot{\phi}\sin\theta + \dot{\phi}\dot{\theta}\cos\theta + \ddot{\psi} \tag{32.22f}$$

The first and second time derivatives of ϕ, θ, and ψ can be approximated using the same techniques (i.e., discrete differentiation, or differentiation of time function approximations) used to estimate translational velocity and acceleration.

Computation of rotational kinematics can be readily applied to a wide range of video movement measurements; however, the occupational ergonomist must provide meaningful interpretation for these angles. For example, the designations "flexion/extension," "abduction/adduction," and "internal/external rotation" could be assigned to the angles describing relative orientation of a femoral LCS with respect to a pelvic LCS. Assignment of relevant designations, however, can be considerably more complicated for some body segments.

Construction of plots of rotation angles versus time are often useful with tasks for which cycles can be defined (e.g., a reach across a workspace). In such plots, time is often expressed as a percentage of the time to complete one cycle. Averages, maximums, minimums, and ranges of rotation angles, as well as angular values at specific instances, are all potential variables of interest to the ergonomist. Selection of the most appropriate variables for particular video movement measurements, however, is a matter of the ergonomist's discretion.

The techniques, described in this discussion, for video movement data collection and processing can be generally applied to a variety of occupational tasks. Ergonomists using these techniques for specific applications will need to adapt them for the particular nuances of the occupational tasks of interest. In doing so, they should consider that any aspect of video movement measurement (calibration, marker placement, image reconstruction, biomechanical modeling, and kinematic computation) can represent a weak link in the process if not appropriately considered.

References

Abdel-Aziz YI, and Karara HM: Direct Linear Transformation from comparator coordinates into object space coordinates in close range photogrammetry. Proceedings of the Symposium on Close Range Photogrammetry, Falls Church, VA, 1-18, 1971.

Antonsson EK, and Mann, RW: Automatic 6-d.o.f. kinematic trajectory acquisition and analysis. *Journal of Dynamic Systems, Measurement and Control*, 111:31-39, 1989.

Asato KT, Cooper RA, Robertson RN, and Ster JF: SMARTWheels: development and testing of a system for measuring manual wheelchair propulsion dynamics. *IEEE Transactions on Biomedical Engineering,* 40(12):1320-4, 1993.

Berguer R, Rab GT, Abu-Ghaida H, Alarcon A, and Chung J: A comparison of surgeons' posture during laparoscopic and open surgical procedures. *Surgical Endoscopy,* 11(2):139-42, 1997.

Boninger ML, Cooper RA, Robertson RN, and Rudy TE: Wrist biomechanics during two speeds of wheelchair propulsion: an analysis using a local coordinate system. *Archives of Physical Medicine & Rehabilitation,* 78(4):364-72, 1997.

Boston JR, Rudy TE, Lieber SJ, and Stacey BR: Measuring treatment effects on repetitive lifting for patients with chronic low back pain: speed, style, and coordination. *Journal of Spinal Disorders,* 8(5):342-51, 1995.

Cooper RA, Robertson RN, VanSickle DP, Boninger ML, and Shimada SD: Projection of the point of force application onto a palmar plane of the hand during wheelchair propulsion. *IEEE Transactions on Rehabilitation Engineering,* 4(3):133-42, 1996.

Garg A, Owen B, Beller D, and Banaag J: A biomechanical and ergonomic evaluation of patient transferring tasks: wheelchair to shower chair and shower chair to wheelchair. *Ergonomics,* 34(4):407-19, 1991.

Gracovetsky S, Kary M, Pitchen I, Levy S, and Ben Said R: The importance of pelvic tilt in reducing compressive stress in the spine during flexion-extension exercises. *Spine,* 14(4):412-6, 1989.

Gracovetsky S, Kary M, Levy S, Ben Said R, Pitchen I, and Helie J: Analysis of spinal and muscular activity during flexion/extension and free lifts. *Spine,* 15(12):1333-9, 1990.

Gracovetsky S, Newman N, Pawlowsky M, Lanzo V, Davey B, and Robinson L: A database for estimating normal spinal motion derived from noninvasive measurements. *Spine,* 20(9):1036-46, 1995.

Kadaba MP, Ramakrishnan HK, Wootten ME: Measurement of lower extremity kinematics during level walking. *Journal of Orthopaedic Research,* 8(3):383-92, 1990.

Kumar S, and Cheng CK: Spinal stresses in simulated raking with various rake handles. *Ergonomics,* 33(1):1-11, 1990.

Kumar S: Lumbosacral compression in maximal lifting efforts in sagittal plane with varying mechanical disadvantage in isometric and isokinetic modes. *Ergonomics,* 37(12):1975-83, 1994.

Lee YH, Cheng CK, and Tsuang YH: Biomechanical analysis in ladder climbing: the effect of slant angle and climbing speed. Proceedings of the National Science Council, Republic of China — Part B, Life Sciences, 18(4):170-8, 1994.

Marzan GT: Rational design for close-photogrammetry. Doctoral dissertation, University of Illinois at Urbana-Champaign, 1975.

Pascarelli EF, and Kella JJ: Soft-tissue injuries related to use of the computer keyboard. A clinical study of 53 severely injured persons. *Journal of Occupational Medicine,* 35(5):522-32, 1993.

Peterson B, and Palmerud G: Measurement of upper extremity orientation by video stereometry system. *Medical & Biological Engineering & Computing,* 34(2):149-54, 1996.

Robinson ME, O'Connor PD, Shirley FR, and MacMillan M: Intrasubject reliability of spinal range of motion and velocity determined by video motion analysis. *Physical Therapy,* 73(9):626-31, 1993.

Roozmon P, Gracovetsky SA, Gouw GJ, and Newman N: Examining motion in the cervical spine. I: Imaging systems and measurement techniques. *Journal of Biomedical Engineering,* 15(1):5-12, 1993.

Roozmon P, Gracovetsky SA, Gouw GJ, and Newman N: Examining motion in the cervical spine. II: Characterization of coupled joint motion using an opto-electronic device to track skin markers. *Journal of Biomedical Engineering,* 15(1):13-22, 1993.

Rudy TE, Boston JR, Lieber SJ, Kubinski JA, and Delitto A: Body motion patterns during a novel repetitive wheel-rotation task. A comparative study of healthy subjects and patients with low back pain. *Spine,* 20(23):2547-54, 1995.

Vaughan CL, Davis BL, and O'Connor JC: *Dynamics of Human Gait.* Human Kinetics Publishers, Champaign, Illinois, 1992.

Whittle MW: Calibration and performance of a three-dimensional television system for kinematic analysis. *Journal of Biomechanics,* 15:185-196, 1982.

Wickstrom G, Laine M, Pentti J, Hyytiainen K, and Salminen JJ: A video-based method for evaluation of low-back load in long-cycle jobs. *Ergonomics,* 39(6):826-41, 1996.

Woltring HJ: Calibration and measurement in 3-dimensional monitoring of human motion by optoelectronic means II. *Biotelemetry,* 3:65-97, 1976.

33

Force Dynamometers and Accelerometers

Robert G. Radwin
University of Wisconsin–Madison

Thomas Y. Yen
University of Wisconsin–Madison

33.1 Introduction

Forces considered in industrial ergonomics are usually classified as external or internal. External forces act against the human body, and they may be produced by an external object or in reaction to the human body exerting forces against an external object. It is possible to directly measure external forces using mechanical or electromechanical force measurement instruments. Internal forces are tension, compression, torsion, or shearing within muscles, tendons, bones, or other anatomical structures. Voluntary motions and exertions are produced through the generation of internal forces through active muscle contraction and passive action of connective tissues. Internal forces produce torque or rotation about the joints. External forces often result from internal force actions. Internal forces are usually not measured directly but by using indirect electrophysiological correlates such as electromyography (EMG).

The means of measuring external force will vary greatly depending on the circumstances of the task and practical considerations such as the accuracy required and the equipment and expertise available. Strength or maximal voluntary exertions represent the maximum force an individual is capable of producing. Forces associated with industrial tasks are usually less than maximal and are sometimes estimated from indirect measurements of the task requirements rather than measuring the exertions of individuals performing the task. These include measuring the weight of objects carried or lifted, or measuring the force necessary to do work, such as pushing or pulling a control. Direct force measurements should consider variability among individuals.

Human vibration is quantified from acceleration of objects that transmit vibrational energy to the body by contact either through the seat or feet (whole-body vibration) or by grasping vibrating objects (hand–arm vibration). Whole-body vibration is associated with vibration from riding in a vehicle or from standing on a moving platform. Hand–arm vibration may be introduced by using power hand tools or operating controls such as steering wheels on off-road vehicles.

Force and acceleration are expressed as a vector having magnitude and direction. The unit for force in the MKS system is the Newton ($1\ \text{N} = 1\ \text{kg m/s}^2$). The corresponding unit for acceleration is meters per second per second (m/s^2). Sometimes acceleration is expressed as multiples of the acceleration of gravity in units of g ($1\ g = 9.8\ \text{m/s}^2$). Force and acceleration measurements should specify both magnitude and direction. Usually force and acceleration vectors are decomposed into orthogonal components in a reference coordinate system.

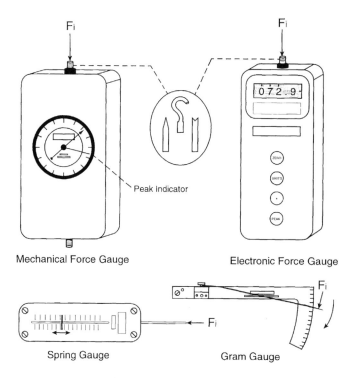

FIGURE 33.1 Hand-held force transducers with a variety of attachments.

Instruments applicable to force and acceleration measurements range from simple mechanical instruments to electromechanical devices. This chapter presents the theory and practice of force and acceleration measurement devices in industrial ergonomics. Mechanical and electromechanical force measurement devices are described, and force measurement applications using specialized dynamometers are discussed. The chapter also describes accelerometer theory and applications in human vibration measurement.

33.2   Mechanical Measurement Devices

Simple mechanical devices such as spring scales are used in many instances for estimating forces and loads, and are adequate for numerous industrial ergonomics applications. An assortment of these instruments is illustrated in Figure 33.1. Spring scales are available in various load levels, ranging from just a few grams to thousands of kilograms. These devices typically have a precision of 1% full scale. Most spring scales actually display units of mass (kg) rather than force because they are usually calibrated against a known mass. Force, which is measured in units of Newtons, can be determined by multiplying kilograms by the acceleration of gravity (9.81 m/s^2). Ergonomics practitioners find that instruments that are threaded for attachments such as various hooks and points are convenient for force measurement applications in the field (see Figure 33.1).

Mechanical force measurement instruments operate like a "fish scale" on the principle that applied force displaces a spring that is mechanically coupled by a spring-lever or a spring-cable system to a pointer and scale display. These simple devices often act as second-order mechanical systems consisting of a single mass, spring, and damping element as illustrated in Figure 33.2. The relationship between the input force and the output displacement is established as a simple second-order differential equation of the following form:

$$F_i - kx_o - c\dot{x}_o = m\ddot{x}_o \tag{33.1}$$

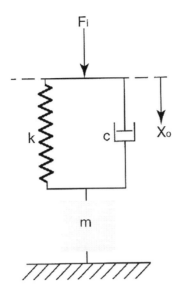

FIGURE 33.2   Second-order mechanical system consisting of a single mass, spring, and damping element.

where F_i is the applied force, k_s is the spring constant, c is the damping factor, m is the mass of elastic member, and x_o is the resulting displacement.

It is usually sufficient to assume that the mass of the elastic member does not affect the spring displacement, so it can considered zero. A damping element is used for damping the mass-spring system and preventing it from being excited into resonance, but in the static or quasi-static case where the excitation frequency is low it can be ignored. Consequently, the spring scale displacement can be simplified to:

$$x_o = F_i/k \qquad\qquad (33.2)$$

indicating that the displacement is proportional to the applied force. A stiffer spring results in a less sensitive instrument, yielding a greater force range.

Many common mechanical force measurement instruments used in ergonomics practice have a continuously moving pointer and a peak force indicator. Peak forces are recorded by the peak indicator, while sustaining forces are read directly from the indicator dial. The peak force indicator in mechanical force transducers is sometimes used for measuring isometric strength. This may be accomplished by anchoring one end of the transducer while exerting force against the free end of the transducer with an appropriate handle. Grip dynamometers such as the Jamar and Smedely dynamometers are spring force measurement instruments specifically designed for pinch and power grip strength.

Mechanical force transducers are most suitable for static force measurements such as determining the weight of a stationary object or for measuring quasistatic, or very slowly changing forces such as the force needed to overcome friction and push or pull a rolling cart along the floor. Although these simple mechanical devices are easy to use and they do not require an external power source, they are somewhat limited in their application and if not used cautiously they can yield erroneous measurements. For example, when measuring the force needed to push a hand truck, forces when starting and stopping the truck are usually greater due to friction and inertia of the truck. But since the measured force is read directly off the display dial, it is difficult to continuously record fluctuating forces, particularly when the force rapidly changes. Care must also be taken to ensure the alignment of the scale or dynamometer with the axis of the exertion. Because springs can become permanently deformed when stretched beyond their elastic limits, their calibration should be periodically tested using known loads.

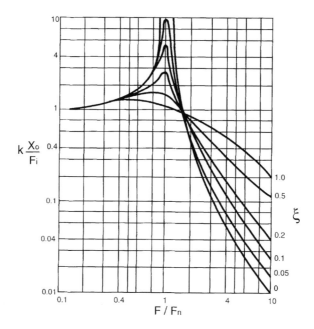

FIGURE 33.3   Second-order system response characteristics over a range of frequencies.

As is the case with any measurement, it is critical that the measurement instrument doesn't affect the system it is measuring. When dynamic or changing force is applied to spring devices, the mechanical spring instrument does in fact become part of the mechanical system. Second-order systems respond to changing inputs as depicted in Figure 33.3. When the system input excitation frequency is near or at resonance, the ratio of output displacement to input force greatly increases and the system becomes unstable. Since spring scales behave as second-order systems with resonant frequencies typically less than 1 Hz, they can be put into resonance and become unstable even by a relative low frequency input or by an impulsive force. Furthermore, if the changing force input has a frequency that is greater than the natural frequency of the instrument, the force measurements may read much less than its actual value due to attenuation by the spring-mass system (See Figure 33.3).

33.3   Electromechanical Measurement Devices

Electronic force transducers overcome many of the limitations of mechanical force transducers. Strain gauge load cells are capable of measuring static force, and they are much better suited for measuring forces that change with time than mechanical spring scales. Since electronic force transducers are much stiffer than mechanical force instruments, their response is much less affected by the system being measured. They also have better accuracy, and they can provide continuous force-time data. There are many self-contained electronic force measurement devices commercially available for use in industrial ergonomics studies (see Figure 33.1). These devices often contain circuitry for recording continuous force, peak force, and average force. It may also be possible to record the force output using an external instrument, and the device may have provisions to connect it directly to a digital computer. A summary of the advantages and disadvantages for each type of force transducers is shown in Table 33.1.

Many electromechanical sensors operate on the principle that mechanical inputs can alter the electrical resistance of a resistive element. The resistance (Ω) of a cylindrical electrical conductor is $R = \rho L/A$, where L is its length (m), A is its cross-sectional area (m^2), and ρ is the resistivity of the particular material (Ωm). The greater resistivity a material has, the more it acts as an insulator. Mechanical actions affect resistive sensors by either changing L or A (for metals) or by changing ρ (for semiconductors), conse-quently changing the sensor's resistance and the voltage across it. Metal strain gauges are made from

TABLE 33.1   Force Transducer Comparison

	Mechanical Force Transducers	Electronic Force Transducers
Advantages	Relatively inexpensive Simple to use Requires no power	Highly linear Continuous recording is possible Large force range with single transducer Small size Durable Fast response time
Disadvantages	Coarse resolution Continuous force recording is difficult Large size Limited dynamic capability	Calibration needed often Expensive transducer and instrumentation

lengths of very fine wire (<25 μm diameter). When the wire is stretched, its resistance changes, mainly due to changes in its cross-sectional area A and length L. The resistivity of metals increases with increasing temperatures because of the increased number of collisions that electrons make, thus increasing their electrical resistance. Consequently, resistive sensor accuracy may be affected by temperature.

Commercial strain gauge transducers of various types are available with a capacity for measuring loads of a few grams up to hundreds of thousands of kilograms and in several accuracy grades. The lowest accuracy grade typically has an overall (combined nonlinearity, hysteresis, nonrepeatability, etc.) error of about 1% of full scale, while the best accuracy grade has about 0.15% of full scale overall error.

Strain gauge transducers are often constructed using a calibrated metal plate or beam that undergoes a very small change (strain) in one of its dimensions. This mechanical deflection, usually a fraction of 1%, causes a small change in electrical resistance in the gauge wire. Low modulus materials such as aluminum are used to increase strain per unit force. Strain gauges are usually used in pairs. A simple load cell consists of a cantilever beam with strain gauges bonded on two opposing sides. The load cell can be made more sensitive and accurate when two gauges are placed on the top of the beam and an additional two gauges are attached to the bottom. The folded cantilever beam permits four gauges to be placed on the top surface, which eases manufacture and increases performance (Doebelin, 1990).

Silicon strain gauges are made of diffused resistors integrated into a silicon substrate. Both silicon and metal strain gauges are used in a similar manner and provide a very linear response within the elastic limits of the material they are fastened to. Silicon strain gauges exhibit even higher temperature effects than metal because deformation affects ρ. These temperature effects can often be controlled using special compensation circuits.

The gauge factor $G = (\Delta R/R_0)/(\Delta L/L_0)$, specifies a strain gauge's sensitivity to mechanical deformation. In this proportion, $\Delta R/R_0$ is the fractional change in resistance due to strain, and $\Delta L/L_0$ is the fractional change in strain gauge length. The gauge factor for metal strain gauges is typically between 2 and 5, while the gauge factor for silicon can be as high as 170. This mechanical deflection, usually a fraction of 1%, causes a small but measurable change in resistance.

When strain gauges are arranged in a Wheatstone bridge circuit, their sensitivity can be increased. The resulting imbalance in voltages in the bridge is proportional to the applied force. Bridge circuits also provide temperature compensation which prevents output shifts due to temperature effects on the transducers. The bridge circuit configuration increases sensitivity by $2(1 + v)$ over a single strain gauge configuration, where v is Poisson's ratio. Bridge circuits also provide temperature compensation which prevents output shifts due to temperature effects on the transducers. A low modulus material such as aluminum is often used to increase strain per unit force. When four strain gauges are used in a Wheatstone bridge they yield an electrical output that is insensitive to bending stresses due to the force being applied off center or at an angle, and they can be temperature compensated (Doebelin, 1990). The circuit in Figure 33.4 uses two strain gauges in adjacent legs of the bridge. The output voltage of the bridge V_o is given by the equation:

$$V_o = V_e \left[\frac{R_2}{R_2 + R_4} - \frac{R_1}{R_1 + R_3} \right] \tag{33.3}$$

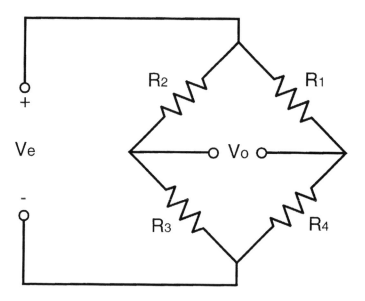

FIGURE 33.4   Strain gauges in Wheatstone bridge circuit.

When the beam bends, gauge 1 elongates and increases ΔR, while gauge 2 is compressed and decreases ΔR. The two fixed resistors in the opposite two legs of the bridge have identical resistance to the undeformed strain gauges, R_0, but they do not deform. This arrangement is known as a half-bridge circuit. If ΔR is much less than R_0 then

$$V_o \approx -\frac{V_e}{2R_0}\Delta R, \tag{33.4}$$

and the output is proportional to the change in resistance. A full bridge configuration in which all four legs contain strain gauges yields even higher sensitivity with an output given by $V_o \approx \Delta R/R_0$. Load cell amplifiers are commercially available that already contain the necessary Wheatstone bridge circuitry and excitation power sources.

Piezoelectric and piezoresistive load cells require minute deformations of their atomic structure within a block of crystalline material. Piezoelectric materials, such as quartz and barium titanate, produce a change in charge distribution when subjected to a mechanical stress. Quartz is a naturally found piezo-electric material, and deformation of its crystalline structure changes the electrical characteristics such that the electrical charge across its surfaces is altered. The charges collect on metal electrodes deposited onto the surface of piezoelectric material. Special amplifiers are used called charge amplifiers that output a voltage proportional to a charge at its input. Since piezoelectric sensors operate on changes in charge distribution, the most notable drawback to piezoelectric sensors is their inability to respond to static loads.

Piezo material can be quite small in size, which allows for easy mounting in many applications. Rapidly changing forces can be measured using piezoelectric force sensors. Because of their small size, many commercial piezo load cells have the necessary electronic circuitry already built into the package. A single piezo load cell is useful over a very wide range of forces because the 1% nonlinearity applies to any calibration range. Piezoelectric load cells can measure forces as great as 16,000 N. Piezo transducers are excellent for dynamic force measurements because of their very fast time constant. Typical piezo trans-ducers have very high stiffness, and natural frequency ranging from 10 KHz to 300 KHz (Doebelin, 1990). Because piezoelectric load cells are very stiff, they are suitable for measuring isometric forces. Since they only respond to changing and impulsive forces, piezoelectric load cells are unsuitable for measuring steady state forces.

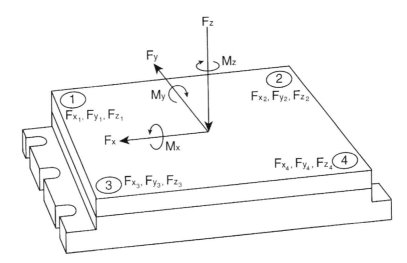

FIGURE 33.5 A force platform for measuring the three orthogonal force components (F_x, F_y, F_z) and three moments (M_x, M_y, M_z) from triaxial force transducers 1 through 4.

Unlike piezo*electric* sensors that produce changes in charge when stressed, piezo*resistive* sensors change resistance when they are stressed. Consequently, piezoresistive sensors are not limited to changing forces and are suitable for static loads. Piezoresistive sensors are usually configured in a Wheatstone bridge, and like strain gauges they require an external excitation voltage and a sensitive instrumentation amplifier.

Force platforms are used for measuring forces in two or more directions, such as ground reaction force acting on the feet during standing or walking. Ground reaction force contains a vertical component plus two shear components acting along the surface (Winter, 1990). This is accomplished using three or more force sensors that are arranged at right angles to each other. Force platforms are available with strain gauge or piezo force transducers. A common force plate configuration contains a flat plate supported by four triaxial force transducers as shown in Figure 33.5. This configuration measures three orthogonal force components in the x, y, and z axes for each of the four transducers (1, 2, 3, 4). Orthogonal force components (F_x, F_y, F_z) and moments (M_x, M_y, M_z) are resolved by using the equation:

$$
\begin{bmatrix} F_x \\ F_y \\ F_z \\ M_x \\ M_y \\ M_z \end{bmatrix} =
\begin{bmatrix}
1 & 1 & 1 & 1 & 0 & 0 & 0 & 0 & 0 & 0 & 0 & 0 \\
0 & 0 & 0 & 0 & 1 & 1 & 1 & 1 & 0 & 0 & 0 & 0 \\
0 & 0 & 0 & 0 & 0 & 0 & 0 & 0 & 1 & 1 & 1 & 1 \\
0 & 0 & 0 & 0 & -1 & -1 & 1 & 1 & 0 & 0 & 0 & 0 \\
0 & 0 & 0 & 0 & 0 & 0 & 0 & 0 & 1 & -1 & 1 & -1 \\
-1 & -1 & 1 & 1 & 1 & -1 & 1 & -1 & 0 & 0 & 0 & 0
\end{bmatrix}
\cdot
\begin{bmatrix} F_{x_1} \\ F_{x_2} \\ F_{x_3} \\ F_{x_4} \\ F_{y_1} \\ F_{y_2} \\ F_{y_3} \\ F_{y_4} \\ F_{z_1} \\ F_{z_2} \\ F_{z_3} \\ F_{z_4} \end{bmatrix}
\tag{33.5}
$$

33.4   Force Measurement Applications in Ergonomics

One way of directly measuring hand and grip force is by installing strain gauge force sensors directly in handles and objects grasped in industrial tasks. For example, Armstrong, et al. (1994) installed strain gauge load cells directly underneath computer keyboards for measuring finger exertions during typing tasks. Grip measurements are sometimes complicated by the fact that forces are unevenly applied and distributed throughout the palmar and finger surfaces and often involve multiple digits. A conventional strain gauge force instrument that measures force using the strain produced from the bending moment of a cantilever beam is extremely limited because the point of application must be controlled in order to know the particular bending moment arm. Furthermore, these instruments cannot linearly sum forces applied at arbitrary locations along the beam. Because of these constraints, the simple cantilever beam strain gauge system will not suffice for practical hand force measurements when using an instrumented handle.

A strain gauge dynamometer was developed that has sensitivity independent of the point of force application and linearly sums forces applied at multiple locations along the length of the active area. It is based on the principle of shearing stress acting in the cross-section of a beam when a transverse force is applied (Pronk and Niesing, 1981; Radwin et al., 1991). Instead of basing the dynamometer on sensing bending stresses produced when an applied force creates a bending moment in a cantilever beam, which is commonly used in many strain gauge instruments and is highly dependent on point of application, this instrument employs the principle of measuring beam shear stresses acting in the cross section of the beam when a transverse force is applied. This is accomplished by measuring shearing stress acting in the cross section of the beam. Strain gauges are mounted on a thin web machined into the central longitudinal plane and aligned at 45° with respect to the long axis (Pronk and Niesing, 1981; Radwin, Masters and Lupton, 1991). By selecting a measurement point at the neutral axis of the beam, the effect of bending stresses are completely removed from the strain gauges and all strain at the measurement point is strictly due to shear stress. Shear strain is totally independent of the point of application. A schematic diagram of the dynamometer design and the critical dimensions are given in Figure 33.6. This instrument is highly linear and has a typical force sensitivity of 2 mV/N.

The strain gauge dynamometer was used in a number of ergonomics investigations involving both maximal and submaximal exertions. A dynamometer for measuring grip strength when grasping a cylindrical handle (Oh and Radwin, 1993) was constructed for an average force of 250 N and a maximum load of 1000 N force. Aluminum caps were attached to both beams for producing a cylindrical surface. The handle length was 145 mm in order to accommodate hands of various sizes. The beams were mounted on a track so that they were capable of being separated arbitrary distances in order to provide a variable grip span.

A similar dynamometer was designed and constructed for measuring submaximal finger forces during five-finger prehension tasks when weights of various sizes were suspended from the dynamometer (Radwin et al., 1992). Another version of this instrument was attached to an electromagnetic shaker stage for directly measuring forces exerted when the handle was set into vibration resembling a vibrating power hand tool (Radwin et al., 1987). A smaller version with greater sensitivity was constructed for measuring pinch strength (Jeng et al., 1994). The pinch force instrument was 85 mm in length and was designed for an average force of 10 N and a maximum force of 200 N. Two opposing active dynamometer beams were used for independently measuring forces exerted by the thumb in opposition to one of the four fingers. Since the dynamometer was insensitive to the point of application, it was useful for clinical evaluations where it may be difficult to control the location where the fingers apply forces against the bars, particularly when using different fingers.

Another version of this instrument was used for investigating the forces involved in operating a pistol grip power hand tool (Oh and Radwin, 1993). The apparatus is shown in Figure 33.7. Two strain gauge instrumented beams were constructed for the handles of the power hand tool and an in-line pneumatic hand tool was mounted perpendicular to the handle to resemble a pistol grip hand tool. The tool was completely operational and capable of measuring palmar feed force and finger exertions during power

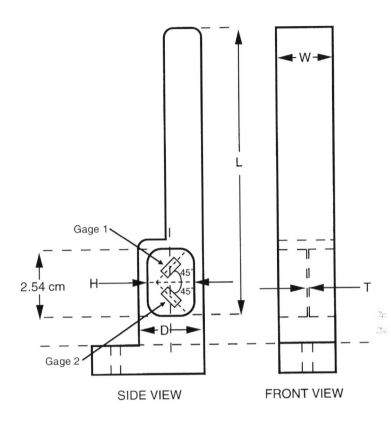

SIDE VIEW   FRONT VIEW

FIGURE 33.6 Schematic diagram containing dimensional variables for general dynamometer design. Dimensions that are varied include length (L), width (W), depth (D), pocket length (H), and web thickness (T). (From Radwin, R. G., Masters, G. P., and Lupton, F. W. 1991. A linear force-summing hand dynamometer independent of point of application, *Applied Ergonomics*, 22(5): 339-345. Copyright (1991) with kind permission from Elsevier Science Ltd., The Boulevard Langford Lane, Kidlington, U.K.)

hand tool operation. Contoured plastic caps were constructed and attached to the handle for simulating the shape of an actual hand tool. A trigger for activating the power hand tool was integrated into the finger side of the handle. The trigger contained a leaf spring switch used for activating a relay and solenoid valve for controlling the tool air supply.

A̲lthough strain gauge load cells can measure force with great accuracy, load cells with sufficient range are sometimes too large and bulky for attaching to handles or to the body, and in many instances it is difficult to use load cells for directly measuring exerted hand force in industrial tasks. A durable and thin conductive polymer force transducer was found useful for measuring external forces on regions of the body where conventional force sensors were far too large (Jensen et al., 1991). Because of its very small size and high durability, the sensor can be easily attached to the skin, which makes fingertip and palmar forces easy to measure. The conductive polymer sensing elements are composed of two conducting interdigitated patterns deposited on a thermoplastic sheet facing against another sheet containing a conductive polyetherimide film (see Figure 33.8). A spacer between the two plastic layers prevents contact, causing the sensor to have infinite impedance. As applied force increases, the two layers compress together, increasing the contact area. This subsequently decreases the electrical resistance of the sensor. An increased force leads to a decrease in resistance.

A̲ dome for distributing force over the active sensing area is necessary for these elements to operate as force sensors (see Figure 33.8). Without the dome, the measurements are erroneous as shown in Figure 33.9. These sensors are very limited; their useful range is up to 30 N with an accuracy of 1 N; however, there are few alternatives available for directly measuring finger and hand forces.

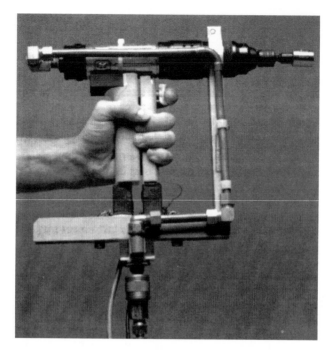

FIGURE 33.7   Dynamometer used for measuring finger and palm forces exerted when operating a completely functional simulation of a pistol-grip pneumatic power hand tool. (From Radwin, R. G., Masters, G. P., and Lupton, F. W. 1991. A linear force-summing hand dynamometer independent of point of application, *Applied Ergonomics*, 22(5): 339-345. Copyright (1991) with kind permission from Elsevier Science Ltd., The Boulevard Langford Lane, Kidlington, U.K.)

The small size and flexibility of the transducer materials, greatly increases the mountability of the transducers on a large variety of objects, and tool handles and grips. Due to the physical properties or the transducer mounting configuration, most electronic force transducers are only sensitive to forces acting normal or perpendicular to the surface plane of the transducer. This allows the identification of directional components of the forces acting on or by an object by using a different force transducer for each force direction of interest.

33.5   Acceleration Measurements

Acceleration is measured directly using devices called accelerometers. Accelerometers consist of a small mass and a piezoelectric element that measures the resulting force when mass accelerates. Recall that piezoelectric force sensors operate on the principle that the electrical charge measured across a piezo-electric material is proportional to its deformation when force is applied. Piezoelectric accelerometers operate the same way except a small mass is mounted on top of the piezoelectric material. The mass weighs usually no more than several grams. When the device is accelerated, the mass exerts a force against the piezoelectric material which produces a signal proportional to its acceleration. A typical accelerometer design is shown in Figure 33.10.

Piezoelectric accelerometer sensitivity is usually expressed in terms of coulomb charge per unit of acceleration (typically pC/g). The outputs are amplified using a charge amplifier that converts charge into voltage. Accelerometers may also be made from piezoresistive material, and their sensitivity is expressed in terms of millivolts per unit of acceleration (mV/g). Like piezoresistive force sensors, piezore-sistive accelerometers require an excitation voltage, and they require an instrumentation amplifier similar to a load cell.

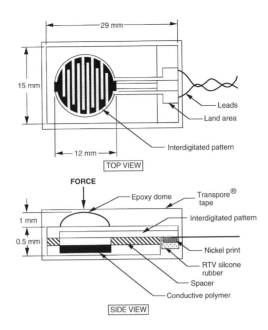

FIGURE 33.8 Schematic diagram of the conductive polymer finger force sensor showing top and side views. After a dome was placed over the conductive polymer sensing area the sensor was encased in Transpore® tape. (Reprinted from Jensen, T. R., Radwin, R. G., and Webster, J. G. 1991. A conductive polymer sensor for measuring external finger forces, *Journal of Biomechanics,* 24 (9): 851-858. Copyright (1991) with kind permission from Elsevier Science Ltd., The Boulevard Langford Lane, Kidlington, U.K.)

Accelerometer sensitivities are proportional to their mass. Consequently, the smaller the accelerometer, the less sensitive it is. Typical accelerometer frequency response characteristics are shown in Figure 33.11. It is important that an accelerometer not be excited by frequencies near or at its resonant frequency. This would produce erroneously large measurements. Accelerometer resonant frequencies vary inversely proportional to the square root of their mass. Therefore, smaller accelerometers have higher resonant frequencies and are usable over a greater frequency range. Accelerometers are also influenced by temperature changes, humidity, and other harsh environmental conditions.

It is important that the total mass of an accelerometer is sufficiently small not to interfere with the measurement by loading the vibrating body. Commercial accelerometers are small enough and light enough to attach directly to the limbs for measuring body motions. Accelerometers weighing more than 15 g are typically unsuitable for vibration measurements made by mounting them on a human body.

Usually accelerometers are sensitive to motion in a single direction. Triaxial accelerometers are available for simultaneously measuring acceleration in three orthogonal directions. Angular acceleration may be measured by mounting an accelerometer tangential to the rotating object. The angular acceleration α is the tangential acceleration a_t, divided by the radius of rotation r: $\alpha = a_t/r$.

When accelerometers are mounted on the body, they are usually located near bony eminences and surfaces. More commonly, accelerometers are mounted on objects that transmit vibration to the body. This may be a seat or a platform. In that case, accelerometer mass is not as critical for measuring whole-body vibration (WBV), although size might be a consideration when mounting accelerometers under a seated operator. Accelerometer resonant frequencies should be greater than 300 Hz and be capable of sustaining instantaneous acceleration levels up to 100 m/s^2 without damage. A triaxial seat disk accelerometer may be inserted between a vehicle seat and a passenger's buttocks. Vibration at the feet of a vehicle passenger can be measured by mounting an accelerometer directly to the floor. SAE J1013 specifies a

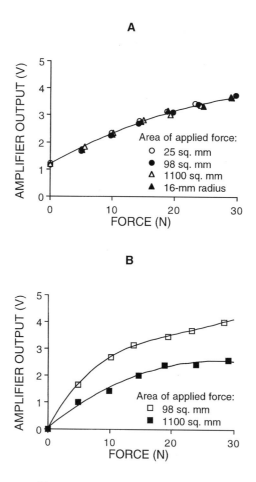

FIGURE 33.9   (A) Sensor response with an epoxy dome for input forces applied using different size surface areas, including a curved surface, showing insensitivity to area of application. (B) Sensor response without an epoxy dome for forces applied using different size surface areas, showing high sensitivity to the contact area of force application when the dome is not included. (Reprinted from Jensen, T. R., Radwin, R. G., and Webster, J. G. 1991. A conductive polymer sensor for measuring external finger forces, *Journal of Biomechanics,* 24 (9): 851-858. Copyright (1991) with kind permission from Elsevier Science Ltd., The Boulevard Langford Lane, Kidlington, U.K.)

transducer mounting for measuring seated vibration using either a 200-mm-diameter by 6-cm-thick disc placed between the operator and the seat cushion, or using a semirigid disc made of 80 to 90 durometer molded rubber or plastic. The semirigid disc is the most practical, and it is recommended particularly for soft or highly contoured cushions. The disc should be placed on the seat so that the triaxial accelerometers are located midway between the ischial tuberosities and are aligned parallel to the ISO basicentric seated operator coordinate axes. The disc should be taped or similarly attached to the cushion to maintain its location.

The frequencies of interest for whole-body vibration are between 0 Hz and 80 Hz. Piezoresistive accelerometers are the most common for WBV because of their low frequency response. Calibration is simplest for piezoresistive accelerometers with a DC response because they can be calibrated by just inverting the accelerometer in the direction of gravity. A 180° tilt represents a 19.61 m/s^2 (2g) peak-to-peak change in acceleration. WBV acceleration measurements are referenced to the ISO biodynamic coordinate system (ISO 2631) as shown in Figure 33.12.

Hand–arm vibration (HAV) usually is transmitted through the handles of manually operated equipment, such as power hand tools. Small piezoelectric accelerometers are the most common type of

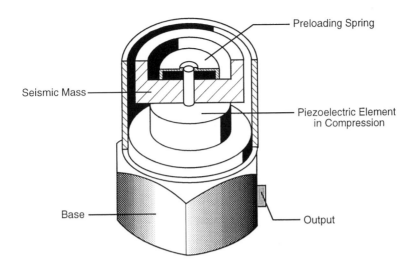

FIGURE 33.10   Piezoelectric accelerometer.

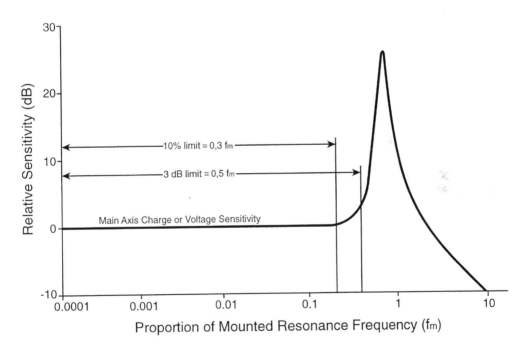

FIGURE 33.11   Representative accelerometer frequency characteristics. The usable frequency range is typically limited to the frequency where the output becomes amplified less than 10% or the frequency where the output becomes amplified no more than 3 dB.

transducer used for HAV measurements. Size and durability are critical considerations since the accelerometers are mounted on tool handles as operators use the tools for work. Accelerometer mass should be as small as possible. Accelerometers weighing 10 to 15 grams or less with a cross-axis sensitivity less than 10% are most suitable for HAV measurements. Piezoelectric accelerometers should be calibrated using a portable vibration exciter (such as the Brüel & Kjær 4294) that can verify accelerometer condition at a study site, because accelerometers are easily damaged during measurements in the field.

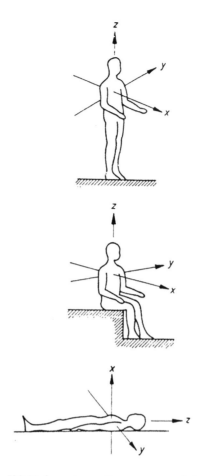

FIGURE 33.12   ISO Biodynamic coordinate system as defined in ISO 2631.

Vibration measurements of the hand are performed using the ISO basicentric or biodynamic coordinate system in ISO 5349 as shown in Figure 33.13. The basicentric coordinate system is usually easier to use than the biodynamic coordinate system and is almost exclusively used for handle vibration measurements. The three linear axes designated as x (motion through the palm), y (motion across the knuckles), and z (motion along the axis of the arm) are measured simultaneously. The accelerometer should be mounted directly on the handle nearest to where the hand grasps the handle. All three linear axes should be simultaneously measured and recorded, with each axis analyzed separately.

Accelerometers are usually mounted directly to the handle using a hose clamp or similar strap, or welding a mounting block with a stud for the accelerometer. If the vibration magnitude varies significantly over different parts of the handle, then the maximum value at the point of contact should be measured. Accelerometers are commonly screwed to a small aluminum block with a radius on the bottom and strapped to the handle using a hose clamp or a Panduit cable tie. If a resilient element is interposed between the hand and the vibrating surface, a thin and suitably shaped piece of metal may be placed between the hand and the handle.

Triaxial accelerometers are the most convenient because remounting an accelerometer in different orientations is often time-consuming and more obtrusive in the field. Single axis accelerometers may be used for measuring vibration as long as the event is relatively repeatable; however, error might be introduced if each successive measurement involves considerably different tool orientations and motions, as is possible in field measurements.

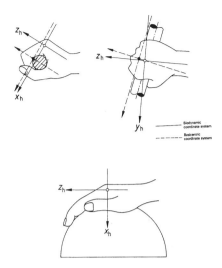

FIGURE 33.13 Biodynamic and basicentric coordinate system as defined in ISO 5349.

Root-mean-squared (RMS) linear acceleration (in meters/second/second or "g," where $1g = 9.81$ m/sec/sec) is the measurement parameter of choice for vibration magnitude. The hand–arm vibration frequency range for most standards cover the one third octave bands having center frequencies from 6.3 to 1250 Hz, octave bands having center frequencies from 8 to 1000 Hz, and frequency weighted measurements covering the frequency range 5.6 Hz to 1400 Hz. A frequency range as high as 5 KHz for unweighted acceleration measurements has been advocated by some researchers.

Vibration signals with very high peak acceleration, such as percussive tools, require a more durable transducer with a resonant frequency above 25 KHz and cross axis sensitivity at least 20 dB below the sensitivity of the axis to be measured. One source of artifact error comes from DC shifts that result when an accelerometer is excited at its resonant frequency. It sometimes may be necessary to mechanically attenuate vibration in order to avoid accelerometer resonances, which should be made as near to the transducer as possible. The preferred method of measurement involves inserting a mechanical low-pass filter with a cut-off frequency no less than 2000 Hz between the transducer and the vibrating surface.

Human occupational vibration exposure is usually assessed by measuring vibration acceleration and determining acceleration magnitude and frequency characteristics, in addition to vibration exposure time. There are no specified methods for obtaining accurate exposure time (duration) data. Exposure time could be based on a time study, work sampling, historic data, or determined directly from the vibration record. Exposure time is critical in most standards. Statistical methods may be useful for determining duration from sampled data if repetitive events are within a specified error for a certain accuracy (e.g., 95% probability that sampling error is less than 5%).

Vibration exposure time is difficult to predict based on production or work standards and may vary significantly depending on the specific operation, individual work methods, and operator experience. A study was undertaken to evaluate individual worker's exposure to hand–arm vibration from power hand tool use in an electric appliance assembly plant where workers used small in-line pneumatic screwdrivers to perform highly repetitive tasks on an assembly line (Radwin and Armstrong, 1985). The average daily exposure was predicted from observed vibration samples recorded for each operator on the assembly line.

Accelerometers were attached to the power hand tools used by each operator, and vibration measurements were recorded for an observation period ranging from 4 to 18 min. The analysis consisted of measuring the average air tool operation time for each operator. Vibration and frequency measurements were made at a later time in the laboratory. There were two phases associated with the screwdriver operation. The first, or "run-down," phase corresponded to running the tool during screw driving. The second "clutch-slippage" phase corresponded to slippage of the clutch mechanism when the screw was tightened to the set torque. Visual inspection of the acceleration waveform clearly revealed that the vibration produced during the clutch-slippage phase was much greater in amplitude and fundamentally lower in frequency than the run-down phase. A computer program was used to recognize the run-down and chutch-slippage phases from data sampled using an analog–digital converter, on the basis of an empirically derived digital moving average filter. Total exposure for each operator was computed by measuring samples of their screwdriver operation times and multiplying the cumulative time by the number of operations per shift.

Current HAV standards frequency weightings were arrived at by consensus and originate from early studies by Miwa (1967, 1968) of subjective perception of vibration and discomfort and tolerance levels. The ISO 5349 and ANSI S3.34 specify similar frequency weighting networks. These weighted curves were revised in the 1980s and issued in 1986 as ISO 5349. Frequency weighting is accomplished by applying specified one third octave band weightings or by passing the data through the specified filter network. Discrete spectra may also be weighted according to the filter network characteristics. Because early HAV standards were based solely on subjective vibration perception and the paucity of epidemiological data available, NIOSH (1989) and others have advocated measuring unweighted HAV.

Use of linear integrating vibration exposure measurement equipment is best suited for measuring vibration exposure when the vibration signal is of short duration and its operation varies substantially with time. When task cycle times are relatively long, lasting more than one minute or longer, linear integrating vibration spectrum analyzers or software having integration periods greater than one minute are not always available. Vibration exposure is difficult to assess directly using many fast Fourier transform (FFT) spectrum analyzers because of long tool operating times. Providing that the vibration signal is stationary (i.e., the signal is time-invariant) during tool operation, linear integration over the entire tool operating period is not always necessary. A study showed that if tool operation time can be accurately determined independently of vibration acceleration, vibration exposure levels can then be computed using transient vibration measurements (Radwin, et al., 1990). The vibration measurements can be taken off-line where conditions can be controlled and not interfere with ongoing operations.

References

American National Standards Institute (1986). *S3.34: Guide for the Measurement and Evaluation of Human Exposure to Vibration Transmitted to the Hand*, New York, NY.

Armstrong, T. J., Foulke, J. A., Martin, B. J., Gerson, J., and Rempel, D. M. 1994. Investigation of applied forces in alphanumeric keyboard work, *American Industrial Hygiene Association Journal*, 55: 30–35.

Doebelin E. O. 1990. *Measurement Systems: Application and Design*, 4th ed. New York: McGraw-Hill.

Eastman Kodak Company 1986. *Ergonomic Design for People at Work: Volume 2*, New York: Van Nostrand Reinhold.

International Organization for Standardization 1985. *ISO 2631: Evaluation of Human Exposure to Whole-body Vibration–Part 1: General Requirements*, Geneva.

International Organization for Standardization 1986. *ISO 5349: Guidelines for the Measurement and Assessment of Human Exposure to Hand-Transmitted Vibration*, Geneva.

Jeng, O. J., Radwin, R. G., and Rodriquez, A. A. 1994. Functional psychomotor deficits associated with carpal tunnel syndrome, *Ergonomics*, 37(6): 1055-1069.

Jensen, T. R., Radwin, R. G., and Webster, J. G. 1991. A conductive polymer sensor for measuring external finger forces, *Journal of Biomechanics*, 24 (9): 851-858.

Miwa, T. 1967. Evaluation methods for vibration effect: Part 3: Measurements of threshold and equal sensation contours on hand for vertical and horizontal sinusoidal vibrations, *Industrial Health*, 5: 213-220.

Miwa, T. 1968. Evaluation methods for vibration effect: Part 6. Measurements of unpleasant and tolerance limit levels for sinusoidal vibrations, *Industrial Health*, 6, 18-27.

National Institute for Occupational Safety and Health 1989. *Criteria for a Recommended Standard: Occupational Exposure to Hand–arm Vibration*, DHHS/NIOSH Publication No. 89-106.

Oh, S. and Radwin, R. G. 1993. Pistol grip power tool handle and trigger size effects on grip exertions and operator preference, *Human Factors*, 35(3), 551-569.

Pronk, C. N. A. and Niesing, R. 1981. Measuring hand-grip force using a new application of strain gauges, *Medical & Biological Engineering and Computing*, 19, 127-128.

Radwin, R. G. and Armstrong, T. J. 1985. Assessment of hand vibration exposure on an assembly line, *American Industrial Hygiene Association Journal*, 46(4), 211-219.

Radwin, R. G., Armstrong, T. J., and Chaffin, D. B. 1987. Power hand tool vibration effects on grip exertions, *Ergonomics*, 30 (5), 833-855.

Radwin, R. G., Armstrong, T. J., and VanBergeijk, E. 1990. Vibration exposure for selected power hand tools used in automobile assembly, *American Industrial Hygiene Association Journal*, 51(9): 510-518.

Radwin, R. G., Masters, G. P., and Lupton, F. W. 1991. A linear force-summing hand dynamometer independent of point of application, *Applied Ergonomics*, 22(5): 339-345.

Radwin, R. G., Oh, S., Jensen, T. R., and Webster, J. G. 1992. External finger forces in submaximal static prehension, *Ergonomics*, 35(3): 275-288.

Winter, D. A. (1990). *Biomechanics and Motor Control of Human Movement*, New York: John Wiley & Sons.

Section III

Cognitive Environment Issues

34

How Complex Human-Machine Systems Fail: Putting "Human Error" in Context

Klaus Christoffersen
The Ohio State University

David D. Woods
The Ohio State University

"Rather than being the main instigators of an accident, operators tend to be the inheritors of system defects… Their part is that of adding the final garnish to a lethal brew whose ingredients have already been long in the cooking"

(Reason, 1990; p. 173).

34.1 Introduction

Research on human error is as old as the field of human factors (Fitts and Jones, 1947). Despite this long record of investigation into the factors that lead to erroneous actions and assessments, the belief remains commonplace that the verdict "human error" is a meaningful statement of the cause of failures. The belief that human errors are random acts constituting a basic category of human performance blocks our ability to understand and therefore control the factors that lie behind failures of complex systems. Human factors has always probed for the systematic factors behind the label "human error" in the nature of the problems people face, in the design of the artifacts that people use, and in the organization that provides resources and sets goals. In other words, the label human error is not a conclusion, but rather a starting point for investigation (Woods et al., 1994).

A fundamental premise of human factors research on errors, therefore, is that human performance is shaped by systematic factors. The scientific study of failure is not possible unless one understands this basic tenet of social and behavioral science. We do not yet understand all of the factors and how they interact. Individual differences are always a prominent fact of human behavior. But there are regularities which shape human cognition, collaboration and performance, and we do understand how these factors can make certain *kinds* of erroneous actions and assessments predictable. Again, finding points where these design-induced and organization-induced errors will occur and developing countermeasures has been among the major topics of human factors from its origins. Our ability to predict the timing and number of erroneous actions is very weak, but our ability to predict the kind of errors that will occur, when people do err, is often good or very good.

A number of serious, widely publicized accidents, such as the Three Mile Island nuclear power accident in 1979, the Bhopal chemical plant accident in 1984, the Challenger space shuttle accident in 1986, the Strasbourg automated airliner crash in 1992, and numerous less celebrated but no less catastrophic events led to a new wave of research into the factors that lead complex systems to fail (Perrow, 1984; Senders and Moray, 1991; Reason, 1990; Hollnagel, 1993). This research has shown how popular beliefs that such accidents are due simply to isolated acts of human error mask the deeper story. In pursuing this story, the research has shown that the processes that lead complex systems to fail are much more complex than a single "error" by a single human. The opportunity to learn from accidents and incidents, and the ability to make systems more reliable and robust, depends on pursuing the underlying factors beyond the label "human error" (Woods et al., 1994). This chapter presents a portion of the deeper story of how complex systems fail.

34.2   Hindsight Bias

Why is human error so persistently viewed as a legitimate explanation for system failures? In part, the problem is that attributions of "human error" occur after the fact, in the aftermath of a failure, through a process of social judgment. After an accident or incident, it is easy for us with the benefit of hindsight to say, "How could they have missed x?" or "How could they have not realized that x would obviously lead to y?" This is because our knowledge of the bad outcome makes it seem that participants failed to account for information or conditions that "should have been obvious" or behaved in ways that were inconsistent with the (now known to be) significant information. (See Woods et al., 1994, Chapter 6 for an extensive discussion of how the hindsight bias degrades our ability to learn from accidents.) Fundamentally, this omniscience is not available to any of us before we know the results of our actions. To react, after the fact, as if this knowledge were available to operators, trivializes the situation confronting the practitioner, and masks the processes affecting practitioner behavior before-the-fact.

Hindsight bias is the tendency for people to "consistently exaggerate what could have been anticipated in foresight" (Fischhoff, 1975). Studies have consistently shown that people have a tendency to judge the quality of a process by its outcome. The information about outcome biases their evaluation of the process that was followed. Decisions and actions followed by a negative outcome will be judged more harshly than if the *same* decisions had resulted in a neutral or positive outcome. Indeed, this effect is present even when those making the judgments have been warned about the phenomenon and been advised to guard against it (Fischhoff, 1975, 1982).

Given knowledge of outcome, reviewers will tend to *simplify* the problem-solving situation that was actually faced by the practitioner. The dilemmas, the uncertainties, the trade-offs, the attentional demands, and double binds faced by practitioners may be missed or under-emphasized when an incident is viewed in hindsight. Because the hindsight bias masks the real dilemmas, uncertainties, and demands practitioners confront, we have a distorted view of the factors contributing to the incident or accident. In this vacuum, we only see human performance after an accident or near miss as irrational, willing disregard (for what is now obvious to us and to them), or even diabolical. This fuels the traditional responses to punish the individuals associated most closely with the outcome in time and space.

34.3 Design and Organizational Factors Shape Human Performance

If we accept human error as a symptom rather than as a cause of problems, then what is human error a symptom of? Three of the underlying factors which most significantly influence human performance and create opportunities for errors by front-line operators are the design of tasks, the design of human–computer interfaces, and organizational characteristics of systems. A comprehensive discussion of the ways in which design and organizational factors can negatively impact human cognition, collaboration, and performance is outside the scope of this chapter, but some examples will serve to illustrate the point.

Task Design: Omissions of Isolated Acts

One example of a kind of slip of action is the omission of an isolated act (often the last action in a prescribed sequence or post-completion slip; Byrne and Bovair, 1997). In slips of action, the user intends to act in the appropriate way, but the process of translating that correct intention into the specific sequence of detailed action needed to carry out that intention is disturbed (cf. Norman, 1981). One example of this class of erroneous actions can occur when one of the actions that makes up an action sequence is unconnected physically or functionally to previous or successive actions. As a result of how the task has been designed, there are no cues in the structure of the task to act as an external memory or remind the actor to carry out the step. This characteristic increases the user's working memory load. When this increased memory burden is combined with the occurrence of other factors that challenge user's working memory (e.g., disruptions, multiple tasks, high work load, fatigue), the isolated step is easily omitted (Byrne and Bovair, 1997). The design of the task influences the memory load on the actor, i.e., it shapes the cognitive activities of the people in the system. If other factors are present that also challenge working memory, a specific kind of erroneous action can result.

This is only one kind of erroneous action that is affected by high memory loads on users. Many tasks and devices are designed in ways that place high demands on user memory; so many in typical human–computer interfaces that Norman (1988) refers to a conspiracy against human memory. Symptomatic of the loads on user memory are adaptations, such as the creation of external memory aids, which users often devise in response to this threat to their performance. For example, people may develop paper reminders or "crib notes" and attach them to a computer workstation to aid in various tasks.

Human–Computer Interface Design: Mode Error

A classic example of how design factors can create the potential for poor human performance is mode error (see Woods et al., 1994, Chapter 5 for a summary). Mode error is one kind of breakdown in the interaction between humans and machines, especially computerized devices. Norman (1988, p. 179) summarizes the source of mode error quite simply by suggesting that one way to create or increase the possibilities for erroneous actions is to "… change the rules. Let something be done one way in one mode and another way in another mode." When this is the case, a human user can commit an erroneous action by executing an intention in the way appropriate to one mode of the device when the device is actually in another mode. Put simply, multiple modes in devices create the potential for mode errors. The consequences of mode errors depending on the context in which they occur. Mode errors in human–computer interaction have been critical contributors to accidents in aviation (Billings, 1996).

Mode error is inherently a human–machine system breakdown. It requires a user who loses track of the system's active mode configuration and a machine that interprets user input differently depending on the current mode of operation. Mode error has been identified and studied as a systematic form or user error created by design factors since at least 1981 (e.g., Norman, 1981; Lewis and Norman, 1986; Monk, 1986; Sellen et al., 1992; Sarter and Woods, 1995; Obradovich and Woods, 1996). Interestingly, characteristics of the computer medium and common pressures on the design process have made it easy for designers to proliferate modes and to create more complex interactions across modes. The result is an epidemic of new opportunities for mode errors to occur and new kinds of mode-related problems in today's computerized devices. These studies provide methods to identify when computerized devices will produce mode errors and suggest several design techniques as countermeasures.

Organizational Factors: Goal Conflicts

Organizational factors also shape human cognition and collaboration in ways that lead predictably to certain forms of erroneous actions (e.g., Reason, 1997). Organizations provide resources but also create or sharpen the dilemmas practitioners face. Organizational pressures to meet some goals (generally throughput or economic goals) without taking into account potential conflicts (typically with goals that provide a safety margin against the potential for failure) can place operators in "double-" or "N-tuple-" binds. (See Woods et al., 1994, Chapter 4 for an in depth treatment of goal conflicts and organizational pressures.) In these situations, any attempt to achieve one goal involves sacrificing achievement of another goal. Multiple conflicting goals result in situations where all of the operator's degrees of freedom for action are consumed by the various demands and there is no course of action left which does not violate some goal. In these situations, operators are essentially forced to choose among the lesser of N evils, thereby committing some "error" by default.

In one tragic aviation disaster, the Dryden Ontario crash (Moshansky, 1992), several different organizational pressures to meet economic goals and organizational decisions to reduce resources created a situation where a pilot faced this kind of double bind. Deciding not to take off in deteriorating weather conditions would strand a full plane of passengers, disrupt schedules, and lose money for the carrier. Such a decision would be regarded as an economic failure with potential sanctions for the pilot. On the other hand, the means to accommodate the weather threat (de-icing equipment) were not available due to organizational choices not to invest or to cut back on equipment at peripheral airports such as Dryden. In the end, the pilot attempted to take off after an overlengthy delay despite the risk of icing, and the aircraft crashed.

The folk models that lead us to think that human error is the cause of accidents mislead us. These folk models create an environment where accidents are followed by a search for a culprit and solutions that consist of punishment and exile for the apparent culprit and increased regimentation or remediation for other practitioners as if the cause resided in defects inside people. However, these countermeasures are ineffective or counterproductive because they completely miss the systematic deeper factors that produced the multiple conditions necessary for failure. Other practitioners, regardless of motivation levels or skill levels remain vulnerable to the same systematic factors. If the erroneous action was an omission of an isolated act, the memory burdens imposed by task mis-design are still present as a latent factor ready to contribute to the same type of error. If a mode error was part of the failure chain, the computer interface design still creates the potential for others to commit the same kind of error. If a double bind was behind the actions that contributed to the failure, that goal conflict remains to perplex other practitioners.

If we examine deeper organizational and design factors and understand how they shape the cognitive and collaborative activities of people who work in the field of practice, then we can predict the kinds of human performance problems that will arise and we can learn where and how to invest to improve the system.

34.4   How Do Complex Systems Fail?

The Nature of Complex Systems

The research on disasters and the role of human performance over the last 20 years have revealed the basic form of failure in complex systems. However, before we can enter into a discussion of the characteristic signature of complex system failures, we must clarify what we mean by a complex system. Specifically, we will outline those features of modern systems which shape and contribute to the nature of the failures we see. These characteristics are becoming increasingly prevalent today as the forces of economy and safety exert ever greater pressure on system designers, managers, and operators.

Complexity

The defining characteristic of complex systems refers to the degree of interconnection and interdependence one finds among the system components (Perrow, 1984). While it is difficult to objectively measure complexity, systems that rate highly in this dimension tend to exhibit a high number of common mode connections, highly interconnected subsystems, numerous feedback loops, and interacting control parameters (Perrow, 1984; p. 88). In part, these characteristics are design responses to pressures to make systems efficient. These qualities can be exploited to make systems highly responsive to changes in demands and to certain classes of internal disturbances.

Highly integrated manufacturing systems offer a prime example of these characteristics. Individual machines and machine cells are informationally linked such that their activities co-determine one another. Thus, the arrival of a new production order or a machine breakdown somewhere in the system can produce a wave of effects in the form of rescheduling production, rerouting part flows, and reconfiguring machines and machine cells, thus allowing the system to dynamically redesign itself in response to changing conditions.

The negative aspect of these qualities is a tendency to result in system dynamics that are highly nonlinear and very difficult for operators to understand or predict accurately, especially in the presence of disturbances (Kugler and Lintern, 1995). Because of the interrelationships among parts of the system, disruptions or anomalies can produce effects that are "distant" (in a physical or functional way) from their source. Single faults can have consequences for multiple system elements. Multiple faults can simultaneously influence individual elements. All of these factors add significantly to the difficulty confronting operators attempting to assess and respond to disturbances in the system.

A typical "error" in complex systems is for operators to miss side-effects of their actions. Because control parameters and systems can interact, operator actions can often have unintended effects on other parts of the system. When these effects are not anticipated or accounted for, operators can be drawn into erroneous situation assessments. For example, an operator may judge that a new and independent fault has occurred when in fact the new indications are a result of the operator's previous actions. In the presence of one or multiple faults, system functioning may be altered such that it becomes impossible for operators to anticipate the full effects of their response actions.

Coupling

Often associated with complexity, coupling represents another dimension of modern complex systems which has direct implications for the way in which these systems fail. Perrow (1984) describes coupling as the amount of "slack" or buffering between system elements. In a tightly coupled system, effects will propagate very quickly between neighboring parts of the system. Again, this is a highly desirable property for systems that need to be sensitive and responsive to changes in demands. Such systems are also more efficient in that they shed superfluous capacity and adaptability in exchange for a more streamlined process. Just-in-time (JIT) inventory management practices represent an example where increasing the coupling in systems can result in demonstrably more efficient performance.

The price of the increased responsiveness and efficiency observed in tightly coupled systems is that the effects of disturbances propagate very quickly, with limited opportunities for intervention, and can rapidly impact overall system functions. The margins between successful system performance and system breakdown become significantly narrowed. For example, the United Parcel Service strike of August 1997 caused rapid, widespread shutdowns in the manufacturing sector due to shortages in parts supplies (a direct result of JIT inventory practices). Under exceptional circumstances (see Latent Failure model below), these characteristics can create a window of opportunity for individual events (e.g., a disturbance or erroneous operator action) to lead quickly and decisively to a large-scale failure of the system.

Uncertainty

There is an irreducible level of uncertainty inherent in complex human–machine systems. For designers, managers, and operators, the complexity of these systems is such that it is difficult if not impossible to be entirely certain of what the effects of certain decisions will be. In the context of operator decision making, uncertainty arises due to the indirect nature of most information. Operators must take into account the possibility of failed sensors, noisy information channels, inaccuracies in human-reported data, stale data, the presence of faults, etc. The complexity of the system also means that there is often a many to one mapping between root causes and the observable symptoms in data. One of the ways these factors manifest themselves is in the form of a trade-off operators must negotiate between their confidence in their assessment of a situation and the value of using resources (time, cognitive effort) to gather more information. Because system elements are highly interdependent and tightly coupled, the effects of a delay in taking action can quickly spread throughout the system. Therefore, operators will often be under considerable pressure to act to preserve the integrity of the system, even if decisions about how to act must be made on the basis of highly uncertain data.

Variability

Ashby's (1956) Law of Requisite Variety states that the possibility for functionally significant variation, or "variety," that the environment presents to a controlled system must be matched by the variety of the control system if effective regulation over the "essential variables" (e.g., productivity, safety) is to be maintained. There are two fundamental ways to ensure that the "essential variables" of a system are unaffected by unwanted variability in the environment. The first is to insulate the system by restricting and controlling the input channels which can transmit variety from the external environment to the system. Examples include the building in which a production system is housed, which serves to block variety due to the weather, to control access by the general public, etc. The second approach to protecting the essential variables is to design variety into the control system to allow it to respond to both externally and internally generated variability. This is most commonly done by attempting to anticipate significant classes of variability and designing preplanned measures specifically to recognize and respond to those types of conditions. Automated safety systems and the development of detailed standard operating procedures (SOPs) are two examples of this method.

However, this second method is necessarily limited because it is impossible, even in principle, for designers to exhaustively foresee every anomaly, every twist of circumstances, and every combination of events that may confront the system. That is, it is impossible to eliminate the potential for *unanticipated variability* in the system. In the extreme case, genuinely novel events can occur, i.e., events for which no planned responses have been developed in advance, for example, the explosion of an oxygen tank on Apollo 13. In these cases, operators must improvise a response with whatever resources are available. At the other extreme, there will be cases that match very well with previously developed plans and that can thus be handled routinely (so called "textbook cases"). However, even in these instances, the presence of SOPs, while reasonable and important, cannot reduce activity to the status of simple rote procedure following, no matter how detailed they are. (See Suchman, 1987 for an extensive discussion of the practical and theoretical limits of plans to specify completely all needed actions in advance.)

Moreover, the most frequent manifestation of unanticipated variability consists of small complicating factors (e.g., multiple faults that mask each other or suggest conflicting responses; see Roth et al., 1992 for a list of examples) that present subtle variations on "textbook cases." These situations challenge operational personnel to consider how previously developed plans and automatic responses are relevant to the unique circumstances they confront and to consider how they should be modified if critical goals are to be achieved (Woods et al., 1990). For example, despite massive efforts by the nuclear power industry to develop comprehensive and fully detailed procedures to guide operators through any conceivable scenario, Roth et al. (1992) were able to show that highly plausible scenarios involving complicating factors could be generated that presented serious challenges to the procedures.

Thus, helping the operational (human–machine) system to confront and absorb the variability of the domain is critical to producing highly robust systems. The notion that a system is reliable in the sense that it performs as designed a high proportion of the time is incomplete — first, because "a high proportion" can never be 100% and second, because circumstances *will* arise that circumvent or challenge the designed features of the system.

The Signature of Complex System Failures

Our usual conception is that a system fails when some single catastrophic event occurs that overwhelms the coping ability of the people on the scene. But because engineers and others are aware of the potential for disaster they develop multiple redundant mechanisms, safety systems, and elaborate policies and procedures to keep them from failing in ways that produce bad outcomes. The results of combined operational and engineering measures make these systems relatively safe from single point failures; that is, they are protected against the failure of a single component or procedure directly leading to a bad outcome.

The scale, complexity, and coupling of these systems create a different pattern for serious failures where incidents develop or *evolve* through a *conjunction* of several small failures, both machine and human (e.g., Turner, 1978; Pew et al., 1981; Perrow, 1984; Wagenaar and Groeneweg, 1987; Reason, 1990). This pattern has been seen in multiple disasters or incidents in a variety of different industries, and despite the fact that each critical incident is unique in many respects.

1. Incidents *evolve* toward failure.

    These incidents evolve through a series of interactions between the people responsible for system integrity and the behavior of the technical systems themselves (the engineered or physiological processes under control). One acts, the other responds, which generates a response from the first and so forth. The incident evolution can be stopped or redirected away from undesirable outcomes at various points.

2. Multiple contributors, each necessary but only jointly sufficient.

    System failures are characterized by a concatenation of several small failures and contributing events rather than a single large failure (e.g., Pew et al., 1981; Reason, 1990). The multiple contributors are all necessary but individually insufficient for the system failure to have occurred. If any of the contributing factors were missing, the failure would have been avoided. Similarly, a contributing disturbance or fault can occur without producing negative outcomes if other potential factors are not present.

3. Human–machine interaction.

    Often the multiple contributing factors include aspects of human–machine interaction.

4. Latent factors.

    Some of the factors that combine to produce a disaster are latent in the sense that they were present before the incident began (Turner, 1978). Reason (1990) uses the term *latent failures* or factors to refer to conditions resident in a system that can produce a negative effect but whose consequences are not revealed or activated until some other enabling condition is met. These conditions are latent or hidden because their consequences are not manifest until the enabling conditions occur. A typical example is a condition that makes safety systems unable to function properly if called on, such as the maintenance problem that resulted in the emergency feedwater system being unavailable during the Three Mile Island incident (The Kemeny Commission, 1979). Latent failures require a trigger, i.e., an initiating or enabling event, that activates its effects or consequences. For example, in the Space Shuttle Challenger disaster, the decision to launch in cold weather was the initiating event that activated the consequences of the latent failure in booster seal design (Rogers et al., 1986).

    Latent failures are typically associated with managers, designers, maintainers, or regulators; people who are generally not directly involved in routine operations and handling incidents and accidents. The latent failure model thus highlights the fact that the causes of system failure are much broader than simply erroneous actions on the part of front-line operators. Organizational activities such as goal-setting, planning, maintaining, and communicating shape and interact with task and environmental conditions, individual unsafe acts, and failed system defenses to produce failures.

An Example: The "Going Sour" Scenario

In the "going sour" class of accidents, an event occurs or a set of circumstances come together that appear to be minor and unproblematic, at least when viewed in isolation or from hindsight. This event triggers an evolving situation that is, in principle, possible to recover from. But through a series of commissions and omissions, misassessments and miscommunications, the human team or the human–machine team manages the situation into a serious and risky incident or even accident. In effect, the situation is managed into hazard (originally, Cook et al., 1991; cf., Woods and Sarter, 1997) Several recent accidents in aviation involving highly automated aircraft show this signature (Billings, 1996).

After the fact, going sour incidents look mysterious and dreadful to outsiders who have complete knowledge of the actual state of affairs (Woods et al., 1994). Since the system is managed into hazard, in hindsight, it is easy to see opportunities to break the progression toward disaster. The benefits of hindsight allow reviewers to comment (Woods et al., 1994, Chapter 6),

- "How could they have missed X; it was *the* critical piece of information?"
- "How could they have misunderstood Y; it is so logical to us?"
- "Why didn't they understand that X would lead to Y, given the inputs, past instructions and internal logic of the system?"

In fact, one test for whether an incident is a going sour scenario is to ask whether reviewers, with the advantage of hindsight, make comments such as, "All of the necessary data was available, why was no one able to put it all together to see what it meant?"

Luckily, going sour accidents are relatively rare even in complex systems. The going sour progression is usually blocked because of two factors:

- The expertise embodied in operational systems and personnel allows practitioners to avoid or stop the incident progression
- The problems that can erode human expertise and trigger this kind of scenario are significant only when a collection of factors or exceptional circumstances come together.

The going sour accident is one kind of latent failure scenario and illustrates how latent failures are a side effect of complexity. The latent failure signature is an evolving process in which there are several points or opportunities to detect that the system is being managed into hazard and to act to recover the situation.

34.5   Adaptation in Human–Machine Systems

Unanticipated variability is a primary example of the factors that can force operational systems to adapt, structurally and behaviorally, away from standard or canonical practices in order to cope with the potential for change. In general, adaptation is a potent concept for describing and interpreting the nature of human–machine systems, particularly with respect to its implications for how systems evolve over time and how we ought to interpret the meaning of "human error" in these systems. Rasmussen (e.g., 1990) has argued convincingly for the need to more fully understand the forces that serve to drive and limit adaptation in complex work environments, both bottom-up through the nature of the technology (e.g., human–machine interfaces) used to support work, and top-down through management practices and organizational structures. The ubiquity and power of adaptation makes it important to understand how it expresses itself in human machine systems, what factors shape the forms of adaptation we see, and particularly how adaptive processes can lead to weaknesses. By examining adaptation more closely we can hope to learn more about how to support it and encourage it in ways that are effective, coherent, and consistent with global goals.

Adaptation can be interpreted as a sort of equilibrium-seeking process, guided by subjective criteria (e.g., speed, quality, robustness against unanticipated variability, etc.), in which people attempt to dynamically match their behavior to the state of their environment. That is, based on their experience and knowledge of their environment, and on available feedback, people will actively try to locate an acceptable balance among their criteria by modifying their behavior, modifying their environment, or both. In any sufficiently complex and dynamic environment, this process will take place continually as demands and constraints shift, appear, and disappear. There will always be forces compelling people to explore new or modified ways of performing tasks. This is especially true in industries such as manufacturing, where the drive to be "flexible" and "agile" (Karwowski et al. 1997) means that the system must constantly reinvent itself to accommodate novel production runs, integrate new process technologies, and dynamically manage breakdowns in the system. Total quality management and continuous improvement strategies attempt to harness people's adaptive tendencies by promoting and capturing positive adaptations.

We take it as a basic premise then that adaptation *does* occur. In fact, it *must* occur to allow complex human–machine systems to function at all. Regardless of how well thought out any system or piece of technology is beforehand, there will always be a gap between the description of system functioning as designed on paper and the complete, situated reality of what must be done to make the system work (cf. the "irremediable incompleteness" of plans; Suchman, 1987). The job of designers and planners is to create a system that *can* work; the function of operators is to resolve the remaining degrees of freedom in ways such that the system *does* work (Rasmussen et al., 1994).

One of the most important ways in which adaptation is influenced is by the nature of the technology with which operators must interact in the performance of their tasks. Embedded within the broader environment of dynamically changing demands and constraints, an ongoing dialogue of co-adaptation between operator and technology continually takes place. This is especially apparent in the context of the introduction of new information technology into the workplace. As Rasmussen (1995) notes, "When the system is put to work, the human elements change their characteristics; they adapt to the functional characteristics of the working system, and they modify system characteristics to serve their particular needs and preferences" (p. 4). Note the dual forms of adaptation implied: not only do operators adapt to the system characteristics, they actively alter the system characteristics to suit their own criteria (Woods et al., 1994, Chapter 5). Cook and Woods (1996) have referred to these co-adaptive processes as *task tailoring* and *system tailoring*. These are fundamental processes in any human–machine work system, and are significant in shaping opportunities for "error" and system failure.

Task and System Tailoring

Operators are not passive in the process of accommodating to changes in technology. Rather, they are actively adaptive. Multiple studies have shown that practitioners adapt information technology provided for them to the immediate tasks at hand in a *locally* pragmatic way, usually in ways not anticipated by the designers of the information technology (Flores et al., 1988; Hutchins, 1990; Cook and Woods, 1996; Obradovich and Woods, 1996). In fact, human adaptive processes can often obscure the effects of poorly designed information technology because humans compensate for the weaknesses of the technology. The point is that the artifacts of information technology that designers introduce to the workplace are shaped by their users until they become useful "tools."

System Tailoring

System-tailoring types of adaptations tend to focus on shaping the technology itself to fit the pre-existing strategies of operators and the demands of the field of activity (e.g., adaptation focuses on the setup of the device, device configuration, how the device is situated in the larger context). For example, in one study (Cook and Woods, 1996), operators set up the new device in a particular way to minimize their need to interact with the new technology during high-criticality, high-tempo periods. This occurred despite the fact that the operators' configurations neutralized many of the putative advantages of the new system (the flexibility to perform greater numbers and kinds of data manipulation). Note that system tailoring frequently results in only a small portion of the "in principle" device functionality actually being used operationally. That is, operators will throw away or alter functionality in order to achieve simplicity and ease of use.

Task Tailoring

In task tailoring, operators adapt their strategies, especially cognitive processing strategies, for carrying out tasks to accommodate constraints imposed by new technology. Thus, task-tailoring types of adaptations tend to focus on how operators adjust their activities and strategies given the constraints imposed by the characteristics of the device. For example, information systems that force operators to access related data serially instead of in parallel result in a proliferation of windows and new window management tasks (e.g., searching for related data, decluttering displays as windows accumulate, etc.). Operators may tailor the device itself, for example, by trying to configure windows so that related data are available in parallel, but they may still need to tailor their activities. For example, the may need to learn when to schedule the new decluttering task (e.g., by devising external reminders) to avoid being caught in a high criticality situation where their first need is to reconfigure the display so that they can "see" what is going on in the monitored process.

Brittle Adaptations

Task and system tailoring represent examples of operators' adaptive coping strategies for dealing with clumsy aspects of new technology, usually in response to criteria such as work load, cognitive effort, ease of use, robustness to common errors, etc. The danger associated with these strategies is that adaptation based on operators' locally defined criteria can lead to "brittle" features in the larger system. In the language of adaptation, local work practices become "overspecialized" with respect to the prevailing conditions, and thus become highly sensitive and prone to failure when these conditions change. For example, data monitoring strategies developed in the context of routine operations may cause critical data to be missed in the context of a fault detection scenario. Such adaptations can thereby become a form of latent failure (Reason, 1990) within the system which can then be triggered by critical changes in the local conditions which originally shaped the adaptation.

The problem is that operator adaptation is a fundamentally local phenomenon, in terms of both time and space. In the absence of influences to the contrary, the criteria which serve to shape adaptation will tend to be applied with respect to *recent* experience and the *current* state of the *immediate* environment. The tendency towards overspecialization occurs when operators (knowingly or not) trade the ability to successfully adapt to novel conditions in exchange for better adaptation to the current, prevailing conditions. That is, there is a tendency to trade long-term adaptability for short-term efficiency and simplicity. Adaptation at higher organizational levels is subject to the same processes.

Thus, while adaptation can be a powerful force for positive change, there is no guarantee that the individually local adaptive activities of operators will result in emergently adaptive changes at a global level. Tailoring can be clever or it can be brittle. Adaptations can conflict, compete, and lead to weaknesses in the system. Therefore, to be successful in a global sense, adaptation at local levels must be guided by the provision of appropriate criteria, informational and material resources, and feedback (Rasmussen et al., 1994). In a sense, this can be seen as a primary function of management. Because operators have privileged access to the dynamic details and demands of their situated work context, management should provide the freedom and resources for operators to adapt to local conditions. But at the same time management must provide ways to constructively constrain and coordinate that adaptation such that global goals such as productivity and safety are protected.

Adaptation and Error

There is a fundamental sense in which the adaptive processes that lead to highly skilled, highly robust operator performance are precisely the same as those that lead to errors. Adaptation is basically a process of exploring the space of possible behaviors in search of stable and efficient modes of performance. Occasionally, exploration may happen unintentionally (e.g., in the case of action slips; Norman, 1981). Other times, the exploration will be a deliberate modification to existing behaviors due to pressures or changes in the environment. Such explorations are not random, but represent "educated guesses," normally based on minor modifications to existing strategies or heuristically guided generation of novel strategies and methods.

"Errors" in this context represent information about the limits of successful adaptation. Adaptation thus relies on the feedback or learning about the system which can be derived from failures, near misses, and incidents. A simple example is the ubiquitous "speed–accuracy trade-off" observed in manual tasks. All other things being equal, it is generally true that as the speed at which a manual task is attempted increases, there comes a point at which the accuracy or quality of the task performance begins to degrade. This point may change as skill increases, but within local bounds this relationship places a practical limit on the speed with which the task can be performed. If the process of adaptation in a manual task is guided by the simultaneous criteria of speed and accuracy, then each individual must attempt to locate an acceptable balance between these two factors. Generally though, the *only* way to locate this point is to continue to increase speed until accuracy begins to suffer; i.e., "errors" begin to occur.

Adaptation with respect to more complex cognitive activities and work practices are subject to the same basic processes. Operators' limited access to the state of the environment, compounded by the inherent variability and uncertainty of the world means that every modification to established behaviors has the potential to result in a negative outcome, particularly if the system is operating at or near acceptable performance limits (see below). If the result is positive, we tend to call it "successful adaptation" and reward the operator. If the result is negative, we tend to call it "human error" and begin remedial action. The point is that in order to reap the rewards of the power of human adaptive processes, *people must be allowed to be wrong.* "Zero tolerance" attitudes toward error, or policies that enforce strict adherence to SOPs in an attempt to eliminate errors, create a double bind by forcing operators to persist in established practices, despite whatever forces of change in the environment may be pushing them to adapt.

Boundary-Seeking Behavior

Rasmussen et al. (1994) have described how complex human–machine systems will naturally tend to migrate toward the boundaries of safe/acceptable performance (i.e., boundaries beyond which significant negative consequences become imminent). In response to pressures of efficiency and work load, managers and operators will naturally adapt in ways that push the system nearer to the edges of its performance envelope. (See Rasmussen et al., 1994, p. 149 for a graphical depiction of this phenomenon.) In a resource-constrained, competitive environment, organizations are rewarded for operating as closely as possible to the edges of system performance limits in pursuit of maximum efficiency, productivity, etc. However, consistent operation near the boundaries of performance means that the system is continually vulnerable to the effects of critical changes in the environment or to erroneous operator actions. In critical situations, events that might otherwise be relatively benign can in fact nudge the system into an unsafe or unacceptable region of performance.

When changes in technology and organizational structures are introduced, the shape of the performance envelope is changed. Some weak points are eliminated; new weak points are created; the nature of weak points can change. The result is not simply a system that is equally safe but more productive. For example, technological changes designed to increase efficiency or safety often do so at the price of increased complexity. This can drastically change the nature and location of the system's performance limits. Moreover, competitive stresses mean that the benefits of changes tend to be taken in productivity gains rather than as increased safety margins. Thus, the system remains vulnerable to failures, but not necessarily the same failures. The system may fail in new, unexpected ways; it may fail more suddenly; the consequences of failure may be more severe. The point is that system performance boundaries are always changing, and that the forces of adaptation will tend to seek out those boundaries. The key to a system that is both efficient and robust in the long run lies not in artificially constraining operators to established work practices, but in allowing them to sense and become familiar with system performance boundaries so that they can recognize when performance limits are being approached and know how to recover when they have been crossed.

34.6   Error Tolerance, Error Recovery, and High Robustness Systems

The implication of the preceding discussion is that systems must be able to cope effectively with the nature of operating at the boundaries of system performance (Rasmussen, 1997). Ultimately, system reliability is measured by outcomes. That is, evaluations of system performance depend not so much on whether the system operated optimally in every detail, but whether the "essential variables" (Ashby, 1956), such as those defined by production or safety goals were satisfactorily kept in their desired ranges. One must accept that systems will experience disturbances, some due to operator actions, which push the system into unacceptable states where negative consequences become imminent. Efforts to improve system reliability cannot focus exclusively on preventing these disturbances because of our limited ability to predict for all eventualities. To achieve a truly robust system, the key is to incorporate mechanisms that allow the system to detect, recover from, and absorb the effects of errors.

Rasmussen (1986) has described the concept of "unkind" work environments, where the effects of errors lead quickly and decisively to failures. Recovery intervals (the period during which the effects of errors can be reversed) are short, errors are difficult to detect, the system degrades quickly, and the consequences are severe. The variabilities in behavior and performance which normally drive adaptive processes become "an unsuccessful experiment with unacceptable consequences" (Rasmussen, 1986). In these sorts of environments, the natural tolerance to errors is very low. What is needed are measures that make the environment "kinder" by supporting not only avoidance, but the detection of and recovery from errors.

"The ultimate error frequency largely depends upon the features of the work interface which support immediate error recovery, which in turn depends on the observability and reversibility of the emerging unacceptable effects. The feature of reversibility largely depends upon the dynamics and linearity of the system properties, whereas observability depends on the properties of the task interface which will be dramatically influenced by the modern information technology"

(Rasmussen, 1985; p. 1188).

Human–Machine Cooperation

Studies of highly reliable, high-performance organizations (e.g., Rochlin et al., 1987; Seifert and Hutchins, 1992; Hutchins, 1990, 1995) have shown that the processes that support error detection and recovery are fundamentally *cooperative* and *distributed*. These studies reveal that reliability *emerges*, not because the individual agents never commit errors, but because the cooperative structures in the system (e.g., shared workspaces, cross-checking strategies) provide mechanisms for catching errors and correcting them before negative consequences begin to accrue. For example, miscommunications between air traffic control and aviation crews are relatively common, but the air transport system has evolved robust cooperative processes such as crew cross-checks and readbacks which help ensure that these errors are revealed and corrected quickly. The reliability of the larger system is not a function of the reliability of the agents but rather of the *ways in which the agents interact*.

The prevalence of advanced automation in human–machine systems makes it important to ask how well automated agents support the cooperative processes which produce highly robust performance in distributed human teams. We must recognize that introducing advanced automation into a system is not simply a matter of substituting machine activity for human activity. Automation changes the cooperative structure of the system and the patterns of communication and coordination that occur among the human and machine agents in the system. Patterns of errors will be changed; some types of error will be eliminated, some will be created. What is needed are automated systems which consider the larger cooperative structure in which they are implemented (Sarter et al., 1997). To make automation support robust systems we must pay close attention to the question of how well such systems support the interactive activities that form the basis for error detection and recovery.

The answer, unfortunately, is often not very well. The properties of advanced automation have tended to hinder cooperation with human partners rather than help it. Woods (1996) has observed that automated systems tend to be:

Strong. Much advanced automation has the ability to act with considerable autonomy. That is, it is capable of performing extended sequences of tasks without any direct intervention from human operators. Closely related to autonomy is the issue of authority. This refers to the automation's ability to *independently* assess when a situation calls for intervention based on its own internal criteria, and to take control of the situation if it finds it to be warranted.

Silent. Many automated systems are "silent" in the sense that they provide little feedback about their activities. This makes it very easy for operators to lose track of the state of the automation. Particularly when combined with the properties of "strong" automation as outlined above, this creates the potential for "automation surprises" (Sarter et al., 1997). These are situations where operators have lost track of the activities of their automated partners and are "surprised" by the actions of the automated system. Operators can be left asking questions of the automation such as "What is it doing?" "Why did it do that?" "What is it going to do next?" (Weiner, 1989). All of these are symptoms of a general breakdown in coordination among the human and the automated system.

Clumsy. One of the putative benefits of the use of automation is an expected decrease in work load for human operators. However, rather than smoothing out the peaks and troughs in operator work load or reducing it in an absolute sense, many times automation has been found to simply amplify patterns of work load in a syndrome termed "clumsy automation" by Wiener (1989). That is, work load reductions tend to occur during periods in which work load was already low, while additional work load tends to appear in periods where work load is already high and the consequences of breakdowns are greatest. It is at these times that the automated system demands the most of the operator in terms of providing input and coordinating activities, leaving the user in the paradoxical position of needing the automated system's help but not having the time or attentional resources to help the automation do so.

Difficult to Direct. Uncooperative automated systems make it difficult for users to interrupt and/or redirect the automation's activities if the operator recognizes a need to do so. If the automation is not designed to support cooperative problem solving, a typical result is that the user's only option for intervening is to essentially "turn off" the automated system and take over the problem in its entirety and full complexity.

Strong, silent, clumsy, difficult to direct automated systems act as uncooperative partners rather than resources adapted to support people as situations vary in tempo and criticality. Such uncooperative interactions in both human–human and human–machine interaction have been recognized as a latent factor in disasters (Billings, 1996). Uncooperative partners lead to miscommunications and misassessments which push situations toward greater hazard, which retard the detection of the deteriorating situation, and impair initiation of recovery strategies.

34.7   The Complexity of Human Error

We have attempted to convey some of the flavor of the results of the past two decades of research into the ways that complex human–machine systems fail and the role of human error. Fundamentally, research has shown that human performance, including errors, is shaped by systematic factors such as task design, human computer interaction, and organizational influences. Reacting to failures as if they were strictly a function of inherent human fallibility prevents us from examining the deeper factors that shape performance and lead to errors and system failures. Although we do not yet understand precisely how all of these factors interact, our ability to predict the sorts of errors that will occur given certain system characteristics is very good.

Pressures of economy and safety have driven up the complexity of modern human–machine systems and have led to a new signature of failure characterized by the presence of multiple contributors, including factors latent in the system. The pressures of the environment, unanticipated variability, and the potential for failure cause people to adapt their behavior and available technology in ways that can be both positive and negative. While adaptation drives much of the behavior we see, it is based on feedback about the limits of successful performance. Moreover, competitive stresses will encourage systems to migrate toward these limits.

A constructive response to failures, near misses, and incidents involves a search for the vulnerabilities, constraints, pressures, and dilemmas behind the label "human error." In the end, improving the system lies in helping people in various roles and at different levels of an organization adapt successfully to these demands of their field of practice. This is best done by making systems error tolerant and by supporting detection and recovery from errors before negative consequences result.

The Paradox of Simultaneous Success and Vulnerability

Ultimately, the difficulties and controversies in dealing with human error arise from seemingly paradoxical effects of our efforts to improve system performance. The highly successful efforts of engineers in protecting systems against catastrophic single-point failures have lead to systems that are "safer" in an actuarial sense, but which at the same time are vulnerable in new ways — the latent failure model and going sour incidents. The efforts to improve systems seem to produce unanticipated side effects on human performance as the systems increase in complexity. The characteristics of complex system failures are perplexing — although individual failures become less frequent, the consequences of failures are more severe. Failures may be fewer, but they more visible and more dreadful to stakeholders.

Effort after Success

It is common for us to think of complex technical systems as inherently safe. When problems occur, we are led to believe that the reason is that the system was somehow interfered with or prevented from operating *as designed*. This belief leads to a search for the (presumably human) guilty parties, and simultaneously blinds us to the tremendous effort invested by human operators in the (normally) successful performance of the system. The truth is that such systems are inherently hazardous. People and organizations are aware of these hazards and actively work to devise defenses against them. This effort after success is needed continuously. When these efforts break down and we see a failure, we gain information, not about the innate fallibilities of people, but about the nature of the threats to complex systems and the limits of the defenses we have put in place.

We can thus begin to see how attributing complex system failures to mere "human error" represents a basic misunderstanding of the nature of complex systems and the relationship between errors and system failure. The story behind human error is every bit as complex as the environments in which it occurs. Human operators behave in ways that are rational given the demands imposed upon them and the combination of cognitive, technological, and organizational resources at their disposal. The typical mismatches between these demands and resources cause operators to develop coping strategies; they structure their environment; they adjust and adapt their behavior to make systems work. Errors are not symptoms of random variability in human performance, but rather represent what is simply the most visible evidence of the mismatches between demands and resources which human ingenuity and diligence hide from us most of the time.

References

Ashby, W. R. (1956). *An Introduction to Cybernetics.* New York, NY: Wiley.

Billings, C. E. (1996). *Aviation Automation: The Search for a Human-Centered Approach.* Hillsdale, NJ: Erlbaum.

Byrne, M. D. and Bovair, S. A working memory model of a common procedural error. *Cognitive Science,* 21(1), 31-61, 1997.

Cook, R. I. and Woods, D. D. (1996). Adapting to new technology in the operating room. *Human Factors,* 38(4), 593-613.

Cook, R. I., Woods, D. D., and McDonald, J. S. (1991). *Human Performance in Anesthesia: A Corpus of Cases*. (CSEL Technical Report 91-TR-03). Columbus, OH: The Ohio State University, Department of Industrial and Systems Engineering, Cognitive Systems Engineering Laboratory.

Fischhoff, B. (1975). Hindsight-foresight: The effect of outcome knowledge on judgment under uncertainty. *Journal of Experimental Psychology: Human Perception and Performance*, 1(3), 288-299.

Fischhoff, B. (1982). For those condemned to study the past: Heuristics and biases in hindsight, in D. Kahneman, P. Slovic, and A. Tversky (Eds.), *Judgment under Uncertainty: Heuristics and Biases*. Cambridge, England: Cambridge University Press.

Fitts, P. M, and Jones, R. E. (1947). *Analysis of Factors Contributing to 460 "Pilot-Error" Experiences in Operating Aircraft Controls* (Memorandum Report TSEAA-694-12). Wright Field, OH: U.S. Air Force Air Materiel Command, Aero Medical Laboratory.

Flores, F., Graves, M., Hartfield, B., and Winograd, T. (1988). Computer systems and the design of organizational interaction. *ACM Transactions on Office Information Systems*, 6, 153-172.

Hollnagel, E. (1993). *Human Reliability Analysis: Context and Control*. London: Academic Press.

Hutchins, E. (1990). The technology of team navigation, in J. Galegher, R. Kraut, and C. Egido (Eds.), *Intellectual Teamwork: Social and Technical Bases of Cooperative Work*. Hillsdale, NJ: Erlbaum.

Hutchins, E. (1995). *Cognition in the Wild*. Cambridge, MA: MIT Press.

Karwowski, W., Warnecke, H. J., Hueser, M., and Salvendy, G. (1997). Human factors in manufacturing, in G. Salvendy (Ed.), *Handbook of Human Factors and Ergonomics* (2ed.), (pp. 1865-1925). New York, NY: Wiley.

Kemeny, J. G. et al. (1979). *Report of the President's Commission on the Accident at Three Mile Island*. New York: Pergamon Press.

Kugler, P. N. and Lintern, G. (1995). Risk Management and the Evolution of Instability in Large-Scale, Industrial Systems, in P. Hancock, J. Flach, J. Caird, and K. Vicente (Eds.), *Local Applications of the Ecological Approach to Human–Machine Systems* (*Vol. 2*), (pp. 416-450). Hillsdale, NJ: Erlbaum.

Lewis, C. and Norman, D. A. (1986). Designing for error, in D. A. Norman and S. W. Draper (Eds.), *User-Centered System Design: New Perspectives on Human–Computer Interaction* (pp. 411-432). Hillsdale, NJ: Erlbaum.

Monk, A. (1986). Mode errors: A user centered analysis and some preventative measures using keying-contingent sound. *International Journal of Man-Machine Studies*, 24, 313-327.

Moshansky, V. P. (1992). *Final Report of the Commission of Inquiry into the Air Ontario Crash at Dryden Ontario*. Ottawa, Canada: Ministry of Supply and Services.

Norman, D. A. (1981). Categorization of action slips. *Psychological Review*, 88, 1-15.

Norman, D. A. (1988). *The Psychology of Everyday Things*. New York: Basic Books.

Obradovich, J. H. and Woods, D. D. (1996). Users as designers: how people cope with poor HCI design in computer-based medical devices. *Human Factors*, 38(4), 574-592.

Perrow, C. (1984). *Normal Accidents: Living with High Risk Technologies*. New York, NY: Basic Books.

Pew, R. W., Miller, D. C., and Feehrer, C. E. (1981). *Evaluation of Proposed Control Room Improvements Through Analysis of Critical Operator Decisions* (NP-1982). Palo Alto, CA: Electric Power Research Institute.

Rasmussen, J. (1990). The role of error in organizing behavior. *Ergonomics*, 33(10/11), 1185-1199.

Rasmussen, J. (1985). Trends in human reliability analysis. *Ergonomics*, 28(8), 1185-1196.

Rasmussen, J. (1986). *Information Processing and Human–Machine Interaction: An Approach to Cognitive Engineering*. New York: North-Holland.

Rasmussen, J. (1995). The concept of human error and the design of reliable human–machine systems. (Invited address) *1st Berliner Workshop on Man-Machine Systems*, Berlin, October 1995.

Rasmussen, J. (1997). Risk management in a dynamic society: A modeling problem. *Safety Science*, 27(2/3), 183-213.

Rasmussen, J., Pejtersen, A. M., and Goodstein, L. P. (1994). *Cognitive Systems Engineering*. New York, NY: Wiley.

Reason, J. (1997). *Managing the Risks of Organizational Accidents*. Aldershot, U.K.: Ashgate.

Reason, J. (1990). *Human Error.* Cambridge, England: Cambridge University Press.

Rochlin, G., Laporte, T. R., and Roberts, K. H. (1987). The self-designing high reliability organization: Aircraft carrier flight operations at sea. *Naval War College Review,* (Autumn), 76-90.

Rogers, W. P. et al. (1986). *Report of the Presidential Commission on the Space Shuttle Challenger Accident.* Washington, D.C.: Government Printing Office.

Roth, E. M., Mumaw, R. J., and Pople, H. E. (1992). Enhancing the training of cognitive skills for improved human reliability: lessons learned from the cognitive environment simulation project, in *Proceedings of the IEEE Fifth Conference on Human Factors in Power Plants.* Monterey, CA: IEEE.

Seifert, C. M. and Hutchins, E. (1992). Error as opportunity: Learning in a cooperative task. *Human-Computer-Interaction, 7,* 409-435.

Sellen, A. J., Kurtenbach, G. P. and Buxton, W. A. S. (1992). The prevention of mode errors through sensory feedback. *Human-Computer-Interaction, 7,* 141-164.

Sarter, N. B. and Woods, D. D. (1995). "How in the world did we get into that mode?" Mode error and awareness in supervisory control. *Human Factors, 37,* 5-19.

Sarter, N. B., Woods, D. D., and Billings, C. E. (1997). Automation surprises, in G. Salvendy (Ed.), *Handbook of Human Factors and Ergonomics (2ed.),* (pp. 1926-1943). New York, NY: Wiley.

Senders, J. and Moray, N. (1991). *Human Error: Cause, Prediction, and Reduction.* Hillsdale, NJ: Erlbaum.

Suchman, L. A. (1987). *Plans and Situated Actions: The Problem of Human–Machine Communication.* New York, NY: Cambridge University Press.

Turner, B. A. (1978). *Man-Made Disasters.* London: Wykeham.

Wagenaar, W. and Groeneweg, J. (1987). Accidents at sea: Multiple causes and impossible consequences. *International Journal of Man-Machine Studies, 27,* 587-598.

Weiner, E. L. (1989). *Human Factors of Advanced Technology ("Glass Cockpit") Transport Aircraft.* (NASA Contractor Report 117528). Moffett Field, CA: NASA-Ames Research Center.

Woods, D. D. (1996). Decomposing automation: apparent simplicity, real complexity, In R. Parasuraman and M. Mouloula, (Eds.), *Automation Technology and Human Performance: Theory and Applications,* (pp. 3-17). Hillsdale, NJ: Erlbaum.

Woods, D. D. and Sarter, N. B. (1997) *Learning from Automation Surprises and Going Sour Accidents.* (CSEL Technical Report 97-TR-05). Columbus, OH: The Ohio State University, Institute for Ergonomics, Cognitive Systems Engineering Laboratory.

Woods, D. D., Roth, E. M., and Bennett, K. B. (1990). Explorations in joint human–machine cognitive systems, in S. Robertson, W. Zachary, and J. Black, (Eds.), *Cognition, Computing and Cooperation.* Norwood, NJ: Ablex.

Woods, D. D., Johannesen, L. J., Cook, R. I., and Sarter, N. B. (1994). *Behind Human Error: Cognitive Systems, Computers, and Hindsight.* CSERIAC State-of-the-Art-Report. Crew Systems Ergonomics Information Analysis Center: Wright-Patterson AFB, OH.

35

Human and System Reliability Analysis

Joseph Sharit
The University of Miami

35.1 The Evolution of Perspectives to Human and System Reliability Analysis

In most scientific applications, the reliability of a system is formally expressed as the probability that the system performs its intended objective. When systems consist only of machine or material components, mathematical methods exist for computing or estimating component or system reliability, either in terms of the probability of functioning normally each time it is used or in terms of the probability that it will not fail during some prescribed time of use (Kapur and Lamberson, 1977; Dhillon and Singh, 1981).

However, when humans are integral to system operation, as is the case in many of today's systems, the problem of reliability assessment is altered. First, there is the need to assess human reliability. Second, the assessment of system reliability must now take into account the effects, both immediate and delayed, of human behaviors on other components of the system, including other humans. Although a reasonable understanding of human reliability is pivotal to evaluating system reliability, as human–system interactions become more complex the balance in emphasis will likely shift to methods that focus on how these interactions can adversely affect system reliability, even in the absence of identifiable human reliability problems or human errors.

With engineering reliability methods having matured much earlier than methods of human reliability analysis (HRA), it is not surprising that approaches to HRA initially emphasized computing probabilities of human error. This perspective enabled computations of system reliability based on probability assessments of all individual system components — mechanical and human components alike. However, improved understanding of cognitive and sociotechnical processes and development of approaches to the analysis of work settings has led to a shift in emphasis, away from computing human error probabilities (HEPs) and toward understanding how and why human errors occur. The underlying premise of this qualitative perspective to HRA is that with this understanding we can better predict the potential for particular types of human errors as well as better understand why accidents or adverse system consequences occur, thereby providing insights into system reliability and the design of countermeasures that are not necessarily apparent from application of quantitative techniques. However, this perspective also places much greater demands on analysts to identify and analyze the complex texturings underlying work contexts that are capable of triggering human errors or propagating human actions into adverse system outcomes.

Paralleling this qualitative or systems perspective was a change in safety philosophy adopted by some industries. The traditional safety perspective emphasized a blame culture whereby accidents were considered the fault of the worker who either violated procedures or was not exercising proper caution. This perspective was gradually being replaced in some corporations by safety philosophies that recognized the role of organizational structure and management in shaping work environments that promote human errors and violations (Reason, 1990, 1995). Although quantitative approaches to HRA also recognized such factors, they were used either to modify or directly assess the probability of human error. In contrast, the qualitative perspective to HRA was primarily interested in understanding the link between these factors and human behaviors that result in errors, violations, and inadequate performance.

35.2 The Current Dilemma Facing Industries

As a result of these developments, there are currently many options available to industries who are interested in human and system reliability, including the consideration of hybrid approaches that combine both quantitative and qualitative perspectives. First, industries must determine the perspective that is most consistent with their goals and interests. If the industry is primarily concerned with meeting regulatory standards that require "quantitative risk assessments" — i.e., that require the derivation of probabilities associated with potentially hazardous system events — then a quantitative perspective to human and system reliability will be required. However, if an industry is concerned about a growing culture of procedural violations that have potentially important implications for worker safety and system productivity, then a qualitative perspective emphasizing sociotechnical considerations will likely be more appropriate.

Even if an appropriate or dominant perspective can be identified, decisions still need to be made concerning the degree of depth to which human (and system) reliability analysis should be pursued. For example, with quantitative approaches to HRA, methods range from relatively quick assessments to very involved analyses. These decisions will be based, in part, on the availability of resources and the expertise of analysts. Decisions also need to be made concerning the degree to which other perspectives are to be incorporated. The current pressures industries face, which include the need to reduce accidents, litigation, and worker compensation costs while increasing both the quantity and quality of the product, often demand a well-balanced, multifaceted perspective to the problem of human and system reliability analysis.

Industries also need to consider issues of training, data collection, and data organization required for maintaining and improving HRA programs that are put in place. For example, if an industry's objective is to improve the design of written work procedures in order to minimize human errors that may occur during execution of these procedures, some form of training program will likely be necessary that enables designers to anticipate the potential for such errors. Such anticipation will necessarily require some understanding of the interrelationships between work contexts and tendencies humans have in processing information that form the basis for various types of errors. Finally, the ability of an industry to determine whether they have adopted an appropriate perspective (or "model") and have the capability to continuously improve human and system reliability through the implementation of design strategies requires having mechanisms in place that dictate what information should be collected, how it should be organized, and how it should be evaluated in order to determine what adjustments may be needed to the human and system reliability programs (Center for Chemical Process Safety, 1994).

35.3 Hazard Analysis Techniques and Quantitative Risk Assessments

It is perhaps misleading to suggest that qualitative and quantitative approaches to HRA are necessarily mutually exclusive. When one considers that HRA generally encompasses human error identification, prediction, and reduction, some degree of qualitative emphasis is necessarily implied. The extent to which a quantitative perspective is adopted will likely depend on the extent to which the HRA will serve as input to a quantitative risk assessment. Depending on the industry, this type of system reliability assessment is typically referred to as a probabilistic safety analysis (PSA) or probabilistic risk assessment (PRA). Most industries that carry out such assessments, such as the chemical processing or nuclear power industries, are concerned about hazards arising from interactions between hardware and software failures, environmental events, and human errors and violations that can result in injuries, fatalities, and damage to the plant and environment.

A variety of well-known hazard analysis techniques have been developed that can be used to identify hazards (Center for Chemical Process Safety, 1992). These methods are usually differentiated based on the stage in the system design life cycle on which the analysis is being performed. At the conceptual and preliminary design stages of system development, Preliminary Hazard Analysis, Failure Mode and Effects Analysis, and HAZOP techniques are often used. These approaches, by and large, enable design solutions to be implemented, including those incorporating human factors and ergonomics design principles, for the various undesirable consequences of hazards that were identified. PRAs, however, are usually more concerned with undesirable events at the later stages of the system's life cycle that cannot be resolved by these methods.

The two primary hazard analysis techniques that have become associated with PSAs and PRAs are Fault Tree (FT) Analysis and Event Tree (ET) Analysis. Both have as starting points undesirable events to which hazards contribute. These events may have been identified through informal methods such as expert opinion, or from previous hazard analysis techniques such as HAZOP which were unable to further resolve the hazard. The FT, however, represents a deductive, top-down decomposition of this event, whereas the ET represents an inductive analysis that determines how this event can propagate, and consequently the ways in which the event can be circumvented. In both cases, human interventions in combination with other system components and environmental factors can be assessed.

Figure 35.1 illustrates an ET for an offshore emergency shutdown scenario. This particular ET addresses the sequence of human actions in response to the initiating event and is therefore often referred to as an Operator Action Event Tree. Each branch of the tree represents either success (the upper branch) or failure (represented in this diagram as a HEP) in achieving the required actions specified along the top. The probability of each failure state on the right is the product of the error and/or success probabilities of each branch leading to that failure state, where the overall probability of failure is the sum of the individual failure states. Dashed lines indicate paths through which recovery from previous errors can occur.

A portion of an FT developed by Ozog (1985) associated with loading a flammable liquid storage tank from a tank truck is illustrated in Figure 35.2 (Center for Chemical Process Safety, 1989). As indicated in this figure, FTs are essentially Boolean logic models that depict the relationships between events in a system — either hardware, human, or environmental — that lead to a final outcome or top event which, in principle, could be either desirable or undesirable. Qualitative applications of FTs emphasize the different combinations of events that could lead to the top event. For many applications, this information is revealing in its own right. As a quantitative method, "basic events" or events for which no further analysis of the cause is carried out, are assigned probabilities which can either be static or in the form of rate measures that take into account dynamic considerations. Methods are then employed to propagate these values into either a probability or rate measure associated with the top event (Dhillon and Singh, 1981). These methods also enable the individual contributions of events to the top event to be computed, making this technique very suitable for performing cost-benefit analyses that can lead to design interventions.

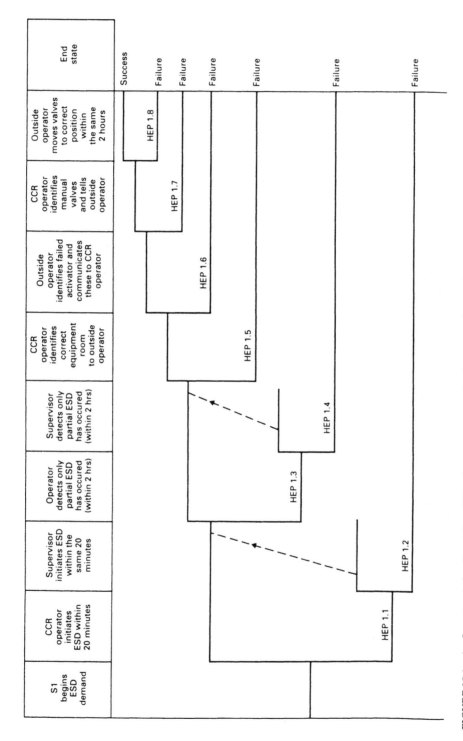

FIGURE 35.1 An Operator Action Event Tree. ESD refers to emergency shutdown procedure, and CCR refers to chemical control room. (From Kirwan, B. (1994). *A Guide to Practical Human Reliability Assessment.* London: Taylor & Francis. With permission.)

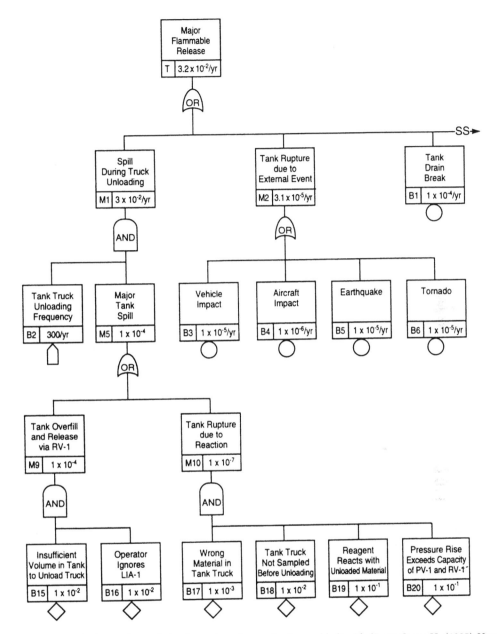

FIGURE 35.2 A fault tree for analysis of a major release of a flammable liquid. (From Ozog, H. (1985) Hazard identification, analysis, and control. *Chemical Engineering*, February 18, 161-170. With permission.)

FTs often differ in their depictions due to subtle distinctions in symbology (Roland and Moriarty, 1983). In Figure 35.2, the circles under the rectangles denote basic events; the house symbol denotes a normal event that either occurs or does not; and the diamond symbol denotes an undeveloped event that the analyst chooses not to analyze further but one that, in theory, is at a sufficiently high level in the decomposition process that it can be further analyzed. Events for which these symbols are not present represent intermediate events that are further analyzed through successive levels of causality as described by logical AND and OR relationships between events.

In those cases where the top event in an FT represents the initiating event potentially leading to accidental conditions in an ET, Cause Consequence Charts which combine FTs and ETs can be used

(CCPS, 1992). Although potentially cumbersome, they are capable of representing both the deductive and inductive analyses associated with a particular undesirable event.

When FTs and ETs are used as a basis for performing PSAs, they necessarily require quantitative solutions. However, regardless of whether FTs or ETs are used there will be a need to quantify the human error contributions to the event of interest. In addition to these solutions, PSAs also involve determining the risks associated with the consequences of these events, and ultimately whether the results of the analysis are consistent with the industry's (or regulatory agency's) acceptable risk criteria.

35.4   The HRA Process

The HRA process as viewed by Kirwan (1994) is presented in Figure 35.3. In this depiction, the qualitative–quantitative emphasis issue is addressed in the problem definition stage, representing the first step in the HRA process. If the HRA is not driven by the need to perform a PSA, the emphasis of the HRA will likely be qualitative. A quantitative component may be useful for prioritizing different human errors, especially in meeting the objectives of safety improvement programs that have limited resources for implementing risk reduction strategies. However, such quantitative components need not invoke many of the assumptions that underlie approaches that emphasize quantification of HEPs.

Clearly, the most important aspect of HRA following problem definition, and which qualitative approaches are extending to increasingly greater levels of depth and sophistication, is the identification and modeling of human errors (Figure 35.3). In general, this aspect remains insufficiently addressed in HRAs driven by PSAs, where the balance in emphasis on human errors is toward quantification. As noted earlier, there is a growing concern in many high-risk systems (such as health care) for understanding adverse system outcomes that may not arise from human error *per se*, but rather from the complex couplings between organizational, individual, collaborative work group, and environmental factors, including human tendencies for errors and violations (Sharit et al., 1996). In these cases, HRAs and not PSAs will determine which situations should be analyzed and how they should be assessed.

For HRAs that are driven by PSAs, a number of issues need to be addressed, including specification of PSA criteria (such as the target criteria for the PSA) and the types of human interactions that will be dealt with (e.g., whether human-initiated accident sequences will be considered in addition to those initiated by hardware failures or environmental events). Many of these issues relate to the scope of the HRA, which will depend on the system's degree of vulnerability to human error. Systems that are complex in terms of the underlying system design processes, highly interactive in terms of the relationships between subsystems and components, and tightly coupled in the sense that such interactions provide minimal flexibility in preventing negative system events to be isolated from system processes, represent high-risk systems (Perrow, 1984) that are likely to require more detailed HRAs. This increase in scope, in turn, implies that the human error analysis component of the HRA (Figure 35.3) be given greater emphasis. It is in these cases where justification would likely exist for investing resources in high-level or hybrid HRAs that emphasize both the qualitative and quantitative aspects of HRA.

Task Analysis, Cognitive Task Analysis, and System Analysis

Most HRAs, whether quantitative or qualitative in emphasis, require some form of organizing structure for describing human involvement with other system components in order to identify human behaviors that could potentially lead to undesirable consequences. This would also provide a basis for suggesting error reduction design interventions and for conducting follow-up assessments of these interventions. Task analysis (TA) represents the most formal approach to this problem. Many TA methods exist that differ primarily on the basis of the types of information acquired and in the ways in which this information is represented. These distinctions, in turn, depend on the types of problems the analyst is interested in. A particular type of TA technique referred to as hierarchical task analysis (HTA) can readily organize large amounts of task-based information (Stammers and Shepherd, 1995), making it particularly suitable for HRAs involving high-risk, but relatively routine tasks. An illustration of the use of HTA in HRA will

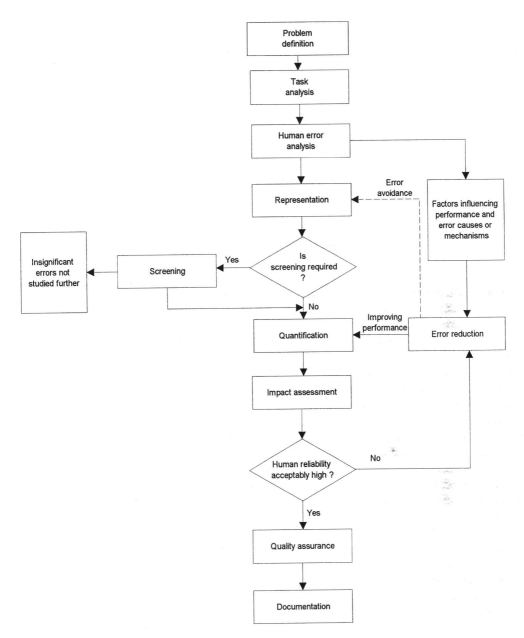

FIGURE 35.3 The HRA process. (From Kirwan, B. (1994). *A Guide to Practical Human Reliability Assessment.* London: Taylor & Francis. With permission.)

be presented in the next section. Reviews of task analysis techniques can be found in Kirwan and Ainsworth (1992), CCPS (1994), and Luczak (1997).

While TA remains essential in almost any study of human reliability, qualitative approaches to HRA benefit significantly from cognitive task analysis and systems analysis. Cognitive task analysis (Rasmussen, 1986; Rasmussen et al., 1994) focuses on cognitive processes that support human planning, problem solving, and decision making and control activities that are often expressed through interactions with computer-based interfaces to complex systems. Examples of such systems include expert system based fault-diagnostic systems, nuclear power systems, intelligent vehicle highway systems, and intelligent information management systems that act as pilot assistants in advanced aviation. Cognitive task analysis often relies on "modeling frameworks" for deriving insights concerning human performance. For example, the distinctions between skill-based, rule-based, and knowledge-based levels of performance (discussed in the next section), and the use of a "stepladder" model of human performance developed by Rasmussen (1986) for depicting the relationships between various stages of information processing that occur in response to process disturbances (Figure 35.4) can, when used in conjunction, enable predictions to be made concerning the types of errors that can be expected.

Cognitive task analysis would generally be preceded by a systems analysis (Rasmussen et al., 1994). In HRA, systems analysis entails describing the overall system, in terms of its various characteristics, in ways that can provide insights into work contexts that are most relevant to the problem of human and system reliability analysis. A brief illustration of systems analysis is provided in the following section. When systems analysis is combined with cognitive task analysis, the tendencies for performing at either the skill-, rule-, or knowledge-based levels can be determined. In addition, analysis at the systems level can reveal tendencies for errors and violations arising due to complex "higher-level" system factors. Cognitive task analysis and the more traditional task analysis would then provide additional layers of texture in the analysis of work contexts that could enhance predictions concerning these errors, including their types and the potential for their recovery, and also provide more detailed design recommendations for their reduction.

35.5   The Qualitative Perspective to Human and System Reliability Analysis

The qualitative perspective to HRA emphasizes system factors — specifically, the contexts that arise from the interplay of system factors. The identification and analysis of system contexts can potentially provide insight not only into how situational factors can trigger particular types of human errors and how these errors can propagate into adverse outcomes, but also how human interventions, even if not considered erroneous, can lead to negative consequences. In addition, the qualitative perspective is also potentially capable of predicting worker violations, depending on the extent to which factors such as organizational structure and work culture are considered.

In adopting this perspective, it is important that the analyst choose an appropriate system description that is consistent with the general problem definition. System descriptions can take many forms (cf. Sharit, 1997, pp. 303-305), and ultimately the analyst may choose to explore the degree to which different descriptions provide different insights. For example, in a large-scale trauma center, system descriptions in the form of: (1) defining relevant subsystems and the links between the different subsystems within which a trauma patient may directly or indirectly receive care; (2) the types of collaborations that exist between health care providers both within and across these subsystems; (3) the temporal constraints governing these collaborations; and (4) the communication channels between these providers, provide a basis for understanding the possibility for corruption of relevant patient-based information, and thus the occurrence of adverse system consequences. This analysis, in turn, would dictate the methods that would be adopted for analyzing human–system interactions in more detail in order to better evaluate human and system reliability. In contrast, in the case of a nuclear power plant control room, a different approach to systems description may be advised. For example, when the concern is for a nuclear power plant operator's capabilities in handling abnormal events through interaction with a control room computer interface, a more appropriate system description may be one based on an abstraction hierarchy (Rasmussen, 1986; Vicente and Rasmussen, 1992).

Unlike the quantitative perspective, which is driven by quantitative risk assessments and design recommendations that are primarily influenced by risk potential, the qualitative perspective is compatible with the "proactive" problem of predicting human and system reliability as well as with the "retrospective" problem of analysis of accidents. In the latter case, a clear starting point will almost invariably result in the identification of accident causes. However, when the objective is prediction, there are many "contextual paths" that could conceivably lead to reduced human and system reliability. These distinct cases are illustrated in Figure 35.5.

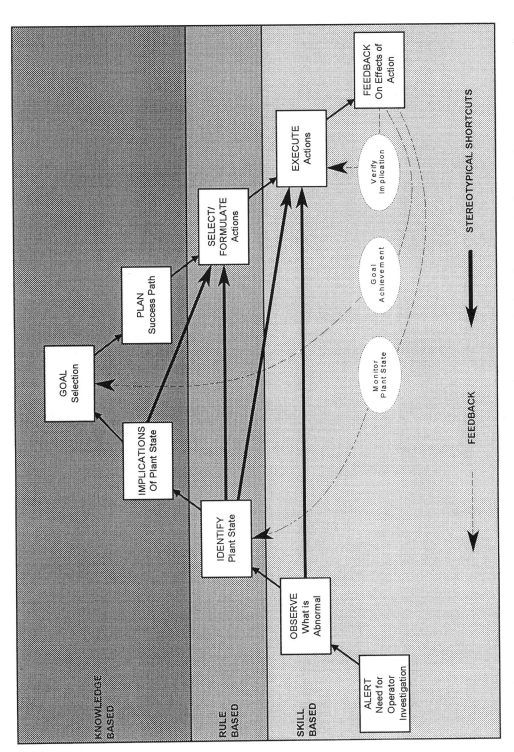

FIGURE 35.4 A "stepladder" model for depicting the relationships between various stages of information processing that occur in response to process disturbances. (From Center for Chemical Process Safety (CCPS). (1994). *Guidelines for Preventing Human Error in Process Safety*. New York: American Institute of Chemical Engineers. With permission. Adapted from Rasmussen, J. (1986). *Information Processing and Human-Machine Interaction*. Amsterdam.)

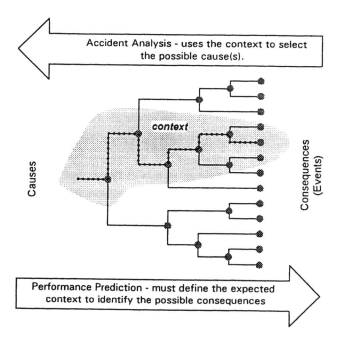

FIGURE 35.5 Accident analysis and performance or human error prediction. (From Hollnagel, E. (1993). *Human Reliability Analysis: Context and Control.* London: Academic Press. With permission.)

Prediction of Human Error

Prediction of human error and violations, and more generally, of adverse system outcomes, requires understanding the interplay between three factors: (1) the general error-inducing environment; (2) particular types of events characteristic to the system that could promote or trigger human behaviors leading to adverse system consequences; and (3) the tendencies humans have for committing errors and violations. These factors are summarized in Figure 35.6 for a trauma care system application (Sharit et al., 1996). As discussed earlier, the particular system analysis framework adopted will influence the characterization of the error-inducing environment. This analysis can be further aided by tools or methods that focus on identifying and elaborating on relevant sociotechnical factors, and by analysis of Performance Shaping Factors (PSFs). Discussion of each of these concepts follows.

An example of a "systems method" that considers sociotechnical factors is the Human Factors Analysis Methodology or HFAM (Pennycook and Embrey, 1993). This methodology is comprised of 20 groups of factors that are subdivided into three broad categories: (1) management-level factors (such as degree of worker participation, effectiveness of communications, and effectiveness of procedures development system); (2) operational-level generic factors (such as work group factors, training, process management, and job aids and procedures); and (3) operational-level job-specific factors (such as computer-based systems, control panel design, and maintenance). HFAM first invokes a screening process to identify the major areas vulnerable to human error (e.g., maintenance operations involving steam generators in nuclear power plant operations); the generic and appropriate job-specific factors are then applied to these areas. The components of each factor that applies can then be evaluated at two levels of detail as illustrated in Figure 35.7. The problems that are identified ultimately reflect failures at the management control level. Corresponding management-level factors would then be evaluated to identify the nature of the management-based error; these types of errors are often referred to as "latent" errors (Reason, 1990). Management-level factors fall into various categories, including: (1) those that can be specifically linked to operational-level factors (e.g., training, procedures); (2) those that are indicators of the quality of the safety culture and therefore can affect the potential for both errors and violations; and (3) those

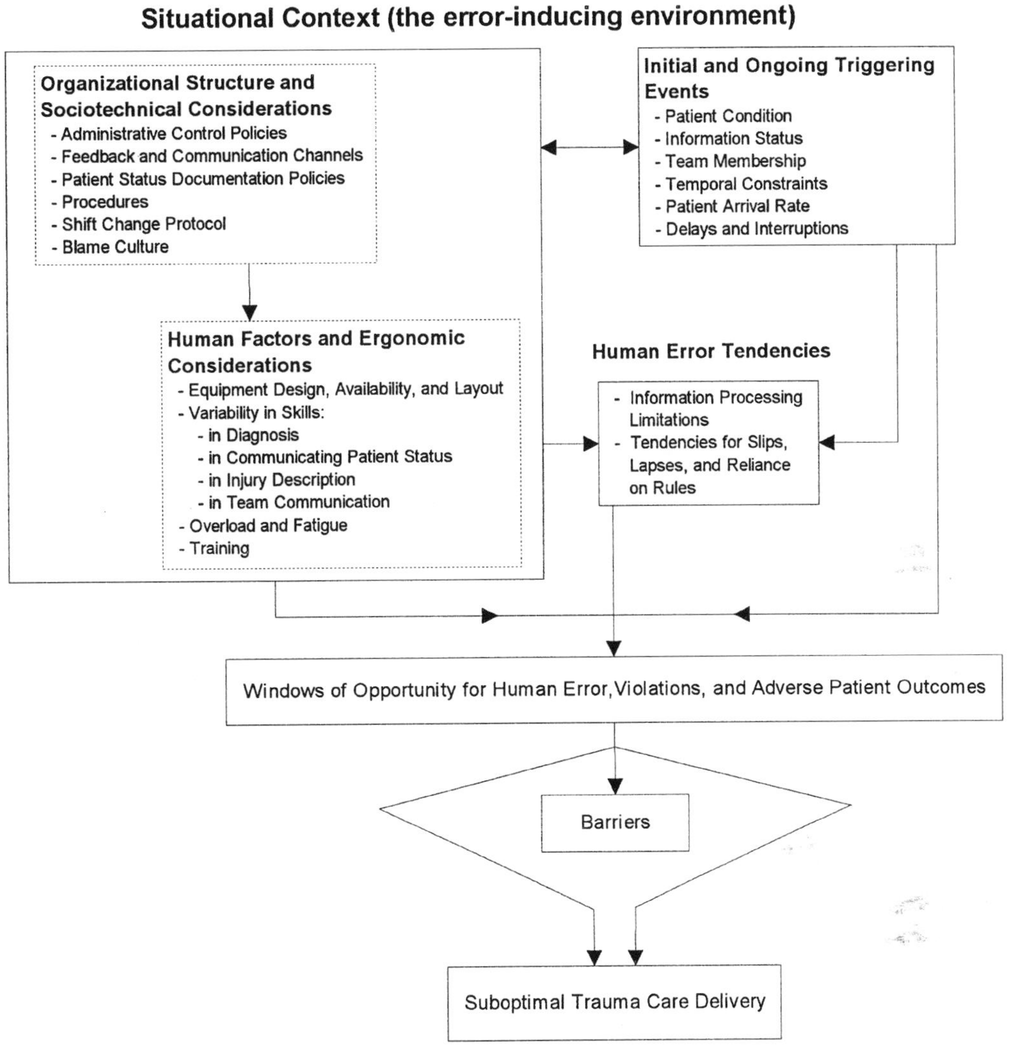

Situational Context (the error-inducing environment)

Organizational Structure and Sociotechnical Considerations
- Administrative Control Policies
- Feedback and Communication Channels
- Patient Status Documentation Policies
- Procedures
- Shift Change Protocol
- Blame Culture

Initial and Ongoing Triggering Events
- Patient Condition
- Information Status
- Team Membership
- Temporal Constraints
- Patient Arrival Rate
- Delays and Interruptions

Human Factors and Ergonomic Considerations
- Equipment Design, Availability, and Layout
- Variability in Skills:
  - in Diagnosis
  - in Communicating Patient Status
  - in Injury Description
  - in Team Communication
- Overload and Fatigue
- Training

Human Error Tendencies
- Information Processing Limitations
- Tendencies for Slips, Lapses, and Reliance on Rules

Windows of Opportunity for Human Error, Violations, and Adverse Patient Outcomes

Barriers

Suboptimal Trauma Care Delivery

FIGURE 35.6 The interplay between situational context and human error tendencies in generating windows of opportunity for human error and adverse system outcomes. (From Sharit, J., Czaja, S. J., Augenstein, J., and Dilsen, K. (1996). A systems analysis of a trauma center: a methodology for predicting human error, in A. F. Ozog and G. Salvendy (eds.) *Advances in Applied Ergonomics*, 996-1101. Indiana: U.S.A. Publishing Corporation. With permission.)

that reflect communication of information throughout the organization, including the capability for learning lessons from operational experience based on various forms of feedback channels.

In principle, Performance Shaping Factors (PSFs), also referred to as Performance Influencing Factors (PIFs), represent any factors that can influence the potential for human error or violations. As will be discussed, in quantitative HRA they are used to modify or directly compute estimates of HEPs. In qualitative human and system reliability assessments, PSFs can serve a number of purposes. For example, their identification and assessment are important components of human factors and ergonomics audits that are used for establishing which design features might be susceptible to human error. This application of PSFs can also be used by workers as part of a participative error-reduction program. When used for *predicting* human error and adverse system outcomes, PSFs essentially shape the system context — their interplay with initiating and ongoing system events and with human error tendencies define contexts conducive to error and adverse outcomes (Figure 35.6).

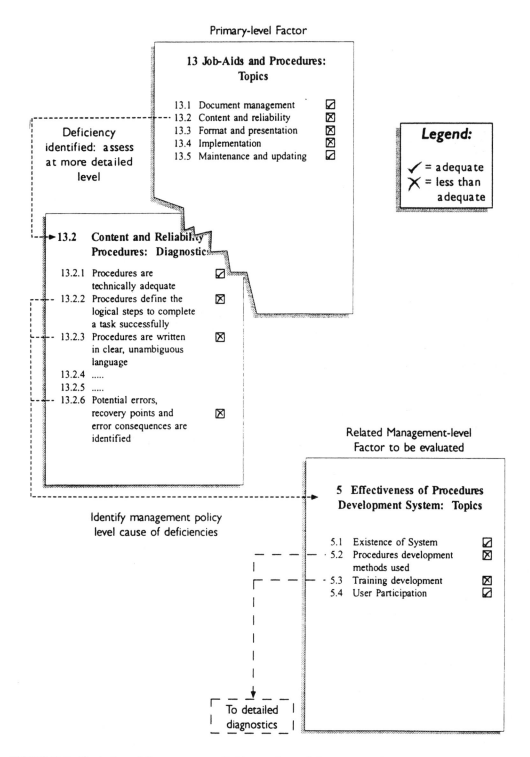

FIGURE 35.7   Illustration of the structure applied in one form of the HFAM tool. (From Pennycook, W. A. and Embrey, D. E. (1993). An operating approach to error analysis, in *Proceedings of the First Biennial Canadian Conference on Process Safety and Loss Management*. Edmonton, Alberta, Canada. Waterloo, Ontario, Canada: Institute for Risk Research, University of Waterloo. With permission.)

Table 35.1 presents a classification structure for PSFs compiled by Swain and Guttmann (1983) for application to the nuclear power industry. It is not uncommon for analysts to devise their own classification schemes based on the characteristics of the particular operating environment or organization. In general, at least three broad categories of PSFs need to be considered: those related to demands (e.g., task and physical work environment factors), resources (e.g., job aids and training), and management policies. The assumption generally adopted by both the quantitative and qualitative perspectives is that when all PSFs relevant to a particular situation are optimal, error likelihood will be minimized. However, errors will still occur due to a phenomenon known as "stochastic variability" in human performance that can derive, for instance, from movement variability or from unique intentions and biases. Note that many of the sets of factors in HFAM are essentially PSFs which are expressed at increasingly finer levels of definition.

While PSFs are essential for constructing a framework for error prediction, as alluded to earlier many critical links between human behavioral tendencies and the situational context will still likely be "under-specified." The additional level of articulation generally required is typically provided by task and cognitive task analysis. As with systems analysis, however, it is important that the analyst choose an appropriate method of task analysis (TA) in order to expose the subtleties that best satisfy the contextual systems model. A number of taxonomies exist for classifying human performance (Fleishman and Quaintance, 1984), and many variations and hybrid schemes can be derived from these fundamental structures. Returning to the example of complex trauma care systems, a TA method that, among other considerations, emphasizes: (1) how teams of workers coordinate activities over time and different locations; (2) the constraints imposed on workers relating to how information is documented; (3) the variability of worker skills for various areas of health care expertise; (4) the types of communication protocol; and (5) the opportunities for interruptions and delays, are examples of the types of descriptions required for elaboration of the error-inducing environment.

Areas where cognitive task analysis could help further establish the potential for error should now be more readily identifiable. For example, it can be used to analyze how the constraints imposed by a computer-based information system on information documentation can affect information processing activities by other workers who must assimilate this documented patient information with other data during information acquisition activities. At this point, an understanding of human error tendencies based on models of human information processing and human error is needed.

Human Error: Information Processing, Classification Schemes, and Models

The Traditional Human Factors Approach

Predicting human error requires some understanding of the relationships between the various attentional processes or components comprising the human's information processing system (Figure 35.8). Even with only a very fundamental appreciation of these processes it may, for example, be possible to predict that: (1) a worker does not have enough time to input information accurately given the design of the interface; (2) the design of displays is likely to evoke control responses that are contraindicated; (3) equipment is positioned in a manner that makes it likely that a poor position will be adopted when performing some activity or that other operations will be interfered with; and (4) decision making will take place without the benefits of complete or unambiguous information.

This perspective to prediction essentially translates the contextually rich system-based and task-based information into task demands. Mismatches between these demands and the human's capabilities for meeting these demands, which are largely reflected in information processing considerations (Figure 35.8), help map out areas with increased vulnerability to human error. Depending on the objectives of the human and system reliability assessment, this type of analysis may be sufficient. It is, however, suboptimal in that it is generally incapable of predicting the *types* of errors that might occur, and in this sense it is likely to underutilize contextually rich descriptions of the error inducing environment.

Systematic Approaches to Error Prediction

Systematic approaches to error prediction provide several advantages, including the ability to more easily rationalize and document the consequences of errors and error reduction strategies. This capability, in

TABLE 35.1 Examples of Performance Shaping Factors

External PSFs	Stressor PSFs	Internal PSFs
Situational Characteristics Those PSFs General to One or More Jobs in a Work Situation	**Psychological Stressors:** PSFs which Directly Affect Mental Stress	**Organismic Factors:** Characteristics of People Resulting from Internal & External Influences

Architectural features
Quality of environment:
  Temperature, humidity, air quality, and radiation
  Lighting
  Noise and vibration
  Degree of general cleanliness
Work hours/work breaks
Shift rotation
Availability/adequacy of special equipment, tools, and supplies
Manning parameters
Organizational structure (e.g., authority, responsibility, communication channels)
Actions by supervisors, co-workers, union representatives, and regulatory personnel
Rewards, recognition, benefits
	Suddenness of onset
	Duration of stress
	Task speed
	Task load
	High jeopardy risk
	Threats (of failure, loss of job)
	Monotonous, degrading, or meaningless work
	Long, uneventful vigilance periods
	Conflicts of motives about job performance
	Reinforcement absent or negative
	Sensory deprivation
	Distractions (noise, glare, movement, flicker, color)
	Inconsistent cueing
		Previous training/experience
		State of current practice or skill
		Personality and intelligence variables
		Motivation and attitudes
		Emotional state
		Stress (mental or bodily tension)
		Knowledge of required performance standards
		Sex differences
		Physical condition
		Attitudes based on influence of family and other outside persons or agencies
		Group identifications

Task and Equipment Characteristics: Those PSFs Specific to Tasks in a Job	**Physiological Stressors:** PSFs which Directly Affect Physical Stress	
Perceptual requirements Motor requirements (speed, strength, precision) Control–Display relationships Anticipatory requirements Interpretation Decision-making Complexity (information load) Narrowness of task Frequency and repetitiveness Task criticality Long- and short-term memory Calculational requirements Feedback (knowledge of results) Dynamic vs. step-by-step activities Team structure and communication Man–machine interface factors: design of prime equipment, test equipment, manufacturing equipment, job aids, tools, fixtures	Duration of stress Fatigue Pain or discomfort Hunger or thirst Temperature extremes Radiation G-Force extremes Atmospheric pressure extremes Oxygen insufficiency Vibration Movement constriction Lack of physical exercise Disruption of circadian rhythm	
Job and Task Instructions: Single Most Important Tool for Most Tasks		
Procedures required (written or not written) Written or oral communications Cautions and warnings Work methods Plant policies (shop practices)		

Source: Swain, A. D. and Guttmann, H. E. (1983). *Handbook of Human Reliability Analysis with Emphasis on Nuclear Power Plant Applications.* NUREG/CR-1278, U.S. Nuclear Regulatory Commission, Washington, D.C.

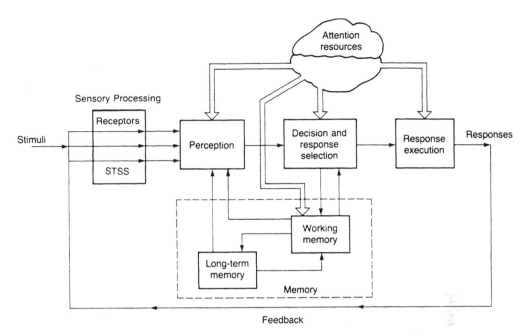

FIGURE 35.8 A model of human information processing (STSS refers to the short-term sensory store). (From Wickens, C. D. (1992). *Engineering Psychology and Human Performance*, Second Edition. New York: HarperCollins. With permission.)

Action		Retrieval	
A1	Action too long / short	R1	Information not obtained
A2	Action mistimed	R2	Wrong information
A3	Action in wrong direction	R3	Information retrieval incomplete
A4	Action too little / too much		
A5	Misalign	**Transmission**	
A6	Right action on right object	T1	Information not transmitted
A7	Wrong action on right object	T2	Wrong information transmitted
A8	Action omitted	T3	Information transmission incomplete
A9	Action incomplete		
A10	Wrong action on wrong object	**Selection**	
		S1	Selection omitted
Checking		S2	Wrong selection made
C1	Checking omitted		
C2	Check incomplete	**Plan**	
C3	Right check on wrong object	P1	Plan preconditions ignored
C4	Wrong check on right object	P2	Incorrect plan executed
C5	Check mistimed		
C6	Wrong check on wrong object		

FIGURE 35.9 An example of an error taxonomy used as part of a systematic methodology for predicting human error for the chlorine tanker filling problem. (From Center for Chemical Process Safety (CCPS). (1994). *Guidelines for Preventing Human Error in Process Safety*. New York: American Institute of Chemical Engineers. With permission.)

turn, increases the chances of management's acceptance of the HRA program. An example of one such approach is SPEAR — System for Predictive Error Analysis and Reduction (CCPS, 1993). SPEAR consists of five steps: task analysis (TA), PSF analysis, predictive human error analysis, consequence analysis, and error reduction analysis. For each step of the TA, and taking the influence of PSFs into consideration, the occurrence of one or more errors derived from some type of "error taxonomy," such as the one presented in Figure 35.9, is evaluated.

An example of this approach taken from CCPS (1993) involves the filling of a storage tank with chlorine from a tank truck. The primary purpose of this analysis was to identify potential human errors that could

0. **Fill tanker with chlorine**
    Plan: Do tasks 1 to 5 in order.

1. **Park tanker and check documents (not analyzed)**

2. **Prepare tanker for filling**
    Plan: Do 2.1 or 2.2 in any order then do 2.3 to 2.5 in order.
    2.1 Verify tanker is empty
    Plan: Do in order.
        2.1.1 Open test valve
        2.1.2 Test for Cl_2
        2.1.3 Close test valve
    2.2 Check weight of tanker
    2.3 Enter tanker target weight
    2.4 Prepare fill line
    Plan: Do in order.
        2.4.1 Vent and purge line
        2.4.2 Ensure main Cl_2 valve closed
    2.5 Connect main Cl_2 fill line

3. **Initiate and monitor tanker filling operation**
    Plan: Do in order.
    3.1 Initiate filling operation
    Plan: Do in order.
        3.1.1 Open supply line valves
        3.1.2 Ensure tanker is filling with chlorine
    3.2 Monitor tanker filling operation
    Plan: Do 3.2.1, do 3.2.2 every 20 minutes on initial weight alarm, do 3.2.3 and 3.2.4. On final weight alarm, do 3.2.5 and 3.2.6.

    3.2.1 Remain within earshot while tanker is filling
    3.2.2 Check road tanker
    3.2.3 Attend tanker during last 2-3 ton filling
    3.2.4 Cancel initial weight alarm and remain at controls
    3.2.5 Cancel final weight alarm
    3.2.6 Close supply valve A when target weight reached

4. **Terminate filling and release tanker**
    4.1 Stop filling operation
    Plan: Do in order.
        4.1.1 Close supply valve B
        4.1.2 Clear lines
        4.1.3 Close tanker valve
    4.2 Disconnect tanker
    Plan: Repeat 4.2.1 five times then do 4.2.2 to 4.2.4 in order.
        4.2.1 Vent and purge lines
        4.2.2 Remove instrument air from valves
        4.2.3 Secure blocking device on valves
        4.2.4 Break tanker connections
    4.3 Store hoses
    4.4 Secure tanker
    Plan: Do in order.
        4.4.1 Check valves for leakage
        4.4.2 Secure log-in nuts
        4.4.3 Close and secure dome
    4.5 Secure panel (not analyzed)

5. **Document and report (not analyzed)**

FIGURE 35.10 Part of a hierarchical task analysis for the chlorine tanker filling problem. (From Center for Chemical Process Safety (CCPS). (1994). *Guidelines for Preventing Human Error in Process Safety*. New York: American Institute of Chemical Engineers. With permission.)

contribute to a major flammable release resulting from either a spill during unloading of the truck or from a tank rupture. Prior to this HRA, an FTA revealed that the frequency of such a release is largely due to human errors. Figure 35.10 illustrates a portion of a HTA — the TA method used in this case — that addresses the operations that could lead to a spill during tanker loading. Note that with an HTA, task operations can be described to whatever level of detail is required. Figure 35.11 illustrates some of the results of this analysis, where errors are coded according to their classification (Figure 35.9). The appeal of such a systematic method should be obvious. Moreover, this procedure does not preclude quantification of human errors considered important for satisfying PSA requirements.

Cognitive Engineering Approaches to Error Prediction

Perspectives that emphasize mental or cognitive processes that underlie human error can potentially reveal the causes of errors. This knowledge, in turn, provides a stronger basis for design countermeasures, and for making use of data on "near misses" — incidents that are precursors of more serious events—that industries would be well advised to keep. These "cognitive engineering system" perspectives, however, require descriptions of error-inducing environments that are generally more detailed and less systematic than approaches that emphasize the external form of the error — i.e., *what* happened (in terms of its observable manifestation) — rather than *how* (from the standpoint of information processing mechanisms) and *why* (from the interplay between behavioral tendencies and the situational context) it happened.

A modeling framework that is consistent with the goal of understanding underlying causes of human error is based on distinguishing fundamentally different categories of human information processing. These categories are referred to as skill-based, rule-based, and knowledge-based levels of performance

STEP	ERROR TYPE	ERROR DESCRIPTION	RECOVERY	CONSEQUENCES AND COMMENTS	ERROR REDUCTION RECOMMENDATIONS		
					PROCEDURES	TRAINING	EQUIPMENT
2.3 Enter Tanker target weight	Wrong information obtained (R2)	Wrong weight entered	On check	Alarm does not sound before tanker overfills	Independent validation of target weight	Ensure operator double checks entered date. Recording of values in checklist.	Automatic setting of weight alarms from unladen weight. Computerize logging system and build in checks on tanker reg. No. and unladen weight linked to warning system. Display differences
3.2.2 Check Tanker while filling	Check omitted (C1)	Tanker not monitored while filling	On initial weight alarm	Alarm will alert the operator if correctly set. Equipment fault, e.g., leaks not detected early and remedial action delayed.	Provide secondary task involving other personnel. Supervisor periodically checks operation.	Stress importance of regular checks for safety	Provide automatic log-in procedure
3.2.3 Attend tanker during last 2-3 ton filling	Operation omitted (O8)	Operator fails to attend	On step 3.2.5	If alarm not detected within 10 minutes tanker will overfill	Ensure work schedule allows operator to do this without pressure	Illustrate consequences of not attending	Repeat alarm in secondary area. Automatic interlock to terminate loading if alarm not acknowledged. Visual indication of alarm.
3.2.5 Cancel final weight alarm	Operation omitted (O8)	Final weight alarm taken as initial weight alarm	No recovery	Tanker overfills	Note differences between the sound of the two alarms in checklist.	Alert operators during training about differences in sounds of alarms	Use completely different tones for initial and final weight alarms
4.1.3 Close tanker valve	Operation omitted (O8)	Tanker valve not closed	4.2.1	Failure to close tanker valve would result in pressure not being detected during the pressure check in 4.2.1	Independent check on action. Use checklist	Ensure operator is aware of consequences of failure	Valve position indicator would reduce probability of error
4.2.1 Vent and purge lines	Operation omitted (O8) Operation incomplete (O9)	Lines not fully purged	4.2.4	Failure of operator to detect pressure in lines could lead to leak when tanker connections broken	Procedure to indicate how to check if fully purged	Ensure training covers symptoms of pressure in line	Line pressure indicator at controls. Interlock device on line pressure.
4.4.2 Secure locking nuts	Operation omitted (O8)	Locking nuts left unsecured	None	Failure to secure locking nuts could result in leakage during transportation	Use checklist	Stress safety implication of training	Locking nuts to give tactile feedback when secure

FIGURE 35.11 Results from the application of a systematic approach to error prediction and analysis for the chlorine tanker filling problem. (From Center for Chemical Process Safety (CCPS). (1994). *Guidelines for Preventing Human Error in Process Safety*. New York: American Institute of Chemical Engineers. With permission.)

(Rasmussen, 1986). Work activities at the skill-based level are highly practiced routines that require little conscious attention. The rule-based level involves the use of rules that the worker invokes from memory, or obtains from other sources such as reference manuals or co-workers. For example, a maintenance worker might conclude that a certain fault is present in the equipment based on symptoms revealed by diagnostic tests. This conclusion may then trigger another rule that addresses actions that need to be taken in response to that fault. Knowledge-based performance typically occurs when workers are attempting to solve problems in relatively unfamiliar situations, and demands the greatest use of information processing resources.

The distinctions implicit to this "SRK framework" can be used in accident analysis to trace the observed or "external error form" to its underlying causes. For error prediction, while models of human error have been proposed that utilize this framework (e.g., Reason, 1990), the usefulness of these models is not likely to be readily apparent to analysts. Much more useful are the distinctions Reason (1990) makes between different types of errors, specifically, between skill-based slips and lapses, rule-based mistakes, and knowledge-based mistakes. In his view, the potential for error begins with "cognitive underspecification," implying that at some point in the processing of task-related information the specification of information is incomplete. This underspecification, which to some extent results from information processing and memory limitations, can promote one of two forms of biases: "frequency bias" or "similarity bias." These biases, respectively, reflect tendencies to process information and act based on the frequency with which a behavior has been performed, or the degree to which information currently being perceived or processed appears similar to patterns of information the person is readily tuned to. The manifestation of these biases in terms of different types of errors will depend largely on the particular level of performance, the situational context, and characteristics unique to the person.

For example, assume a worker is performing a series of operations on a machine that are fairly well-practiced, requiring little attention. A modification is made to the machine that requires the operations ADB to be performed instead of ABCD. Following the modification, the worker immediately proceeded to perform operation B after performing operation A, even though the worker was aware of the new sequence and intended to perform it. This error is referred to as a "double-capture slip." More generally, this error is classified as a "skill-based slip" due to "inattention" (Table 35.2): had the worker invested more attention at the critical point where, due to the sheer frequency with which the previous routine was performed, one would expect the worker to slip into the old routine, then this error would likely have been avoided. An example of a skill-based lapse resulting from this same absence of attentional control is when a worker intends to initiate a sequence of operations but is interrupted by an alarm. After addressing the source of the alarm the worker goes on to other activities, perhaps because initiating the intended sequence of events has not in the past been generally associated with corrective actions in response to alarms. This type of error is referred to as "an omission following an interruption" (Table 35.2). Excessive attentional control can also lead to skill-based slips, as when a worker disrupts activities being performed in order to analyze the situation. Disruption of the "preprogrammed" or automatic sequence of activities typical of skill-based work can result in the worker picking up the task at a point further along than it is (an "omission due to overattention"), or repeating steps already taken ("repetition due to overattention").

Unlike errors at the skill-based level, errors at the rule-based level represent actions that are intentional; they just turn out to be wrong, and thus are viewed as "mistakes." As with the skill-based level, cognitive underspecification serves to set in motion frequency and similarity biases that shape the form errors assume at the rule-based level. Consider a job-shop situation where machine setup operations are based on factors such as part type and processing operations. The current work part has features that do not completely match the conditions associated with the IF or antecedent portion of any of the rules available to the worker. However, despite these inconsistencies (i.e., cognitive underspecification) that result in only partial matching of the rule, the worker may choose to use a rule that has been successful many times before, but one that is incorrect for the current situation. This form of error or "failure mode" is referred to as "rule strength" (Table 35.2), and represents the misapplication of a good rule in that under the appropriate conditions this rule would have been successful.

TABLE 35.2   Human Error Failure Modes at Each of Three Levels of Human
Performance Associated with the SRK Framework

Skill-based performance

Inattention	*Overattention*
Double-capture slips	Omissions
Omissions following interruptions	Repetitions
Reduced intentionality	Reversals
Perceptual confusions	
Interference errors	

Rule-based performance

Misapplication of good rules	*Application of bad rules*
First exceptions	Encoding deficiencies
Countersigns and nonsigns	Action deficiencies
Informational overload	Wrong rules
Rule strength	Inelegant rules
General rules	Inadvisable rules
Redundancy	
Rigidity	

Knowledge-based performance

Selectivity	*Problems with causality*
Workspace limitations	*Problems with complexity*
Out of sight out of mind	Problems with delayed feedback
Confirmation bias	Insufficient consideration of processes in time
Overconfidence	Difficulties with exponential developments
Biased reviewing	Thinking in causal series not causal nets
Illusory correlation	Thematic vagabonding
Halo effects	Encysting

Source: Reason, J. (1990). *Human Error.* Cambridge: Cambridge University Press.

As indicated in Table 35.2, there are many forms of knowledge-based mistakes, which is not surprising given the enormous variance in human behavioral tendencies manifest during performance at the knowledge-based level. The challenge for the analyst is to recognize the potential for these situations and to recommend design solutions such as decision and memory aids. The use of principles of ecological interface design (Vicente and Rasmussen, 1992) can provide a means for handling abnormal or unanticipated situations characteristic of this level of performance. Through these methods and principles, system representations on human–computer interfaces can potentially be devised that can transform knowledge-based problem solving to the rule-based level, where the worker has a more realistic opportunity to succeed.

Rasmussen (1982) has provided, in flow-chart form, a guide for answering questions concerning *what*, *how*, and *why* an error occurred. Figure 35.12 illustrates the flow chart corresponding to identification of the internal error mechanism — i.e., *how* an error occurred. Although this procedure is more easily applied to the analysis of accidents, it can also be used to predict errors if work contexts have been analyzed to sufficient detail. In that case, the links between various factors comprising the error-inducing environment and the possible internal error mechanisms will represent the underlying causes of the error. Note that the end points in Figure 35.12 related to information processing can, in principle, be further resolved based on Reason's (1990) analysis of failure modes. However, to extend the analysis of causality to this level of detail requires methods for: (1) identifying cognitive underspecification in work situations; and (2) translating this underspecification into particular failure modes.

For a cognitive-based analysis of human error that still maintains the appeal of systematic methods, one can substitute the predictive human error analysis stage of SPEAR with a classification scheme that

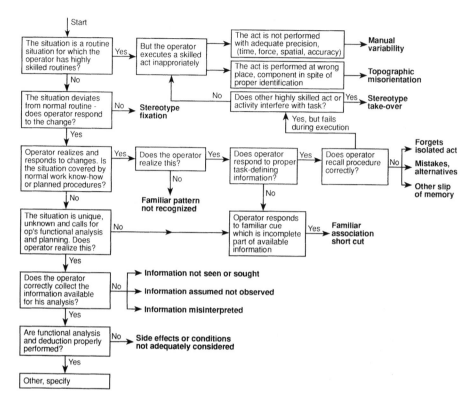

FIGURE 35.12   Guide for identifying the mechanisms concerning "how" a human error occurred. (From Rasmus-sen, J. (1982). Human errors: a taxonomy for describing human malfunction in industrial installations, *Journal of Occupational Accidents*, 4, 311-333. With permission.)

adds considerable depth to the one illustrated in Figure 35.9. An example of such a scheme is the "human error identification in systems tool" (HEIST). Based on an analysis of PSFs, this tool first identifies the external error mode. A listing of external error modes classified according to eight stages of human information processing is presented in Table 35.3. The first column in a "HEIST table" would then consist of a code whose first letter (or first two letters) refers to one of these eight stages, and whose next letter refers to one of six general or "global" PSFs: time (T); interface (I); training/experience/familiarity (E); procedures (P); task organization (O); and task complexity (C).

These external error modes are then linked to underlying psychological error mechanisms. Many of these mechanisms are consistent with modes of failure or types of errors listed in Table 35.2. Table 35.4 presents an extract from a HEIST table corresponding to two of the eight stages of human information processing referred to above — specifically, "Activation/detection" and "Observation/data collection." The linking to psychological error mechanisms is depicted in column four of the HEIST table (Table 35.4). Table 35.5 presents a listing of these mechanisms for the two stages of information processing included in Table 35.4. Note that the error-reduction guidelines contained in the HEIST table can be used for making specific error reduction recommendations in methods such as SPEAR (Figure 35.11). A complete HEIST table and associated listing of psychological error mechanisms can be found in Kirwan (1994).

35.6   Simulators and Computer Simulation

A simulator is a facility consisting of a mock-up of a system and its human interface, and containing a mathematical representation of the system. Although typically used for training purposes, the potential exists for investigating various types of errors on the part of workers and worker crews, depending on

TABLE 35.3   A Listing of External Error Modes Classified According to Stages of Human Information Processing

1. Activation/detection
   1.1   Fails to detect signal/cue
   1.2   Incomplete/partial detection
   1.3   Ignore signal
   1.4   Signal absent
   1.5   Fails to detect deterioration of situation

2. Observation/data collection
   2.1   Insufficient information gathered
   2.2   Confusing information gathered
   2.3   Monitoring/observation omitted

3. Identification of system state
   3.1   Plant-state-identification failure
   3.2   Incomplete-state identification
   3.3   Incorrect-state identification

4. Interpretation
   4.1   Incorrect interpretation
   4.2   Incomplete interpretation
   4.3   Problem solving (other)

5. Evaluation
   5.1   Judgment error
   5.2   Problem-solving error (evaluation)
   5.3   Fails to define criteria
   5.4   Fails to carry out evaluation

6. Goal selection and task definition
   6.1   Fails to define goal/task
   6.2   Defines incomplete goal/task
   6.3   Defines incorrect or inappropriate goal/task

7. Procedure selection
   7.1   Selects wrong procedure
   7.2   Procedure inadequately formulated/short-cut invoked
   7.3   Procedure contains rule violation
   7.4   Fails to select or identify procedure

8. Procedure execution
   8.1   Too early/late
   8.2   Too much/little
   8.3   Wrong sequence
   8.4   Repeated action
   8.5   Substitution/intrusion error
   8.6   Orientation/misalignment error
   8.7   Right action on wrong object
   8.8   Wrong action on right object
   8.9   Check omitted
   8.10 Check fails/wrong check
   8.11 Check mistimed
   8.12 Communication error
   8.13 Act performed wrongly
   8.14 Part of act performed
   8.15 Forgets isolated act at end of task
   8.16 Accidental timing with other event/circumstance
   8.17 Latent error prevents execution
   8.18 Action omitted
   8.19 Information not obtained/transmitted
   8.20 Wrong information obtained/transmitted
   8.21 Other

Source: Kirwan, B. (1994). *A Guide to Practical Human Reliability Assessment.* London: Taylor & Francis.

the apparatus available for tracking human performance. When properly run, simulator trials can also be used as a basis for estimating HEPs for quantitative PRA-driven HRAs.

Computer simulation attempts to capture the interaction between the human and environment through software. When applied to work contexts, one can define worker tasks, environmental events, and performance variables, and, through the simulation's software logic, determine various outcomes of interest. An example of such a simulation is MicroMAPPS (Gertman and Blackman, 1994), a PC-based version of the Maintenance Personnel Performance Simulation (Siegel et al., 1984). Sponsored by the Nuclear Regulatory Council, this simulation enables the user to define various subtasks for particular maintenance crew tasks and rate a number of variables that impact performance (e.g., time since task was last performed, time available, crew's initial ability). Given its underlying probabilistic basis, over many simulation runs MAPPS can provide the average HEP for any given subtask. When using this simulation model, analysts have the opportunity to evaluate changes in worker and work crew reliability as a function of changes in a variety of relevant parameters. In addition, HEP estimates can be used as inputs into PRAs.

The Cognitive Environment Simulation (CES) is a computer simulation model that is more consistent with the objective of qualitative approaches to HRA that emphasize causality of error (Woods et al., 1987). CES uses as inputs the data produced by a nuclear power plant (NPP) simulator. It responds to those inputs through an array of human information processing mechanisms and a knowledge base derived from models of what humans would do in facing the circumstances unfolding over time in the CES model. This simulation model primarily focuses on modeling the processes by which humans assess

TABLE 35.4   Extract from a HEIST Table

Code	Error-identifier prompt	External error mode	System cause/ psychological error-mechanism	Error-reduction guidelines
		1. Activation/Detection		
AT1	Does the signal occur at the appropriate time? Could it be delayed?	Action omitted; performed too early or too late	Signal timing deficiency; failure of prospective memory	Alter system configuration to present signal appropriately; generate hard copy to aid prospective memory; repeat signal until action has occurred
AI1	Could the signal source fail?	Action omitted or performed too late	Signal failure	Use diverse/redundant signal sources; use a higher-reliability signal system; give training and ensure procedures incorporate investigation checks on 'no signal'
AI2	Can the signal be perceived as unreliable?	Action omitted	Signal ignored	Use diverse signal sources; ensure higher signal reliability; retrain if signal is more reliable than it is perceived to be
AI3	Is the signal a strong one, and is it in a prominent location? Could the signal be confused with another?	Action omitted; or performed too late; or wrong act performed	Signal-detection failure	Prioritize signals; place signals in primary (and unobscured) location; use diverse signals; use multiple-signal coding; give training in signal priorities; make procedures cross-reference the relevant signals; increase signal intensity
AI4	Does the signal rely on oral communication?	Action omitted or performed too late	Communication failure; lapse of memory	Provide physical back-up/substitute signal; build required communications requirements into procedures
AE1	Is the signal very rare?	Action omitted or performed too late	Signal ignored (false alarm); stereotype fixation	Give training for low-frequency events; ensure diversity of signals; prioritize signals into a hierarchy of several levels
AE2	Does the operator understand the significance of the signal?	Action omitted or performed too late	Inadequate mental model	Training and procedures should be amended to ensure significance is understood
AP1	Are procedures clear about action following the signal or the previous step, or when to start the task?	Action omitted or performed either too early or too late	Incorrect mental model	Procedures must be rendered accurate, or at least made more precise; give training if judgment is required on when to act
AO1	Does activation rely on prospective memory (i.e., remembering to do something at a future time, with no specific cue or signal at that later time)?	Action omitted or performed either too late or too early	Prospective memory failure	Proceduralize task, noting calling conditions, timings of actions, etc...; utilize an interlock system preventing task occurring at undesirable times; provide a later cue; emphasize this aspect during training
AO2	Will the operator have other duties to perform concurrently? Are there likely to be distractions? Could the operator become incapacitated?	Action omitted or performed too late	Lapse of memory; memory failure; signal-detection failure	Training should prioritize signal importances; improve task organization for crew; use memory aids; use a recurring signal; consider automation; utilize flexible crewing
AO3	Will the operator have a very high or low work load?	Action omitted or performed either too late or too early	Lapse of memory; other memory failure; signal-detection failure	Improve task and crew organization; use a recurring signal; consider automation; utilize flexible crewing; enhance signal salience

TABLE 35.4 (continued)   Extract from a HEIST Table

Code	Error-identifier prompt	External error mode	System cause/ psychological error-mechanism	Error-reduction guidelines
AO4	Will it be clear who must respond?	Action omitted or performed too late	Crew-coordination failure	Emphasize task responsibility in training and task allocation among crew; utilize team training
AC1	Is the signal highly complex?	Action omitted, or wrong act performed, or act performed either too late or too early	Cognitive overload; inadequate mental model	Simplify signal; automate system response; give adequate training in the nature of the signal; provide on-line, automated, diagnostic support; develop procedures which allow rapid analysis of the signal (e.g., use of flow charts)
AC2	Is the signal in conflict with the current diagnostic "mindset"?	Action omitted or wrong act performed	Confirmation bias; signal ignored	Procedures should emphasize disconfirming as well as confirmatory signals; utilize a shift technical advisor in the shift-structure; carry out problem-solving training and team training; utilize diverse signals; implement automation
AC3	Could the signal be seen as part of a different signal set? Or is, in fact, the signal part of a series of signals which the operator needs to respond to?	Action performed too early or wrong act performed	Familiar-association shortcut/stereo-type take-over	Training and procedures could involve display of signals embedded within mimics or other representations showing their true contexts or range of possible contexts; use fault-symptom matrix aids, etc.
		2. Observation/Data Collection		
OT1	Could the information or check occur at the wrong time?	Failure to act; or action performed too late or too early; or wrong act performed	Inadequate mental model/ inexperience/ crew coordination failure	Procedure and training should specify the priority and timing of checks; present key information centrally; utilize trend displays and predictor displays if possible; implement team training
OI1	Could important information be missing due to instrument failure?	Action omitted or performed either too late or too early; or wrong act performed	Signal failure	Use diverse signal sources; maintain back-up power supplies for signals; have periodic manual checks; procedures should specify action to be taken in event of signal failure; engineer automatic protection/action; use a higher-reliability system
OI2	Could information sources be erroneous?	Action omitted or performed either too late or too early; or wrong act performed	Erroneous signal	Use diverse signal sources; procedures should specify cross-checking; design system-self-integrity monitoring; use higher-reliability signals
OI3	Could the operator select a wrong but similar information source?	Action omitted or performed either too late or too early; or wrong act performed	Mistakes alternatives; spatial misorientation; topographic misorientation	Ensure unique coding of displays, cross-referenced in procedures; enhance discriminability via coding; improve training

TABLE 35.4 (continued)   Extract from a HEIST Table

Code	Error-identifier prompt	External error mode	System cause/ psychological error-mechanism	Error-reduction guidelines
OI4	Is an information source accessed only via oral communication?	Action omitted or performed either too late or too early; or wrong act performed	Communication failure	Use diverse signals from hardwired or softwired displays; ensure back-up human corroboration; design communication protocols
OI5	Are any information sources ambiguous?	Action omitted or performed either too late or too early; or wrong act performed	Misinterpretation; mistakes alternatives	Use task-based displays; design symptom-based diagnostic aids; utilize diverse information sources; ensure clarity of information displayed; utilize alarm conditioning
OI6	Is an information source difficult or time-consuming to access?	Action omitted or performed too late; or wrong act performed	Information assumed	Centralize key data; enhance data access; provide training on importance of verification of signals; enhance procedures
OI7	Is there an abundance of information in the scenario, some of which is irrelevant, or a large part of which is redundant?	Action omitted or performed too late	Information overload	Prioritize information displays (especially alarms); utilize overview mimics (VDU or hard wired); put training and procedural emphasis on data-collection priorities and data management
OE1	Could the operator focus on key indication(s) related to a potential event while ignoring other information sources?	Action omitted or performed too late; or wrong act performed	Confirmation bias; tunnel vision	Training in diagnostic skills; enhance procedural structuring of diagnosis, emphasizing checks on disconfirming evidence; implement a staff-technical-advisor role; present overview mimics of key parameters showing whether system integrity is improving or worsening or adequate
OE2	Could the operator interrogate too many information sources for too long, so that progress toward stating identification or action is not achieved?	Action omitted or performed too late	Thematic vagabonding; risk-recognition failure; inadequate mental model	Training in fault diagnosis; team training; put procedural emphasis on required data-collection time frames; implement high-level indicators (alarms) of system-integrity deterioration
OE3	Could the operator fail to realize the need to check a particular source? Is there an adequate cue prompting the operator?	Action omitted or performed either too late or too early; or wrong act performed	Need for information not prompted; prospective memory failure	Procedural guidance on checks required; training; use of memory aids; use of attention-gaining devices (flash; alarms; central displays and messages)
OE4	Could the operator terminate the data collection/observation early?	Action omitted or performed either too early or too late; or wrong act performed	Overconfidence; inadequate mental model; incorrect mental model; familiar-association short-cut	Training in diagnostic procedures and verification; procedural specification of required checks, etc.; implement a shift-technical-advisor role

TABLE 35.4 (continued)   Extract from a HEIST Table

Code	Error-identifier prompt	External error mode	System cause/ psychological error-mechanism	Error-reduction guidelines
OE5	Could the operator fail to recognize that special circumstances apply?	Action omitted or performed either too late or too early; or wrong act performed	Fail to consider special circumstances; slip of memory; inadequate mental model	Ensure training for, as well as procedural noting of, special circumstances; STA; give local warnings in the interface displays/controls
OP1	Could the operator fail to follow the procedures entirely?	Action omitted or wrong act performed	Rule violation; risk-recognition failure; production–safety conflict; safety-culture deficiency	Training in use of procedures; operator involvement in development and verification of procedures
OP2	Could the operator forget one or more items in the procedures?	Action omitted or performed either too early or too late; or wrong act performed	Forget isolated act; slip of memory; place-losing error	Ensure an ergonomic procedure design; utilize tick-off sheets, place keeping aids, etc.; team training to emphasize checking by other team member(s)
OO1 (AO2)	Will the operator have other duties to perform concurrently? Are there likely to be distractions? Could the operator become incapacitated?	Action omitted or performed too late	Lapse of memory; memory failure; signal-detection failure	Training should prioritize signal importances; develop better task organization for crew; use memory aids; use a recurring signal; consider automation; use flexible crewing
OO2 (AO3)	Will the operator have a very high or low work load?	Action omitted or performed either too late or too early	Lapse of memory; other memory failure; signal-detection failure	Better task and crew organization; utilize a recurring signal; consider automation; use flexible crewing; enhance signal salience
OO3 (AO4)	Will it be clear who must respond?	Action omitted or performed too late	Crew-coordination failure	Improve training and task allocation among crew; team training
OO4	Could information collected fail to be transmitted effectively across shift-hand-over boundaries?	Failure to act; or wrong action performed; or action performed either too late or too early; or an error of quality (too little or too much)	Crew-coordination failure	Develop robust shift-hand-over procedures; training; team training across shift boundaries; develop robust and auditable data-recording systems (logs)
OC1	Does the scenario involve multiple events, thus causing a high level of complexity or a high work load?	Failure to act; or wrong action performed; or action performed too early or too late	Cognitive overload	Emergency-response training; design crash-shutdown facilities; use flexible crewing strategies; implement shift-technical-advisor role; develop emergency operating procedures able to deal with multiple transients; engineer automatic information recording (trends, logs, printouts); generate decision/diagnostic support facilities

Source: Kirwan, B. (1994). *A Guide to Practical Human Reliability Assessment*. London: Taylor & Francis.

TABLE 35.5   Listing of Psychological Error Mechanisms for the Two Stages of Information Processing in Table 35.4

Activation/detection

1. Vigilance failure: lapse of attention.
   Ergonomic design of interface to allow provision of effective attention-gaining measures; supervision and checking; task-organization optimization, so that the operators are not inactive for long periods, and are not isolated.
2. Cognitive/stimulus overload: too many signals present for the operator to cope with.
   Prioritization of signals (e.g., high-, medium-, and low-level alarms); overview displays; decision-support systems; simplification of signals; flow chart procedures; simulator training; automation.
3. Stereotype fixation: operator fails to realize that situation has deviated from norm.
   Training and procedural emphasis on range of possible symptoms/causes; fault-symptom matrix as a job-aid; decision support system; shift technical advisor/supervision.
4. Signal unreliable: operator treats signal as false due to its unreliability.
   Improved signal reliability; diversity of signals; increased level of tolerance on the part of the system, or delay in effects of error, which allows error detection and correction (decreases "coupling"); training in consequences associated with incorrect false-alarm diagnosis.
5. Signal absent: signal absent due to a maintenance/calibration failure or a hardware/software error.
   Provide signal; redundancy/diversity in signaling-design approach; procedures/training to allow operator to recognize when signal is absent.
6. Signal-discrimination failure: operator fails to realize the signal is different.
   Improved ergonomics in the interface design; enhanced training and procedural support in the area of signal differentiation; supervision and checking.

Observation/data collection

7. Attention failure: lapse of attention.
   Multiple signal coding; enhanced alarm salience; improved task organization with respect to back-up crew and rest pauses.
8. Inaccurate recall: operator remembers data incorrectly (usually quantitative data).
   Non-reliance on memorized data, which would necessitate better interface design — as data are received, they can either be acted on while still present on a display (controls and displays are co-located) or at least be logged onto a "scratch pad"; sufficient displays for presenting all information necessary for a decision/action simultaneously; printer usage; training in non-reliance on memorized data.
9. Confirmation bias: operator only selects data that confirms given hypothesis, and ignores other disconfirming data sources.
   Problem-solving training; team training (including training in the need to question decisions, and in the ability of the team leader(s) to take constructive criticism); shift technical advisor (diverse, highly qualified operator who can "stand back" and consider alternative diagnoses); functional procedures; high-level information displays; simulator training; high-level alarms for system-integrity degradation; automatic protection.
10. Thematic vagabonding: operator flits from datum to datum, never actually collating it meaningfully.
    Problem-solving training; team training; simulator training; functional-procedure specification for decision-timing requirements; high-level alarms for system-integrity degradation.
11. Encystment: operator focuses exclusively on only one data source.
    Problem-solving training; team training (including training in the need to question decisions, and in the ability of the team leader(s) to take constructive criticism); shift technical advisor; functional procedures; high-level information displays; simulator training; high-level alarms for system-integrity degradation.
12. Stereotype fixation revisited: need for information is not prompted (by either memory or procedures).
    Emergency procedure enhancements, and emphasis of key symptoms and indicators to be checked; team training; problem-solving training; alarm re-prioritization; simulator training.
13. Crew-functioning problem: allocation of responsibility or priorities is unclear, with the result that data collection/observation fails.
    Improved crew coordination, and allocation of responsibilities; team training; emergency training; accident-management procedures; remote-incident-monitoring/back-up center; high-level displays; "crash-shutdown" facilities.
14. Cognitive/stimulus overload: operator too busy, or being bombarded by signals, with the result that effective data collection/observation fails. See 2 above.

Source: Kirwan, B. (1994). *A Guide to Practical Human Reliability Assessment.* London: Taylor & Francis.

situations and form intentions to act in emergency NPP operations; inappropriate intentions therefore define human error. CES allows the analyst to investigate situational characteristics conducive to intention failures and the form and consequences of these failures. It does so by allowing the analyst to make various adjustments in the processing mechanisms and knowledge base underlying CES's problem-solving model, as well as in the input stream of plant data. A relatively sophisticated understanding of human information processing and situational modeling is obviously required of the analyst in utilizing this model.

Data from both simulators and simulations must be interpreted with caution (Sharit, 1993). With simulators the particular application may determine the usefulness of data on human error. For example, in NPP simulators real-world conditions of stress can rarely be duplicated, whereas in flight simulators pilots can not only experience "simulator sickness," but, unlike NPP operators, are also likely to have experienced many of the emergency conditions being simulated, and therefore may respond with different strategies. In any simulator application, however, it is unlikely that many of the more subtle factors that contribute to establishing the context within which the human performs can be captured. With simulation models, the requirements for verifiability and validity, fundamental to any simulation modeling application, must be satisfied. In human and system reliability applications, validating simulation outcomes based on assumptions concerning human performance variables and their multiple interactions is far from trivial.

35.7   Accident and Incident Analysis

While the primary goal in human and system reliability should be on preventing adverse outcomes, the occurrence of accidents demands that mechanisms be in place for determining the causes of such events, and these mechanisms will necessarily be qualitative in emphasis. As indicated in Figure 35.5, tracing paths from accidents to causes is much more constrained, and therefore feasible; not surprisingly, these endeavors tend to be successful, though often expensive. Depending on the method used, however, these backward paths can be traced to levels of causality that are relatively shallow (e.g., the worker forgot to check protective equipment properly prior to entering a hazardous area), to those that are much more revealing. A modeling framework consisting of an error causation chain (Figure 35.13) and corresponding flow charts corresponding to different stages in this chain (e.g., Figure 35.12) is, in principle, more effective at analyzing adverse consequences such as an injury, near miss, or poor product quality that directly result from the *external* error or mode of malfunction (Figure 35.13), than when used for the purpose of error prediction. A hypothetical case study illustrating the use of this modeling framework in a process control scenario is presented in CCPS (1994, p. 100).

An intuitively appealing and cost-effective method for investigating accidents is "change analysis." Change analysis techniques are based on the well-documented general relationship between change and increased risk. These techniques were developed at the Rand Corporation and improved by Kepner and Tregoe (1981). The primary objective of these techniques is to establish accident-free reference bases and then systematically identify changes or differences relative to the accident/incident situation. A worksheet is typically used to explore potential changes contributory to adverse outcomes. The various factors that the analyst feels should be included in the analysis are listed under the first column. These are designed to ask the questions: who (e.g., operator, fellow worker, supervisor), where, what (e.g., protective equipment), when, how, and why in terms of task factors, working conditions, initiating events, and management control factors. The next columns, respectively, address each of these factors in terms of the present (accident/incident) situation, prior situation, a comparison of these two situations in order to identify changes or differences, and a listing of all the differences. Finally, differences are analyzed for their effect on the accident in terms of both their independent and interactive contributions, and the resulting analysis is integrated into the overall accident system evaluation process.

If integrated with a contextual modeling framework, change analysis can contribute to identification of underlying causes of human error and/or adverse system consequences. For example, a change in a work procedure may result in workers violating the new procedure due to a perceived increase in risk or work load associated with the task. An organizational structure characterized by poor feedback channels between workers and management, and by policies that emphasize productivity, reinforce continuation of these violations which, under other changes instituted in the system, can result in increased exposure to hazards. Note that change analysis techniques can also be used *proactively* to predict adverse consequences

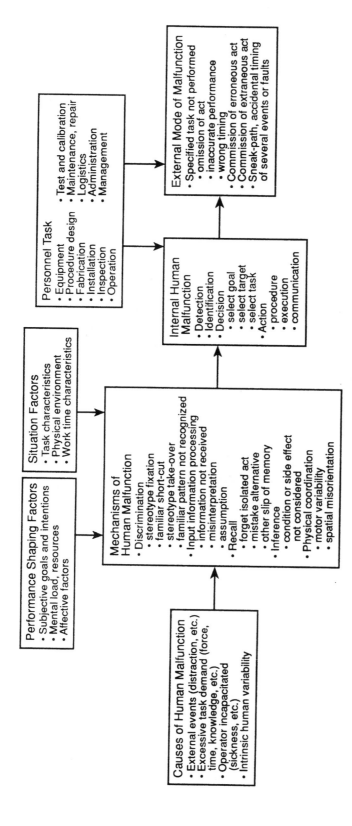

FIGURE 35.13   A model of an error causation chain tracing human error in a backward sequence from its consequences, through "what" happened (the external error mode), "how" it happened (the internal error mode), and "why" it happened (an interplay between PSFs, error-inducing factors, and the internal error mechanisms or error tendencies). (From Rasmussen, J. (1982). Human errors: a taxonomy for describing human malfunction in industrial installations, *Journal of Occupational Accidents*, 4, 311-333. With permission.)

by investigating potential problems associated with proposed changes in normal or stable functioning systems. In this "change control" application, mechanisms for detecting and monitoring change are essential.

The Management Oversight and Risk Tree (MORT) relies on a logic diagram for investigating the various factors contributing to an accident (Johnson, 1980). Factors considered by MORT include lines of responsibility, barriers to unwanted energy, and management factors. The emphasis on barriers essentially constitutes a "barrier analysis" which, like change analysis, is an accident analysis technique. In barrier analysis, the various energy sources to be considered (e.g., the energy associated with heights as derived from scaffolds or the energy associated with gamma radiation) and barriers to these energy sources (e.g., lanyards or lead-shielded body suits) are analyzed to determine the availability and adequacy of barrier safeguards.

By reasoning backward through a sequence of contributory factors, posing "yes" and "no" to questions along the way, and through the availability of accompanying text that aids the analyst in judging whether a factor is adequate or less than adequate (LTA), MORT assists the analyst in detecting omissions, oversights, or defects. MORT is especially powerful in identifying organizational root causes, which makes it useful for PRAs as well as accident analysis (Gertman and Blackman, 1994).

A generic MORT diagram is illustrated in Figure 35.14. The three flow charts in Figures 35.15 through 35.17 correspond to elaboration of the "OR" gate in Figure 35.14 labeled "S" arising from "Specific Control Factors." Specifically, they expand on the breakdown of the "Accident" event resulting from specific control factors being LTA, into "Barriers" (Figure 35.15), "Persons or Objects in Energy Channel" (Figure 35.16), and "Incident" (Figure 35.17). The codes in these diagrams correspond to further elaborations depicted in a series of MORT flow charts that can be found in Gertman and Blackman (1994).

Other relatively well-known accident analysis techniques include Events and Causal Factors Charting (Ontario Hydro, 1978), the Sequentially Timed Events Plotting Procedure or STEP (Hendrick and Benner, 1987), and Root Cause Coding (Armstrong et al., 1988).

35.8   The Quantitative Perspective to Human and System Reliability Analysis

The primary objective of all quantitative approaches to HRA is to derive estimates of the probability of error associated with various facets of performance. While some behavioral scientists may accept the validity of such estimates for relatively simplistic and repetitive behaviors, they would not be inclined to attach any significance to these probability estimates for behaviors that are even moderately complex. Such behaviors are considered too multidimensional to be summarized by single numerical estimates. However, with this information the opportunity exists for PRAs to be performed that enable system reliability to be evaluated, and ultimately, for the most cost-effective design interventions to be determined from the standpoint of reducing system risks.

The most classical approach to quantitative human reliability assessment is generally associated with engineering reliability techniques (Kapur and Lamberson, 1977). When applying these techniques, a fundamental distinction exists between static and dynamic reliability models. In the static case, a prescribed time period of operation is implied, whereas in the dynamic case the reliability of the component is expressed as a function of time. Some of the early quantitative approaches to modeling human reliability closely paralleled the classical paradigm (Askren and Regulinski, 1969, 1971). In the dynamic case, this involved deriving an instantaneous error rate, $h_e(t)$ (analogous to the hazard function in classical reliability modeling), from which the human reliability function, $R_e(t)$, corresponding to the probability that the human will not have committed an error by time t, could be derived.

This distinction between the static and dynamic cases also exists in many of the current quantitative perspectives to HRA. However, in contrast to approaches to modeling human error that have adopted classical engineering reliability methods, these perspectives address real-world tasks and a variety of task-related behavioral issues. The models corresponding to the dynamic case are often referred to as "time

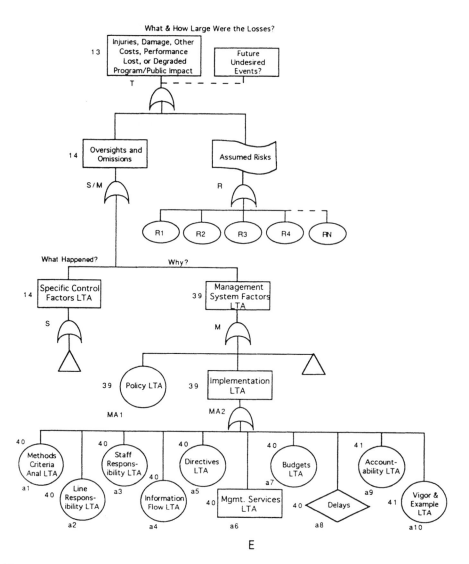

FIGURE 35.14   A Generic MORT diagram. (From Gertman, D. I. and Blackman, H. S. (1994). *Human Reliability & Safety Analysis Data Handbook*. New York: John Wiley & Sons. With permission.)

reliability correlations" (TRC). These models constitute mathematical relationships between the failure of an operator or crew to appropriately deal with some compelling situation and the time t (in minutes) after the initiation of this situation, and are usually depicted in the form of probability vs. time plots. An example of a TRC is the human cognitive reliability model (Hannaman and Spurgeon, 1988). This model is characterized by a three-parameter Weibull distribution (a type of probability distribution often used in reliability modeling) whose parameter values depend on whether the crew response to a problem can be best described by either the skill-, rule-, or the knowledge-based level of performance. Figure 35.18 illustrates the predictions of this model for a normalized time variable (that controls for contributions to crew response times that are unrelated to human activities). More general TRC models capable of reflecting a wide variety of influences are discussed by Dougherty and Fragola (1988).

In the remainder of this section, the focus will be on two techniques used to determine HEPs. Although each of these techniques relies heavily on the concept of PSFs (as do most other currently used quantitative approaches to HRA), they illustrate the different types of approaches that can be taken to predicting human error.

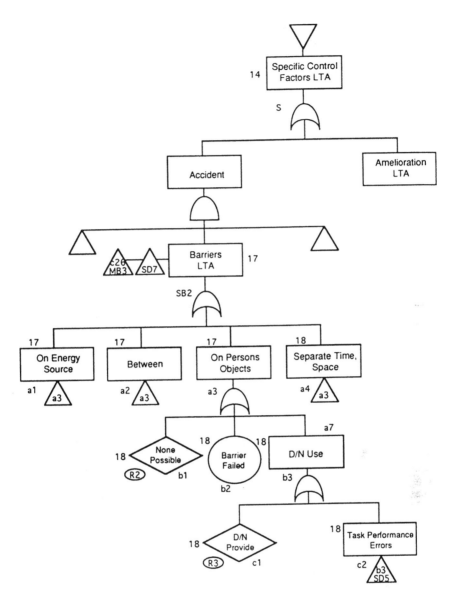

FIGURE 35.15 The Specific Control Factors MORT tree deriving from Figure 35.14. (From Gertman, D. I. and Blackman, H. S. (1994). *Human Reliability & Safety Analysis Data Handbook*. New York: John Wiley & Sons. With permission.)

THERP

The Technique for Human Error Rate Prediction, generally referred to as THERP, is perhaps the most well-known of all HRA techniques. The motivation for its development was twofold: (1) to derive estimates of human error that could be used in system fault trees in combination with other failure event data in order to evaluate system reliability; and (2) to aid in making decisions involving design trade-offs by evaluating the contribution of human error to subsystem or system reliability. The basic steps comprising THERP are outlined in Figure 35.19 and detailed in a work by Swain and Guttmann (1983) sponsored by the U.S. Nuclear Regulatory Commission. The overall objective of the method is to (1) *decompose* human tasks to a level of description whereby human error probabilities (HEPs) can be

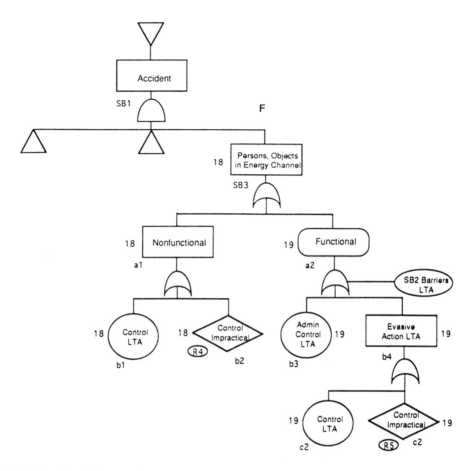

FIGURE 35.16 The Accidents with Persons, Objects in Energy Channel MORT tree derived from Figure 35.15. (From Gertman, D. I. and Blackman, H. S. (1994). *Human Reliability & Safety Analysis Data Handbook.* New York: John Wiley & Sons. With permission.)

estimated for the constituent sequential subtasks, and then to (2) *aggregate* these estimates to derive probabilities of task failure.

The first four steps of this method establish which work activities will be emphasized, the concerns for human error associated with these operations, time and skill requirements, and factors related to error detection and the potential for error recovery. The results of this effort are represented by a type of event tree referred to as a "probability tree." Each relevant subtask in a probability tree is characterized by two limbs representing either successful or unsuccessful performance, as illustrated in Figure 35.20. In this hypothetical task, a worker checks the calibration of a series of set points consisting of three comparators that operate by OR logic to detect abnormal pressure of a plant stream. This task requires that test equipment be set up correctly; an uncorrected error would result in miscalibration of all three comparators and consequently a failure (denoted by F_1 and F_2). Note that path "a" has been arbitrarily designated a success path (S_1) under the assumption that if the test equipment was set up correctly, adopting a conservative estimate of 10^{-2} for the probability of miscalibration would result in a probability of 10^{-6} of miscalibrating all three comparators, a value considered negligible in most PRAs for which this type of analysis would serve as input.

The next five steps (Figure 35.20) constitute the quantitative assessment stage. First, HEPs are assigned to each of the limbs of the tree corresponding to incorrect performance. These probabilities are referred to as *nominal* HEPs and, in theory, are presumed to represent medians of lognormal distributions.

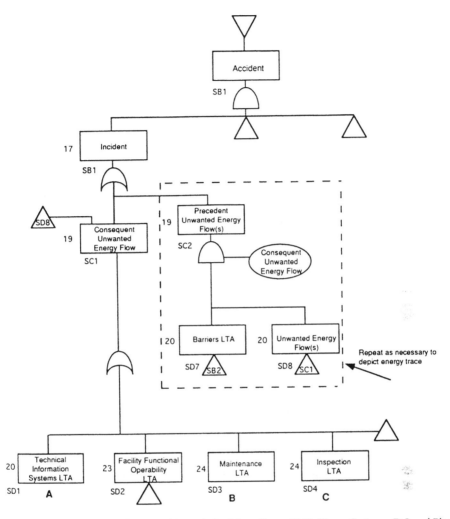

FIGURE 35.17   The Accident, Incident MORT tree derived from Figure 35.15. (From Gertman, D. I. and Blackman, H. S. (1994). *Human Reliability & Safety Analysis Data Handbook*. New York: John Wiley & Sons. With permission.)

Associated with each nominal HEP are upper and lower uncertainty bounds (UCBs) that reflect the variance associated with any given error distribution. The square root of the ratio of the upper to the lower UCB defines the *error factor*, the choice of which reflects the variability associated with the distribution for a particular error. Large error factors reflect variance arising from the assignment of nominal HEPs in addition to the variance associated with individual differences in worker performance. Swain and Guttmann (1983) provide a variety of nominal HEPs and their associated error factors for a variety of NPP tasks. Tables 35.6 and 35.7 illustrate these values for two different tasks; the values in Table 35.7 refer to *joint* HEPs in that the performance of a crew rather than an individual worker is being evaluated. In general, however, the absence of existing data from the operations in question will require that nominal HEPs be derived from other sources such as: (1) expert judgment using a variety of techniques such as absolute probability judgment and paired comparisons (Kirwan, 1994); (2) simulators (Gertman and Blackman, 1994); and (3) data from other jobs similar in psychological content to the operations of interest.

In order to account for more specific individual, environmental, and task-related influences on performance, nominal HEPs are subjected to a series of refinements (Figure 35.19). First, nominal HEPs are modified based on the influence of PSFs, resulting in *basic* HEPs (BHEPs). In some cases, guidelines are

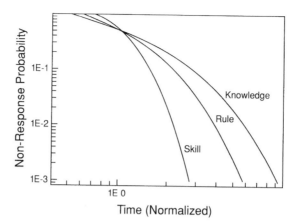

FIGURE 35.18 An example of a time reliability correlation model corresponding to skill-based, rule-based, and knowledge-based processing for work crews. (From Hannaman, G. W. and Worledge, D. H. (1988). Some Developments in Human Reliability Analysis Approaches and Tools. *Reliability Engineering and System Safety*, 22, 235-256. With permission.)

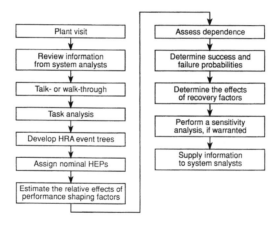

FIGURE 35.19 The steps comprising THERP. (From Swain, A. D. and Guttmann, H. E. (1983). *Handbook of Human Reliability Analysis with Emphasis on Nuclear Power Plant Applications*. NUREG/CR-1278, U.S. Nuclear Regulatory Commission, Washington, D.C. With permission.)

provided in the form of tables indicating the direction and extent of influence on nominal HEPs of particular PSFs. For example, Table 35.8 illustrates the influence of stress on nominal HEPs as a function of type of task and worker experience. Next, a nonlinear *dependency model* is incorporated that considers *positive* dependencies that exist between adjacent limbs of the tree, resulting in *conditional* HEPs (CHEPs). In a positive dependency model, failure on a subtask increases the probability of failure on the following subtask, and successful performance of a subtask decreases the probability of failure in performing the subsequent task element. Instances of negative dependence can be accounted for, but require the discretion of the analyst. In the case of positive dependence, THERP provides equations for modifying BHEPs to CHEPs based on the extent of dependence assumed by the analyst.

For example, in Figure 35.20, for the case of a small setup error (α), *complete* dependence was assumed between the setup task element and the calibration of the first set point, as well as between the subtasks of calibrating the second and third set points. The .9 probability that the operator will be alerted by the misalignment of the second setup and therefore recheck the test setup represents an instance of negative

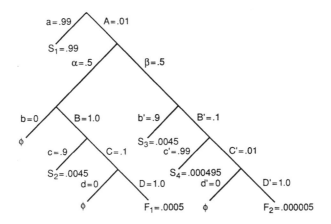

A = Failure to set up test equipment correctly

α = Small miscalibration of test equipment	β = Large miscalibration of test equipment
B = For a small miscalibration, failure to detect miscalibration for first setpoint	B' = For a large miscalibration, failure to detect miscalibration for first setpoint
C = For small miscalibration, failure to detect miscalibration for second setpoint	C' = For a large miscalibration, failure to detect miscalibration for second setpoint
D = For a small miscalibration, failure to detect miscalibration for third setpoint	D' = For a large miscalibration, failure to detect miscalibration for third setpoint

FIGURE 35.20 An example of a probability tree used in THERP for a hypothetical calibration task. (From Swain, A. D. and Guttmann, H. E. (1983). *Handbook of Human Reliability Analysis with Emphasis on Nuclear Power Plant Applications.* NUREG/CR-1278, U.S. Nuclear Regulatory Commission, Washington, D.C. With permission.)

TABLE 35.6 Estimated HEPs for Errors of Commission in Reading and Recording Quantitative Information from Unannunciated Displays

Item	Display of Task	HEP[a]	EF
1	Analog meter	0.003	3
2	Digital readout (less than four digits)	0.001	3
3	Chart recorder	0.006	3
4	Printing recorder with large number of parameters	0.05	5
5	Graphs	0.01	3
6	Values from indicator lamps that are used as quantitative displays	0.001	3
7	Recognition that an instrument being read is jammed, if there are no indicators to alert the user	0.1	5
8	Less than three	—[b]	—
9	More than three	0.01	3
10	Simple arithmetic calculations with or without calculators	0.01	3
11	Detect out-of-range arithmetic calculations	0.05	5

[a] Multiply HEPs by 10 for reading quantitative values under a high level of stress if the design violates a strong populational stereotype (e.g., a horizontal analog meter in which values increase from right to left). In this case, "letters" refer to those that convey no meaning. Groups of letters such as motor-operated value (MOV) do convey meaning, and the recording HEP is considered to be negligible.

[b] Negligible 0.001 (per symbol).

Source: Gertman, D.I. and Blackman, H.S. (1994). *Human Reliability & Safety Analysis Data Handbook.* New York: John Wiley & Sons.

dependence; for most PRA applications, however, the assumption of *zero* dependence as opposed to negative dependence will lead to more conservative estimates of HEPs. In addition to zero and complete dependence, the dependency model also accounts for low, medium, and high levels of dependence between adjacent task elements.

TABLE 35.7   Nominal Model of Estimated HEPs and EFs for Diagnosis within Time (T) by Control Room Personnel of Abnormal Events Annunciated Closely in Time[a]

Item	T (minutes)[b][c] after T_0	Median joint HEP[d] for diagnosis of single or first event	EF	Item	T (minutes)[b] after T_0	Median joint HEP[d] for diagnosis of single or second event	EF	Item	T (minutes)[b] after T_0	Median joint HEP[d] for diagnosis of single or third event	EF
1	1	1.0	—	7	1	1.0	—	14	1	1.0	—
2	10	0.1	—	8	10	1.0	—	15	10	1.0	—
3	20	0.01	10	9	20	0.1	10	16	20	1.0	10
4	30	0.001	10	10	30	0.01	10	17	30	0.1	10
5	60	0.0001	30	11	40	0.001	10	18	50	0.01	10
6	1500	0.00001	30	12	70	0.0001	30	19	50	0.001	10
				13	1510	0.00001	30	20	80	0.0001	30
								21	1520	0.00001	30

[a] Closely in time refers to cases in which the annunciation of the second event occurs while the control room personnel are still actively engaged in diagnosing and/or planning the responses to cope with the first event. This is situation specific, but for the initial analysis, use within 10 minutes as a working definition of closely in time. Note that this model pertains to the control room crew rather than to one individual. Note that this nominal model for diagnosis includes the activities listed in Table 12-1 of NUREG/CR-1278, as perceive, discriminate, interpret, diagnosis, and the first level of decision making. The modeling includes those aspects of behavior included in the Annunciator Response Model in NUREG/CR-1278, Table 20-23; therefore, when the nominal model for diagnosis is used, the annunciator model should not be used for the initial diagnosis. The annunciator model may be used for estimating recovery factors for an incorrect diagnosis.

[b] For points between the times shown, the medians and EFs may be chosen from NUREG/CR-1278, Figure 12-4.

[c] T_0 is a compelling signal of an abnormal situation and is usually taken as a pattern of annunciators. A probability of 1.0 is assumed for observing that there is some abnormal situation.

[d] NUREG/CR-1278, Table 12-5, presents some guidelines to use in adjusting or retaining the nominal HEPs presented above.

Source: Gertman, D.I. and Blackman, H.S. (1994). *Human Reliability & Safety Analysis Data Handbook.* New York: John Wiley & Sons.

TABLE 35.8 Modifications of Estimated HEPs for Step-by-Step and Dynamic Processing as a Function of Stress

		Modifiers for nominal HEPs[a]	
Item	Stress level	Skilled[b]	Novice[b]
1	Very low (very low task load)	×2	×2
	Optimum (optimum task load)		
2	Step-by-step[c]	×1	×1
3	Dynamic[c] moderately high (heavy task load)	×1	×2
4	Step-by-step[c]	×2	×4
5	Dynamic[c] extremely high (threat stress)	×5	×10
6	Step-by-step	×5	×10
7	Dynamic[d]	0.25 (EF = 5)	0.50 (EF = 5)
	Diagnosis[d]	These are the actual HEPs to use with dynamic tasks or diagnosis. They are NOT modifiers.	

[a] The nominal HEPs are those in the data tables in Part III and in Chapter 20 of NUREG/CR-1278. Error factors are listed in NUREG/CR-1278, Table 5-20.

[b] A skilled person is one with 6 months or more experience in the tasks being assessed. A novice is one with 6 months or less experience. Both levels have the required licensing or certificates.

[c] Step-by-step tasks are routine, procedurally guided tasks, such as carrying out written calibration procedures. Dynamic tasks require a higher degree of man–machine interaction, such as decision making, keeping track of several functions, controlling several functions, or any combination of these. These requirements are the basis of the distinction between step-by-step tasks and dynamic tasks, which are often involved in responding to an abnormal event.

[d] Diagnosis may be carried out under varying degrees of stress, ranging from optimum to extremely high (threat stress). For threat stress, the HEP of 0.25 is used to estimate performance of an individual. Ordinarily, more than one person will be involved. NUREG/CR-1278, Tables 5-6 and 5-8, lists joint HEPs based on the number of control room personnel presumed to be involved in the diagnosis of an abnormal event for various times after annunciation of the event and their presumed dependence levels, as presented in the staffing model in NUREG/CR-1278, Table 5-9.

Source: Gertman, D.I. and Blackman, H.S. (1994). *Human Reliability & Safety Analysis Data Handbook.* New York: John Wiley & Sons.

At this point, success and failure probabilities for the entire task are computed. Various approaches to these computations can be taken. The simplest approach is to multiply the individual CHEPs associated with any path on the tree leading to failure and then to sum these individual failure probabilities to arrive at the probability of failure for the total task, and then assign UCBs to this probability. More complex approaches to these computations take into account the variability associated with the combinations of events comprising the probability tree (Swain and Guttmann, 1983).

Following these computations, the analyst can choose to consider ways in which errors can be recovered. Common "recovery factors" include: (1) alerting the operator to the occurrence of an error through an annunciator, in which case the HEP associated with correctly responding to the annunciator would also have to be considered in the HRA; (2) the presence of co-workers who can potentially catch a fellow worker's errors, especially during work crew performance; and (3) various types of walk through inspections that are scheduled. As with event trees, these "recovery paths" can easily be represented on the original HRA probability tree. In the case of annunciators or inspectors, the relevant failure limb is extended into two (one failure and one success) additional limbs, with the probability of the operator being alerted to the annunciator or the inspector spotting the error, respectively, feeding back into the success path of the original tree. In the case of recovery through fellow workers, BHEPs are modified to CHEPs by considering the degree of dependency between the operator and one or more fellow workers who are in a position to notice the error. The computations for total task failure can now be repeated to determine the effects of recovery factors.

In addition to considering error recovery factors, the analyst can choose to perform sensitivity analysis. An example of one approach to sensitivity analysis is to identify the most probable errors on the tree and determine the degree to which design modifications corresponding to those task elements, which would reduce the magnitudes of those errors accordingly, affect the total failure probabilities previously computed. Finally, the results of the HRA are incorporated into system risk assessments such as PRAs.

An obvious deficiency of THERP, even among proponents of this approach, is its limitations in addressing decision-based errors, or what some have referred to as "errors of commission." One approach proposed for identifying these types of errors in situations, such as NPP applications, where the system's operations are flow oriented, is SNEAK analysis (Hahn et al., 1991), which is based on methods used to identify faults in electrical circuits. A computer-interactive "Sneak Analysis Tool" is available to the analyst for identifying these errors or "sneak conditions." These errors can then be used as starting points (i.e., initiating events) in event trees to determine whether these errors could be recovered (Blackman, 1991). Although these tools potentially provide valuable qualitative information to HRA analysts, if probabilities for these errors of commission were available the modification of this error based on the use of event trees for determining recovery paths would also provide useful inputs into PRAs. A method proposed for estimating HEPs for these types of errors is referred to as INTENT (Gertman et al., 1992). This method is based on subjective assessments from experts and borrows heavily from a method for deriving HEPs discussed below.

SLIM-MAUD

SLIM refers to the Success Likelihood Index Methodology (Embrey et al., 1984), a procedure for deriving HEPs; MAUD (Multi-Attribute Utility Decomposition) refers to a computer-interactive implementation of SLIM. In contrast to THERP, SLIM allows the analyst to focus on any human action or task, including those that can lead to highly infrequent events. Consequently, this method can provide inputs into PRAs at various system levels; that is, the HEPs can reflect relatively low-level actions that cannot be further decomposed as well as more broadly defined actions that encompass many of these lower-level actions. This increased flexibility, however, comes at the expense of a greatly reduced emphasis on task analysis and an increased reliance on subjective assessments.

SLIM assumes that the probability a human will carry out a particular task successfully depends on the combined effects of a number of PSFs that can be identified and appropriately evaluated through expert judgment. Task domain experts are assumed to be capable of assessing the relative importance (or weights) of each PSF with respect to the likelihood of human error in the task being evaluated and, independently of this assessment, rating how good or bad each PSF actually is for each of these tasks. The likelihood of success for each human action under consideration is determined by summing the product of the weights and ratings for each PSF, resulting in numbers (SLIs) that represent a scale of likelihood of success. These SLIs are useful in their own right, especially when the actions under consideration represent alternative modes of response in an emergency scenario, and the analyst is interested in determining which types of responses are least or most likely to succeed. However, for the purpose of conducting PRAs, SLIM converts the SLIs to HEPs.

The basic procedures for implementing SLIM are summarized as follows. First, an appropriate group of task-domain experts are identified. These experts help identify the potential error modes associated with the human actions of interest, and the set of PSFs most relevant to performance of these actions. The identification of all possible error modes is essential, and is generally arrived at through in-depth analysis and discussions that could include task analyses and reviews of documentation concerning emergency operating procedures. Relative importance weights for the PSFs are then derived by asking each "judge" to assign a weight of 100 to the most important PSF, and then assign weights to the remaining PSFs as a ratio of the one assigned the value of 100. Each individual weight is then divided by the sum of the weights for all the PSFs, resulting in normalized weights. The judges then rate each PSF on each task, with the lowest scale value indicating that the PSF is as poor as it is likely to be under real operating conditions, and the highest value indicating that the PSF is as good as it is likely to be in terms of promoting successful task performance. The ranges of values associated with the rating scale will dictate the range of possible SLI values that are subsequently computed.

SLIs are computed for each task or action by summing the product of the normalized weights with the ratings for each PSF. An estimate of the HEP, which equals one minus the probability of success ($P(S)$), can then be derived using the relationship $log\ P(S) = a \times SLI + b$. To derive the constants a and b, the probabilities of success must be available for at least two tasks taken from the cluster of tasks for which the relevant set of PSFs were identified. However, even if information on such "reference" tasks is not available, methods exist for deriving HEPs for the tasks of interest. Methods also exist for deriving upper and lower uncertainty bounds for these HEPs that PRAs typically require.

MAUD represents a user-friendly computer-interactive environment for implementing SLIM. This feature ensures that many of the assumptions that are critical to the theoretical underpinnings of this methodology are met. For example, MAUD ensures that the ratings for the various PSFs by a given judge (or analyst) are independent of one another, and that the relative importance weights elicited for the PSFs are consistent with the judge's preferences. In addition, MAUD provides procedures for assisting the expert in identifying the relevant PSFs. Further details concerning SLIM-MAUD can be found in Embrey et al. (1984) and Kirwan (1994).

35.9   Summary

In summary, the task facing a human and system reliability analyst has become increasingly more complex. The analyst must first understand the problem from the standpoint of determining whether a qualitative or quantitative perspective should be emphasized. Given the various conceptual and methodological tools at the analyst's disposal, a variety of options exist in adopting either perspective. In applying qualitative approaches, the analyst may have to consider the extent to which effort is to be invested in generating appropriate system descriptions, and in identifying or developing appropriate sociotechnical systems analysis, task analysis, and cognitive task analysis techniques. At the next level, the analyst must decide on the extent to which scenarios or contexts, and the propagation of both human errors and "reasonable actions" given these contexts, are to be modeled and analyzed. Contextual analysis requires that the sociotechnical systems analysis and task and cognitive task analyses be linked; this process, in turn, is governed by the relevant system descriptions. Furthermore, this contextual modeling and analysis process implies that a suitable characterization or model of human error (and of human violations) be available and integrated into the systems model.

In adopting a quantitative emphasis, the analyst must have a clear understanding of the motivation for deriving quantitative estimates of human error. Often, this requirement will be equivalent to understanding the motivation and objectives associated with performing quantitative risk assessments such as PRAs and PSAs. The analyst must then choose from a number of methods available for deriving estimates of human error probabilities (HEPs). In addition to understanding the assumptions and inner workings of each of these methods, the analyst must also be able to address issues related to criteria such as cost, practicality, usefulness, face validity, and training requirements. Moreover, the analyst must decide to what extent qualitative techniques need to be integrated into the HRA, giving rise to a potentially large number of hybrid HRA methods.

Finally, a more complete treatment of human and system reliability analysis, which is beyond the scope of this chapter, entails consideration of issues related to data collection and information systems organization in order to establish a reliability program dedicated to continuous improvement. However, understanding what data to collect, how to organize data so that it answers the questions of interest, and how to update these efforts based on ongoing human and system reliability analysis, requires an understanding of many of the fundamental qualitative and quantitative considerations discussed to this point. Finally, risk management or error reduction strategies need to be integrated into any human and system reliability analysis program. This requires understanding the effects of various design interventions, both in terms of how well they serve as barriers to human error and violations and to the adverse system outcomes they can induce, and in terms of how the design interventions can contribute to new contexts that can give rise to adverse outcomes.

This summary, which essentially defines the boundaries within which the ideal human and system reliability analyst operates, is presented in Figure 35.21. In the coming years we can expect many of these concepts and methods to become further refined and new techniques to be introduced, especially as analysts continue to gain insights into this problem.

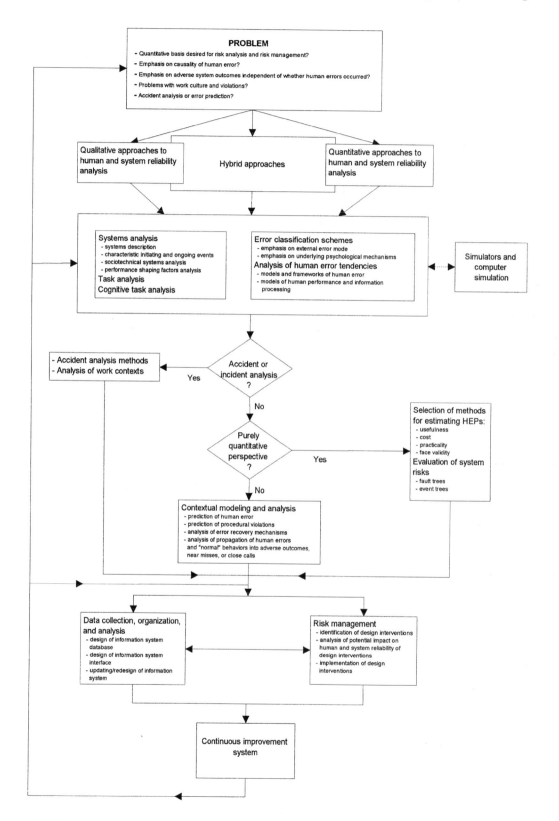

FIGURE 35.21 Defining the boundaries within which the human and system reliability analyst operates.

References

Armstrong, M. E., Cecil, W. L., and Taylor, K. (1988). *Root Cause Analysis Handbook*. Report No. DPSTOM-81, E.I. Dupont De Nemours & Company, Savannah River Laboratory, Aiken, SC.

Askren, W. B. and Regulinski, T. L. (1969). *Mathematical Modeling of Human Performance Errors for Reliability Analysis of Systems*. Aerospace Medical Research Laboratory, AMRL-TR-68-93.

Askren, B. W., and Regulinski, T. L. (1971). *Quantifying Human Performance Reliability*. Technical Report AFHRL-TR-71-22, Air Force Systems Command, Brooks Air Force Base, Texas.

Blackman, H. S. (1991). Modeling the influence of errors of commission on success probability, *Proceedings of the Human Factors Society 35th Annual Meeting*, pp. 1085-1089.

Center for Chemical Process Safety (CCPS). (1989). *Guidelines for Chemical Process Quantitative Risk Analysis*. New York: American Institute of Chemical Engineers.

Center for Chemical Process Safety (CCPS). (1992). *Guidelines for Hazard Evaluation Procedures, Second Edition, with Worked Examples*. New York: American Institute of Chemical Engineers.

Center for Chemical Process Safety (CCPS). (1994). *Guidelines for Preventing Human Error in Process Safety*. New York: American Institute of Chemical Engineers.

Dhillon, B. S. and Singh, C. (1988). *Engineering Reliability*. New York: John Wiley & Sons.

Dougherty, E. M. and Fragola, J. R. (1988). Foundations for a time reliability correlation system to quantify human reliability. *4th IEEE Conference on Human Factors and Power Plants*, pp. 268-278.

Embrey, D. E., Humphreys, P., Rosa, E. A., Kirwan, B., and Rea, K. (1984). *SLIM-MAUD: An Approach to Assessing Human Error Probabilities Using Structured Expert Judgment*. NUREG/CR-3518, U.S. Nuclear Regulatory Commission, Washington, D.C.

Fleishmann, E. and Quaintance, M. (1984). *Taxonomies of Human Performance: The Description of Human Tasks*, Orlando, FL, Academic Press.

Gertman, D. I., Blackman, H. S., Haney, L. N., Seidler, K. S., and Hahn, H. A. (1992). INTENT: A method for estimating human error probabilities for decision-based errors. *Reliability Engineering and System Safety*, 35, 127-137.

Gertman, D. I. and Blackman, H. S. (1994). *Human Reliability & Safety Analysis Data Handbook*. New York: John Wiley & Sons.

Hahn, A. H. and deVries II, J. A. (1991). Identification of Human Errors of Commission Using Sneak Analysis, *Proceedings of the Human Factors Society 35th Annual Meeting*, pp. 1080-1084.

Hannaman, G. W. and Worledge, D. H. (1988). Some Developments in Human Reliability Analysis Approaches and Tools. *Reliability Engineering and System Safety*, 22, 235-256.

Hendrick, K. and Benner, L. Jr. (1987). *Investigating Accidents with STEP*. New York: Marcel Dekker.

Hollnagel, E. (1993). *Human Reliability Analysis: Context and Control*. London: Academic Press.

Johnson, W. G. (1980). *MORT Safety Assurance Systems*. New York: Marcel Dekker.

Kepner, C. H. and Tregoe, B. B. (1981). *The New Rational Manager*. Princeton, NJ: Kepner-Tregoe Inc.

Kapur, K. C. and Lamberson, L. R. (1977). *Reliability in Engineering and Design*. New York: John Wiley & Sons.

Kirwan, B. (1994). *A Guide to Practical Human Reliability Assessment*. London: Taylor & Francis.

Kirwan, B. and Ainsworth, L. K. (1992). *Guide to Task Analysis*. London: Taylor & Francis.

Luczak, H. (1997). Task analysis, in G. Salvendy (ed.) *Handbook of Human Factors and Ergonomics, Second Edition*. New York: John Wiley & Sons.

Ontario Hydro (1977). *Events and Causal Factors Charting*. U.S. Department of Energy 76-45/14, SSDC-14, Ontario Hydro Toronto, Canada.

Ozog, H. (1985) Hazard identification, analysis, and control. *Chemical Engineering*, February 18, 161-170.

Pennycook, W. A. and Embrey, D. E. (1993). An operating approach to error analysis, in *Proceedings of the First Biennial Canadian Conference on Process Safety and Loss Management*. Edmonton, Alberta, Canada. Waterloo, Ontario, Canada: Institute for Risk Research, University of Waterloo.

Perrow, C. (1984). *Normal Accidents: Living with High-Risk Technologies*. New York: Basic Books.

Rasmussen, J. (1982). Human errors: a taxonomy for describing human malfunction in industrial installations, *Journal of Occupational Accidents*, 4, 311-333.

Rasmussen, J. (1986). *Information Processing and Human-Machine Interaction: An Approach to Cognitive Engineering*, New York: North-Holland.

Reason, J. (1990). *Human Error*. Cambridge: Cambridge University Press.

Reason, J. (1995) A systems approach to organizational error. *Ergonomics*, 38, 1708-1721

Roland, H. E. and Moriarty, B. (1983). System Safety Engineering and Management. New York: John Wiley & Sons.

Sharit, J. (1993). Human reliability modeling, in K.B. Misra (ed.) New Trends in System Reliability Evaluation. Amsterdam: Elsevier, 369-410.

Sharit, J. (1997). Allocation of functions, in G. Salvendy (ed.) *Handbook of Human Factors and Ergonomics*, Second Edition. New York: John Wiley & Sons.

Sharit, J., Czaja, S.J., Augenstein, J., and Dilsen, K. (1996). A systems analysis of a trauma center: a methodology for predicting human error, in A.F. Ozok and G. Salvendy (eds.) *Advances in Applied Ergonomics*, 996-1101. Indiana: U.S.A. Publishing Corporation.

Siegel, A. I., Bartter, W. D., Wolf, J. J., Knee, H. E., and Haas, P. M. (1984). *Maintenance Personnel Performance Simulation (MAPPS) Model: Summary Description*. NUREG/CR-3626, U.S Nuclear Regulatory Commission, Washington, D.C.

Stammers, R. B. and Shephard, A. (1995). Task analysis, in J. R. Wilson and E. N. Corlett (eds.) *Evaluation of Human Work, Second Edition*. London: Taylor & Francis, 144-168.

Swain, A. D. and Guttmann, H. E. (1983). *Handbook of Human Reliability Analysis with Emphasis on Nuclear Power Plant Applications*. NUREG/CR-1278, U.S. Nuclear Regulatory Commission, Washington, D.C.

Vicente, K. J. and Rasmussen, J. (1992). Ecological interface design: theoretical foundations, *IEEE Transactions on Systems, Man, and Cybernetics*, 22, 589-606.

Wickens, C.D. (1992). *Engineering Psychology and Human Performance*, Second Edition. New York: HarperCollins.

Woods, D. D., Roth, E., and Pople, H. (1987). *Cognitive Environment: System for Human Performance Assessment*. NUREG-CR-4862, U.S. Nuclear Regulatory Commission, Washington, D.C.

36

Some Developments in Human Reliability Assessment

Barry Kirwan
University of Birmingham

36.1   Introduction

This chapter deals with the subject of human reliability assessment (HRA). HRA may be considered a subdiscipline of ergonomics or human factors (these terms are used interchangeably in this chapter), but it emanates also from the fields of reliability engineering and risk assessment, and is therefore a hybrid discipline. HRA is fundamentally the analysis of human failures. Unlike accident analysis, however, HRA is prospective or predictive — it is concerned with determining what can go wrong, before it happens. This is no trivial task. HRA also not only tries to determine what can go wrong (i.e., human errors), but also how likely it is to go wrong, i.e., it predicts the probabilities of different errors and failures occurring. Furthermore, since HRA has become more linked to psychology and ergonomics over the last decade and a half, it has focused on how human failures occur, and what factors cause them or increase their likelihood of occurrence. Therefore, based on such analysis, it then becomes possible to determine how to prevent such errors from occurring at all, or at least to decrease their likelihood. HRA, broadly speaking, can therefore be seen to have three interlinked functions:

1. Determination of what can go wrong (human error identification)
2. Quantification of the probabilities of errors (human reliability quantification)
3. Reduction of error likelihood (error reduction analysis)

HRA is most commonly used in a risk assessment format, essentially determining how frequently accidental outcomes (e.g., fatalities) will occur in a given period of operation of a system (usually such predicted frequencies are very small, e.g., once in one hundred thousand years of operation). When utilized within risk assessment, HRA is effectively assessing the human contribution to risk. This contribution is integrated within the overall risk assessment framework, so that the human contribution to risk can be seen in conjunction with other contributions to risk: hardware and software failures, and environmental events. Therefore, when total risk is estimated for a system such as a chemical plant or an offshore platform, the relative contribution of human error (and human recovery capabilities) to risk can be judged by the owners, designers, and/or regulators of such a system. Sometimes human error will be seen as a major contributor to risk, and other times its role may be negligible, or at least tolerable. If, however, risk assessment and HRA do show that human error is of significant concern, there will be the need for more human factors effort to improve the designed operator support systems (interfaces, training, procedures, etc.). HRA can therefore lead to the determination of the adequacy, from a safety perspective, of the human factors considerations designed into a system.

A typical question that HRA might be used to address, therefore, is the following:

Is the human error contribution to risk for an offshore drilling system (or nuclear power plant, or chemical plant, or transportation system, etc.) acceptable?

Such a question might be posed by a regulatory body, and the oil and gas company would then be obliged to carry out an HRA/risk assessment to answer it. Sometimes, however, HRAs are carried out to compare two different designs, e.g., an automated process vs. a semiautomated process, and then the risks of the two systems are compared, in order to determine which design to utilize.

Whether HRA is used for risk assessment purposes or for design evaluations and comparisons, it is clearly relevant to human factors on two major counts. First, the prime content or subject matter of HRA is human performance (and particularly human error in actions or decisions) in the working environment, leading to system failures, often as a function of poor original attention to ergonomics factors in the design of the system. Second, the impact of many HRAs leads to the determination that more human factors input into the design process is required. Therefore, HRA is obviously related to human factors by its content and can act as a mechanism for enhancing the incorporation of more human factors into system designs.

A number of questions arise concerning the approach of HRA, and among them are the following, which will be at least partly addressed within the confines of this chapter:

- How does HRA work? (the HRA process)
- What is the scope of real HRAs, i.e., how big are they, how long do they take, what do they achieve, etc.?
- What contemporary developments are occurring in HRA?
- Is there evidence to support the validity of HRA approaches, given that they ultimately have a rather ambitious aim of predicting human performance in non-trivial industrial tasks?

This chapter therefore attempts to define the HRA process, and briefly outline some recent practical HRAs and research initiatives,[1] which show in more detail how HRA may be applied and how it is developing as an approach. A recent validation study is also summarized, which supports the quantitative part of the HRA process, thereby at least in part supporting the empirical validity of the approach of HRA.

---

[1]Much of the work reported in this chapter is of U.K. origin. This is due to the accessibility of this work to the author and, as the referee for this chapter noted, due to a current decrease in work in HRA in the U.S. at the present time. Although there is therefore a slight bias to U.K. work, these studies are still relatively representative of work ongoing in other countries.

36.2   Objectives

The objectives of the chapter can be stated formally as follows:

1. To give an overview of the HRA process
2. To outline the scope and some of the impacts of two recent large-scale (U.K.) HRAs, and a U.K. HRA assessment program
3. To consider recent developments in two major areas of HRA
   A new human error analysis system
   A human error probability (HEP) data bank
4. To summarize a recent validation exercise

36.3   Scope

In the limits of this chapter, a good many short-cuts are necessary. For the more interested reader, there are a number of texts on HRA which describe the overall process, how it can be used, and its impacts (Swain and Guttmann, 1983; Dhillon, 1986; Park, 1987; Dougherty and Fragola, 1987; Kirwan et al., 1988; Swain, 1989; Embrey et al., 1994; Gertman and Blackman, 1994, and Kirwan, 1994). There are also two other texts of interest which take a more pessimistic view of HRA (Reason, 1990; Hollnagel, 1993), but which are also important for their discussion of the nature of human error, as is a recent review of the nature of human error (Woods et al., 1994). The reader interested in finding out more about risk assessment may consult a number of texts (e.g., Henley and Kumamoto, 1981; Green, 1983; Cox and Tait, 1991).

Since this chapter is written by a U.K.-based author, its contents are biased to developments in HRA approaches in the U.K. However, it is believed that these developments are also relevant to other international spheres of HRA activity.

36.4   Human Reliability Assessment — The HRA Process

HRA usually progresses through a number of stages, embodied in the HRA process as shown in Figure 36.1. These stages are described below.

Problem Definition

This refers to scoping the HRA, i.e., deciding what tasks or human involvements will be addressed by the HRA (very few HRAs can address all human involvements, due to resource constraints). In some cases, many tasks will be analyzed in detail, whereas in other HRAs only a representative sample of tasks will be assessed, and many of these will be assessed only using basic rather than detailed methods. There are no fixed criteria for determining the depth and breadth of an HRA, but the following encapsulate some of the major considerations:

- The nature of the plant being assessed and the cost of failure
- The criticality of the role of the operator
- The novelty of the plant design
- The system life cycle stage

The more hazardous the plant, the more critical the role of the operator, and the more novel the plant design, then the deeper or more exhaustive the assessment must be (as with novel systems, there will inevitably be more unpredicted events or situations that the operator will have to deal with). Therefore, a hazardous, human-critical, novel plant design will warrant full error analysis of all types of errors (and a corresponding and interrelated human factors assessment). Alternatively, a well-known and tested plant system that has already been operational for 30 years, has a good (i.e., low) accident record, and has been

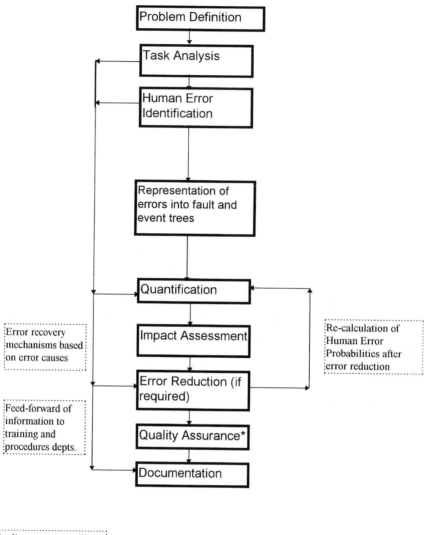

FIGURE 36.1   The HRA process. (Adapted from Kirwan, B. (1990) — A resources flexible approach to human reliability assessment for PRA, Safety and Reliability Symposium, Altrincham, September. London: Elsevier Applied Sciences, pp. 114-135.)

analyzed several times previously will warrant much less assessment effort. However, even an old plant may warrant special attention, e.g., if rule violations are starting to appear in incident records, or if new control systems are being installed, etc. Last, the earlier the life cycle stage, the more the plant can be favorably influenced as a result of error reduction, but the less powerful are the error identification and prediction techniques, and the more difficult task analysis (see below) becomes due to lack of detailed information. Furthermore, some analyses will be difficult at an early design stage, e.g., rule violation assessment (see below) in particular.

The assessment team must also decide upon the scope in terms of which operational stages are to be addressed. Typically, the major focus will be on emergency events, such as post-trip recovery actions in a nuclear power plant following a reactor trip, or the equivalent recovery actions following an event in a chemical plant. However, a number of phases of operation may be considered, namely the following:

- **Normal operation** — Actions and errors may occur in normal operation which lead to hazardous events or situations
- **Abnormal operation** — Conditions which are off-normal must be detected by operators and prevented from developing into incidents, or if the conditions are intentional, special provisions will have been made to maintain safety integrity, and the operators must act accordingly
- **Emergency operation** — An event or incident has happened, and the operators must maintain or restore safety barriers, or mitigate accident consequences (this is often the focus of HRA)
- **Beyond design basis accident** (BDBA) — A severe accident has occurred and operators must limit the accidental consequences (this will involve usage of emergency provisions and procedures, and will probably involve offsite organizations and resources)
- **Maintenance, test, and calibration** — Errors may occur during these phases which may then lead to initiating events or to the unavailability or erroneous operation of safety-related systems or instrumentation during a later event or incident (a so-called *latent failure*)
- **Outage/shutdown** — Certain operations will occur during outages or shutdown (or partial shutdown) conditions, when the usual interlocks and protective systems may be in a partly disabled state. Operations during these phases may still be hazardous, and safety integrity may rely more on administrative controls (e.g., both *ad hoc* and formal procedures), which are ultimately a human barrier form, rather than the usual physical protective systems
- **Start-up/shut-down** — Errors may occur during start-up/shut-down which will cause complications later on, or which could lead directly to events during start-up or shut-down
- **Other** (specific events related to a specific plant)

Once the scope of the analysis has been determined, the next question becomes what can go wrong in the identified spheres of activities, and this is the area of human error analysis. However, before identifying what can go wrong, it is (logically) necessary to determine how tasks and operations *should* proceed, i.e., to determine what is correct or normative performance. This is the human factors subject area of task analysis, discussed next.

Task Analysis

Task analysis is a fundamental approach describing and analyzing how the operator interacts with a system and with other personnel in that system (Kirwan and Ainsworth, 1992). In particular, task analysis describes what an operator is required to do, in terms of actions and/or cognitive processes, to achieve a system goal. Task analysis methods can also detail the displays which cue the operator to perform or cease an operation, and the controls with which such operations are achieved. The first primary aim of task analysis is to create a detailed picture of human involvement, with all the necessary information for analysis of the adequacy of that involvement. Thus, although all task analysis techniques aim to describe the task, they differ in the information that is encoded, depending on the aim of the evaluation.

Prior to representing tasks in some format, data on the operator's task must be collected in some way. There are a number of data collection approaches, the most commonly used ones in HRA being the following (see Kirwan and Ainsworth, 1992; and also Sinclair, 1995):

Observation
Interviews
Critical incident technique (CIT)
Operational experience review (OER) (of incident and near-misses, etc.)
Walk-through and talk-through (WT/TT)

The most frequently used formal approach in task analysis for HRA purposes is *hierarchical task analysis* (HTA) (see Shepherd, 1989; Kirwan and Ainsworth, 1992). This is a useful way of eliciting the *goals* driving behavior, the *tasks* used to achieve the goal or goals, the tasks' subordinate *operations* (actions and behaviors), and the operational sequence options, called *plans,* in a human operator task. HTA is a powerful and flexible method and can represent the task either graphically or in a tabular fashion, and both representations are concise and amenable to usage for the next stage of the HRA process (error identification). There are a number of other more detailed task analysis techniques, which can be useful in the support of error identification, as follows:

Tabular Task Analysis (TTA) — To consider interface-related errors, i.e., associated with problems in knowing when to act or whether an act has been performed correctly, as a function of the feedback available to the operators via the interface.

Tabular Scenario Analysis (TSA) — To define the sequence of tasks in an abnormal/emergency event-driven situation (see Kirwan, 1994).

Timeline Analysis (TLA) — To determine overall task feasibility with respect to time constraints (usually used for emergency response tasks — *horizontal timeline analysis*), and the overall coordination of personnel required (including communications and personnel locations) and individual personnel involvement in tasks (thus, problems of work load, in a crude sense, can be identified and addressed — *vertical timeline analysis*).

Link Analysis (LA) — To be used if it is required to consider detailed movements in a confined location, e.g., to determine if task interference will occur, or if signals will be missed due to poor location of instruments etc., or to consider accidental activation of critical controls, or to consider information flow-paths.

Decision-Action-Diagrams (DADs) — These are used to model decision making sequences and are a useful representation method for all but simple/straightforward decisions.

The usage of these various approaches for HRA purposes is shown in Figure 36.2.

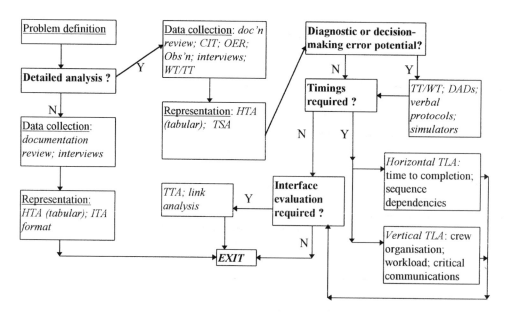

FIGURE 36.2 HRA task analysis technique selection. (Adapted from Kirwan, B. (1994) A guide to practical human reliability assessment. London: Taylor & Francis.)

Human Error Analysis

Once task analysis has been completed, human error identification may then be applied to consider what can go wrong. At the least, this error identification process will consider the following types of error (adapted from Swain and Guttmann, 1983):

- Omission error — Failing to carry out a required act
- Action error — Failing to carry out a required act adequately:
  - Act performed without required precision, or with too much/little force
  - Act performed at the wrong time
  - Acts performed in the wrong sequence
- Extraneous act — Unrequired act performed instead of or in addition to the required act (also called error of commission, or EOC)

While this is a basic and useful set of error modes, a more refined, recent, and cognitive process control-related set is shown in Table 36.1 (Meister, 1995). The human error identification phase can identify many errors, and there are many techniques available for identifying errors (see Kirwan, 1992a; 1992b; 1995). Not all of these errors will be important for the study, as can be determined by reviewing their consequences on the system's performance. This is done during the error identification stage, usually by completing a human error analysis table, as shown in Table 36.2. This simplified table shows the task step (or operation, or goal, or plan) from the task analysis, and then the error(s) associated with that step. Consequences are determined (often in conjunction with reliability engineers assessing hardware, software, and environmental failures), and possible recovery steps (e.g., checks by supervisor, etc.) are considered. The assessor may also at this stage identify how the error could be prevented, or how its effects could be mitigated.

The errors which most clearly contribute to a degraded system state, alone or in conjunction with other hardware/software failures and/or environmental events, must next be *represented* in the risk analysis framework.

Representation

Having defined what the operators should do (via task analysis) and what can go wrong (at a detailed error level or simply at the overall task execution level), the next step is to represent this information in a form in which quantitative evaluation of human error impact on the system, in conjunction with other failure events, can take place. It is usual that the human error impact be seen in the context of all other potential contributions to system risk, such as hardware and software failures, and environmental events. This enables total risk to be calculated from all sources, and enables interactions between different sources of failure to be assessed. It also means that when risk is calculated, human error will be seen in its proper perspective, as one component factor affecting risk. Sometimes human error will be found to dominate risk, and sometimes it will have less importance than other failure types (e.g., hardware, software, or equipment failures).

Risk assessments typically use logic trees, called fault and event trees, to determine risk, and human errors and recoveries are usually embedded within such logical frameworks (see Henley and Kumamoto, 1981; Green, 1983; Cox and Tait, 1991; Kirwan, 1994). Fault trees look at how various failures and combinations of failures can lead to a "top event" of interest (e.g., loss of core cooling in a nuclear power reactor). Event trees are more sequential in nature, and look at how an event, once occurred, develops and proceeds toward accidental circumstances (e.g., loss of cooling can eventually lead to core melt if certain hardware and human functions fail, which in turn could lead to leakage of radioactivity into the biosphere, and to the need for public evacuation, etc.). Both fault trees and event trees can be very large, and so are usually developed and mathematically evaluated using computerized methods and tools.

TABLE 36.1   Possible Cognitive Errors in Process Control

Information Gathering

1. Failure to detect a disturbance
2. Misreading system status information
3. Misinterpretation of system status information: a general term for instances not covered by one of the more specific interpretation errors below
4. Failure to associate two or more items of information, when their combined effect should be noted
5. Wrongly associate two or more items of information
6. Interpret the situation as having changed, when in fact it hasn't
7. Failure to realize that the situation has changed, as has been indicated by new information
8. Over/under-estimation of situation severity
9. Use excessive time acquiring information
10. Spend too little time acquiring information about system status
11. Great difficulty in interpreting symptoms of system status
12. Concentration of attention on one deviation to the exclusion of a second concurrent deviation

Stabilization Requirements

1. Failure to attempt stabilization
2. Incomplete stabilization requirements
3. Incorrect stabilization requirements
4. Over/under-estimate basic requirements for stabilization

Hypothesis Generation

1. Inability to develop a hypothesis
2. Generate too few possible hypotheses
3. Difficulty in deciding between competing hypotheses
4. Selection of an incorrect hypothesis
5. Considers correct hypothesis of fault cause, but rejects it without even testing
6. Perseveres with working hypothesis of fault cause, despite contrary evidence

Hypothesis Testing

1. Failure to test hypothesis
2. Select incorrect test of hypothesis
3. Perform selected test incorrectly, due to skill-based/procedural errors
4. Misinterpretation of test result

Performing Corrective Action

1. Selection of corrective action inconsistent with information gathered
2. Skill-based/procedural errors during execution of the corrective action
3. Failure to complete execution of corrective action
4. Reacts to system status information in excessively rapid manner when the need to do so is not apparent
5. Operator continues with corrective action after it is clear it has not rectified the situation
6. Proceed to initiate selected action, even though it is no longer applicable because the situation has changed
7. Failure to monitor the effects of corrective actions
8. Inability to decide whether or not the corrective action has been effective, and has subsequent problems in deciding whether the action should continue
9. Misinterpretation of effects of corrective action

From Meister, D. (1995) Cognitive behaviour of nuclear reactor operators. *International Journal of Industrial Ergonomics,* 16, 109-122. With permission.

TABLE 36.2   Simple HEA Tabular Format

Task Step (from the HTA)	Error	Consequence	Recovery	Error Reduction
4.1.3 Close valve	Omits task step	Line will rupture leading to release	Pressure build up alarm in CCR	Emphasis in procedures; spring-returned valve

A further representation issue is that of dependence between two or more errors or tasks, where for example, failure on one task will increase the likelihood of failure on a subsequent task (e.g., misdiagnosis may affect the successful outcome of a number of subsequent tasks). It is important that such dependencies are included in the representation, and that their effects are given appropriate mathematical weighting, as otherwise, risk may be seriously under-estimated. Currently only one HRA technique really deals with dependence (THERP — see later), and most other techniques borrow from this method when required.

Human Error Quantification

Once the human error potential has been represented, the next step is to quantify the likelihood of the errors to determine the overall effect of human error on system safety or reliability. Human reliability quantification techniques all quantify the human error probability (HEP), which is the metric of human reliability assessment. The HEP is defined as follows:

$$\text{HEP} = \frac{\text{number of errors occurred}}{\text{number of opportunities for error to occur}}$$

Thus, if when buying a cup of coffee from a vending machine, on average one time in a hundred tea is accidentally purchased, the HEP is taken as 0.01 (it is somewhat educational to try and identify HEPs in everyday life with a value of less than once in a thousand opportunities, or even as low as once in ten thousand). In an ideal world, there would be many studies and experiments in which HEPs were recorded, (e.g., operator fails to fully close a valve once every five thousand times (s)he is required to close a valve). In reality there are few such recorded data. The ideal source of human error "data" would be from industrial studies of performance and accidents, but at least three reasons can be deduced for the lack of such data:

- Difficulties in estimating the number of opportunities for error in realistically complex tasks (the so-called denominator problem)
- Confidentiality and unwillingness to publish data on poor performance
- Lack of awareness of why it would be useful to collect in the first place (and hence lack of financial incentive for such data collection)

There are other potential reasons (see Kirwan et al., 1990) but the net result is a scarcity of HEP data. HRA therefore uses quantification techniques, which either rely on expert judgment, or a mixture of data and psychologically based models which evaluate the effects of major influences on human performance.

Below are listed the major techniques in existence in the field of human reliability quantification, arguably the most developed field within human reliability assessment today. These are categorized below into four classes, depending on their data sources, and mode of operation.

1. Unstructured Expert Opinion Techniques
   - Absolute Probability Judgment or Direct Numerical Estimation (APJ or DNE: Seaver and Stillwell, 1983)
   - Paired Comparisons (PC: Hunns, 1982; Comer et al., 1983)
2. Data-Driven Techniques
   - Human Error Assessment and Reduction Technique (HEART: Williams, 1986; 1988; 1992)
   - Technique for Human Error Rate Prediction (THERP: Swain and Guttmann, 1983)
   - Human Reliability Management System (HRMS: Kirwan and James, 1989; Kirwan, 1994)
   - Justification of Human Error Data Information (JHEDI: Kirwan 1990; 1994)
3. Structured Expert Opinion Techniques
   - Success Likelihood Index Method using Multi Attribute Utility Decomposition (SLIM-MAUD: Embrey et al., 1984)
   - Socio-technical Approach to Assessing Human Reliability (STAHR: Phillips et al., 1983)

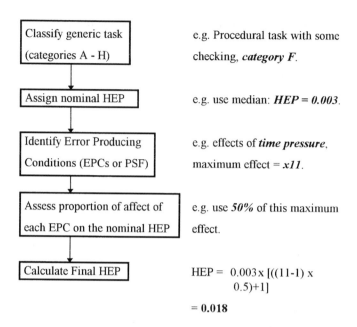

FIGURE 36.3   HEART quantification process.

4. Accident Sequence Data Driven Techniques
   • Human Cognitive Reliability (HCR: e.g., Hannaman et al., 1984)
   • Accident Sequence Evaluation Program (ASEP: Swain, 1987)

All of these techniques (and others) generate human error probabilities. Swain (1989) and Kirwan et al. (1988) discuss the relative advantages and disadvantages of these techniques, and Kirwan (1988a) gives the results of a comparative evaluation of five techniques. Kirwan et al. (1988) and Kirwan (1994) also give selection guidance to help practitioners decide which one(s) to use for a particular assessment problem.

To give a simplified indication of how data-driven techniques work, the THERP technique uses a database of "nominal" human error probabilities, e.g., failure to respond to a single annunciator alarm. Performance shaping factors (PSF) such as the *quality of the interface design* (e.g., whether alarms are prioritized, adequately color-coded, near to the operator and in the normal viewing range, etc.), or *time pressure*, are then considered with respect to this error. If such factors are indeed evident in the scenario under investigation, then the nominal human error probability may be modified by the assessor (e.g., in this case increased to reflect poor quality of interface) by using an error factor (EF) of, say, 10. Thus, if an initial nominal HEP is 0.001, an EF of 10 can be used to increase the actual estimated HEP to a value of 0.01. This is then the HEP that could be inserted into the fault or event tree.

Expert judgment techniques, (1) and (3) above, on the other hand, use personnel with relevant operational experience (e.g., more than 10 years) to estimate HEPs. The rationale is that these personnel will have had significant opportunities for error and will have also committed certain errors (and seen others commit errors), and hence have information in their memories which can be used to generate HEPs. Such expert opinion methods may either ask experts directly for such estimates, or may use more subtle and indirect methods, to avoid the various biases associated with human recall which can occur (see Tversky and Kahneman, 1974).

Another technique, currently popular in the U.K., is the HEART approach. The HEART quantification process is shown in Figure 36.3. The first stage is to consider what category the task in question belongs to, from a range of nine categories. Each category has an associated nominal or baseline error probability range, for example as follows:

TABLE 36.3 HEART Calculational Formula and Example Calculation
(task = carry out valve closure sequence)

| Type of Task F | Generic Error Probability — 0.003 | | |
Error Producing Conditions (EPCs)	Maximum effect	Assessed Proportion of Effect	Calculation
Inexperience	×3	0.4	((3-1).0.4) +1 = 1.8
Opposite technique	×6	1.0	((6-1).1.0) +1 = 6.0
Low morale	×1.2	0.6	((1.2-1).0.6)+1 = 1.12

HEP = 0.003 × 1.8 × 6.0 × 1.12 = 0.036

Category A: Totally unfamiliar task, performed at speed with no real idea of likely consequences. HEP range = 0.35–0.97, median HEP = 0.55.

Category F: Restore or shift a system to original or new state following procedures, with some checking. HEP range = 0.0008–0.007, median HEP = 0.003.

Once the assessor has decided which task category to use, the next stage is to consider whether there are any negative factors that could lead to an increased HEP. More generally known as performance shaping factors (PSF), in HEART these factors are called error producing conditions (EPCs). There are over 30 EPCs, though most assessors use a range of approximately 20 EPCs. Each EPC has a derived associated maximum effect it can have on the nominal HEP, for example, as follows:

EPC	Maximum Effect on Nominal HEP
Unfamiliarity	x17
Time pressure	x11
Operator inexperience	x3

The next step with HEART, the calculation of the final HEP, is fairly straightforward (see example in Table 36.3). It should be noted that the formula uses a mathematical "fix" to avoid low-maximum-effect EPCs inadvertently resulting in decreasing the HEP, rather than increasing it. The effectiveness of this fix can be seen in Table 36.3 with respect to the EPC of motivation: if the assessed proportion of effect was applied directly to the maximum effect for this EPC, then the resulting multiplier would be less than unity (0.72 in this example), and this would then reduce the final HEP instead of increasing it.

In summary, therefore, HEART begins the quantification process by determining which of a number of generic error probabilities is appropriate to the scenario under investigation. The assessor then considers PSF effects on the task, though these are called error producing conditions (EPCs) in HEART terminology. Having selected EPCs from a range of 32 EPCs, which are weighted in terms of the maximum effect they can each have on performance, the assessor then judges how much of that maximum effect should be applied in the scenario. The overall HEP is then calculated as a function of the original generic probability, the number of EPCs, and each EPC's weighting factored by the assessor's assessed rating of the EPC's level in the scenario.

One important advantage of techniques like HEART (and SLIM, and HRMS/JHEDI) is that they can give insights into avenues for error reduction. This facility is available because of their utilization of performance shaping factors in the HEP estimation process. Thus, for example, if the above HEP was considered unacceptably high, it can be seen that the EPC of *opposite technique* is contributing significantly to the final derived HEP. This EPC may have been used by the assessor because one of the valves closes in the opposite direction to the others (also called a stereotype violation). If the plant designers/owners were to change that valve to make it consistent with others in the plant, the EPC of opposite technique could be removed from the HEART calculation. This would lead to a revised HEP of 0.006 (an error reduction factor of 6). This facility of some of the HRA techniques is useful for determining risk reduction measures and priorities.

Impact Assessment, Error Reduction Assessment, Quality Assurance, and Documentation

Following such quantifications, system risk will be calculated. It will then be determinable whether the installation's overall risk is acceptable. If not, it may be necessary or desirable to try to reduce the level of risk. If human error is contributing significantly to risk levels, then it will be desirable to attempt to find ways of reducing the human error impact. This may or may not utilize the quantification technique (as shown above) in devising error reduction mechanisms (ERMs) which will reduce the system vulnerability to human error. If ERMs are derived (whether by the quantification means or other qualitative means, or both), then the quantification technique will be used to recalculate the HEPs, for the system as it would perform with ERMs in place. Following this stage (which can run through several iterations), the results will be documented and quality assurance systems should ensure that ERMs are effectively implemented, and that assumptions made during the analysis remain valid throughout the lifetime of the system (see Kirwan, 1994). This completes the HRA process. The next section gives a brief overview of some applications of HRA.

36.5   Human Reliability Assessment in Practice

The following examples are aimed at giving insight into large HRAs in practice. There are three examples as follows (this section is adapted from Kirwan, 1996a):

1. Two recent plant design HRAs, discussing briefly their scope and impact
2. An HRA program for a number of existing and aging nuclear power plants, showing the impact of HRA on the human factors in the system design

HRA for THORP (Thermal Oxide Reprocessing Plant, BNFL, Sellafield)

THORP is a very large nuclear fuel reprocessing plant, sometimes described as a number of conventionally sized plant modules joined together under one roof, and has a staff complement approaching 800 personnel. It has process complexity similar to that of a nuclear power plant. The HRA approach started in earnest in the detailed design stage for THORP, and was predicated upon a large human factors assessment exercise (Kirwan, 1988b; 1989), amounting to approximately 15 person-years of effort. This assessment addressed the safety adequacy (from a human factors perspective) of the central control room, local control rooms, control and instrumentation panels local in the plant, staffing and organization issues, training, and emergency preparedness. The human factors reviews were used to determine the effect of certain performance-shaping factors on performance (e.g., the adequacy of the interface in the central control room during emergency conditions). Such information significantly influenced the two computerized quantification systems developed for the THORP HRA, namely the Human Reliability Management System (HRMS: Kirwan and James, 1989; Kirwan, 1990) and the Justification of Human Error Data Information system (JHEDI: Kirwan, 1990).

JHEDI was designed to rapidly but conservatively assess all identified human involvements in the THORP risk assessment, and assessed over 800 errors. JHEDI requires a simplified task analysis, error identification via keyword prompting, quantification according to PSF, and noting of training or procedural implications to be fed forward to the respective THORP operations departments.

HRMS is a more intensive system and is used for those errors that are found to be risk significant in the PSA (e.g., for the errors that have a potentially major risk impact on the THORP system: see Kirwan, 1990). HRMS requires detailed task analysis, error identification and quantification, and computer-supported error reduction is also carried out based on the PSF assessment. Approximately 20 tasks were the subject of the more detailed HRMS assessment approach.

All assessments have been rigorously documented, and information arising out of the assessments is fed forward to the operational departments that are now running THORP and assessing its safety performance. JHEDI is still being applied to other plant designs at BNFL Risley, U.K., the design center for BNFL.

Sizewell B Pressurized Water Reactor HRA[2]

The Sizewell B risk assessment and HRA (Whitworth, 1987; Whitfield, 1991; 1995) for the U.K.'s first and only PWR were also to an extent predicated upon extensive human factors assessment of the design of the interface and other systems (e.g., see Umbers and Reiersen, 1995; Ainsworth and Pendlebury, 1995), although the linkage between the human factors assessments and the HRA inputs was less formalized than for THORP. The HRA approach involved a very large amount of initial task analysis and error analysis. The Human Error Assessment and Reduction Technique (HEART: Williams, 1986) was used as the main quantification tool, supplemented by the Technique for Human Error Rate Prediction (THERP: Swain and Guttmann, 1983), and error reduction was carried out as required.

One of the early impacts of the human reliability assessment work for Sizewell B was the recognition of the need for more automation to support the operator in a particular accidental event scenario (Fewins et al., 1992).

In the early HRA phase, a great deal of human error analysis and task analysis was carried out, whereas later on, as the design became more detailed and construction started, the focus shifted to particular significant scenarios and the quantification of HEPs for these scenarios. The balance of effort in the HRA was therefore largely qualitative in the early design stages, and more quantitative later on. This is a sensible approach, since early in the design phase is when most design impact can be easily achieved, and the numbers (HEPs and risk calculations) may be seen as secondary when compared to the goal of achieving a good (safe) and operable (ergonomic) design. Later on, having a strong safety case so that operation will be allowed by the regulators, becomes the primary driving force behind assessment. Therefore, the HRA effort becomes more focused on quantitative predictions and fault and event trees, as the resultant risk estimates will determine when the plant may become operable. This shift of emphasis within the life cycle of a large HRA for a novel and complex plant is probably typical of such projects.

The Sizewell B detailed design and assessment program spanned over a decade, following an exhaustive public inquiry, during which one of the key recommendations was for a high degree of human factors support for the station design and development. (This inquiry spawned the Advisory Committee for the Safety of Nuclear Installations, ACSNI, which still advises the nuclear power industry in the U.K.). As a consequence, the project received a good deal of guided human factors and HRA effort, as it was seen as a high-profile project.

Continued Operation Risk Assessments/HRAs

These risk assessments and associated HRAs are part of a required program of work to determine whether the aging gas-cooled reactor plants are safe to continue operating beyond their original sanctioned lifetimes (e.g., whether the Magnox stations can operate beyond 30 years). So far, two Magnox plants have been shut down (due probably more to economic reasons than safety concerns), and the results of the continued operation HRAs for the other stations are still being reviewed by the relevant U.K. regulatory body, HM Nuclear Installations Inspectorate.

These HRAs have used a significant amount of task analysis and have generally each followed a similar basic format: detailed task analysis for a small number of key scenarios, and less detailed analysis of the remainder of the scenarios. No detailed error analysis is utilized, and task failure likelihood is calculated by using the HEART quantification method. Error reduction measures are identified either based on the HEART calculations or on the task analysis. An example of the methodology from one of the continued operation HRAs is given in Kirwan, Scannali, and Robinson (1995). In this particular study, which was carried out over a period of two years, four tasks were assessed in exhaustive depth using a range of task analysis approaches (hierarchical, tabular, and timeline analyses), and approximately 40 other tasks were

---

[2]It should be noted that the insights into the Sizewell "B" PWR nuclear power plant project are based largely on the author's observations as an outsider to that program of HRA activities, and discussions with colleagues working on the project at the time. These views may therefore not be representative of the owners' or project team's views.

analyzed in less (though still substantial) depth. The HEART approach was used for quantification, with THERP used to independently corroborate the quantifications for three of the tasks (the results using the two techniques agreed within a factor of three of each other, and this was considered a positive result). Fault trees were the predominant representation method used, and THERP's dependence model was adapted for the HRA. The HRA led to a number of impacts on the existing plant, which have since been implemented. Some examples are as follows:

- Procedural recommendations (i.e., changes to existing procedural documentation)
- Training recommendations (concerning simulator training)
- Emergency lighting recommendations
- Recommended changes to the VDU interface
- Recommendations for certain local alarms being made more central

As a result of the risk assessment and HRA, and the implementation of certain human factors and other risk assessment-identified design recommendations, the plant is still currently operating and producing power.

36.6   Developments in HRA

This section discusses two recent developments in HRA in the U.K. The first is the development of a HEP database to support HRA activities, and the second is to support error identification activities. Both of these areas were seen as primary and urgent areas for development of practical tools in the U.K. The first was necessary to lend credibility to the whole HRA quantification process (i.e., so that at least some real and robust HEPs were in evidence, among all the expert judgment that enshrouded the area of HRA), and to provide at least enough HEPs to be useful for validation and training efforts. The second was to provide much needed support to assessors in the difficult area of determining what can go wrong (error identification). This latter project on error identification was also pursued since industry and academia alike realized that, whereas prediction of HEPs was reasonably robust and a range of tools existed for quantifying HEPs, the logical precursor to quantification, namely deciding what needed to be quantified, was an immature area in terms of technique development and was probably a large source of inconsistency in practical HRA. Although there have been other developments in HRA, therefore, these two are seen as significant in enhancing this applied methodology.

Human Error Database (CORE-DATA) Development

Although much human reliability assessment (HRA) is carried out internationally, in nuclear power and other industries (e.g., chemical and offshore), there has been a paucity of HEP data useful for predicting how often errors will occur. This lack of data has been despite a number of attempts to develop such a database over the past 30 years (e.g., see Topmiller et al., 1984). However, in the late 1980s, certain developments in the theory of human error suggested that a database could be constructed. In the U.S. this led to the NUCLARR database (e.g., see Gertman and Blackman, 1994).

Following an Advisory Committee for the Safety of Nuclear Installations (ACSNI: 1991) report in the U.K. recommending the construction of a HEP database, a three-year project was set up to develop the Computerised Operator Reliability and Error Database (CORE-DATA: see Taylor-Adams and Kirwan, 1995; Taylor-Adams, 1995). The work was funded by the U.K. Nuclear Power and Reprocessing Generic Nuclear Safety Research (GNSR) Programme.

The main objectives of the project were as follows: to aggregate existing HEP data and to structure the database to be theoretically satisfactory yet usable for assessors; to collect new data for the database; and to produce a computerized prototype of the database for demonstration purposes. There was also a more long term objective, namely the investigation of the feasibility of developing extrapolation rules to render the technique usable as a human reliability assessment tool in its own right, i.e., so that data could be manipulated to be applicable to new situations, scenarios, and even new industrial contexts.

The development of a sound theoretical structure for the database relies on the development of a number of taxonomies or classification systems, which enable categorization of the data in psychologically meaningful, robust, and mutually exclusive terms. This was not a trivial task, and five taxonomies were developed, analyzed, and evaluated in the context of the current dominant models in human factors and human reliability, namely the Information Processing Model (Wickens, 1992), and the Skill, Rule and Knowledge Model (Rasmussen et al., 1981). Once these taxonomies had been developed, existing data could be incorporated into a working database.

CORE-DATA currently contains about 250 HEPs, and a further 900 or so, of varying pedigree, have been put into hard copy CORE-DATA format. There have also been two recent and successful data collection exercises (one offshore evacuation study [Basra and Kirwan, 1996], and one manufacturing study), which have produced new data for the CORE-DATA system. An example of the CORE-DATA interface is shown in Figure 36.4.

Human Error Analysis

A recent research program has produced a prototype human error analysis system for dealing with the following types of error:

- Skill- and rule-based error forms (slips and lapses, and rule-following errors)
- Cognitive errors (diagnosis and decision-making errors)
- Errors of commission
- Rule violations
- Teamwork and communication errors

This subsection gives some insights into the human error analysis tool-kit that has been developed, called the Human Error and Recovery Analysis (HERA) system (Kirwan, 1998), with respect to the most detailed and computerized element, the skill and rule-based error analysis section.

The skill and rule-based error analysis section has seven sets of error identification prompts or questions, as follows:

1. Mission — High level mission-oriented questions, e.g., Could the task fail to be achieved in time?
2. Operations — Still fairly high-level questions, e.g., Could a previous latent maintenance error lead to errors or difficulties in the current task?
3. Goals — Concerned with high-level planning of the team, e.g., Could the team fail to realize the need to shift to a higher goal?
4. Plans — Concerned with the timing and sequencing of the task, e.g., Will the pre-conditions for the plan be met?
5. Error — External error mode prompts, e.g., omission; right action on wrong object; etc.
6. Performance shaping factors (PSF) — macro human factors considerations, e.g., Is the alarm reliable? How often do operators expect to see this event?
7. Psychological error mechanisms — The internal mechanisms of failure, e.g., manual variability; substitution error; information transmission error, etc.

It is up to the assessor as to which of the above seven modules are utilized, and several or even all can be applied to the hierarchical task analysis that is entered in a tabular format into the computerized system. Once errors are identified, they are automatically encoded into a human error analysis table. The assessor can then add the consequences of the errors and any likely error recovery possibilities. The tabulated results can then be printed out, ready for the representation phase of the HRA process.

An example of the usage of the computerized HERA system is shown in Figures 36.5 and 36.6, and Table 36.4. In this example, goals analysis is being applied to a (UK) Nuclear Power Plant Loss of Offsite Power (Loss of Grid) scenario. Figure 36.5 shows the goals analysis questions in full, with their abbreviated short-form highlighted in bold. Figure 36.6 shows a screen dump from the HERA system, showing a goals analysis in progress. Table 36.4 shows part of the resultant human error analysis table based on goals analysis of this scenario.

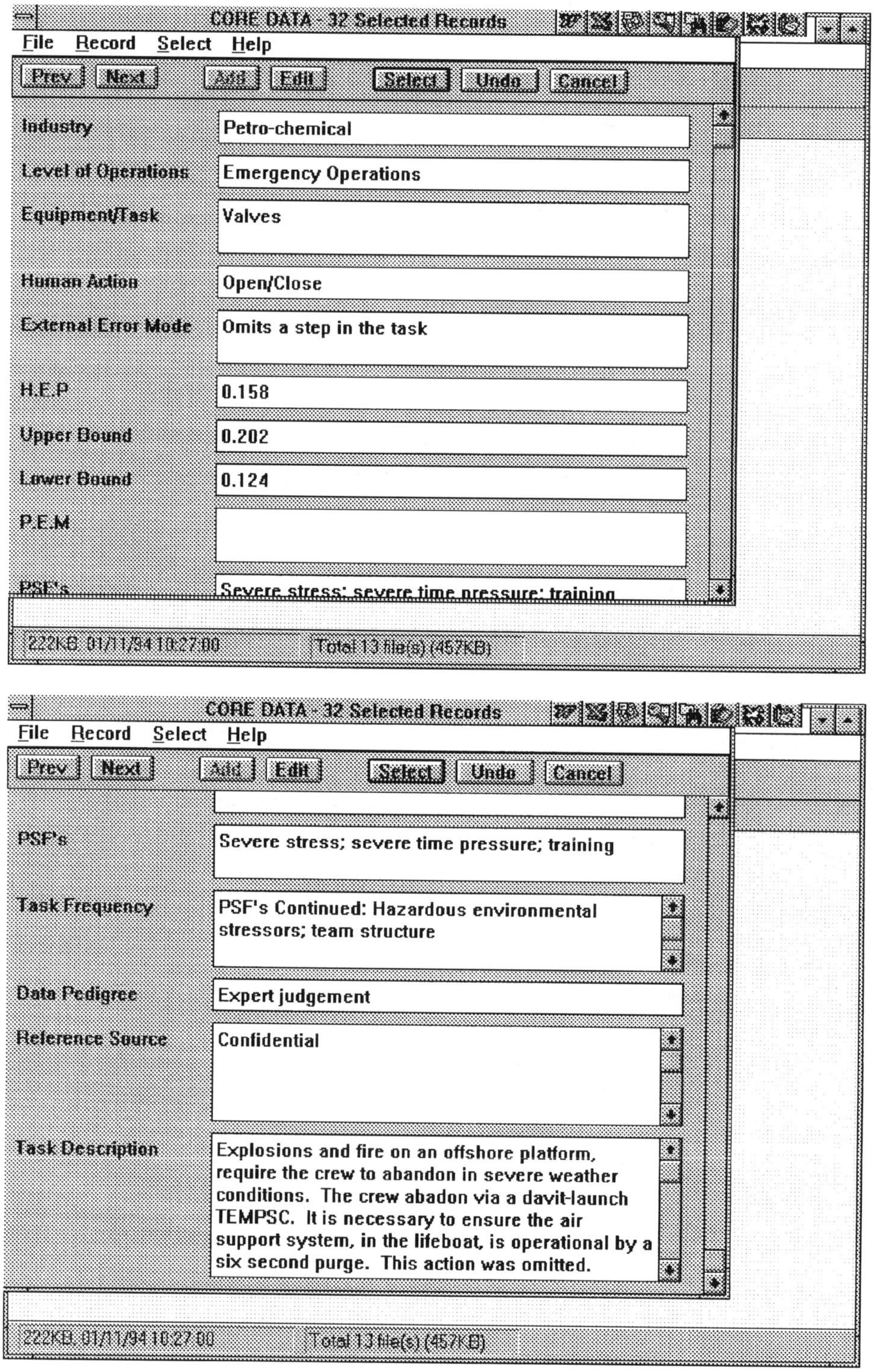

FIGURE 36.4   Screen of first part of CORE-DATA data screen (split here into two halves).

GOALS ANALYSIS
1. Could the operators have no goal, e.g. due to a flood of conflicting information; the sudden onset of an unanticipated situation; a rapidly evolving and worsening situation; or due to disagreement or other decision-making failure to develop a goal ? **[no goal]**
2. Could operators find themselves without any clear strategy, i.e. they would be 'outside' the procedures ? {as in a Beyond Design Basis Accident or BDBA} **[outside procedures]**
3. Could the key decision-maker(s) panic or become unavailable, and hence obstruct goal-setting? {chain of command is broken - also known as 'decapitation' if the task leader is lost} **[no goal]**
4. Could the operators have the wrong goal, e.g. due to a misperception of the circumstances, or due to over-familiarity with that goal (e.g. due to training bias)? **[wrong goal]**
5. Could the operators have the wrong goal due to production pressure, or due to practical difficulties in implementing the stated goal, leading to apparent violations ? **[wrong goal]**
6. Could there be a goal conflict, e.g. between production and safety goals (e.g. shutting down plant when the plant safety margin is narrow), or between competing safety goals (e.g. rescuing operational personnel from fire versus protecting the public from contamination) **[goal conflict - note: this may result in goal delayed, wrong goal, no goal, or goal inadequate]**
7. Could the operators fail to realise the need to shift to a higher goal, as the scenario worsens, e.g. due to detection failure or disbelief of indications, or due to operators getting 'locked in' to a lower level goal or task (e.g. diagnosis), ignoring the deterioration of events, and persevering with a goal when it should be aborted in favour of another? **[wrong goal]** ?

FIGURE 36.5 HERA Goals Analysis.

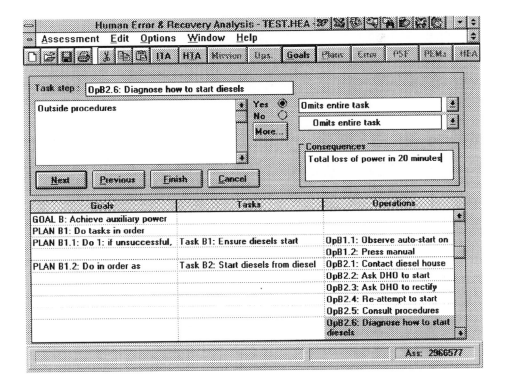

FIGURE 36.6 HERA Goals Analysis Screen.

TABLE 36.4 Goals Analysis: Skill and Rule Identified Errors for Loss of Offsite Power (Grid) Scenario

Identifier from HERA	Task Step	Error Identified	Consequence	Recovery	Comments
1. No goal	2.6:. Diagnose how to start diesels	Fail to derive plan	Total loss of power and forced cooling in 20 minutes.	Restoration of grid within 20 minutes (unlikely).	Overall failure to diagnose.
2. Outside procedures	2.6: Diagnose how to start diesels	No *ad hoc* procedures derived	Total loss of power and forced cooling in 20 minutes.	Restoration of grid within 20 minutes (unlikely).	
3. Decapitation		None identified			
4. Wrong goal	2.6: Diagnose how to start diesels	Switch off all pony motors	Undesirable to switch off all pony motors. Cooling will suffer.	When diesels started, will restart pony motors.	
5. Production pressure		N/A			
6. Goal conflict	2.6: Diagnose how to start diesels	Unwilling to turn off pony motors	Total loss of power and forced cooling in 20 minutes.	Restoration of grid within 20 minutes (unlikely).	Also failure to start diesels, a form of reluctance.
7. Goal shift failure	2.6: Diagnose how to start diesels	Fail to realize they must go beyond procedures	Total loss of power and forced cooling in 20 minutes.	Restoration of grid within 20 minutes (unlikely).	
8. Premature shift	2.6: Diagnose how to start diesels	Switch off cooling when not required (i.e, before off-loading non-essential supplies)	Loss of cooling when not necessary	Diesels start and pony motors restarted.	A partial diagnostic failure — it is safer to do this than do nothing, but will compromise short-term cooling.
9. Too many goals		None identified			
10. Goal delayed	2.6: Diagnose how to start diesels	Delay opening of circuit breakers until too late (>20 minutes)	Total loss of power and forced cooling in 20 minutes.	Restoration of grid within 20 minutes (unlikely).	
11. Incomplete goal		None identified			
12. Violating goal		None identified			

36.7   Validation of Human Reliability Quantification Techniques

A recent large-scale validation has taken place in the U.K. to test the three quantification techniques THERP, HEART, and JHEDI. Thirty U.K. practitioners took part in the exercise, and each assessor independently used only one of the three techniques (very few assessors were practitioners in more than one technique). Each assessor quantified HEPs for 30 tasks, and for each task there was the following information:

- General description of the scenario
- Inclusion of relevant PSF information in the description
- Provision of simple linear task analysis
- Provision of diagrams where necessary and relevant
- Statement of exact human error requiring quantification

Each assessor had two days to carry out the assessments, and experimental controls were exercised, so that the assessors were working effectively under invigilated examination conditions. For each of the 30 tasks the HEP was known to the experimenter, but not to any of the assessors. Tasks were chosen to be relevant to nuclear power and reprocessing industries, since all of the assessors were currently working in these two areas. A large proportion of the data were from real recorded incidents,[3] with the data range spanning five orders of magnitude (i.e., from 1.0 to 1E-5). The results are summarized below. (For a fuller analysis and presentation/discussion of results, see Kirwan et al., 1996.)

Predictive Validity

The analysis of all the data (i.e., all 895 estimated HEPs — there were five missing values) showed a significant correlation between estimates and their corresponding true values (Kendall's coefficient of concordance: $Z = 11.807$, $p < 0.01$). This supports the validity of the HRA quantification approach as a whole, especially as no assessors or outliers have been excluded from these results. The analysis of all data for individual techniques shows a significant correlation in each case (using Kendall's Coefficient of Concordance): THERP $Z = 6.86$; HEART $Z = 6.29$; JHEDI $Z = 8.14$; all significant at $p < 0.01$.

Individual correlations for all subjects are shown in Table 36.5. There are 23 significant correlations (some significant at the $p < 0.01$ level) out of a possible 30 correlations. This is a very positive result, again supporting the validity of the HRA quantification approach.

Precision

Table 36.6 shows that there is an overall average of 72% precision (estimates within a factor of 10) for all assessors, irrespective of whether they were significantly correlated or not. This figure includes all data estimates, even the apparent outliers that have been identified in the study. This is therefore a reasonably good result, supporting HRA quantification as a whole. Furthermore, no single assessor dropped below 60% precision in the study. The precision within a factor of 3 is approximately 38% for all techniques. This is a fairly high percentage given the required precision level of a factor of 3. The degree of optimism and pessimism is not too large, as also shown in the histogram in Figure 36.7, with only a small percentage of estimates at the extreme optimistic and pessimistic ends of the histogram (i.e., greater than a factor of a hundred from the actual estimate). Certainly there is room for improvement, but the optimism and pessimism are not in themselves dominating the results, and estimates were more likely to be pessimistic (17.5% of the total number of estimates) than optimistic (9.7% of the total number of estimates).

The highest and lowest precision values for the techniques within a factor of 3 and within a factor of 10 are shown in Table 36.7.

---

[3]Real data came largely from actual recorded incidents in the nuclear chemical and other heavy industry sectors for which the number of opportunities for error could be calculated robustly. See Kirwan et al., 1990 for similar data collection and generation activities.

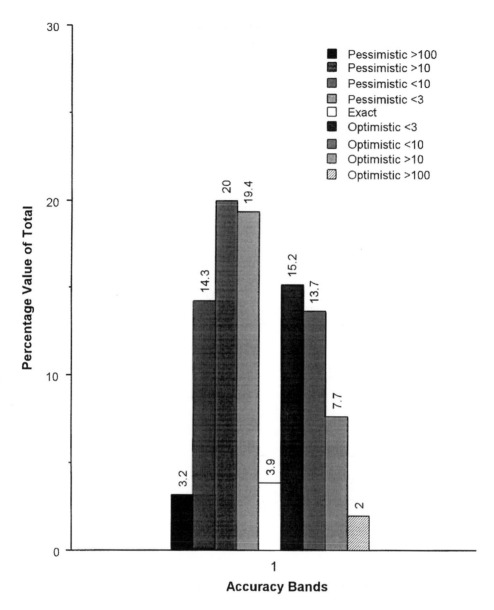

FIGURE 36.7 Overall HRA optimism and pessimism.

The overall results were therefore positive, with a significant overall correlation of all estimates with the known true values, with 23 individual significant correlations, and a general precision range of 60 to 87%, the average precision being 72%. These results lend support for the HRA quantification part of the HRA process.

36.8   Concluding Comments

This chapter has attempted to outline the subject of Human Reliability Assessment, by describing the HRA process, and citing some recent large-scale applications in the U.K. The two recent developments and the validation exercise are also encouraging for HRA practitioners, since they represent improvements in HRA technology, and also demonstrate that such projects themselves are considered worth funding by industry. In the U.K. HRA is also seen as a means of evaluating the human factors adequacy of a

TABLE 36.5   Correlations for Each
Subject for the Three Techniques

	THERP	HEART	JHEDI
1	.615**	.577**	.633**
2	.581**	.558**	.551**
3	.540**	.473**	.533**
4	.521**	.440**	.452**
5	.437**	.389*	.436**
6	.311*	.370*	.423*
7	.298	.351*	.418*
8	.297	.347*	.401*
9	.254	.217	.386*
10	.078	.124	.275

* SIGNIFICANT P < 0.05.
** SIGNIFICANT P < 0.01.

TABLE 36.6   Precision for the Three Techniques for all HRA Assessors

	Factor 3	Factor 10	Optimistic	Pessimistic
THERP	38.33%	72%	13.67%	13.33%
HEART	32.67%	70.33%	11%	18.33%
JHEDI	43.67%	75%	3.33%	21.33%

TABLE 36.7   Precision for the Three Techniques

	Highest Precision		Lowest Precision	
	Factor of 3	Factor of 10	Factor of 3	Factor of 10
THERP	56.67%	80%	10%	63.33%
HEART	46.67%	76.67%	16.67%	60%
JHEDI	66.67%	86.67%	26.67%	70%

system, albeit from a safety or risk perspective, and many HRA assessors, particularly on large HRA projects, achieve significant human factors improvements in systems via the HRA process. Human factors and HRA are therefore not only clearly related, but can operate synergistically, with benefits to both camps. It is hoped that in the future such collaboration between HRA and Human Factors will continue and grow. Human error and human performance, after all, are simply two sides of the same coin.

References

ACSNI (1991) *Human Reliability Assessment — a Critical Overview.* Advisory Committee on the Safety of Nuclear Installations, Health and Safety Commission, London: HMSO.

Ainsworth, L. and Pendlebury, G. (1995) Task-based contributions to the design and assessment of the man-machine interfaces for a pressurised water reactor. *Ergonomics,* 38, 3, pp. 462-474.

Basra, G. and Kirwan, B. (1998) Collection of offshore human error probability data. *Reliability Engineering & System Safety,* 61, 77-93.

Comer, M. K., Seaver, D. A., Stillwell, W. G. and Gaddy, C. D. (1984) *Generating Human Reliability Estimates Using Expert Judgment.* Vols. 1 and 2. NUREG/CR-3688 (SAND 84-7115). Sandia National Laboratory, Albuquerque, New Mexico, 87185 for Office of Nuclear Regulatory Research, U.S. Nuclear Regulatory Commission, Washington, D.C. 20555.

Cox, S.J. and Tait, N.R.S. (1991) *Reliability, Safety and Risk Management.* Oxford: Butterworth-Heine-mann.

Dhillon, B.S. (1986) *Human Reliability with Human Factors.* Oxford: Pergamon.

Dougherty, E.M. and Fragola, J.R. (1988) *Human Reliability Analysis: a Systems Engineering Approach with Nuclear Power Plant Applications,* New York: John Wiley & Sons.

Embrey, D.E., Humphreys, P.C., Rosa, E.A., Kirwan, B. and Rea, K. (1984). *SLIM-MAUD: An Approach to Assessing Human Error Probabilities Using Structured Expert Judgment.* NUREG/CR-3518, Volumes 1 and 2, U.S. Nuclear Regulatory Commission, Washington, D.C. 20555, 180 pages.

Embrey, D.E., Kontogiannis, T. and Green, M. (1994) *Preventing Human Error in Process Safety.* Centre for Chemical Process Safety (CCPS), American Institute of Chemical Engineers. New York: CCPS.

Fewins, A., Mitchell, K., and Williams, J.C. (1992) Balancing automation and human actin through task analysis, in Kirwan, B., and Ainsworth, L.K. (Eds.) *A Guide to Task Analysis.* London: Taylor & Francis, pp. 241-251.

Gertman, D.I. and Blackman, H. (1994) *Human Reliability and Safety Analysis Data Handbook.* Chichester: John Wiley & Sons.

Green, A.E. (1983) *Safety Systems Reliability.* Chichester: John Wiley & Sons.

Hannaman, G.W., Spurgin, A.J., and Lukic, Y.D. (1984) *Human Cognitive Reliability Model for PRA Analysis.* Report NUS-4531, Electric Power Research Institute, Palo Alto, California.

Henley, E.J. and Kumamoto, H. (1981) *Reliability Engineering and Risk Assessment.* New Jersey: Prentice-Hall.

Hollnagel, E. (1993) *Human Reliability Analysis: Context and Control.* London: Academic Press.

Hunns, D.M. (1982) The method of paired comparisons, in Green, A.E. (Ed.) *High Risk Safety Technology.* Chichester: John Wiley & Sons.

Kirwan, B. (1988) Integrating human factors and reliability into the plant design and assessment process, in *Contemporary Ergonomics,* Megaw, E.D. (Ed.). London: Taylor & Francis, pp. 154-162.

Kirwan, B. (1988a) A comparative evaluation of five human reliability assessment techniques, in *Human Factors and Decision Making.* Sayers, B.A. (Ed.). London: Elsevier, pp. 87-109.

Kirwan, B., Embrey D.E., and Rea, K. (1988b) *The Human Reliability Assessors Guide,* Report RTS 88/95Q, NCSR, UKAEA, Culcheth, Cheshire, 271 pages.

Kirwan, B. (1989) A human factors and reliability programme for the design of a large nuclear chemical plant. *Human Factors Annual Conference,* Denver, Colorado, October, pp. 1009-1013.

Kirwan, B. and James, N.J. (1989) Development of a human reliability assessment system for the management of human error in complex systems. *Reliability '89,* Brighton, June 14–16, pp. 5A/2/1-5A/2/11.

Kirwan, B. (1990) A resources flexible approach to human reliability assessment for PRA, *Safety and Reliability Symposium,* Altrincham, September. London: Elsevier Applied Sciences, pp. 114-135.

Kirwan, B., Martin, B.R., Rycraft, H., and Smith, A. (1990) Human error data collection and data generation, in *International Journal of Quality and Reliability Management,* 7.4, pp. 34-66.

Kirwan, B. (1992a) Human error identification in human reliability assessment. Part 1: overview of approaches. *Applied Ergonomics,* 23, 5, 299-318.

Kirwan, B. (1992b) Human error identification in HRA. Part 2: Detailed comparison of techniques. *Applied Ergonomics,* 23, 6, 371-381.

Kirwan, B. (1992c) A task analysis programme for THORP, in Kirwan B and Ainsworth L K (eds.), *A Guide to Task Analysis,* pp. 363-388. London: Taylor & Francis.

Kirwan, B. and Ainsworth, L.K. (Eds.) (1992) *A Guide to Task Analysis.* London: Taylor & Francis.

Kirwan, B. (1994) *A Guide to Practical Human Reliability Assessment.* London: Taylor & Francis.

Kirwan, B. (1995) Current trends in human error analysis technique development, in *Contemporary Ergonomics,* Robertson, S.A. (Ed.), London: Taylor & Francis, pp. 111-117.

Kirwan, B., Scannali, S., and Robinson, L. (1995) Practical HRA in PSA — a case study. *European Safety and Reliability Conference,* ESREL '95, Bournemouth, June 26–28. Institute of Quality Assurance.

Kirwan, B. (1996) Human Reliability Assessment in the U.K. Nuclear Power and Reprocessing Industries, in Stanton, N (Ed.) *Human Factors in Nuclear Safety.* London: Taylor & Francis.

Kirwan, B. (1998a) Human error identification techniques for risk assessment of high risk systems — Part 1: Review and evaluation of techniques. *Applied Ergonomics,* 29, 3, 157-177.

Kirwan, B. (1998b) Human error identification techniques for risk assessment of high risk systems — Part 2: Towards a framework approach. *Applied Ergonomics,* 29, 5, 299-318.

Meister, D. (1995) Cognitive behaviour of nuclear reactor operators. *International Journal of Industrial Ergonomics,* 16, 109-122.

Park, K.S. (1987) *Human Reliability: Analysis, Prediction, and Prevention of Human Errors.* Oxford: Elsevier.

Phillips, L.D., Humphreys, P., and Embrey, D.E. (1983) *A Socio-Technical Approach to Assessing Human Reliability.* London School of Economics, Decision Analysis Unit, Technical Report 83-4.

Rasmussen, J., Pedersen, O.M., Carnino, A., Griffon, M., Mancini, C., and Gagnolet, P. (1981) *Classification System for Reporting Events Involving Human Malfunctions,* RISO-M-2240, DK-4000, Riso National Laboratories, Roskilde, Denmark.

Reason, J.T. (1990) *Human Error.* Cambridge: Cambridge University Press.

Seaver, D.A. and Stillwell, W.G. (1983) *Procedures for Using Expert Judgment to Estimate Human Error Probabilities in Nuclear Power Plant Operations.* NUREG/CR-2743, Washington, D.C. 20555.

Shepherd, A. (1989) Analysis and training of information technology tasks, in Diaper, D. (Ed.) *Task Analysis for Human–Computer Interaction.* Chichester: Ellis Horwood, pp. 15-54.

Sinclair, M. (1995) Subjective assessment, in Wilson, J.R., and Corlett, N.E. (Eds.) *Evaluation of Human Work.* London: Taylor & Francis, pp. 69-100.

Swain, A.D. and Guttmann, H.E. (1983) *Human Reliability Analysis with Emphasis on Nuclear Power Plant Applications.* NUREG/CR-1278, USNRC, Washington, D.C. 20555.

Swain, A.D. (1987) *Accident Sequence Evaluation Program Human Reliability Analysis Procedure.* NUREG/CR-4722. Washington, D.C.-20555: USNRC.

Swain, A.D. (1989) *Comparative Evaluation of Methods for Human Reliability Analysis.* Gessellschaft fur reaktorsicherheit, GRS-71. Schwertnergasse 1, 5000 Koln.

Taylor-Adams, S.E. (1995) The use of the Computerised Operator Reliability and Error Database (CORE-DATA) in the Nuclear Power and Electrical Industries, in the *IBC Conference on Human Factors in the Electrical Supply Industries,* Copthorne Tara Hotel, London, 17/18 October. 16 pages.

Taylor-Adams, S.T., and Kirwan, B. (1995) Human reliability data requirements. *International Journal of Quality and Reliability Management,* 12, 1, 24-46.

Topmiller, D.A., Eckel, J.S., and Kozinsky, E.J. (1984) *Human reliability databank for nuclear power plant operations.* USNRC Report NUREG/CR-2744, Washington, D.C.-20555

Tversky, A. and Kahneman, D. (1974) Judgment under uncertainty: heuristics and biases. *Science,* 185, 1124-1131.

Umbers, I. and Reiersen, C.S. (1995) Task analysis in support of the design and development of a nuclear power plant safety system. *Ergonomics,* 38, 3, pp. 443-454.

Whitfield, D. (1991) An overview of human factors principles for the development and support of nuclear power station personnel and their tasks, in the Conference *Quality management in the nuclear industry: the Human Factor.* London: Institute of Mechanical Engineers.

Whitfield, D. (1995) Ergonomics in the design and operation of Sizewell B nuclear power station. *Ergonomics,* 38, 3, pp. 455-461.

Whitworth, D. (1987) Application of operator error analysis in the design of Sizewell "B", in *Reliability '87,* NEC, Birmingham, April. 14–16. London: Institute of Quality Assurance, pp. 5A/1/1-5A/1/14

Wickens, C. (1992) *Engineering Psychology and Human Performance.* New York: Harper-Collins.

Williams, J.C., 1986, HEART — A proposed method for assessing and reducing human error, *Proceedings of the 9th "Advances in Reliability Technology"* Symposium, University of Bradford.

Williams, J.C. (1988) A data-based method for assessing and reducing human error to improve operational performance, in *IEEE Conference on Human Factors in Power Plants,* pp. 436-450. Monterey, California, June 5–9.

Williams, J.C., 1992, Toward an improved evaluation analysis tool for users of HEART, *Proceedings of the International Conference on Hazard Identification, Risk Analysis, Human Factors & Human Reliability in Process Safety,* Orlando.

Woods, D.D. et al. (1994) *Behind Human Error: Cognitive Systems, Computers and Hindsight.* CSERIAC state of the art report. Ohio: Wright Patterson Air Force Base. October.

ACRONYMS

ACSNI	(UK) Advisory Committee for the Safety of Nuclear Installations
APJ	Absolute Probability Judgment
BDBA	Beyond Design Basis Accident
BNFL	British Nuclear Fuels plc
CIT	Critical Incident Technique
CORE-DATA	Computerised Operator Reliability and Error Database
DADs	Decision Action Diagrams
DHO	Diesel House Operator
EEM	External Error Mode
ERM	Error Reduction Mechanism
GNSR	Generic Nuclear Safety Research
HEART	Human Error Assessment and Reduction Technique
HEP	Human Error Probability
HERA	Human Error and Recovery Assessment
HF	Human Factors
HRA	Human Reliability Assessment
HRMS	Human Reliability Management System
HRQ	Human Reliability Quantification
HTA	Hierarchical Task Analysis
JHEDI	Justification of Human Error Data Information
LA	Link Analysis
MAGNOX	Magnesium Oxide (cladded fuel reactor)
NUCLARR	Nuclear Computerised Library for Assessing Reactor Reliability
OER	Operational Experience Review
PC	Paired Comparisons
PEM	Psychological Error Mechanism
PSFs	Performance Shaping Factors
SLIM-MAUD	Success Likelihood Index Method using Multi-Attribute UtilityDecomposition
THERP	Technique for Human Error Rate Prediction
TLA	Timeline Analysis
TSA	Tabular Scenario Analysis
TTA	Tabular Task Analysis
WT/TT	Walk-through/talk-through
UK	United Kingdom

37

DIALOG:
A Computer-Based
System for
Development of
Experimental Data on
Human Reliability

Heiner Bubb
Technische Universität München

Iwona Jastrzebska-Fraczek
Technische Universität München

37.1 System Ergonomics and Human Error

Most of the conventionally applied human reliability assessment (HRA) methods are based on an event-orientated categorizing system, as it is given by, for example, every accident statistic. A most important system of this category is THERP (Technique for Human Error Rate Prediction) prepared by Swain and Guttmann (1983). Systems of this kind provide reliable prognosis for so-called skill-based actions, which are often enough repeated. In the case of cognitive tasks they cannot support sufficient data. The HCR (Human Cognitive Reliability, Hannaman and Spurgin, 1984) methods classify human error as belonging only to the ratio of time available to time required. This assumption is often challenged. As system ergonomics deals with the information flow in a man–machine system (Figure 37.1) as well as with the information flow between different systems, this would serve as a base to assess human reliability independent of the level of information processing.

By the information flow of a singular man–machine system, the quality of human work can be calculated as the relation of the result of work to the task. Performance in this context is: Quality per time. Limits of quality are explicitly or implicitly defined in every system. If any human action results in a jumping over such predefined limits, we call that *human error* (HE, Rigby, 1970). The probability of nonoccurrence of such HEs defines *human reliability* (HR). This consideration is important not only for a simple singular man–machine system but also for any complicated combinations of different singular man–machine systems, which describes in its totality any arbitrary system, as for example a power plant, an assembly plant, or a traffic system. Apart from technical defects, human errors influence the reliability

Man-machine-system (MMS): Combination and entirety of interaction between man and working materials

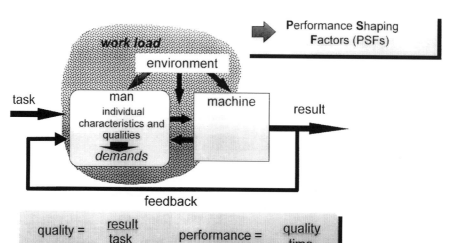

FIGURE 37.1 Structural block diagram of human work.

and dependability of such systems. Primary technical, often unimportant, failures occur and initiate inadequate human reactions. Because such effects cannot be treated only by measurement in the technical area, more profound treatments of human operation in the technological context and its possibilities of deviation are necessary. System ergonomics can be used as a base.

37.2   The Idea of System Ergonomics

System ergonomics starts with a general description of properties of every task and assigns experimental experience, and from that, ergonomic recommendations to the partial aspects of the task can be made. The fundamental idea is that by the knowledge of the information transfer by the subsystems man and machine, the tasks to be performed by the operator may be designed. For this designing of the task that results from the system mission and from the specifically chosen lay-out of the system and the system components (e.g., the machine), the following fundamental rules are to be considered, formulated as questions (see also Figure 37.2):

1. Function: "What has the operator in view and how far is he assisted by the technical system?"
2. Feedback: "Is the operator allowed to recognize if he has effected something and what was the success of it?"
3. Compatibility: "How much effort has the operator to make in order to convert the code system between the different technical information channels?"

The *function* may be separated into the intrinsic task contents and the influencable task design. Under this aspect the *task contents* are essentially defined by the *temporal and spatial order* of the activities which are to be carried out to perform the total task. It may be described by the terms operation (= *temporal* organization), dimensionality (= *spatial order* of the task), and manner of control (= the *time and location window* within which the task must be finished (see Bubb, 1988). The *task design* refers to the system structure which may be chosen to a large extent by the system planner. In this area the degree of difficulty may be influenced by the specifically designed layout. It can be distinguished between the manner of presenting task and result to the operator — the so-called *display* (*compensatory* or *pursuit*

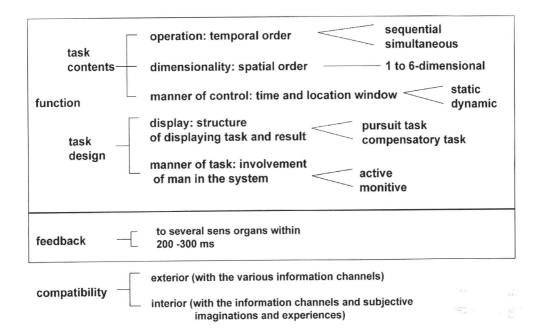

FIGURE 37.2 System ergonomic elements.

task), and the manner of involving the operator in the total system — the so-called *manner of task* (*active* or *monitive task*).

Feedback calls for a certain kind of redundancy which utilizes a *number of sensory organs*. A further aspect is the *time* that elapses between the input of information on the control element and the reaction of the system on the output side. Considering the relevant recommendations a well-designed feedback allows the operator to answer the questions:

- What have I done?
- In what a state is the system?

Compatibility describes the facility that enables the operator to convert the code system between the different information channels that he has to handle. It characterizes the obvious between different areas of information, as indication by instruments, control elements, and internal models of the operator. Neglecting the rules of compatibility decreases human reliability immensely (Spanner, 1993).

37.3   The Software Tool DIALOG

With DIALOG, a great part of system ergonomic aspects can be investigated. For instance, tasks of differing difficulties are realized by entering figures which appear as a combination of figures. Additionally, continuous tasks in the form of simple tracking experiments are possible, simultaneous tasks can be simulated, and time delay, concerning feedback, can also be presented. With another feature of the program, the subject can be put under time pressure.

The Task "Entering Figures"

In the "entering figures" mode the task for the subject is to rewrite a combination of several numbers between 1 and 20. This is an example for *sequential operation* of different degree of difficulty. Simultaneously further tasks can be displayed in form of traffic lights (with the color sequence green, yellow,

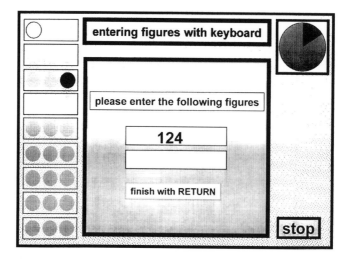

FIGURE 37.3 Example for displayed tasks by DIALOG. In this case: Sequential task of 3 numbers, 2 simultaneous tasks (rewriting of numbers and switching off of the traffic lights), restricted time window (indicated by the green/red disk right above).

red), which are to be switched off by a mouse click before the appearance of the color red. Up to 9 traffic lights can be presented. That is an example for *simultaneous tasks* of different degrees of difficulty. By an additional indicated disk different time windows can be investigated. This disk is originally green. During the task a red segment appears increasing with time. That is a representation of a *static task* (manner of control). Figure 37.3 shows an example of the video screen as it is presented to the subject during the experimental run.

The designer of the experiment can adjust the subject's task by a special input mask (see Figure 37.4). As it can be seen from this figure, the software tool DIALOG allows us to choose the number of test runs (here 30) and to adjust the time limit in steps of milliseconds. By a mouse click the time limit may be activated (indicated by a cross). The box "feedback" allows us to choose a feedback delay and to activate or deactivate optical feedback (= indication of the numbers) and/or acoustic feedback (= a clicking sound). In the box "determination of figures" the minimum and maximum number of figures may be determined. If the same number is not taken for both, by arbitrary numbers with different figures appear in the task window in the chosen range. The box "simultaneous operation" allows us to determine the number of control actuators (= traffic lights) and independent of it the number of actual activated operation during a test run. In the field "speed" the time between the appearance of two traffic lights is determined. Of course, this time must be chosen in such a manner, that the amount of "number of operations" can run within the general chosen "time limit."

The program modus "entering of figures" offers an example of a one-dimensional, static task presented as a pursuit task. The manner of task is active. In this combination the following elements of system ergonomics can be treated by experiment: Sequential and simultaneous operations of different degrees of difficulty and different conditions of feedback.

The Task "Following a Line"

In a second modus, DIALOG allows us to investigate continuous tasks. This is a kind of tracking task. In this case the subject has to redraw a line with the mouse, which is presented under a defined angle. This task can be demanded with and without time pressure and with and without optical feedback. By changing the allowed tolerance field, the difficulty of task can be varied. The described task is presented to the subject by a screen picture as shown in Figure 37.5.

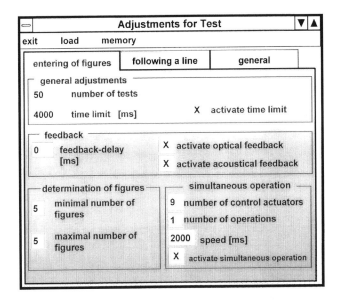

FIGURE 37.4   Possibilities of adjustment for the test "entering of figures."

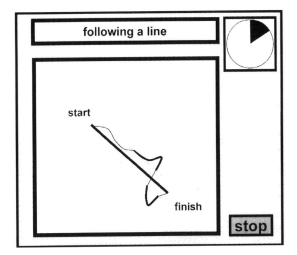

FIGURE 37.5   Example for displayed tasks by DIALOG. In this case: Sequential (Following the line), optical feedback, restricted time window (indicated by the green/red disk right above).

In order to adjust the different experimental conditions the user surface "following a line" of Figure 37.6 is applied. Also in this case by "general adjustments" the number of tests (here: 30) is chosen and the time limit, which determines the time pressure, is adjusted and in a given case selected ("activate time limit"). The box "description of range" allows us to determine the width of tolerance (radius). This can be done even asymmetrically (tolerance: positive, negative, or symmetric). By the box "feedback" the optical replay of the drawn line can be switched on and off. The box "general adjustments" allows us to define the starting position of the displayed line, the angle under which it is inclined, and its length.

The program modus "following a line" represents a two-dimensional static task with a location window given by the width of tolerance. Under the view point of operation it is a singular task. The display again creates a pursuit task.

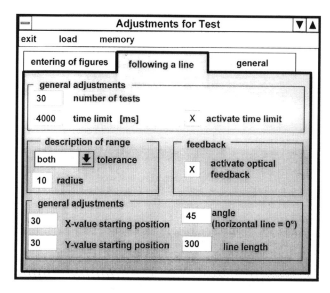

FIGURE 37.6 Possibilities of adjustment for the test "following a line."

37.4   Results

Using the described equipment, experiments were carried out. Thirty-two subjects had to accomplish different tasks presented by DIALOG during a session of about 1.5 hours. All reactions of the subjects were recorded by DIALOG. In the following only a part of the possible analysis is reported as an example. All results are statistically safe on a significance level of 0.95.

Entering Figures

In a first investigation the areas "operation" (sequential task of the difficulty "5 numbers" and "7 numbers" in combination with and without simultaneous switching off of traffic lights) and manner of control (with and without a temporal window of 4s or 8s, within which the tasks had to be accomplished) have been the objective. In this case the decision "task correctly accomplished" or "not" can be made, and the usual definition of human error probability (HEP) can be used:

$$\text{HEP} = \text{number of observed errors/number of opportunities for error}$$

We found out that time pressure results in more increased error probability than additional simultaneous tasks. In the case of recording 5 numbers, the average error probability was about 0.03. This value increased with the factor 14 to 0.43 under the condition of time pressure (Figure 37.7).

The observation shows that a more difficult (sequential) task under the condition of time pressure has an under-proportional ascent of error probability in relation to the behavior of easier tasks (Figure 37.7).

The simultaneous task of switching off traffic lights had no additional effect to the error probability, with a time window of 8s. Without time pressure, the increase of the difficulty of a sequential task of 5 to 7 numbers had an effect of the factor 2.5. Under the condition of time pressure of 4s, the same increase had only an effect of the factor 1.3. A more difficult task seems to stimulate the attention (see Figure 37.8).

This suggestion is substituted by a more detailed investigation of errors, so we could separate the following kinds of error (see also Swain and Guttmann, 1983):

- Error of quality, z_1 (writing a wrong number, e.g., 123 instead of 124)
- Error of part omission, z_2 (omitting one number, e.g., 12 instead of 124)
- Error of omission, or negligence, z_3 (the task is totally neglected)
- Error of addition, z_4 (writing an additional number not asked for, e.g., 1234 instead of 124)

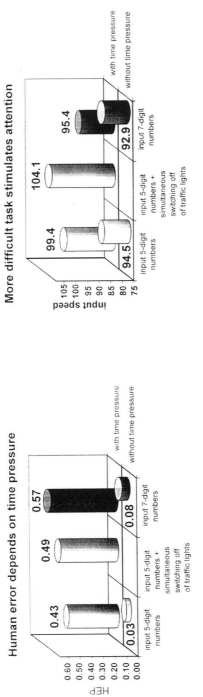

FIGURE 37.7 Human error depends on time pressure.

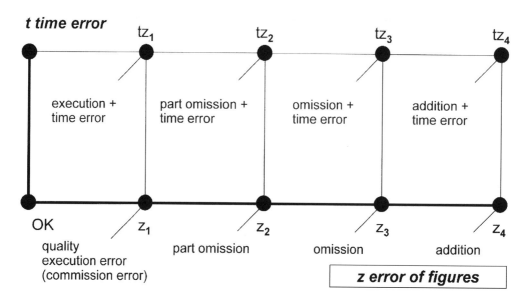

FIGURE 37.8 Errors made at "entering of numbers" with time pressure.

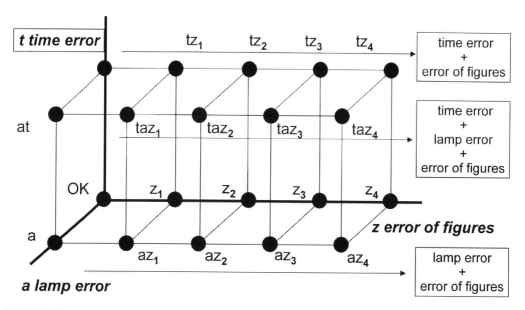

FIGURE 37.9 Errors made at "entering of figures, simultaneous switching off of the traffic lights" under time pressure.

In the case of time pressure, all the types of error can occur in combination with a time error, i.e., the task is not finished within the time window, t (Figure 37.9).

Whereas the error of negligence does not appear without time pressure, this type crops up with time pressure and gets a value of 0.076 (see Figure 37.10). Under the condition of time pressure the time error appears with a probability independent of the difficulty of the task of 0.07 to 0.09. Of course, all types of error are increasing with time pressure. There is one exception: the error of quality disappears under this condition. The probability of all kinds of error becomes smaller when these appear in combination with time error. There is again one exception: the error of omission increases in connection with time error with the factor 16 for 5 numbers and 12 for 7 numbers.

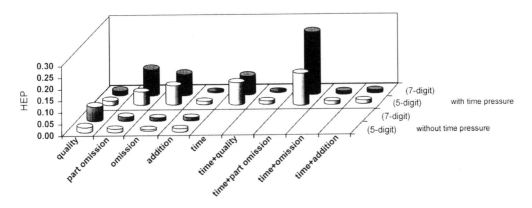

Errors at "manual input of numbers", sequential operation without and with time pressure

FIGURE 37.10   In the case of time pressure, all the types of errors can occur in combination with the time error.

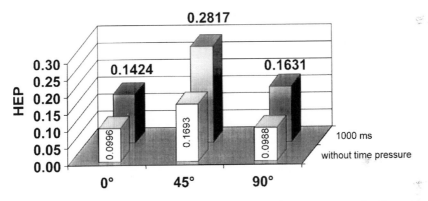

FIGURE 37.11   HEPs during "following a line" under the condition of time pressure and without time pressure and the orientation of the line of 00, 45° and 900.

Following a Line

As during continuous tasks the number of tasks and thereby the number of errors cannot be defined, the definition was carried out considering the time:

$$HEP = \text{total of time quota status "error"/total time}$$

In the case of these results, the tolerance width was adjusted to 10 pixels. Figure 37.11 shows the results of the HEP values without time pressure and time pressure of 1s. Different orientations of the line were used. We observed that under the condition of 0° and 90° the same error probability occurs, whereas in connection with an inclination of 45° the error probability is nearly doubled. Under this condition the time pressure plays a more important role. However, as Figure 37.12 shows, the jump of error probability from without time pressure to any time pressure is much higher than the increase from time pressure of 3s over 2.25s to 1.5 s. Figure 37.12 shows also that the inclination of the line is altogether unimportant.

The results reported here only show examples of the possibilities of experiments that can be done by DIALOG. In future, by using this program, a more systematic investigation in the area of system ergonomics will be possible in regard to error probability. There is a certain hope, that experiences observed by DIALOG will complement and add to observations of errors during real situations. In future, there might be a system in which a prediction of human error probability will be possible in cognitive tasks as well.

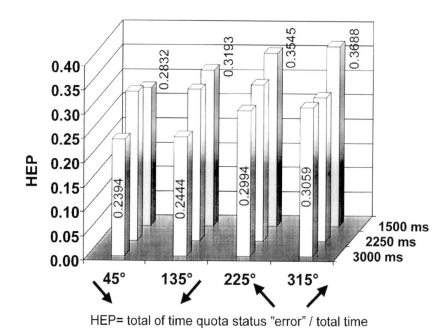

HEP= total of time quota status "error" / total time

FIGURE 37.12   HEPs during "following a line" under the condition of different time pressure and different inclination of the line.

References

Bubb, H. (1988): System ergonomics as an approach to improve human reliability. *Nuclear Engineering and Design* 110, S. 233-245.

Hannaman, G.W. and Spurgin, A. J. (1984): *Systematic Human Action Reliability Procedure (SHARP).* ERPI NP-3583, Electric Power Research Institute, Palo Alto, CA, June 1984.

Rigby, L. V. (1970): The nature of human error, in: *Amer. Soc. for Quality Control, Annual Technical Conference,* 24th, 457-466.

Spanner, B. (1993): Einfluß der Kompatibilität von Stellteilen auf die menschliche Zuverlässigkeit. Fortschritts-Berichte VDI, Reihe 17 "Biotechnik," VDI-Verlag Düsseldorf.

Sträter, O. (1997): *Beurteilung der menschlichen Zuverlässigkeit auf der Basis von Betriebserfahrungen.* Dissertation an der Technischen Universität München.

Swain, A.D. and Guttmann, H.E. (1983): *Handbook of Human Reliability Analysis with Emphasis on Nuclear Power Plant Applications.* NUREG/CR-1278, Scandia Laboratories, Albuquerque, NM, 97185.

38

Managing the Speed–Accuracy Trade-Off

Colin G. Drury
*State University of New York
at Buffalo*

38.1   Introduction: Speed, Accuracy and Performance

This chapter examines two aspects of task performance, speed and accuracy, and shows how they interact with each other. The best-known interaction is a speed/accuracy trade-off (SATO) — as speed increases accuracy decreases. But this is not the only interaction between these two performance measures, so managing their interaction needs more detailed knowledge. Speed and accuracy are first defined, and their modes of interaction shown as three levels. Finally, a practical way of managing this interaction is presented with a worked example.

Under conditions of increasingly global competition, industry must remain focused on performance in order to survive and prosper. We use many measures of "performance" based on our different perspectives and on the level of aggregation of our data. Thus at a low level, a supervisor might define performance as labor hours to produce the week's output. At a somewhat higher level, a production manager may define it as percentage of orders shipped on time. At the board level, an appropriate performance measure may be how the stock price has increased in the past quarter.

While all are useful performance measures, at the plant level an augmented set may be more appropriate, for example:

Quality: e.g., fraction of acceptable product shipped
Timeliness: e.g., fraction of product shipped on time
Efficiency: e.g., resources used to achieve the quality and timeliness levels.
Plant Health: e.g., how fit the plant remains to achieve its performance in subsequent time periods

In fact, these are all examples of the three basic measures of performance — effectiveness, efficiency, and well-being — which can be applied to an individual employee, a production unit, or the whole plant:

1. *Effectiveness:* How well does the outcome match expectations? Was the refrigerator well designed? Was the error rate in hospital prescriptions low? Effectiveness is measured by accuracy or quality, or by their complement: errors.

2. *Efficiency:* How much resource is consumed to achieve the effectiveness? Did the design of the refrigerator take too many designer-months? What staffing level was needed to achieve the low prescription error rate? Efficiency is measured traditionally by throughput or speed, but in fact includes other resource use, such as capital and lead-time.

3. *Well-being:* How did the achievement of *that* level of effectiveness with *that* efficiency affect the worker and the organization? What was the operator's stress level? How many injuries were sustained? Did job satisfaction measures increase? Well-being can be measured by the cost to the operator and the organization of achieving the desired performance.

Note that the first two, effectiveness and efficiency, are those typically measured and used by business people and together define "performance." Effectiveness, particularly customer satisfaction, has been promoted as the key measure. Indeed, customer satisfaction is seen by many as the price of entry into the competitive global marketplace. Efficiency is closely scrutinized by both managers and market analysts to determine whether the costs are outstripping the sales base. Corporate well-being is a more recent consideration, but is becoming more readily measured by satisfaction surveys. Most companies' annual reports talk about "Our people are our greatest asset" and "We put safety before all other considerations." Some companies, such as chemical and continuous process operations, have espoused the philosophy that if a process is understood well enough to make it safe, then it will also be effective and efficient. Thus the modern view is that effectiveness, efficiency, and well-being are all required to define "performance."

These three aspects can, of course, co-vary. Under some conditions, high efficiency can mean low effectiveness. This is the classic speed/accuracy trade-off where lack of time reduces quality. Under others, a plant-wide change can give both high effectiveness and high efficiency. Finally, some means of achieving high effectiveness and efficiency (for example, piece-rate payment systems) can adversely affect the well-being of plant and workers. Figure 38.1 shows some of the possible interactions between efficiency, effectiveness, and well-being applied to the individual human performing a task. It is these covariations, specifically between effectiveness and efficiency, that constitute the speed–accuracy trade-off (SATO). This chapter explains how different SATO functions can arise, and how to manage processes that may have SATO characteristics.

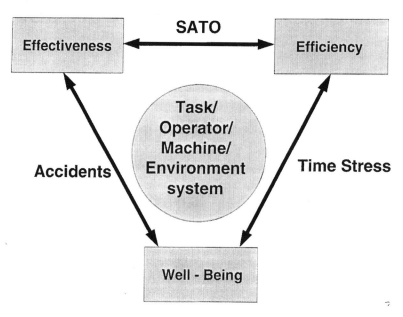

FIGURE 38.1 Interactions between effectiveness, efficiency, and well-being for the individual operator.

To understand SATO, we need to make the connections of speed with efficiency and accuracy with effectiveness. Speed is the rate at which a task is performed. At times it is measured as a rate, for example throughput of a manufacturing cell in parts per day or speed of a conveyor or forklift truck in meters per second (ms⁻¹). Often, however, speed is only implied and its reciprocal, time per unit, is measured. Thus a sewing task has standard allowed hours for each garment, or an aircraft engine maintenance shop has a throughput time measured in person days per engine. In research studies, speed or time is a basic parameter measured or controlled. A very early study of human motions (Woodworth, 1899) had people make rapid movements at different rates from 20 to over 200 per minute, while accuracy was measured to give one of the earliest examples of SATO. Since that date, studies of reaction times, decision times, movement times, search times, and so on have provided the basis for many of our quantitative models of human performance. Time and rate measurements have an equally long history in industry, stemming from scientific management, and evolving into time study and methods study (e.g., Konz, 1990; Mundel and Danner, 1994).

Accuracy is measured as "freedom from error in discrete tasks" (Drury, 1994), with analogous measures for continuous (e.g., tracking) tasks. Thus to define accuracy, we must also define error. Rasmussen (1986) considered errors as unsuccessful experiments in an unkind environment (p. 150). In a major compilation of current positions on error, Senders and Moray (1991) suggest that an error is a human action that fails to meet an implicit or explicit standard. More formally they characterize errors as (p. 25):

Actions not intended by the actor
Actions not desired by a set of rules or an external observer
Actions that led the task or system outside acceptable limits

Responding to costly and visible errors, such as the Bhopal disaster, the study of human errors is undergoing something of a renaissance in ergonomics (e.g., Reason, 1990; Norman, 1981), with numerous classification schemes being proposed. However, for the purpose of this chapter a simple classification into slips and mistakes (Norman, 1981) is appropriate. While both can be affected by speed, it is the skill-based slips that most obviously show this effect. Rule-based and knowledge-based mistakes represent incorrect intentions, which could be the result of insufficient time for information processing, but correct intentions incorrectly executed (slips) are the most typical speed-affected errors.

At the beginning of this introduction, we emphasized that modern manufacturing requires both speed *and* accuracy to survive. Now that speed and accuracy have been defined, we can proceed to see how they interact with each other, i.e., SATO. We are not forgetting well-being, merely holding it constant. We will only consider SATO which leaves well-being unchanged, meaning that we do not allow the plant to deteriorate, or its workforce to develop chronic injuries in our pursuit of speed and accuracy.

38.2 The Speed–Accuracy Trade-Off

Not all tasks exhibit a speed–accuracy trade-off. In some tasks, the accuracy attainable is an inherent function of the quality of the data used to perform the task, rather than a function of how long the operator can consider that data. For example, in a difficult quality-control discrimination task, such as judging the noisiness of electric motor bearings, more time to decide does not produce increased accuracy. Such tasks are called *data-limited* (Norman and Bobrow, 1979). "Data limited" means that the inherent quality of the data limits performance, no matter how much time is devoted to the task. Tasks which *do* exhibit a SATO are ones where evidence in favor of a correct response is accumulated over time, for example, searching a circuit board for defects. These tasks are called *resource-limited* in that the limitation on accuracy is how much resource (in this case, time) is expended on the task.

Examining the shape of the speed–accuracy trade-off function of many tasks suggests that most are resource-limited at very short times and data-limited at very long times. Figure 38.2, for example, shows a task that is resource limited at short durations (500 ms or so), but data limited for durations longer than about 500 ms. In the remainder of this chapter, we shall consider tasks that do exhibit SATO, i.e., those that are resource-limited over a practical range of times.

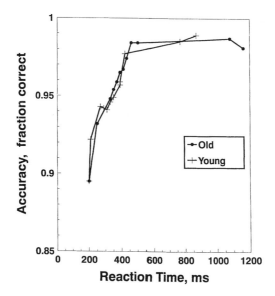

FIGURE 38.2   Level 1 speed–accuracy trade-off in reaction times for two age groups. (Adapted from Smith, G.A. and Brewer, N. 1995. Slowness and age: speed–accuracy mechanisms. *British Journal of Psychology*, 76:199-203.)

Woodworth's early movement study showed that as speed increased, accuracy worsened. This classic inverse relationship between speed and accuracy is what most consider "SATO." Perhaps the first comprehensive SATO studies were Garrett's (1922) inquiries into SATO for human perception and coordination. Since then, major advances were made by Pachella (1974) and Pew (1969). The latter coined the term "Speed–Accuracy Operating Characteristic," or SAOC, to describe the plot of speed *versus* accuracy for a particular task or experiment. It is one type of isoperformance curve (Jones and Kennedy, 1996) showing a limit on possible combinations of performance. We will examine at a "task" level where SATO arises and the different forms it can take.

A common-sense view of human performance suggests that people will choose levels of speed and accuracy that do them the most good. Such ideas have indeed been influential in the study of SATO, under the general title of economic maximization models. They postulate that human operators consider the rewards for speed and the penalties for error, finally choosing an operating point that maximizes their expected payoff given a fixed level of ability. Examples of this thinking include:

1. A model of choice reaction time (Fitts, 1966) that postulates that the operator gradually builds up evidence for one or the other alternative until the amount of evidence passes some threshold that the operator considers enough to start a response. Because the evidence also contains noise, the operator's path toward the final alternative is a "random walk." Time to respond can be reduced by reducing the threshold amount of evidence needed for a response, but the result will also mean more errors.

2. Signal Detection Theory Models of the decision process as a choice between alternatives based on inherently noisy data. In, for example, a quality control task, the inspector has to decide on whether or not to reject an item of product based upon whether it is judged to meet a standard. Choice of the criterion of what to accept depends on the relative costs and probabilities of the various outcomes of the decision (McNichol, 1972).

3. Optimal stopping models of visual search (e.g., Karwan et al., 1995) where a visual inspector is modeled as having to choose when to stop searching one item of a product for possible defects and move on to the next item. Stopping earlier on each item will save time but lead to the inspector missing some defects. Stopping time is optimum when the costs of increased search time begin to outweigh the value of finding a defect.

TABLE 38.1 Possible Interventions by SATO Level

SATO Level	Interventions	Intervention Label
1. Micro-SATO	Keep strategy at constant point on the SAOC	1. Reduce SATO variability
2. Macro-SATO	Choose the best operating point on the existing SAOC	2. Optimize SATO point
3. Performance intervention	(a) Reduce the fixed time associated with task	3. (a) Reduce overhead
	(b) Improve the slope of the SAOC, i.e., improve efficiency of process	3. (b) Improve process capability

4. Driving along a roadway of fixed width (Montazer et al., 1987). During a fixed visual sampling interval, a driver covers a distance proportional to road speed. There is value to covering the distance as quickly as possible, but large costs associated with deviating into a fixed boundary, e.g., drifting off the road. The optimization model correctly predicts the speed and accuracy as a function of roadway width.

While such models have been useful and influential within laboratory tasks, their use in shaping long-term human behavior at work may be limited. People tend to be poor at estimating the costs and probabilities needed in everyday life, and often act as "satisficers" who select a solution that is good enough rather than as "optimizers" who seek the best solution. In particular, using rewards and punishments to control the speed–accuracy trade-off may well distort the very system it seeks to optimize. Even Deming (1986) has removal of speed incentives as one of his 14-points for improving quality.

In parallel, we are being urged by modern management techniques to stress the achievement of quality as a way to improve speed of performance. This is taking place at a higher level of aggregation than the task or its detailed components. We are here considering the introduction of dedicated cells, just-in-time manufacturing, statistical process control procedures, or even plant-wide total quality management programs.

There is a clear disparity between the known negative SATO at a task level and the known positive SATO at a more highly aggregated level. This disparity is mainly the result of the multiple ways in which SATO can take place. We can classify where SATO occurs at three levels as a micro-SATO (Level 1), where speed and accuracy vary from trial to trial in a repetitive task, a macro-SATO (Level 2), where the task may be performed under different speed and accuracy reward structures or a performance-intervention-SATO (Level 3), where deliberate changes are made to improve both the speed and accuracy of task performance. We shall later examine the possible interventions at each level, as shown in Table 38.1. First, we consider each level in turn.

Level 1: Micro-SATO: In a repeated task, the repetition-to-repetition performance changes can show a speed–accuracy trade-off. In laboratory reaction-time tasks, where the criteria are for high speed with high accuracy, both speed and accuracy vary from trial to trial. Trials with the shortest reaction times tend to exhibit the highest error rates. This was demonstrated by Pew (1969) and has since been a consistent finding in reaction time research. For example, Smith and Brewer (1995) found very similar speed/accuracy relationships for two groups of subjects 18 to 30 years old and 58 to 75 years old (Figure 38.2), despite the older subjects generally responding more slowly. Note that in these carefully controlled laboratory tasks, the time from stimulus to response (reaction time) may be very short compared with typical industrial task completion times.

Studies of human performance within a sequence of repetitions of a task (e.g., Drury and Corlett, 1975; Rabbitt, 1981) have shown that people control their level of effort on a trial-to-trial basis. For control over speed and accuracy, people try to maintain speed performance within a narrow band which gives a sufficiently low error rate. When speeds increase beyond this band, errors occur. An error (at least a detected error) is followed by a slowing of performance for a time until the desired speed band is achieved once again. This effect can be seen in the Smith and Brewer study of SATO and aging. Figure 38.3 shows mean reaction times in the trials before and after an error occurred.

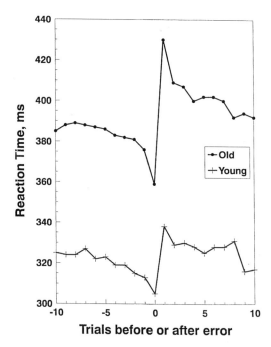

FIGURE 38.3 Speed-up prior to error and slow down following error. (Adapted from Smith, G.A. and Brewer, N. 1995. Slowness and age: speed–accuracy mechanisms. *British Journal of Psychology*, 76:199-203.)

Level 2: Macro-SATO: When the same task is repeated under different reward conditions for speed and accuracy, the mean speed and accuracy over each reward condition show a SATO. In a classic experiment, Hick (1952) rewarded subjects in his choice reaction time experiment either for speed or for accuracy. He found that when encouraged to react faster, more errors occurred. In fact the increase in mean errors balanced the decrease in mean reaction times exactly as predicted from information theory, i.e., the subjects' rate of transmission of information remained constant across all reward conditions.

This macro-SATO level is the most typical lay interpretation of SATO: if you are encouraged to increase speed, you may be expected to decrease accuracy. Perhaps because of this popularity, there have been many studies that have demonstrated a macro-SATO. Wickelgren (1977) reviews earlier studies and examines how the SAOC can be measured. Wickens (1994) reviews SATO in an information theory context, while Meyer, Smith, Abrams and Wright (1990) provide a comprehensive historical view of SATO applied to accurate movements. Measurement of macro-SATO in visual search is described in Drury and Forsman (1996).

Much of the SATO literature comes from the study of accurate movements, e.g., Gan and Hoffmann (1988). For low accuracy movements, known as ballistic movements, the movement time is a function only of the distance traveled. For high accuracy movements, time depends both on the distance traveled and the accuracy demanded. This is known as Fitts' Law, with the formula given in Table 38.2.

In an industrial context, the same effects are observed. Omissions and errors increase with pacing speed in industrial assembly (Dudley, 1958) and in textile machine servicing (Conrad, 1954). Accurate movement of a computer's mouse obeys Fitts' Law (Card et al., 1983). Forklift trucks are driven more slowly as aisle width decreases (Drury and Dawson, 1972). Inspection of minted coins for defects shows decreased accuracy at higher pacing speeds (Fox, 1977).

The macro-level of SATO has involved a number of models of human performance. These predict the levels of both speed and accuracy that are possible, provide some explanation of *why* the SATO effects occur, and give a rationale for measuring and plotting the SAOC. It is not the intention of this chapter to present these models in any detail, nor to continue the arguments for and against alternative models.

TABLE 38.2   Speed–Accuracy Functions for Selected Tasks

Name	Task	SAOC Form	Time Range, s
Target detection (Bloch's Law)	Detect dim targets of intensity I on a dark field with exposure time T	$I\,T = a$	<0.1
Visual search (random strategy)	Search for and locate a target in an extended, often cluttered, background	$p(det) = 1 - \exp(-a\,T)$	<100
Signal detection (SD theory)	Detect signal in the presence of noise. p(det) = prob. of detection; p (FA) = prob. of false alarm. z(p) is normal deviate for prob. p	$z(p(det)) +$ $z(p(FA)) = a\ \sqrt{T}$	<2
Choice reaction time (Hick's Law)	Respond correctly to one of a number N of different events	$\mathrm{Log}\left(\dfrac{p(correct)}{p(error)}\right) = a\,T$	<2
Accurate movements (Fitts' Law)	Move the hand or an object a distance A to a target of width W	$T + a + b\,\mathrm{Log}\left(\dfrac{2A}{W}\right)$	<2
Driving/path following	Control body or vehicle along a path with lateral boundaries giving clearance = W	$\mathrm{Speed} = \dfrac{k}{T} = a + bW$	—

T = time allowed or taken; a, b are constants

Rather, some models of specific tasks are given in Table 38.2 to indicate how much assistance is available in the literature in understanding SATO.

A single example is instructive, however. Figure 38.4 shows a SAOC for the task of automobile driving, from DeFazio et al. (1992). Here, the issue was the effect of an accuracy restriction on chosen speed. Subjects drove an automobile around a circular course of radius 8.13 m as fast as possible without exceeding the course boundaries. A model of this process (Drury, 1971; Montazer et al. 1987) showed that if the driver adjusts the speed to maintain a constant (low) probability of leaving the road, then a linear relationship would be expected between speed and car/boundary clearance. The model tells us what to control and measure in this experiment (control clearance, measure speed), and what shape of SAOC to expect. In this experiment, the aim was to determine whether car width and boundary width have the same effect on clearance, and hence speed. They did not, and the authors were able to calculate effective car width which differed from the actual car width.

Level 3: Performance Intervention SATO: When a change (other than just a change of reward structure) has been made to a task, then both speed and accuracy can be affected. The particular relationship between speed and accuracy seen in Levels 1 and 2 will probably not hold when changes are made to the task. Thus an improved task may well give:

1. Improved accuracy at the same speed
2. Improved speed at the same accuracy
3. Simultaneous improvements in speed and accuracy

The point is that changes to the task move us to an entirely new SAOC, rather than merely moving *along* a single SAOC as was the case for Levels 1 and 2. For example, Figure 38.4 shows SAOCs for two tasks in circuit board inspection: finding a wrong IC chip and finding a bent component lead (Drury and Chi, 1995). Both SAOCs are examples of systematic visual search, but at any chosen search time the probability of detection is always higher for the Wrong IC condition.

Change of conditions can be due to changing the operator (e.g., selection, motivation training), the task (as in the above example), the equipment (e.g., hybrid automation in inspection: Hou et al., 1993) or the environment (e.g., lighting, Drury, 1987). Such changes have costs attached, but also payoffs for increased performance. For example, Drury (1991a) extended the data from Hasslequist (1981) on a before-and-after study of ergonomics changes at an assembly plant. Quality improved — 50% fewer errors — at the same time as productivity was up 6%.

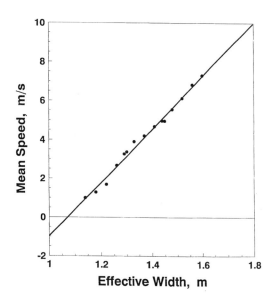

FIGURE 38.4   Speed–accuracy operating characteristic for car driving along narrow road. Chosen speed varies with the effective width of clearance. (Adapted from DeFazio, K., Wittman, D., and Drury, C.G. 1992. Effective vehicle width in self-paced tracking, *Applied Ergonomics*, 23(6): 382-386.)

Much of the data supporting the concept that improving quality will also improve productivity comes from popular publications rather than scientific journals. From Crosby's classic book, *Quality Is Free* (1979) comes the idea that the true cost of quality is no greater than that of producing defects, but productivity is not addressed. In other similar books, executives from successful quality companies, such as Motorola, do make the connection, saying "Quality is quantity; quality is low cost" (Dobyas and Crawford-Mason, 1991, p. 253). In the same source an executive at Siemens shows 30,000 fuel injectors being produced in a month (1985), in a week (1987), and now in a day (1990). Plans were for them to be produced in one shift in 1992. "That has been and is being done through quality improvement" is the quote.

An excellent special issue of *Business Week* on quality in 1991 provided many more examples. Toyo Ltd., a maker of bathroom fixtures, started a quality program in 1984, and credits it with halving inventories and lead time, and also with a 50% increase in productivity at the same time as a 25% decrease in customer complaints (p. 23). Similarly, at NEC's Tohoku printer plant a quality function deployment program reduced defect rates from 26% to 1.2%, while doubling productivity (p. 61). As a final example, a quality vice president at IBM is quoted as saying "The longer you work on something, the more time you have to interject defects." (p. 68).

Some examples have more traditional references. The compilation of interventions in *Work in America* (Anon., 1973) showed many that reported both throughput and quality increases. A more recent example (Drury, 1991a) introduced a just-in-time manufacturing cell for machining end-caps for automotive alternators. In just-in-time (JIT), the aim is to improve performance by reducing inventory levels between processes. When one process fails, it brings the whole production system to a halt, which focuses management attention on finding permanent cures for process failures. This eventually improves quality by reducing many of the failure modes. With a higher-quality product and system, the efficiency can also increase, thus giving simultaneous speed and accuracy benefits. In the manufacturing cell (Drury, 1991a), cycle time decreased from 7 days to 2.5 hours, while quality improved so much that no defects were measured during the test period of several months after the change.

Incidentally, changing to a different person performing the task can change both speed and accuracy, but rarely in any predictable manner. Because people differ in their overall capabilities, they can display better or worse speed and accuracy. Thus a plot of speed against accuracy where each point is the average performance of an individual will, in general, produce just a cloud of points with no correlation. We cannot expect that faster individuals will be any more, or any less, accurate than slower individuals. This concept has been used to classify people on the two dimensions of speed and accuracy using a standard test, the Matching Familiar Figures Test (MFFT). The four groups are:

Slow, inaccurate
Slow, accurate ("Reflective")
Fast, inaccurate ("Impulsive")
Fast, accurate

We have used this categorization in two inspection tasks (Schwabish and Drury, 1984; Chi and Drury, 1996) without much success. In fact, performance on an inspection task was best predicted in both cases by using only the accuracy dimension of MFFT.

38.3 Practical Intervention

Corresponding to each level of SATO, there are appropriate interventions to manage those jobs where SATO has an influence. Drury (1994) developed four possible interventions, provided a methodology for managing SATO, and gave an example of its use. In the current chapter, his interventions, methodology, and example will be adapted to the three levels of SATO defined in the previous section.

The micro-SATO of Level 1 showed how trial-to-trial variability in SATO can arise. For example, in Figure 38.2, if the mean reaction time had been chosen for each reaction, the error would probably not have occurred. A suitable intervention for Level 1 is thus to ensure that a consistent strategy is used, i.e., the operator does not drift along the SAOC toward either very slow performance or frequent errors.

Level 2 introduced macro-SATO, showing how the operating point chosen on the SAOC moves as the rewards for speed and accuracy are manipulated. Here, the intervention strategy should be to ensure that operators choose the "best point" on the SAOC to operate. "Best" has to be defined in systems terms depending upon the shape of the underlying SAOC and the relative value to the company of efficiency and effectiveness.

At the performance intervention level (Level 3) the aim should be to move to a better SAOC, rather than merely choosing and sticking to the best point on the existing SAOC. Figure 38.5 shows two SAOCs that illustrate the ways in which a SAOC can be "better." First, the two fitted lines cut the time axis at different points. This point is where effectiveness (here, probability of detection) is zero, and represents the time to perform the task when no searching is taking place. A suitable name for this would be the overhead time. Reduction of overhead time is thus one aspect of a Level 3 intervention. The other difference between the SAOCs of Figure 38.5 is in their slopes. The "better" SAOC has a steeper slope, i.e., probability of detection increases more rapidly with every second spent searching. In general, this represents improved performance efficiency, or improved process capability.

To summarize the interventions, Table 38.1 shows the levels, interventions and appropriate labels. Note that most of the simple views on SATO are at Levels 1 and 2 which emphasize trading off speed for accuracy (e.g., haste makes waste). However, most of the quality-oriented management philosophies emphasize the simultaneous improvement of quality *and* productivity, largely by using Level 3 strategies.

The objective of an ergonomic analysis of a task is not just to understand the task, but to generate beneficial interventions. How can we improve speed and accuracy aspects of productivity? If a task is expected to exhibit SATO effects, the four generic intervention strategies can be employed to generate specific interventions. All that is required is a task description, and a task analysis which determines whether SATO is likely. Such a task analysis should be the first step in almost any ergonomic project, so that its production does not represent additional ergonomics work load. Thus a four-step process is recommended (Drury, 1994):

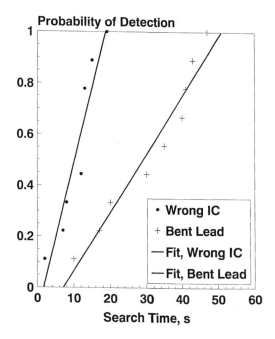

FIGURE 38.5 Speed–accuracy operating characteristics for two circuit board search tasks. (Adapted from Drury, C.G. and Chi, C.-F. 1995. A test of economic models of stopping policy in visual search. *IIE Transactions* (1995), 27: 382-393.)

1. *Perform Task Description and Task Analysis*: Use standard methods (e.g., Kirwan and Ainsworth, 1992; Drury et al., 1987) to produce a list of tasks, with the ergonomic models or database underlying each.
2. *Select SATO Tasks*: Determine whether each task is potentially resource-limited. Prototypical tasks are those given in Table 38.2.
3. *Form Task/Intervention Matrix*: List all SATO-prone tasks and, for each, list the four generic intervention types.
4. *Use Matrix to Generate Specific Interventions*: For each cell in the Task/Intervention matrix, use standard techniques (e.g., group processes) to allow the design team to determine specific interventions.

This will now be applied to a task about which much has been written recently (Drury et al., 1990), the inspection of aircraft structures.

The world's fleet of airliners has built up over several decades. With excellent safety records and a shortage of new aircraft in the 1980s, older aircraft are often replaced at much longer time intervals. The mean age of several aircraft types in the U.S. is now over 20 years (Shepherd, 1990). Continuing airworthiness has demanded increased inspection load to discover faults (e.g., cracks, corrosion) before they affect flight safety. While the system has several beneficial redundancies, people (inspectors) are still required to detect small targets in complex structures, visually or with electronic aids.

As the first part of a study of improvements to aircraft inspection, task analyses were undertaken of many inspection activities in several airlines (Drury et al., 1990). From those came a generic list of inspection functions which can serve here as the required task description. Although this same list covers both visual and non-destructive inspection (electronically aided), only the former will be considered here because it represents the majority of inspection activities in the airline practice.

It should be noted that for this task, errors of failure to detect defects would have enormous cost. This is recognized by inspectors and management whose philosophy represents the extreme accuracy end of the SAOC. However, there are time pressures involved, and (more commonly) adverse working conditions

TABLE 38.3 Generic Tasks in Aircraft Inspection, with Simple Task Analysis, Task Type, and SATO Potential

Task	Description	Type	SATO Potential
1. Initiate	Get workcard, read and understand area to be covered. Get equipment.	Accurate, movement Reading	Small
2. Access	Locate area on aircraft. Move to position to inspect.	Self-paced movement	Large
3. Search	More field of view (FOV) across area. More fixation across FOV. Stop if any indication.	Search	Large
4. Decision	Examine indication against standards. Reach conclusion.	Discrimination Response selection	Medium
5. Respond	Mark defect on aircraft. Record data onto repair record.	Accurate movement	Medium

of cramped spaces and awkward postures which can create the same effects as time pressures. The four step process is as follows:

Perform Task Description and Task Analysis. Table 38.3 shows the generic functions, with a brief task description for each and a note about the type of task.

Select SATO Task. The major tasks where SATO is to be expected in Table 38.3 are in the continuous movement to the area inspected (access) and the serial search for defects (search), although all five tasks have some SATO potential. Search is the most time consuming of the tasks and potentially the most error prone (Drury, 1991b), so that it is the obvious place to concentrate efforts at intervention. However, other tasks cannot be neglected. For example, although the SATO for the response selection task (in decision) is over periods of less than 1.0 s from the literature, interventions to improve the task should lead to higher decision reliability. Equally, although the recording of defects (respond) is relatively brief, it will necessarily interrupt search activities. Thus any improvement here will decrease the probability of forgetting the current search point.

Form Task/Intervention Matrix. Table 38.4 shows this matrix, with all tasks listed. To save space, a completed matrix (step 4) is presented.

Use Matrix to Generate Specific Interventions. Examples of relatively obvious interventions are given in Table 38.4. They were generated by considering the form of the SAOC for each of the tasks, and postulating practical methods of affecting each aspect of the SAOC. In a design (or redesign) setting, a team would generate ideas, whether practical or not, within this matrix for later careful evaluation.

Some of the interventions given in Table 38.4 are already being implemented. More readable workcards (Patel et al., 1992), better lighting (Reynolds et al., 1992), and training interventions for search strategy (Drury and Gramopadhye, 1992) have all been evaluated.

38.4 Summary and Suggestions for Implementation

This chapter has defined how speed and accuracy can be measured and the various ways in which speed and accuracy can be traded off. Level 1, or micro-SATO, is where successive repetitions of a task can either be performed slowly to gain accuracy or rapidly, in which case accuracy suffers. An appropriate strategy to manage Level 1 is to train operators to keep a consistent strategy from trial to trial. At the level of macro-SATO (Level 2) operators respond to overall pressures by changing their mean speed, and hence accuracy. Here, management should ensure that operators have a clear idea of the costs of both slow and erroneous performance. Management should also not insist on either very high speeds, which have a disproportionate effect on errors (Figure 38.1), or on unrealistic error rates, which lead operators to become overly slow (Figure 38.1 again). Level 3 SATO is performance intervention where the task, operator, machinery, or environment is changed to move to a new and better SAOC. This is the level at which most quality-oriented management interventions take place.

TABLE 38.4   Tasks/Interventions Matrix for Aircraft Structural Inspection

Task	Intervention Level			
	Level 1 1. Reduce SATO Variability	Level 2 2. Optimize SATO Point	Level 3 3.a. Reduce Overhead	Level 3 3.b. Improve Capability
1. Initiate	Train to complete understanding before accessing area		Accessible workcard Accessible equipment	More readable workcard Checklist for equipment
2. Access	Provide performance feedback on using safe access practices	Make walkways as wide as possible Train not to hurry	Use fixed walkways If not, use easily movable equipment Ensure equipment available	Design walkways for easy walking, e.g., clear boundaries, good surface Ensure correct footwear
3. Search	Train to use correct overlap of field of view (FOVs)	Determine the correct degree of overlap of fixations/FOVs Minimize awkward posture constraints to discourage early terminations	Allow easy FOV moves (e.g., lightweight flashlight) Well-defined boundaries to search area	Train to increase size of visual lobe Improve lighting for indication detection Workcard should list expected fault types
4. Decision		Minimize space and posture constraints as above	Clear listing of possible defects and their standards Accessible comparison standards	Correct lighting to discriminate faults Training in judgment against standards
5. Respond	Training to complete information	Standards for information completeness	Accessible recording device	Voice recording at inspection site Easy transcribing of voice recording to repair records (e.g., barcodes)

The common factor in all of these interventions is that they cost money, at least more money than the occasional talk to the workforce. Most have a single one-time cost, although others, such as training, have a continuing cost element. Replacing ongoing costs by one-time costs is, however, the basic strategy of industry. Much innovation, and practically all automation, has a single investment which produces long-term savings. Standard engineering texts have chapters on return-on-investment calculations, while management texts elevate these simple formulae to an impressive level of real-world complexity. It is not that industry has no precedents for such interventions of substance, but the will has been somewhat lacking. Presumably, the "new logic" of insistence on high quality, better design, and continuous improvement will allow more scope in the future for implementation.

But to follow the methodology outlined here places the onus on both the ergonomist and the manager to interpret tasks in terms of models so that predictions of benefits can be made. Practicing managers will not have to dismiss models as "academic," and the ergonomics research community will have to consider that their models may be used to evaluate investment decisions. The benefits of such changes in approach are that we utilize detailed, quantitative knowledge of human performance to make more rational use of the time and material resources of society.

References

Anon 1973. *Work in America, Report of a Special Task Force to the Secretary of Health, Education, and Welfare,* MIT Press, Cambridge, MA.

Brewer, N. and Smith, G.A. 1989. Developmental changes in processing speed: influence of speed–accuracy regulation. *Journal of Experimental Psychology: General,* 118(3): 298-310.

Business Week 1991. Special Issue, October 1991, The Quality Imperative.

Card, S.K., Moran, T.P. and Newall, A. 1983. *The Psychology of Human-Computer Interaction,* Lawrence Erlbaum Associates, Hillsdale, NJ.

Chi, C.-F. and Drury, C.G. 1998. Do people choose an optimal response criterion in an inspection task? *IIE Transactions,* 30, 257-266.

Conrad, R. 1954. Speed stress, in *Human Factors in Equipment Design,* eds. W.F. Floyd and A.T. Welford, Lewis Publishers, Chelsea, Michigan.

Craig, A. and Condon, R. 1985. Speed–accuracy trade-off and time of day. *Acta Psychologica* 58, p. 115-122, Elsevier Science Publishers B.V., North-Holland.

Crosby, 1979. *Quality is Free. The Art of Making Quality Certain.* McGraw-Hill, New York.

Dar-el, E.M. and Vollichman, R. 1996. Speed v/s accuracy under continuous and intermittent learning, in *Advances in Applied Ergonomics, Proceedings of the 1st International Conference on Applied Ergonomics (ICAE'96),* ed. A.F. Ozok, Istanbul, Turkey, 676-682.

DeFazio, K., Wittman, D., and Drury, C.G. 1992. Effective vehicle width in self-paced tracking, *Applied Ergonomics,* 23(6): 382-386.

Deming, W.E. 1986. *Out of the Crisis,* Massachusetts Institute of Technology, Cambridge, Mass.

Dobyns, L. and Crawford-Mason, C. 1991. *Quality or Else. The Revolution in World Business,* Houghton Mifflin Company, Boston.

Drury, C.G. 1971. Movements with lateral constraint. *Ergonomics,* 14: 293-305.

Drury, C.G. 1987. Inspection performance and quality assurance, Chapter 65, *Job Analysis Handbook,* John Wiley & Sons, New York.

Drury, C.G. 1991a. Ergonomics practice in manufacturing. *Ergonomics,* 34: 825-839.

Drury, C.G. 1991b. Errors in aviation maintenance: taxonomy and control, In Proceedings of the 35th Annual Meeting of the Human Factors Society, San Francisco, CA, 42-46.

Drury, C.G. 1994. The speed–accuracy trade-off in industry. *Ergonomics,* 37: 747-763.

Drury, C.G. and Chi, C.-F. 1995. A test of economic models of stopping policy in visual search. *IIE Transactions* (1995), 27: 382-393.

Drury, C.G., and Corlett, E.N. 1975. Control of performance in multi-element repetitive tasks, *Ergonomics,* 18: 279-298.

Drury, C.G., and Dawson, P. 1974. Human factors limitations in fork-lift truck performance. *Ergonomics,* 17: 447-456.

Drury, C.G. and Forsman, D.R. 1996. Measurement of the speed accuracy operating characteristic for visual search. *Ergonomics,* 39(1): 41-45.

Drury, C.G. and Gramopadhye, A. 1992. Training for visual inspection: controlled studies and field implications, in *Meeting Proceedings of the Seventh Federal Aviation Administration Meeting on Human Factors Issues in Aircraft Maintenance and Inspection,* Atlanta, GA, 135-146.

Drury, C.G., Paramore, B., Van Cott, H.P., Grey, S.M. and Corlett, E.N. 1987. Task Analysis, Chapter 3.4, in *Handbook of Human Factors,* ed. G. Salvendy, p. 370-401, John Wiley & Sons, New York.

Drury, C.G., Prabhu, P. and Gramopadhye, A. 1990. Task analysis of aircraft inspection activities: methods and findings, in *Proceedings of the Human Factors Society 34th Annual Conference,* Santa Monica, California, 1181-1185.

Dudley, N.A. 1958. Output pattern in repetitive tasks, *Institute of Production Engineers' Journal,* 37: 303-313.

Fitts, P.M. 1966. Cognitive aspects of information processing: III. Set for speed vs. accuracy. *Journal of Experimental Psychology,* 71(6): 849-857.

Fox, J.G. 1977. Quality control of coins, in *Case Studies in Ergonomics Practice,* eds. H.G. Maule, and J.S. Weiner, Vol. 1, Taylor & Francis, London.

Gan, K.C. and Hoffmann, E.R. 1988. Geometrical conditions for ballistic and visually-controlled movements. *Ergonomics,* 31: 829-840.

Garrett, H.E. 1922. A study of the relation of accuracy to speed, in *Archives of Psychology,* ed. R.S. Woodworth, No. 56, G.E. Stechert & Co., London.

Hasslequist, R.J. 1981. Increasing manufacturing productivity using human factors principles. *Proceedings of the Human Factors Society,* 25th Annual Conference, Santa Monica, CA, 204-206.

Hick, W.E. 1952. On the rate of gain of information, *Quart. J. Exp. Psychol.,* 4: 11-26.

Hou, T.-S., Lin, L., and Drury, C.G. 1993. An empirical study of hybrid inspection systems and allocation of inspection function. *International Journal of Human Factors in Manufacturing,* 3: 351-367.

Jones, M.B. and Kennedy, R.S. 1996. Isoperformance curves in applied psychology. *Human Factors,* 38(1): 167-182.

Karwan, M., Morawski, T.B., and Drury, C.G. (1995). Optimum speed of visual inspection using a systematic search strategy. *IIE Transactions (1995),* 27: 291-299.

Kirwan, B. and Ainsworth, C.K. 1992. *A Guide to Task Analysis,* Taylor & Francis, London.

Konz, S. 1990. *Work Design: Industrial Ergonomics,* p. 219-236, Publishing Horizons, Inc., Worthington, Ohio.

Meyer, D.E., Irwin, D.E., Osman, A.M., and Kounios, J. 1988. The dynamics of cognition and action: mental processes inferred from speed–accuracy decomposition. *Psychological Review,* 95(2): 183-237.

Meyer, D.E., Smith, J.E.K., Abrams, R.A. and Wright, C.E. 1990. Speed–accuracy tradeoffs in aimed movements: toward a theory of rapid voluntary action, in *Attention and Performance XIII, Motor Representation and Control,* ed. M. Jeannerod, Chapter 6, 173-226.

McNichol, D. 1972. *A Primer of Signal Detection Theory,* Allen and Unwin, Sydney.

Montazer, M.A., Drury, C.G., and Karwan, M. 1987. Self-paced path control as an optimization task, *IEEE Transactions,* SMC-17.3: 455-464.

Mundel, M.E. and Danner, D.L. 1994. *Motion and Time Study, Improving Productivity,* Seventh Edition, Prentice Hall, London.

Myerson, J., Hall, S., Wagstaff, D., Pon, L.W., and Smith, G.A. 1990. The information-loss model: a mathematical theory of age-related cognitive slowing. *Psychological Review,* 97(4): 475-487.

Norman, D.A. 1981. Categorization of action slips. *Psychological Review,* 88(1): 1-15.

Norman, D.A. and Bobrow, D.J. 1979. On data limited and resource limited processes. *Cognitive Psychology,* 5: 44-64.

Pacella, R.G. 1974. The interpretation of reaction time in information processing research, in *Human Information Processing: Tutorials in Performance and Cognition,* ed. B.H. Kantowitz, p. 41-82, Lawrence Erlbaum Associates, Hillsdale, NJ.

Patel, S., Prabhu, P., and Drury, C.G. 1992. Design of work control cards, in *Meeting Proceedings of the Seventh Federal Aviation Administration Meeting on Human Factors Issues in Aircraft Maintenance and Inspection,* Atlanta, GA, 163-172.

Pew, R.W. 1969. The speed–accuracy operating characteristic, in *Acta Psychologica,* 3: 16-26.

Rabbit, P.M.A. 1981. Sequential reactions, in *Human Skills,* ed. D.H. Holding, John Wiley & Sons, Chichester.

Rasmussen, J. 1986. *Information Processing and Human-Machine Interaction, An Approach to Cognitive Engineering,* North-Holland, New York.

Reason, J. 1990. *Human Error,* Cambridge University Press, Cambridge, U.K.

Reynolds, J.L., Gramopadhye, A., and Drury, C.G. 1992. Design of the aircraft inspection/maintenance visual environment, in *Meeting Proceedings of the Seventh Federal Aviation Administration Meeting on Human Factors Issues in Aircraft Maintenance and Inspection,* Atlanta, GA, 151-162.

Salvendy, G. and Smith, M.J. (eds.) 1981. *Machine Pacing and Occupational Stress,* Taylor & Francis, London.

Schwabish, S.D. and Drury, C.G. 1984. The influence of the reflective-impulsive cognitive style on visual inspection, *Human Factors,* 26.6: 641-647.

Senders, J.W. and Moray, N.P. 1991. *Human Error: Cause, Prediction and Reduction,* Lawrence Erlbaum Associates, Hillsdale, New Jersey.

Shepherd, W.T. 1990. Human factors in the maintenance and inspection of aircraft. *Proceedings of the Human Factors Society 34th Annual Meeting,* Santa Monica, CA, 1167-1170.

Smith, G.A. and Brewer, N. 1995. Slowness and age: speed–accuracy mechanisms. *British Journal of Psychology,* 76:199-203.

Wickelgren, W.A. 1977. Speed–accuracy tradeoff and information processing dynamics. *Acta Psychologica* 41: 67-85.

Wickens, C.D. 1994. *Engineering Psychology and Human Performance,* Harper Collins, New York.

Woodworth, R.S. 1899. The accuracy of voluntary movement. *Psychol. Rev. Monogr. Suppl.,* 3, No. 3: 1-114.

For Further Information

For an excellent review of SATO in accurate movements, see Meyer, Hall, Wagstaff, Pon, and Smith (1990).

For a discussion of the methodology of measuring the SAOC, see Wickelgren (1977).

A new way of measuring the SAOC in the laboratory has been developed by Meyer, Irwin, Osman, and Kounios (1988).

The issue of paced performance has not been covered in this chapter. A useful book on the subject is Salvendy and Smith (1981).

For an example of current studies of SATO and learning, see Dar-el and Vollichman (1996).

Time-of-day effects on SATO are covered in Craig and Condon (1985), while development and aging processes affecting SATO can be found in Brewer and Smith (1989) and Meyerson, Hall, Wagstaff, Pon, and Smith (1990).

39

Receiver Characteristics in Safety Communications

Stephen L. Young
Liberty Mutual Research Center for Safety & Health

Kenneth R. Laughery
Rice University

Michael S. Wogalter
North Carolina State University

David R. Lovvoll
Rice University

39.1 Introduction

An interesting aspect of the warning process occurs when people pick up a prescription drug at a pharmacy. In many cases, these medications are accompanied by a patient package insert (PPI). PPIs contain detailed information about the nature of a drug, potential side effects, prescriptions and proscriptions for use, and a wealth of other details about the chemical makeup of the drug. PPIs are similar in scope and detail to the information contained in drug reference books (e.g., *Physician's Desk Reference*). One look at these documents will tell you that they are not designed for the layperson, but rather they are designed to provide the kinds of information and the level of detail that would primarily benefit an individual with substantial medical training. Because of the level of sophistication required to acquire relevant information from PPIs, pharmacies will often provide a briefer and simpler summary of the relevant information for use by the lay customer. The purpose of the summary is to provide the end user with the most important information necessary to use the drug properly.

While a central tenet of warning theory is that it is important to provide people with information so they can make informed choices regarding their behavior, it is not necessarily true that more information is better. Table 39.1 shows the information provided in a PPI and in a pharmacy summary for the same drug. It is clear that the information is targeted for two different audiences. Physicians are provided detailed information, because they need it to make proper prescribing decisions. However, patients (for the most part) will not find much of the detail helpful, and they could find it difficult and confusing. It may actually make the extraction of relevant information *more* difficult.

This example demonstrates, in a very basic way, that one must consider who is the audience when designing, producing, and delivering safety-related information. Other examples might include the presentation of information in material safety data sheets (MSDSs) and in OSHA regulations.

TABLE 39.1   Example of Pharmaceutical Information Provided to Physicians and Patients

Information Category	Physicians	Patients
Dosage	Adult: 3-4 g/day in evenly divided doses Child (>2): 40–60 mg/kg/day in 3–6 evenly divided doses	Take two tablets twice daily. Take this medicine with meals or a snack. Try to space your doses evenly over each 24 hour period. If you miss a dose of this medicine, take it as soon as possible. If it is almost time for your next dose, skip the missed dose and go back to your regular dosing schedule. Do not take 2 doses at once.
Adverse Reactions	Cardiovascular — Vasculitis, pericarditis with or without tamponade CNS — Headache, transverse myelitis, convulsions, meningitis, transient lesions of the posterior spinal column, cauda equina syndrome, Guillain-Barre syndrome, peripheral neuropathy, mental depression, vertigo, hearing loss, insomnia, ataxia, hallucinations, tinnitus, drowsiness Genitourinary — Oligospermia, infertility Gastrointestinal — Anorexia, nausea, vomiting, gastric distress, hepatitis, pancreatitis, diarrhea, stomatitis, abdominal pains, neutropenic enterocolitis Hematological — Heinz body anemia and hemolytic anemia, aplastic anemia, agranulocytosis, leukopenia, megaloblastic (macrocytic) anemia, purpura, thrombocytopenia, hypoprothrombinemia, methemoglobinemia, congenital neutropenia	Side effects, that may go away during treatment, include headache, nausea, loss of appetite, or indigestion. If they continue or are bothersome, check with your doctor. CHECK WITH YOUR DOCTOR AS SOON AS POSSIBLE if you experience sore throat, fever, rash, tightness of chest or difficulty breathing. If you notice other effects not listed above, contact your doctor, nurse or pharmacist.
Cautions	Drug-induced hypersensitivity reactions, blood dyscrasias, neuromuscular and CNS reactions, hepatotoxicity, nephrotoxicity, or fibrosing alveolitis may result in death. Watch for clinical signs suggesting a serious blood dyscrasia; obtain a CBC frequently. To prevent crystalluria and lithiasis, perform a urinalysis, including a careful microscopic examination, frequently and instruct patients to maintain an adequate fluid intake. Caution patients that their skin or urine may turn orange-yellow during therapy. Contraindications include hypersensitivity to sulfasalizine, sulfapyridine, or other sulfonamides or to 5-aminosalicylic acid or other salicylates.	TELL YOUR DOCTOR OR PHARMACIST if you are allergic to sulfa drugs. Keep all medical appointments while receiving this medicine. Drinking extra fluids while you are taking this medicine is recommended. Check with your doctor or nurse for instructions. This medicine may cause increased sensitivity to the sun. Avoid exposure to the sun or sunlamps until you know how you react to this medicine. Use a sunscreen or protective clothing if you must be outside for a prolonged period. This medicine may cause a harmless, yellow-orange discoloration of the urine or skin.

FIGURE 39.1   Basic communication model.

The task of communicating warning information to an individual, whether through product warnings, safety signs, auditory warnings, etc., can be described in terms of communication theory (Lehto and Miller, 1986). Using this theoretical context, McGuire (1980) defined warnings as communications designed to influence behavior with respect to a product. Figure 39.1 shows a simple, generic communications model, which includes four primary components: the sender, the message, the medium, and the receiver.

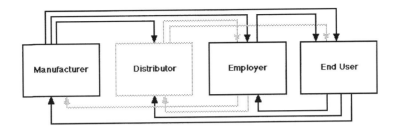

FIGURE 39.2   Complex warning communication system.

The sender, or source, represents the originator of the communication. With respect to warnings, the message is the relevant information that is to be transmitted via some medium. The message could (and preferably would) contain information about the nature of the hazard, consequences of exposure to the hazard, and/or instructions on how to avoid the hazard. The medium refers to the channel or route by which information moves from the source to the receiver. Media for warnings can include MSDSs, on-product labels, package inserts, signs, oral instructions, and so forth. The receiver refers to any and all persons who are at risk and to whom the warning should be directed. Characteristics of each of these components may, and often do, play a critical role in the effectiveness of a warning.

In the most simple application of this model to the warning process, a manufacturer of a product (the source) attempts to relay some warning message using one or more media to an end user of the product (the receiver). However, the process of conveying warning information is not always so straightforward. For example, Figure 39.2 represents the elements of a more complex warning communication system for a product being used in an industrial setting. Here the product might be marketed through a distributor (or a series of distributors) to an employer (the business and its management) who in turn communicates in various ways with the end user (employee). Communication from the manufacturer to the end user may be direct, such as labels on the product, or indirect, through various intermediaries (e.g., distributors). The media through which the information is communicated may also be quite varied. Feedback between various components may be involved, such as an employer notifying the manufacturer about a safety problem associated with the use of the product.

Even more complex warning systems could have several receivers, including distributors, employers, and end users. These receivers might differ markedly in several important respects. For example, an employer's industrial toxicologist who may be a receiver in the communication process will probably have a great deal more technical knowledge than a laborer working in the plant who is the end user of the product. This knowledge difference may have implications for the warning system associated with the product. A parallel example, is that given at the outset of the chapter, where there are at least two kinds of users with respect to medications manufactured by a drug company (e.g., a prescribing physician, pharmacists, and a patient who are all targeted receivers of safety information).

Whether the circumstances are simple or complex, the success of a warning communication system depends on accounting for the properties of the various system components. Previous research has examined issues related to the source, the medium, and the message. In this chapter, we focus on the receiver. We review the literature on the most commonly addressed receiver characteristics and present warning-design implications that stem from these characteristics. This chapter is organized into four sections: demographic variables, competence, familiarity, and risk perception. Finally, we offer observations that warning designers should consider with respect to the receiver.

39.2   Demographic Variables

Demographics are statistical characteristics of individuals that can be used for the purpose of grouping. It is easy to collect such data and many warning-related studies have done so in the past. Two common demographic variables are gender and age.

Gender

Research indicates that there are gender differences in the perception of product hazards and in willingness to read and comply with warnings. Where gender differences are found, it appears that females report being more likely to look for and read warnings than are males (Godfrey et al., 1983). However, in many instances, the results do not provide a clear picture regarding the exact nature and cause of gender differences. Much of this confusion may result from the fact that many of the sex differences reported in the literature come from *post hoc* analyses. That is, gender effects are not manipulated *a priori* but are simply analyzed after the data are collected.

Post Hoc Analyses of Gender Effects

For example, Donner (1990) manipulated warning modality (written versus oral versus both) and formality (informal versus formal) for two different products (fabric protector versus a bench grinder). Compliance with the warning information, as well as noticing and reading the warning, was reported. The results showed a strong product effect, with no significant effect of warning modality or formality. After these analyses (which were the primary focus of the study), an evaluation of gender effects was conducted. Donner found gender differences in previous experience with the product (e.g., females had greater experience with the fabric protector), but this variable did not affect compliance with the product warning. This type of analysis is common among studies that report gender effects. After the primary analyses are performed, a *post hoc* analysis is conducted to determine how gender interacted with or influenced the findings.

Studies employing *post hoc* analyses of gender have produced mixed results. Dorris and Tabrizi (1978) had male and female subjects rate the hazardousness of different products and found very small effects of gender. Godfrey et al. (1983) showed that females were more likely to look at warnings than were males. Silver and Braun (1993) found no gender differences in preference for font size and style in warning labels. Green and Pew (1978) and Laux, Mayer, and Thompson (1989) found that females produced lower accuracy in safety symbol comprehension than males, but only under certain conditions (e.g., gender effects interacted with previous training) and with certain symbols. Leonard, Hill, and Karnes (1989) reported that females were more likely to wear seatbelts than were males. Other research shows that females would be more likely to report complying with warnings than would males (Desaulniers, 1991; Viscusi et al., 1986).

Note that in these studies, gender was not a formal variable under study, nor was gender crossed with the other experimental variables. In such *post hoc* analyses, there is often little reasoning provided as to why gender might influence the results. Gender analyses are presented because the authors thought to do them and because they could be done. It is possible that many studies do not report *post hoc* analyses of gender because of a lack of observed differences. Thus, it is not really surprising that gender has produced mixed results.

A Priori Analyses of Gender Effects

Several studies have examined sex differences as a primary focus of the study. These studies have, for the most part, demonstrated differences between males and females with regard to several safety-related variables. For example, Goldhaber and deTurck (1988a; 1988b) reported that, while females were less likely to recall a "No Diving" warning sign posted around a swimming pool, they were significantly less likely to actually dive than were males. LaRue and Cohen (1987) had males and females rate different consumer products according to several questions. Reported familiarity and perceptions of product hazardousness were statistically similar between males and females in this study. However, females reported being significantly more likely to read warnings for the products than males. Young, Martin, and Wogalter (1989) had males and females rate consumer products according to several questions. The products in this study were classified from the ratings as being either masculine or feminine. Product ratings were then examined as a function of subject gender. The results demonstrated a main effect of product's masculinity/femininity and an interaction between this variable and the subject's gender. According to the interaction, males and females judged the hazardousness of feminine products similarly, but females rated the masculine products as significantly more hazardous than did the males.

Although gender was the variable of interest in these studies, it remains unclear as to whether gender is truly responsible for the observed results. Specifically, gender may simply act as a proxy or alias for another variable or group of variables (e.g., familiarity). Young et al. (1989) demonstrated that greater hazard ratings were given by females to products that they used infrequently. Other studies have shown that familiarity and product hazard ratings are negatively correlated (Otsubo, 1988; Young, 1996; Wogalter et al., 1987; Young et al., 1990). It may be that females and males would rate product hazards similarly if females only used the products more frequently and/or had greater knowledge about them generally or their hazards specifically. However, other studies suggest that this hypothesis may not be true. For example, Godfrey and Laughery (1984) showed that females underestimated the risks associated with tampon use due, in part, to their familiarity with the product. Karnes and Leonard (1986) examined male and female knowledge of hazards associated with a contraceptive device (an IUD) based on a safety-related pamphlet. Males were included in the subject sample specifically because it was assumed (whether true or not) that they would serve as a low-familiarity control to the females. Females in this study perceived the risks to be significantly higher than did the males.

Conclusions: Gender

These studies suggest that gender effects may or may not be observed in research that does not specifically address gender issues. Evidence tends to suggest that females are more willing to act safely with products (e.g., read or comply with warnings) than are males. However, it is unclear whether this trend would be observed in all situations and with all products. It is also unclear whether gender is the true source of observed differences in product perceptions or whether those differences are related to more basic considerations (e.g., knowledge of the hazards, familiarity, frequency of use, etc.). Further systematic research, which accounts for confounding variables (e.g., familiarity, etc.), is needed before design considerations with regard to warnings can be provided.

Age

When considering age, it is generally believed that (a) older people tend to be more risk averse and (b) younger people (especially younger males) tend to be predisposed to taking greater risks. Smith and Watzke (1990) demonstrated that older people (30 to 59 years, 60 to 75 years, and over 75 years old) more carefully consider the risks and are more cautious than younger adults (under 30 years of age). The authors suggest that cautiousness may be a characteristic of older adults and that this characteristic may start to exhibit itself in the middle years of life (between 30 and 59 years of age). If it is true that people are more or less risk averse depending on their age, then this should manifest itself in safety-related behaviors (e.g., looking for, reading, and complying with warnings). For example, older people (> 25 years old) are more likely to wear seatbelts than younger individuals (< 25 years old) (Leonard et al., 1989). Desaulniers (1991) found that older people, 40 and above, reported being more likely to take precautions in response to warnings.

There are many hypotheses as to why age may influence risk perceptions and/or safety-related behaviors, but the most reasonable ones surmise that younger people do not actually take risks in a formal sense. Specifically, they do not consider an action with a conscious view toward the costs and benefits of acting one way over another (see Lehto, 1991; Wagenaar, 1992). As such, their behavior may appear to be nonrational and more dangerous than the types of judgment-based behavior exhibited by older individuals. While a great deal of evidence for risk-taking in younger people can be found in the literature on traffic accidents (see Edwards and Ellis, 1976; Leonard et al., 1989), there is somewhat less evidence of this phenomenon in day-to-day behaviors. Thus, it is not surprising that the literature is not entirely consistent in demonstrating age effects.

For example, Mazis, Morris, and Gordon (1978) showed that a sample of premenopausal females preferred longer and more detailed information regarding the risks associated with oral contraceptives, but that this trend was more pronounced for younger subjects (e.g., college-age students) than for older subjects. Purswell, Schlegel, and Kejriwal (1986) demonstrated that older people (>30 years old) exhibited safer behaviors than younger subjects (<30 years old) with a router, but that the opposite pattern was observed with an electric knife. Leonard, Ponsi, Silver, and Wogalter (1989) found minor differences between older (M = 37 years) and younger (M = 18 years) subjects on willingness to read warnings for pest-control products. Wright, Creighton, and Threlfall (1982) observed no effect of age (ages in the sample ranged from under 30 years old to over 50) with regard to willingness to read instructions.

As with the evidence regarding gender effects, findings associated with age in warnings research have not been entirely consistent. Much of the conflicting data may result from the fact that a majority of analyses are *post hoc* evaluations. There is an additional problem associated with research on age: defining "younger" and "older" categories in research. These terms can vary widely in meaning from one study to the next. The "older" group of subjects in one study could be the "younger" group in another. Because of these issues, the effect of age on risk perceptions and safety-related behaviors is inconclusive. It appears as though younger adults may be a more difficult group to warn because of their lack of (formal) consideration of risks and benefits in the decision-making process. Relative to younger people (and possibly younger males in particular) older adults may be more likely to comply with safety-related information. However, this hypothesis needs additional testing, with systematic research that controls for confounding variables (e.g., knowledge and familiarity with products, etc.).

39.3   Competence

Competence can be defined as possessing the capacity to meet the demands of a particular task. There are many dimensions of receiver competence that may be relevant to the design of warnings. We discuss three here: sensory, physical, and cognitive capabilities.

Sensory Capabilities

It is obvious that the blind person cannot see a written warning, nor would the deaf person receive an auditory warning. Although these extreme examples are obvious, we also know that sensory capabilities lie along a continuum. Consider that many older adults, who use more medications as a group, cannot read medication labels because of age-related visual decrements. Yet many over-the-counter pharmaceutical labels have print that is too small for older adults to read. One way to solve this problem is to design product labels to accommodate larger type (see Wogalter and Young, 1994). Another way is to provide relevant safety information in the form of pictorial symbols. However, several studies have demonstrated that older adults have greater difficulty in interpreting safety-related symbols (Collins and Lerner, 1982; Easterby and Hakiel, 1981; Ringseis and Caird, 1995).

Physical Capabilities

This topic deals with the extent to which the user will be physically capable of carrying out a task. For example, older adults may not have the dexterity to grab hold of a three-point manual seat belt or they may not be able to generate the torque necessary to open small medicine containers. Wogalter and Young (1994) demonstrated that different label designs, while increasing the size of warning information (and thereby making it easier to read), could provide the user with a greater surface area on which to apply force. If special equipment is required to comply with the warning, it must be available or obtainable. For example, some hair dyes contain warnings that direct the use of gloves during application. Rather than imposing on the user to find and/or purchase gloves separately, plastic gloves are generally included in the package with the dye. If special skills are required, they must be present in the receiver population. To some extent, as with the sensory limitations of receiver populations, the behavioral limitations that may be involved could be considered rather obvious, although we are constantly amazed at the number of warnings that violate such considerations — especially in the behavioral domain where basic product instructions (e.g., for assembly/installation) are often difficult to carry out.

Cognitive Capabilities

Examples of cognitive competence include requisite technical capacity, language, and reading ability.

Technical Capacity

One of the primary issues in warning design with respect to competence concerns the level of technical information to be communicated. Comprehension of such information is generally a function of the receiver's existing technical knowledge of the domain. Here we are referring to conceptual knowledge that includes both factual information and process understanding (the receiver's mental model). Some examples include: (a) medications where knowledge of physiology may be relevant, (b) chemical reactions that require an understanding of what not to mix with what, and (c) mechanical properties where knowledge is needed to understand the hazards of handling certain kinds of equipment. In formulating warnings, it is important to take into account the relevant technical knowledge of the receiver. Further, the problem may be more complicated in the sense that warnings regarding a particular product hazard may be directed to multiple groups (or receivers) differing in knowledge.

The point to be emphasized here is that the level or levels of knowledge and understanding must be considered. Of course, it is also a valid concern that variability in knowledge about facts and processes exists within the target audience for a particular product warning. There may be various approaches to address these concerns. One approach is to construct a single warning system that will be understood at a range that reaches as many people in the target audience as possible. Another approach is to develop a multiple-component warning system where different components are directed at subgroups varying in technical knowledge. The second approach, as shown in the example provided at the beginning of the chapter, is the one selected for presentation of drug-related information by pharmacies. Physicians receive detailed information about prescription drugs and patients receive summaries of that information.

Language

A second cognitive issue with respect to competence is language. The target audience may know a language different from the majority. A warning printed in only one language is much less likely to be accessible to all potential users. Attempts to deal with this problem include the use of pictorial symbols and printing the message in multiple languages. The latter technique is commonly employed in instruction booklets that accompany various electronic products such as watches and calculators. Signs printed in multiple languages must be either increased in size to accommodate the extra material or, if the size is held constant, they must be more cluttered or dense. Neither of these alternatives is desirable. Also, selection of languages to appear on these signs may not be so straightforward. How many languages does one need in order to cover all potential users? The number could be prohibitively high.

Symbols, on the other hand, provide the promise of non-verbal communication — a method of conveying safety-related information regardless of the language spoken in the population. Research has demonstrated that symbols are effective in attracting user attention (Young, 1991; Laughery and Young, 1991; Wogalter et al., 1996; Young and Wogalter, 1990) and in conveying safety information (Collins, 1983; Collins and Lerner, 1982; Laux et al., 1989). However, the promise of completely non-verbal communication has not been and may not be fully realized. Symbols necessarily involve an abstraction of some message. This method of information display is easier for certain safety-related concepts (e.g., slippery floor) than with others (e.g., biohazard, cancer). Designing symbols to convey information that can be interpreted accurately under many different circumstances can be difficult.

Reading Ability

Many warnings require high levels of reading ability on the part of the receiver. The usual recommendation for general target audiences is a reading level near the elementary school range. An exception to this rule is found in Leonard et al. (1989), who found that college students and other highly educated individuals reported being more likely to read complex warnings than simple ones. The complex warnings in this study were used primarily for more hazardous products. Perceived hazardousness is a factor that (as we will see in the next section) has a strong relationship to willingness/likelihood of reading warnings. Obviously, if comprehension of a warning is to be achieved, the material should be written at the level that accommodates the readership. One way to evaluate the readability of safety-related information is to conduct some type of readability analysis of the materials. A discussion of reading level measures and their application in the design of instructions and warnings can be found in Duffy (1985). It should be noted that readability formulas are only indications of comprehensibility, and they may be less suitable with short messages like those that commonly appear in warning signs. Readability scores and comprehension measures are not always highly correlated. Thus, readability formulas should be used with caution and probably as only a first step in determining the material's understandability.

The problem of warning readability may require more than simply keeping reading levels to a minimum. There are a very large number of functionally illiterate adults in the population who cannot read written (verbal) warnings at any level. We offer no simple solutions to this problem, but certainly pictorial symbols, oral warnings, special training programs, etc. may be important ingredients of warning systems for such populations.

Conclusions: Competence

There are many factors that influence the capacity of individuals to meet the task of acting safely around products. Sensory, physical, and cognitive capabilities can influence whether users are capable of accessing, understanding, and using safety-related information. These characteristics of potential users must be considered when designing warnings.

39.4   Familiarity and Experience

One of the issues that has received substantial attention in research concerns the familiarity and/or experience that people have with products and how such factors influence the effectiveness of warnings. Familiarity has been defined in many ways, but we define it here as a state of being intimate or closely acquainted with a product and its hazards. Familiarity is not a dichotomous state. People may be unfamiliar with a product. They may be familiar with a product generally (i.e., they have heard *about* the product) or they may be familiar with it specifically (i.e., they have used the product). Users can become familiar with a product indirectly (i.e., through the acquisition of knowledge about it) or directly (i.e., through direct use). Familiarity is a belief and it is most commonly measured through ratings where people express their familiarity with a product on a Likert-type scale (e.g., 0 = "not at all familiar" to 7 = "extremely familiar"). Experience, on the other hand, can be operationally defined in terms of time and/or frequency of use. A distinction between familiarity and experience has been noted by Wogalter et al. (1986, 1987). We do not consider the familiarity and experience to be synonymous, but we will discuss them together here for the purpose of dealing with users' knowledge-based product perceptions.

Familiarity

Numerous studies have explored the effect of familiarity on safety-related product perceptions and behavior. In general, higher levels of product familiarity or experience are associated with decreases in the probability that warnings will influence user behavior. The reasons for this conclusion are varied, but they tend to revolve around the notion that as people use a product and become more familiar with and knowledgeable about it, they perceive it to be less dangerous. Desaulniers (1989) showed that people perceived more familiar products as less hazardous. Karnes and Leonard (1986) showed that subjects with greater experience riding ATVs considered them to be less dangerous than did subjects with less experience.

The utility of warning information may be reduced as people come to see the products as less hazardous. Thus, users may not seek out or read relevant information. Godfrey and Laughery (1984) showed that females reported being less likely to read warnings for products with which they are familiar (e.g., tampons). Johnson (1992) showed that willingness to look for and read warning information for scaffolds was negatively related to the number of times workers had previously used scaffolding. Morris, Mazis, and Gordon (1977) showed that about 78% of females read a PPI for oral contraceptives the first time they used the drug, but that less than 11% read the insert when it accompanied subsequent prescriptions. However, other research suggests that unwillingness to look for and read warnings is not exclusively related to familiarity. Leonard et al. (1989) demonstrated that willingness to read warnings for a pest-control product was unrelated to familiarity with that product. In addition, Godfrey et al. (1983) demonstrated that familiarity was not related to subjects' reported willingness to look for warning labels on products perceived as hazardous. However, they also reported being more willing to look for warnings on unfamiliar products when the hazard level was perceived as being low. Thus, familiarity may be one factor, along with other perceptions, that influence the extent to which users may seek information.

Behavioral effects of familiarity have also been demonstrated. Goldhaber and deTurck (1988a, b) showed that previous experience with diving into pools was related to lower likelihood of noticing a "No Diving" sign, a higher likelihood of diving into shallow water, and a lower perception of the risks associated with such activities. Otsubo (1988) showed that people with less experience were more likely to read the warnings for two types of saws.

The above review suggests that the more people become familiar with a product, the less likely they will be to engage in safe behaviors (and vice versa). While this relationship may be linear (or at least monotonic), there is some evidence to suggest that the relationship is nonlinear. Bettman and Park (1980) found that subjects with a moderate level of knowledge or experience relied most on external information when making a product-related decision. People with low and high levels of previous knowledge relied to a greater extent on this external information. The authors suggested that users with high levels of experience did not need the information and that users with low levels did not have the capacity to use it properly. Johnson and Russo (1980) demonstrated that both the linear and nonlinear ("inverted-U") functions were observed in different decision-making tasks.

Conclusions: Familiarity

Research generally suggests that lower levels of familiarity are associated with higher levels of perceived hazard and greater reported willingness to act with caution (and vice versa). The most common explanation for this finding is that greater familiarity is associated with greater knowledge of and appreciation for the product's hazards. However, familiarity with a product is not synonymous with knowledge of the hazards associated with it. People may report being familiar with a product and yet have little or no knowledge of its hazards. Familiarity lies along a continuum — people can have indirect, general familiarity with a product or they can have more direct, specific familiarity (i.e., from lower to higher forms of familiarity). Subjects who provide ratings of familiarity in research studies may not make the distinction between the two types. Thus, they might report a high degree of familiarity with a product that they have very little personal knowledge of or experience with (e.g., they may have heard a lot about a product, but have no direct experience with it).

High levels of perceived familiarity may lull people into thinking that they have greater knowledge of and control over product hazards than they actually have and/or that the products are less hazardous. This perception, coupled with the fact that familiarity may reduce information-seeking behaviors on the part of users, can produce a dangerous situation and a special challenge to safety professionals. That challenge is to make warning information salient (i.e., to attract the attention of familiar users) and to make the information seem relevant. A considerable body of research has addressed various forms of salience. However, there is much less research dealing with relevance issues. Ways to make warnings more relevant can include prioritizing warning information based on users' needs and presentation of information to specific users at intermittent schedules.

39.5   Risk Perception

Risk perception in the present context refers to the way people understand and consider the hazards associated with products and the ways in which these perceptions influence people's behavior when using them. A consistent finding in warning research is that people's perception of the risk associated with a product or situation is an important determinant of warning effectiveness — the greater the risk, the more likely people will look for, read, and comply with warnings (Donner and Brelsford, 1988; Friedmann, 1988; Godfrey et al., 1983; LaRue and Cohen, 1987; Leonard et al., 1986; Otsubo, 1988; Wogalter et al., 1991). While most of the research on risk perception has evaluated the nature of the products themselves, some research has examined subject characteristics.

One study (Young, 1996) had subjects rate a list of consumer products according to several different rating questions:

- How hazardous is this product?
- How frequently do you use this product?
- How likely are you to be injured while using this product?

As in other studies of risk perception, the results demonstrated that subjects' risk perceptions varied as a function of the product being evaluated, with some products being considered more hazardous (as a whole) than others. However, the risk ratings also varied as a function of the subject. Specifically, there were differences in the way different subjects perceived the hazard for individual products (e.g., chain saw).

Based on the ratings in this study, Young (1996) partitioned the subjects into three distinct groups which varied in terms of how they perceived the products in general and in terms of the information they accessed when evaluating product risks. The first group of subjects was labeled *Fearful*, because they subjects considered the products as a whole to be quite hazardous while having only average knowledge of the risks and average familiarity with the products. The second group of subjects, labeled *Fearless*, considered the products as a whole to be nonhazardous despite having little knowledge of the products' risks and little familiarity with them. The third group of subjects was labeled the *Informed* group. These subjects considered the products as a whole to be nonhazardous, but they also knew a great deal about the risks associated with the products and they were very familiar with them as well. This group is similar to the internal locus-of-control subjects reported in Laux and Brelsford (1989) in that these subjects believe they are capable of controlling the hazards. One interesting demographic relationship was the finding that subjects in the *Fearless* group were significantly younger than were subjects in the *Fearful* group.

Young (1996) also demonstrated that these subject groups considered different kinds of information when forming risk perceptions. The *Fearful* group considered products to be risky if they had severe potential injury consequences, if an injury was likely, if the number of different risks associated with the product was high, and/or if the subject had been injured or had known someone who had been injured with the product in the past. The *Fearless* group considered products to be dangerous only if they had the potential to injure or kill many people at a time and if the product hazards were encountered involuntarily. When considering product hazards, the *Informed* group not only looked at the potential for catastrophe, but they also weighed information about the benefits provided by the product and the degree of control they exercised over the hazards. Thus, subjects not only perceived the products differently, but they accessed different information when forming perceptions of the risk associated with consumer products.

The results of this work demonstrated that at least some variance of risk perceptions could be attributed to how people perceive consumer products as a whole. The results suggest that information in warnings could be designed to suit the informational needs of the targeted audience. For instance, with the *Fearful* group, one could provide information about the nature of the hazard and the potential severity of injury associated with it. One way to accomplish this is to provide explicit information regarding injury consequences. Research has demonstrated that the explicitness with which the consequence information is expressed is an important determinant of perceived hazard and of recall of warning information (Laughery and Stanush, 1989; Sherer and Rogers, 1984). As expected, the more explicit the consequence information, the greater the perceived hazard and the more information recalled. This would hold true for the *Fearful* group of subjects. However, different information may be needed for the other groups. For the *Fearless* and *Informed* groups, information about the potential catastrophic nature of the hazard, about the extent to which exposure to the hazard is voluntary, and/or about the degree of personal control over the hazard may be necessary for these subjects to develop a proper appreciation of the risks.

39.6   Conclusions and Recommendations

In this chapter we have focused on characteristics of receivers that are important in the design of warnings. There are several principles or guidelines that appear warranted on the basis of the analyses presented.

Principle #1 — Know thy receiver. This statement may seem trivial and obvious; yet, as noted earlier, warnings are often designed with little or no regard for characteristics of the people to whom they are directed. Examples include warnings that require reading levels greater that the receiver's capability and that contain unfamiliar, technical terminology. Gathering knowledge and data about relevant characteristics of target audiences may require time, effort and money, but without such information, the warning designer and ultimately the receiver will be at a serious disadvantage. Analyzing existing data, such as demographic information, or collecting new data by conducting surveys may be necessary.

Principle #2 — When variability exists in the target audience, design for the low end of that audience. Whether the variability exists in competence, technical knowledge, familiarity, perception of hazardousness, or other receiver characteristics, it is important that warnings not be designed for the average. While it would be desirable to choose a criterion for warning designs that would include up to 99% of the population, there are several instances in which this may not be possible. For example, warning about such hazards as radon gas will necessarily involve information that may not be understood by all people. It is inappropriate to suggest that warning information should not be provided simply because the information may not be understood by 100% of the population. The point is to consider the variability in the target audience and to design the safety information so that it can be used by as many people in the target audience as is practical.

Principle #3 — When the target audience consists of subgroups that differ in relevant characteristics, consider employing a warning system that includes different components designed for the different subgroups. As in the prescription drug example provided at the beginning of the chapter, different types of information and different levels of detail are provided to different groups of receivers. This information is tailored to the needs and capabilities of these receivers.

A corollary to this principle is: do not try to accomplish too much with a single warning. Consider the current OSHA guidelines regarding the variety of subgroups in the target audience for material safety data sheets (MSDSs). These subgroups include toxicologists, safety engineers, managers, physicians, and end users (such as the laborer using the product). It is unlikely that one warning or pamphlet will be sufficient to meet the informational needs and capabilities of all these users. If the warning system does not include communications designed for the capabilities (both the strengths and weaknesses) of each group, it is probably destined to fail.

Principle #4 — Warnings should be tested using samples of potential receivers. Warning design guidelines (e.g., ANSI, 1991; FMC, 1985; Westinghouse Electric Corporation, 1985) can be used to develop candidate warnings for testing, thereby limiting the number of items that need to be tested. However, it is not always possible to use these guidelines to design a perfect warning system. The guidelines presented here can enable one to develop a preliminary warning. Testing of the warning system could assist the designer in refining and developing an effective system by providing information on ways to modify and improve the warnings. These tests might consist of "trying it out" on a target audience sample to assess comprehension and/or behavioral intentions. Our experience indicates that even such *minimal efforts* are seldom part of the warning design process, but would benefit the produced warning had they been taken.

Last, warnings should be viewed within the context of a communication *system* that includes the message, the medium, and the receiver. This chapter has sought to demonstrate that it is important to consider the capabilities and limitations of the receiver when designing the warning message. The most well-researched receiver characteristics were discussed in this chapter. Other receiver characteristics have been reported in the literature (e.g., locus of control, risk taking as a personality trait), but thus far they have received much less attention. The essential point of the chapter is that receiver characteristics should be considered in the warning design process in order to maximize effectiveness for their intended target audience.

References

ANSI (1991). *American National Standard for Product Safety Signs and Labels, ANSI Z535.4-1991*. Washington, D.C.: National Electrical Manufacturers Association.

Bettman, J.R. and Park, C.W. (1980). Effects of prior knowledge and experience and phase of the choice process on consumer decision processes: A protocol analysis. *Journal of Consumer Research, 7*, 234-248.

Collins, B.L. (1983). Evaluation of mine-safety symbols, in *Proceedings of the Human Factors Society 27th Annual Meeting* (pp. 947-949). Santa Monica, CA: The Human Factors Society.

Collins, B.L. and Lerner, N.D. (1982). Assessment of fire-safety symbols. *Human Factors, 24*, 75-84.

Desaulniers, D.R. (1989). Consumer product hazards: What will we think of next? In *Interface '89: The Sixth Symposium on Human Factors and Industrial Design in Consumer Products* (pp. 115-120). Santa Monica, CA: The Human Factors Society.

Desaulniers, D.R. (1991). *An Examination of Consequence Probability as a Determinant of Precautionary Intent*. Doctoral dissertation, Rice University, Houston, TX.

deTurck, M.A. and Goldhaber, G.M. (1991). A developmental analysis of warning signs: The case of familiarity and gender. *Journal of Products Liability, 13*, 65-78.

Donner, K.A. (1990). *The Effects of Warning Modality, Warning Formality, and Product on Safety Behavior*. Masters thesis, Rice University, Houston, TX.

Donner, K.A. and Brelsford, J.W. (1988). Cueing hazard information for consumer products, in *Proceedings of the Human Factors Society 32nd Annual Meeting* (pp. 532-535). Santa Monica, CA: The Human Factors Society.

Dorris, A.L. and Tabrizi, M.F. (1978). An empirical investigation of consumer perception of product safety. *Journal of Products Liability, 2*, 155-163.

Duffy, T.M. (1985). Chapter 6: Readability Formulas: What's the Use? In T.M. Duffy and R. Waller (Eds.), *Designing Usable Texts*. (pp. 113-140). Orlando: Academic Press, Inc.

Easterby, R.S. and Hakiel, S.R. (1981). The comprehension of pictorially presented messages. *Applied Ergonomics, 12*, 143-152.

Edwards, M.L. and Ellis, N.C. (1976). An evaluation of the Texas driver improvement training program. *Human Factors, 18*, 327-334.

FMC (1985). *Product Safety Sign and Label System*. Santa Clara, CA: FMC Corporation.

Friedmann, K. (1988). The effect of adding symbols to written warning labels on user behavior and recall. *Human Factors, 30*, 507-515.

Godfrey, S.S., Allender, L., Laughery, K.R. and Smith, V.L. (1983). Warning messages: Will the consumer bother to look? In *Proceedings of the Human Factors Society 27th Annual Meeting* (pp. 950-954). Santa Monica, CA: Human Factors Society.

Godfrey, S.S. and Laughery, K.R. (1984). The biasing effect of familiarity on consumer's awareness of hazard, in *Proceedings of the Human Factors Society 28th Annual Meeting* (pp. 483-486). Santa Monica, CA: Human Factors Society.

Goldhaber, G.M. and DeTurck, M.A. (1988a). Effects of consumers' familiarity with a product on attention to and compliance with warnings. *Journal of Products Liability, 11*, 29-37.

Goldhaber, G.M. and deTurck, M.A. (1988b). Effectiveness of warning signs: Gender and familiarity effects. *Journal of Products Liability*, 11, 271-284.

Green, P. and Pew, R.W. (1978). Evaluating pictographic symbols: An automotive application. *Human Factors*, 20, 103-114.

Johnson, D. (1992). A warning label for scaffold users, in *Proceedings of the Human Factors Society 36th Annual Meeting* (pp. 611-615). Santa Monica, CA: The Human Factors Society.

Johnson, E.J. and Russo, J.E. (1980). Product Familiarity and Learning New Information, in K.E. Monroe (Ed.) *Advances in Consumer Research*.

Karnes, E.W. and Leonard, S.D. (1986). Consumer product warnings: reception and understanding of warning information by final users, in W. Karwowski (Ed.) *Trends in Ergonomics/Human Factors III, Part B: Proceedings of the Annual International Industrial Ergonomics and Safety Conference* (pp. 995-1003).

LaRue, C. and Cohen, H. (1987). Factors influencing consumers' perceptions of warning: an examination of the differences between male and female consumers, in *Proceedings of the Human Factors Society 31st Annual Meeting* (pp. 610-614). Santa Monica, CA: The Human Factors Society.

Laughery, K.R. and Stanush, J.A. (1989). Effects of warning explicitness on product perceptions, in *Proceedings of the Human Factors Society 33rd Annual Meeting* (pp. 431-435). Santa Monica, CA: The Human Factors Society.

Laughery, K.R. and Young, S.L. (1991). An eye scan analysis of accessing product warning information, in *Proceedings of the Human Factors Society 35th Annual Meeting* (pp. 585-589). Santa Monica, CA: The Human Factors Society.

Laux, L.F. and Brelsford, J.W. (1989). Locus of control, risk perception and precautionary behavior, in *Interface 89: The Sixth Symposium on Human Factors and Industrial Design in Consumer Products* (pp. 121-124). Santa Monica, CA: Human Factors Society.

Laux, L.F., Mayer, D.L. and Thompson, D.B. (1989). Usefulness of symbols and pictorials to communicate hazard information, in *Interface 89: The Sixth Symposium on Human Factors and Industrial Design in Consumer Products* (pp. 79-83). Santa Monica, CA: Human Factors Society.

Lehto, M.R. (1991). A proposed conceptual model of human behavior and its implications for design of warnings. *Perceptual and Motor Skills*, 73, 595-611.

Lehto, M.R. and Miller, J.M. (1986). *Warnings, Volume 1*. Ann Arbor, MI: Fuller Technical Publications (p. 18).

Leonard, S.D., Hill, G.W. and Karnes, E.W. (1989). Risk perception and use of warnings, in *Proceedings of the Human Factors Society 33rd Annual Meeting* (pp. 550-554). Santa Monica, CA: The Human Factors Society.

Leonard, S.D., Matthews, D., and Karnes, E.W. (1986). How does the population interpret warning signals? In *Proceedings of the Human Factors Society 30th Annual Meeting* (pp. 116-120). Santa Monica, CA: Human Factors Society.

Leonard, D.C., Ponsi, K.A., Silver, N.C. and Wogalter, M.S. (1989). Pest-control products: Reading warnings and purchasing intentions, in *Proceedings of the Human Factors Society 33rd Annual Meeting* (pp. 436-440). Santa Monica, CA: The Human Factors Society.

Mazis, M., Morris, L.A., and Gordon, E. (1978). Patient attitudes about two forms of printed oral contraceptive information. *Medical Care*, 16, 1045-1054.

McGuire, W.J. (1980). The communication-persuasion model and health-risk labeling. (pp. 99-122), in L.A. Morris, M.B Mazis and I. Barofsky (Eds.) *Product Labeling and Health Risks: Banbury Report 6*, Cold Spring Harbor Laboratory.

Morris, L.A., Mazis, M.B. and Gordon, E. (1977). A survey of the effects of oral contraceptive patient information. *The Journal of the American Medical Association*, 238, 2504-2508.

Otsubo, S.M. (1988). A behavioral study of warning labels for consumer products: Perceived danger and use of pictographs, in *Proceedings of the Human Factors Society 32nd Annual Meeting* (pp. 536-540). Santa Monica, CA: The Human Factors Society.

Purswell, J.L., Schlegel, R.E. and Kejriwal, S.K. (1986). A prediction model for consumer behavior regarding product safety, in *Proceedings of the Human Factors Society 30th Annual Meeting* (pp. 1202-1205). Santa Monica, CA: The Human Factors Society.

Ringseis, E.L. and Caird, J.K. (1995). The comprehensibility and legibility of twenty pharmaceutical warning pictograms, in *Proceedings of the Human Factors & Ergonomics Society 39th Annual Meeting* (pp. 974-978). Santa Monica, CA: The Human Factors & Ergonomics Society.

Sherer, M. and Rogers, R.W. (1984). The role of vivid information in fear appeals and attitude change. *Journal of Research in Personality*, 18, 321-334.

Silver, N.C. and Braun, C.C. (1993). Perceived readability of warning labels with varied font sizes and styles. *Safety Science*, 16, 615-626.

Smith, D.B.D. and Watzke, J.R. (1990). Perception of safety hazards across the adult life-span, in *Proceedings of the Human Factors Society 34th Annual Meeting* (pp. 141-145). Santa Monica, CA: The Human Factors Society.

Viscusi, W.K., Magat, W.A. and Huber, J. (1986). Informational regulation of consumer health risks: an empirical evaluation of hazard warnings. *Rand Journal of Economics*, 17, 351-365.

Wagenaar, W.A. (1992). Risk taking and accident causation, in Yates, J.F. (Ed.) *Risk-Taking Behavior.* (pp. 257-281). New York: John Wiley & Sons.

Westinghouse Electric Corporation (1985). *Product Safety Label Handbook.* 2nd Edition. Trafford, PA: Westinghouse Printing Division.

Wogalter, M.S. and Silver, N.S. (1990). Arousal strength of signal words. *Forensic Reports*, 3, 407-420.

Wogalter, M.S. and Silver, N.C. (1995). Warning signal words: Connoted strength and understandability by children, elders, and non-native English speakers. *Ergonomics*, 38, 2188-2206.

Wogalter, M.S. and Young, S.L. (1994). The effect of alternative product-label design on warning compliance. *Applied Ergonomics*, 25, 53-57.

Wogalter, M.S., Desaulniers, D.R., and Brelsford, J.W. (1986). Perceptions of consumer product hazards: Implications for the need to warn, in *Proceedings of the Human Factors Society 30th Annual Meeting* (pp. 1197-1201). Santa Monica, CA: Human Factors Society.

Wogalter, M.S., Desaulniers, D.R. and Brelsford, J.W. (1987). Consumer products: How are the hazards perceived? In *Proceedings of the Human Factors Society 31st Annual Meeting* (pp. 615-619). Santa Monica, CA: The Human Factors Society.

Wogalter, M.S., Sojourner, R.J. and Brelsford, J.W. (1997). Comprehension and retention of safety pictorials. *Ergonomics*, 40, 531-542.

Wogalter, M.S., Brelsford, J.W., Desaulniers, D.R., and Laughery, K.R. (1991). Consumer product warnings: The role of hazard perception. *Journal of Safety Research*, 22, 71-82.

Wright, P., Creighton, P. and Threlfall, S.M. (1982). Some factors determining when instructions will be read. *Ergonomics*, 25, 225-237.

Young, S.L. (1991). Increasing the noticeability of warnings: effects of pictorial, color, signal icon and border, in *Proceedings of the Human Factors Society 35th Annual Meeting* (pp. 580-584). Santa Monica: Human Factors Society.

Young, S.L. (1996). Subject differences in the perception of risk, in *Proceedings of the Human Factors and Ergonomics Society 40th Annual Meeting* (pp. 503-507). Santa Monica, CA: The Human Factors and Ergonomics Society.

Young, S.L. and Wogalter, M.S. (1990). Comprehension and memory of instruction manual warnings; Conspicuous print and pictorial icons. *Human Factors*, 32, 637-649.

Young, S.L., Brelsford, J.W., and Wogalter, M.S. (1990). Judgments of hazard, risk and danger: Do they differ? In *Proceedings of the Human Factors Society 34th Annual Meeting* (pp. 503-507). Santa Monica, CA: Human Factors Society.

Young, S.L., Martin, E.G. and Wogalter, M.S. (1989). Gender differences in consumer product hazard perceptions, in *Interface '89: The Sixth Symposium on Human Factors and Industrial Design in Consumer Products* (pp. 73-78). Santa Monica, CA: The Human Factors Society.

40

Design of Industrial Warnings

David R. Clark
*GMI Engineering & Management
Institute*

Susan A. H. Benysh
Purdue University

Mark R. Lehto
Purdue University

40.1 Introduction

The design of an effective warning sign or label is a complex and difficult task. For the working professional in ergonomics, it can be perplexing to decipher the vast amount of research that has been, and is being, performed in the field, as well as the regulations and standards that affect the design and use of warning signs or labels. The purpose of this chapter is to provide industrial professionals with a succinct, but comprehensible, overview of the design of warning signs and labels for application in their own work environment.

This chapter is divided into four main sections: (1) the hierarchy of hazard control, where the role and place of warnings in a comprehensive program of hazard control is discussed, (2) generic guidelines for warning signs and labels, which covers the underlying human factors principles that guide the development of warnings, (3) effectiveness of warning signs and labels, which provides guidance on assessing whether or not the labels or signs are affecting behavior as desired, and (4) specific requirements and guidelines for warning signs and labels, where sources and provisions of existing design standards are outlined.

40.2 The Hierarchy of Hazard Control

Due to the increase of product liability litigation in the United States, especially those lawsuits citing "the failure to warn" as grounds for seeking damages, preventive criteria may be developed by the litigators of such cases that are not based on research. One example of this is the preventive criteria for companies to warn against ALL hazards with explicit warning signs and labels. The dangers of overwarning have been noted by several authors, including Lehto and Miller (1986). Overwarning might reduce the overall effectiveness of warnings, and thus increase product accidents. Too many warnings may cause habituation. Information overload is another possible problem. Therefore, warnings should be considered a supplement to, but not a substitute for, engineering of the product itself. The placement of warning signs and labels to products should be a final step in the attempt to control hazards associated with a product (Lehto and Salvendy, 1995). The engineering approaches to increase safety in the industrial environment related to product use that will be discussed in this section are product design (Norman, 1988; Rouse, 1992), job design (Campion and Medsker, 1992), personnel selection (Conoley and Kramer, 1989), training (Phillips, 1983), and supervision (Peterson, 1975; 1984).

Figure 40.1 (Lehto and Salvendy, 1995) illustrates the requirements for safe product design and use within workplaces in terms of the approaches discussed above. The most crucial stage for safe product design and use is product and process design. Essentially, this stage is fundamental for ensuring safe use of the product. During this stage, designers should be considering not only the intended use of the product, but also the many different ways in which the product might be misused. For example, an employee might use a chair as a ladder. Consideration of potential incorrect product-usage scenarios is vital to the design of a safer product. It is up to the employer to ensure that the employee uses the product for what it is designed to do, thus minimizing accidents associated with it.

Personnel selection is the next critical stage. This stage is important due to the individualistic nature of workers and the variety of skills and abilities they may possess. The selection process must ensure that employees have the prerequisite abilities to effectively and safely use a product. Training may be essential to ensure that the workers are qualified to perform tasks involving the product. Job design is also important since it determines exactly how, when, and for how long the operator will be using or exposed to the product. Through job design, the correct product will be chosen for the job, ensuring that misapplication does not occur. Forms of job performance aids, such as written instructions and operator manuals, perform a vital function at this stage.

Employee supervision is an essential method of ensuring that safe behavior is followed during employee training and job design. As well, supervision is a necessary deterrent when the employee can rationalize unsafe behavior or protocol, i.e., the behavior is an inconvenience or is unnecessary.

Warning signs and labels are the final resort, when all else fails, to produce reasonably safe behavior. The purpose of the warning is to alert operators to the presence of hazard, in hopes of avoiding unsafe or risky behavior. Warnings can enter the employee's field of attention in a variety of modes, verbal or nonverbal, through the auditory channel, the tactile channel, making use of the olfactory sense, or through the visual senses. Also, they can be presented through many different types of communications, including product packaging, the instruction manual, and the product itself.

From this discussion of accident prevention comes the question: At what stage is the most effective intervention performed? Lehto and Salvendy (1995) suggest starting by determining the error scenarios that involve the product. From there, determine intervention strategies for each stage mentioned in Figure 40.1. Brainstorming will suggest many options, and the preferred intervention strategy can then be determined and initiated. Generally, error scenarios are performed case by case, but intervention strategies for generic error categories can also be determined. This approach can be seen in Table 40.1.

Employee errors can be classified into three broad categories: errors that occur when performing routine behavior, when performing nonroutine behavior, and when the employee commits an error through an intentional violation (Reason, 1990). According to Rasmussen (1986), experienced workers performing routine tasks make errors when they fail to notice changes from ordinary conditions. When employees perform nonroutine actions, errors usually occur because of the user not knowing the proper procedure. Intentional violations occur when the user knowingly performs an unsafe act.

From the earlier discussion of Figure 40.1, product design is the preferred intervention strategy for any error scenario, especially with respect to errors occurring from routine behavior. One approach using product design to reduce potential routine behavior errors is to have the product provide warnings signals when the product is in a hazardous state (Lehto, 1991; 1992). An example of such a warning signal is when hazardous chemicals have a distinctive odor. This type of warning ensures that the hazard is immediately presented to the worker, so an effective action may be taken to possibly reduce the severity of the accident. Another approach is to design the products so that they do not provide affordances for, or trigger, inappropriate responses (Gibson, 1979). An example of this type of intervention would be to design an oven door out of glass to discourage users from standing on it. A final approach is to design

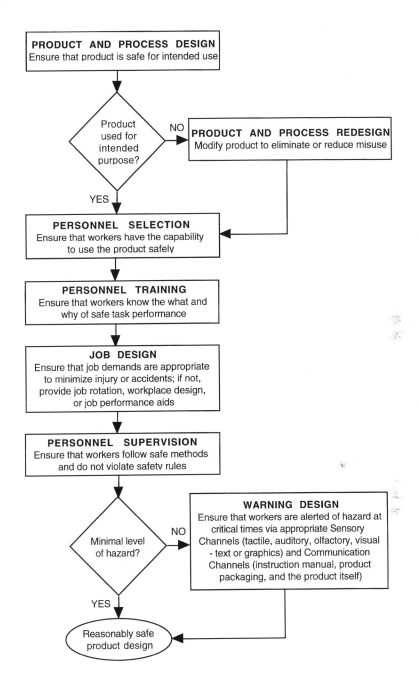

FIGURE 40.1 Intervention strategies and requirements for safe product design and use. (From Lehto, M.R. and Salvendy, G. 1995. Warnings: A supplement not a substitute for other approaches to safety. *Ergonomics*. 38 (11):2155-2163. With permission.)

the product to tolerate a minor range of errors, allowing for variability of work behavior. An example would be to have a knob click to certain values, instead of a free spinning dial which requires the worker to dial to an exact value.

Personnel selection is the next preferred strategy for error prevention. Again, this strategy seems to only be applied to errors associated with routine behavior. The only approach that these authors can foresee for the reduction of accidents is to ensure that the employees have adequate abilities to use the

TABLE 40.1   Intervention Strategies Based on Specific and Generic Error Scenarios

Error Scenario	Preferred Intervention Strategies	Objectives	Examples
ROUTINE BEHAVIOR Fail to perceive hazard condition or forget hazard condition	**PRODUCT DESIGN** hazard signals interruptive features **WARNINGS** interactive	Provide signal at time of hazard. Trigger shift away from routine. Combine warning text with product interruptive feature.	Bumps marking border of roadway. Barricade between street curb and exit of building. Taping the top drawer of a file cabinet shut with a warning label. Placing a sign on barricade.
	selective, nonvisual **TRAINING** on product signals	Provide signal at time of hazard. Develop necessary skill to interpret product hazard signals.	Auditory warning triggered when person enters hazard area. The odor of ethyl mercaptan indicates a gas leak.
Forget intended action	**JOB DESIGN** checklists	Confirmation of each action taken.	Flight start up and shut down checklist.
	PRODUCT DESIGN hazard signal	Response relevant at time of signal.	Vibrations from speed bumps.
	WARNINGS reminders	Stimulate error recovery.	Seat belt buzzer/chimes.
Activate incorrect script	**PRODUCT DESIGN** affordances and constraints	Eliminate design features that afford inappropriate responses. Deactivate product features.	Eliminate stair-like ladder features that encourage users to face away instead of toward rungs. Use of interlocks and lockouts.
Psycho-Motor variability	**PRODUCT DESIGN** hazard signals error tolerance **TRAINING** skill development **EMPLOYEE SELECTION**	Increase signal strength. Tolerates normal response range. Reduce response variability. Ensure adequate abilities.	Ethyl mercaptan (odor) added to natural gas. Adequate separation distance between controls. Driver training on simulator. Testing driver visual acuity.
NON-ROUTINE BEHAVIOR	**TRAINING** impart knowledge **JOB DESIGN** written procedures **WARNINGS** procedures	Teach safe procedures, safety rules, etc. Change task to routine following of instructions. Change task to routine following of instructions.	HAZCOM program in U.S. Step-by-step maintenance or repair procedures. Instruction manuals or step-by-step procedures on product labels.
INTENTIONAL VIOLATIONS	**SUPERVISION** enforcement information gathering **JOB DESIGN** compliance and noncompliance costs and benefits	Make sure safety rules followed. Identify extent of violations. Reduce the cost of compliance. Eliminate benefits of non-compliance.	Visible enforcement and supervision. Work sampling, employee monitoring, etc. Provide personal protective equipment (PPE) near point of hazard. Make PPE comfortable. Make sure PPE doesn't interfere with task.
	TRAINING impart knowledge	Teach about hazards, reasons for safety rules, etc.	HAZCOM program in U.S.

Adapted from Lehto, M.R. and Salvendy, G. 1995. Warnings: A supplement not a substitute for other approaches to safety. *Ergonomics*. 38 (11):2155-2163. With permission.

product. This would include selecting workers for both current abilities and their potential for learning new ones. The next stage for accident intervention is personnel training. Training can be employed for all three types of behavior — routine, nonroutine, and intentional violations — to reduce errors. Training can be used to ensure that the worker has the necessary skills to interpret hazard signals and understand the reasons for safety rules. Safe behavior can also be taught to the worker. Finally, training to reduce response variability by employees can lead to a minimization of accidents by the worker.

Designing the job to minimize errors is the next level of priority from Figure 40.1. As with employee training, this level can be applied to all three categories of behavior. For routine behavior, a set of checklists can be created to ensure the employee is following a set routine. A checklist will also provide for validation of a correctly completed task. For nonroutine behavior, written procedures will alter the task from a nonroutine task to a routine of following instructions. Finally, job design can be employed to reduce intentional violations by reducing the compliance costs and increasing the benefits associated with safe behavior. An example of this would be to provide personal protective equipment near the point of a hazard.

Supervision is useful when attempting to reduce errors that occur due to intentional violations. This reduction of errors is a result of enforcing safety rules. Supervision is also useful in reminding employees of the negative repercussions of improper behavior. The final step in accident intervention is to use warning signs or labels. Warnings are useful by interrupting the task, making certain that the employee is reminded to not perform incorrect behavior before it occurs (as found by Frantz and Rhoades, 1993). As well, warnings can stimulate recovery from errors. One example of this type of warning is to have a seatbelt buzzer reminding users to secure their seatbelts before driving.

Warnings, though, should be used with care, especially when considering the level of experience of the employees to which they are directed. Dorris and Purswell (1977) and others (Frantz and Rhoades, 1993; Otsubo, 1988) found that experienced users, who were familiar with a task and the product used to complete that task, often failed to read warning labels. One potential reason may be that the user is highly focused on task-related goals, only looking for information that will satisfy the end-goal of task completion. Interruptions may also take the user out of an automatic form of behavior and encourage him or her to enter a less routine information-seeking mode of behavior.

40.3   Generic Guidelines

From the research regarding the effectiveness of warning signs and labels, many generic guidelines can be inferred. These guidelines are: "(#1) conform to standards and population stereotypes when possible; (#2) make warning signs and labels conspicuous and each component legible; (#3) simplify the syntax of text and combinations of symbols; (#4) make symbols and text as concrete as possible; (#5) make sure that the cost of compliance is within a reasonable level; (#6) be selective; (#7) integrate the warning into the task and hazard related context; and (#8) match the warning to the level of performance at which critical errors occur for the relevant population" (Lehto, 1992, p. 108). These general guidelines are expanded upon and classified into three broad categories: format, content, and mode of presentation. Though much research has been performed on the format of warnings, content and mode of presentation are still, if not more, important to the overall effectiveness of warning signs and labels.

Format

The format of a warning sign or label refers to the physical aspects of the warning, which include text, iconic, and/or graphic information, both in individual and in collective form. Other format aspects of warning signs and labels are the size of the label, methods of color coding, and the arrangement of the warning components. The two generic guidelines that are of the greatest relevance to format specification are (#1) conform to standards and population stereotypes when possible; and (#2) make warning signs and labels conspicuous and each component legible. Usually, the normal variations in typography are acceptable, but there exist at least four circumstances (according to Sanders and McCormick, 1993) in which it may be preferred to have set guidelines for labels, one being when working with warning messages. There are twelve specific guidelines that have surfaced from the literature. These will be presented, with examples when appropriate, below.

Provide a signal word or words in capital letters at the top of a warning sign or label to indicate the sign or label is a warning. Examples of signal words are NOTICE, WARNING, CAUTION, and DANGER (ANSI Z535.2, 1996; ANSI Z535.4, 1996; and ANSI Z129.1, 1994). A study by Bresnahan and Bryk (1975) addressed the perception of the terms DANGER and CAUTION by industrial workers. They found that greater levels of hazard were associated with the term DANGER. Lirtzmann (1984) also found these results, but when he tested WARNING and CAUTION, there was found no significant difference between perceived danger levels. Hadden (1986) provides support for a new system of signal words, EXTREME-DANGER, SERIOUS-DANGER, and MODERATE-DANGER.

Use lowercase for text within a warning sign or label. Poulton (1967) found that lowercase is easier to read and faster to comprehend than when the text is ALL UPPERCASE. This seems to be the general consensus of the literature in this area (Sanders and McCormick, 1993).

For text within smaller labels, a visual angle of at least 10 minutes of arc for the intended viewing distance should be provided (Heglin, 1973; Woodson, 1963). For text within signs, a visual angle of a minimum of 25 minutes of arc should be present for the intended viewing distance (Heglin, 1973; Woodson, 1963). Other factors, such as luminance and the population, should be factored into determining the proper visual angle (Sanders and McCormick, 1993). The National Bureau of Standards (Howett, 1983) developed a formula for determining the size of alphanumeric characters to be read at various distances by persons with various Snellen acuity scores. The following formulas should be used:

$$W_S = 1.45 \times 10^{-5} \times S \times d \tag{1}$$

$$H_L = W_S/R \tag{2}$$

where W_s, d, and H_L are in the same units (inches or mm) and:

W_s = stroke width
S = denominator of Snellen acuity score (i.e., if Snellen acuity = 20/40, then S = 40)
d = reading distance
H_L = letter height
R = stroke width-to-height ratio of the font, expressed as a decimal proportion
     (i.e., for a ratio of 1:10, R = 1/10 = 0.10)

For short messages or signal words, text characters should be sans serif (ANSI Z35.1, 1972; SAE Recommended Practice J115a Safety Signs, 1979; Westinghouse, 1981; FMC Guidelines, 1980).

Test characters should be dark against a light background. This recommendation is based on research that shows that the contents of a label or sign need to be larger when they are presented as white on a dark or black background (Heglin, 1973). Also, irradiation may occur, which is the phenomenon in which white features on a black background tend to "spread" into adjacent dark areas (though the reverse is not true).

The stroke width-to-height ratio of text characters should be between 1:6 and 1:10 (Heglin, 1973). The stroke ratio can also be expressed as a proportion (i.e., 1:10 = 0.10).

The width-to-height ratio of alphanumeric characters should be between 1:1 and 1:3.5 (Heglin, 1973; Sanders and McCormick, 1993). Alphanumeric characters are a mix of both text and numeric characters. A ratio that is often used is 3:5. This ratio has its roots in the fact that most letters of the English alphabet have five elements in height and three elements in width. This can easily be seen in the numeric 8 (two vertical strokes plus the space in-between: three horizontal strokes plus two spaces between these).

Provide for a brightness contrast of at least 50% between the warning text and its background. The perceived brightness of a warning sign or label depends on the energy spectrum of the emitted or reflected light and the type of color that is used on the warning. One should not depend on color contrasts, since printing inks and fading may cause the contrast to lessen. The brightness of a particular stimulus can be described using photometric units. Data regarding the energy spectrum of various forms of lighting on different types of signs and labels is available in the *IES Lighting Handbook* (1981).

FIGURE 40.2 Example of a stroke height-to-width ratio.

Consider color coding schemes consistent with ANSI Z35.1 (1996). If color is used, consider the energy spectrum of foreseeable lighting sources when specifying the color mix. Avoid using the extremes of the color spectrum, such as reds and blues (Matthews, 1987), since some subjects with various types of color blindness may not be able to read these signs or labels.

Avoid crowding of components on the sign or label, assuring that each component is legible. Increased density of the sign or label (even when the necessary information is present) increases both reaction times and errors (Tullis, 1988). Avoid crowding by eliminating unnecessary information and use concise wording.

The presence of adverse viewing environments, such as the presence of dirt, grease, and other contaminants, can degrade the legibility of warning components. Compensation should take place to counteract these variables by testing the legibility under realistic conditions and replacing damaged signs and labels.

Consider the use of symbols/pictographs when users are performing routine behavior. Consider using text when employees need to make decisions or are performing non-routine behavior. FMC (1980) and Westinghouse (1981) encourage the use of symbols/pictographs. Conversely, the ANSI Z535 (1996) standards recommend using symbols as a supplement to words.

Content

The content of the warning sign or label refers to the warning message itself, its level of abstraction, and its syntactic structure. The generic guidelines applicable here are (#3) simplify the syntax of text and combinations of symbols, (#4) make sure that the cost of compliance is within a reasonable level, (#5) be selective, and (#8) match the warning to the level of performance at which critical errors occur for the relevant population. Some specific guidelines that can be derived from these generic guidelines and from the research follow.

Messages should focus on critical errors that cause a significant safety problem. When choosing warnings, one should be selective to avoid habituation and warning overload, and thus reduce the potential for ignoring the warning. Therefore, avoid long lists of hazards and messages that describe trivial hazards or hazards that are obvious to the intended audience. Such information is best provided in media other than warning signs or labels (Lehto and Miller, 1986; Scammon, 1977).

Focus on developing messages for the following two types of error situations: (1) forgetting to perform an action ordinarily performed (i.e., the sign or label reminds), and (2) not knowing the consequences of performing or failing to perform some action.

When a user's performance is skill-based, meaning they are performing an automatized set of procedures, and they commit an error based on their failure to perceive a condition or motor variability, provide a warning signal and consider training. An example of this type of error is when a person is walking in a dry area and does not change his or her gait before stepping on a wet spot. Rhoades et al. (1990) provide many case studies of such errors.

When performance is rule-based, meaning the user is following a set of rules and the behavior is not yet automatized, and the error is caused by a incorrect or inadequate rule, determine whether the rule was originally developed on the basis of knowledge or judgment-based behavior. If it seems to be judgment-based, like when people speed on the highway, focus on enforcement. If it seems to be knowledge-based, determine whether a warning sign or label can be used to interrupt the task (i.e., to place its message into short term memory at the time it is relevant). Frantz and Rhoades (1991) showed that interrupting a task increased the compliance in a task from 13% to 73%. Other studies have also concluded that interrupting a task increases warning sign or label effectiveness (Dingus et al., 1991).

When performance is knowledge-based, meaning the user is problem-solving, and the error is caused by inadequate knowledge, the amount of knowledge necessary to prevent the error should be determined. If the knowledge can be described with a small number of rules, consider a warning sign or label containing these rules in the form of step-by-step instructions. Otherwise, focus on training, instruction manuals, or other forms of education.

When performance is judgment-based, meaning the user is experiencing an affective reaction of some sort, and the error is caused by inappropriate priorities, evaluate the user's behavior pattern. If the undesired behavior pattern appears to have significant value to the user (i.e., pleasure, comfort, convenience, etc.) or is likely to be entrenched, focus on enforcement through supervision.

Regardless of the level of a user's performance, consider messages that minimize the cost and increase the benefits of compliance. Therefore, the behavior most desired by the user will be the safer and more effective behavior.

If a large number of potential warnings are present after applying these guidelines, increasing the probability of overloading the user, other means of providing the information (such as instruction manuals or training courses) should be considered to reduce warning overload.

Given that a message satisfies the preceding rules for content, further recommendations pertain to the *level of abstraction* and are based on the generic guidelines (#8) match the warning to the level of performance at which critical errors occur for the relevant population and (#4) make symbols and text as concrete as possible. Specific recommendations inferred from these guidelines are:

When subjects are inexperienced, consider pictographs (having a more detailed design) instead of symbols. The abstractness of the symbols have been found to cause a decrease in warning comprehension (Jacobs et al., 1975; Lerner and Collins, 1980).

When performance is at a skill-based or rule-based level, consider brief messages that describe conditions or actions. Also, consider symbols or pictographs instead of text, since these are found to make the warning more memorable (Young and Wogalter, 1988).

When performance is at a knowledge-based level, that is the user needs to understand "why" an action or behavior needs to be performed, consider more detailed messages that describe both conditions and actions.

When performance is at a judgment-based level, in that the incoming information is processed as values that evoke an effective reaction, consider messages that describe the hazard and the benefits of compliance. Also, consider citing highly credible sources. Craig and McCann (1981) found that compliance to warnings was higher when the message came from a government source versus a public utility company.

When the hazard is complex or occurs in different manifestations, consider abstract text, which better covers hazard contingencies, rather than a long list of concrete examples or symbols. It was found by Lehto and DeSalvo (1992) and Morris and Kanouse (1980) that these longer messages created problems for some subjects. Also, increasing the number of items on a label has been shown to decrease overall recall performance (Scammon, 1977; Reder and Anderson, 1982). It was found in numerous studies (Cahill, 1976; Johnson, 1980; Cairney and Sless, 1982) that symbols and pictographs were poorly comprehended when describing a complex hazard (as opposed to reading and understanding text). However, it was also found that concrete text may be better comprehended than abstract text (Lehto and DeSalvo, 1992; Laughery et al., 1991; Leonard and Matthews, 1986). An example of concrete text is "Using this product without the use of a mask will give you lung cancer." An example of abstract text is "Using this product incorrectly will increase your chances of illness."

When knowledge or understanding of a product or task is low, consider concrete text instead of abstract symbols or pictographs. Concrete text is easier to comprehend, and the interpretation of symbols has been found to vary across different cultures (Easterby and Zwaga, 1976; Sinaiko, 1975; Cairney and Sless, 1982).

Use text and symbols that people in the intended user population can comprehend. Consider language, reading level, and cultural effects.

The *syntactic structure* of warning signs and labels is addressed by generic guideline (#3) simplify the syntax of text and combinations of symbols. Specific recommendations derived from this guideline and research in the area are:

Use short simple sentences; complex conditional sentences, particularly those containing negations, should be avoided. Longer messages are not necessarily comprehended better (Morris and Kanouse, 1980; Lehto and DeSalvo, 1992). The latter study showed that action statements ALONE consistently received the highest comprehensibility ratings.

Symbolic signs or labels should focus on describing conditions (i.e., flammable). With few exceptions (i.e., a slash/bar to indicate negation) they should not combine multiple meanings or be used to describe complicated sequences of actions. This should lead to an increase in the comprehension of the warning (Young and Wogalter, 1988).

Mode of Presentation

The mode of presentation of a warning sign or label refers to the location and task-specific timing of contact with a warning sign or label. The following generic guidelines apply here: (#5) make sure that the cost of compliance is within a reasonable level, (#7) integrate the warning into the task and hazard related context, and (#8) match the warning to the level of performance at which critical errors occur for the relevant population. Specifically, the following guidelines should be followed:

The warning sign or label should be presented at a location and time in which the danger is still avoidable. It was shown by Wogalter et al. (1987) that the location of warnings in instructions can be a major determinant of effectiveness. Another example is the use of a seatbelt buzzer that alerts the driver of an automobile to secure his or her seatbelt when starting the automobile.

The location and timing of presentation should minimize the cost or difficulty of compliance (i.e., a sign to wear goggles should be close to an available source of goggles). Wogalter et al. (1989) found that subjects were more likely to use goggles and gloves in a chemistry lab task when they were provided with the equipment. Several other studies have shown that the likelihood of following a warning is influenced by the perceived cost of compliance with respect to cost to the worker, effort, time, comfort, productivity, and more (Godfrey et al., 1985).

Make an attempt to present the warning sign or label at a time when the person has available attentional capacity. Several studies have shown that people have trouble remembering warnings shortly after completing a task (Wright, 1979; Strawbridge, 1986; Ursic, 1984).

When performance is skill-based (Rasmussen, 1986), determine if the task can be interrupted to bring attention to the label. If this can be done, consider a warning sign or label which describes the condition and prescribes an action. If this cannot be done, consider providing a warning signal, training, or modifications of the product.

Avoid embedding the sign or label in a cluttered background. Several studies have shown the adverse effects of visual clutter on the perception of signs (Holahan, 1977; Boersema and Zwaga, 1985). Holahan was able to show that traffic accidents increased at a stop sign where the presence of commercial signs increased.

40.4 Evaluating Effectiveness

To measure the effectiveness of a warning, one must first understand the steps that must occur before a sign or label can prevent an accident. These include: (1) attending to the warning, (2) comprehending the warning, (3) deciding to perform the proper action to avoid an accident, and (4) taking the appropriate behavior. Finally, (5) the action decided upon must be sufficient to avoid the accident. Since all the steps must be taken in order to avoid an accident, the probability of achieving correct behavior, or avoiding an accident, can never be greater than the probability of successfully completing a single step.

A simple accident illustrates this argument. Consider the hypothetical situation in which (1) 50% of the population will read the warning, (2) 80% of those individuals understand the warning once they have read it, (3) 95% of those individuals decide to perform the proper action to avoid an accident, (4) 100% of those individuals take the appropriate behavior, and (5) that action is sufficient to avoid the accident 75% of the time. When the probability of successfully completing the entire sequence of events is defined as the effectiveness of the warning, then effectiveness is the multiplication of the conditional probabilities of completing each step. This example results in an effectiveness of an 0.285, or the warning will be 28.5% effective for the population of users.

Recent research has confirmed the basis of this model, that the end result of compliance cannot be greater than any probability of completing a single step. Otsubo (1988) observed that warning labels created to encourage workers to wear gloves when operating a circular saw were complied with by 38% of the workers, though 74% of the subjects noticed the warning and 52% read the warning. The same results were found when Otsubo performed the same experiment with a jigsaw. It was found that 54% of the subjects noticed the warning, 25% read it, and only 13% complied with the warning to wear gloves.

There are many dependent variables related to the effectiveness of warnings. A majority of these dependent variables can be grouped into three areas, perceptual factors, comprehension levels, and those factors associated with behavior patterns. The evaluation of the perception of warnings should give insight as to the conspicuity, or attention-attracting ability of the sign. There are numerous variables related to the perception of warning signs or labels. Some of these measures include reaction time, accuracy of task completion, attention to different elements of the warning (accomplished through the study of eye movement), the use of tachistoscopic procedures, and measures of legibility distance. Table 40.2 gives the approach taken for each procedure and their respective advantages and disadvantages.

When evaluating the comprehensibility of a warning, four variables are often used: symbol recognition or matching, message recall, psychometric (rating scales), and readability indexes. Symbol recognition is the most commonly applied technique used to measure warning sign and label comprehension. It is usually applied via open-ended questions, in which the user is asked to describe the symbol. When symbol matching is used, the user is asked to match the symbol to possible meanings. Recognition gives a more accurate understanding of comprehension, though matching allows for more quantifiable and, thus, analytical results. Message recall is an approach in which subjects are asked to remember the message after a task has been completed. Obviously, the user's memory may introduce error, but this measure provides the best indication as to how the contextual material is related to perception of the warning. Psychometric scales are often used when requesting users to rate their perceptions of different aspects of the warning. From this rating, comprehension can be inferred using statistical approaches such as factor analysis or cluster analysis. A lot of pre-experiment work should be performed to ensure that the scales give the information that is desired by the experimenter. Finally, readability indexes are applied to the warning signs and labels to describe the difficulty of written material in terms of word length, sentence length, or other variables. Many indexes have been developed, but due to the nature of most warnings, being in terse fragments vs. prose, readability indexes have little value.

Evaluating behavior patterns also offers some insight as to the effectiveness of warning signs and labels. When evaluating a subject's behavior, this allows the warning to be evaluated under realistic conditions as to how it affects overall behavior. Since the ultimate objective of a warning sign or label is to change behavior, this technique will offer insight into the realization of this goal. The most common approach is to set up an experiment in which two groups of subjects are tested on the same task. One group is the control without the warning, and the other group receives the warning. The prevalence of safe behavior can then be compared between the groups. Field observations are also useful in determining warning effectiveness. This will allow even more realistic conditions, and should reduce potential error introduced due to the unrealistic situation of a controlled experiment.

TABLE 40.2 The Evaluation of Perceptual Factors for Warning Signs and Labels

Procedure	Approach	Advantages	Disadvantages
1. Reaction time	Measures the reaction time to different warnings. Assumes that a symbol quickly reacted to is more salient than one reacted to more slowly	Useful when time to react is needed for the task	Reaction time not always related to sign or label quality
2. Accuracy of product use	Conversely, errors can be measured to determine possible perceptual problems associated with the sign or label. A confusion matrix can be created to establish the correct and incorrect responses	Shows areas that need to be investigated further to decrease errors An absolute measure is obtained and can be used for comparison to other warning designs	May not give insight into the area that has caused accuracy to decrease
3. Attention to items within the warning	Usually, measuring attention is achieved through the measurement of the placement and time of eye movements. This gives insight into the elements that capture a user's attention	This allows the specific aspects of the warning that is causing trouble or decreasing accuracy to be identified	The amount of data produced by this method makes it difficult to analyze Equipment used to gather this data can be quite intrusive and not allow for an accurate setting for the study
4. Tachistoscopic procedures	A sign or label is presented to the subject with a tachistoscope for precisely timed intervals. After viewing the sign or label, the subjects are asked questions regarding it's content	Allows for the amount of time that is needed to perceive a warning to be correctly measured Gives an accurate measure of the attention needed to perceive a warning	This measure not based on a realistic setting and therefore, may confound errors The results of this study rely on the user's memory, which may introduce error
5. Measures of legibility distance	Determines the distance in which a warning can be perceived and read	Gives an absolute values as to the distance required to read a warning	Questionable measure of symbol effectiveness when quick perception is not important Studies have found this measure to not be related to measures of comprehension or even other measures of legibility

There are four methods commonly used to measure the dependent variables needed to evaluate the effectiveness of warning signs and labels. These methods include interviews, questionnaires, behavioral observations, and ratings of warning effectiveness. Each method has advantages and disadvantages. Therefore, the best method will depend upon the objectives of the study.

Interviewing is a good method for understanding the reasoning behind user actions. The interview is normally performed some time after the task is performed. Therefore, the basic errors that can occur in knowledge elicitation will be compounded with memory errors. A structured interview begins by the interviewer having a transcript of questions to ask the subject. These questions are general, such as "What did you do to reach this goal?" and "What did you do next?" The interviewer will then ask these questions over and over again until all relevant information is extracted from the subject. Obviously, this process can take a very long time. The main concern with applying this technique is that the subject can only produce what can be verbalized, so automatic actions are not given as much detail as the processes that have non-proceduralized actions (Wilson and Corlett, 1990). Also, subjects tend to justify their decision when in an interview situation, creating false confidence (Wilson and Corlett, 1990).

The advantages of questionnaires and rating scales are often the same as for an interview, but the amount of knowledge obtained from the questionnaire and rating scales will be limited to the amount of forethought by the researcher. Since a questionnaire can contain either open- or closed-ended questions, one must decide on this as well. Closed-ended questions (such as "On a scale of 1 to 10, with 10 being the highest, how safe do you view this product?") are easy to analyze, but may not have the construct validity that is needed. Open-ended questionnaires can be difficult to analyze and can be open to interpretation error by the researcher. Also, while some subjects will explain their reasoning, others will give very short and, quite often, uninformative answers.

Behavioral observations can be used to describe the use of the product. In some cases, interpretation on the part of the researcher may need to be employed, possibly introducing error. Also, it may not be apparent from the user's actions what mental actions were taken to perform a consequential action. Therefore, sometimes a verbal protocol is performed along with the observation. A verbal protocol is a knowledge elicitation technique in which subjects are requested to "think aloud" as they perform a task (Shadbolt and Burton, 1990; Belkin et al., 1987; Kuipers and Kassirer, 1983). During this time, the analyst records the process (usually on video). After the task is completed, the video is reviewed by the analyst and the subject, with the subject giving details about how the task was performed or why certain actions were taken. Also, a review of the video by experts could offer insights regarding certain focal areas that the subject did not, or could not, elaborate upon. The first step to performing a verbal protocol involves some preliminary work, so that the analyst has basic knowledge of the task for best time utilization (Wilson and Corlett, 1990). Also, a few training sessions in "thinking aloud" should be held so as to avoid embarrassment to the subject. The main problem that may occur due to verbal protocol technique is that the verbalization may interfere with the subject's performance (Wilson and Corlett, 1990). But, with enough training sessions, that problem should be minimized.

The foregoing provides a basis for the use of warnings within the hierarchy of hazard control, and outlines general guidelines for the construction and evaluation of warning signs and labels based on research findings and principles of human factors. As a practical matter, the designer of warnings must also consider an array of government and consensus standards that can either (1) provide explicit guidelines on how to accomplish that which has been generically discussed above, or (2) require compromise between incompatible generic guidelines and explicit requirements (and not infrequently, between different explicit requirements). The remainder of this chapter identifies sources of some of the most important or commonly used standards and illustrates some of the provisions of those standards.

40.5   Government and Consensus Standards

Consider the typical industrial signs shown in Figure 40.3. Are these warning signs or are they not? Are they instructions instead? Are they combinations of both? A warning may be defined as something that (1) gives or serves to give (a) notice beforehand, especially of danger or evil, or (b) admonishing advice, or (2) calls or serves to call one's attention; an instruction is a direction calling for compliance (Webster 1988). Consequently, warnings do not need to include instructions; or conversely, instructions do not constitute a warning. On the other hand, certain guidelines for warning signs and labels recommend that provisions be made for information on how to avoid the hazard, that is, instructions. Depending on the nature of the hazard and the target audiences of the warning, information of differing types may be warranted. Therefore, in the following review of specific warning standards, it is not possible to totally separate the giving of instructions in the construction of warnings. Perhaps it would be better to think of the combination of elements as hazard communication. Indeed, the current American National Standards Institute warning sign and label standards (ANSI Z535.1 *et seq.*, 1996) refer to themselves as a Hazard Communication System.

FIGURE 40.3   Warnings or instructions?

Sources of Standards

It is natural to divide warnings standards into two categories: mandatory requirements and voluntary guidelines. Each is discussed below.

Mandatory Requirements

Sources of mandatory design guidelines include government regulatory agencies at all levels. The primary source of interest to industry is the U.S. Department of Labor's Occupational Safety and Health Administration, although other important sources include the U.S. Department of Transportation, the Environmental Protection Agency, and even the Consumer Product Safety Commission. Many states have analogs of these agencies whose regulations supplement or supersede the federal guidelines, although they must be at least as rigorous as the federal versions.

In many cases, the information available in these regulations to the designer or specifier of an industrial warning is generally not of a systems nature. This means that certain components of warnings designed to address certain specific hazards are given, but guidelines on how to address new or unanticipated hazards are not.

Occupational Safety and Health Administration (OSHA). The primary OSHA regulations on warnings are found in 29 CFR 1910.144 (color coding) and .145 (accident prevention signs and tags), and in 29 CFR 1910.1200 (hazard communication standard, or "right to know"); these will be covered in more detail later in this chapter. Other provisions address specific hazards such as egress (29 CFR 1910.36 *et seq.*), radiation (.96 *et seq.*), and flammables or explosives (.103 *et seq.*), while others focus on processes/industries, such as sawmills (.265) or telecommunications (.268). Similar provisions are found in OSHA's construction (29 CFR 1926) and maritime standards (29 CFR 1915 *et seq.*).

Department of Transportation (DOT). The DOT has extensive labeling, marking, and placarding regulations that cover materials determined to be hazardous when transported. Inasmuch as these regulations provide cradle-to-grave coverage, they are applicable to all parts of industry, whether in the preparation, transportation, or use of these materials.

Environmental Protection Agency (EPA). The hazard labeling authority of the EPA arises out of the Federal Insecticide, Fungicide and Rodenticide Act of 1978 (FIFRA), the Toxic Substances Control Act of 1976 (TSCA), and the Resource Conservation and Recovery Act of 1976 (RCRA). FIFRA requires that pesticide labels include information on product identification, ingredients, warnings, and use.

Consumer Product Safety Commission (CPSC). Although not directly applicable to the industrial workplace, many products that find their way into the workplace would be considered consumer products under other circumstances. Indeed, the general availability of Material Safety Data Sheets (MSDS) through the OSHA Right-to-Know Standard has been extended to consumer products. Therefore, the role of the Consumer Product Safety Commission should be appreciated, although no specific CPSC regulations will be treated herein.

The Courts. Also bear in mind that the courts have influenced the role and design of warnings, although the decisions typically give very little in the way of engineering design criteria.

Voluntary Guidelines

Several voluntary systems-based warning sign and label guidelines are provided by private sector standards-making organizations. These systems promote consistency in design and allow designers to leverage the considerable experience of other organizations in the development of warning signs and labels.

Consensus Organizations. These organizations develop voluntary requirements that represent a consensus, or agreement, among its membership and the affected industries or public. This process is somewhat similar to that which the federal government follows under the rules of the Administrative Procedures Act. Such organizations include: (1) international groups, such as the United Nations, the European Community, the International Standardization Organization (ISO), and the International Electrotechnical Commission (IEC), (2) national organizations, such as the American National Standards Institute (ANSI), the British Standards Institute, the Canadian Standards Association, the German Institute for Normalization (DIN), and the Japanese Industrial Standards Committee, and (3) independent agencies which often submit their standards to ANSI for approval as consensus standards, such the National Fire Protection Association (NFPA). Table 40.3 identifies examples of various U.S. consensus

TABLE 40.3 Examples of Warnings or Warnings-Related U.S. Consensus Standards

ANSI A13.1	Piping
ANSI C95.2	Radio Frequency Radiation
ANSI D6.1, 6.1b, 10.1	Traffic Control
ANSI N12.1	Fissile Material
ANSI N2.1	Radiation
ANSI Z129.1	Industrial Chemicals
ANSI Z138.2	Color
ANSI Z241.1	Sand Foundry Industry
ANSI Z244.1	Lock Out/Tag Out
ANSI Z35.1, 535.2	Accident Prevention Signs
ANSI Z35.2, 535.5	Accident Prevention Tags
ANSI Z35.4	Informational Signs
ANSI Z35.5	Biological Hazard
ANSI Z53.1, 535.1	Color Code
ANSI Z535.2	Environmental and Facility
ANSI Z535.3	Symbols
ANSI Z535.4	Consumer Product
ANSI/ASAE S338.1	Towed Equipment
ANSI/ASTM D1535	Color
ANSI/ASTM D56, 93	Flash Point
ANSI/ISA S5.5	Process Display Symbols
ANSI/MH11.3	Powered Industrial Trucks
ANSI/NEMA ICS1, 6	Industrial Control & Systems
ANSI/NFPA 30	Flammable and Combustible Liquids
ANSI/NFPA 70	Electrical
ANSI/NFPA 101	Safety to Life from Fire
ANSI/NFPA 178	Fire Fighting Operations
ANSI/NFPA 1901	Automotive Fire Apparatus
ANSI/SAE J1116	Off-Road Work Machines
ANSI/SAE J115	Safety Signs
ANSI/SAE J1164	ROPS and FOPS
ANSI/SAE J137	Agricultural Equipment on Highways
ANSI/SAE J1500	Operator Controls
ANSI/SAE J208, 389, 841, 1170	Agricultural Equipment
ANSI/SAE J284	Agricultural, Construction, and Industrial Equipment
ANSI/SAE J298	Industrial Equipment
ANSI/SAE J575	Motor Vehicle Lighting
ANSI/SAE J594	Reflex Reflectors
ANSI/SAE J674	Motor Vehicle Glazing
ANSI/SAE J725, 943	Slow-Moving Vehicles
ANSI/SAE J99	Industrial Equipment on Highways
ASAE S441	Safety Signs
ASTM C1023, ES6	Ceramic Art Material
ASTM D1014	Paints on Steel
ASTM D1729, 2244, E308	Color
ASTM D1788	ABS Plastic
ASTM D2794	Organic Coatings
ASTM D3278	Flash Point
ASTM D4086	Metamerism
ASTM D4257	Coatings & Lining Industry
ASTM D4267	Parenteral Drug Containers
ASTM E239	Paint, Varnish, Lacquer, and Related Products
ASTM E42, 188, 822, G23, 26	Nonmetallic Material
ASTM E991	Fluorescent Color
ASTM ES9, F926	Kerosine Containers
ASTM F406	Play Yards
ASTM F839	Gasoline Containers
EIA RS257	Mercury
NEMA 260	Switchgear & Transformers

TABLE 40.3 Examples of Warnings or Warnings-Related U.S. Consensus Standards (continued)

NEMA EW6	Arc-Welding and Cutting
NEMA IB1	Lead-Acid Batteries
NFPA 291	Fire Hydrants
NFPA 704	Fire Hazards
SAE J1048	Motor Vehicles
SAE J107	Motorcycles
SAE J1150	Agricultural Equipment
SAE J179	Truck Wheel Rims
TAPPI UM586	Label & Tape Aging Testing

standards. Two sources of particular interest to U.S. industry are the ANSI Z535 Committee on Safety Signs and Colors and the ANSI Z129 Committee on Labeling of Hazardous Industrial Chemicals.

The ANSI Z535 Committee on Safety Signs and Colors committee was formed in 1979 by combining the previous Z35 Committee on Safety Signs with the Z53 Committee on Safety Colors. The Z35.1 (Specification for Accident Prevention Signs) and Z53.1 (Safety Color Code for Marking Physical Hazards) standards (along with Z129.1) existing at that time were the primary standards for industrial warning systems. The new Z535 Committee was "to develop standards for the design, application, and use of signs, colors, and symbols intended to identify and warn against specific hazards and or other accident prevention purposes" (ANSI Z535.1). In 1991, a series of five standards were published by the Z535 co-secretariat National Electrical Manufacturers Association after being approved by ANSI. These addressed color coding (Z535.1, which replaced Z53.1), environmental and facility safety signs (Z535.2, which replaced Z35.1 and Z35.4), safety symbols (Z535.3), product safety signs and labels (Z535.4), and accident prevention tags for temporary hazards (Z535.5, which replaced Z35.2). A 1996 version of each of these standards is now in the approval process. Note, however, that Z35.1-1968 and Z53.1-1967 were used in 1972 as the basis for the current OSHA regulations on accident prevention signs, and most sign manufacturers still provide Z35.1/Z53.1-compliant signs. New design efforts should normally conform to the Z535 series of standards.

Notable among the exceptions in the scope of the ANSI Z535 standards is that of chemical products and mixtures. These are addressed by the ANSI Z129 Committee on Labeling of Hazardous Industrial Chemicals and the provisions of a standard that was first constituted in 1946 as an industry guide under the Manufacturing Chemists Association and later approved as ANSI Z129.1 in 1976, now sponsored by the Chemical Manufacturers Association. The current standard was approved in 1994.

Professional Organizations. This includes such organizations as the Society of Automotive Engineers (SAE), the American Society of Mechanical Engineers (ASME), the National Safety Council (NSC), and the American Society of Safety Engineers (ASSE).

Industrial Trade Associations. Typically based on a product or service focus, these organizations include such organizations as the National Electrical Manufacturers Association (NEMA), the Chemical Manufacturers Association (CMA), and the Material Handling Institute, (MHI). They are often the initial source of standards approved by consensus organizations.

Companies. Individual companies often develop internal systems for warning design, some quite extensive. And these efforts are sometimes made available outside the company. Two notable examples of this type are the *Product Safety Sign and Label System* developed by the FMC Corporation (1980) and the *Product Safety Label Handbook* developed by the Westinghouse Electric Corporation (1981). Elements of the FMC system were the basis of the current ANSI Z535 labeling standards.

Commercial. There are a number of "sign" companies that provide warning signs and labels that are marketed as conforming to various and specific standards set by many of the organizations mentioned above.

Other Sources. There are a number of reference books, handbooks, and textbooks that cover warning issues to varying degrees. Most books on safety, human factors, and ergonomics include requisite sections on warnings as a hazard control measure, and often discuss general issues about design and effectiveness.

A few books focus on warnings, albeit from a theoretical or research perspective, including Miller, Lehto, and Frantz (1994), Lehto and Miller (1986), and Edworthy and Adams (1996), and offer little in terms of specific design guidelines. Then there are examples of books designed to address the needs of specific users, such as the chemical industry (O'Connor and Lirtzman, 1984).

Provisions Within Selected Standards

According to ANSI Z535.2 (1996), a safety sign is "a visual alerting device in the form of a sign, label, decal, placard, or other marking which advises the observer of the nature and degree of the potential hazard(s) which can cause injury or death…." The key here is that a sign (or label) is intended to be a permanently mounted device. The difference is normally in terms of where the device is placed. If it is on the hazard-producing or -containing object, especially as manufactured or distributed, it is usually referred to as a label. On the other hand, if it is mounted not on the object but nearby or in the general environment, or refers to hazards not specifically "attached" to an object, it is usually referred to as a sign. Temporary devices (or tags) used in either place are also used in industry for similar purposes and are addressed by their own standard (i.e., Z535.5). There are a fairly large number of existing warning sign and label standard "systems," not all of which are compatible. These differ in the choice of words, symbols, colors, and layout (see Table 40.4). The following is based on a selection of the most important of these standard systems. The reader should note that, due to space limitations, not all of the provisions of these standards can be included here. Further, standards are time-sensitive entities. While every effort has been made to use the latest information, it is a certainty that parts of these standards will change over time. Consequently, one should always consider obtaining the latest version of the standard(s) in their full text.

ANSI Z535.2-1996 Environmental and Facility Safety Signs

ANSI Z535.2 establishes requirements for a uniform system of visual identification of potential hazards in the environment, such as industrial facilities. It specifically excludes applicability to product labels (see ANSI Z535.4) and chemicals (see ANSI Z129.1). As ANSI Z535.2 signs include the use of safety colors and symbols, it incorporates by reference the Safety Color Code (ANSI Z535.1) and Criteria for Safety Symbols (ANSI Z535.3). As noted in Table 40.5, there are seven types of safety signs in ANSI Z535.2: (1) danger, (2) warning, (3) caution, (4) notice, (5) general safety, (6) fire, and (7) directional. The selection of the appropriate type of sign for a given hazard and the design of its panel are also illustrated in Table 40.5, and further information on the colors used is shown in Table 40.6. A typical ANSI Z535.2 sign consists of multiple panels for signal words, messages, and symbols (see Table 40.7). Finally, the use of symbols is encouraged, provided that they are understood. ANSI Z535.3 describes methodology for the development of such symbols. In general, symbols must be correctly identifiable 85% of the time, with no more than 5% critical confusions (i.e., results in an action that is opposite of the desired response). ANSI Z535.3 provides 22 symbols that meet this criteria (see Table 40.8) along with 6 appropriate surround shapes (see Table 40.9).

OSHA 29 CFR 1910.145 Specification for Accident Prevention Signs and Tags, and 29 CFR 1910.144 Safety Color Code for Marking Physical Hazards

For all practical purposes, one can follow either the old ANSI Z35.1/Z53.1 standards or the new ANSI Z535 standards to meet the requirements of OSHA for hazard warning signs. Consequently, there is no need to replace existing compliant signs. Also, specific reference is made within the regulations to conforming with the ANSI Z53.1 color code. However, conformance with the new ANSI Z535.1 color code will result in the same design. Therefore, when designing or specifying new signs, follow the new ANSI standards.

ANSI Z535.5–1996 Accident Prevention Tags

These tags are intended to be used to identify temporary hazards. Therefore, they are to be used only until such time as the hazard is eliminated. They should never be used as a permanent substitute for

```
┌─────────────────────────────────────────────────┐
│           PRODUCT NAME OR IDENTIFICATION          │
│   (Identity of Hazardous Component(s), Where      │
│                  Appropriate)                     │
│                                                   │
│                  SIGNAL WORD                      │
│            IMMEDIATE HAZARD(S) STATEMENT          │
│          DELAYED HAZARD(S) LABEL STATEMENT        │
│                                                   │
│               Precautionary Measures             │
│                                                   │
│       Instructions in Case of Contact or Exposure│
│    (First Aid Statements and Antidotes Where      │
│                  Appropriate)                     │
│                                                   │
│                  Fire Instructions                │
│                                                   │
│              Spill or Leak Instructions           │
│                                                   │
│      Container Handling and Storage Instructions  │
│                                                   │
│                   References                      │
│                                                   │
│             Additional Useful Statements          │
│                                                   │
│              Name and Address of Company          │
│                                                   │
│                 Telephone Number                  │
└─────────────────────────────────────────────────┘
```

FIGURE 40.4 ANSI Z129.1-1994 general example of label format.

ANSI Z535.2 signs. The format follows that of the two-panel designs of ANSI Z535.2 (see Table 40.7) with the lower panel reserved for word messages and/or symbols.

ANSI Z129.1-1994, Hazardous Industrial Chemicals – Precautionary Labeling

This standard is applicable to chemical products and mixtures and is complementary in scope to ANSI Z535.2 in the environmental or facility area, and to ANSI Z535.4 in the product area. It is directed toward manufacturers, distributors, and employers who wish to alert persons to hazards inherent with chemicals. The required contents of an ANSI Z129.1 label is shown in Table 40.10, with an example arrangement shown in Figure 40.4.

OSHA 29 CFR 1910.1200 Hazard Communication Standard

This section of the OSHA General Industry Standards, also known as the Right-to-Know Standard, requires employers to provide information to employees about the hazardous chemicals they are or may be exposed to during normal conditions, or during any foreseeable emergency. This is to be done through hazard communication including on-container labels and Material Safety Data Sheets (MSDS). The information necessary for such communication must be provided by the chemical manufacturers or importers through their distributors.

Labels — Manufacturers' Responsibility. Manufacturers' container labels (or tags, other markings) must include: (a) the identity of the chemical, (b) appropriate hazard warnings, and (c) the name and address of the manufacturer, importer, etc.

Labels — Employers' Responsibility. Employers must ensure that all labels on incoming containers of hazardous chemicals are maintained. Further, it is the employer's responsibility that all containers in the workplace are labeled, tagged, or marked with: (a) the identity of the chemical, and (b) appropriate hazard warnings. As an alternative, and provided that the same information is provided and is readily accessible to all employees, the employer may use signs, placards, or other written materials. Where material is being transferred from labeled containers to portable containers, and for immediate use by the same employee who performed the transfer, labeling is not required.

TABLE 40.4　Summary of Recommendations Within Selected Warning Systems

System	Signal Words	Color Coding	Typography	Symbols	Arrangement	Hazard ID
NEMA Guidelines: NEMA 260	Danger Warning	Red Red	Not specified	Electric shock symbol	Defines signal word, hazard, consequences, instructions, symbol. Does not specify order.	Not specified
SAE J115 Safety Signs	Danger Warning Caution	Red Yellow Yellow	Sans serif typeface, upper case	Layout to accommodate symbols; specific symbols/pictographs not prescribed	Defines 3 areas: signal word panel, pictorial panel, message panel. Arrange in order of general to specific.	Provides guidance
ISO R557, 3864	None. 3 kinds of labels: Stop/prohibition Mandatory action Warning	Red Blue Yellow	Message panel is added below if necessary	Symbols and pictographs	Pictograph or symbol is placed inside appropriate shape with message panel below if necessary	Not specified
OSHA 1910.145 Specification for Accident Prevention Signs and Tags OSHA 1926.200 Accident Prevention Signs and Tags	Danger Warning (tags) Caution Biological Hazard, BIOHAZARD, or symbol (safety instruction) (slow-moving Vehicle)	Red Yellow Yellow Fluorescent orange/orange-red Green Fluorescent yellow-orange & dark red per OSHA 1910.144 and ANSI Z53.1	Readable at 5 feet or as required by task	Biological hazard symbol. Major message can be supplied by pictograph (tags only). Slow-Moving Vehicle (SAE J943)	Signal word and major message (tags only)	Provides guidance
OSHA 1910.1200 Hazard Communication Standard			In English		Only as Material Safety Data Sheet	Provides guidance
		Per applicable requirements of EPA, FDA, BATF, and CPSC; not otherwise specified.				
Westinghouse Handbook; FMC Guidelines	Danger Warning Caution Notice	Red Orange Yellow Blue	Helvetica bold and regular weights, upper/lower case	Symbols and pictographs	Recommends 5 components: signal word, symbol/pictograph, result of ignoring warning, avoiding hazard	Provides guidance about how to select signal words

Standard	Signal Words	Colors	Typeface/Lettering	Symbols	Format/Arrangement	Selection Guidance
ANSI Z35.1 Specifications for Accident Prevention Signs	Danger Caution	Red Yellow	Sans serif typeface. All upper case or upper and lower case.	Symbols only as supplement to words.	Defines signal word, message, symbol panels (optional, attached to side of label).	Not specified
(replaced by ANSI Z535.2, but 1968 version is still basis for OSHA regulations)						
ANSI Z129.1 Precautionary Labeling of Hazardous Chemicals	Danger Warning Caution Notice* Attention* * optional words for "delayed" hazards	Not specified	Not specified	Exclamation point may follow signal word for emphasis. Skull-and-crossbones as supplement to words.	Label arrangement not specified; examples given.	Provides guidance about how to select signal words and precautionary text.
ANSI Z535.2 Environmental and Facility Safety Signs	Danger Warning Caution Notice (general safety) (arrows)	Red Orange Yellow Blue Green as above when combined; black & white otherwise per ANSI Z535.1	Sans serif, upper case only for signal word, acceptable typefaces, letter heights	Symbols and pictographs per ANSI Z535.3; safety alert symbol	Defines signal word, word message, symbol panels in 1-3 panel designs. 4 shapes for special use.	Provides guidance
(use ANSI Z129.1 for chemical hazards)						
ANSI Z535.4 Product Safety Signs and Labels	Danger Warning Caution	Red Orange Yellow per ANSI Z535.1	Sans serif, upper case only for signal word, acceptable typefaces, letter heights	Symbols and pictographs per ANSI Z535.3; safety alert symbol	Defines signal word, message, pictorial panels in order of general to specific. Can use ANSI Z535.2 for uniformity.	Provides guidance
(use ANSI Z129.1 for chemical hazards)						
ANSI Z535.5 Accident Prevention Tags (for Temporary Hazards)	Danger Warning Caution	Red Orange Yellow per ANSI Z535.1	Sans serif, upper case only for signal word, acceptable typefaces, letter heights	Symbols and pictographs (to replace or supplement word messages) per ANSI Z535.3; safety alert symbol	Defines signal word and message panels.	Provides guidance

TABLE 40.5   ANSI Z535.2-1996 Signal Word Panels

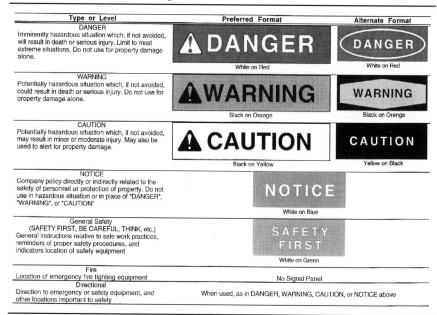

TABLE 40.6   ANSI Z535.1-1996 Safety Colors, Meanings, and Examples of Use

Color	Meaning	Commonly Used Examples
Red	Identification of DANGER or STOP.	Flammable liquid containers Emergency machine-stop bars/buttons/switches Fire protection equipment
Orange	Identification of WARNING. Identification of hazardous parts of machines.	Machine parts that may cut, crush, or otherwise injure, and emphasizing such when guards are open Inside of moveable guards Exposed edges of power transmission devices
Yellow	Identification of CAUTION. May be striped or checkered with black for enhancing contrast with background.	Hazards of striking against, stumbling, falling, tripping, or being caught in-between Flammable material storage cabinets Corrosive or unstable material containers
Green	Identification of SAFETY, emergency egress, and location of first aid and safety equipment	Gas masks First aid kits or dispensary Stretchers Safety deluge showers Safety bulletin boards and signs Emergency egress routes
Blue	Identification of SAFETY INFORMATION used on informational signs and bulletin boards	Mandatory action signs for wearing of personal protective gear such as hard hats
Other	Unassigned	

Labels — Format. The only specifications for label format and placement is that the labels must be: (a) legible, (b) in English, and (c) prominently displayed or readily available. Other standards with more specific design guidelines that can be applied include: (a) OSHA 29 CFR 145 *et seq.*, (b) ANSI Z535.2, and (c) ANSI Z129.1, all of which are covered elsewhere herein.

Material Safety Data Sheets — Manufacturers' Responsibility. Chemical manufacturers and importers must obtain or develop an MSDS for every hazardous chemical produced or imported by them.

Material Safety Data Sheets — Employers' Responsibility. Employers must obtain, maintain, and make readily accessible MSDSs for every hazardous chemical they use.

TABLE 40.7 ANSI Z535.2-1996 Multi-Panel Sign Layouts by Type Classification

Material Safety Data Sheets — Format. While there is no required format for OSHA MSDSs, they must be in English and contain the information outlined in Table 40.11.

The physical layout of an MSDS is not specified by OSHA. However, two guidelines exist. The first is a nonmandatory form by OSHA itself, shown in Figure 40.5. The second is a standard proposed by the Chemical Manufacturers Association and ANSI, Z400.1. An example from this draft standard is shown in Figure 40.6.

ANSI Z535.4 Product Safety Signs and Labels

Part of the ANSI Z535 series, this standard addresses product signs and labels, with their inherent characteristics of generally being smaller and observed at a closer distance than environmental or facility signs or labels. The other primary differences between Z535.2 and Z535.4 signs and labels are that there are fewer types for products, with fewer signal words and panel layouts (see Table 40.12 and Figure 40.7).

DOT 49 CFR 170 Hazardous Materials Transportation Regulations

The probably familiar "diamond" signs on the outside of trucks, especially tankers, is due to U.S. Department of Transportation regulations. The primary purpose is to alert emergency response personnel to potential hazards present within and in the vicinity of the vehicle, especially when there has been an accident. These signs are comprised of nine classes of hazards, with corresponding colors, symbols, and text, with one version using standardized four-digit code numbers to identify the hazard (see Figure 40.8).

NFPA 704 Standard System for the Identification of Fire Hazards

The National Fire Protection Association developed the NFPA 704 system to aid fire emergency response personnel in the identification of hazards and the appropriate procedural response. It addresses fire,

TABLE 40.8   ANSI Z535.3-1996 Symbol Examples

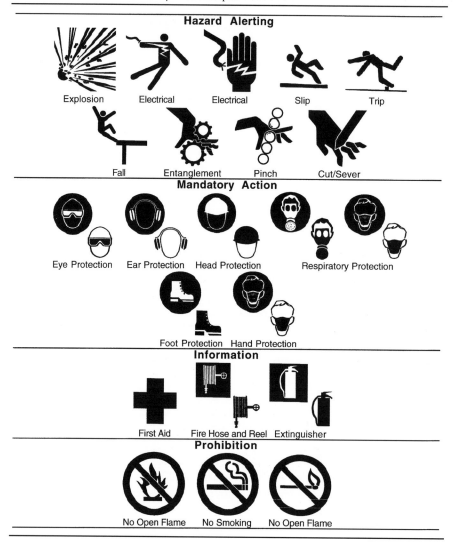

health, reactivity levels, and specific hazard concerns (see Figure 40.9). These signs or labels may be commonly found both in the environment and on specific product containers.

Hazardous Material Identification System

The HMIS® was developed by the National Paint and Coatings Association. It is similar to another system, the ASTM Safety Alert System (ASTM D 4257, 1984). In both systems (and like NFPA 704), coded indices indicating health, flammability, and reactivity are given. In addition, they both indicate the types of personal protective equipment that should be used. Figure 40.10 illustrates the sign or label format and the meanings of the four indices.

40.6   Summary

This chapter has endeavored to give the reader some background on the fundamentals of warnings research and practice. The conclusions reached give guidance on how to construct warnings that will hopefully have a better chance at achieving the goals of eliciting appropriate behavioral response for

TABLE 40.9   ANSI Z535.3-1996 Safety Symbol Surround Shapes and Use

Symbol Type and Use	Shape
Hazard Alerting Usually not recommended. Reduces symbol size on on Z535.2, .4, and .5 formats. Preferred colors: Black on white. Red on white for some symbols. White may be replaced by orange or yellow if used with Z535.2	(triangle and diamond shapes)
Mandatory Action Generally used on Z535.2 Notice signs, although optional. Colors: White symbol on solid blue/black circular surround, in turn on white symbol panel background. Reverse when no surround.	(solid black circle)
Information Generally used on Z535.2 General Safety or Fire Safety signs. Colors: Green or black image on white background. For fire-related symbols, red image on white background, or reverse.	(solid black square and outlined square)
Prohibition Use is mandatory. Colors: Black symbol on white background with red or black circle and slash. White border on slash if both symbol and slash are black.	(circle with slash)

accident avoidance. This was followed by a review of the major warning sign and label standards and their provisions that are applicable or likely to be found in common industrial environments.

The reader is cautioned that both knowledge from warnings research and mandatory requirements and voluntary guidelines are constantly evolving, resulting in changing criteria for the design and assessment of warning signs and labels. Since most standards are revised every five years or less, one should always try to obtain the latest versions of the references given for the standards in this chapter.

Finally, bear in mind that, in a court of law, compliance with standards is generally not a sufficient sole defense because standards are often viewed as minimum requirements (often due to a "one size fits all" strategy in their development). Therefore, it is necessary to fully understand the potential hazards of any system being analyzed (i.e., environment, facility, product, etc.), apply the hierarchy of hazard control to achieve the most effective protection for those who use or are exposed to the hazards, and apply warnings as a sole remedy sparingly and in a rigorous manner.

TABLE 40.10   ANSI Z129.1-1994 Precautionary Label Content

Category	Criteria or Guidelines
Identification of chemical product or its hazardous components	Adequate to allow proper action in case of exposure. Chemical name(s) of substances contributing substantially to hazards (if mixture proprietary, informative not required on label, but procedure to obtain such information in an emergency situtation must be provided). Nondescriptive code or trade name only not permitted
Signal word	Based on greatest immediate hazard. DANGER, WARNING, or CAUTION for immediate hazards. NOTICE or ATTENTION for delayed hazards. May be followed by exclamation mark (!) for emphasis.
Immediate hazards	Reasonably foreseeable immediate physical and health hazards associated with reasonably foreseeable handling, use, or misuse. Order by seriousness of hazard. Standard provides guidelines for selection of statement of hazard., 6.2, 7.1, Table 1
Delayed hazards	Reasonably foreseeable delayed physical and health hazards associated with reasonably foreseeable handling, use, or misuse. Order by seriousness of hazard. No need to repeat hazards included as immediate hazards., 6.2, 6.3, 7.1, Table 2
Precautionary measures	Brief supplement to hazard statement(s). Measures to avoid injury., 7.2, Table 1
Contact or exposure instructions (first aid or antidotes)	Section captioned "FIRST AID": Use when immediate treatment warranted and simple remedial measures may be taken; limit to procedures that do not require special training. Section captioned "ANTIDOTE": Use when specific antidoes known and administration does not required special training. Section captioned "NOTES TO PHYSICIANS": Recommended medical practices or antidotes to be administered by a physician., 7.2, Table 1
Fire instructions	Intended for persons who handle containers during shipment and storage. Instructions for confining and extinguishing fire. Simple and brief as possible. Advise suitable control material. When appropriate and personnel and property not at risk, may advise to allow fire to burn out rather than risk contamination., 6.7, 7.1, 7.2, Table 3
Spill or leak instructions	Methods to use, in the absence of fire, to contain spills or leaks to minimize exposures and prevent personal injury and environmental contamination., 6.8, 7.1, 7.2, Table 4
Container handling and storage instructions	Special or unusual handling and storage procedures. Where flash point of flammable or combustible liquids determines storage requirements, include this information as pertinent to storage code (ANSI/NFPA 30)., 6.9, 7.1, 7.2
References	Reference to MSDS, if available.
Additional useful statements	5.2, 6.9, 7.1, 7.2
Name and address of company	Manufacturer, importer, or distributor. Actual corporate or business name. Include street address, city, state, and Zip code. Country and mail code if foreign.
Telephone number	For additional information on product. Indicate hours or type of information restrictions.

TABLE 40.11   Information Required on Material Safety Data Sheet (MSDS) by OSHA (29 CFR 1910.1200(g)(2))

Identity of chemical as used on the container label
Physical and chemical charteristics
Physcial hazards, such as fire, explosion, or reactivity
Health hazards, signs and symptoms, and medical conditions that may be aggravated by exposure
Primary routes of entry
OSHA permissible exposure limit (PEL), ACGIH Threshold Limit Value (TLV), and common or recommended exposure limits
Potential as carcinogen
Precautions for safe handling and use
Control measures
Emergency and first aid procedures
Date of preparation or last revision to MSDS
Name, address, and telephone number of party responsible for MSDS who can provide additional information

Material Safety Data Sheet	**U.S. Department of Labor**
May be used to comply with OSHA's Hazard Communication Standard, 29 CFR 1910.1200. Standard must be consulted for specific requirements.	Occupational Safety and Health Administration (Non-Mandatory Form) Form Approved OMB No. 1218-0072
IDENTITY *(As Used on Label and List)*	*Note: Blank spaces are not permitted. If any item is not applicable, or no information is available, the space must be marked to indicate that.*

Section I

Manufacturer's Name	Emergency Telephone Number
Address *(Number, Street, City, State, and ZIP Code)*	Telephone Number for Information
	Date Prepared
	Signature of Preparer *(optional)*

Section II – Hazardous Ingredients/Identity Information

Hazardous Components (Specific Chemical Identity: Common Name(s))	OSHA PEL	ACGIH TLV	Other Limits	% *(optional)*

Section III – Physical/Chemical Characteristics

Boiling Point		Specific Gravity (H_2O = 1)	
Vapor Pressure (mm Hg)		Melting Point	
Vapor Density (AIR = 1)		Evaporation Rate (Butyl Acetate = 1)	
Solubility in Water			
Appearance and Odor			

Section IV – Fire and Explosion Hazard Data

Flash Point (Method Used)	Flammable Limits	LEL	UEL
Extinguishing Media			
Special Fire Fighting Procedures			
Unusual Fire and Explosion Hazards			

FIGURE 40.5a & 5b OSHA nonmandatory Material Safety Data Sheet (MSDS) format.

Section V – Reactivity Data

Stability	Unstable		Conditions to Avoid
	Stable		

Incompatibility *(Materials to Avoid)*

Hazardous Decomposition or Byproducts

Hazardous Polymerization	May Occur		Conditions to Avoid
	Will Not Occur		

Section IV – Health Hazard Data

Route(s) of Entry	Inhalation?	Skin?	Ingestion?

Health Hazards *(Acute and Chronic)*

Carcinogenicity	NTP?	IARC Monographs?	OSHA Regulated?

Signs and Symptoms of Exposure

Medical Conditions
Generally Aggravated by Exposure

Emergency and First Aid Procedures

Section VII – Precautions for Safe Handling and Use

Steps to Be Taken in Case Material is Released or Spilled

Waste Disposal Method

Precautions to Be Taken in Handling and Storing

Other Precautions

Section VIII – Control Measures

Respiratory Protection *(Specify Type)*

Ventilation	Local Exhaust		Special
	Mechanical *(General)*		Other

Protective Gloves	Eye Protection

Other Protective Clothing or Equipment

Work/Hygienic Practices

FIGURE 40.5a & 5b   (continued)

1. **CHEMICAL PRODUCT AND COMPANY INFORMATION**
   Company name and address   **PRODUCT NAME:**
   Emergency Phone   **PRODUCT CODE:**
   Effective Date: Print Date:

2. **COMPOSITION/INFORMATION ON INGREDIENTS**
   Chemical Ingredients (% by wt.)
   Component A   CAS#   xx–xx%
   Component B   CAS#   xx–xx%
   Impurity C   CAS#   xx ppm max

3. **HAZARDS IDENTIFICATION**

   ┌───┐
   │ EMERGENCY OVERVIEW │
   │ [short description of characteristics and hazards] │
   └───┘

   POTENTIAL HEALTH EFFECTS

   EYE:
   SKIN CONTACT:
   SKIN ABSORPTION:
   INGESTION:
   INHALATION

   CHRONIC EFFECTS/CARCINOGENICITY:

4. **FIRST AID MEASURES**

   EYE:
   SKIN:
   INGESTION:
   INHALATION

5. **FIRE FIGHTING MEASURES**

   FLAMMABLE PROPERTIES
   FLASH POINT:
   METHOD USED:
   FLAMMABLE LIMITS
   LFL:
   UFL:
   EXTINGUISHING MEDIA:
   FIRE & EXPLOSION HAZARDS:
   FIRE-FIGHTING EQUIPMENT:

6. **ACCIDENTAL RELEASE MEASURES**

7. **HANDLING AND STORAGE**

8. **EXPOSURE CONTROLS/PERSONAL PROTECTION**

   RESPIRATORY PROTECTION:
   SKIN PROTECTION:
   EYE PROTECTION:
   EXPOSURE GUIDELINE(S):
   ENGINEERING CONTROLS:

9. **PHYSICAL AND CHEMICAL PROPERTIES**

   APPEARANCE:
   ODOR:
   BOILING POINT:
   VAPOR PRESSURE:
   VAPOR DENSITY:
   SOLUBILITY IN WATER:
   SPECIFIC GRAVITY:
   FREEZING POINT:
   pH:
   VOLATILE:

10. **STABILITY AND REACTIVITY**

    STABILITY: (CONDITIONS TO AVOID)
    INCOMPATIBILITY: (SPECIFIC MATERIALS TO AVOID)
    HAZARDOUS DECOMPOSITION PRODUCTS:
    HAZARDOUS POLYMERIZATION:

11. **TOXICOLOGICAL INFORMATION**

12. **ECOLOGICAL INFORMATION**

13. **DISPOSAL CONSIDERATIONS**

14. **TRANSPORT INFORMATION**

15. **TRANSPORTATION AND HAZARDOUS MATERIALS DESCRIPTION:**

    REGULATORY INFORMATION

    OSHA HAZARD COMMUNICATION RULE, 29 CFR 1910.1200:
    CERCLA/SUPERFUND, 40 CFR 117, 302:
    SARA HAZARD CATEGORY:
    SARA 313 INFORMATION:
    TOXIC SUBSTANCES CONTROL ACT (TSCA):
    CALIFORNIA PROPOSITION 65:

16. **OTHER INFORMATION**

    MSDS STATUS:

FIGURE 40.6   Example of Material Safety Data Sheet (MSDS) format per proposed ANSI/CMA Z400.1.

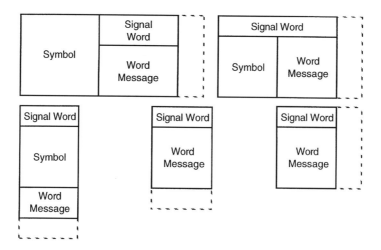

FIGURE 40.7   ANSI Z535.4-1996 multi-panel sign layouts.

TABLE 40.12   ANSI Z535.4-1996 Signal Word Panels

Type or Level	Format
DANGER Imminently hazardous situation which, if not avoided, will result in death or serious injury. Limit to most extreme situations. Do not use for property damage alone.	⚠ **DANGER** White on Red
WARNING Potentially hazardous situation which, if not avoided, could result in death or serious injury. Do not use for property damage alone.	⚠**WARNING** Black on Orange
CAUTION Potentially hazardous situation which, if not avoided, may result in minor or moderate injury. May also be used to alert against unsafe practices.	⚠ **CAUTION** Black on Yellow
CAUTION without Safety Alert Symbol Propery-damage-only accidents.	**CAUTION** Black on Yellow

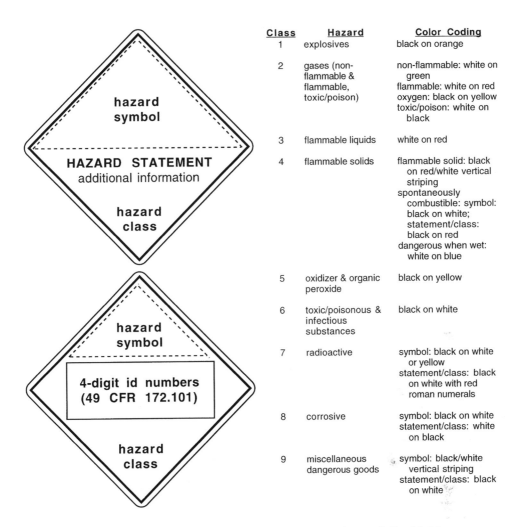

Class	Hazard	Color Coding
1	explosives	black on orange
2	gases (non-flammable & flammable, toxic/poison)	non-flammable: white on green flammable: white on red oxygen: black on yellow toxic/poison: white on black
3	flammable liquids	white on red
4	flammable solids	flammable solid: black on red/white vertical striping spontaneously combustible: symbol: black on white; statement/class: black on red dangerous when wet: white on blue
5	oxidizer & organic peroxide	black on yellow
6	toxic/poisonous & infectious substances	black on white
7	radioactive	symbol: black on white or yellow statement/class: black on white with red roman numerals
8	corrosive	symbol: black on white statement/class: white on black
9	miscellaneous dangerous goods	symbol: black/white vertical striping statement/class: black on white

FIGURE 40.8   U.S. Department of Transportation (49 CFR 172) Hazard Class label formats.

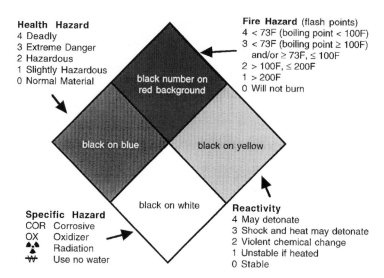

FIGURE 40.9   Label format from NFPA 704-1990, Standard System for the Identification of the Fire Hazards of Materials.

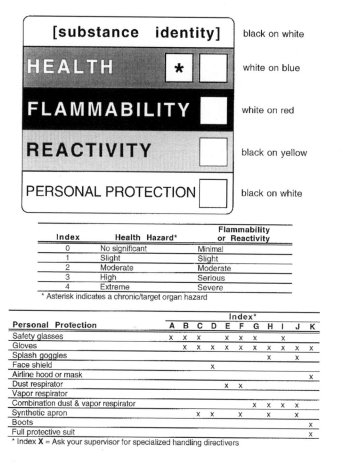

FIGURE 40.10   Hazardous Materials Identification System (HMIS®) (From National Paint and Coatings Association. *Hazardous Materials Identification System*. With permission.)

References

American National Standards Institute/Chemical Manufacturers Association. 1994. *American National Standard for Hazardous Industrial Chemicals – Precautionary Labeling*, ANSI Z129.1.

American National Standards Institute/Chemical Manufacturers Association. 1995, *American National Standard for Material Safety Data Sheets*, ANSI Z400.1 (draft version).

American National Standards Institute/National Electrical Manufacturers Association. 1996. *American National Standard for Accident Prevention Tags (for Temporary Hazards)*, ANSI Z535.5 (draft revision).

American National Standards Institute/National Electrical Manufacturers Association. 1996. *American National Standard for Criteria for Safety Symbols*, ANSI Z535.3 (draft revision).

American National Standards Institute/National Electrical Manufacturers Association. 1996. *American National Standard for Environmental and Facility Safety Signs*, ANSI Z535.2 (draft revision).

American National Standards Institute/National Electrical Manufacturers Association. 1996. *American National Standard for Product Safety Signs and Labels*, ANSI Z535.4 (draft revision).

American National Standards Institute/National Electrical Manufacturers Association. 1996. *American National Standard Safety Color Code*, ANSI Z535.1 (draft revision).

American National Standards Institute/National Bureau of Standards. 1979. *American National Standard Safety Color Code for Marking Physical Hazards*, ANSI Z53.1.

American National Standards Institute/National Electrical Manufacturers Association/National Safety Council. 1972. *American National Standard Specifications for Accident Prevention Signs*, ANSI Z35.1.

Belkin, N.J., Brooks, H.M., and Daniels, P.J. 1987. Knowledge elicitation using discourse analysis. *International Journal of Man-Machine Studies*. 27:127-144.

Boersema, T. and Zwaga, H. 1985. The influence of advertisements on the conspicuity of routing information. *Applied Ergonomics*. 16(4):267-273.

Bresnahan, T.F. and Bryk, J. 1975. The hazard association values of accident-prevention signs. *Professional Safety*. 20(1):17-25.

Cahill, M.C. 1976. Design features of graphic symbols varying in interpretability. *Perceptual and Motor Skills*. 42(2):647-653.

Cairney, P.T. and Sless, D. 1982. Communication effectiveness of symbolic safety signs with different user groups. *Applied Ergonomics*. 13(2):91-97.

Campion, M.A. and Medsker, G.J. 1992. Job design, in *Handbook of Industrial Engineering*, 2nd edition, ed. G. Salvendy, p. 845-881. John Wiley & Sons, New York, NY.

Conoley, J.C. and Kramer, J.J., (Eds.) 1989. *The Mental Measurements Yearbook*. Buros Institute of Mental Measurements of the University of Nebraska, Lincoln.

Craig, C.S. and McCann, J.M. 1978. Assessing communication effects on energy conservation. *Journal of Consumer Research*. 5:82-88.

Dorris, A.L. and Purswell, J.P. 1977. Human factors in the design of effective product warnings. *Proceedings of the Human Factors Society*. 22:343-346.

Easterby, R.S. and Hakiel, S.R. 1981. Field testing of consumer safety signs: The comprehension of pictorially presented messages. *Applied Ergonomics*. 12(3):143-152.

Easterby, R.S. and Zwaga, H.J.G. 1976. Evaluation of public information symbols ISO tests: 1975 Series, A.P.Report No. 60. University of Aston in Birmingham, UK.

Edworthy, J. and Adams, A. 1996. *Warning Design: A Research Perspective*. Taylor & Francis, London, UK.

FMC Corporation. 1980. *Product Safety Sign and Label System*, 2nd ed. Santa Clara, CA.

Frantz, J.P. and Rhoades, T.P. 1993. A task analytic approach to the temporal and spacial placement of product warnings. *Human Factors*. 35:719-730.

Gibson, J.J. 1979. *The Ecological Approach to Visual Perception*. Houghton-Mifflin, Boston, MA.

Godfrey, S.S., Rothstein, P.R., and Laughery, K.R. 1985. Warnings: Do they make a difference. *Proceedings of the Human Factors Society*, 29th Annual Meeting. 669-673.

Hadden, S.G. 1986. *Read the Label*. Westview Press, Boulder, CO.

Heglin, H. 1973, July. NAVSHIPS display illumination design guide, Volume 2: Human Factors (NELC-TD223), Naval Electronics Laboratory Center, San Diego, CA.

Holahan, C.J. 1977. Relationship Between Roadside Signs and Traffic Accidents: A Field Investigation, Research Report 54. Council for Advanced Transportation Studies, Austin, TX.

Illuminating Engineering Society. 1981. *IES Lighting Handbook: Application Volume*. New York, NY.

Jacobs, R.J., Johnston, A.W., and Cole, B L. 1975. The visibility of alphabetic and symbolic traffic signs. *Australian Road Research*. 5(7):68-86.

Johnson, D.A. 1980. The design of effective safety information displays. *Proceedings of the Symposium: Human Factors and Industrial Design in Consumer Products*. 314-328.

Kuipers, B. and Kassirer, J.P. 1983. How to discover a knowledge representation for causal reasoning by studying an expert physician. *IJCAI-83: Proceedings of the 8th International Conference on Artificial Intelligence*.

Laughery, K.R., Rowe-Hallbert, A.L., Young, S.L., Vaubel, K.P., and Laux, L.F. 1991. Effects of explicitness in conveying severity information in product warnings. *Proceedings of the Human Factors Society, 35th Annual Meeting*. 481-485.

Lehto, M.R., and Miller, J.M. 1986. *Warnings: Volume I: Fundamentals, Design, and Evaluation Methodologies*. Fuller Technical Publications, Ann Arbor, MI.

Lehto, M.R. and Papastavrou, J.D. 1993. Models of the warning process: Important implications towards effectiveness. *Safety Science*. 16:569-595.

Lehto, M.R. and Salvendy, G. 1995. Warnings: A supplement not a substitute for other approaches to safety. *Ergonomics*. 38 (11):2155-2163.

Lehto, M.R. 1991. A proposed conceptual model of human behavior and its implications for the design of product warnings. *Perceptual and Motor Skills*. 73:595-611.

Lehto, M.R. 1992. Designing warning signs and warning labels: Scientific basis for initial guidelines. *International Journal of Industrial Ergonomics*. 10:115-138.

Leonard, S.D. and Matthews, D. 1986. How does the population interpret warning signals? *Proceedings of the Human Factors Society, 30th Annual Meeting*. 116-120.

Lerner, N.D. and Collins, B.L. 1980. The assessment of safety symbol understandability by different testing methods, PB81-185647. National Bureau of Standards, Washington, D.C.

Lirtzman, S.I. 1984. Labels, perception, and psychometrics, in *Handbook of Chemical Industry Labeling*, Ed. C.J. O'Connor and S.I. Lirtzman. Noyes Publications, Park Ridge, NJ.

Matthews, M. 1987. The influence of color on CRT reading performance and subjective comfort under operational conditions. *Applied Ergonomics*. 18:259-271.

Miller, J.M., Lehto, M.R., and Frantz, J.P. 1994. *Warnings & Safety Instructions*. Fuller Technical, Ann Arbor, MI.

Morris, L.A. and Kanouse, D.E. 1981. Consumer reaction to the tone of written drug information. *American Journal of Hospital Pharmacy*. 38(5): 667-671.

National Electronic Manufacturers Association. 1982. *Safety Labels for Padmounted Switchgear and Transformers Sited in Public Areas*, NEMA 260.

National Fire Protection Association. 1990. *Standard System for the Identification of the Fire Hazards of Materials*, NFPA 704.

National Paint and Coatings Association. *Hazardous Materials Identification System*.

Norman, D.A. 1988. *The Psychology of Everyday Things*. Basic Books, NY.

O'Connor, C.J. and Lirtzman, S.I. 1984. *Handbook of Chemical Industry Labeling*. Noyes Publications, Park Ridge, NJ.

Otsubo, S.M. 1988. A behavioral study of warning labels for consumer products: Perceived danger and use of pictographs. *Proceedings of the Human Factors Society, 30th Annual meeting*. 1202-1205.

Peterson, D. 1975. *Safety Management — A Human Approach*. Aloray, Deer Park, NY.

Peterson, D. 1984. *Analyzing Safety Performance*. Aloray, Deer Park, NY.

Phillips, J.J. 1983. *Handbook of Training Evaluation and Measurement Methods*. Gulf Publishing Company, Houston.

Poulton, E. 1967. Searching for newspaper headlines printed in capitals or lower-case letters. *Journal of Applied Psychology*. 51:417-425.

Rasmussen, J. 1986. *Information Processing and Human-Machine Interaction*. North-Holland.

Reason, J. 1990. *Human Error*. Cambridge University Press, Cambridge, UK.

Reder, L.M. and Anderson, J.R. 1982. Effects of spacing and embellishment on memory for the main points of a text. *Memory and Cognition*. 10(2):97-102.

Rhoades, T.P., Frantz, J.P., and Miller, J.M. 1991. Emerging strategies for the assessment of safety related product communications. *Proceedings of the Human Factors Society*, 35th Annual Meeting, San Francisco, CA. 998-1002.

Rouse, W.B. 1992. Human-centered product planning and design, in *Handbook of Industrial Engineering*, 2nd edition, Ed. G. Salvendy. John Wiley & Sons, New York, NY. 1220-1240.

Sanders, M.S., and McCormick, E.J. 1993. *Human Factors in Engineering and Design*, 7th edition. McGraw-Hill, New York, NY.

Scammon, D.L. 1977. Information load And consumers. *The Journal of Consumer Research*. 4:148-155.

Shadbolt, N. and Burton, M. 1990. Knowledge elicitation, in *Evaluation of Human Work*, Ed. J.R. Wilson and E.N. Corlett, Taylor & Francis, London, UK.

Sinaiko, H.W. 1975. Verbal factors in human engineering: some cultural and psychological data, in *Verbal Factors in Human Engineering*. The Smithsonian Institution, Washington, D.C. 159-177.

Strawbridge, J.A. 1986. The influence of position, highlighting, and imbedding on warning effectiveness. *Proceedings of the Human Factors Society*, 30th Annual Meeting. Human Factors Society. Santa Monica, CA. 716-720.

Tullis, T. 1988. Screen design, in *Handbook of Human-Computer Interaction*, Ed. M. Helander. Elsevier Science, Amsterdam. 377-411.

U.S. Consumer Product Safety Commission. Hazardous Substances and Articles, 16 CFR 1500.

U.S. Department of Labor, Occupational Health and Safety Administration. Accident Prevention Signs and Tags, 29 CFR 1926.200, in *Construction Standards*, 29 CFR 1926.

U.S. Department of Labor, Occupational Health and Safety Administration. *Hazard Communication Standard*, 29 CFR 1910.1200, in *General Industry Standards*, 29 CFR 1910.

U.S. Department of Labor, Occupational Health and Safety Administration. *Safety Color Code for Marking Physical Hazards*, 29 CFR 1910.144, in *General Industry Standards*, 29 CFR 1910.

U.S. Department of Labor, Occupational Health and Safety Administration. *Specification for Accident Prevention Signs and Tags*, 29 CFR 1910.145, in *General Industry Standards*, 29 CFR 1910.

U.S. Department of Transportation. *Hazardous Materials Transportation Regulations*, 49 CFR 170 *et seq.*

Ursic, M. 1984. The impact of safety warnings on perception and memory, *Human Factors*. 16(6):677-682.

Webster's Ninth New Collegiate Dictionary. 1988. Merriam-Webster, Springfield, MA.

Westinghouse Electric Corporation. 1981. *Product Safety Label Handbook*. Westinghouse Printing Division. Trafford, PA.

Wilson, J.R. and Corlett, E.N. 1990. *Evaluation of Human Work*. Taylor & Francis, London, UK.

Wogalter, M.S., Allison, S.T., and McKenna, N.A. 1989. Effects of cost and social influence on warning compliance. *Human Factors*. 31:133-140.

Wogalter, M.S., Godfrey, S.S., Fontenelle, G.A., Desaulniers, D.R., Rothstein, P.R., and Laughery, K.R. 1987. Effectiveness of warnings. *Human Factors*. 29(5):599-622.

Woodson, W. 1963. Human engineering design standards for spacecraft controls and displays (General Dynamics Aeronautics Report GDS-63-0894-1). National Aeronautics and Space Administration, Orlando, FL.

Wright, P. 1979. Concrete actions plans in TV messages to increase reading of drug warnings. *Journal of Consumer Research*. 6:256-259.

Young, S.L. and Wogalter, M.S. 1988. Memory of instruction manual warnings: Effects of pictorial icons and conspicuous print. *Proceedings of the Human Factors Society*, 32nd Annual Meeting. Human Factors Society, Santa Monica, CA. 905-909.

41

Ergonomics Methods
in the Design of
Consumer Products

Neville A. Stanton
University of Southampton

Mark S. Young
University of Southampton

41.1   Introduction

There is an immense variety in the range of methods that can be applied to the design of consumer products. The methods differ in what they address (i.e., the human element, the device element, or the interaction) and what they produce (e.g., task descriptions, predicted errors, and performance times). Ergonomics has a practical role to play in device design, on the basis that use of the methods should improve design by: reducing device interaction time, reducing user errors, improving user satisfaction, and improving device usability. Although there are many different design processes, most can be reduced to six main phases:

Concept: in which the idea for the device is considered in a largely informal manner, many implementations are considered, and many degrees of freedom remain.

Flowsheeting: in which the ideas for the device become formalized and the alternatives considered become very limited.

Design: in which the design solution becomes crystallized and blueprints are devised.

Prototyping: in which soft- and hard-built prototype devices are developed for evaluation.

Commissioning: in which the final design solution is implemented and the product enters the marketplace.

Operation and maintenance: when the device is supported in the marketplace.

We believe that ergonomics methods may have the greatest impact at the prototyping stage, particularly *analytic prototyping* (i.e., when the device exists as a paper-based or computer-based model). Although in the past, it may have been costly to alter design at structural prototyping, and perhaps even impossible, with the advent of computer-aided design such retooling is made much simpler. It may even be possible to compare alternative designs at this stage with such technology. These ideas have yet to be proved in practice; however, given the nature of most ergonomics methods, it would seem most sensible to apply the methods at the analytic prototyping stage. We have selected 12 methods for consideration, based upon our analysis that these are a representative spread of methods that are currently being used to evaluate human–machine performance and assess the demands and effects upon people (Diaper, 1989; Kirwan and Ainsworth, 1992; Kirwan, 1994; Corlett and Clarke, 1995; Wilson and Corlett, 1995; Jordan et al., 1996). They were also chosen because of the appropriateness to the assessment of consumer products and user activity. Methods selected were as follows:

- Heuristics
- Checklists/Guidelines
- Observation
- Interviews
- Questionnaires
- Link analysis

- Layout analysis
- Hierarchical task analysis
- Systematic human error reduction and prediction approach
- Task analysis for error identification
- Repertory grids
- Keystroke level model

In terms of the analytic prototyping of human interfaces, we feel that there are three main forms: functional analysis (i.e., consideration of the range of functions the device supports), scenario analysis (i.e., consideration of the device with regard to a particular sequence of activities), and structural analysis (i.e., nondestructive testing of the interface from a user-centered perspective). We have classified the methods in this chapter into each of these types, as follows:

Functional Analysis	Scenario Analysis	Structural Analysis
Interviews	Link analysis	KLM
Questionnaires	Layout analysis	PHEA
Checklists	HTA	TAFEI
Repertory grids	Heuristics	Observation

We hope this chapter will help to highlight the different contributions each of the methods makes to analytic prototyping.

41.2   Radio-Cassette Players

Given the number of methods to be covered, emphasis will be given to providing an example of each approach with some accompanying text and reference to source material on the approach. The review is based upon the application of the methods to the evaluation of two radio-cassette machines taken from a car, as shown in Figures 41.1 and 41.2.

Although these analyses were undertaken as part of a research project concerned with operation of in-car devices, we believe that the methods would be equally appropriate to all kinds of consumer products, e.g., lawn mowers, hairdryers, kettles, videocassette recorders, cookers, washing machines, freezers, power tools, and vacuum cleaners. A fuller evaluation of devices may be found in Stanton (1998).

Ford 7000 RDS-EON

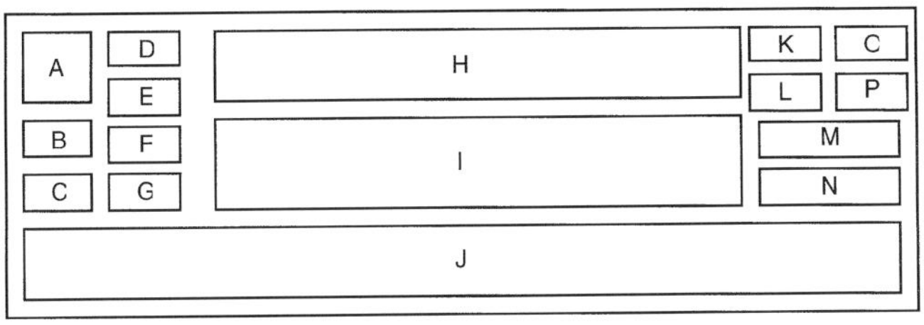

A - On/Off/Vol.
B - Bass/Treb.
C - Fade/Bal.
D - Eject
E - Dolby
F - News
G - TA
H - Cassette Door

I - Display
J - Presets
K - Tape
L - PTY
M - Menu
N - Seek
O - CD
P - AM/FM

This is a schematic diagram of the Ford radio cassette referred to in the examples. It is quite an advanced device, with RDS facilities, automatic volume control, automatic music search and programme type tuning amongst its functions. It is generally very mode-driven, with most of the more advanced functions being hidden in a menu. It is about twice the height of a standard radio, and apart from the On/Off/Volume control, all of the controls are buttons. This means that simple tasks such as adjusting the bass or treble involve more than one element (i.e., press Bass button, then turn Volume control).

FIGURE 41.1   The Ford car radio-cassette player (7000 RDS-EON).

41.3   Examples of Ergonomics Methods

Heuristics (Nielsen, 1992)

Heuristics require the analyst to use judgment, intuition, and experience to guide product evaluation. This method is wholly subjective and the output is likely to be extremely variable. In favor of the heuristic approach is the ease and speed with which it may be applied. Several techniques incorporate the heuristic approach (e.g., checklists, guidelines, SHERPA) but serve to structure heuristic judgment (see Figure 41.3).

An heuristic analysis was applied to the Sharp car radio. Needless to say, the analysis was very quick to execute and did not require any special knowledge on the analyst's part. Indeed, in many ways it resembled a simple walkthrough. Problems encountered in the analysis were few; the only real dissatisfaction resided in the fact that most of the output pertained to very similar faults with the device. Regarding the output, it may be observed that it is largely similar to the checklist approach in form and content, particularly as here we also see many items concerned with anthropometrics (e.g., button sizes). In addition, one of the remedial suggestions was also proposed in the SHERPA analysis (although, this particular section of SHERPA is also subjective). Thus there is some overlap between these approaches, which can only be encouraging for heuristics.

Checklists/Guidelines (Ravden and Johnson, 1989; Woodson et al., 1992)

Checklists and guidelines would seem to be a useful *aide memoir*, to make sure that the full range of ergonomics issues have been considered. However, the approach may suffer from a problem of situational sensitivity, i.e., the discrimination of an appropriate item from a nonappropriate item largely depends

Sharp RG-F832E

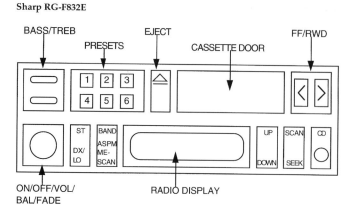

This is a schematic diagram of the Sharp radio cassette referred to in the examples. It is a rather standard radio and has no RDS facilities. Some of the controls may need further elaboration.

On/Off:	This is a knob-twist control - turn clockwise for On, then further to increase volume. Push and turn to adjust fade; a collar adjusts balance.
ST:	Push the top of this button to toggle stereo/mono radio reception.
DX/LO:	Toggles local or distance reception when scanning for radio stations.
BAND:	Switches between wavebands.
ASPM ME-SCAN:	Scans the preset stations.
UP/DOWN:	Manual radio tuning.
SCAN:	Scans the current waveband for radio signals; continues until interrupted by user.
SEEK:	Looks for next radio signal on current waveband and locks onto it.
CD:	CD/auxilliary input socket.
BASS/TREB:	Sliding controls for bass and treble.

FIGURE 41.2   The Sharp car radio-cassette player (RG-F832E).

upon the expertise of the analyst. Nevertheless, checklists offer a quick and relatively easy method for device evaluation (see Figure 41.4).

Woodson, Tillman, and Tillman (1992) Transport Checklist turned out to be more of a set of guidelines, so could not be used as an example in this analysis. A cursory inspection of the vehicular-related guidelines revealed that such recommendations are largely concerned with anthropometric improvements and safety (e.g., ingress and egress; protruding units etc.). Thus, these can only deal with issues such as consequentiality of accidents, comfort, and satisfaction, rather than error prediction and usability. (Although comfort and satisfaction are elements of usability, there are also more cognitive components.) The Human Engineering Design Checklist proved more useful. Only one section was deemed relevant to assessing a car radio; this was the section on console and panel design. Even so, it was clear that this was constructed for control room assessments, and many of the items were simply irrelevant. However, it would not be an arduous task to extract the relevant items and thus construct a checklist for assessing in-car devices. Furthermore, some items that were not relevant to a car radio (e.g., those concerned with CRTs) may be applicable to other devices (e.g., navigation aids). Again, though, the checklist was, for the most part, concerned with anthropometric issues.

**Example of Heuristic Output
(Sharp RG-F832E)**

- On/Off/Volume control is a tad small and awkward, combined with difficult Balance control
- Pushbutton operation would be more satisfactory for On/Off, as Volume stays at preferred level
- Fader Control is particularly small and awkward
- Both of the above points are related to the fact that a single button location has multiple functions - this is too complex
- Treble and Bass Controls also difficult and stiff; although these functions are rarely adjusted once set
- Station Preset Buttons are satisfactory; quite large and clear
- Band Selector Button and FM Mono-Stereo Button should not have 2 functions on each button - could result in confusion if wrong function occurs. These buttons are the only buttons on the radio which are not self-explanatory - the user must consult the manual to discover their function
- Tuning Seek and Tuning Scan Buttons are easier to understand and use, although there are still two functions on the same button. These are probably used more than the aforementioned buttons
- Cassette FF, RWD and Eject Buttons are self-explanatory; the same accepted style that is on all car radio designs. FF and RWD Buttons could be a little larger
- Auto-reverse function is not so obvious, although it is an accepted standard (pressing FF and RWD Buttons simultaneously)
- Illumination - is daytime/nighttime illumination satisfactory? A dimmer control would probably aid matters

FIGURE 41.3   Heuristic analysis of the radio-cassette for Sharp RG-F832E.

Observation
(Drury, 1990; Kirwan and Ainsworth, 1992; Baber and Stanton, 1996a)

Observation is perhaps the most obvious way of collecting information about a person's interaction with a device; watching and recording the interaction will undoubtedly inform the analyst of what occurred on the occasion observed (see Figure 41.5). Observation is also a deceptively simple method. One simply watches, participates in, and/or records the interaction. However, the quality of the observation will largely depend upon the method of recording and analyzing the data. There are concerns about the intrusiveness of observation, the amount of effort required in analyzing the data, the objectivity of the analysis, and the comprehensiveness of the observational method. Despite these concerns, it is difficult to manage without some form of observational data, as most ergonomics methods rely upon it, e.g., hierarchical task analysis and link analysis.

The observational studies show data from 30 participants performing a range of tasks on two occasions. These data on actual performance show errors and response times. These data may seem to be highly credible in the eyes of designers, but observational studies are very resource-intensive and provide little output regarding cognitive mechanisms. They are applicable only late in the design process. The wide use of the observational technique suggests that it is both reliable and valid, as well as useful.

Interviews (Cook, 1988; Sinclair, 1990; Kirwan and Ainsworth, 1992)

Like observation, the interview has a high degree of ecological validity associated with it. If you want to find out what people think of a device you simply ask them (see Figure 41.6). Interviewing has many forms, ranging from highly unstructured (free-form discussion) through focused (a situational interview), to highly structured (an oral questionnaire). For the purposes of device evaluation, a focused approach would seem most appropriate. The interview is good at addressing issues beyond direct

**Example of Checklist Output
(Sharp RG-F832E)**

The following are selected items from the Human Engineering Design
Checklist (Woodson, 1981) which are relevant and/or marginal
(unsatisfactory) for the car radio under analysis.

4. Console and panel design
4.1 Displays
4.1.1 Principles
     e. Crucial visual checks identified by attention-getting devices (e.g.
     visual or aural signals).
     g. Probability of confusion among instruments is minimal.
4.1.2 Labeling
     a. Trade names and other irrelevant information deleted.
     b. Easy to read under expected conditions of illumination.
4.1.4 Scales, dials, counters
     a. Numbers and letters are large enough for accurate reading at
     normal distance.
     b. Reflected light does not create illusion warning is "ON" or
     obscure reading.
4.1.6 Indicator and legend lights
     k. Displays are arranged in relation to one another to reflect the
     sequence of use or the functional relations of the components they
     represent, in that order of preference.
     l. Distinct, functional areas set apart for purposes of ready
     identification are outlined by black lines...
     w. Button surfaces are concave to fit the finger, or provide a high
     degree of frictional resistance to prevent slipping.
     x. Buttons provide "snap feel" or an audible click to indicate that the
     control has been activated.
     y. A channel or cover guard is provided when prevention of
     accidental activation is imperative.
4.2 Control/Display Relationships
4.2.1 Arrangements
     a. All controls having sequential relations, or having to do with a
     particular function or operation, or which are operated together, are
     grouped together, along with the associated displays.
     d. If a control knob is adjacent to the instrument it controls, it is
     located so that the control or the hand normally used for setting
     does not obscure the indicator.
4.2.2 Precautions
     a. The control is located or oriented so that the operator is not likely
     to hit it or move it accidentally in the normal sequence of control
     movements.
     d. Interlocks are provided so that extra movement of the prior
     operation of a related or locking control is required.
     e. Resistance is built into the control so that definite or sustained
     effort is required to actuate it.

FIGURE 41.4   Checklist applied to the radio-cassette for Sharp RG-F832E.

interaction with devices, such as the adequacy of manuals and other forms of support. The strengths of the interview are the flexibility and thoroughness it offers. For the purposes of this review we undertook the interview within the Ravden and Johnson (1989) framework of 11 areas of usability. This served as an interview agenda to focus the interviewer and respondent on usability issues associated with the operation of the radio-cassette.

The most striking aspect about the interview was its speed of administration — the whole process lasted around 30 minutes. As its structure was based on the sections of an HCI checklist, we can be quite confident that it thoroughly covered all aspects of device interaction. Admittedly, some aspects of the checklist were simply inapplicable to a car radio, but this just affirmed one advantage of the interview — its flexibility in adapting to changing scenarios. The output of the interview was on the whole unsurprising; much of it resembling the output of the heuristic analysis. Again, though, there are a number of advantages to the structured approach. Thorough coverage has been mentioned; in addition, it was possible to relate

Task	Errors Observed	F1	F2	T1 (s) mean (sd)	T2 (s) mean (sd)
1				4.64 (4.38)	4.05 (3.02)
2	Didn't turn knob enough to adjust	3	1	10.5 (7.67)	6.14 (1.88)
	Pressed Seek	1			
3	Adjusted Treble	1		20.4 (13.3)	10.1 (3.75)
	Adjusted Volume	2			
	Pressed On/Off	1			
4				15.7 (15.5)	8.55 (4.50)
5	Adjusted Fade	2	5	18.1 (9.61)	11.9 (5.53)
	Adjusted Bass	1			
	Didn't attempt - forgot how		1		
6	Used Seek	1		7.14 (9.88)	3.86 (1.73)
7	Pressed Preset and Seek together	1	1	31.5 (24.7)	23.6 (10.7)
	Held Seek button down	2	1		
	Interrupted Seek by pressing Preset		1		
	Pressed preset		1		
	Used Manual tuning		1		
	Failed to store - didn't know how	10			
	Didn't hold preset long enough	10	4		
	Pressed Seek to store	1			
8	Didn't know function	4		44.2 (23.8)	29.5 (13.6)
	Used Seek	15	3		
	Held Seek button down	2			
	Pressed Preset	1			
	Failed to store - didn't know how	7			
	Didn't hold preset long enough	10	4		
	Hit 2 Presets and storage failed		1		
9				3.64 (1.87)	3.18 (1.33)
10	Failed to stop FF/RWD	1	2	36.6 (19.2)	30.9 (16.1)
	Pressed Seek instead of autoreverse	3			
	Pressed wrong direction	3	2		
	Turned tape over manually	6			
	Failed to Seek	1	1		
11	Pressed twice		1	4.55 (2.77)	3.91 (1.69)
12				2.86 (1.21)	2.05 (0.576)

Task list:

1. Switch On
2. Adjust Volume
3. Adjust Bass
4. Adjust Treble
5. Adjust Balance
6. Choose a new Preset station
7. Choose a new station using Seek and store it
8. Choose a new station using Manual search and store it
9. Insert cassette
10. Find next track on other side of cassette
11. Eject cassette
12. Switch Off

FIGURE 41.5   Observation of user activity with the radio-cassette for Ford 7000 RDS-EON.

the responses to psychological issues of design (e.g., cognitive compatibility) as a domain expert was present. It could be argued that the subjective responses, combined with professional wisdom, make this a very strong technique. In particular, one aspect emerged from this analysis which had not been covered by any of the other techniques — the usability of the instruction manual.

Questionnaires (Brooke, 1996)

There are few examples of standardized questionnaires appropriate for the evaluation of in-car devices. However, the Software Usability Scale (SUS) may, with some minor adaptation, be appropriate. SUS was developed as part of the usability engineering program in integrated office systems developed at the Digital Equipment Company. SUS comprises 10 items that relate to the usability of the device. Originally conceived as a measure of software usability, it has some evidence of proven success. The distinct advantage of this approach is the ease with which the measure may be applied. It takes less than a minute to complete the questionnaire, and no training is required.

Example of Interview output (Sharp RG-F832E)

SECTION 1: VISUAL CLARITY
Information displayed on the screen should be clear, well-organised and easy to read.
• There is a certain amount of visual clutter on the LCD
• Writing (labelling) is small but readable
• Ambiguous abbreviations (e.g., DX/LO; ASPM ME-SCAN)
SECTION 2: CONSISTENCY
The way the system looks and works should be consistent at all times
• Tuning buttons (especially Scan and Seek functions) present inconsistent labelling
• Moded functions create problems in knowing how to initiate the function
SECTION 3: COMPATIBILITY
The way the system looks and works should be compatible with user expectations
• 4 functions on 'On/Off' switch makes it somewhat incompatible
• Auto-reverse function could cause cognitive compatibility problems
SECTION 4: INFORMATIVE FEEDBACK
Users should be given clear, informative feedback on where they are in the system
• Tactile feedback is poor, particularly for the 'On/Off' switch
• Operational feedback poor when programming a preset station
SECTION 5: EXPLICITNESS
The way the system works and is structured should be clear to the user
• Novice users may not understand station programming without instruction
• Resuming normal cassette playback after FF or RWD is not clear
SECTION 6: APPROPRIATE FUNCTIONALITY
The system should meet the needs and requirements of users when carrying out tasks
• Rotating dial is not appropriate for front/rear fader control
• Prompts for task steps may be useful when programming stations
SECTION 7: FLEXIBILITY AND CONTROL
The interface should be sufficiently flexible in structure, information presentation and in terms of what the user can do, to suit the needs and requirements of all users
• Users with larger fingers may find controls fiddly
• Radio is inaudible whilst winding cassette - this is inflexible
SECTION 8: ERROR PREVENTION AND CORRECTION
The system should be designed to minimise the possibility of user error, users should be able to check their inputs and to correct errors
• There is no 'undo' function for stored stations
• Separate functions would be better initiated from separate buttons
SECTION 9: USER GUIDANCE AND SUPPORT
Informative, easy-to-use and relevant guidance and support should be provided
• Manual is not well structured, relevant sections are difficult to find
• Instructions in the manual are matched to the task
SECTION 10: SYSTEM USABILITY PROBLEMS
• Minor problems in understanding function of 2 or 3 buttons
• Treble and bass controls are tiny
SECTION 11: GENERAL SYSTEM USABILITY
• Best aspect: This radio is *not* mode-dependent
• Worst aspect: Ambiguity in button labelling
• Common mistakes: Adjusting balance instead of volume
• Recommended changes: Substitute pushbutton operation for 'On/Off' control

FIGURE 41.6   Interviewing users about the radio-cassette for Sharp RG-F832E.

The SUS score is determined by taking 1 from all the scores on items with odd numbers and subtracting the scores from 5 for all the items with even numbers. The resultant value is multiplied by 2.5 to give an overall SUS rating of between 0 (extremely poor usability) and 100 (excellent usability). In the example shown in Figure 41.7, this resulted in the value 29 multiplied by 2.5. giving a SUS rating of 72.5. On its own, this just offers a subjective value, but the rating could be of most use when a dozen or more participants are rating two or more products. Statistical analysis of the results could be used to indicate real differences between the products.

Given the brevity of the approach, it is likely to serve as a useful adjunct to other methods. Brooke (1996) reports that SUS is a reliable measure and correlates well with other subjective measures.

Example of SUS output and scoring
(Ford 7000 RDS-EON)

	Strongly disagree				Strongly agree
1. I think that I would like to use this system frequently	1	2	3	4	5 ✓
2. I found the system unnecessarily complex	1	2 ✓	3	4	5
3. I thought the system was easy to use	1	2	3	4	5 ✓
4. I think that I would need the support of a technical person to be able to use this system	1 ✓	2	3	4	5
5. I found the various functions in this system were well integrated	1	2	3	4 ✓	5
6. I thought there was too much inconsistency in this system	1 ✓	2	3	4	5
7. I would imagine that most people would learn to use this system very quickly	1	2	3	4 ✓	5
8. I found the system very cumbersome to use	1	2 ✓	3	4	5
9. I felt very confident using the system	1	2	3	4	5 ✓
10. I needed to learn a lot of things before I could get going with this system	1	2	3 ✓	4	5

Scoring SUS

Odd-numbered items	Even-numbered items
score = scale position - 1	score = 5 - scale position
1. 5 - 1 = 4	2. 5 - 2 = 3
3. 5 - 1 = 4	4. 5 - 1 = 4
5. 4 - 1 = 3	6. 5 - 1 = 4
7. 4 - 1 = 3	8. 5 - 2 = 3
9. 5 - 1 = 4	10. 5 - 3 = 2

Total for odd-numbered items = 18   Total for even-numbered items = 16
Grand Total = 34 (multiply this by 2.5 to obtain usability score)
SUS overall usability score = 34 * 2.5 = **85**

FIGURE 41.7   Rating the radio-cassette using SUS for Ford 7000 RDS-EON.

Link Analysis
(Stammers et al., 1990; Kirwan and Ainsworth, 1992; Drury, 1995)

Link analysis represents the sequence in which device elements are used in a given task or scenario. The sequence provides the links between elements of the device interface. This may be used to determine if the current relationship between device elements is optimal in terms of the task sequence. Time data recorded on duration of attentional gaze may also be recorded in order to determine if display elements are laid out in the most efficient manner. The link data may be used to evaluate a range of alternatives before the most appropriate arrangement is accepted. The following diagrams represent the relevant link diagrams and tables for the Ford 7000 RDS EON. The analyses are abbreviated and based on a standard subset of tasks (see below). Redesign is offered on the basis of the analyses. As can be seen, very little is changed on the Ford radio, suggesting that the original satisfied the principles of link analysis well (see Figure 41.8).

A task list follows:

1. Switch On
2. Adjust Volume
3. Adjust Bass
4. Adjust Treble
5. Adjust Balance
6. Choose New Preset
7. Use Seek, then store station
8. Use Manual Search, then store station
9. Insert cassette
10. Autoreverse, then Fast Forward
11. Eject cassette and Switch Off

An initial dilemma was encountered in link analysis in whether to study hand or eye movements. For simplicity, the analysis was restricted to hand movements. A basic walk-through was used for the data collection. One particular problem here was concerned with the fact that operating a radio is far from being a set procedure, leading to a possibly infinite set of links. This was circumvented by analyzing a single run of a typical task.

Layout Analysis (Easterby, 1984)

Layout analysis builds upon link analysis to consider functional groupings of device elements (see Figure 41.9). Within functional groupings, elements are sorted according to optimum trade-off of three criteria: frequency of use, sequence of use, and importance of element.

Layout analysis was undoubtedly easier to execute. Functional groupings on a radio are obvious, and their importance, sequence, and frequency of use were also easily determined. The analysis also maintains something of a hierarchical structure, as it progresses from general categories to specific functions. Overall, this was a very straightforward and seemingly effective technique. Both techniques (link and layout analysis) lead to suggested improvements for interface layout.

Hierarchical Task Analysis (Annett et al., 1971; Stammers and Shepherd, 1995)

Hierarchical task analysis (HTA) has been a technique central to the discipline of ergonomics in the U.K. for over 2 decades. Application of the technique breaks tasks down into goals, plans, and operations in a hierarchical structure (see Figure 41.10). While the technique offers little more than a task description, it serves as the input into other predictive methods, for example SHERPA and KLM. The concepts of

Link Analysis - Ford 7000 RDS EON

Initial design:

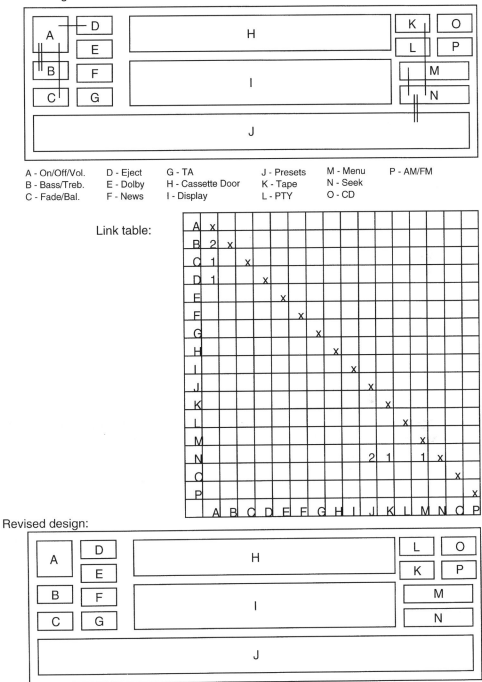

A - On/Off/Vol.   D - Eject   G - TA   J - Presets   M - Menu   P - AM/FM
B - Bass/Treb.   E - Dolby   H - Cassette Door   K - Tape   N - Seek
C - Fade/Bal.   F - News   I - Display   L - PTY   O - CD

Link table:

Revised design:

FIGURE 41.8   A link analysis for Ford 7000 RDS-EON.

HTA are relatively straightforward, but the approach requires some practice and reiteration before HTA can be applied with confidence.

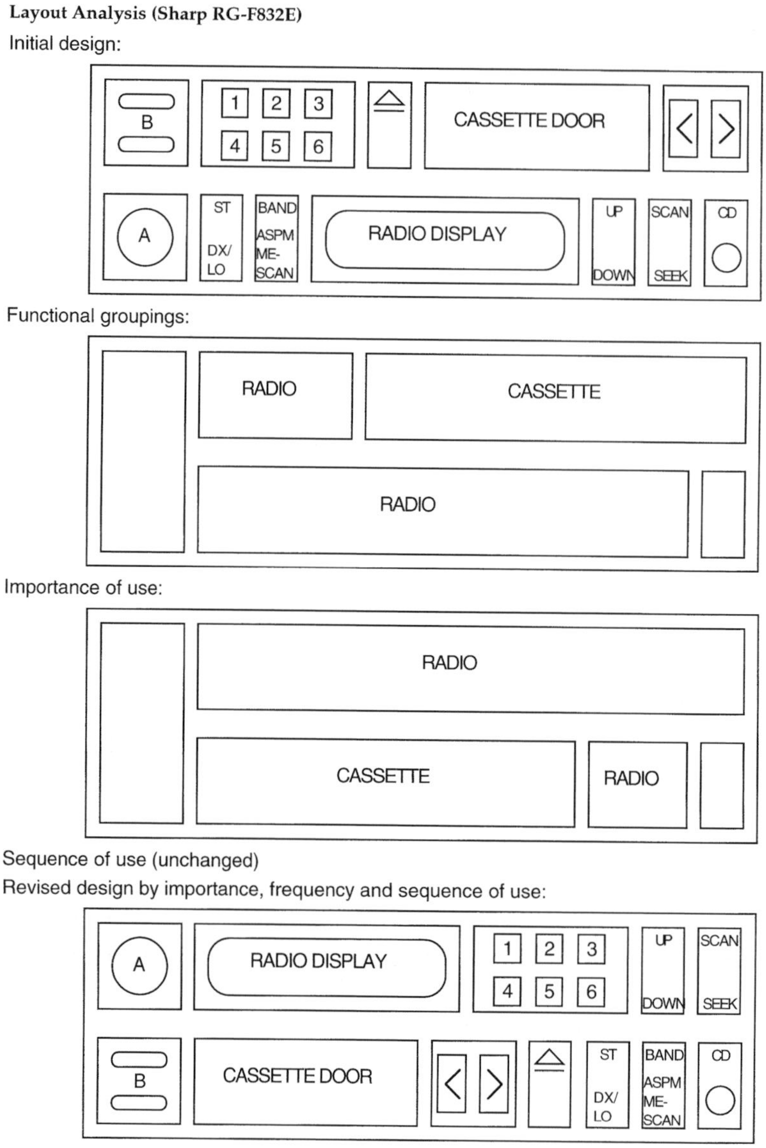

Layout Analysis (Sharp RG-F832E)

Initial design:

Functional groupings:

Importance of use:

Sequence of use (unchanged)

Revised design by importance, frequency and sequence of use:

FIGURE 41.9   A layout analysis for Sharp RG-F832E.

The structure of the task presented itself immediately; however, some detailed aspects later in the analysis (such as the logic involved in decisions and plans) were slightly more problematic. This, though, merely illustrates one of the characteristics (and some may say benefits) of HTA — that it is an iterative technique. This was certainly borne out in the analysis conducted.

Systematic Human Error Reduction and Prediction Approach (Embrey, 1993; Stanton, 1995; Baber and Stanton, 1996b)

Systematic human error reduction and prediction approach (SHERPA) is a semistructured human error identification technique. It is based upon hierarchical task analysis (HTA) and an error taxonomy. Briefly, each task step from the bottom level in HTA is taken in turn and potential error modes associated with that activity are identified (see Figure 41.11). From this the consequences of those errors are determined.

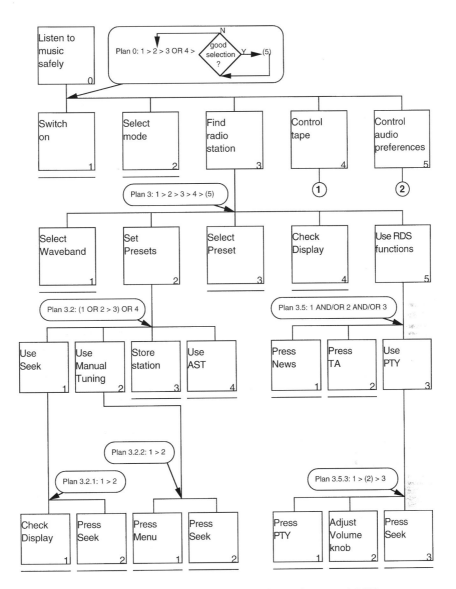

FIGURE 41.10 Hierarchical task analysis for Ford 7000 RDS-EON.

SHERPA appears to offer reasonable predictions of performance but may have some limitations in its comprehensiveness and generalizability.

While the human error taxonomy incorporated in SHERPA is certainly a handy prompt in identifying errors, it does have limitations in generalization across tasks. As SHERPA was originally designed for process control room tasks, some sections of the taxonomy are quite inappropriate for product design and evaluation. Thus, as stated above, there is room for improvement in this area. The error predictions and remedies offered by SHERPA are certainly of applied value if proved to be valid. Indeed, there are some recent studies (e.g., Baber and Stanton, 1996b; Stanton and Stevenage, 1997) that suggest such predictions are, to a large extent, accurate and reliable. However, the credibility and salience of some of the predicted errors may be questioned. Even after a thorough SHERPA, error reduction strategies must be evaluated along dimensions such as practicability and cost-effectiveness to determine whether they are worth applying. Moreover, cognitive aspects of errors, which may aid in the above determination,

Abridged example of HTA tabular format (Ford 7000 RDS-EON)

Super-ordinate	Subtask/plan	Notes
0	LISTEN TO IN-CAR ENTERTAINMENT P0: 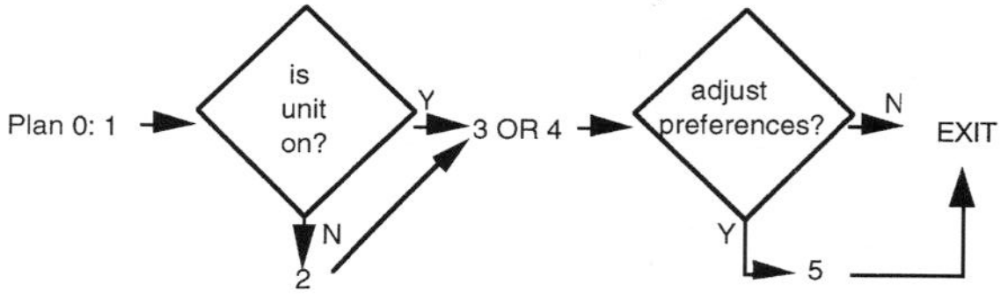 *[flow diagram: Plan 0: 1 → is unit on? Y → 3 OR 4; N → 2; → adjust preferences? Y → 5; N → EXIT]* 1. Check unit status // 2. Press on/off button // 3. Listen to radio 4. Listen to cassette 5. Adjust audio preferences	User checks whether radio is already on. If not, user switches on by **pressing** on/off knob. User may then decide to listen to radio or cassette, and may wish to adjust their audio preferences.
5	ADJUST AUDIO PREFERENCES P5: Plan 5: 1 AND/OR 2 AND/OR 3 AND/OR 4 AND/OR 5 1. Adjust volume 2. Adjust bass 3. Adjust treble 4. Adjust fade 5. Adjust balance	User may decide to adjust any or all of these preferences.
etc.	...	HTA table continues in this manner.

FIGURE 41.10 (continued)

**Abridged example of PHEA output
(Sharp RG-F832E)**

Task Step	Error Mode	Description	Consequence	Recovery	P	C	Remedies
1	A4	Volume level adjusted inappropriately	Volume is at undesirable level	Immediate /4.2	M		Separate vol/on/off Preset startup volume
	A7	Balance adjusted instead of on/off/vol	Unit is not switched on; balance settings altered	Immediate /4.3	H		Separate balance/on/off Lockout mechanism
2.1	C1	Omit check of station	Listening to wrong station	2.2	L		Untuned on startup Reminder to check
	C2	Misidentify station	Listening to wrong station	2.2	M		RDS
2.2.1	S2	Select wrong preset	Listen to wrong station	2.2.2 on	H		Label buttons clearly Aide-mémoire
2.2.2	A1	Preset button held too long	Undesired station storage	2.3	L		Confirm before storage
	A6	Press wrong button	Desired station not found	Immediate	H		Label buttons clearly Aide-mémoire
2.3.1	C1	Omit check of wavelength	Unnecessary retuning	2.3.2	L		Display conspicuity (e.g., RDS)
	C3	Check wrong display	Wavelength not identified	2.3.2	L		Display conspicuity (e.g., RDS)
etc...							

FIGURE 41.11   SHERPA for Sharp RG-F832E.

are largely avoided by SHERPA (although differing error mechanisms were intimated when multiple error modes lead to similar consequences). Finally, an observation worthy of note is that executing the SHERPA served to highlight deficiencies in the HTA — undoubtedly a bonus for task analysis.

Task Analysis For Error Identification
(Baber and Stanton, 1994; Stanton and Baber, 1996b)

Task analysis for error identification (TAFEI) is an approach for modeling the interaction between device and user (see Figure 41.12). TAFEI is based upon hierarchical task analysis (HTA) and state space diagrams (SSDs), both established techniques with a pedigree of over 25 years. By mapping HTA onto SSDs and employing a transition matrix, it is possible to start to consider what may go wrong in the interaction. Essentially, TAFEI is a human error identification method (like SHERPA).

Once one has understood HTA and SSDs, executing a TAFEI analysis is not difficult, although it is still a little time consuming. It is certainly an advantage to possess either or both HTA and SSD before beginning, as this saves a great deal of time. The most difficult part of this trial analysis was in constructing the SSDs for a car radio, as these have to be quite accurate for the remainder of the analysis to be effective. Completing the transition matrix is then a straightforward affair.

Repertory Grids (Kelly, 1955; Baber, 1996)

Repertory grids may be used to determine people's perception of a device. In essence, the procedure requires the analyst to determine the elements (the forms of the product) and the constructs (the aspects of the product that are important to its operation) (see Figure 41.13). Each version of the product is then rated against each construct. This approach seems to offer a way of gaining insight into consumer perception of the device, but does not necessarily offer predictive information.

The repertory grid is not a difficult technique to execute, once the concept has been grasped. Constructing a thorough grid should take no more than an hour. However, analysis is a different story. Initially in this example, the revised analysis technique of Baber (1996) was attempted. While it certainly seemed an easier method than the usual factor analysis, there also appeared to be weaknesses in its approach. These weaknesses were particularly borne out when it came to the factor extraction stage, for no constructs were significantly related to be grouped as a factor. Thus while the process of constructing the repertory grid was useful in itself, no useful quantification of the grid could be gleaned. Thus, the analysis turned

Abridged example of TAFEI

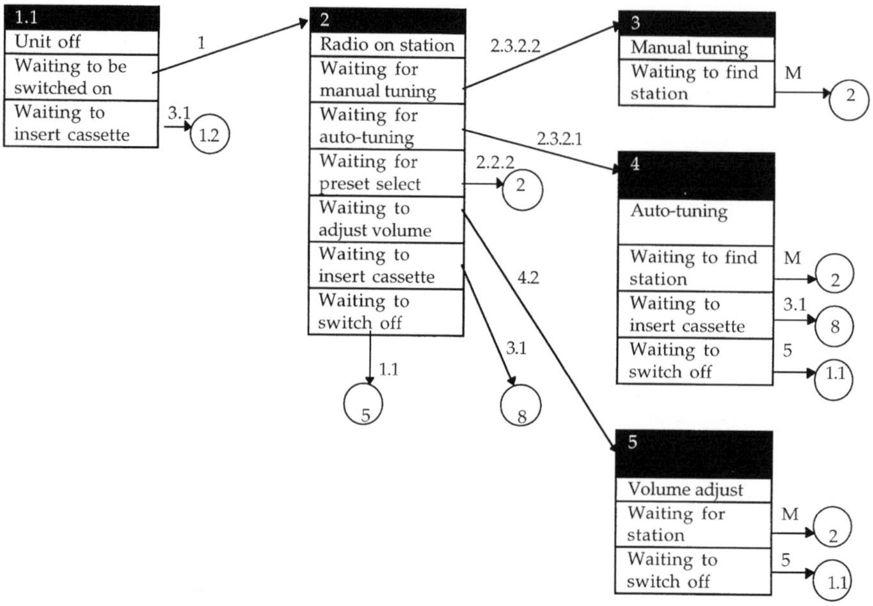

1.1
Unit off
Waiting to be switched on
Waiting to insert cassette

2
Radio on station
Waiting for manual tuning
Waiting for auto-tuning
Waiting for preset select
Waiting to adjust volume
Waiting to insert cassette
Waiting to switch off

3
Manual tuning
Waiting to find station — M → (2)

4
Auto-tuning
Waiting to find station — M → (2)
Waiting to insert cassette — 3.1 → (8)
Waiting to switch off — 5 → (1.1)

5
Volume adjust
Waiting for station — M → (2)
Waiting to switch off — 5 → (1.1)

Transitions: 1 → 2; 3.1 → (1.2); 2.3.2.2 → 3; 2.3.2.1 → 4; 2.2.2 → (2); 4.2; 3.1; 1.1 → (5); (8)

Transition matrix

	1.1	1.2	1.3	1.4	2	3	4	5	6	7	8	9	10	11	12	13	14
1.1	-	I	-	-	L	-	-	-	-	-	-	-	-	-	-	-	-
1.2	L	-	I	I	-	-	-	-	-	-	L	-	-	-	-	-	-
1.3	-	I	-	-	-	-	-	-	-	-	-	L	-	-	-	-	-
1.4	-	I	-	-	-	-	-	-	-	-	-	-	L	-	-	-	-
2	L	-	-	-	-	L	L	L	L	L	L	-	-	-	-	-	-
3	-	-	-	-	L	-	-	-	-	-	-	-	-	-	-	-	-
4	L	-	-	-	L	-	-	-	-	-	L	-	-	-	-	-	-
5	I	-	-	-	L	-	-	-	-	-	-	-	-	-	-	-	-
6	-	-	-	-	L	-	-	-	-	-	-	-	-	-	-	-	-
7	-	-	-	-	L	-	-	-	-	-	-	-	-	-	-	-	-
8	-	I	-	-	L	-	-	-	-	-	-	L	L	L	L	L	L
9	-	-	I	-	-	-	-	-	-	-	L	-	-	-	-	-	-
10	-	-	-	I	-	-	-	-	-	-	L	-	-	-	-	-	-
11	-	-	-	-	-	-	-	-	-	-	L	-	-	-	-	-	-
12	-	I	-	-	-	-	-	-	-	-	L	-	-	-	-	-	-
13	-	-	-	-	-	-	-	-	-	-	L	-	-	-	-	-	-
14	-	-	-	-	-	-	-	-	-	-	L	-	-	-	-	-	-

FIGURE 41.12　TAFEI for Sharp RG-F832E.

to more conventional methods. It is quite possible to execute a number of analyses on a repertory grid, such as factor analysis, multidimensional scaling, and even analyses of variance (although none of these methods were actually carried out in the present example). In summary, the repertory grid did provide a useful insight into perception of this product, such that the output could be useful in design. However, the constructs elicited also seemed to mirror somewhat the output of some of the other techniques reviewed so far, such as checklists and heuristics.

Keystroke Level Model (Card et al., 1983)

The keystroke level model (KLM) is a technique that is used to predict task performance time for error-free operation of a device. The technique works by breaking tasks down into component activities, e.g.,

**Example of Repertory Grid Output
(Ford 7000 RDS-EON and Sharp RG-F832E)**

Constructs	Rover	Ford	Vaux-hall	New Rover	Worst	Best	Opposites
Mode dependent	1	5	4	1	5	1	Separate functions
Pushbutton operation	2	5	4	2	1	5	Knob-turn operation
Bad labelling	2	5	4	2	5	1	Clear labelling
Easy controls	1	5	5	2	1	5	Fiddly controls
Poor functional grouping	4	5	2	2	5	1	Good functional grouping
Good illumination	2	4	5	2	1	5	Poor illumination

5 = left side very much applicable (right side not applicable at all)
4 = left side somewhat applicable (right side not really applicable)
3 = in between
2 = left side not really applicable (right side somewhat applicable)
1 = left side not applicable at all (right side very much applicable)
0 = characteristic irrelevant

FIGURE 41.13   Repertory grids for Sharp RG-F832E and Ford 7000 RDS-EON.

mental operations, motor operations, and device operations, then determining response times for each of these operations and summing them. The resultant value is the estimated performance time for the whole operation. While there are some obvious limitations to this approach (such as the analysis of cognitive operations) and some ambiguity in determining the number of mental operations to be included in the equation, the approach does appear to have some support. There are four motor operators in KLM — keystroking, pointing, homing and drawing; one mental operator; and one operator for system response. Each of these operators has an associated nominal time, derived by experiment (although drawing and response times are variable). It is thus a simple matter of determining the components of the task in question and summing the times of the associated operators to arrive at an overall task time prediction (see Figure 41.14).

Although KLM is indeed a simplistic method, it was designed for human–computer interaction (HCI), and this is evident when attempting to apply it. With a car radio, an immediate stumbling block was encountered because there is no operator accounting for "turning a knob." Thus, while some operators have no place outside HCI (e.g., pointing, drawing), there are others that are not foreseen within HCI. A consequence of this is that we are immediately limited as to the tasks we can analyze in the automobile. Further restrictions are imposed regarding the fixed times associated with most operators, leading to some errant predictions — a time of 2 seconds simply for selecting a preset station seems generous. A final limitation of KLM is that it does not predict anything over and above task execution time, nor does it claim to.

41.4   Conclusions

The detailed review of ergonomics methods led to a greater insight into the demands and outputs of the methods under scrutiny. A study by Stanton and Young (1997) indicated that link analysis, layout analysis, repertory grids, and KLM appear to offer good utility when compared with other, more commonly used methods. These data also seem to reinforce the reason for the popularity of questionnaires, interviews, observations, checklists, and heuristics, as they take relatively little time to apply when compared with HTA, TAFEI, and SHERPA. Perhaps it is surprising that link and layout analysis are not more popular given that they are also relatively quick to use. Similarly, repertory grids and the keystroke level model

Worked example for comparing two alternative car radio designs using the Keystroke Level Model (KLM)

Task	Time - Design 1 (s)	Time - Design 2 (s)	Difference +/-
Switch On	MHKR=2.65+1=**3.65**	MHKR=2.65+1=**3.65**	0
Adjust Volume	MHKR=2.65+0.1=**2.75**	MHKR=2.65+0=**2.65**	+0.1
Adjust Bass	MHKR=3.95+0.2=**4.15**	MHKR=2.65+0=**2.65**	+1.5
Adjust Treble	MHKKHKR=4.15+0.3=**4.45**	MHKR=2.65+0=**2.65**	+1.8
Adjust Balance	MHKKHKR=4.15+0.3=**4.45**	MHKKR=2.85+0.1=**2.95**	+1.5
Choose new Preset	MHKR=2.65+0.2=**2.85**	MHKR=2.65+0.2=**2.85**	0
Use Seek	MHKR=2.65+1=**3.65**	MHKR=2.65+1=**3.65**	0
Use Manual search	MHKHKR=3.95+1=**4.95**	MHKR=2.65+1=**3.65**	1.3
Store station	MHKR=2.65+1=**3.65**	MHKR=2.65+3=**5.65**	-2
Insert Cassette	MHKR=2.65+1=**3.65**	MHKR=2.65+1=**3.65**	0
Autoreverse and FF	MHKRHKRKR=4.15+5=**9.15**	MHKRKRK=3.05+5=**8.05**	1.1
Eject Cassette	MHKR=2.65+0.5=**3.15**	MHKR=2.65+0.3=**2.95**	0.2
Switch Off	MHKR=2.65+0.5=**3.15**	MHKR=2.65+0.7=**3.35**	-0.2
Total time	**53.65**	**48.35**	**5.3**

This table represents the calculations for execution times of standard tasks across the two different radio designs. Design 1 is the Ford 7000 RDS EON, design 2 is a Sharp RG-F832E. As is evident, on this set of tasks, the Ford design takes around 5 seconds longer to complete. Analysing the operators suggests that this extra time is very much taken up by the moded nature of the device.

N.B.: System response times have been largely estimated in the above table. For purposes of equality, standard response times (e.g., for radio tuning) were applied to both designs.

FIGURE 41.14   KLM for Sharp RG-F832E and Ford 7000 RDS-EON.

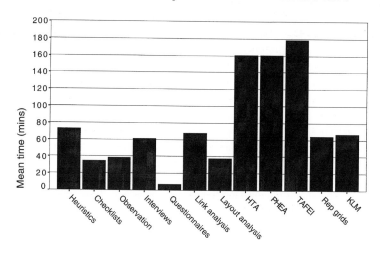

FIGURE 41.15   Time to apply each method to the evaluation of a radio-cassette (Ford 7000 RDS-EON).

seem to be no more time consuming to use than the focused interview. However, these techniques are rather more specialized in their output, like link and layout analysis (see Figure 41.15).

These analyses seem to suggest that some methods will be more acceptable than others because of the time required to apply them to an evaluation. However, we hope that ergonomists and designers would explore the utility of other methods rather than always relying on three or four of their favorite approaches.

Acknowledgments

The research reported in this chapter was supported by the EPSRC under the LINK Transport Infrastructure and Operations Programme in the U.K. The authors are grateful to Taylor & Francis for allowing them to reproduce some material from *Human Factors in Consumer Products*.

References

Annett, J., Duncan, K. D., Stammers, R., and Grey, M. J. (1971) *Task Analysis.* Department of Employment Training information paper 6. HMSO, London.

Baber, C. (1996) Repertory grid theory and its application to product evaluation, in P. W. Jordan; B. Thomas; B. A. Weerdmeester and I. L. McClelland (eds.) *Usability Evaluation in Industry* Taylor & Francis, London, pp. 157-165.

Baber, C. and Stanton, N. A. (1994) Task analysis for error identification: a methodology for designing error-tolerant consumer products. *Ergonomics,* 37, 11, 1923-1941.

Baber, C. and Stanton, N. A. (1996a) Observation as a usability method, in P. W. Jordan; B. Thomas; B. A. Weerdmeester and I. L. McClelland (eds.) *Usability Evaluation in Industry* Taylor & Francis, London, pp. 85-94.

Baber, C. and Stanton, N. A. (1996b) Human error identification techniques applied to public technology: predictions compared with observed use, *Applied Ergonomics,* 27 (2) pp. 119-131.

Brooke, J. (1996) SUS: a "quick and dirty" usability scale, in P. W. Jordan; B. Thomas; B. A. Weerdmeester and I. L. McClelland (eds.) *Usability Evaluation in Industry* Taylor & Francis, London, pp. 189-194.

Card, S. K., Moran, T. P., and Newell, A. (1983) *The Psychology of Human-Computer Interaction* Erlbaum, Hillsdale, NJ.

Corlett, E. N. & Clarke, T. S. (1995) *The Ergonomics of Workspaces and Machines.* 2nd Edition Taylor & Francis, London.

Diaper, D. (1989) *Task Analysis in Human Computer Interaction.* Ellis Horwood, Chichester.

Drury, C. G. (1995) Methods for direct observation of performance, in J. Wilson and N. Corlett (eds.) *Evaluation of Human Work.* 2nd Edition Taylor & Francis, London, pp. 45-68.

Easterby, R. (1984) Tasks, processes and display design, in R. Easterby and H. Zwaga (eds.) *Information Design* Taylor & Francis, London.

Embrey, D. (1983) Quantitative and qualitative prediction of human error in safety assessments, in *The Institution of Chemical Engineers symposium Series* 130 pp. 329-350.

Jordan, P. W., Thomas, B., Weerdmeester, B. A., and McClelland, I. L. (1996) *Usability Evaluation in Industry* Taylor & Francis, London.

Kelly, G. A. (1955) *The Psychology of Personal Constructs* Norton, New York.

Kirwan, B. and Ainsworth, L. (1992) *A Guide to Task Analysis* Taylor & Francis, London.

Kirwan, B. (1994) *A Guide to Practical Human Reliability Assessment* Taylor & Francis, London.

Nielsen, J. (1992) Finding usability problems through heuristic evaluation, in *Proceedings of the ACM Conference on Human Factors in Computing Systems* ACM Press, Monterey, CA, pp. 373-380.

Ravden, S. J. and Johnson, G. I. (1989) *Evaluating Usability of Human-Computer Interfaces: A Practical Method.* Ellis Horwood, Chichester.

Sinclair, M. (1995) Subjective assessment, in J. Wilson and N. Corlett (eds.) *Evaluation of Human Work.* 2nd Edition Taylor & Francis, London, pp. 69-100.

Stammers, R. B. Carey, M., and Astley, J. A. (1990) Task analysis, in J. Wilson and N. Corlett (eds.) *Evaluation of Human Work* Taylor & Francis, London, pp. 134-160.

Stammers, R. B. and Shepherd, A. (1995) Task analysis, in J. Wilson and N. Corlett (eds.) *Evaluation of Human Work.* 2nd Edition Taylor & Francis, London, pp. 144-168.

Stanton, N. A. (1995) Analysing worker activity: a new approach to risk assessment, *Health and Safety Bulletin,* 240 pp. 9-11.

Stanton, N. A. (1998) *Human Factors in Consumer Products.* Taylor & Francis, London.

Stanton, N. A. and Baber, C. (1996a) Factors affecting the selection of methods and techniques prior to conducting a usability evaluation, in P. W. Jordan; B. Thomas; B. A. Weerdmeester and I. L. McClelland (eds.) *Usability Evaluation in Industry.* Taylor & Francis, London pp. 39-48.

Stanton, N. A. and Baber, C. (1996b) A systems approach to human error identification, *Safety Science,* 22 (1-3) 215-228.

Stanton, N. A. and Stevenage, S. (1998) Learning to predict human error: issues of reliability, validity and acceptability. *Ergonomics,* (in press)

Stanton, N. A. and Young, M. (1995) *Development of a Methodology for Improving Safety in the Operation of In-Car Devices* EPSRC/DOT LINK Report 1. University of Southampton, Southampton.

Stanton, N. A. and Young, M. (1997) Is utility in the mind of the beholder? A study of ergonomics methods, *Applied Ergonomics,* 29 (1) 41-54.

Wilson, J. (1995) A framework and context for ergonomics methodology, in J. Wilson and N. Corlett (eds.) *Evaluation of Human Work.* 2nd Edition Taylor & Francis, London pp. 1-39

Wilson, J. and Corlett, N. (1995) *Evaluation of Human Work.* 2nd Edition. Taylor & Francis, London.

Woodson, W. E., Tillman, B., and Tillman, P. (1992) *Human Factors Design Handbook.* 2nd edition McGraw-Hill, New York.

Index

load measurement devices for, **32**-3

local coordinate system, **32**-4, **32**-11 to 12

object space calibration, **32**-4 to 8

overview of, **32**-1

procedure for, **32**-3 to 4

rotational kinematics, **32**-14

tasks, **32**-3 to 4

virtual markers for, **32**-8

Videotaping for job analysis

checklists for, **30**-8 to 13

data sheets for, **30**-8 to 13

description of, **30**-1

equipment for, **30**-1 to 3

flow chart for, **30**-4 to 5

objective of, **30**-3 to 4

preparation for, **30**-1 to 3

procedure, **30**-3 to 4

quality-related problems, **30**-4

results of, **30**-6

supplemental data gathering, **30**-4 to 5

Vigilance decrement, **16**-6

Virtual reality, **29**-2, **29**-9

Visual acuity, **16**-6

Visual angle, **16**-6

Visual display units

advantages of, **8**-15 to 17

cathode ray tubes, **8**-12 to 14

description of, **8**-10

design of, **8**-10 to 12

digital presentation, **8**-11 to 12

disadvantages of, **8**-15 to 17

equipment, **8**-11

screen, **8**-26

screens, **8**-16

Visual instruments, **8**-2 to 4

Voice identification instruments, **8**-38

VO$_{2max}$, *see* Maximal oxygen uptake

W

Warning signals, **8**-9

Warning signs and labels

color coding schemes for, **40**-7, **40**-20

competence considerations, **39**-6 to 8

comprehensibility of, **40**-10

content guidelines for, **40**-7 to 9

demographic considerations, **39**-3 to 6

effectiveness evaluations, **40**-9 to 12

employer's responsibilities, **40**-17

familiarity issues, **39**-8 to 9

format of, **40**-5 to 7, **40**-30

guidelines for, **40**-5 to 9

hazardous industrial chemicals, **40**-17

material safety data sheets, **39**-1, **39**-3, **39**-11, **40**-20 to 21, **40**-24 to 27

methods of, **39**-2

mode of presentation, **40**-9

model of, **39**-2

nonverbal communication, **39**-7

patient package inserts, **39**-1 to 2

perceptual factors for, **40**-11

principles of, **39**-11 to 12

questionnaires for evaluating, **40**-11 to 12

receiver considerations, **39**-3 to 11

recommendations for, **39**-11 to 12

standards for

ANSI Z535 committee, **40**-15, **40**-20, **40**-24

Consumer Product Safety Commission, **40**-13

Department of Transportation, **40**-13, **40**-29

description of, **40**-12

Environmental Protection Agency, **40**-13

fire hazards, **40**-21 to 22

hazard communication, **40**-17, **40**-20 to 21

hazard materials transportation, **40**-21

hazardous industrial chemicals, **40**-17

hazardous materials identification system, **40**-22, **40**-30

mandatory requirements, **40**-13

Occupational Safety and Health Administration, **40**-13

recommendations, **40**-18 to 19

sources of, **40**-13 to 16

voluntary guidelines, **40**-13 to 16

summary overview of, **40**-22 to 30

symbols, **40**-23

symbols used in, **39**-7

wording of, **40**-6, **40**-28

Warning theory, **39**-1

Web page, **5**-8

Web site, **5**-8

Weibull distribution, **35**-30

Weight, **9**-12, **9**-14

Wet bulb globe temperature, **23**-12

Wheatstone bridge circuit, **33**-5

Wheels, design considerations, **8**-35

White finger syndrome, **27**-, **27**-11

Work factors

health effects of, **15**-5

physical, **15**-5

psychosocial, **15**-2 to 7

control of, **15**-7

Work organization, **15**-3

Work time, **12**-17

Workers

cumulative trauma disorder training for, **7**-11

strength evaluations

definition of, **21**-2

description of, **11**-3, **11**-7

dynamic, **21**-4

factors that affect, **21**-4 to 7

isoinertial testing, **21**-11 to 13

isokinetic tests, **21**-13 to 14

isometric, **21**-4, **21**-10 to 11

job design and, **21**-7 to 9

measurement of

description of, **21**-2 to 3

purposes, **21**-7 to 10

progressive inertial lifting evaluation, **21**-12 to 13

psychophysical approach, **21**-9

purposes of, **21**-1 to 2

strength aptitude test, **21**-11 to 12

types of, **21**-3 to 4

worker selection and placement based on, **21**-9 to 10

training of, **7**-11

Working posture

discomfort, **12**-2 to 4

hand positions, **12**-4 to 10

maximum holding time

acceptable, **12**-17

definition of, **12**-2, **12**-17, **31**-20

description of, **12**-2 to 3, **31**-20

discomfort and, **12**-3

muscle effort and, **12**-3

standards developed using, **12**-14 to 16

overview of, **12**-1 to 2

standards, from maximum holding time data, **12**-14 to 16

static, **13**-2

static load evaluations, **12**-4

work to rest model, **12**-10 to 14

Workload analysis, **17**-14 to 15

Workplace

design of, **13**-10

seated, **10**-18 to 19

standing, **10**-18 to 20